Thomas E. Dillinger

VLSI ENGINEERING

PRENTICE HALL
Englewood Cliffs, New Jersey 07632

Library of Congress Cataloging-in-Publication Data

Dillinger, Thomas E.,
  VLSI engineering.

  Bibliography: p. 853
  Includes index.
  1. Integrated circuits—Very large scale integration.
2. Microcomputers. I. Title.
TK7874.D55 1988   621.395   87-11395
ISBN 0-13-942731-7

Cover design: Photo Plus Art
Manufacturing buyer: Gordon Osbourne

Cover photo courtesy of the Integrated Technology Laboratory, IBM-Rochester; photo was
taken with a Leitz Metalloplan microscope, Vario-Orthomat II camera.
    Photo caption: Macro circuit layout (after polysilicon patterning). Note the large area
nodes allocated for *n*-well (VDD) and substrate (GND) contacts adjacent to circuit layouts.

© 1988 by Prentice Hall
A division of Simon & Schuster
Englewood Cliffs, New Jersey 07632

Printed in the United States of America

10 9 8 7 6 5 4 3 2 1

ISBN 0-13-942731-7 025

Prentice-Hall International (UK) Limited, *London*
Prentice-Hall of Australia Pty. Limited, *Sydney*
Prentice-Hall Canada Inc., *Toronto*
Prentice-Hall Hispanoamericana, S.A., *Mexico*
Prentice-Hall of India Private Limited, *New Delhi*
Prentice-Hall of Japan, Inc., *Tokyo*
Prentice-Hall of Southeast Asia Pte. Ltd., *Singapore*
Editora Prentice-Hall do Brasil, Ltda., *Rio de Janeiro*

In memory of my father, Joseph R. Dillinger,
Professor of Physics, University of Wisconsin–Madison

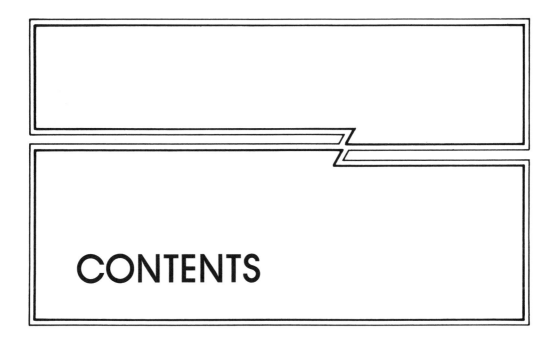

CONTENTS

CHAPTER 10   DESIGN SYSTEM LIBRARY DEVELOPMENT   462

CHAPTER 11   SPECIAL CIRCUIT TYPES   473

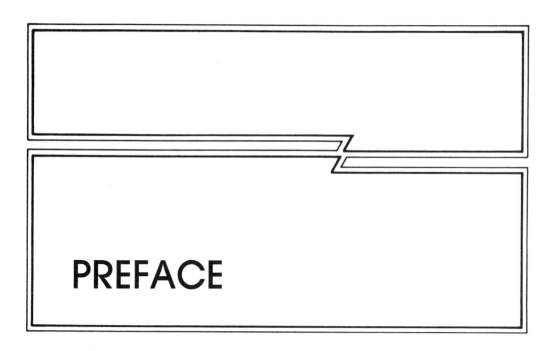

PREFACE

This book describes the steps associated with the design, modeling, and verification of silicon integrated circuitry and the tools developed to assist with managing the complexity and volume of data that supports those tasks. The era of technological development pertinent to these discussions encompasses very large scale integration of microelectronic circuitry (VLSI); the set of application programs and database structures that supports VLSI design efforts is collectively referred to as a *design system*. The computational resource constraints and algorithmic trade-offs that are part of application program development are described in considerable detail; these engineering judgments imply that some design flexibility will be forsaken to improve the efficiency of the task relegated to the design system and, likewise, to ensure the manufacturability of the resulting part. The implementation of design structure rules and the associated checking procedures used to ensure that those rules are observed are recurring themes throughout the design and verification tasks described. The use of a design system technology model library is also highlighted, as this database provides the link between process engineering, logic/circuit design, and system architecture.

This text is intended for senior-level undergraduates and first-year graduate students in the fields of electrical engineering and computer science and as a reference for electronic engineers and design system application programmers. The order in which the topics are presented is perhaps a bit unusual: the first six chapters concentrate on the design process itself and specifically on the design system tools used to assist with that process. With this background, the succeeding six chapters discuss circuit design, modeling, and performance evaluation specifically pertaining to MOS technologies. Chapter 7 provides a review of solid-state device concepts; for those readers with an electronics background, this discussion can be skipped. Chapters 13 and 14 describe the back-end design verification tasks of testability evaluation and final design checking prior to manufacture; although referred to as back-end steps in the design flow, their influence on the complete design cycle will be evident throughout previous chapters. The final chapter discusses integrated-circuit process development and characterization, with particular emphasis on manufacturing reliability. Most of the chapter problems included with the discussion are major algorithm development and/or circuit analysis projects. The motivation for including such broad development goals is to encourage further research in computer-aided design algorithms and techniques; by analyzing the circuit design efforts of others, new techniques for enhancing performance, density, and reliability measures are likely to be uncovered.

Portions of this material have been used for teaching internal courses at IBM for several years. In addition, this material was used for a two-semester course taught at the University of Wisconsin–Madison in 1982–1983. Several of the development objectives alluded to were assigned as small group or graduate student projects during that time. The design description languages introduced in Chapters 4 and 5 have their origins in those efforts.

A number of people at IBM–Rochester were instrumental in helping this text reach its final form. In particular, I would like to thank Joe Jackson and Miller Ness for their reviews of the material and their insightful comments. The enthusiastic teaching and exceptional class notes provided by Joe and Dick Donze were the springboard for this effort. I owe a great deal of thanks to Mike Sheehan, whose confidence and persistence enabled both the teaching opportunity at Wisconsin and this writing project to get off the ground.

I would like to thank the students of UW ECE 601, 1982–1983, for their participation in a new and rather unconventional class. The contributions of Randy Schnier, John Wardale, Jeff Jackson, Bob Karau, Mahmood Azhar, and Jeff Koehler deserve special mention.

The faculty at Wisconsin who have helped to shape this text are too numerous to mention—the discussions with Professors D. Dietmeyer, J. Wiley, C. Kime, A. Scidmore, and J. Nordman have been extremely helpful throughout the years. There is one very special faculty member at Wisconsin whose teachings, ideas, and dedication I appreciate more and more each day—to Professor Henry Guckel, my most heartfelt thanks.

To Mike Gruver, Jim Worisek, and Roy Ames, thanks for always being around to help. To Bill O'Brien, thanks for sharing your perspectives on this industry and your amazing sense of humor.

Finally, thanks to my family for their encouragement, especially my wife Suzi for her patience and tolerance of all those late hours.

Tom Dillinger
Rochester, Minnesota

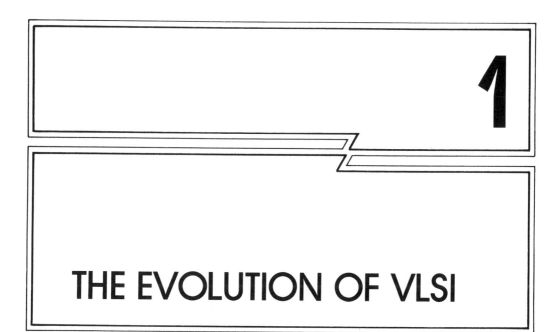

THE EVOLUTION OF VLSI

*If all the computer companies continue to buy all their logic devices from the
merchants, the uniqueness of their computer products would disappear.*

—Peter R. Tierney
Sperry–Univac

1.1 MOORE'S LAW

The influence of integrated-circuit technology in the past few years on our society
has been pervasive, in areas ranging from consumer products to business management to manufacturing control. The driving force behind this pervasiveness
is that the functional capability of modern integrated circuitry has increased in
scope and complexity *exponentially* with time over the past 20 years; in short,
the designers of modern integrated circuitry have continually endeavored to provide more computational speed with less dissipated electrical power and less
circuit board area, while maintaining a low failure rate and an aggressive cost.
This text describes the design steps and the engineering constraints and trade-
offs that are currently being addressed in the production of monolithic, silicon
integrated circuits, commonly referred to as *chips*. An underlying theme present
throughout the material is the need to address economic constraints as well, with
an emphasis on designer productivity and design reliability. A manufacturable,
reliable product developed on an aggressive schedule with a minimum amount of
development cost is required to keep pace with a fast-changing applications market.

The exponential growth pattern in integrated circuit function over time was first described in the late 1960s by Gordon Moore (then with Fairchild Corporation), and the projections he made based on this pattern have become known as Moore's law (reference 1.1). The extrapolation of Moore's law has remained surprisingly accurate over the past decade, in the sense that this exceptional growth required similar advancements in a myriad of related fields and disciplines (Figure 1.1). Indeed, the rapid increase in technological capacity was in large part made possible by the continuing refinement of our understanding of the physical, electrical, and chemical properties of materials, particularly at very small dimensions. This progress was also enhanced by advances in the design and manufacture of the equipment used to fabricate modern integrated circuits, especially in the area of photolithographic patterning of active device areas and their interconnections. When existing manufacturing techniques were not adequate to continue to support this growth, new procedures and apparatus were rapidly conceived and developed. Yet, for reasons to be discussed in this chapter, this growth pattern has recently abated to a significant degree. The technological and economic hurdles in the path of further advancements are larger than ever. Nevertheless, the economic reward for overcoming these obstacles remains inviting; the key is incorporating the critical mass of engineering, manufacturing, financial, and computer resources.

The criterion used to develop Moore's law was the capacity of the densest production memory storage integrated circuit at that time. This particular function has in the past been the forerunner of each new technology. Due to the continual demand for increased on-line computer storage capacity and the resulting performance enhancement, the large sales volume of memory integrated circuits provides the greatest return on the initial nonrecurring design and engineering

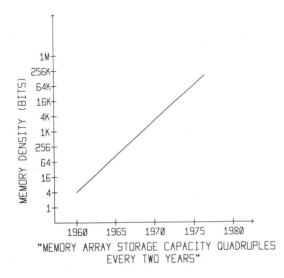

Figure 1.1. Plot of *Moore's law* depicting the exponential growth in the function available on a single integrated circuit over time. The criterion used for illustration is the memory array storage for production hardware.

costs. The development of a new memory product has therefore been the natural motivating factor behind technological strides. Other functions (processor, controller, and interface circuitry) have also progressed in circuit density and performance in a similar fashion, using the engineering knowledge base gained during early memory technology development as a guide toward designing for manufacturability and testability.

Other benchmarks besides memory storage capacity have been used to illustrate Moore's law, the most common being the minimum linewidth dimension used in the technology for patterning a conductor. The two specifications are very strongly correlated: the finer the lithography of interconnection, the greater the achievable circuit and memory array density.

This text will use yet a third benchmark for describing integrated circuit complexity: the number of logic circuits that have been collectively placed and wired on the final integrated circuit. It was mentioned earlier that the growth in functional complexity predicted by Moore's law has slowed in recent years. This trend particularly applies to the logic-intensive functions: microprocessors, microcontrollers, communications and graphics hardware. In addition to technology obstacles, other major factors impeding the development of logic function parts are as follows:

1. The number (and diversity) of high-volume product applications for logic hardware has not grown at the same pace as the technological capability.
2. The expense and quantity of resource required to verify the design prior to manufacture and to test the manufactured product have increased tremendously.
3. The complexity of current designs is becoming considerably more difficult to manage *conceptually* among the design and engineering group.

The last reason is undoubtedly the most significant; the design group must have the *tools* available that will permit the total integrated-circuit function to be partitioned for manageability and then verified collectively. These tools must span the range of design tasks, including partitioned design entry, design simulation and verification, circuit design (including the physical placement of circuits and routing of interconnections as per the design schematic), and the analysis of design testability and manufacturability. These tools are commonly implemented as computer-aided design (CAD) programs, which the design group uses to describe and analyze the logical and circuit design data. The evolution of integrated-circuit development has become heavily dependent on the development of CAD resources for design support.

In addition to CAD programs and algorithms, a key element in the ability to effectively partition and verify the design is the design *database*, that is, the computer storage means for representing the various models of a design element (logical, physical, and test) and the interconnection of those elements into *higher-order* functions. This text will also be concerned with the organization of the design database and the development of tools to be exercised against that data for design verification. In addition, CAD tools are indispensable for enhancing the productivity of the design group, particularly for such error-prone tasks as circuit placement and interconnection. The text material stresses the design/engineering trade-offs faced when implementing these productivity-related tools.

The partitioning of the design will be discussed in terms of a design *hierarchy*; the data structures illustrated for storing the design data will be hierarchical, using *pointers* to the related elements in the design description.

The technical literature contains much debate over the question of the *optimum* methodology for building the design database in terms of designer productivity and design evaluation; the terms *top-down* and *bottom-up* refer to the direction followed when constructing the design hierarchy. This text will also discuss to some extent the merits and disadvantages of each of these approaches.

Before leaving this section, it is worthwhile to note that the realization that Moore's law would be slowed by economic constraints and design complexity (as opposed to more fundamental technological limits) was also initially made by Gordon Moore (reference 1.2).

1.2 THE DEFINITION OF VLSI

With each successive *generation* of logic function complexity available on an integrated circuit, a new category was created to describe the improvement. The acronym VLSI (very large scale integration) was introduced about 1977 to describe the commercial products being developed at that time. Figure 1.2 illustrates the evolution of logic complexity and includes *rough* estimates of the dates and function that define each category. In the figure, the logical function is defined in terms of the number of logic blocks (circuits) implemented on a representative part. As alluded to in Section 1.1, the era of VLSI integrated-circuit designs is best characterized by the logic/system design being implemented, rather than the memory storage density.

ERA	DATE	COMPLEXITY (NUMBER OF LOGIC BLOCKS PER CHIP)
SINGLE TRANSISTOR	1959	< 1
UNIT LOGIC	1960	1
MULTI-FUNCTION	1962	2 - 4
COMPLEX FUNCTION	1964	5 - 20
MEDIUM SCALE INTEGRATION	1967	20 - 200 (MSI)
LARGE SCALE INTEGRATION	1972	200 - 2000 (LSI)
VERY LARGE SCALE INTEGRATION	1977	2000 - ? (VLSI)

Figure 1.2. Evolution of logic complexity in integrated-circuit hardware; the criterion used is the number of logic gates implemented on a chip versus the year when such function was available.

A flow chart of the steps in the design phase of a VLSI part is presented in Figure 1.3. In particular, the iterative nature of the design process should be noted. Constraints imposed on the physical design may require modifications to the original logic design. Likewise, the generation of test patterns may require restructuring the logic to improve the testability of embedded functions. The utilization of a *design system* can reduce the number of iterations required in the design phase by imposing rigid constraints at each step in the design flow. In short, a VLSI design system consists of (1) a set of logical (and circuit) design *standards* and (2) the CAD program tools used to verify that the design satisfies these standards. The structure imposed by these standards enables the development of the CAD programs and design database to assist with the more difficult and error-prone design and verification tasks. Some of these standards can be interpreted as being independent of the particular fabrication technology, whereas others will vary in magnitude and significance from technology to technology.

Referring again to Figure 1.3, it is unlikely that any practical design system would eliminate design iterations completely, for the following two reasons:

1. The overall system/circuit performance is not known until the physical design is complete. To ensure that the overall performance goals of the design are met, the logic designer must be involved in the evaluation of the final physical design.
2. Enhancing the testability of the design (i.e., the controllability and observability of all internal nets by the input test pattern set) may require modifications to the logic design.

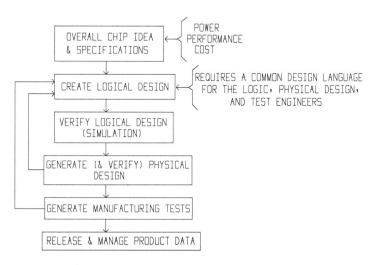

Figure 1.3. Flowchart of the steps in the design of a VLSI part. In particular, note the iteration indicated in the design flow between the logical description, the physical design, and the test file development.

In reference to item 1, during logic simulation the logic designer will likely use delay *estimates* for the logic gates and networks in the design. These estimates are typically based on the number of logic gates in a signal path, the specific gate/ circuit type, and the circuit loading (fan-out and estimated interconnection capacitance). After the physical design is complete and the actual interconnection loading is more accurately known, the design must be resimulated for verification. Delay paths that do not meet performance objectives will require modifications to the logic design. Area and power trade-offs may also require contemplating changes to the logic design after the physical design is complete.

Regarding item 2, the model used for assessing the testability of the design is based to a large degree on the need for diagnostic information about the parts that fail manufacturing test. If the goal of testing is to assess the *functionality* of the part on a go/no go basis, the influence of testability on the logic design is reduced and multiple design iterations for test purposes are unlikely. In this case, the same input patterns used by the logic designer for simulation are exercised against each part during manufacturing test; the measured outputs are compared against their expected logic simulation values. Alternatively, the design can be regarded for test-modeling purposes more in terms of its failure modes than with respect to its functionality. The internal signal nets are regarded as either valid or faulty representations of their logic model. The test pattern set must try to isolate a good network from a faulty one in as detailed a manner as possible, down to individual nets or logic paths. This diagnostic information (the correlation of measured fails to the physical failing location) is of considerable importance in pinpointing marginal circuit designs, isolating manufacturing defects, and enhancing manufacturing yield.

To illustrate the concept of design system standards, two examples are presented here (others will be introduced throughout the remainder of the text). The first standard is technology independent and pertains to all logic designs; the impact of the second will vary from technology to technology. The first also assumes a testing methodology that strives for diagnostic information.

Example 1

> If the integrated-circuit design contains a memory array as a subportion of the total design, that array must be able to be isolated for the purposes of testing. In other words, the inputs to the array must be *directly controllable* from the chip inputs and the outputs of the array must be *directly observable* at the chip outputs (Figure 1.4). Thus, the correct function of the memory array may be verified at manufacturing test in a more efficient manner.

Directly controllable means that the embedded array inputs (all clocks, data inputs, and memory addresses) may be directly toggled by toggling an appropriate subset of the chip inputs; all intervening logic gates must therefore be transparent during array test. Likewise, *directly observable* implies that the value of the array outputs can be captured at the outputs of the integrated circuit, again through

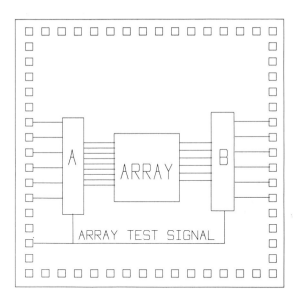

Figure 1.4. Technology-independent design system standard regarding the isolation of internal storage arrays for more efficient testing. Networks *A* and *B* are designed to be transparent logic networks upon application of the Array Test signal to permit direct controllability and observability of the array inputs/outputs. In addition, these logic networks must be able to be verified independently.

transparent logic. The intervening logic networks are typically controlled by an Array Test signal. This signal could be applied to a chip input pad during test, yet need not be accessible at higher levels of system packaging (unless higher-level diagnostics are also required).

The logic networks illustrated in Figure 1.4 must also be able to be tested independently of the memory array. Note that the requirement of transparent logic networks precludes the use of any flip-flop in the signal path; if sequential functions are required, latches should be used.[1] Adopting this design system standard can result in reduced tester time (*per* integrated circuit) after manufacture and therefore reduced manufacturing cost. In addition, the development of the test pattern set for verifying the array is considerably simplified. Indeed, a CAD tool could likely be used to add the memory array test patterns to the tester pattern file. The disadvantages of this design system standard, however, are the constraints imposed on the overall logic design and the additional circuitry required to implement the Array Test mode. This standard would apply to all logic designs, irrespective of technology.

Example 2

A technology-dependent standard requires that the logic circuits of a particular technology have unique limits on the circuit *fan-out* (the collective number of logic gate inputs to which the output of the driving circuit is connected).

[1] The terminology used in this text regarding flip-flops and latches are that (1) flip-flops are by design *edge-triggered* circuits, that is, the data value is captured and the output(s) may only change in response to a *transition* on the clock input; and (2) latches are *level-sensitive* circuits, that is, changes at the data input will be observed at the latch output(s) as long as the clock signal remains at its *active* level. This distinction is illustrated in Figure 1.5.

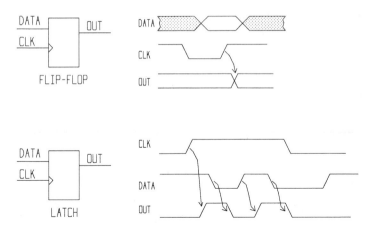

Figure 1.5. Terminology differentiating *latches* and *flip-flops*; latches are *level-sensitive*, while flip-flops are *edge triggered*.

The fan-out limits are commonly imposed to ensure that the total current from the driving logic gate will not substantially degrade the output voltage level (Figure 1.6). A fan-out limit may also be imposed to bound the output capacitive loading on the driving logic gate (and the subsequently poor performance that would result). Rather than exceed this limit, the logic design should contain instead a parallel implementation of the function of the network. For example, widely distributed internal clock signals must often be implemented in a carefully designed *powering tree* with balanced loading to minimize clock skew (Figure 1.7).

Whatever the reason for the implementation of this standard, a CAD tool is indispensable for examining the interconnections in the logic design description

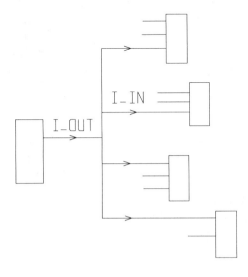

Figure 1.6. Technology-dependent design standard regarding limits on the block *fan-out*, the sum of the number of driven-block inputs. A technology rule must be developed to assess the output current capability of the sourcing block against the sum of the input current requirements of the fan-out pins.

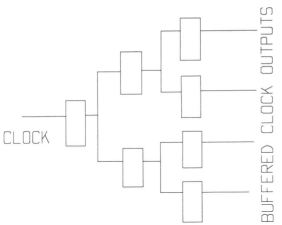

Figure 1.7. A design system fan-out limit may be imposed to contain the interconnection capacitance of heavily loaded nets. A *clock powering tree* is illustrated, where the loading on parallel buffering circuits is balanced to minimize skew.

and reporting fan-out errors. This tool must have available the technology-specific database of output current capability and input current/capacitive loading for the logic circuits of the technology. The associated technology-dependent files contain the *rules* that are applied against the design by the design system tools. Once the technology choice is made for the implementation of a particular design, that design must satisfy the rules of the technology; some logic design iteration may be required to satisfy the technology *rules check*. The fabrication of the design is predicated on the successful completion and audit of all the rules-checking procedures, giving an increased measure of confidence in the manufacturability of the part.

Given that some iteration may indeed be required in the design phase, a design system should also address how to best provide for the fastest detection and correction of problems with the design. Throughout this text, the following terminology will be used:

Front-end tools will refer to the algorithms and programs used to assist the designer(s) with completing the original portion of the design phase.

Back-end tools will refer to the algorithms and programs used to detect (and potentially correct) problems found after the chip physical design is completed.

The benefits of using a design system and the associated standards do not necessarily pertain to increased circuit density and enhanced performance; indeed, circuit density and/or performance may suffer. Rather, the motivation is to ensure a manufacturable, testable design. CAD tools are commonly used to reduce the time required to perform the design verification tasks. Recall that one of the obstacles to VLSI technology development was the need for more specialized applications of integrated-circuit technology to provide increased revenue; to enter the market for a new application in a timely, cost-effective manner, multiple hardware design passes and long design cycles must be avoided. Often, maximizing circuit density and performance are only secondary goals. Only after initial market entry may subsequent product enhancements require addressing more aggressive circuit density and performance measures to maintain product attractiveness, using an existing design as a foundation. These improvements can also be realized in a design system environment by

1. implementing an existing logic design in a newer technology, and/or
2. customizing and combining the physical circuit design of individual logic gates into a single physical implementation, optimizing the circuit sizes for each specific hand-wired interconnection loading; this handcrafted physical design is then treated as a whole for placement and interconnection on the chip (Figure 1.8).

The customizing process in item 2 may be viewed as providing a higher-level function in the design hierarchy as a single physical circuit design, replacing the individual lower-level elements.

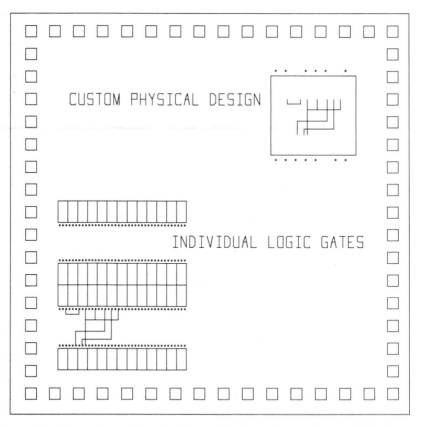

Figure 1.8. Customizing of the physical circuit designs of several logic primitives into a single physical design, where the constituent circuit sizes are optimized for the specific, local hand-wired interconnection capacitance.

PORTLAND COMMUNITY COLLEGE
LEARNING RESOURCE CENTERS

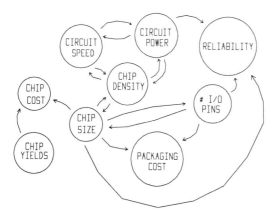

Figure 1.9. VLSI physical design trade-offs in evaluating an integrated-circuit technology.

One area where current design systems provide little support is with respect to front-end tools to assist with the selection of the technology offering best suited for the physical chip design. The electrical and environmental specifications of a VLSI part (and the technology chosen for its implementation) must be compatible with the goals and constraints of the products that utilize the part. The physical design trade-offs (and their interdependencies) in evaluating one technology versus another are illustrated in Figure 1.9. In addition, each technology will differ significantly in the number and complexity of their design system rules. The resource required to complete the physical design (and generate manufacturing tests) will vary widely among technologies and is difficult to quantify.

In short, the problem of technology choice is multifaceted; it is indeed difficult to incorporate into the design flow the tools that would encompass all aspects of the decision. As a result, this text will use a single family of VLSI technologies for illustrating design system concepts when discussing physical design; very little discussion will be given to the merits of one technology family versus another as it pertains to the problem of technology choice. The silicon MOS device family of technologies with two levels of metal interconnect will be discussed and evaluated in some detail; this family is at the forefront of VLSI technology development. Alternative technologies will not be treated with regard to the physical design of logic and/or memory storage circuits or the technology rules required to be incorporated into a design system. The discussion given should serve as a basis for evaluating technology alternatives and making the technology choice.

Summary

The era of VLSI is characterized by the ability to provide a large and diverse function on an individual integrated circuit and the need for aggressive design cycles with manufacturability and testability as major concerns. The key to addressing these needs is the adoption of design standards and the development of design system tools to maximize designer productivity and verify the resulting

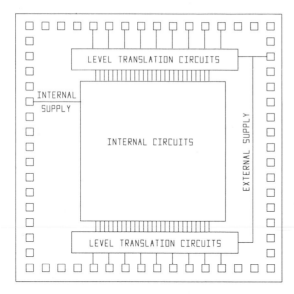

Figure 1.10. Use of multiple supply voltages on a chip; a reduced supply voltage for the internal circuits, and an external supply for interface voltage levels at the chip inputs/outputs.

design against those standards. Yet, as technologies evolve, the technology rules can rapidly increase in number and intricacy, making the enforcement of these rules more difficult. For example, some current technologies require multiple power supply voltages to be distributed internally; some fraction of the circuits may use a reduced supply voltage to reduce the overall power dissipation. A technology rule will therefore require the correct use of logic voltage level translation circuits between networks connected to different supply voltages (Figure 1.10). The development of design system standards is complicated further by the diversity of function that is desirable to implement as a physical element in the overall chip design. An example given earlier was the incorporation of a memory array into the design. When performance and circuit density in a network are of primary importance, it was also indicated that a collective, custom physical design would be optimal. These functions, or *macros*, must be placed and wired on the integrated circuit in conjunction with (potentially) many thousands of single logic gate circuit designs. In addition, models must be developed for these macros for the purposes of logic simulation and manufacturing test. Section 1.3 presents some of the economic considerations driving design system development, specifically with regard to macro-level functions.

1.3 U.S. INDUSTRY AND VLSI: A MATTER OF ECONOMICS

The cost of introducing a VLSI part includes the initial nonrecurring development engineering expense (NRE), as well as the ongoing costs of manufacturing and process engineering, sales and marketing, and field engineering support. Estimates of development costs are typically made in units of person-years, the product of the number of people involved in the development project times the duration of their involvement; a reasonable (perhaps conservative) expense multiplier for project planning would be $250K per person-year. This figure includes both direct and overhead expenses associated with the development group:

Salary, benefits, overtime

Computer and data processing (DP) charges: CPU time, storage

Depreciation on capital equipment

Facility costs: heat, lights, maintenance, office and laboratory space

Administrative overhead: management, secretarial support, nonincome-producing functions (e.g., personnel, finance, payroll, legal and patent operations, cafeteria, recreation)

Taxes

The development effort for a VLSI part can range from less than 1 to in excess of 100 person-years, depending on the uniqueness of function, the performance constraints, and the availability and applicability of design system tools for support. It should be evident that unique, highly optimized, resource-intensive VLSI designs have been pursued only when the expected volume of sales would justify the necessary development expenditure. However, the market for these high-volume, universal part numbers is starting to erode. Instead, application-specific designs are emerging as new markets are being explored; new products, incorporating unique VLSI parts, are being introduced into the automotive, medical, business, and consumer industries, from data/voice communications to graphics animation. The effect of this shift in the electronics market from high-volume, few part numbers to low-volume, multiple-part numbers has had a significant impact on the U.S. integrated-circuit industry; the design cycle time and design resource for new part numbers must be economized, implying increased design system support.

In the past, U.S. companies have been categorized as either captive suppliers (who develop and manufacture parts for use in their own higher-level products) or merchant vendors (who develop and manufacture parts for direct sale to other equipment and product manufacturers). Lately, a change has occurred among both captive and merchant suppliers in an attempt to create additional markets for new part numbers to offset development costs. Captive suppliers are beginning to offer lower-level components (subassemblies, boards, integrated circuits) for sale to other manufacturers; merchant vendors are starting to *vertically integrate*, that is, to incorporate their parts into higher-level components and/or systems (and to provide the software and software development support for those systems). In this regard, the U.S. integrated-circuit industry is starting to more closely resemble the Japanese and European industry structure, where neither pure captive nor pure merchant supplier is common.

In the meantime, a third category of U.S. integrated-circuit manufacturer has begun to evolve, which specifically addresses the low-volume, application-specific IC marketplace—the silicon *design center*. These design centers are geared toward providing small-to-moderate volumes of unique part numbers to outside product groups on an aggressive manufacturing turnaround schedule, with a range of technology and packaging options available. Prospective designers are able to license the technology and circuit design information necessary to be able to submit a part for manufacture. The design centers are primarily interested in highly manufacturable parts, customer designs whose performance objectives are conservative and whose test requirements are straightforward. Little emphasis is placed on failure analysis and yield optimization for a particular part; these are contradictory to the goals of providing functional, single design cycle hardware in low volumes. To enhance this manufacturability, the design systems offered by most design centers tend to restrict circuit design flexibility. In addition, those that offer CAD tools for physical placement and interconnection of circuits often permit only limited customer interaction with the physical design database.

The design centers are striving to support more aggressive and diverse design requirements to attract a wider customer base; the captive and merchant suppliers are striving toward reducing the development resource required to implement unique, optimized functions for their product markets. As a result, design system development is being driven to provide chip designs with near-custom function and performance, as well as a high confidence level for the resulting manufac-turability. Design systems are therefore evolving to support greater utilization of medium- and large-scale macros; the initial resource invested in the logic design, circuit design, and modeling of these higher-level functions is returned when the macros are later incorporated into multiple chip designs. The key to this evolution is developing the design system tools to be able to verify the correct use of a variety of macro offerings to maintain overall manufacturability.

1.4 SYSTEM DESIGN FOR VLSI

VLSI parts are truly system-level parts; the function incorporated onto a VLSI chip will undoubtedly have a major impact on the features of the final product that uses that part. Indeed, the availability of VLSI hardware has fostered in-creased interest in a variety of computer system architectures, from parallel to pipelined execution of instructions, from on-chip high-level programming lan-guage support to reduced instruction set implementations. This section will not discuss any of these topics; rather, it will describe some of the system product design considerations that must now be evaluated at the integrated-circuit design level. The two areas to be highlighted are (1) design hierarchy, expanding on the introduction in Section 1.1, and (2) product reliability and serviceability.

Design Hierarchy

The decomposition of a design into hierarchically organized constituent components was described earlier as a means of more easily managing a complex design task both conceptually and for the purposes of design verification (Figure 1.11). There are three aspects to the hierarchical design database of a VLSI part, and they are closely interrelated.

The *logic model* of the part describes its function for the purposes of logic simulation. The simulation procedure *expands* the logic design representation down to the interconnection of logic primitives, nodes in the logic hierarchy where output values can be calculated directly from input stimuli (Figure 1.12). Asso-

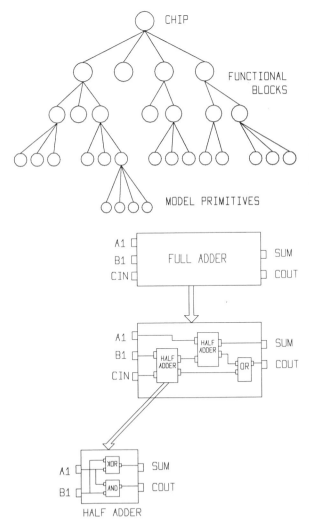

Figure 1.11. Decomposition of a design into a hierarchically organized database; the chip, functional block, and model primitive designations apply to the logic and test models of the design.

Figure 1.12. Logic model expansion of full-adder and half-adder logic functions into their logic primitives.

ciated with each of these primitives is a delay value (or values) for scheduling changes in future simulation time, corresponding to changes in the input condition at the current simulation time. (Techniques for describing the logic design hierarchy and representing the logic design database for simulation are described in Chapter 4.)

The *test model* of the part describes the model of nets internal to the part for test pattern evaluation. If logic simulation patterns are to be used for manufacturing test, the test and logic models of the part are identical; if test diagnostics are required, the test model may be considerably more elaborate than the logic model, expanding a logic simulation primitive further into a decomposition suitable for isolating a faulty from a good network (Figure 1.13).

The third model is the *physical model* of the part, that is, the description of individual physical circuit designs, their location, and the wires that interconnect them. Each circuit design is realized by a collection of two-dimensional shapes of particular geometries and orientations. These shapes define the devices, the contacts to the device terminals, and their local interconnection to form logic circuits. The collection of shapes that constitutes a physical circuit design may have its own internal hierarchical decomposition; this structure is used during *layout* to more efficiently develop and describe the complete physical design. The entire collection of shapes (i.e., the physical macro) is regarded as a unit when placing the circuit design on the chip in correspondence to a logic design function (Figure 1.14). The physical design hierarchical representation is continued above the macro level to encompass the entire chip design. The individually placed physical circuits (with their associated chip locations) are combined with the *global* interconnection wiring and via shapes to complete the chip physical design. A means of representing the chip physical design database is described in Chapter 5.

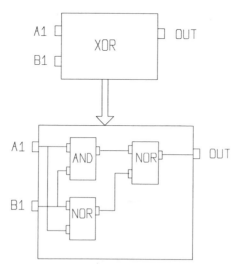

Figure 1.13. The test model may contain further expansions from the logic model. For example, the logic simulation program may interpret the XOR function as a logical primitive; however, the test model may require a further decomposition of the function into the constituent physical circuits.

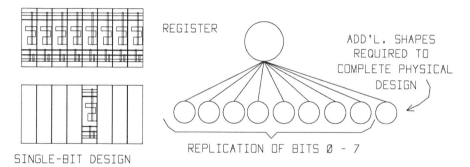

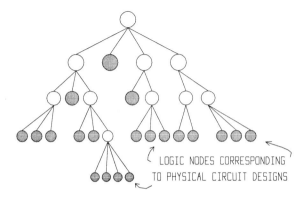

Figure 1.14. The physical design of a logic function may also be organized hierarchically; that is, the collection of shapes that defines the devices, contacts, and wires may be comprised of smaller units (which are then translated and replicated to realize the full logical function).

Each physical circuit design on the chip therefore has a corresponding node in the logic design hierarchy. The logical-to-physical agreement between the simulation model and the resulting physical design requires that each branch of the logic model, as traversed from the top node, eventually encounters a node corresponding to a circuit placed on the chip (Figure 1.15). In no case should more than one physical circuit design be encountered in traversing a branch of the logic design hierarchy.

The function of a design system CAD circuit placement tool is to traverse the logic design database, calling forth the appropriate individual circuit designs and optimally locating them relative to one another to facilitate the efficient routing of the interconnection wires between them; a wiring tool finds those routes and provides the coordinates for adding wires to the physical design database.

Figure 1.15 is a bit misleading in indicating that all physical circuit designs on the chip correspond to logic simulation primitives. The function of the physical circuit may require a further logical expansion into primitives that the simulator can evaluate (Figure 1.16). In this case, the physical circuit design has a logical

Figure 1.15. The logic functions present in the logic model must agree exactly with the physical circuit designs placed on the chip. The shaded nodes in the logical description correspond to equivalent physical designs. Each traversed branch of the logic model should encounter one and only one physical design.

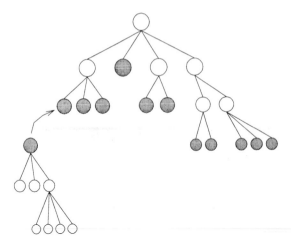

Figure 1.16. The physical circuit design may include a further logical expansion model to provide a description that the logic simulator can interpret. Commonly, physical *macros* include an expanded logic model.

model directly associated with it that must be appended to the logical design database in order to be able to simulate the function of the network. The assignment of simulation delays to the expanded primitives of the logic model must be done in a judicious manner so as to agree with the actual circuit delays. Note that this situation will commonly arise with the incorporation of macros, physical circuit designs whose function may not be directly interpretable by the simulation program. Note also that it is critical when developing the macro circuit design that the physical circuit and its logical model be in agreement with regards to function.

      Figure 1.17 illustrates the case where the logic function being simulated corresponds to more than one physical circuit design. If the simulator has the capability to accept a higher-level description of a network, this description may

Figure 1.17. Hierarchical structure where a logic simulation model may correspond to multiple physical circuit designs. The logic simulator in this case must be sophisticated enough to incorporate a *behavioral* description of a network, in addition to the direct interpretation of logic model primitives. Design system support is often provided to construct the hierarchy below the behavioral model automatically (including the physical circuit technology assignments). The resulting physical circuits and their interconnection as determined from the behavioral model are said to be *correct by construction.*

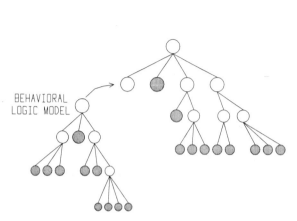

BEHAVIORAL
LOGIC MODEL

supersede the interpretation of its constituent primitives. As before, the delays used by the simulator must be verifiable against the actual circuit design; nevertheless, the benefits in simulation productivity can be substantial. In the case depicted in Figure 1.16, the physical macro circuit design is used to create an expanded model for simulation; this model must be verified logically-to-physically against that circuit design. For the example in Figure 1.17, however, it is necessary to build a valid higher-level *behavioral* description from the interconnection of a number of physical designs, typically a much more difficult task to accomplish. Rather than develop a high-level behavioral simulation model from individual circuit designs in a bottom-up fashion, it may be preferable to generate the physical circuit description *from* an initial high-level logic description in a top-down manner. Design systems are beginning to incorporate physical design *synthesis* tools to produce an equivalent decomposition. The simulation productivity is maintained by using a powerful logic model, and the check for validity of the resulting physical network can potentially be incorporated into the synthesis tool. Such a top-down design flow is often deemed to be *correct by construction*. This correctness is maintained only if the performance of the resulting physical network agrees with that of the behavioral model and if no changes are made in the physical design once decomposition has been performed.

Developing a hierarchical design database permits greater manageability of the total design, specifically enabling correspondence among the logical, physical, and test models to be more easily verified. It encourages the development of tools to generate detailed physical circuit implementations from productive behavioral-level simulation models. It also supports the incorporation of macro designs with logic model expansions.

Should the hierarchical representation of a product design terminate at the chip level? For the same reasons just described, the answer is no. System simulation also requires a logic model expansion of the constituent parts; this expansion uses the detailed logic models of the individual components and is also hierarchical (Figure 1.18a). (Ideally, if any VLSI hardware representing a component of the system is already available, the part can perhaps be incorporated *electrically* into an otherwise *software-driven* system simulation. Design systems are indeed beginning to provide simulation capability to electrically configure physical hardware into the system model.) Also, as newer, denser technologies become available, individual parts may be redesigned to incorporate more of the overall system function, reducing the component count and cost. A hierarchical system representation provides the database to support this component integration.

Increased levels of component integration, circuit density, and function in current VLSI chips have also accentuated the need for good design documentation. In this regard, a hierarchical database and engineering graphics go hand in hand. The decomposition of a design element can easily be illustrated graphically (as in Figure 1.11); more importantly, functional-level block diagrams are an inherent part of the design database and therefore the design documentation (Figure 1.18b).

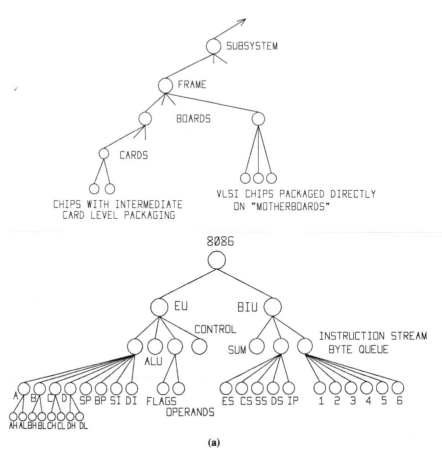

Figure 1.18. (a) The hierarchical representation can continue to be used throughout the system description, from the individual chip designs to the highest levels of system function. (b) The hierarchical description of the design database easily lends itself to graphical documentation, as illustrated here for the Intel 8086 microprocessor. Such documentation is often provided with the assistance of engineering graphics workstations. (Block diagram from *Intel 8086 User's Manual*, p. 2-1. Copyright © INTEL Corp. Reprinted with permission.)

Reliability and Serviceability

The most unreliable aspect of any system design is the physical connection. Component integration has replaced printed circuit cards full of MSI and LSI logic components with a single VLSI part. The connections between circuits are now implemented in patterns of thin film wires on the chip and are not subject to problems with mechanical vibration, incomplete insertion, and dirty contacts that plague the circuit card packaging level of the system. As such, the VLSI part *should* be considerably more reliable and, in many cases, that is indeed true. Yet,

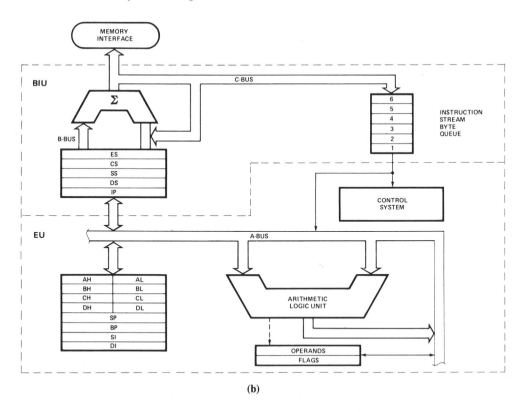

(b)

as VLSI parts have replaced cards full of logic, the parts are now being incorporated directly onto the system motherboard. A failing VLSI part can result in an expensive, difficult service problem to isolate and fix, particularly if the lowest-level *field-replaceable unit* (FRU) is the motherboard itself. Unlike the logic on the card, the VLSI part cannot be probed internally to isolate and correct a problem. The individual part may be more reliable than the card, but the cost of failure detection and correction is considerably more severe, not to mention the consequences of system downtime. As a result, the failure modes of a VLSI part (corrosion, electrical stress and breakdown, thermal and environmental stress) are extremely important to isolate, characterize, and quantify for each new technology.

The direction in system design, motivated by the availability of VLSI hardware, is to concentrate on increasing product reliability: higher percentage system up-time, lower service and maintenance costs, greater dependency on the product user (and the product itself) to employ diagnostic tests for failure isolation. Integrated-circuit designs are now beginning to incorporate redundant networks and diagnostic circuitry for chip-in-place self-test. Redundancy implies multiple, identical networks whose outputs are compared; if a disagreement is detected, the correct value may be determined by a majority vote of the networks and/or an error signal is raised. Chip-in-place self-test requires additional circuitry to be able to internally sequence through a number of states with known and fixed network inputs, compressing the network output values into a means suitable for making a final judgment as to the validity of the function of the network. Design system standards for redundant and self-test networks typically have not been well defined to support these unique design criteria. Four key areas yet to be addressed (in a design system environment) are as follows:

1. Developing a measure of testability and effectiveness of test pattern coverage for redundant networks.
2. Providing standards for implementing self-test data compression.
3. Providing simulation support for redundant networks with faults (manually or automatically) introduced.
4. Providing test diagnostics for failure analysis (for both redundant and self-test networks).

One last comment about reliability: the reliability of a new VLSI technology is extremely difficult to quantify. The effects that accelerate device and circuit degradation over time increase in significance at smaller lithography dimensions and increased circuit densities. A VLSI technology fabrication process is a dynamic set of manufacturing steps that change as a reliability *learning curve* develops. Part of that learning curve involves isolating the important failure mechanisms; the other is modifying the process steps and/or the testing procedures to prevent *infant* failing parts from reaching the final product. A severe failure mechanism may require modification to the technology rules, which may significantly ripple back through the design system, affecting circuit designs, models, and CAD tools. During this ongoing churning of manufacturing process and technology rules, the product designer must choose when to commit to this technology the design of a new part or the component integration of an existing design (while maintaining an aggressive development schedule). The benefits of early entry into a new technology are product performance and function; the major disadvantage is uncertainty in the resulting product reliability. Opting for a more *mature* technology may result in a smoother design flow and a more reliable product, while possibly sacrificing performance and circuit density. Another major factor influencing this decision is the manufacturing cost learning curve, which decreases with time; as a technology matures and yield (and reliability) measures increase, the manufacturing costs decrease. Recall the point made in Section 1.2 about the difficulty in providing tools to support the technology choice; these factors make that decision all the more difficult to "automate."

VLSI technologies will continue to have a major impact on system and product design. To date, this era has been characterized by the decentralization of the function and performance provided by VLSI hardware into a wider variety of more *intelligent* products with which a product user interacts. As a result, new applications markets are evolving. This trend is likely to continue for some time.

1.5 FUTURE DIRECTIONS

Many future technology and product directions have already been described in previous sections of this chapter. From the product viewpoint, increased emphasis on application-specific markets will emerge. From the technology arena, increased component integration will result from the capability to achieve better circuit densities and performance. Design systems will provide wider CAD tool support to accommodate greater diversity of function and more intricate technology rules. To date, these trends have been occurring primarily in the logic design environment. In many product applications, this component integration trend has reached the interface between the digital and analog functions of the system. A key future direction is to coordinate the technology, the design system, and the product definition to provide combined analog and digital function on the same chip. This involves providing the macro-level support for functions such as A/D and D/A converters, comparators, single-ended and differential amplifiers, oscillators, and filters. The incorporation of analog function on a logic part also requires the design system standards to ensure correct use of the function, as well as the ability to create valid simulation and test models of the logic.

The unique design requirements of analog circuits will also require substantial design system support, for example, additional supply voltage distribution, small signal device modeling, and emphasis on tight parameter tolerances and strong correlation between the characteristics of *matched* devices and circuits. The roadblock to the increased utilization of analog-logic parts is the manufacturing process modifications required to provide a device set optimized for analog design considerations; these modifications imply greater resulting cost. Nevertheless, combining analog and digital function in the same part or package can physically permit the electronics to be closer to the transducer and thereby reduce susceptibility to noise in the analog signals (Figure 1.19). The ultimate goal is to

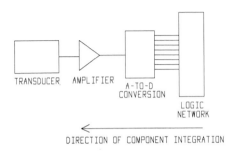

Figure 1.19. One of the future directions in component integration is the development of VLSI hardware containing both analog and digital functions.

be able to integrate electronics and transducer on the same monolithic part in an optimized package.

1.6 IMMEDIATE AND FUTURE PROBLEMS

The previous section indicated some future directions for VLSI product and technology development. One item was the goal of synergism of chip, package, and product. A more immediate problem is to be able to ensure that the technology and design system can accommodate a wide variety of packaging options. VLSI package types that may have unique leverage in a particular product design include multichip and hybrid packages, surface mount packages (no pin-through-plated via hole in a circuit board required), and direct chip-on-board attach. Each of these types has requirements on the chip pad design for the chip-to-carrier connection; some of these restrictions are with regard to minimum pad size, pad spacing, minimum and maximum spacing from the edges and corners of the chip, and the minimum distance from pad to adjacent circuits. The design system tools should ensure that the physical design adheres to these rules for the selected package type. In addition, the system designer may require a particular logic signal-to-pin number assignment at the package level; a design system tool should verify the assignment of the signal to associated pad at the chip level.

Closely related to the problem of supporting the use of a growing number of package options is the problem of the large electrical power dissipation of VLSI chip designs. To ensure reliable operation, the temperature at the device level must be maintained below an operating range maximum value. The system designer must perform a thermal analysis of the product environment, available airflow (or some other means of heat extraction), and a model of the thermal resistance of the chip, chip-to-package attach, and the package. A significant problem is the mismatch in thermal expansion coefficients between the chip, package, and board at operating versus ambient temperatures. This mismatch results in mechanical stresses and a potential long-term reliability problem. The VLSI packaging development effort is striving toward better matching of expansion coefficients, reduced thermal resistances, and more efficient cooling methods to permit greater power dissipation density as VLSI parts are closely packed together on system motherboards. The growing applications market for portable and battery-operated products is also forcing VLSI technology development and the circuit design techniques used to pursue very low power implementations.

Packaging development is also actively pursuing increased pin count packages, a significant problem for increased component integration. The system designer must partition the logic design to ensure that the number of required input/output connections and the resulting chip size do not exceed the limitations of the package. Often, to remain at or under the pin count limit, the logic design and system architecture must be substantially altered. Techniques such as time multiplexing of address and data signals on a set of pins and *common input/output pads* are used (Section 11.8), with considerable impact on the overall performance of the system.

As always, designing for manufacturing reliability remains a high goal and an elusive problem. A high measure of testability for VLSI designs is becoming more difficult to achieve as the number of internal circuits increases exponentially while the observability (as represented by the number of I/O pads) does not. This problem, as it affects the development of design system standards, logic design, and circuit design, will be addressed repeatedly throughout this text.

PROBLEMS

1.1. The means used for describing the evolution of integrated-circuit technologies in Figure 1.2 was the number of logic blocks present on the part. Yet, as described later in the chapter, a greater diversity of functions (logic and memory) is being incorporated onto VLSI hardware.

    (a) Develop an improved scheme for categorizing and comparing hardware complexity other than those mentioned in the chapter; include a combination of the criteria given in the chapter, the physical design characteristics of Figure 1.9, and/or criteria of your own. (Another criterion to consider, which has also been used in the technical literature, is the number of *hand-drawn* transistors in the design, with reference to the complexity of incorporating macros into the physical design.)

    (b) Research the technical literature for descriptions of recent VLSI hardware designs and evaluate the complexity of those designs using the method developed in part (a). What was the minimum lithography dimension used in fabrication for those designs?

    (c) Enhance the scheme or method developed in part (a) to include the possibility of analog macros present on the part.

1.2. Research the currently available VLSI packaging options, specifically including pin-grid arrays and leaded or leadless chip carriers. Compare and rate these options against the following criteria:

I/O count

Ability to automate assembly

Package-to-board attach techniques

Soldering techniques for package-to-circuit board trace connection

Ability to probe and rework boards containing the package

Size

Cost

1.3. Using mail-order advertisements from back issues of popular magazines in the computing and electronics fields, plot the purchase cost over time for several VLSI part numbers; include a wide diversity of hardware types: memory, microprocessor, and microcontroller. Also include several analog parts of moderate complexity. Indicate any overall trends and differences between the cost curves of the various types of parts.

1.4. Research and develop definitions for the following terms:

Lead time
Book-to-bill ratio
Second source
OEM
JEDEC standard
JAN (or MIL) spec
Radiation hard

1.5. (a) Research and develop a list of the ten largest U.S. semiconductor manufacturers for the most recent year for which data are available. Classify these manufacturers as captive suppliers, merchant vendors, or both.

(b) Develop a similar list for the top ten semiconductor corporations in both the Far East and Europe.

(c) For each captive supplier from parts (a) and (b), determine the range of product markets in which these companies are participating. Compare the diversity of products offered by U.S., Far East, and European captive suppliers.

(d) How many of the captive and merchant suppliers on the lists in parts (a) and (b) are also providing *design center* services?

REFERENCES

1.1 Moore, Gordon, "VLSI: Some Fundamental Challenges," *IEEE Spectrum* (April 1979), 30–37.

1.2 Moore, Gordon, "Are We Really Ready for VLSI²?" *IEEE International Solid State Circuits Conference* (1979), 54–55.

Additional References

Allan, Roger, "VLSI: Scoping Its Future," *IEEE Spectrum* (April 1979), 30–37.

Evanczuk, Stephen, "Special Report: Workstations—Integrating the Engineer's Environment," *Electronics* (May 17, 1984), 121–139.

Faggin, Federico, "How VLSI Impacts Computer Architecture," *IEEE Spectrum* (May 1978), 28–31.

Gossen, R. N., and Heilmeyer, G. H., "100 000+ Gates on a Chip: Mastering the Minutia—(I) The 64 Kbit RAM: A Prelude to VLSI, and (II) Needed: A 'Miracle Slice' for VSLI Fabrication," *IEEE Spectrum* (March 1979), 42–47.

Heller, William, "Contrasts in Physical Design between LSI and VLSI," *18th Design Automation Conference* (1981), 676–683.

Intel Corp., "HMOS-III: Continuing the Legacy of Economy, Performance, and Availability," *INTEL Solutions* (May/June 1982), 10–11.

Iversen, Wesley, "Captive Semiconductor Facilities Are Gearing up to Compete against Established Merchant Suppliers," *Electronics* (May 19, 1982), 133–136.

Keyes, Robert, "The Evolution of Digital Electronics towards VLSI," *IEEE Journal of Solid-State Circuits*, SC-14:2 (April 1979), 193–201.

———, "Fundamental Limits in Digital Information Processing," *Proceedings of the IEEE*, 69:2 (February 1981), 267–277.

Ousterhout, J. K., and Ungar, D. M., "Measurements of a VLSI Design," *19th Design Automation Conference* (1982), 903–908.

Posa, John, "Superchips Face Design Challenge," *High Technology* (January 1983), 34–42, 86.

Puthuff, Steven, "Technical Innovations in Information Storage and Retrieval," *IEEE Transactions on Magnetics*, MAG-14:4 (July 1978), 143.

Robson, Gary, "Benchmarking the Workstations," *VLSI Design* (March/April 1983), 58–61.

Stapper, Charles, "Defect Density Distribution for LSI Yield Calculations," *IEEE Transactions on Electron Devices*, Correspondence (July 1973), 655–657.

Texas Instruments, Inc., VLSI Laboratory, "Technology and Design Challenges of MOS VLSI," *IEEE Journal of Solid-State Circuits*, SC-17:3 (June 1982), 442–449.

Verhofstadt, Peter, "VLSI and Microcomputers," *IEEE International Computer Conference* (Spring 1978), 10–12.

Werner, Jerry, "Sorting out the CAE Workstations," *VLSI Design* (March/April 1983), 46–55.

2

Design System Concepts

A library is thought in cold storage.

—Lord Samuel

2.1 DEFINITIONS

This section will provide brief definitions and descriptions of the terminology used in subsequent discussion of design system development.

Methodology. The dictionary definition of *methodology* is "a system of principles or rules." In the context of VLSI design system development, a methodology is provided to guide the design flow of the part in proper sequence and using the appropriate tools, subject to the design system standards and technology rules.

Rules. The term *rules* has been used previously to refer to limitations imposed on the design by a particular technology. Many different types of rules govern the design of a VLSI part, and the term will be used rather broadly. For example, there are technology *shapes* rules that provide the dimensional requirements (minimum widths, spaces, and overlaps) for the orientation of the shapes that form the circuit designs. There are circuit fan-out rules, as described earlier. Indeed, a number of *usage* rules may need to be asserted, including the following:

1. Limitations on *dotting* of logic block outputs to effectively increase fan-in and/or provide a logical function with a simple interconnection (Figure 2.1).
2. Logic voltage interface level checking between the values provided at circuit outputs and required at circuit inputs (if the circuit designs used in a particular technology include multiple interface levels).

From the previous examples, it is evident that the greater the need to use the features of the technology to the fullest extent (to maximize performance and circuit density), the greater the number and intricacy of the rules required to assure those features are not incorrectly implemented and hence the broader the support required from the design system tools.

Pin.   A *pin* usually refers to the physical interconnection on the chip package; it will also be used in a very general sense to refer to an input or output connection of a circuit design (i.e., internal circuits also have input and output "pins" to be interconnected). The connection is implemented with shapes on wiring and via levels of the design; these shapes must suitably overlap the *pin shapes* of the circuit design to form a continuous electrical connection (Figure 2.2).

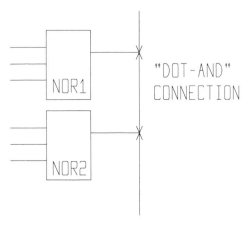

Figure 2.1.   *Usage rules* must be developed to define the possibilities and/or limitations on the *dotting* of outputs of different circuits (used to effectively increase fan-in or implement a logical function with a simple interconnection). In the example illustrated, the effective fan-in of two three-input NOR gates is increased to six by the *dot-AND* connection:

$$\overline{A + B + C} * \overline{D + E + F} = \overline{A + B + C + D + E + F}$$

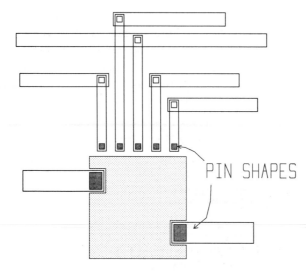

PIN SHAPES

Figure 2.2. Pin shapes included with each physical circuit design must be suitably overlapped by the interconnection wires between circuits. This ensures that the pins of the individual circuits and the global wires make a continuous electrical connection.

Net.   A *net* refers to the collective set of all pins and their interconnections forming a single node that can be assigned a logical value during simulation (Figure 2.3).

Block.   A *block* is a term commonly used to describe primitive components of the logic simulation (or test) model of a design; the term is synonymous with that of a logic *gate*. In a more general sense, it is used to refer to *any* node in the hierarchical logic model, as in a "functional block diagram" of the highest levels of the design.

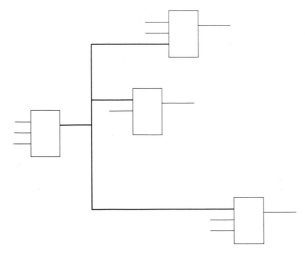

Figure 2.3. A *net* is the collective set of all interconnections between circuits that can be assigned a unique logical value during simulation.

Image.   To facilitate the development of design system tools to assist with the placement and wiring of physical circuit designs, it is necessary to develop a *chip image* for the technology. The image is the description of how the chip area is to be allocated for the location of circuits, wires, and input/output pads. Most design systems (and their tools) use relatively structured images; the area allocated for circuits, wires, and pads is all fixed, as is the total chip size. Other, less structured design methodologies provide more flexible utilization of chip real estate, permitting the wiring area to expand and contract around circuit designs as required (Figure 2.4).

A key consideration in designing the chip image is the technique used for power supply distribution to the circuits. For the structured image with fixed area allocation, the power supply rails may be included as part of the *background* design (Figure 2.5). The physical circuits can then be designed with knowledge of the locations where the rails may be contacted; when the circuit is assigned a proper coordinate location and the background image and circuit design shapes are merged, connection to the supply voltage is provided. The predefined rails in the image prevent (block) the use of wires for interconnection in the circuit design from utilizing that area (Figure 2.6).

A less structured image, with more flexible circuit and wiring area allocation, does not typically contain predefined power supply rails throughout the image. Rather, the routing of the power supply distribution among circuits is included in the responsibilities of the CAD wiring tool. The physical circuit designs may contain multiple access points for the entry (and possibly exit) of the supply voltage; the design may provide a *wire-through* for the continuity of the supply voltage connection (Figure 2.7). A major difficulty in providing this flexibility is the requirement that the supply interconnections be of adequate size to minimize

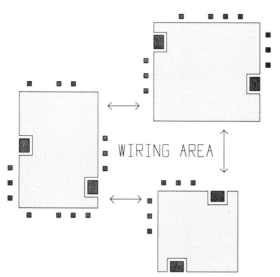

Figure 2.4. A flexible chip image may tentatively place individual circuit designs, allowing the wiring area between circuits to expand and contract as required to accommodate the necessary global wires.

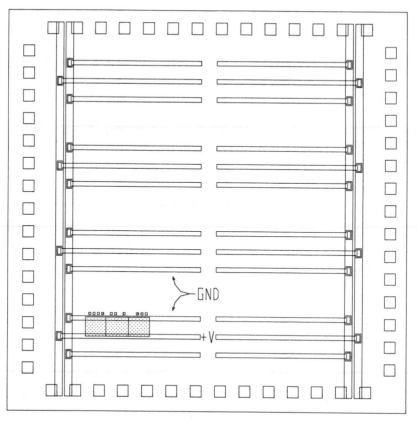

Figure 2.5. A *background image* for a structured design, with fixed supply rail locations, wiring area, and pad I/O. The triplet of rails permits the circuit designs to be placed back to back, sharing the middle power supply rail.

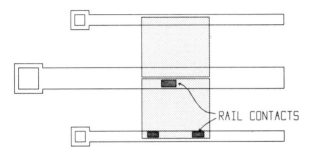

Figure 2.6. The predefined power supply rails in the structured image are properly contacted by the circuit designs; they also block the use of circuit wires.

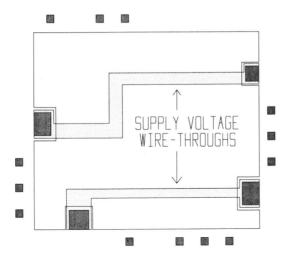

Figure 2.7. The physical design in an unstructured image does not use predefined supply rails, but rather pin shapes for the routing of power supply connections. The design may also contain *wire-throughs*, permitting multiple entry and exit points for the supply connection.

resistive and inductive losses. The wiring tool would therefore have to accommodate generating wires of different widths for signals and supply voltages; another tool would be required to evaluate the number (and dissipative power) of the circuits connected to a particular branch of the supply voltage to determine what that specific width should be. Of special concern is the sufficiency of the power supply distribution in the case of a wire-through. Since the connection is part of the circuit design, the wiring program has no control over that branch of the connection when the design is placed. For a structured image, the worst-case number of circuits and switching current on each branch can be estimated, the appropriate supply line widths can be calculated, and a single (albeit pessimistic) power supply distribution image designed. This single image would then be used with all design system designs for the particular technology.

Other benefits of structured images can be derived from the fixed pad locations; for example, the required number of different test fixtures for probing at manufacturing test is minimized, thereby reducing the tester setup and debug time. Likewise, the ability to automate the chip-to-package bonding operation is enhanced. The sacrifice made is the ability to better utilize the allocated pad area for circuits and interconnections, if pads can be displaced (or deleted).

Chip images will be described further in Section 3.2.

Cell.   The term *cell* will be used in two contexts, one pertaining to image design and one pertaining to physical circuit design. With regard to image design, a cell is the unit of area allocated for placing circuits. In other words, the size of a circuit design is usually specified as a number of cells in the *x*-direction by a number in the *y*-direction. For structured images with fixed power supply distribution, a limit of one or two is typically placed on the extent of the circuit design in one of the dimensions (Figure 2.8). The chip capacity is commonly specified in terms of the number of cells available for placing circuits. The multiple-cell

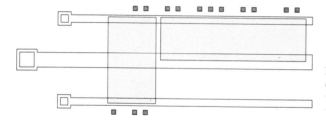

Figure 2.8. Multiple-cell circuit designs, one and two cells high in the vertical direction, with a variety of widths in the horizontal direction.

macro design will likely provide greater circuit density than the equivalent inter-connection of single-cell (logic primitive) designs. Figure 2.9 illustrates a logic function of five gates realized in a two-cell physical design. As a result, it is also common (and somewhat confusing) to specify the design system chip capacity in terms of an "equivalent number of logic primitives," where this equivalent is *greater* than the number of cells. For example, a structured image chip capacity of 5000 cells may be specified as capable of containing a design of "10,000 two-input NAND gates." The ability to actually realize that quantity of logic on the design system chip depends to a large extent on the percentage of multiple-cell macros used. The entire area of the structured image is usually not subdivided into cells; the dedicated wiring area for interconnection is described in different units, as will be discussed shortly.

One might wonder why the quantization of circuit area is made in the first place, why the cell granularity is imposed. Circuit designs that implement different functions (or the same function at different performance levels) do not all occupy the same area. Why not try to maximize circuit density, foregoing the units of cells in favor of the more precise physical coordinates that describe the design? The answer is in order to achieve a high level of designer productivity. Recall from Chapter 1 that an ultimate goal of a design system is to lessen the impact of iterations in the design cycle, occasionally at the expense of circuit density and/or performance. Consider the example of a chip design that has completed the physical design step, after which the situation develops that a minor logic design change is required; the time and resource impact of implementing the change can be minimized if the flexibility exists to *locally* make changes to the

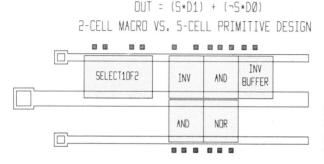

OUT = (S*D1) + (¬S*D0)

2-CELL MACRO VS. 5-CELL PRIMITIVE DESIGN

Figure 2.9. The multiple-cell macro can provide a greater circuit density than the equivalent implementation using logic primitives.

existing physical design without beginning anew. Quantizing the area of different circuits into identical units often permits minor changes to be embedded without disrupting surrounding designs. In Figure 2.10, a circuit is replaced by another with a different pin count; for this case, the *x*-dimension of the cell area was chosen to accommodate up to four pins. It is common to define one of the cell dimensions in terms of the number of pins to be wired. (Refer to the definition of *wiring pitch* later in this section.)

The other commonly used definition of the term cell pertains to the physical design of the shapes that realize a circuit design. In Chapter 1, it was indicated that the physical design data for a chip are also organized hierarchically, in much the same manner as the logic design description. The collection of shapes for each circuit may also be structured as a hierarchical composition of groups of shapes. Each node in the *physical* design hierarchy is commonly referred to as a cell, from the smallest unique group of shapes to the complete chip design shapes data file.

The common use of the term cell in the two instances can be confusing. Each image cell has a unique location; each physical design cell has a unique cell name and, if used multiple times, possibly many cell locations. A node in the

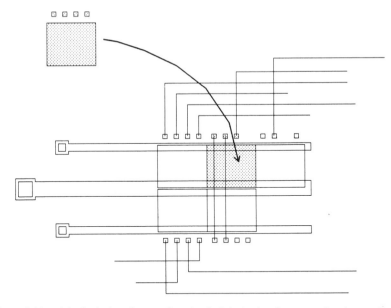

Figure 2.10.   A logic design change after physical design has been completed may often be implemented with only minor modifications to the physical design. In this situation, a single design is to be replaced with one of a different pin count. The allocated area is the same for both designs, as the cell width was defined to contain up to four pins. The changes to the wiring necessary to satisfy the logic design change should be contained locally, without a significant impact to the performance. An engineering graphics workstation can be extremely useful in making the cell replacement and wiring modifications.

physical design hierarchy may correspond to a circuit design with an area greater than or equal to one image cell or may only be a constituent part of a circuit design.

Wiring Bay, Wiring Channels, and Wiring Pitch.   On a structured image, areas are allocated strictly for wiring; these regions are commonly not divided into cells, as are the placement locations for circuit designs. Instead, the area should be denoted in terms of the maximum number of wires available for interconnecting pins. Figure 2.11 illustrates an area between rows of cells to be used for routing interconnection wires (on the first metal level); this area is denoted as the *wiring bay*. The capacity of the wiring bay is specified in terms of the number of *wiring channels*. A key dimension in evaluating and comparing technologies is the *wiring pitch*, that is, the sum of the minimum wire width and spacing permitted by the technology fabrication rules (Figure 2.12).

If the manufacturing technology provides multiple metal interconnection planes (two will be assumed), both interconnection planes have channels available for routing. A grid of wiring lines can therefore be developed (Figure 2.13). There are several features to note about this grid:

1. The wires on the two planes are assumed to run orthogonally.
2. The grid need not be square. (More than likely, the fabrication rules for the wiring pitch on the two planes will differ; the higher-level planes will commonly have a larger pitch.)

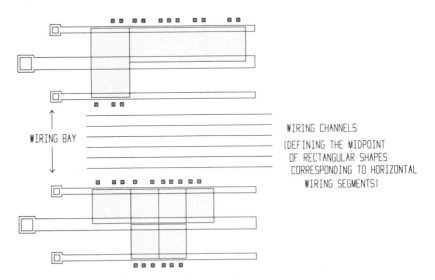

Figure 2.11.   The area between rows of cells in the structured image is called the *wiring bay*; in the horizontal (first metal) direction, it can contain a limited number of wires, fixed by the image definition.

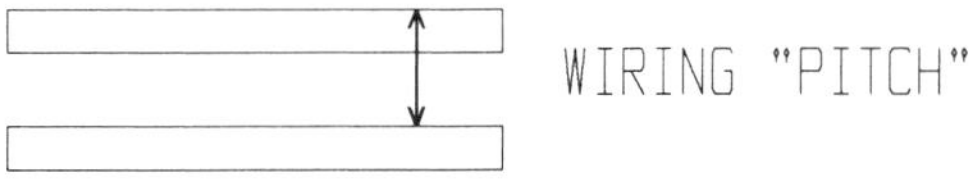

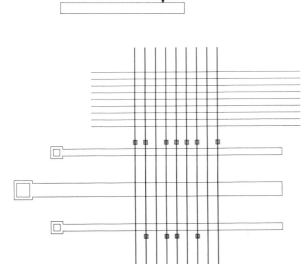

Figure 2.12. The width of the wiring bay is the number of *channels* times the wiring pitch; the pitch is the minimum width plus space as defined by the technology rules. It is a common dimension used in comparing technologies.

Figure 2.13. A two-level metal technology permits the construction of a wiring grid, with wires on the two planes running in orthogonal directions. Note that the vertical (second metal) pitch is greater than the horizontal pitch, from the technology design rules. There may also be technology rule restrictions on the via pitch, which will add complexity to the routing tool.

3. The manufacturing technology must provide for vias to be added at grid points. (When the routing program wants to change direction at a particular point, a via between the two wiring levels must be added. If the fabrication rules do not permit a via pitch equal to the finer grid pitch, additional constraints are typically imposed on the routing algorithms.)

4. Placed circuit designs will typically block the routing of wires on the first wiring plane; there will also be blockages due to the image power supply distribution on both planes. (The routing tool must be able to read in the blocked grid points that describe the image. In addition, if any of the placed circuit designs have constraints on metal lines crossing over, those blockages must also be added.)

The routing tool must also interpret different *costs* assigned to wires on the different planes and to the vias between them; a minimal cost for the sum total of all interconnections between pins is the overall goal. The wiring costs may be based on channel availability and per unit length capacitance. The via cost adder will discourage multiple jogs in routes and, by reducing their number, may tend to enhance the manufacturing yield.

The image definition is closely related to the wiring pitches, the number of circuits to be placed and wired, and the average circuit's number of pins. Enough wiring channels must be included to permit a high percentage of cell occupation with a high confidence that *any* design system logic design will be *wireable*. The ideal situation from a productivity standpoint would be that any design with a cell occupation percentage all the way up to 100% would be fully wireable using a CAD routing tool; a more realistic goal would perhaps be for 90% to 100% occupation, with only a few uncompleted routes requiring intervention. The cell locations, image wiring blockages, and circuit pin locations must all be defined in wiring grid coordinates.

Book.   A *book* is a collection of three things: a logic model, a test model, and the physical circuit design that implements the logic function. All three must satisfy their pertinent design system constraints and specific technology rules. In essence, the logic designer in a design system environment constructs the final chip design in the chosen technology using *only* the previously defined and verified books for that technology. Both macro-level and gate-level circuit designs are referred to as books. The nodes of the logic design hierarchy that correspond directly to the technology's circuit designs call out the books of that technology from a repository of models and shapes data (Figure 2.14). The repository is collectively referred to as the *library* of technology-specific designs.

One point should be made: most (if not all) of the top of the logic design structure to define the chip's function can be developed without reference to a particular technology's books. Indeed, the goal of design productivity encourages a technology-independent design to be developed and simulated to as large an extent as possible. (The circuit performance parameters used for technology-independent simulation are quite coarse, yet they can still be extremely beneficial in the initial functional evaluation of a system/logic design; refer to the discussion in Section 4.1.) Once the technology choice is made, the books from the technology library can be assigned to the logic structure, allowing physical design, test model development, and more detailed simulation to proceed.

The conversion of a technology-independent logic design to a technology-specific implementation is not a straightforward procedure. The technology books will have usage rules that must be observed, constraints that were not applicable when developing a technology-independent logic design structure. Also, it is difficult to develop a logic design that is both technology independent and that utilizes

TECHNOLOGY LIBRARY

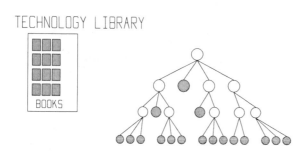

Figure 2.14. The nodes of the logic design hierarchy that refer to physical circuit designs to be placed on the chip extract those designs from the design system technology database, referred to as the *library*. The individual circuit designs are denoted as the *books* of that library and contain shapes data, a logic simulation model, and a test model for the function realized.

macro-level functions effectively; specific macro designs (e.g., memory arrays) are truly only feasible in a limited number of technologies. The decision to incorporate such a function on-chip in many ways dictates the technology choice early in the development cycle. Despite (or perhaps because of) these difficulties, design systems are beginning to incorporate the tools to assist with this assignment or *transformation* process. The goal of the logic transformation procedure is to accept a technology-independent description and assign the books of the chosen technology to the nodes of the logic structure. If the function of the logic node agrees with that of a particular book, the assignment is one for one; otherwise, the function of that node must be *synthesized* by the equivalent interconnection of the books (and their logic models) available in the technology. The technology usage rules must be observed when performing this transformation or synthesis process.

These definitions will be used in subsequent discussions on design system development and in comparison among different approaches to chip physical design in a design system environment.

2.2 STEPS IN THE DESIGN OF A VLSI PART

This section will elaborate on the steps in the design cycle of a VLSI part (see Figure 1.3). Specifically, some of the procedures and options faced in completing each step will be described.

Planning

Before delving into a chip design cycle, a considerable effort must be placed into formulating the plans, schedules, checkpoints, and estimates to assess whether the necessary development resource is available and what the development cost will be. Concurrent with this planning phase, a market analysis should be undertaken to assess what the possible revenue from the part will be (based on the part's projected availability schedule). For a merchant vendor, much of the planning is confined to the chip level; for the captive product supplier, the design cycle for new chip designs must be merged with the plans and schedules for the entire product. In both cases, the timing of market entry is extremely critical; being too late means revenue lost to competitors, while being too early may mean the users of the part or product may not be able, ready, or willing to incorporate it into their application.

The planning phase and market analysis must also share four common specifications, in addition to schedule: performance, function, environment, and cost. The environment specifications should specifically include reliability measures. The performance and function specified may be motivated by the corresponding characteristics of competing hardware or by the need for compatibility with existing hardware (as a cost-reduced replacement). These specifications will serve as an entry to logic design and will help to determine the technology choice for fabrication.

Logic Design and Simulation

Prior to developing a detailed logic structure for the chip design, a common initial step is to perform a *design sizing*, an estimate of the final circuit and pad count, and an assessment of unique macros that may be required to implement the desired chip function. This information can then be used to estimate the chip and package costs, as well as to provide some performance evaluation for the technologies under consideration. Although much of the logic structure can be initiated in a technology-independent manner, it is nevertheless important to make an early design sizing and technology comparison. If a unique macro function is required and is not available in the library for the optimum technology, a design effort to add that book to the library needs to be initiated. If adding that macro function to the technology library is not plausible, the impact of this deadlock should be addressed early in the development cycle.

If the new chip design is based on existing hardware and requires only a technology transfer for component integration and/or cost reduction, much of the logic structure will already have been defined. Logic transformation tools are rapidly being enhanced to be able to discard the old physical design and assign the new technology's books to the structure.

If the design is being developed anew, the logic structure will likely evolve in a combination of top-down and bottom-up fashions. A variety of tools exists within design systems to assist with the task of building the description of the logic structure. It is common to define a text language syntax within the design system to be used for the purposes of describing the logic structure; an example of such a language will be given in Chapter 4. The textual description can then be parsed to create the internal storage database representation of the design. An alternative means for entering the logic structure is through the use of graphic symbols, as supported by most engineering workstations. The graphics files depict the schematics and block diagrams for documentation; an internal representation still serves as the database upon which the design system tools are exercised.

As the top-down and bottom-up structures develop, it is desirable to simulate the nodes in the logic structure for which the behavioral or primitive description is complete. This incremental and independent simulation, although resource intensive, is useful in detecting design errors (at the system and/or block level) relatively early, when such errors are more readily uncovered. The delay information used during this early simulation, prior to physical design, is estimates and may take a variety of forms. One approach is to assume that the delays in logic signal paths between sequential elements are always less that the system clock period; in other words, all the states of the logic simulation output change *synchronously* with the system clock (Figure 2.15). Delays are measured in units of system clock cycles. Another approach, useful for bottom-up structures, is to

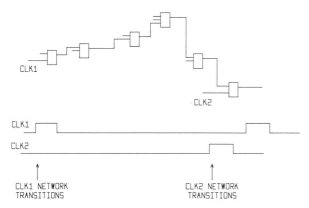

Figure 2.15. Logic simulation may proceed by assuming that the design is fully synchronous; that is, the changes in the output values of all blocks in the network are observed only at transition times of the system clock input.

assign each logic block in a signal path an equal delay. The length of a signal path between sequential elements and the delay of the sequential element are specified as a number of *block delays* (Figure 2.16). (A representative value for a block delay can be used to assign timing information to the simulation output). It is most common to assign specific delays to different nodes in the logic design hierarchy. For a top-down design, the logic designer may wish to use delay estimates for different nodes; for a bottom-up description, block delays with delay adders based on fan-out can be calculated. The detail of timing information used in these early simulation models also translates to the specification of the simulation input stimuli.

Physical Design: Placement and Wiring

In a productivity-driven environment, design system CAD tools are commonly used to assign locations to the collection of technology books called out by the logic design and subsequently to determine routes for the interconnection wires required. On a structured image, those locations are the cell locations of the image, and the wires (and the vias between them) are confined to the wiring grid.

The placement tool can use a number of different criteria in deciding where

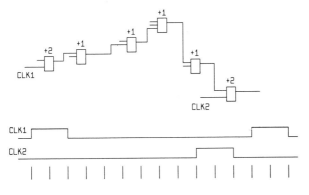

Figure 2.16. Another simulation approach is to assign each gate in a signal path an equal (unit) delay. The system clock period must not be less than the longest delay path in a network; a physical time value may be used for the block delay to provide more detailed timing information.

each book should optimally reside. First and foremost is to minimize the total length of all interconnections, thereby minimizing the average capacitive loading on circuit outputs to enhance the overall performance. Other criteria include the following:

1. Avoiding subsequent wiring congestion.
2. Accepting the designation of some signal paths as *performance critical*, absolutely minimizing these interconnection lengths at the expense of others.
3. Disregarding nonperformance critical signal path interconnection lengths from the scoring method for assigning a placement location.

In addition, the placement tool may accept some preplaced books, circuits affixed by the logic designer to influence the automatic placement of the remaining books. Specific pad locations to which some (or all) of the chip I/O signals are to be assigned may also be provided.

Placement tools usually contain two distinct algorithms, a *constructive* placement and an *iterative* step. The constructive step typically subdivides the chip area into regions corresponding to the higher-level nodes of the logic structure; the size and relative location of the regions are determined by the number of cells required within and the number of wires between the nodes. The iterative placement step attempts to swap and interchange the locations of books assigned during constructive placement in an attempt to further reduce the overall score. The logic designer can also potentially view and assess the merits of individual iterations in the placement procedure.

The wiring tool must input the grid blockages (and any prewired nets) before initiating its path-finding algorithms. As with the placement tool, two steps can be incorporated, *constructive* and *rip-up and reroute*, to strive for 100% wireability. If routes for particular nets (or branches of nets) cannot be found, the wiring output file must so indicate; manual rip-up and reroute of the existing segments must then be addressed. If the design is subsequently deemed to be unwireable, either a different placement, a different chip image, and/or modifications to the logic design are required.

Once the physical design is completed, a key step in proceeding to manufacture is to use a design system tool to calculate more accurate block delays based on circuit type, fan-out, and the *actual* interconnection wiring capacitance. These delays must be *back-annotated* to the existing logic simulation model. This accurate delay information can then be used in place of the assumed delays for logic simulation to confirm the timing behavior. The amount of simulation done after the physical design is complete will vary with the number of performance-critical signal paths in the design and with the accuracy of the original delay assumptions. As illustrated in Figure 2.17, a design system tool can often help to reduce the amount of simulation done after physical design by calculating the slowest and fastest signal path delays in a logic network between sequential elements and comparing those delays against a specified system clock period or

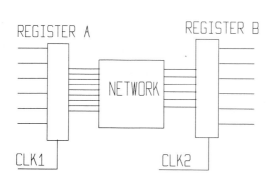

Figure 2.17. A design system *timing analysis* tool may be used in order to reduce the amount of simulation done after physical delays have been calculated and back-annotated to the logical model. This tool would determine the slowest and fastest signal path delays in a combinational network and compare those values against specified timing limits of the system clock distribution. If the register-to-register network delays are not within their bounds, a power/performance optimization step can be initiated. (Refer also to the discussion in Section 4.8.)

assumed network delay value. (Refer also to the discussion of *timing analysis* in Section 4.8.)

If the subsequent analysis indicates that one or more signal paths evaluated with physical delays do *not* meet their timing goals, a means should be provided for enhancing the circuit performance with minimal impact on the existing physical design. In Section 2.1, it was indicated that the quantum of area called the cell was designed to accommodate a variety of book types at the expense of circuit density. In addition, the cell area is chosen and the technology library is enlarged to provide the same book type in a range of *power/performance* measures. Signal path delays may possibly be reduced by replacing the book at a particular location with a higher performance book of the same function. Likewise, the power dissipation should be reduced in paths where the performance goals are being met by a wide margin. To minimize the impact on the existing physical design, each power/performance level of a book type should occupy the same number of cells and have the same pin locations. Indeed, much of the power/performance optimization can be initiated prior to calculating physical delays by directly assigning a performance level from the range available based on interconnection length and specified critical paths. The delay calculation tool then includes the level of the book type as part of its input parameters.

The specification of critical paths by the logic designer prior to placement also typically includes a request for a higher performance level of the book type. Path delay and/or chip power goals that still cannot be met after power/performance optimization will require modifications to the physical design, image, or logic design, much like the case for an unwireable part.

Testing

Although the design cycle flow chart of Figure 1.3 illustrates the generation of test patterns and test specifications as a sequential step after physical design, much of the procedure may proceed in parallel with the physical design. The test pattern file may contain a combination of automatically generated patterns and manual patterns (the latter commonly consisting of setup conditions and logic simulation vectors). The manufacturing test engineer and logic designer must concur that (1) the set of patterns is sufficiently exhaustive to determine that a part that passes has a very low likelihood of being faulty, and (2) good parts are not being misinterpreted as failing due to a mismatch between the input stimuli, the expected output response, and the function of the physical hardware. A severe constraint on item 1 is the cost of tester time.

The test file should also contain the technology-specific *electrical* test parameters. Prior to exercising a part against the set of logic test patterns, it is possible to cull out a significant percentage of failing parts with a simple set of electrical *screening* tests. These tests can be used to detect such obvious defects as supply voltage shorts and opens, input and output pad shorts, and the like. Only if a part passes the brief sequence of electrical screening tests will the logic pattern stimuli be applied.

The set of logic test patterns is not usually applied at functional speed during manufacturing test; the availability and cost of VLSI testers for functional speed evaluation tend to make such testing prohibitive. Rather, to evaluate the performance of a part during test, it is more common to measure only one or two long signal path delays, pad to pad, transition to transition (Figure 2.18). An evaluation of this measure against the calculated or simulated delay value is usually sufficient to assess the performance capabilities of the part. In other words, the correlation of an individual measured path delay to all path delays on a chip is quite high. In this manner, the *sorting* of different parts that pass manufacturing test into different performance measures can be easily accomplished.

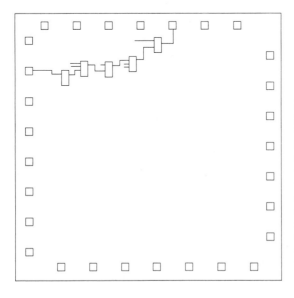

Figure 2.18. A measure of the overall performance of the part may be evaluated at manufacturing test by measuring one or two signal path delays, pad to pad, transition to transition. The measured delays may then be compared to the simulation delay values calculated for the paths in order to assess the part's performance capabilities. The correlation between one or two measured delays and the overall performance is quite high, and thereby permits efficient sorting and screening of parts.

Design Database

The hierarchical database description of a VLSI part is used extensively throughout the design cycle by the designer and by the design system CAD tools. Both
are required to make modifications to the parameters associated with each node
in the structure and with the structure itself. Throughout the design cycle, the
integrity of this database must be maintained. In a multiuser environment, several
members of the design group may request access or modifications to the database
simultaneously. The design system methodology must address this requirement
for maintaining a cohesive design description. One such approach would be to
assign an *owner* to the database; others may copy parts or all of the data into
individual working files. However, only the owner is permitted to make alterations
and/or copy working files back to the main structure. Another approach is to use
the concept of *open* and *closed* files to permit only a single user at a time to access
the data, prohibiting another user from accessing an open file. In any case, a
methodology for keeping a log of access and modifications to the database is
extremely beneficial; this history may help to resolve concerns about data integrity
and will help to document the design evolution.

Documentation

The release of the physical design (shapes) file and the test pattern file to manufacturing is a key checkpoint to achieve in any design cycle, yet it does not
signify the end of the cycle. The design documentation must be prepared for
release; this includes functional block diagrams, timing waveforms, programming
information, electrical requirements, the signal-to-pad correspondence, and so
on. Depending upon the part, the functional description in particular may be
concise or may occupy several manuals.

   The hierarchical database description of the part contains much of the information about the design, which is incorporated into the documentation. The
direction in tool development for documentation is for increased utilization of
engineering graphics and combining graphics with text documentation. A growing
percentage of the documentation is being stored electronically for ease of distribution and maintenance.

2.3 CHECKING

In addition to the tools developed for assisting with design entry and simulation,
a significant percentage of design system support is applied to checking the design
for correctness throughout the design cycle. The set of checking tools provided
by a design system can be divided into three categories based on the criteria being
checked. The first category consists of the tools to verify that the *design system*

standards are indeed being met. The second group checks the design against the *technology rules*, while the third confirms the *correspondence* between the *logical* design hierarchy and the *physical* design.

As design system standards and technology-specific rules are introduced in subsequent chapters, it will be implied that the design will have to be checked for compliance with these constraints. The following list indicates the nature of these standards, some of which have already been mentioned:

1. *Testability:* array isolation, clock signal distribution, limitations on logic feedback loops, restrictions on the use of latches and flip-flops
2. *Usage rules*
   a. Correct reference to book types from the design system library
   b. Fan-out
   c. Voltage interfaces between circuits (including proper noise margins)
   d. Ability to dot circuit outputs together (for increased fan-in/function)
   e. Proper use of special book types (e.g., output pad drivers, input signal receivers, synchronizers)
3. *Fabrication rules*
   a. Adherence to manufacturing lithography (shape) requirements for physical design

The requirement of checking fabrication rules deserves additional mention because it is a very resource intensive task. The development of a technology library of books considerably reduces the magnitude of the task of shapes verification. Each book design is verified against the fabrication rules individually, prior to incorporating the design into the library. It is not necessary to again verify the shape orientations inside each book when subsequently placed and wired on a chip. Rather, the only verification required is to ensure that *globally* the book designs are used correctly; that is, the books are appropriately spaced from one another and global wires make a continuous connection to the book's input/output pins.

To check the global usage of the book, all that is required is (1) a shape defining the outline (or *shadow*) of the design, (2) shapes defining wiring blockages, and (3) shapes defining the location, size, and wiring level for the pins (*pin shapes*). A significant compaction of design data can be achieved when performing global checking of fabrication rules for a chip; the hundreds or thousands of shapes defining each circuit can be equally well represented by a handful of shapes added to the book design. Thus, the global shapes checking can be made considerably more efficient, reducing the design data to shadows, pin shapes, vias, and wires.

There are also checking tools specific to the particular design system image that has been chosen. Prior to executing the wiring program, it is necessary to verify the following:

1. The wireable cell capacity of the image is not exceeded.
2. The placement locations of all the books are valid (i.e., no books extend outside designated cell areas).
3. All pad assignments are valid (e.g., no conflicting pad assignments, proper number and location of supply pads).

The design system placement and wiring tools, the chip image, and technology library are all developed to provide a final physical design that is free of fabrication rule violations. However, if manual intervention into the physical design step is used, the need for these checking tools increases. For example, if manually placed books are accepted as input to placement, it should be verified that two preplaced books do not overlap. If the design requires the manual rip-up and rerouting of wires that overflowed the wiring program, it may also be necessary to verify that the final routing does not violate any technology or image rules. (In both cases, an engineering graphics workstation can be extremely beneficial in ensuring that these manual procedures are done correctly; the necessary checking may be incorporated into the application program menu used on the workstation.) Most importantly, the *quality* of the design may be enhanced considerably if such checking tools are made available; the logic designer is encouraged to participate in optimizing the final physical design, confident that such participation will not introduce errors that go undetected.

There is one type of error, however, that the tools that verify technology rules will not find. Consider the case where the shapes in the global description of the physical design (wires, vias, shadows, and pin shapes) satisfy all the lithography and image rules, yet during manual rip-up and reroute the wrong logic signal wire is connected to a pin shape. The last category of checking tool mentioned at the beginning of this section checks for the correct *logical-to-physical* (L/P) correspondence among the global interconnections. This tool must input the logical description and determine the physical pin locations for all global nets. Then, for each net, the L/P tool traces out in the physical design the branches of the net to ensure that they terminate at the correct locations. (Note that, in addition to shadows and pin shapes, it is also necessary to distinguish individual pins by *name* for books in the global physical design data.) As before, an engineering graphics workstation may provide the software tool to check for L/P wiring errors while manual wiring is being completed on the workstation.

Consistent with the need to maintain the integrity of the design database, it is also necessary to provide an audit trail of the design checking that has been successfully completed. The checking procedures must all be exercised against the "latest and greatest" version of the (logical and physical) design prior to releasing the design for manufacture. After the logic is frozen, the physical design may go through multiple iterations in an attempt to meet performance goals. The design standard and usage rule checks would remain valid, but the fabrication rule and, in particular, the L/P checking will have to be verified against the modified physical design. The time and date *stamp* that a supervisory operating system assigns to closed files is key to determining the latest versions of the various design representations.

There is one final comment about the variety of checking tools required to support a design system. To this point, the discussion has focused on the steps followed by the design system *user*. The chip design develops from a logical description with the aid of design system tools; the resulting physical design was said to have been verified globally. Alas, what of the design system *developer* or the user who wishes to add a new book design to the library? It is evident that a set of *macro-level* checking tools is also required. The same logic and test standards and usage rules will apply directly to the macro-level model. However, the fabrication rule and the L/P checking procedures must be altered to perform circuit-level checking before making additions to the technology library. The characteristics of these *local* checking procedures are discussed in Chapter 5.

2.4 DIRECT-RELEASE DESIGN SYSTEMS

Normally, each new VLSI part number goes through a reliability evaluation before the design is *qualified* to enter the marketplace. A statistically significant number of parts are extracted from several different manufacturing lots and subsequently stressed under extreme environmental conditions (voltage, temperature, and humidity). These conditions are designed to accelerate the aging of the part in order to efficiently determine end-of-life failure rates versus time. In the VLSI design environment, characterized by a proliferation in the number of parts, qualifying each design separately would require a substantial investment in qualification resource and possibly severe adders to the schedule for market entry. In addition, for a VLSI product application where the total anticipated sales volume is small (say 50,000), the wisdom of subjecting a significant fraction of that total to accelerated stressing (say 2000) should be questioned as a matter of economics.

The use of a design system library can eliminate these concerns regarding part number qualification. Rather than qualifying each part developed through the design system individually, the design system library of books can be initially and collectively qualified instead; any user design that draws on the library can then be regarded as *directly releasable*. In short, the qualification is completed once, and the results assumed to be applicable to all subsequent parts.

To qualify the *system*, instead of the parts it produces, two criteria must be satisfactorily met: (1) the design system tools and checking procedures need to be adequately tested, and (2) each distinct book in the technology library should be realized in hardware and subsequently evaluated. The design system qualification procedure therefore requires one or more *test vehicles* to evaluate the system as a whole. Criterion 2 could be satisfied by ensuring that a test chip contained at least one of every book in the library placed somewhere on the image;

ideally, such a design would include an extensive set of test diagnostics for failure analysis. Satisfying criterion 1 may not be as straightforward. Any tools that are new need to be exercised, yet no two designs will likely exercise them in the same manner. In addition, the extent of manual intervention into the physical design step will vary significantly from design to design. For a physical design that contains preplaced books, preassigned pads, or manual routing, the checking programs must also be *qualified* to some degree. To satisfy criterion 1, it may be most expedient simply to turn the design system "loose" to a specific part number, one that will likely approach the limits of database complexity, cell occupation density, and wireability confidence. Assuming this effort meets with manufacturing success, the design system may be released to the user community as a whole. Having the first user part number "debug" the system may be especially frustrating for the designer, but it will likely discover more inconsistencies and errors in the rules and the library than would an artificially concocted example. An added benefit is the feedback available to the design system developers on the overall user-friendliness of the program interfaces.

But how are subsequent additions to the library to be qualified? If the new book is unlike others in the library, a separate hardware test chip for evaluation may be required. Alternatively, if the new book is being developed for a particular part number, it may be just as expedient to subject the part to qualification directly, avoiding the intermediate test chip. Otherwise, if the new book is sufficiently similar to existing members of the library, the successful execution of the set of macro-level checking programs against the new design may be all that is required for qualification. In any case, before requesting an addition to the technology library, the design system user should be cognizant of the resulting qualification costs.

The concept of qualifying a set of CAD tools and a library of design information as opposed to individual part numbers provides a significant productivity boost to a large system design. With the knowledge that the design system parts will be directly releaseable, the partitioning of the overall system function into VLSI hardware may be done with regards to performance, function, and cost, without great concern for the qualification effort.

PROBLEMS

2.1. Selecting a representative collection of commercial SSI and MSI logic parts from a particular technology family, develop a comprehensive set of *usage rules* for that hardware; for each rule, indicate how that information could be represented in the technology library database. Example rules to consider are the following:

Pin locations
Pin names

Pin types (e.g., input, output, bidirectional)

Voltage interface requirements (may be incorporated as part of the pin-type assignment)

Input pin current requirements; output pin fan-out limits

Output and bidirectional pin types that may be dotted with other circuit outputs

Delay values (rising and falling)

Delay adders on input (as a function of input signal transition time)

Delay adders on outputs (as a function of fan-out and/or capacitive load)

Data setup and hold times

Develop a file organization for the rules developed.

2.2. Using technology data books and/or mail-order advertisements from back issues of electronics journals, determine how the SSI and MSI functional libraries for TTL 7400 series and CMOS 4000 series logic technologies have expanded over time. Plot the number of new SSI and MSI part numbers introduced each year; indicate on the plot a breakdown of each year's total by package pin count. Summarize any trends that are evident from this data.

2.3. The objective of this problem is to determine the optimum design system chip image cell width by trying to most efficiently realize a collection of MSI functions.

  (a) Choose a representative set of MSI functions from a particular logic technology library; assume that these functions are to be implemented in a VLSI design system library that contains the following SSI logic primitives: NOR, NAND, INVERT, AOI, and OAI, as well as sequential elements. The *maximum pin utilization percentage* among cell widths can be used as the evaluation criterion for the optimum image definition. For example, with a cell width of six pins, a two-input NAND gate leaves three pins unused. For a cell width of two pins, a two-input NAND gate would require two cells, with one pin unused. Using the logic models given in the technology data book for the MSI functions (with any appropriate modifications for the logic primitive library), determine the number of cells required and the pin utilization percentage of these functions for cell widths ranging from two to six pins. Assume that sequential circuit elements require a circuit area eight pins wide, regardless of the fact that the actual number of pins used may be less (Figure 2.19). An example of the pin utilization analysis for a 7485 4-bit magnitude comparator is given in Figure 2.20. The eight inverters denoted in the figure by an asterisk are not included in the analysis; although present on the MSI part, they would typically not be included in an integrated implementation. Each XOR gate has been replaced by a two-input NOR gate and a two-high, one-high AOI logic gate. [A XOR B = NOT ((A NOR B) OR (A AND B))]. The 7485 contains no sequential elements. Summarizing the results for the set of MSI functions, what is the overall optimum cell width?

  (b) Repeat the calculation of pin utilization percentage for a cell width of four pins, assuming now that the library also contains a *double-inverter* book (Figure 2.21). In other words, each pair of inverters can occupy one four-wide cell, with no unused pins.

  (c) What are the benefits of small cell width, say two or three pins wide (or one pin wide, for that matter)? What are the disadvantages?

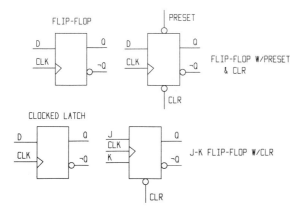

EXAMPLES OF SEQUENTIAL ELEMENTS

Figure 2.19. Examples of sequential elements for Problem 2.3; assume all these different types of circuits require an area eight pins wide.

(d) How does the number of sequential elements present in the function cause the pin utilization to vary?

(e) If a pin utilization greater than 70% is considered to be unwireable, what is the resulting optimum cell width? [Note that, in the examples of images given in this chapter, the pin locations assumed for use in designing the library and wiring grid used only *every other* vertical grid location; a free and clear channel is therefore always present on the vertical (second) wiring plane for long-distance global wires to be directed between pin locations. The 70% figure of merit given therefore translates to a vertical channel density of 35% for routes with local vertices in addition to global interbay connections.]

2.4. The objective of this problem is to define a logic description language (for combinational networks only) and then develop a technique for checking the networks against design system standards.

(a) Suppose that a technology logic library contains only NANDs and INVERTERs. Develop a simple (nonhierarchical) syntax for inputting a NAND network description; assume that the network is strictly combinational (i.e., no feedback paths are present). Internal signal names can be equated to the name of the sourcing block (which could be denoted either by an integer or character string, depending on the complexity incorporated into the syntax). A means for specifying the global input and output signals to the network must also be indicated, including how these names are to be assigned to the appropriate pins of the gates in the network.

(b) Develop a CAD program tool to input this syntax and check for the validity of the network description. Criteria to verify include the following:

   1. No signals are "hanging."

   a. All global input signals are assigned to one or more block inputs (and no block outputs).

   b. All global output signals are assigned to one and only one output (and possibly other block inputs).

   c. All internal block outputs fan-out to other block inputs.

   2. No feedback paths are present in the description. This requires *levelizing* the

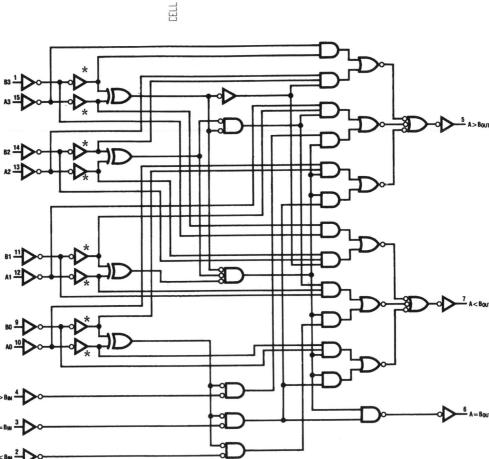

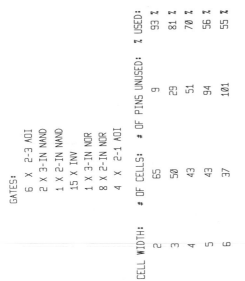

GATES:

    6 X 2-3 AOI
    2 X 3-IN NAND
    1 X 2-IN NAND
   15 X INV
    1 X 3-IN NOR
    8 X 2-IN NOR
    4 X 2-1 AOI

CELL WIDTH:	# OF CELLS:	# OF PINS UNUSED:	% USED:
2	65	9	93 %
3	50	29	81 %
4	43	51	70 %
5	43	94	56 %
6	37	101	55 %

Figure 2.20. Pin utilization analysis for a 7485 4-bit magnitude comparator TTL MSI function. (Schematic diagram from *High Speed microCMOS™ Logic Family Databook*, p. 4–64. Copyright © 1983 National Semiconductor Corp. Reprinted with permission.)

ADDITIONAL WIRING CHANNELS BETWEEN PINS

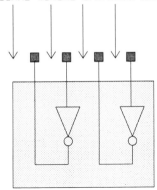

Figure 2.21. A *double-inverter* book; a pair of inverters occupy one four-pin wide cell, with no unused pins.

network; assume the network inputs are at level zero and that the output of a block is one greater than the highest level number among its input signals. If it is not possible to levelize all the signals in the network, a feedback path is present.

3. Block fan-out limits are not exceeded.

(c) Develop a CAD program tool to transfer the NAND description into an equivalent NOR implementation, assuming that another technology library contains only NORs and INVERTERs.

REFERENCES

Breuer, Melvin (ed.), *Design Automation of Digital Systems—Theory and Techniques: Volume 1.* Englewood Cliffs, NJ: Prentice-Hall, Inc., 1972.

Donath, W. E., "Equivalence of Memory to 'Random Logic'," *IBM Journal of Research and Development* (September 1974), 401–407.

Engl, W. L., Dirks, J. K., and Meinerzhagen, B., "Device Modeling," *Proceedings of the IEEE*, 71:1 (January 1983), 10–33.

Keyes, Robert, "A Figure of Merit for IC Packaging," *IEEE Journal of Solid-State Circuits*, SC-13:2, Correspondence, (April 1978), 265–266.

Lohstroh, Jan, "Devices and Circuits for Bipolar (V)LSI," *Proceedings of the IEEE*, 69:7 (July 1981), 812–826.

Raymond, T. C., "LSI/VLSI Design Automation," *IEEE Computer*, 14:7 (July 1981), 89–101.

Ruehli, A. E., and Ditlow, G. S., "Circuit Analysis, Logic Simulation, and Design Verification for VLSI," *Proceedings of the IEEE*, 71:1 (January 1983), 34–48.

Soukop, Jiri, "Circuit Layout," *Proceedings of the IEEE*, 69:10 (October 1981), 89–101.

Sudo, R., Ohtsuki, T., and Goto, S., "CAD Systems for VLSI in Japan," *Proceedings of the IEEE*, 71:1 (January 1983), 129–143.

Szirom, Steve, "Custom-Semicustom IC Business Report," *VLSI Design* (January/February 1982), 32–38.

Werner, Jerry, "Progress toward the 'Ideal' Silicon Compiler—Part 1: The Front End," *VLSI Design* (September 1983), 38–41.

3

EVOLUTION OF INTEGRATED–CIRCUIT DESIGN APPROACHES

3.1 INTRODUCTION

VLSI technology development is constantly striving to reduce the minimum lithographic dimension rules for the shapes on the device and interconnection levels, in an effort to achieve increased measures of circuit density and performance. Much of this development hinges on enhancements in the capabilities of the fabrication equipment used for pattern transfer. In addition to reducing minimum linewidths, VLSI technology evolution can also be characterized by the addition of interconnection levels (such as a second or third metal layer) and a broader offering of device types available in the technology. The former enhancement is required to maintain a high degree of wireability to support the increased circuit density; the latter is provided for a variety of reasons, typically power, performance, circuit density, noise margin enhancements, or a unique application for which a special device type may be required.

In short, newer VLSI technologies are adding more lithography levels at finer dimensions to provide the necessary devices, interconnections, and vias. Whereas MSI and LSI technologies may have used as few as five or six lithography levels, current VLSI technologies are using fifteen or more. A consequence of these additional steps in the fabrication process is higher manufacturing cost. Another related consequence is increased manufacturing time, commonly referred to as the *turn-around time* (TAT). It is crucial to any product development schedule that *prototype* hardware be available for evaluation in system breadboards as early as possible in the development cycle. Once the physical design is complete and final simulation evaluations have been exercised, the manufacturing TAT is a delay to the product development flow; if the TAT is long (e.g., several months), much of the design group's *momentum* can dissipate. In addition, it is likely that a single hardware *pass* of the design may not suffice; in evaluating the first-pass prototypes, design and/or performance problems may be uncovered. Frequently, the system and/or part definition may have changed to some degree during the manufacturing TAT interim, necessitating a logic design revision. Many product development groups allocate time and resource for a *two-pass design*, which makes the TAT even more critical in meeting aggressive product schedule requirements.

The increased number of design and manufacturing levels in current VLSI technologies therefore affects both the initial development of prototype hardware and the final system integration, the former primarily through TAT and the latter primarily in manufacturing cost. The various VLSI design approaches to be discussed in this chapter address these problems from different perspectives. The TAT for fabricating hardware for a particular part can be shortened if the processing sequence is divided into design-dependent (or *personalized*) lithography steps at the back end and design-independent (or *standardized*) levels initially. A standard for the number, size, and orientation of the devices on the chip is adopted, permitting the device-related lithography levels to be design independent; the interconnection of specific devices into circuits (and the global interconnection of those circuits) remains design dependent, obviously.

Figure 3.1 illustrates this technique. Hardware is fabricated up to (but not including) the first personalized level and then stockpiled; the TAT *provided to a design group* is reduced to that required for processing only the personalized levels. The number of different books available in this technology library is limited by the number, size, and type of devices located in each cell area; some circuit designs will leave unused devices in the cell. This design and fabrication alternative is referred to as the *gate array* approach (Section 3.5); the attractive TAT has made this approach the fastest-growing segment of the application-specific VLSI design market. At the other end of the VLSI design spectrum is the *full custom* approach (Section 3.3), which can reduce production-level costs by achieving a minimal chip area. The resource typically required to complete the chip physical design varies considerably between these two approaches; the gate array will commonly use a structured chip image, a limited technology library, and design system placement and wiring tools for support, while the full custom chip design discards the structured image (and the tool support) in favor of fully handcrafted physical designs to completely optimize area and/or circuit performance.

Between these two extremes lies the approach that strives for high design system support for chip physical design and the capability to locally optimize circuit designs using handcrafted cell and macro circuit layouts. This technique is currently referred to as a *standard cell* approach (Section 3.4); although the fabrication TAT is not reduced, the reduction in chip area afforded by using handcrafted circuits will reduce costs as compared to an equivalent gate array.

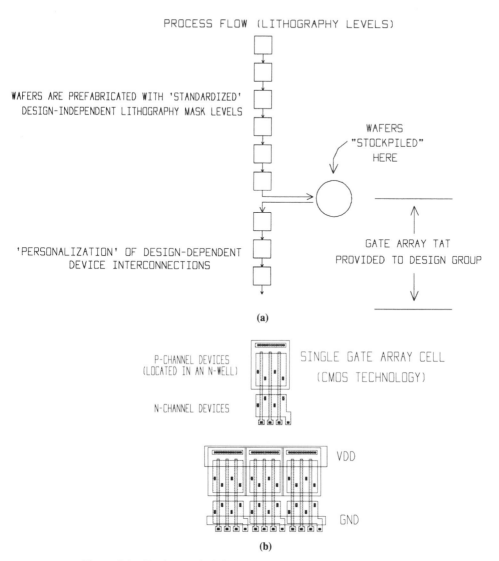

Figure 3.1. Design methodology and cell layout of a *gate array* chip. (a) Process flow diagram illustrating the technique of standardizing the device lithography levels and stockpiling the wafers; unique part numbers are then personalized by specific interconnection level lithography, providing considerably reduced TAT to a design group for prototype parts. (b) Example layout of the standard device and contact shapes present in a *cell* of the gate array chip; a single- and a three-cell layout (with supply and ground metal bus interconnections) are illustrated.

As the design system tools can accommodate a wider variation in chip images and a greater diversity among macro designs, a more descriptive name for this technique might be a *macro image* approach (Section 3.7).

It was mentioned earlier that product development cycles commonly adopt a two-pass design philosophy for the VLSI hardware to be used in the final system. The first pass might be in a gate array design environment to take advantage of the reduced TAT and obtain an early evaluation of the product prototype as a whole. A second pass (with any necessary modifications) could remain with the same design approach or may strive for a smaller area (lower cost) design by choosing a standard cell implementation instead. To make this transition, the logic description of the part needs to be modified to refer to the standard cell technology books, utilizing macro offerings wherever possible.

3.2 DEFINITIONS AND CHIP IMAGES

This section provides specific definitions for the terminology used in discussing VLSI design approaches and their related chip images.

Wafer, Mask, Photoresist.   The starting material for fabricating integrated circuits is a *wafer* of crystalline semiconductor material (e.g., silicon), typically on the order of 20 mils thick (0.020 inch) and from 2 to 6 inches (or more) in diameter. In the initial fabrication steps, impurities such as boron or phosphorus are incorporated locally into the semiconductor crystal; the regions of various impurity types define the device nodes and the electrically isolating regions between adjacent devices. Thin-film dielectric and interconnection layers are deposited onto the surface of the wafer and photolithographically patterned in subsequent processing steps. As a result, the background crystalline wafer is commonly referred to as the chip *substrate* after the individual chip sites on the wafer are separated.

The wafer is obtained by slicing a crystalline *ingot* of the same diameter, an ingot that was pulled from a liquid melt of the semiconductor material. (A carefully measured quantity of impurity is added to the melt and will be distributed uniformly throughout the crystalline material.) The ingot growth and slicing operation is done so that the *active* device surface is of a particular crystalline orientation; the active surface of the wafer is mechanically and chemically polished to a specified flatness. The wafer also contains an alignment reference, a *flat* and/or *notch*, for coarse orientation when beginning the procedure of aligning a lithography level to previous levels (Figure 3.2). The fine alignment requires

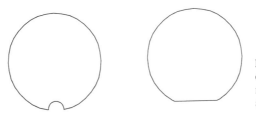

Figure 3.2.   Two typical means of coarse alignment for successive lithography levels on the silicon wafer: a *flat* and a *notch* in the wafer circumference.

reference to high-contrast physical shapes previously patterned on the wafer surface through a high magnification optical system.

The diameter of the wafer is a key parameter in any fabrication process: the greater the diameter, the greater the number of individual chip sites that may be repeated over the wafer surface and the greater the manufacturing throughput. A quantity of wafers for an individual part number is processed collectively in each manufacturing *run*. Monitor wafers are inserted and removed at specific process steps to obtain an isolated (and rapid) evaluation of critical steps prior to the completion of the entire fabrication sequence. Many of the thin-film deposition steps can be performed using a wafer carrier containing the entire run, thereby enhancing manufacturing TAT.

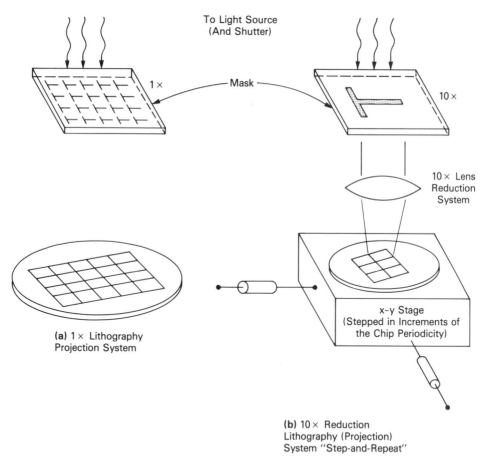

Figure 3.3.   Two means of imaging the chip design shapes of a particular lithography level onto the wafer surface: (a) The case where an array of chip shapes data is contained on the mask plate; (b) A step-and-repeat technique, with a single copy of the design on the mask plate repeated over the wafer surface with multiple exposures.

The level-to-level photolithography alignment step prior to patterning is performed a wafer at a time; thus, the total manufacturing TAT and final costs are strong functions of the number of photolithography levels. Increasing the wafer diameter to increase throughput also has the disadvantage of increasing the sensitivity to wafer flatness when aligning, focusing, and exposing a lithographic pattern over the wafer field. (In addition, migrating to a larger wafer diameter may incur major expenditures for equipment upgrades.) Options currently exist for *blanket exposure* of the entire wafer to an array pattern of chip sites or exposing each chip site on the wafer individually. This latter technique permits the focusing and alignment of the pattern on a site-by-site basis, but at the expense of wafer throughput. Figure 3.3 illustrates these two lithographic techniques. The design shapes are represented as transparent and opaque areas on a glass plate called a *mask*; the photosensitive coat on the wafer surface for pattern transfer is *photoresist*. (Mask manufacture and photoresist exposure are discussed in more detail in Chapters 5 and 15.)

Kerf.   The *kerf* is the narrow wafer area that separates individual chip sites (Figure 3.4). The kerf must be of sufficient width to permit the reliable separation of individual *die* after fabrication and test. The two general classes of separation techniques are *sawing* (e.g., using a diamond-tipped saw blade) and *scribing* (e.g., using a diamond-tipped stylus). Both techniques have restrictions on the surface materials through which a clean cut or scribe can be made. The scribe has the more severe constraints, requiring that the kerf contain no thin-film layers and that the crystalline wafer surface be exposed. This precludes placing alignment targets and/or test structures in the kerf area, which is permissible with a sawing technique.

In both techniques, a piece of tape is typically applied to the wafer backside before sawing or scribing. The saw cuts (or scribe marks) only penetrate a fraction

Figure 3.4. The *kerf* is the area between active chip sites on the wafer; when a sawing separation technique is used, this area may be occupied with parametric and defect structures for process evaluation.

of the wafer surface; the separation between die is completed by breaking the wafer in both directions along these lines. The tape holds all chips in place so that the particular sites that tested *good* on the wafer can be easily extracted. Alternatively, *bad* sites may be visibly marked at manufacturing test so that they are not selected.

Periodicity.   The *periodicity* of the chip sites on the wafer surface is the dimension in the *x*- and *y*-directions between adjacent sites; it is equal to the sum of the chip plus kerf dimensions. For the lithographic technique of Figure 3.3a, it is the distance between chip patterns on the mask plate for each lithography

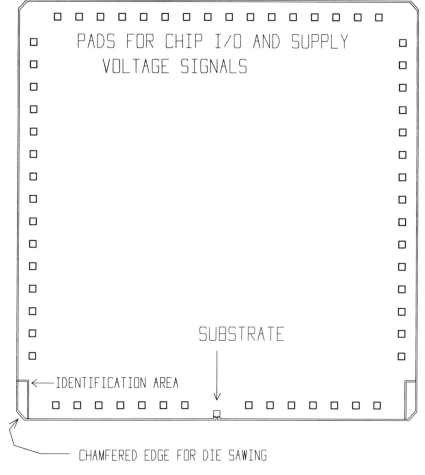

Figure 3.5.   Chip *footprint* for the input/output pads to connect to the chip package; a single row of chip I/O pads is depicted.

level; for the site-by-site exposure technique of Figure 3.3b, it is the translation of the wafer between exposures of the individual pattern.

Chip Footprint.   The *footprint* of the chip is the coordinate information of the input/output pads. This information is used to build probe cards for testing wafer sites after manufacture. It is also used in the design of the package and assembly operation for bonding the chip I/O pads to the external package leads. Chip footprints vary widely, depending on the design approach adopted and the packaging technology used. Gate array (and many standard cell) VLSI design approaches use a fixed footprint to simplify the algorithms of the physical design tools, as well as to save testing and packaging start-up costs. Full custom designs (and some standard cell designs) may use a variable footprint to be able to op-timally trade off the utilization of the I/O area between pads and chip circuits.

Packages that require the active area to be face up in the chip carrier use distinct bonding wires between I/O pad and package pin; the footprint will typi-cally consist of a single row of pads around the chip perimeter (Figure 3.5). Al-ternatively, face-down or *flip-chip* packaging uses wire traces deposited and pat-terned on the ceramic carrier itself; in this case, multiple rows of pads are feasible

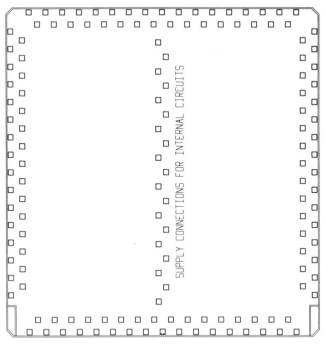

Figure 3.6.   A chip footprint with multiple rows of pads; the internal pads can be used for power supply connection to internal circuits. This footprint is used for the technology de-scribed in reference 3.2.

(Figure 3.6). Section 15.18 discusses chip-to-package bonding techniques in greater detail.

3.3 FULL CUSTOM APPROACH

To develop a fully custom VLSI design, engineering groups are assembled to cover the wide range of skills required to design the part essentially from scratch. These groups span the areas of process engineering, device modeling, circuit design, physical circuit layout, logic design, and system architecture. The design is optimized for the best density and performance as a result of the *close* and *continued* interaction between the engineers in the various groups.

The design methodology of a full custom part includes an extensive set of required tasks to complete the physical design of the part. Initial estimates of circuit and device sizes are made corresponding to the chip logic functions, performance goals, and fan-out. These device sizes are then used to commence circuit (and subsequently macro) layouts. The circuit, macro, and final chip layouts are all handcrafted; the circuit layouts are not confined by any rigid bounds (i.e., cells), and the interconnection wiring is not confined to any grid or directionality. Typically, few design system aids for placement or wiring are used due in large part to the irregular nature of the resulting macro layouts. Figures 3.7 and 3.8 illustrate the physical *floor plans* of two custom chips. In these figures, note that no differentiation is made between the chip area used for circuit implementations and for interconnection wiring; the wiring area has been divided among and incorporated into the individual (abutting) macros. This is in contrast to the other design system approaches, where the distinction between circuit and wiring area is very evident.

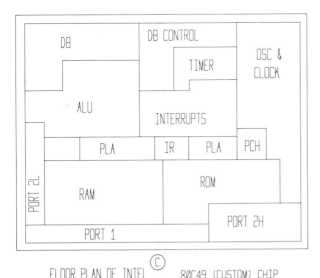

FLOOR PLAN OF INTEL Ⓒ 80C49 (CUSTOM) CHIP

Figure 3.7. *Floor-plan* of a custom chip, illustrating the flexibility in the area and topology of various macro layouts; in addition, note that no distinction is made between circuit and interconnection wiring area (adapted from reference 3.7).

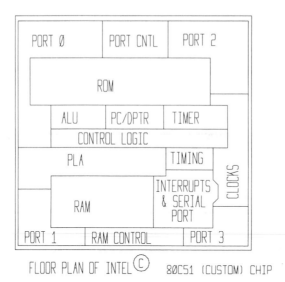

Figure 3.8. Floor-plan of a custom chip; note that the macro topology need not be restricted to 90°, as with the Interrupts and Serial Port macro (adapted from reference 3.7).

As the physical design nears completion, the actual load resistances and capacitances on circuit outputs are better defined. These loading parameters are added to circuit simulation models, and the macro performance is simulated and compared against performance goals. The accuracy of the initial sizing estimates is very important, since the ability to alter a circuit design internal to a tightly packed macro layout is quite limited. Also, it is extremely important to check *frequently* the physical layout against the technology lithography rules during the development of the chip physical design. It can be very difficult to change or correct shapes internal to a macro layout without requiring a major redesign effort.

Considering the large design effort required, a custom design is only warranted for high-volume parts, where the initial engineering expense can be amortized over a long, active product life. Examples of such parts are microprocessor and microcontroller designs; their programmability allows broad use across a wide range of applications. However, in Chapter 1 it was stated that much of the future product market for VLSI designs is for low-volume parts *customized* for different and specific applications; the initial costs and time of the required (physical) design phase for a full custom chip have become an inhibiting factor in meeting the needs of these product markets. Alternative design approaches, like the standard cell approach discussed next, were developed to address this problem.

3.4 MASTER IMAGE (STANDARD CELL) APPROACH

The various VLSI design approaches make different trade-offs and impose different constraints on the chip physical design in an attempt to make this step more manageable, while maintaining sufficient design flexibility. The full custom design approach imposes few constraints on the circuit or physical design, forsaking the use of design system placement and wiring tools. In comparison to full custom designs, the flexibility, circuit density, and performance of the final standard cell design are sacrificed somewhat to achieve an increased level of automation in the logical and physical design steps.[1] Increased automation should result in faster design time.

The standard cell or master image approach typically adopts many of the design system standards described in Chapters 1 and 2. Specifically, the most prevalent characteristics of this approach are:

(−)   A limited set of logic circuits is available.

(+)   The circuit design and physical design for the chosen set of logic blocks has been previously completed (and checked against the technology lithography rules); the physical design is constrained to reside within an integral multiple of fixed area cells.

(+)   To facilitate the automatic wiring of the interconnections of the chip logic functions, the cell I/O pin locations are assigned to a wiring grid.

(+ +) Due to the incorporation of tool support for placement and wiring, the final physical design may be easily (and thoroughly) checked for:

    Logical-to-physical correspondence
    Technology lithography rule violations
    Power supply electrical connection

(Without manual intervention in the physical design step, these will be correct by construction.)

The +'s and −'s associated with these items indicate the attractiveness and sacrifices adopted by this approach. The sophistication of the design system placement tool will dictate the variety of logic circuits available in the master image library, corresponding to the physical size of the circuit design. For a single-cell restriction, the set of circuits would typically consist of unit combinational logic blocks; for multiple-cell designs, more function would be available, and more flexibility would be required of the placement algorithm.

In the master image design approach, a division has been struck between the tasks of handcrafting circuit designs (to fit within the cell area) and placing and wiring those circuits together (with the benefit of design system CAD tools). This separation is based on the assumption that the time-consuming task of handcrafting a custom layout is best restricted in scope to "small" circuit designs only. This layout task will be done *once* and the resulting shapes stored in a technology library for repeated use across many designs. Another deviation from the full custom approach is the distinction made between circuit area (the cells, which include segments of power supply rails, typically) and the interconnection area.

---

[1] Standard cell designs are also denoted as *master image* designs; the two terms will be used interchangeably.

Circuit designs are confined by the cell bounds, and their I/O signals must terminate on a wiring grid point in the wiring bay.

The master image cell area is chosen so as to provide some measure of power/performance optimization after the initial physical design is complete; as described in Section 2.2, a number of different power/performance levels of the same function may be provided in the library, each occupying the same number of cells.

The master image approach does not reduce the manufacturing TAT over that of a full custom design. Each cell area in the final physical design contains handcrafted circuit layouts that are unique at each of the device lithography levels. The master image chip footprint may or may not be standardized, depending on the capabilities of the tools to place pads and I/O circuits and to wire signals (and power) to and from the pads. To a large degree, the flexibility in the number and location of I/O pads is also dictated by the packaging technology.

Examples of industry master image design systems are given in references 3.1 and 3.2. The first describes an older (single-level metal technology) master image system in use at RCA since the mid-1970s; the second describes a master image design system developed within IBM around 1982, in a two-level metal technology. The two design methodologies share some features:

1. Both physical design systems use a high-resistivity (polycrystalline silicon) semiconductor layer to complete the circuit interconnections. For the single-level metal technology, this interconnection layer was (necessarily) used for segments orthogonal to the metal layer. A signal interconnection crossing a row of cells requires a wire-through cell between wiring bays (Figure 3.9). (The placement and wiring tools for this technology strive not only to minimize the total interconnection length but also to emphasize the use of metal,

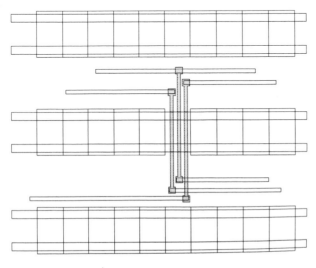

Figure 3.9. A *wire-through* may be used to continue a signal interconnection between wiring bays; the shaded shapes are higher-resistivity polysilicon segments.

instead of the higher-resistivity segments.) The two-level metal technology master image system uses higher-resistivity interconnects for some of the intrabay segments, leaving the second-level metal wiring channel available for global signal routing (Figure 3.10).

2. Both physical design systems permitted the incorporation of larger macro layouts; these designs ranged in size from two cells (placed within a row of cells) to those which spanned several rows of cells (Figure 3.11). The layouts spanning several rows of cells block the wiring bay area from use for first metal segments and sever the through-the-cell first metal power supply rails. The macro sizes fall on cell boundaries in both the horizontal and vertical directions, and the input/output signal pins must fall on wiring grid points. The multiple-row macro layouts are manually placed by the designer prior to executing the placement program. The function incorporated into the larger macro layouts includes static RAMs, *register stacks*, and programmable logic arrays (PLAs).

3. Both design systems implement a means of performance optimization based on the specification of performance-critical signal paths by the designer. In one case, if a net in a signal path is denoted as critical, the highest power/performance level of the logic circuit driving that net will automatically be selected from the technology library; in the other case, the critical paths are interpreted by the placement program to imply that the logic blocks in that

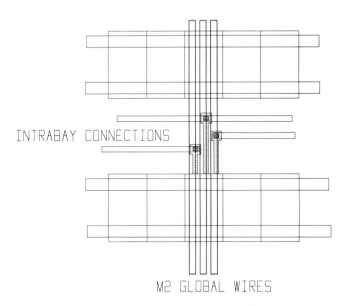

Figure 3.10.   Use of nonmetal interconnection segments for intrabay wiring in a two-level metal technology; this leaves the second metal wiring channels free and clear for interbay wiring.

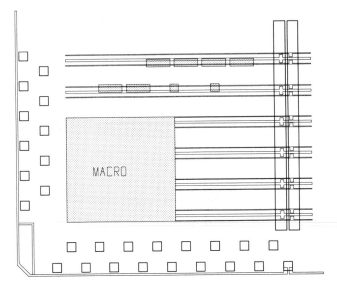

Figure 3.11. A master image chip design incorporating a large macro, which spans several rows of cells; the macro layout truncates the power supply rails, necessitating macro placement at the sides of the chip image.

path should be closely clustered to reduce interconnection length (Figure 3.12).

4. Both chip images locate the chip input/output pad circuits around the chip perimeter, and the I/O circuit size is distinct from that of the internal cells. (The packaging technology developed for the chip image in reference 3.2

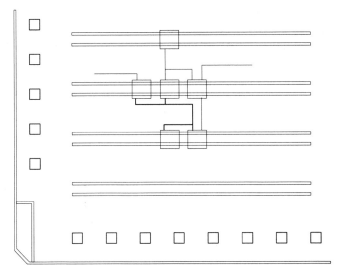

Figure 3.12. The specification of a *critical net* can influence the placement program to provide clustering of the related logic circuits to reduce the net interconnection capacitance.

permits the use of pads in the interior of the image, which were used for power supply distribution to internal circuits.)

In developing a master image design system, trade-offs in area, performance, flexibility, usability, and tool development must all be assessed. The design system developers must have an understanding of the applications, specifications, and productivity needs of the logic designers who will be using the design system library. This is also the case for the next design system approach to be discussed, the *gate array* design system.

3.5 MASTER SLICE (GATE ARRAY) APPROACH

In the master image approach, *all* the layout (shapes) information is merged and sent to manufacturing for fabrication after placement, wiring, and checking. The physical design time was reduced over the full custom design approach by using existing circuit designs (from the design system library) and defining a chip image for automated placement and wiring tools. As described briefly in Section 3.1, the manufacturing turn-around time can be further reduced by adopting a *gate array* approach.[2] In this design methodology, several of the lithographic patterning levels are standardized, ideally, all but the interconnection and via geometries. The devices used to implement individual circuit designs are *prefabricated* in each master slice cell area; that is, the devices and the contacts to the device nodes are fixed in location, but left unconnected, after the initial processing steps. A circuit design/logic block from the technology library is *placed* at a specific cell location by assigning the appropriate pattern of wires to coordinates inside that cell area; these wires connect the devices to form the selected circuit type (Figure 3.13). Device nodes that are not incorporated into the assigned circuit are wired to a particular supply voltage so as not to disrupt the logic function realized by devices that *are* connected. Different logic blocks require different internal cell connections, all of which (for the entire chip) are typically implemented at the first metal interconnection level.

The chip image for the master slice approach is often quite similar to that of the master image designs (Figure 3.11), with the exception that macro layouts spanning multiple rows of cells can no longer be accommodated (unless the shapes for the devices in the macro layout are *also* included in the standardized levels). The cell dimensions will likely differ between the two design system chip images in the same fabrication technology, for reasons to be explained shortly. Also, the number of wiring channels in the wiring bay may be different. The process of placing the appropriate interconnection pattern at each cell location on the master slice chip image is referred to as *personalizing the array*.

---

[2] Gate array designs are also referred to as *master slice* designs; the two terms will be used interchangeably.

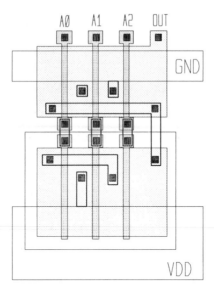

Figure 3.13. A specific logic circuit is *placed* at a gate array cell location by assigning the appropriate pattern of interconnection wires inside the cell to the device contacts; for connections to the supply rails, a shape is included to *merge* with the chip image voltage distribution.

The advantages of the master slice approach are as follows:

1. Wafers can be prefabricated and stockpiled using a single mask set, reducing mask making costs since fewer masks are generated for each new part number.

2. Quick TAT can be provided for prototype hardware. Since the number of prototype pieces required is typically quite small, a full manufacturing lot may not be required. Multiple part numbers may be combined into a single run, even to the extent of sharing the same wafers in the run.

3. In conjunction with reduced TAT, process data and device characteristics are available soon after the personalization processing begins; indeed, prior runs from the same master slice stock may be used to provide yield and performance data for future estimates.

A major disadvantage of the master slice approach pertains to the development of the master slice library. To support a wide variety of logic functions in the library, a large number of transistors is typically required, resulting in an increase in cell area over an equivalent master image library. Alternatively, a relatively sparse number of devices may be offered in a smaller master slice cell, with a limited logic primitive library; however, this results in a greater pin utilization percentage and a corresponding loss in image wireability. These two alternatives are illustrated in Figure 3.14. For either reason, a larger cell area or an inability to wire a densely occupied part, a master slice design typically requires a larger chip area than an equivalent master image design. The resulting production level costs of the larger master slice chip design will be higher. It is not uncommon that the initial design hardware may be developed using a master slice approach

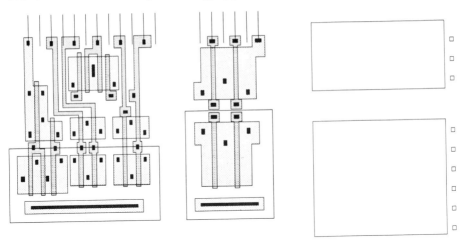

Figure 3.14. The master slice cell may be designed to implement a wide variety of logic circuits in the technology library, requiring a larger number of available devices (and device contacts). Conversely, a simpler, smaller cell with fewer devices may be used, with a correspondingly smaller logic library (and a higher percentage of pin utilization, requiring more available wiring bay channels).

for quicker market entry, with a subsequent cost reduction by phasing over to a smaller master image design.

In addition to the chip designers, the master slice approach appeals to the manufacturing process group as well. Manufacturing lines for an older process technology will find the volume of part orders fluctuating with time, eventually declining as newer technologies are brought on line. These mature technologies will have been well characterized for performance and reliability and are at the bottoming-out portion of their cost curve. Considering the large investment in equipment and people, keeping a manufacturing line at full capacity for as long as possible is a desirable (i.e., profitable) goal. When reduced demand is anticipated, the fabrication of master slice background wafers for stockpiling can be initiated, keeping the manufacturing line full. Personalizing these wafers requires a smaller resource and can provide an economical solution to potential users not in need of the performance and density measures of a newer technology. If the manufacturing facility does not wish to support the test and packaging requirements and the scheduling overhead of frequent low-volume part numbers, the master slice wafers may be offered for sale to other, smaller *satellite* or *pilot* lines that will provide the necessary test and packaging (and possibly the wafer personalization) support.

A master slice approach may also be offered in a newer technology being brought on line. Given the long design time described earlier for custom chip development, master slice (or master image) designs may precede any custom designs. It may be advantageous for the newer technology line to initiate pro-

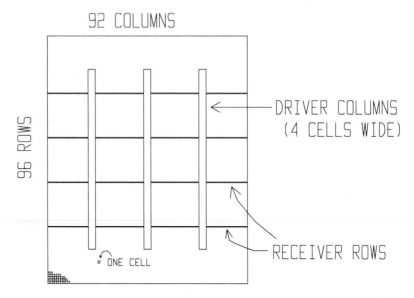

Figure 3.15. A bipolar technology master slice chip image, with cells occupying the entire chip area in a *sea of gates* fashion. Valid circuit placement locations for chip input/output signals (through receiver and driver cells) are interspersed throughout the image.

duction with master slice lots in anticipation of the technology's first users. It is key that the technology-dependent design system tools are developed simultaneously with the final stages of process development so that the design system is available when the manufacturing capability for the new technology is offered.

Examples of master slice design approaches used in industry are given in references 3.1 and 3.3. In reference 3.3, the technology is a bipolar device technology, and the chip image differs considerably from that of the master image chip in Figure 3.11. Instead, the *entire* chip area contains prefabricated devices and cells; that is, the master slice chip contains a *sea of gates* (Figure 3.15).[3] The circuit placement program now has the flexibility to choose from a much less restricted set of locations for assigning circuit coordinates. An occupied cell does not completely block interconnection wires on the first metal layer, but the number of available (horizontal) channels is nevertheless reduced over that of an unoccupied cell (five versus nine); no second-level metal is used for circuit personalization, so the number of available (vertical) wiring channels per cell remains the same (ten). A maximum cell occupation of 65% is specified to maintain overall wireability.

The packaging technology used for this master slice permits the use of I/O

---

[3] The processing technology for this master slice uses three levels of metallization; the first two (orthogonal) levels are used for interconnection and the third primarily for power redistribution from the package pins.

pads internal to the chip perimeter; as a result, unique cells for driver and receiver circuits were interspersed with the internal cells, the receiver cells in rows (occupying four internal rows) and the driver cells in columns (occupying twelve internal columns). The first personalized process lithography layer is the contact etch mask for the device nodes; contacts are not opened in cells that are unoccupied, and the (bipolar) device nodes in unoccupied cells are simply left unconnected.

The gate array described in reference 3.1 is for a single-level metal (MOS device) technology and consists of rows of cells separated by wiring bays, as depicted earlier. As with its master image counterpart in the same technology, high-resistivity passive components are used for vertical interconnection segments; in the master slice image, however, these segments are predefined in location and prefabricated as part of the standardized process. Unlike earlier illustrations, the cells in this image are *not* placed back to back in a double-high row; each row of cells has its own set of supply rails. This permits each cell to contain *two* electrically equivalent access points for each input and output pin (Figure 3.16).

For a design group without access to a fabrication resource, master slice design systems offered by technology houses are a low-cost means of entering the VLSI marketplace. These houses (or *foundries*) often make arrays in mature technologies available to outside firms as an additional source of revenue. In conjunction, engineering workstation and CAD tool support for the array are also commonly provided. The design group is able to specify the schematic and simulate the design using logic and performance models from the design library provided by the foundry. The design is then placed and wired, either by the foundry or by the design group. Physical wiring length net capacitances are measured; then (accurate) delays are calculated and back-annotated to the simulation model for final verification. Figure 3.17 illustrates a design methodology for working with the master slice foundry and gives some of the options available. The design group may choose not to participate in the physical design steps, leaving that responsibility to the foundry's design support group and the foundry's tools.

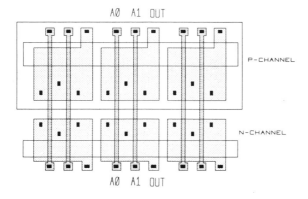

Figure 3.16. A master slice cell design that permits multiple access points for circuit I/Os. Cells are not placed back to back; rather, the access points are in the first wiring channel of adjacent wiring bays.

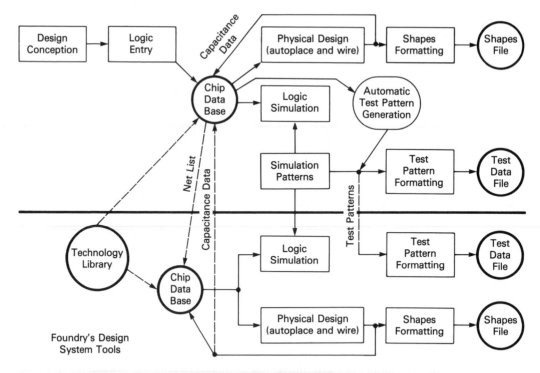

Figure 3.17. Various design methodologies for interfacing with a gate array foundry, illustrating the necessary transfer of data required if the design group uses the foundry's design system (simulation and/or physical design) tools.

Throughout the design flow, several checkpoints are commonly established between foundry and design group to ensure a high confidence on the performance measures and testability of the resulting design.

3.6 COMPARISON OF FULL CUSTOM AND MASTER IMAGE APPROACHES

The economic appeal of both master image and master slice approaches has led one electronics consultant to refer to *logic arrays* as the "fourth component revolution in electronics design," behind the milestones of the development of the transistor, integrated circuit, and microprocessor.[4] Although it is perhaps a bit premature to elevate these design system alternatives to that level of esteem, the master image and master slice approaches have indeed found widespread appeal among VLSI users. The development of engineering workstations and enhancements to design system tools have been key to that acceptance.

[4] Mackintosh Consultants, as reported in *Electronic Engineering Times*, June 21, 1982, p. 4.

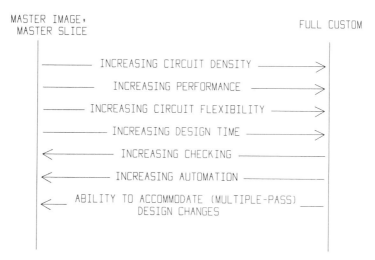

Figure 3.18. Engineering trade-offs between full custom and master image/master slice design system approaches.

Nevertheless, full custom design advocates have remained unconvinced, arguing that the resulting designs do not have the density and performance of full custom chips. In a text devoted to full custom VLSI design, logic arrays are described in the following manner:[5]

> . . . . [this] technique provides the logic designer who has limited knowledge of integrated systems with a means for implementing modest integrated circuit designs directly from logic equations. However, a heavy penalty is paid in area, power, and delay time. Such techniques, while valuable expedients, do not take advantage of the true architectural potential of the technology and do not provide insights for further progress.

It is evident that there exist strong trade-offs between these approaches, some of which are illustrated in Figure 3.18. Of particular significance in the figure is the ability of the array-based systems to more easily accommodate engineering changes relatively late in the design cycle or in a subsequent hardware pass.

Another key consideration in choosing among design alternatives is the knowledge base and/or experience of the system architect or logic design group that requires a VLSI design solution. With the rapidly increasing complexity of integrated-circuit technologies (and system designs), the scope of engineering responsibilities among members of a design group has typically become narrower, thus permitting each member to focus on a more complex problem (or aspect of a design). This evolution is diagrammed in Figure 3.19. As a result, it is becoming

[5] C. Mead and L. Conway, *Introduction to VLSI Systems*, Addison-Wesley, Reading, Mass., 1980.

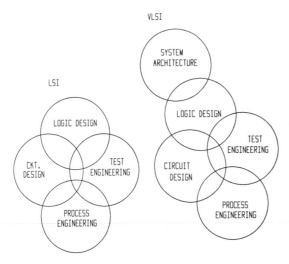

Figure 3.19. Narrowing of engineering responsibilities for a VLSI design group, as compared to the relatively large overlap in job function for an LSI group. The complexity of VLSI design tasks has necessitated the specialization of engineering departments; as a result, a design system can enable a system architecture or logic design group to utilize a new technology in a productive manner, sharing the technology library database with the circuit, test, and process development groups.

less and less likely that the necessary architectural, logic, and physical design experience can all be assembled and dedicated to each new full custom part. A more typical scenario is for the technology library circuit design and technology rules programming to be completed in conjunction with the final stages of fabrication process development. As the system architects and logic designers are now utilizing that technology in their product design, the scarce process development, model development, and circuit design resource must shift attention to technology enhancements. The design system library of models, circuits, and rules acts as the repository of design information, the link between the development and manufacturing capability of a VLSI technology.

With regard to the design system limitations on circuit implementation flexibility, density, and performance, the incorporation of macros into the (master image) design system library is the direction in design system development that addresses the weakness. The goal of the last design system approach to be discussed, the *macro image*, is to provide a highly automated, low resource, and easily verifiable design methodology with performance and density measures comparable to a full custom solution.

3.7 MACRO IMAGE APPROACH

A macro is a logic network that implements a function usually associated with a MSI or LSI part [e.g., a register, counter, multiplexer, decoder, or arithmetic and logic unit (ALU)]. The term macro has been utilized in the integrated-circuit industry in a number of different contexts and meanings; the two most general categories are *logical* macros and *physical* macros. A logical macro is a particular specification of the interconnection of logic primitive cells chosen by the design system developer and added to the design system library. In other words, the design system contains an implementation of a macro-level function using the single-cell books of the library; when called forth by a logic design description, the schema for the interconnection of cells is appended to the hierarchical logic design (Figure 3.20). No single physical layout has been developed for this function; the logical macro is provided as a shorthand convenience to the designer in developing and maintaining the chip logic description. A physical macro (or just macro) will be used to depict the single, dense shapes layout that implements the particular MSI or LSI function. Each macro's physical design is developed in detail (custom); additionally, corresponding simulation and test models must be included in the design system library.

The similarities between integrated circuit development and computer program development are many, and several useful analogies are available. Master image (and master slice) design systems utilize the design library at a level equivalent to programming in the assembly language of a machine. The logic designer using a master image system must have a thorough understanding of the function of each design component, just as the assembly language programmer must have an intimate knowledge of a machine's architecture and the results of each instruction's execution. However, programming at an assembly-language level becomes an increasingly formidable task as the size and complexity of a program grow. A productivity enhancement is provided to the designer (programmer) by using circuit (assembly language) macros, without much loss of understanding of the workings of the design (program) and with potential enhancements in the overall design (program) performance.

Carrying the analogy to a higher level of abstraction, it is also possible to use a *high-level language* description with a corresponding *silicon compiler* to produce the final chip design; several integrated-circuit companies and research communities are involved with this effort. The precise definition of a silicon compiler, however, is subject to various interpretations. The most general one is that a silicon compiler consists of a set of programs that translates a high-level language description (e.g., a register-transfer language description) through a Boolean representation and minimization directly into layouts and mask level data; no pre-

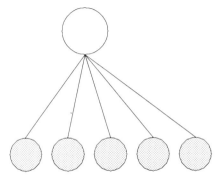

Figure 3.20. A *logical* macro is the hierarchical description of the interconnection of smaller physical technology library circuit designs to implement a higher-order function.

defined array image or library of layouts is used. The shapes defining the Boolean circuits are generated as per the circuit type and then *compacted* to minimal area designs using the particular process lithography rules for minimum widths and spaces. This compaction is also performed globally to reduce the area between circuits while maintaining the overall wireability. As is the case with programming compilers, the designer (programmer) typically has little control over the final area (program size) and/or performance (execution speed); the resulting area and performance would to a large extent be dependent on the transformation, minimization, and compaction (optimization) algorithms of the compiler. For example, the compiler may be able to translate a particular network description into either unit gate combinational logic or a programmable logic array (PLA), with size and delay measures that differ considerably; the criteria used for the choice of implementation may be too complex for the compiler and may indeed require the designer's specification. In addition, changes to the abstract high-level chip description will alter the size and performance of the compiled result in a manner that may not always be predictable.

The custom macro design approach resembles the master image design approach in that the input description is still *structural*, as opposed to procedural. That is, the logic description still consists of the hierarchical structure of blocks, each with its associated pins. Specific pins are interconnected in the description by assigning to each pin a common net name. The physical design of the part requires assigning the books of the technology library to nodes in the structural description; these books now include a wider offering of MSI and LSI functional layouts. The final physical realization of a particular logic network is the decision of the system architect and logic designer. If the requirements of a particular function are not met by existing designs in the technology library, the logic designer must work in conjunction with the physical designer (who has an expertise in the technology and a familiarity with the chip image and constraints) to handcraft a macro layout to address those objectives.

The key difference between custom macro and master image design systems is with regard to the chip image and placement tools for locating the macro and single-cell layouts. The macro layouts will vary in size, power dissipation, and pin count; a flexible chip image is required to accommodate this variation, while ensuring the correct power supply distribution to the macro circuits and a globally wireable part. In addition, the number of macros used in a particular design could range from dozens to none, with a corresponding variation in the number of logic blocks (from few to many thousands). The particular case of *one* large macro plus a few thousand logic blocks is an interesting one, because it is both a difficult image design problem and an attractive design option. A common design requirement is to interface an intelligent system peripheral to the system bus; the microprocessor or microcontroller architecture of the peripheral must be able to communicate to and from a system bus of a differing architecture. If the processor or controller were available as a design library macro, this single large layout

Figure 3.21.   A single large physical macro used on a macro image chip design, with a number of *glue logic* cells.

could be combined with the necessary interface logic into a single chip solution. The resulting *core plus glue* logic design is illustrated in Figure 3.21.[6]

    With regard to flexible image design, macros present the following characteristics:

1. Macro layouts may have outlines (*shadows*) that differ widely in form, although most chip images accept only rectangular descriptions of the area spanned by the layout.

2. The number of pins along the edges of the macro layout is typically relatively small; that is, the macro usually implements a logic function partitioned such that the number of external macro I/Os is small relative to the number of internal nets. With a low percentage of wiring grid points occupied by macro pins, the number of required wiring channels outside the macro border is reduced over a row of unit logic cells.

---

    [6] *Glue logic* is the term commonly used to refer to the small number of gates that in LSI system designs have remained as discrete SSI parts. The component integration characteristic of VLSI design is striving to incorporate that glue logic into the chip definition and off the printed circuit card.

3. Macro pins for logic signals may be restricted by the image to just two sides of the macro shadow, with power connections on the orthogonal sides. Conversely, the physical design placement and wiring tools may permit more flexible pin locations *internal* to the macro shadow (Figure 3.22). Internal macro pins may ease the task of macro layout; it may be difficult (and costly in terms of performance) to route an embedded signal to the macro boundary.

4. In conjunction with item 3, the macro may have some limited wiring channel availability on the interconnection planes. This may include either or both of (a) available wiring channels on a normally blocked wiring plane, and/or (b) blockages on a normally free and clear wiring plane (Figure 3.23). Typically, the macro layout represents a fully blocked area on the layer used for internal signal wires; item a indicates that the macro layout may contain channels on this plane for use in global routing. These channels are inten-

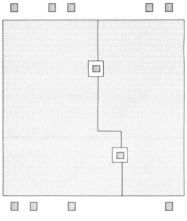

Figure 3.22. Macro signal pin locations, internal to the general blockage area of the macro; the only accessible wiring level in this case would be second-level metal.

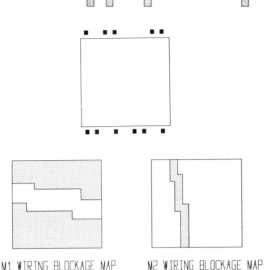

M1 WIRING BLOCKAGE MAP   M2 WIRING BLOCKAGE MAP

Figure 3.23. The design system technology library database may be used to store wiring blockage information on normally free-and-clear wiring levels (e.g., M2) and/or available wiring channels internal to the macro outline on a normally blocked interconnection level (e.g., M1).

tionally left free and clear as part of the original physical design in anticipation of the penalties incurred in routing a number of signals around a large,
fully blocked area. In reference to item b, it is possible that the macro layout
may be reduced in area (and enhanced in performance) if an internal connection is implemented using segments on a (lower capacitance) wiring plane
normally utilized strictly for global routes. In both cases, when the wiring
plane blockages differ from the norm, it is necessary to specifically store
this unique information in the technology library. Prior to routing the placed
design, the wiring program must input the blockage information for the
macros and modify the wiring grid map appropriately.

As custom macro images and their related physical design tools are still
embryonic, macro-intensive physical design systems described in the technical
literature differ considerably (references 3.4, 3.5, and 3.6). Specific features of
these systems include the following:

1. The macro layouts in the technology library are constrained in one dimension
   to a limited number of cell heights; in other words, all macro layouts must
   be h_i cells high, where h_1, h_2, . . . , h_n is the possible set of heights, and h_1
   = 1 for the normal, nonmacro layouts. The number of different macro
   heights, n, should be limited by the design system developer such that the
   probability of multiple macros in a particular design occupying the same
   height is large; the number n should nevertheless be big enough to maintain
   some degree of layout flexibility.

2. Macros of common height may be combined into a single *tier* of layouts.
   This collection of macros can be powered with simple supply rails, obviating
   the need to include power distribution in the global routing step (Figure 3.24).
   Interconnection wiring channels are available adjacent to the power supply
   rails, much like the wiring bay design of the cellular images. This grouping
   technique is particularly well suited for logic functions related to the width
   of a chip data path, such as registers, counters, an ALU, and parity generators and checkers (Figure 3.25).

3. The macro tiers may be developed hierarchically in a manner similar to the
   chip logic structure; that is, a collection of macros is grouped into a *subchip*,
   and the subchips of common width are combined into higher-level structures
   (Figure 3.26). However, the routing of power supply rails now becomes more
   complex (Figure 3.27).

4. An option exists in the means of supply routing to macro tiers; Figures 3.24
   through 3.27 have all illustrated a *treelike* set of segments for providing power
   to the macro circuits. It is also possible to design a *gridlike* image using
   additional global rails (Figure 3.28). The necessary width of the rails depends
   on the current requirements of the circuits and should be chosen to ensure
   low resistive losses and long-term reliability. The more frequent grid rails

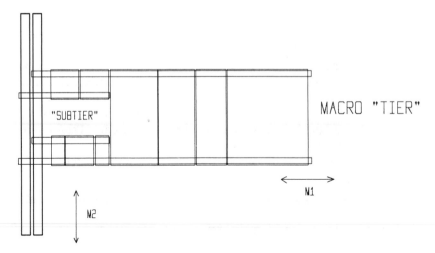

Figure 3.24. A *tier* of macro layouts may be formed to share power supply distribution rails. The figure also illustrates the presence of a *subtier* of macros of smaller dimension, which share part of the power distribution.

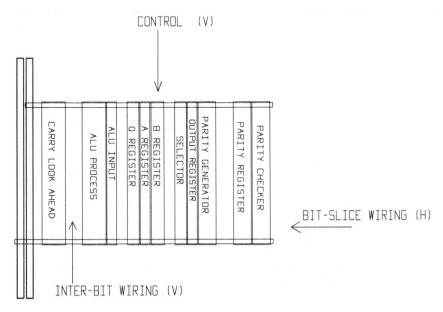

Figure 3.25. The tier of macro layouts is well suited to system designs where a number of different system-level functions share a common bit width; the macros may be developed in *bit-slice* fashion (adapted from reference 3.4).

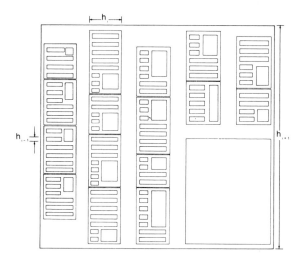

Figure 3.26. Macro tiers may be developed hierarchically. A collection of macros is grouped into a *subchip*; these subchips are then placed on an image based on their width (From E. M. Reingold and K. S. Supowit, "A Hierarchy-Driven Amalgamation of Standard Macro Cells." *IEEE Transactions on Computer-Aided Design*, Vol. CAD-3, No. 1 (January 1984), p. 8. Copyright © 1984 by IEEE. Reprinted with permission).

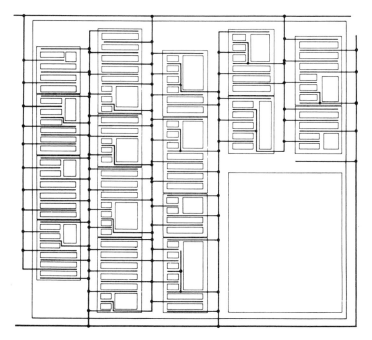

Figure 3.27. The subchips of Figure 3.26 result in a more complex power supply distribution routing problem (From E. M. Reingold and K. S. Supowit, "A Hierarchy-Driven Amalgamation of Standard Macro Cells." *IEEE Transactions on Computer-Aided Design*, Vol. CAD-3, No. 1 (January 1984), p. 8. Copyright © 1984 by IEEE. Reprinted with permission).

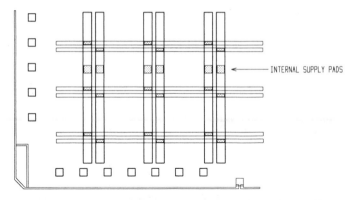

Figure 3.28.   Additional global power supply distribution rails may be provided to produce a *gridlike* distribution; the more frequent (vertical) second-level metal supply rails and supply pads enable the first-level metal cell supply rails to be narrower, as less current-carrying capability is required to service the reduced number of circuits between supply pads.

will carry proportionately less current and may therefore be narrower. Macro pins located under the grid rails are less accessible, however (Figure 3.29).

5. The more flexible macro image may use variable-width wiring bays between macro tiers. The wiring bay can expand or contract to accommodate just the required number of channels to contain the global interconnection segments (Figure 3.30). There may also be some areas where the wiring bays

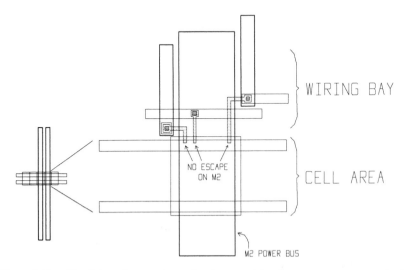

Figure 3.29.   Circuit input/output pins under the second-level metal grid power distribution rails are not accessible to second-level metal interconnection wires, necessitating some additional wiring algorithm complexity.

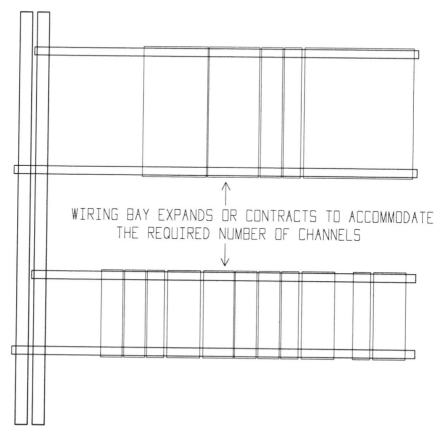

Figure 3.30. The macro image may use variable-width wiring bays between macro tiers to enhance the overall chip wireability.

are *fixed* in width due to the relative location of layouts within structured power supply distribution rails (Figure 3.31). The computational complexity of the physical design algorithms increases considerably for the flexible image. The placement tool must estimate the channels separating circuits to calculate the total interconnection length; the wiring tool must maintain the coordinate information for the circuits and wires as the final routing is completed and the circuit locations are adjusted to accommodate each wiring bay.

6. In conjunction with the flexible (internal) wiring bay widths, the macro image design system must also address the extent to which the total chip size and pad locations will be variable. In addition to the required algorithmic complexity, the variability in design system chip footprint affects the test and packaging functions as well. Each new chip size and pad footprint requires a unique wafer test probe card and possibly substantial tester setup and

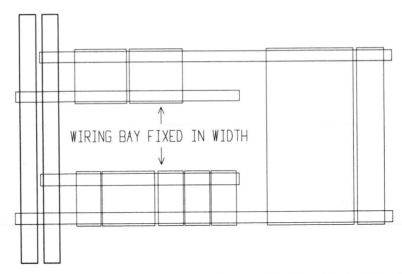

WIRING BAY FIXED IN WIDTH

Figure 3.31. The macro image may also contain areas where the wiring bay is fixed in width, such as with the subtiers of circuits illustrated here.

debug time. Likewise, each new footprint requires modifications in the design of the bonding pattern for chip attach.

Developing a flexible macro image that lends itself easily to automated placement and routing is a difficult task, complicated further by variation in the number and size of macros that may be present on design system parts. The designs will range from the *core plus glue* example given earlier, with one large macro and a small number of gates, to a macro-dominated implementation, with many macros of varying sizes.

Perhaps the most typical (and, in some ways, the most difficult for which to develop an image) is the case where the design contains only a handful of macros and several thousands of gates; the problematic characteristics of this example include the following:

1. The macros may not be strongly interconnected in the logic design; therefore, grouping these layouts into tiers may introduce additional routing difficulty.

2. Locating a large macro toward the chip's center may be optimal for performance reasons: the wiring path lengths from the macro to its fan-outs will likely be reduced. Yet global interconnections that connect pins on opposite sides of the macro may become significantly longer as the macro blockage diverts segments around the layout (Figure 3.32). The resulting wiring channel congestion may cause the chip dimensions to grow considerably.

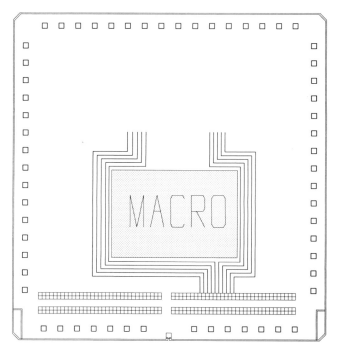

Figure 3.32.   Macro blockages for global interconnections can result in longer signal wires and channel congestion for centrally placed macro layouts.

3. Given the characteristics mentioned in item 2, a flexible image design with macros spaced around the perimeter might appear to be preferable. However, unlike the unit logic circuits, the macros in the technology library will typically not contain multiple power/performance levels; little optimization is available for performance-critical paths through macro circuits. Placing macro layouts toward the chip boundaries increases the likelihood that a high fan-out macro output net in a critical path will have a larger capacitive load and may not meet its performance goals. One possible solution is to incorporate high-drive-capability buffer circuits into the design on all macro outputs. Another solution would be to bias the placement algorithm to locate macros more centrally and multiple power/performance books peripherally, with the associated routing problems described previously.

An alternative approach to macro image design is to adopt the philosophy that an optimum placement of macros (and an ultimately wireable design) can best be achieved if a number of interconnection segments are preassigned in location, *prior* to beginning the placement task. Critical nets and/or nets with a large fan-out are assigned to those segments *initially*, and the placement of circuits with high interconnectivity to those nets is the first step of the algorithm. In essence, those segments act as the *seed* for macro and circuit placement. Figure 3.33 is an example of a *bus-oriented* image; wiring segments are located in the chip center, and the bus and related control signals of the design are assigned to those segments.

The goal of the macro image approach is to provide a high-productivity design path, with many of the area and performance benefits of full custom designs. However, it is a common tendency not to widely utilize an existing macro set in a technology library. It can be difficult and time consuming to become familiar with the function, performance, and usage rules associated with a library's macros; it may also be the case that none of the library's available medium- and/ or large-scale functions are well suited to a unique design or architecture. A design system methodology should be provided for making additions to the macro library for unique functions; this methodology includes the procedures for defining the structure of the layout, pin locations, wiring blockages, and power supply distribution of the physical design, as well as the format of the macro's simulation, test, and performance models. The system architect must realize that a significant

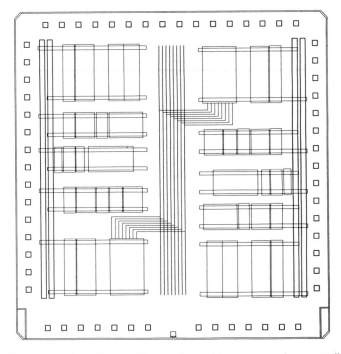

Figure 3.33.   A *bus-oriented* image, with predefined wiring segments; these centrally located routes would commonly be assigned to the bus and primary control signals of the chip architecture.

resource may be required to develop and verify the new layout and models and must assess the cost of that resource against that of alternative design solutions.

PROBLEMS

3.1. Research the architectural definition of the components of a particular VLSI microprocessor/microcontroller family. Develop a common functional-level macro set for the various part numbers of that architecture.

3.2. Develop a flexible macro image for use with a macro-rich library that lends itself to automated physical design tools. Specific points to consider include the following:

Allowed variability in macro sizes

Variability in chip sizes and/or pad footprint

Means for power supply routing

Constraints on macro pin locations and power supply interconnections

Variability in wiring channel areas between layouts

Availability of preassigned wiring segments (i.e., a bus-oriented image)

Assume two levels of metal interconnect are available for global wiring and that macro layouts present a complete blockage to first metal segments.

3.3. The CMOS technology master slice layouts illustrated in Figure 3.16 utilize a device node-to-node spacing for isolation between adjacent cells. Alternatively, the electrical isolation between cells could be provided by n-channel and p-channel devices that are nonconducting; the nonconducting device input is hard-wired to the appropriate voltage rail to hold the device off. In this manner, the chip image may consist of continuous rows of p-channel and n-channel devices, with no distinct cell boundaries. Book circuit designs are composed of one or more n-channel and p-channel device pairs.

(a) Modify the layouts in Figure 3.16 to form a continuous row of devices; show how electrical isolation between adjacent cells can be provided with nonconducting devices. Note that distinct n-channel and p-channel device inputs are now required (like the cell layout of Figure 3.13).

(b) Discuss the advantages and disadvantages of this device isolation approach in comparison to node isolation. Features to consider include the following:

Circuit density (What physical layout design rules are significant in comparing the two approaches?)
Book and macro placement flexibility
Macro- and global-level design checking
Ability to accommodate engineering changes

(c) Show how this technique could be extended throughout the chip image to provide a sea of gates, like the bipolar master slice illustrated in Figure 3.15. How would wiring channels for global routes be accommodated?

REFERENCES

3.1 Feller, A., and others, "LSI and VLSI Automatic Layout Techniques," *1981 Proceedings of the University/Government/Industry Microelectronics Symposium* (May 1981), VIII-1 to VIII-24.

3.2 Donze, R., and others, "PHILO—A VLSI Design System," *19th Design Automation Conference* (June 1982), 163–169.

3.3 Dansky, A. H., "Bipolar Circuit Design for VLSI Gate Arrays," *IEEE International Conference on Circuits and Computers* (October 1980), 674–677.

3.4 Bauge, M., Richarme, M., and Vergnieres, B., "A Highly Automated Semi-Custom Approach for VLSI," *IEEE Journal of Solid-State Circuits*, SC-17:3 (June 1982), 465–472.

3.5 Reingold, E. M., and Supowit, K. J., "A Hierarchy-Driven Amalgamation of Standard Macro Cells," *IEEE Transactions on Computer-Aided Design of Integrated Circuits and Systems*, CAD-3:1 (January 1984), 3–11.

3.6 Hassett, J., "Automated Layout in ASHLAR: An Approach to the Problems of 'General Cell' Layout for VLSI," *19th Design Automation Conference* (June 1982), 777–784.

3.7 Bursky, Dave, "CHMOS Microcontrollers Cut Power Needs by 90% +," *Electronic Design* (November 11, 1982), 240.

Additional References

Beyers, J. W., and others, "A 32-bit VLSI CPU Chip," *IEEE International Solid-State Circuits Conference* (1981), 104–105.

Cushman, Robert, "Arithmetic Chips Assume Greater Importance as Microcomputer Users Demand Faster Response," *EDN*, (April 14, 1982), 61 +.

Feller, A., "Automatic Layout of Low-Cost Quick-Turnaround Random-Logic Custom LSI Devices," *Proceedings of the 13th Design Automation Conference* (June 1976), 79–85.

Kang, S. M., and others, "Gate Matrix Layout of Random Control Logic in a 32-Bit CMOS CPU Chip Adaptable to Evolving Logic Design," *IEEE Transactions on Computer-Aided Design*, CAD-2:1 (January 1983), 18–29.

Mikkelson, J. M., and others, "An NMOS VLSI Process for Fabrication of a 32-Bit CPU Chip," *IEEE International Solid-State Circuits Conference* (1981), 106–107.

Payne, Michael, "An Integrated VLSI Design System," *VLSI Design*, (January/February 1982), 46–50.

Shima, Masatoshi, "Demystifying Microprocessor Design," *IEEE Spectrum* (July 1979), 22–30.

4

LOGIC ENTRY AND VERIFICATION TOOLS

4.1 APPLICATIONS OF A LOGIC DESCRIPTION LANGUAGE

This chapter discusses the means by which the chip designer provides the functional description of the part to the design system; subsequently, the techniques for verifying the behavior and performance of that description through logic simulation will be described. The options currently available for entering the part description are many and varied; engineering workstations commonly offer a graphical interface, while a myriad of logic entry textual *languages* have been developed. Furthermore, these languages can be classified into a number of different groups based on the scope and intent of their usage.

Register-transfer Languages

Register-transfer languages incorporate constructs for describing the behavior of synchronous state machines in terms of logical and/or arithmetic operations and state transitions. The simulation of the system model proceeds by evaluating the network behavior in successive clock cycles. For example, the description of a Memory Add instruction in DDL is given in Figure 4.1.[1] The system consists of global signals (and storage), as well as a local finite state machine (i.e., *automaton*). The global facilities are available to each automaton, which are comprised of independent control and information transfer functions (Figure 4.2).

---

[1] J. R. Duley and D. L. Dietmeyer, "A Digital System Design Language (DDL)," *IEEE Transactions on Computers*, C-17 (September 1968), 850–861.

```
< SYSTEM >  SYSTEM_NAME:

  < REGISTER >  INSTR_REG [16]  , ACC [16], MAR [10] .
     /* DEFINES INSTRUCTION REGISTER, ACCUMULATOR,
        AND MEMORY ADDRESS REGISTER */
  < TI > CLK1 (200E-09).
     /* DEFINES A CLOCK SIGNAL TO WHICH
        A FINITE STATE MACHINE CAN BE SYNCHRONIZED */
  < MEMORY >   M [0:1023,16] .
     /* DEFINES 1024 WORD MEMORY */
  < AUTOMATON > CPU: CLK1:
     /* SIGNAL CLK1 IS THE SYNCHRONIZING SIGNAL
        FOR THE AUTOMATON CPU */
  < STATES >
        ⋮

     MEMADD:   ACC ← ACC + M [MAR] ,→IFETCH.
        ⋮
```

Figure 4.1. A register-transfer language description defines the behavior of a synchronous state machine. State MEMADD will be reached after an instruction fetch and decode of the fields in INSTR_REG. The register operation to be performed is described, along with the succeeding state to be reached (unconditionally). A means for conditional state branching based on signal values is also provided.

SYSTEM

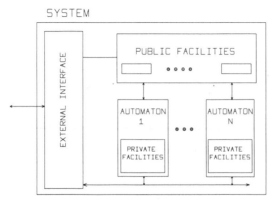

Figure 4.2. A register-transfer language system description consists of independent state machines and public (global) facilities for their intercommunication.

Register-transfer language descriptions are useful in describing and simulating system behavior at an *architectural* level, as opposed to a gate (or primitive) level. Having developed this architectural description, logic synthesis tools can be exercised to develop a technology-independent gate-level description; a logic transformation algorithm can then *map* the technology-independent network into the books of a particular design system technology library. The resulting technology-dependent description is the basis for additional timing verification and, ultimately, physical design.

Yet register-transfer language descriptions are deficient in some respects, due to the synchronous modeling approach of system behavior. Specifically, the ability to define asynchronous state machines, the ability to describe specific system or chip performance measures, and the ability to merge individual chip descriptions hierarchically into a larger system-level model are all typically quite limited. In addition, the methodology by which a technology-dependent description is produced, synthesis and logic transformation, is faced with several difficult tasks:

1. The minimization of logic function within each automaton is difficult to achieve. The conditions that specify the information transfer to each register are used to generate the Boolean description of the required control network; minimizing the collective set of control networks for the total automaton data flow is a formidable effort. To assist with this task, the designer may select to define intermediate Boolean signals in the automaton description and then describe the conditions for data flow in terms of these intermediate variables.

2. An automaton description consists of a collection of data flow operations in each state that must be designed to execute simultaneously (Figure 4.3). This *compatible set of operations* requires synthesis of discrete combinational networks to produce the required manipulations. However, to what extent can a synthesis tool recognize that the function of a control network can be incorporated into similar operations in other states? What are the performance implications of sharing such a network? To direct the synthesis program, the designer will commonly choose to define a single *operation* logic network, which can then be referenced in different automaton states with different input signal sets.

3. The technology-dependent design resulting from synthesis and logic transformation may not effectively utilize the macros available in the technology library. The task of recognizing where higher-function books or macros could (or should) be incorporated is very difficult; the final chip description will likely contain a large concentration of primitive functions. The designer may need to direct the synthesis algorithm to a specific macro for some subset of the design description: register, operation network, or an entire automaton.

4. The flexibility available in describing an operation to a register-transfer level simulator may exceed the capabilities of the synthesis tool to produce the resulting manipulations; such may indeed be the case for arithmetic operators such as binary addition and subtraction.

5. The number and sequence of state transitions in each automaton can be used to attempt to synthesize the appropriate state sequencer function; however, it is likely that the designer will choose to describe the sequencer explicitly, including the assignment of specific Boolean vectors to machine states. The design of the optimum counter implementation for the state sequencer is a difficult task, typically best left to the designer.

6. Perhaps the most frustrating aspect of this design methodology lies not with the synthesis task itself, but with the extreme difficulty present in attempting

```
< STATES >
                /* COMPATIBLE SET OF OPERATIONS */
    EXEC1:  M [MAR]←—INPUT,  ACC ←—ACC+1,  IN ←—0,
              —→ EXEC2.
                /* (UNCONDITIONAL) STATE TRANSITION */
```

Figure 4.3. In state EXEC1, a *compatible set of operations* is listed; these assignments are to execute simultaneously.

the inverse process; that is, if design changes are incorporated into the synthesized gate-level description, to what extent can the architectural-level description be updated to reflect the current function? How can the two descriptions be maintained so as to always be in concurrence?

This last point is undoubtedly the most difficult to contain. Register-transfer language descriptions offer a very productive path for system-level description and verification, but often are of lesser use after synthesis and chip-level simulation.

Netlist Descriptions

A bare-bones language for describing the *structural* interconnectivity between books and macros of a particular chip design is provided by a *netlist* language. Figure 4.4 gives a brief example of a typical netlist format. Two options commonly exist for assigning interconnectivity between pins and net names in a language definition: the netnames listed in each record can either be *positional* or by *explicit assignment*. The positional syntax requires knowledge of the implicit pin order assumed by the design system for each book type.

Netlist descriptions do not typically include any behavioral or performance-modeling information; rather, the chip function is specified at the book or macro level, whereupon models for simulation are extracted directly from the design system technology library. The netlist description may be produced as a result of logic synthesis or graphical schematic entry, or by the designer directly.

Hierarchical Logic Description Languages

Recently, a considerable amount of resource has been invested in the development of a new language type, one that would facilitate the design and verification of higher-level functions prior to logic synthesis, yet one that is robust enough to

```
/*  POSITIONAL FORMAT:
      BLOCK_NAME (OUTPUT_NET_NAME(S)) =
          BOOK/MACRO_NAME (INPUT_NET_NAMES)
      E.G.,  (Q, ¬Q) = DLATCH ( DIN, CLK )  */
REG1 (DBUS1, ) = DLATCH (ALUOUT1,ICLK);
    /*  EXPLICIT ASSIGNMENT:
      BLOCK_NAME(OUTPUT_PIN:NET_NAME) =
          BOOK/MACRO_NAME(INPUT_PIN:NET_NAME, .... );  */
N1 (OUT:N1) = AOI(A0:DEC1,B1:DMAOP,A1:'1',B0:DMACTL);
    /*  NOTE THAT BLOCK NAMES AND NET NAMES HAVE
        DIFFERENT ASSOCIATIONS AND THEREFORE CAN BE
          IDENTICAL  */
```

Figure 4.4. A *netlist* is a structural description of the chip function; the interconnectivity between books is provided by netname. Both positional and explicit assignment formats are illustrated for describing netname to book pin correlation.

represent the full structural interconnectivity of the final technology-dependent description. Rather than attempting to maintain concurrence between a register-transfer language description and an associated technology-dependent netlist, a single description would serve as the design reference, evolving over time to encompass the results of logic synthesis, macro or book selection, and performance analysis. This system or chip description language is no longer just a set of syntactical definitions for logic entry, but rather more of a means for containing that portion of the design database that directly pertains to the system architect or logic designer. As such, the use of the term design *language* is perhaps misleading; the design text file is merely a representation of a substantially more complex data structure. Nevertheless, the term design language will be used, with the implicit knowledge that associated data structures and files are provided.

It is key for VLSI design support that this design reference be able to incorporate a *hierarchical* chip design structure. An example of a hierarchical logic description language is presented in Section 4.3.

The major problem faced in developing a central means for describing a VLSI chip or system is containing the diverse database to support the variety of applications and tools that utilize the information. It is imperative that the language definition be sufficiently open ended so as to be able to expand to accommodate the design data required by new applications as they develop. Some of the *current* applications that must be supported by the design description are described next.

(1) Design documentation.

Typically, the chip or system design documentation [system definition, functional (block diagram) descriptions, performance specifications, physical pin and package information, etc.] is not prepared during the design cycle; preparing documentation inevitably gets postponed until after the design is complete (as it tends to be regarded as closer to drudgery than design). However, much of the necessary information to be provided in the design documentation is indeed resident in the system logic description and can be readily extracted. To as large an extent as possible, the design description should accommodate documentation tools that generate block diagram or schematic plots, summarize physical information, and associate text or graphics files with design nodes.

(2) Simulation modeling.

The design description must include or point to the models used to verify the functionality and performance of the chip or system during simulation. As will be discussed in Section 4.5, the options available for describing the function of a network are many and varied, from truth tables and Boolean equations to *procedural* routines. All these simulation modes can potentially be mixed in a single system verification model.

(3) Physical design.

The chip placement and wiring tools utilize the design description at the netlist level as the basis for their algorithms. However, as will be described in Chapter 6, these applications can be developed to utilize additional design information to influence or give direction to the execution of these procedures. Examples of the physical design information that may be provided include preplaced books and macros (especially chip I/Os) and performance critical nets. In addition, it would be beneficial if the placement and wiring procedures can append related information back to the design description, such as the following:

1. Physical placement locations (and orientations) for all books and macros
2. Net wiring capacitance
3. Any physical *pin swapping* among logically equivalent book inputs (used by the wiring algorithm as an alternative for increasing wireability)

(4) Technology transformation, or *mapping*.

The results of logic synthesis are usually provided in technology-independent form; this generic description must then be mapped into the books and macros of the chosen technology library. The design database must be able to easily accommodate alternative representations of the same network; the representation to be used in building a system simulation model (or beginning physical design) would be selected by the designer's specification of an appropriate keyword or parameter value. In this manner, system performance with different technology implementations of the same chip design can be readily compared, using different targets for the technology transformation.

(5) Technology usage checking.

The design description must contain the technology information for checking of chip capacities and interconnection interfaces; specifically, this includes the following:

    physical pin types (input, output, bidirectional)

    book and macro area (in units of internal or I/O cells)

    Voltage interface logic levels

    Loading parameters (capacitive load plus current sourcing and sinking values)

(6) Test modeling.

Just as a particular node of the system design may point to one of a multiple number of technology implementations for simulation, a similar situation exists with respect to test modeling. The test models for different technology representations will differ in terms of associated circuit or network faults.

This list of applications is by no means complete; the design language and associated data structures used by the design system must be able to grow as new, custom applications are developed.

The design system supervisory software must provide the library (and version) management to individual VLSI descriptions as they evolve. The technology-specific information extracted from the library and appended to a design description may become invalid if changes to that technology library are subsequently introduced. An audit trail of library and application tool usage needs to be maintained to assist with the task of maintaining data integrity between library and system designs.

4.2 HIERARCHICAL LOGIC DESIGN

This brief section is intended to emphasize the importance of hierarchical logic design, as well as describe some of the related pitfalls. The primary motivating factor for developing and maintaining a hierarchical system description is resource management; the design group can make more efficient use of their time and computing power if the design entry and verification task is structured in a hierarchical manner. A secondary, yet not insignificant, reason for adopting this approach is the containment of system design changes; the impact of logic changes (or a change in technology choice) can be better understood and localized in a hierarchically structured system description. Finally, this structure accommodates both top-down and bottom-up design methodology alternatives, supporting the particular approach that may be best suited to different aspects of the overall design task.

The broad use of the term *hierarchy* can be misleading; a hierarchical design structure encompasses both *partitioning* of the design and *nesting* of design functions (Figure 4.5). Design partitioning entails the division of all or part of the function contained within a node into distinct networks; the signal interfaces between partitions (and any logic function explicitly described at the node) are defined and subsequently assigned to that node of the hierarchical description. Partitions of partitions can then be further described, building a design structure in a top-down manner. The structure ends in *leaf* nodes when the network function is entered strictly in terms of library primitives. The bottom of the structure can be reached by the synthesis of a functional description, continuing with a top-down approach. Alternatively, the description of a network can be entered in bottom-up fashion until the function contained (and the external interface) agrees precisely with that of a design partition resident in the top-down structure. Partitioning is used to try to best manage the complexity of a VLSI system design; by dividing the design into smaller functional networks, the tasks of optimization and verification can proceed somewhat independently and with significantly less computer resource. Often, higher-level partitions correspond directly to the design responsibility of different individuals in an engineering group. Partitioning is not

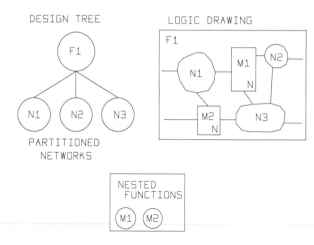

DESIGN TREE

LOGIC DRAWING

PARTITIONED
NETWORKS

NESTED
FUNCTIONS

Figure 4.5. A hierarchical design methodology includes both *partitioning* and *nesting* of design functions. In the figure, networks N1, N2, and N3 are partitions of the function encompassed by node F1; nodes M1 and M2 are also constituent networks of F1, but are nested references (i.e., they are not contained within the design tree). In this manner, these nested functions can be used a number of times within the overall system description.

of itself a great enhancement to design productivity; indeed, a substantial effort may be required to develop the top-down structure, which is often viewed as bothersome. However, the time invested is usually well spent in terms of the design documentation and specifications produced.

A significant enhancement in overall design productivity *can* be realized by utilizing *nested* design functions, where a nonprimitive network is used in multiple instances in the overall description. The associated network model is stored separately; each instance of the function in the design description points to that model, identifying the intended signal interconnections for that particular occurrence. This nested model should be available for all designers to use; thus, individual designers avoid developing, optimizing, and verifying identical networks. The design productivity enhancements afforded by the use of nested functions can be maximized by initial consideration of parallelism present in the overall system; again, the time invested initially can reap significant dividends later.

When building the system simulation model, the design description is *unnested* by copying the nested network model to the appropriate signal interconnections for each instance. These nested functions need not be implemented on the chip as a large physical macro layout; that is, the nested functions may be simply logical macro descriptions of the interconnection of primitive technology books.

The hesitancy to make the initial investment in defining a top-down structure relates to the most common pitfall in using a hierarchical description strategy. If the partitioned functions are subject to considerable change and redefinition, the signal interfaces between different partitions (and designers) will likely be affected. Without good communication between designers, a discrepancy would go undetected while independent design verification tasks are being pursued, until a system-level model is constructed. Another common impediment to hierarchical design methodologies is the reluctance among designers to accept predefined function definitions for nesting in their partitions, opting instead to create unique handcrafted networks for each application. Yet, as VLSI technologies evolve to provide increasing chip circuit counts, a hierarchical design methodology will become increasingly attractive as the best means to apply the available design and verification resource to the problem; overcoming the real and presupposed obstacles to implementing such a methodology is the immediate task at hand.

4.3 LANGUAGE FOR HIERARCHICAL LOGIC ENTRY

There are currently several description languages in various stages of development as *standards* for VLSI system design and verification (e.g., CONLAN, EDIF, and VHDL[2]). Rather than describing one of these languages in detail, this section will illustrate some of the concepts and constructs that are commonly provided in all. Absent from this discussion of design languages is a formal syntactical definition; it should be understood that any such language implementation contains an exhaustive set of rules for ensuring proper syntax and semantics. Syntax refers to the relative positioning of character strings that serve as identifiers, names, separators, and statement terminators; the semantic rules of a language indicate the correct use of names in operations and assignment statements, based on a set of *type* parameters associated with that name. Most design languages (and current programming languages, for that matter) are strongly *typed*, which eases the task of checking the validity of operations and value assignments. Although this section deals strictly with textual design entry, this representation need not be the form in which the design is initially developed; rather, a graphical input means is often used. Ultimately, the full top-down textual description and the network logic drawings should be equivalent; the design system documentation support tools should provide translation between the two representations.

Structure of Hierarchical Input Language Descriptions

The system description language to be discussed in the remainder of this section resembles a structured progamming language in many ways. Both structured programs and hierarchically constructed logic descriptions share the need for the declaration of identifiers used in the description, the *typing* of variables and signal names, and, in particular, the implementation of a set of rules for determining the *scope* of an identifier (to determine the extent of applicability when the same name is used in multiple declarations). Rules of applicability are required in system design, since a number of designers commonly share the development task. Each designer is free to define and use identifiers within the scope of individual partitions; no ambiguities are introduced if the same identifier is declared and used

---

[2] EDIF: Electronic Design Interchange Format; VHDL: VHSIC Hardware Description Language; CONLAN: Consensus Language. For a review of some of the current research in hardware description languages, see *IEEE Computer*, February 1985, special issue.

elsewhere. Two cases arise: a name can be reused in disjoint nonoverlapping scopes, or a name can be redefined within the scope of a previous declaration. As is the case with structured programming languages, the innermost definition will pertain within its specific range of validity.

The basic structure for partitioning the system function into networks will be denoted as a *block*; the exterior of a block can be regarded as a *black box* with pins. To continue the analogy with structured programming languages, a block would be the counterpart of a programming *procedure*. No restrictions are placed on the number of pins or the complexity of the function contained within the scope of a block description; indeed, the entire system is contained within the scope of the top-level declaration.

The framework of a system description is depicted in Figure 4.6. Prior to the collection of block descriptions, a declarative section is provided. Each block description contains signal interfacing and modeling information. The top-level block description encompasses the entire system design; the partitions of this function that define the design hierarchy are contained within the scope of the top-level description. Nested blocks (those referenced by a global name) are outside the scope of the top-level description and are included separately.

The declarative section is used to define the *global* identifiers that are to be encountered by the parser in the subsequent block descriptions; these identifiers include block names, variable names, and constants. The declarative section would commonly be of the form diagrammed in Figure 4.7. It is typically assumed

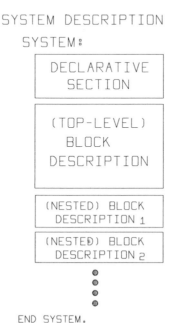

Figure 4.6. Overall structure of a hierarchical system description. The top-level block description contains the entire system partition, exclusive of the external nested functions.

that (1) blanks cannot be embedded in identifier strings (but are suitable anyplace else), (2) commas serve as separators, and (3) semicolons serve as statement terminators. (Comments can be embedded anywhere in a design description and should be used frequently to aid clarity and readability. One means of including comments is to identify a single character that begins a comment; the comment then terminates at the end of a line. Another is to use a complementary pair of delimiters, one to initiate and the other to terminate the comment string. In this manner, a comment section can be located anywhere in the text stream, potentially spanning many lines.) The declarative section in the figure includes some constructs that are similar to those of structured programming languages: constants, variables (with specific typing for enhanced semantic checking), and functions. Of particular note in this initial section is the *UNITS* group, which provides additional possible variable types. A value of this type therefore contains a numeric and units descriptor to be parsed (e.g., 450 FF). The assignment of a value to a variable defined as containing a unit of measure requires normalization of that value to the default measure. For the calculation of variable values to be mean-

```
SYSTEM:
    BLOCK__DECLARATION block₁, block₂, . . . ., blockₙ ;
    /* defines all block names of global scope -- i.e., top-level and
            nested block names */
    UNITS
        time :  nsec,       /* default  */
                psec = 1.0E-3 nsec,
                usec = 1000 nsec,
                msec = 1.0E6 nsec,
                sec = 1000 msec;
        voltage :        . . . .     ;
        current :        . . . .     ;
        power :          . . . .     ;
        frequency :          . . . . ;
        capacitance :        . . . ;
        energy :             . . . . ;
    CONSTANTS
            ⋮
        identifier = integer/real/Boolean/string ;
            ⋮
    TYPE
        sim__model = (LOGIC, TABLE, PARTS, BEHAVIORAL);
        designer =      RECORD
            name :  packed array [1 . . 20] of char ;
            date :  . . . . ;
            revision :  integer ;
            status :  . . . . ;
            end ;
```

(*continues*)

Figure 4.7.   Declarative section of a system description.

```
duty_cycle = 0 . . 100 ;  /*  an integer subrange denoting a percentage */
range = (nom, min, max);
propagation = RECORD
     rising :   array [range] of time;
     falling :  array [range] of time;
   END;
technology = (NMOS_2UM, CMOS_2UM, CMOS_1UM, ECL );
VAR
   sysclk :  frequency;
   supply :  voltage ;
   parameter simulatable :  Boolean ;
   parameter placed :  Boolean ;
   parameter delay :  propagation ;
   parameter default_block_delays :  Boolean;
   /*  default block delays are only sufficient for combinational networks */
   parameter combinational :  Boolean ;
/* function declarations    */
   function f₁ ( var₁:type₁, var₂:type₂, . . . ) : result_type;
   var
   begin
   end;
```

Figure 4.7.   (*Continued*)

ingful when compiling the system description, the default set of units should be defined in a consistent manner.

Another unique attribute of this description format is the designation of some variables as *parameters*. The intent of this definition is to allow *each* block to have an associated set of parameters that can be uniquely assigned to that block. Parameters are not to follow the rules of scope normally applied to variable names; rather, they are global identifiers that can be defined and assigned to individual nodes in the design hierarchy. Most of these block parameters are of type *Boolean* and can be used to describe the *tree* properties of a hierarchically structured description. The parameter can define the block to be a node or a leaf of the design hierarchy to a particular design system application tool. For example, if a node of the design description is incompletely defined when a simulation model is required, the block parameter "simulatable : = FALSE;" could be assigned. During model build, the expansion of this block (and all partitions thereof) could be inhibited; the block outputs would then be given *undefined* values during simulation.

The *TYPE* section is similar to that of structured programming languages in that subranges, arrays, records, and enumerated lists can be described and given to variable definitions.

Recalling the discussion on netlist languages in Section 4.1, it should be evident that a significant fraction of the constructs of future languages does not pertain to the interconnectivity of logic networks at all. Indeed, most of these additional constructs or features are not required. However, the current emphasis in design data management is to be able to contain as much of the user- and design-system-generated information as possible in a single encompassing database. As new types of information are required or produced by design system application tools, the means for design description should be able to expand and absorb those data. The complexity of the compilers required to extract, calculate, and manipulate design data from such a diverse description is considerable. This raises the following questions: precisely what should the system design database consist of? Exactly how encompassing should the design description language be? This chapter and chapter 5 will introduce two more *languages* aside from the current description: one is for specifying the simulation stimulus test cases against which the system network logic model is to be exercised (Section 4.9), and the other is a shapes description language for physical design (Section 5.2). In both cases, it would certainly be feasible to try to contain these data in a more general system description language definition; however, the nature of this information is sufficiently distinct from logic definition (and the quantity of data so extremely large) that attempting to merge these languages (and their associated databases) may be unwise.

The structure of a block description is depicted in Figure 4.8. The block heading defines the pins that are used to connect the block to the exterior environment (Figure 4.9). A *bus* is a collection of related pins, which is specified as an identifier followed by a range of integers; individual pins in the bus are denoted by concatenating the integer to the bus name. The block heading also includes an indication of the signal directionality of each pin or bus, much like assigning a *type* to each signal. The set of signal directionality types could include: *Input, Output, Bidirectional, High impedance,* and open drain/collector (*Dottable*). The compilation of the design description can therefore perform initial checking of the validity of interconnections (e.g., no two outputs driving the same net). The simulator can use these designations to determine signal values when resolving the case of multiple sources in a net, which arises when bidirectional, high-impedance, and dottable circuits are used.

After the block heading comes the bulk of the block description, which resembles the structure of a programming procedure, yet can contain several unique sections.

The *PIN_EQUIVALENCE* section defines logic equivalence relations between pins; this information may be of use during physical design if a difficult interconnection to wire can be simplified by *swapping* pins.[3] Figure 4.10 illustrates a possible syntax for specifying pin equivalence; expressions may be nested to best describe the relationships. Two delimiters are used: square brackets signify that the enclosed list of identifiers is an unordered set, while parentheses indicate that the pin names enclosed constitute a vector or ordered set.

---

[3] If logically equivalent pins are indeed swapped by a wiring program, that information must be written back to a block parameter in the design description database (Section 4.6).

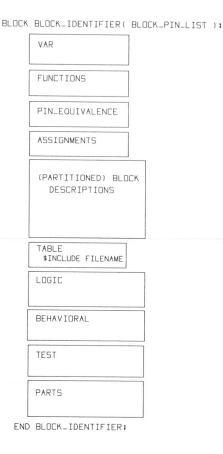

```
BLOCK BLOCK_IDENTIFIER( BLOCK_PIN_LIST );

    VAR

    FUNCTIONS

    PIN_EQUIVALENCE

    ASSIGNMENTS

    (PARTITIONED) BLOCK
        DESCRIPTIONS



    TABLE
      $INCLUDE FILENAME

    LOGIC


    BEHAVIORAL


    TEST


    PARTS


END BLOCK_IDENTIFIER;
```

Figure 4.8. Overall structure of a block description; note that the descriptions of partitioned blocks are contained within the *parent* block description.

```
              BUS DEFINITION
                    /
                   /
                  ∟
BLOCK 256KDRAM ( A(8:0):I, DIN:I, CAS¬:I, RAS¬:I,
                          R_¬W:I, DOUT:O);
```

Figure 4.9. A block heading; signals A0 through A8 are collected into a *bus* definition.

```
BLOCK 2-2AOI (A1:I,A2:I,B1:I,B2:I,OUT:O);

   PIN_EQUIVALENCE [[A1,A2] , [B1,B2]] ;

BLOCK 4-BITREG ( D(0:3):I, CLK:I, Q(0:3):O );

   PIN_EQUIVALENCE [(D0,Q0),(D1,Q1),(D2,Q2),(D3,Q3)] ;
```

Figure 4.10. The PIN_EQUIVALENCE section of the block description indicates which pins (or groups of pins) may be swapped by the physical design wiring program to potentially enhance the overall chip wireability.

The *ASSIGNMENTS* section is the counterpart to the executable section of a procedure; this section contains the statements that assign variable and block parameter values. As alluded to earlier, these block parameters can be defined to contain a great amount of information, from performance data to design status. As the design evolves and additional block information is available, this section will be enhanced; block descriptions that have changed can be recompiled to reflect these data.

The function of the block for simulation and test modeling purposes may be prepared in a number of different representations; often, more than one will be coded for the design. Some of the possible representations are depicted in Figure 4.8. The simplest functional representation, suitable for combinational networks, would be to provide a *truth TABLE*. (Problem 4.3 requests that a truth table syntax and file format be defined for this means of representation.) This technique is particularly well suited for describing ROM contents and PLA functionality. The block delay parameter value would be assigned to all output transitions.

A *LOGIC* section would describe the network function in terms of logic equations and simulation primitives. An example is given in Figure 4.11. (Section 4.5 discusses simulation primitives in further detail.) A logic equation assigns the result of evaluating a Boolean expression to a signal name. A simulation primitive is instantiated by assigning an identifier to the primitive and signal names to the primitive's pins. A netlist-like record is shown; an explicit assignment format is required for primitives with nonsymmetric inputs. Each output pin of the block must be referenced as either an equation or simulation primitive output. Internal delays can be assigned to equations and primitives within the LOGIC model sec-

```
/*    LOGIC section for a dual 2-to-4  decoder  --  74LS155    */

LOGIC
        INTERNALS    NOT1C, NOTA, NOTB, SEL1, SEL2 ;
        G1:  (NOTA) = INV (A)  [3   7] ;
        G2:  (NOTB) = INV (B)  [3   7] ;
        G3:  (NOT1C) = INV (C)  [8   4] ;
        G4:  (SEL1) = NOR ( 1G , NOT1C );
        G5:  (SEL2) = NOR ( 2C , 2G );
        1Y0 = NOT( NOTB AND NOTA AND SEL1 )  [10   18] ;
        1Y1 = NOT( NOTB AND  A  AND  SEL1 )  [10   18] ;
        1Y2 = NOT(    B  AND NOTA AND SEL1 )  [10   18] ;
        1Y3 = NOT(    B  AND  A  AND  SEL1 )  [10   18] ;
        2Y0 = NOT( NOTB AND NOTA AND SEL2 )  [10   18] ;
        2Y1 = NOT( NOTB AND  A  AND  SEL2 )  [10   18] ;
        2Y2 = NOT(    B  AND NOTA AND SEL2 )  [10   18] ;
        2Y3 = NOT(    B  AND  A  AND  SEL2 )  [10   18] ;
```

Figure 4.11.   A network simulation model consisting of logic equations and simulation primitives; unique delays can be defined for individual equations and primitives of the model, to be used in lieu of the default block output delay parameter values.

tion; these can be used in lieu of the default block delay parameter to provide unique path delays to the outputs of the block. A list is provided of internal signal names present in the logic definition but that are not pins of the block.

The *TEST* section depicts the network model as the interconnection of test primitives, in much the same manner as the LOGIC section (only without the option of Boolean equations). Each additional signal name identifier in the test model would typically have an associated stuck-at-1 and stuck-at-0 fault condition pair, which must be isolated by the set of test patterns. (Refer to Chapter 13 for a more detailed discussion of testing methodology.)

The *BEHAVIORAL* section is unlike any other; a behavioral model of a network is a true programming procedure written for simulation. The intent of coding behaviorals for complex networks is to provide simulation efficiency and power, on the order of register-transfer language descriptions. A full range of programming constructs are available (e.g., local variables, conditional expressions, and the ability to output messages to an error file). Unique to a behavioral modeling approach over other techniques are the following capabilities:

1. Reading and locally storing the current simulation time (a global variable) when the procedure is called.
2. Representing signal, bus, and array values in a variety of forms (e.g., binary, octal, decimal, hexadecimal) and performing logical and/or arithmetic operations on those values.
3. *Scheduling* a signal, bus, or array value change in future simulation time from within the procedure (in lieu of using a block delay).

The behavioral procedure is called by the simulator whenever a transition on an input pin to the block has occurred.

The *PARTS* section is the primary means of describing how the function contained within a design hierarchy node has been partitioned. This section lists all subblocks and the nature of their interconnections. An example of a PARTS section is illustrated in Figure 4.12a. The section consists of a NETS declaration list, followed by one or more part declarations. The NETS subsection is simply a list of identifiers that define the net names used within the block description, that is, the names for the interconnections at this node of the design hierarchy that are not pins of the block. (If no additional signal nets are present in the partition, aside from block pins, the NETS section is not required.) Directly following each net identifier, additional delays can be specified; these net-specific values would be added to the rising and falling delay parameters associated with the block(s) driving that net (Figure 4.12b). These data are intended to accommodate more precise net delays after the chip physical design has been completed by adding delay blocks to individual circuit outputs (refer to Section 9.5 on delay calculation). Each subblock declaration begins with a *short name*, followed by the block name of a subblock in the partition. A list of signal names follows the block name; these nets represent the connections to the pins of the subblock. A

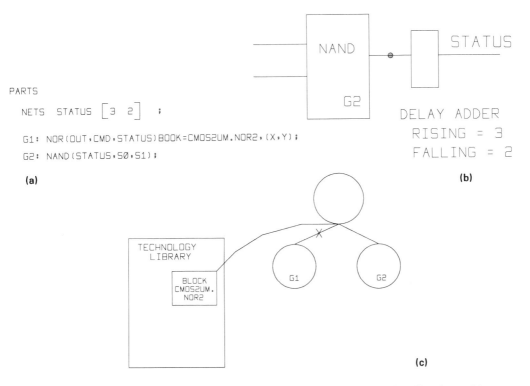

PARTS

  NETS STATUS [3 2] ;

  G1: NOR(OUT,CMD,STATUS)BOOK=CMOS2UM.NOR2,(X,Y);

  G2: NAND(STATUS,S0,S1);

(a)

NAND

G2

STATUS

DELAY ADDER
RISING = 3
FALLING = 2

(b)

TECHNOLOGY
LIBRARY

BLOCK
CMOS2UM.
NOR2

G1   G2

(c)

Figure 4.12.   (a) The PARTS section of a block description, which describes the partitions of the block and their interconnections. The NETS section lists the internal signal names used in the description. (b) Delay values can be assigned to internal nets to add to the delay from the driving block. (c) The field "BOOK = identifier" following the connection list of the subblock is used to point to a specific technology library circuit or macro.

positional format is illustrated; the pin and net names are listed in order so as to correspond with the pin list in the definition of the subblock. An explicit pin assignment format could likewise be used. The reserved identifiers 0, 1, and U can be used in this connection list to indicate an input pin tied to a fixed logical 0 or 1 value or an output pin to be left unconnected. Following the connection list, the optional field BOOK = identifier; can be included. This is an indication that the block name at the beginning of this record is *not* to be used for modeling, but rather is to be supplanted with the description of a technology-specific circuit or macro design. This is the means for selecting a particular book from the design system technology library, ultimately to be placed and wired during chip physical design (Figure 4.12c). An (x, y) coordinate pair describing the physical placement location can also be associated with the book identifier.

   Referring again to Figure 4.8, it should be highlighted that a block description contains within it all the descriptions of partitions beneath it in the design hierarchy. Just as with structured programs, this means of structuring the logic description is necessary to implement rules regarding the scope of block names and signal names. As illustrated further in Figure 4.13, the block names used in the PARTS section of a declaration are defined as part of the next partitioning level or defined as nested blocks. In the case of nested blocks, these globally unique identifiers are declared at the beginning of the system description. Net names and short part names only have significance within the block description node itself; their scope does not extend above or below the block description in the design hierarchy.

An alternative technique would be to simply concatenate individual block descriptions, both nested and partitioned descriptions. In this case, all block names would have to be unique.

In Chapter 5, this alternative approach is indeed used when creating the chip's shapes description using design system placement and wiring tools. Circuit layout descriptions are concatenated to the chip wiring and personalization data in the shapes file. Effectively, all circuit layouts are to be regarded as nested, rather than as partitions of the chip physical design. Whereas system logic partitions are developed independently among members of the design group, the design system circuit or macro layouts are the result of close interactions among the technology physical design team. As a result, layout names will be unique and will not require scope rules; a straightforward concatenation of physical macro descriptions can be used without ambiguity.

A system is fully described for logic simulation when traversal of the design tree encounters a TABLE, LOGIC, or BEHAVIORAL section to cover all branches (i.e., no leaf node of the tree is reached without one of these sections).

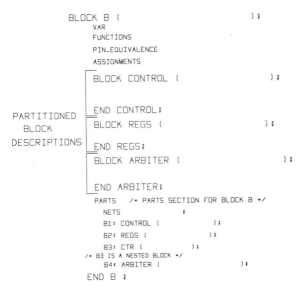

Figure 4.13. The descriptions of block partitions are contained within the parent block definition, just as subprocedures are defined within procedures in structured programs; the exception to this structure is that nested blocks are defined globally, outside the scope of the parent block description.

Similarly, the system is fully described for physical design when traversal yields a BOOK = identifier; reference to cover all nodes.

This section has briefly presented a language for describing system behavior for logic and test modeling simulation; the information necessary to initiate the chip physical design is also provided. This presentation is lacking in several regards. No formal language definition was provided, and the semantic checking involved with compiling such a description was only mentioned in passing. The features described are a composite of several languages currently in use and under development; no implementation of this specific collection of features exists, per se. No design data structure has been constructed to store the associated information contained within this description; no tools for translating a description into a corresponding database (or vice versa) have been developed. Rather, the intent is to provide an indication of the directions in description languages, without selecting one particular implementation over another. The following strategies are clear:

1. A hierarchically constructed system description consisting of partitioned and nested networks can be used to effectively manage a complex design task.

2. A single language or database is attractive for representing a wide variety of design information, above and beyond signal interconnectivity.

3. A description language should be capable of facilitating design implementation changes with a minimum of modifications. These changes could include technology selection, estimated or calculated block delays, and pin swapping during physical design.

4. A description language should support a number of modeling alternatives: Boolean, structural (logic primitives), and behavioral.

The last two points deserve further consideration. As discussed in Section 4.1, design productivity can be gained by developing technology-independent descriptions, subsequently using logic synthesis and technology transformation tools. The modeling options available for network descriptions should be consistent with the capabilities of these tools. Synthesis and transformation algorithms are being enhanced to incorporate specific attributes or goals provided by the designer into their procedures. Block parameters can be used to directly influence the resulting technology implementation. Behavioral modeling also offers a gain in productivity for system simulation tasks. Yet the possibility of multiple representations of a block's simulation model raises the question of consistency among model predictions. It is difficult to verify that a behavioral model coded for a large network concurs with the structural and Boolean expansion of that network. Considerable design system development effort is currently being invested in the tools to assist with comparing *equivalent network* logic simulation results, given that differences in timing resolution and/or detail may be present in the various models.

4.4 DATA STRUCTURES FOR LOGIC DESIGN DATA

When a design description is parsed, the resulting design database becomes the source of reference information for design system application tools. These programs access and update this database. A *hierarchical* data structure can be designed to store these data in a manner so as to correspond exactly with the hierarchical partitioning of the system design. A discussion of data structures is far beyond the scope of this text; this section will only briefly introduce some of the terminology and features of data structures that could be used as a foundation for a design database.

Some pertinent terms used in describing hierarchically constructed designs are illustrated in Figure 4.14. The *relationships* between nodes use references to genealogy; the *structural* picture of the design uses references to the inverted design tree.

The tasks involved in working with a system design description include the following:

Parsing

A parser receives a text file description and interprets the language to fill the basic data structure. Whereas the text file contains a single copy of each nested block description, the database contains separate nodes for each instance of a nested block.

Database Storage and Retrieval

A database management tool facilitates the storage and retrieval of the data structure between main memory and permanent files. An additional application tool

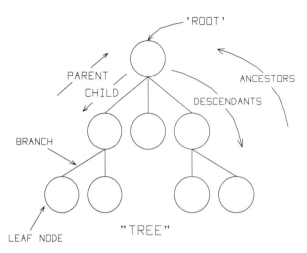

Figure 4.14. Terminology associated with hierarchically constructed designs.

set is required to extract simulation and test modeling information from the system-resident data structure.

Database Editor

Design system application programs need to access and modify the block variable and parametric data resident in the database. An editor assists with the search for specific blocks and block values. To the editor, there will always be a reference node, often referred to as the *current context*. Traversing the tree to locate another node of interest, be it in the direction of ancestors or descendants, utilizes the concept of *pointers*, which are followed until the desired context is reached. Within the current context, pointers are also used to reach block variable and parameter data fields.

A database editor must also perform a global search (or search and replace) for specified block parameter values. In addition, the editor should be able to accept a batch file of edit commands. Both these features would be of use to design system application programs.

Pointers are extremely useful in accessing hierarchically structured data (as well as in defining the contents of a list of records that may be dynamic in length). In the case where the design database is resident in main memory, these pointers are equivalent to memory addresses; they are allocated by the database retrieval program when the design description is loaded from file. Memory address pointers lose significance when the database is written to file; rather, an alternative format for permanent storage is required. Typically, to simplify the task of data storage and retrieval for editing, the complete design database is broken down into many files, which are accessed by file directory pointers. An example of how the data for different blocks might be broken down into separate file directories is given in Figure 4.15. Each block name is used as a file directory name; the concatenation

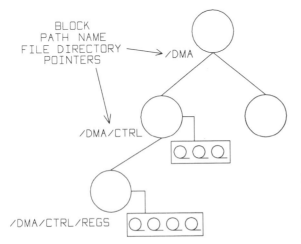

Figure 4.15. Each block's database files are located at the directory defined by the block *pathname*, constructed by the concatenation of file directory pointers.

of the block name to that of its ancestors provides a directory *path name* from the design root. The database file(s) for a specific block is resident at that block's path name.

Database Supervisor

A very important aspect of managing a design database is the ability to exert control over the accessibility to that information as the design evolves. Specifically, a database supervisor must be involved to oversee the following tasks:

1. Assigning file security levels to limit access to only those users and tools with sufficient authority.
2. Ensuring that multiple users and tools do not have the opportunity to simultaneously edit (and thereby corrupt) database files.
3. Tracking the users who have personal design versions *checked out* from the overall database.
4. Keeping file *time-date/user* information current for updates to database files.
5. Making archival copies of design data, enabling recovery to a previous design point.

All these tasks are complicated by the trends that characterize VLSI designs: a potentially large design group participating in the evolution of the design description, and a wider range of design system application tools requiring access to a more encompassing database of design information.

4.5 LOGIC MODELING FOR SIMULATION

Alternatives for describing the logic model of a block were described briefly in Section 4.3; specifically, BEHAVIORAL, TABLE, and LOGIC formats were introduced. The BEHAVIORAL model was described as a procedural programming interface, with access to the internals of the simulator. The TABLE format provides a means of inputting a truth table for a combinational network. This section will expand on the means for describing a network in the LOGIC format, which uses a combination of Boolean equations and simulator primitives.

Before beginning the discussion, it should be mentioned that logic simulation algorithms are complicated considerably by the possibility of circuit outputs dotted together. Recently, simulation algorithms have been enhanced to accommodate multiple driving circuits in a net, with an associated set of *contention* rules. Each block description requires that a parameter value be assigned indicating the specific technology-dependent contention rules to be used for the pins and internals of that block. Figure 4.16 lists the characteristics adopted by the simulator for different technology parameter values. In the figure, a *Forcing* signal strength implies that an active device provides the logic value, while a *Resistive* strength indicates the presence of a passive load device. A high impedance strength (denoted as Z in the figure) indicates that neither an active nor passive device is present in the circuit for a signal transition. When determining the value to be assigned during simulation to a net with multiple drives, the strengths of the sources are compared, in addition to their logic values. The order in which signal strengths of these outputs dominate is Forcing, Resistive, and high impedance (Z); for example, a forcing 0 contending with a resistive 1 results in a net assignment of 0.

The Boolean equations present in a block's LOGIC model section would typically allow the use of the operators NOT, AND, OR and XOR. A priority of evaluation is used by the parser for these operators in case the nesting of parenthetical expressions does not clearly define the intended order (Figure 4.17). The Boolean equation requires less memory in the simulation model (and commonly executes more quickly) than an equivalent representation using logic primitives.

A logic primitive statement is depicted in Figure 4.18; a unique identifier is followed by the output interconnection(s), the primitive type, and an input interconnection list.

A conventional set of simulation primitives would include INV, AND, NAND, OR, NOR, XOR, XNOR, and DOTAND. The last item in this list is for resolving contention, as described earlier. It would be used inside a model when it is desirable for the signals involved to maintain different names. All these examples, except INV, could conceivably accommodate an indefinite number of inputs. A richer simulation primitive set may incorporate the following additional

TECHNOLOGY PARAMETER	DRIVING PIN TYPE	1	0	DISABLED
CMOS	OUTPUT	F	F	-
	HIGH IMPEDANCE	F	F	Z
	BIDIRECTIONAL	F	F	Z
	DOTTABLE	Z	F	Z
BIPOLAR	OUTPUT	F	F	-
	HIGH IMPEDANCE	F	F	Z
	BIDIRECTIONAL	F	F	Z
	DOTTABLE	Z	F	Z
NMOS	OUTPUT	R	F	-
	HIGH IMPEDANCE	F	F	Z
	BIDIRECTIONAL	F	F	Z
	DOTTABLE	Z	F	Z

F - FORCING; R - RESISTIVE; Z - HIGH IMPEDANCE

Figure 4.16. Each technology parameter–pin type combination has an associated *signal strength* used in resolving contention when multiple circuit outputs are dotted together.

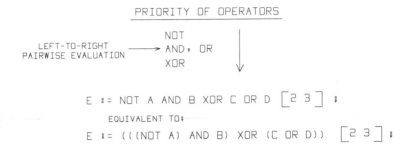

Figure 4.17. When parsing a logic equation, a priority of operators is assumed, unless overridden by the use of parenthetic expressions.

```
IDENTIFIER: (MODEL_NET_NAME) =

              PRIMITIVE (MODEL_NET_NAME, .... ) [        ] ;

    OR

IDENTIFIER: (PRIMITIVE_PIN_NAME:MODEL_NET_NAME) =

              PRIMITIVE (PIN_NAME:MODEL_NET_NAME, ... ) [      ] ;
```

Figure 4.18. The LOGIC simulation model may also consist of primitive definitions. For primitives with logically equivalent inputs, an explicit input pin assignment is not required; for other simulation primitives, an explicit assignment is required.

members: LATCH, DFF, JKFF, RAM, and a high-impedance driver (DRVR, IDRVR).[4] The two primitives DRVR and IDRVR are depicted in Figure 4.19. When the enable input is active, the output value is a function of the input value; when the enable is off, the output goes to a high-impedance condition. This primitive is useful in constructing the logic model for high-impedance, bidirectional, and open drain circuits (Figure 4.16).

Aside from Boolean expressions and primitives, the power of the LOGIC model section can be enhanced substantially by the incorporation of timing check functions within the description. Figure 4.20 illustrates three timing check functions that may prove to be useful: minimum pulse width, setup and hold time, and minimum signal-to-signal transition delay. If a simulation run detects that a signal relationship specification has not been met, a message can be sent to an error file indicating the nature of the violation.[5]

---

[4] Some of these logic *primitives* would ultimately be replaced with a behavioral routine to accurately model their sequential characteristics.

[5] As with sequential logic primitives, these checking functions will ultimately be implemented with small behavioral routines.

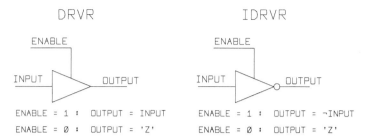

Figure 4.19. The simulation primitives DRVR and IDRVR are used in constructing the LOGIC model for high-impedance, bidirectional, and dottable circuit types.

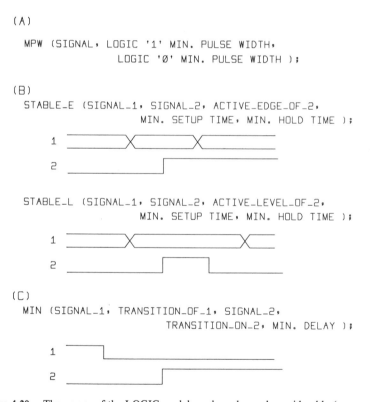

Figure 4.20. The power of the LOGIC model can be enhanced considerably (approaching that of the BEHAVIORAL) if timing check functions are supported. These functions include (a) minimum pulse width (1 and 0), (b) setup and hold times (relative to a signal transition or level), and (c) signal transition–to–signal transition minimum delay.

A recent trend in design system simulation tools provided for engineering workstation equipment is the capability to use a *physical* implementation of the block to replace a software simulation (LOGIC or BEHAVIORAL) model. For example, if one of the blocks in a hierarchical system description is an existing VLSI component (e.g., a microprocessor or microcontroller), the difficult and error-prone task of composing a software model for the part can be eliminated by using the physical hardware instead. The execution efficiency of the simulation model will also be substantially improved. The workstation with this feature would include an electrical interface that can be programmed to force or measure a logic voltage at each pin of that interface. When the workstation simulator detects that an input to the physical model has changed, the electrical equivalent of the new stimulus pattern is applied to the interface connector. After a suitable time interval, the response is measured and appropriate 1 and 0 block output values are returned to the software simulator. The actual performance of the physical electronics is not measured; rather, simulation model delays for the hardware are still used. The intent is to capture only the functional behavior of the electronics; this simplifies the physical breadboarding and, more importantly, allows advanced (preliminary) performance numbers to be used when the corresponding hardware is not yet available. In general, any digital electronics could be exercised—chip, card, or even electromechanical hardware (with a D/A interface). Since the actual time interval between simulation patterns to the interface will typically be "long," dynamic electronics requires special consideration; dynamic logic values will likely have decayed between patterns, rendering the previous state of the logic as *unknown*. Problem 4.6 requests a study of the physical hardware simulation capabilities of current engineering workstation products, with specific emphasis on the modeling of dynamic hardware. To indicate that a physical implementation is to be used for simulation, a new type of section will be required in the block description, one that contains the model-to-electrical interface pin correspondence (Problem 4.7).

4.6 PIN EQUIVALENCE

The *PIN_EQUIVALENCE* section of the block description is optional; it is included as a means of specifying to the design system physical design tools (automated wiring, in particular) which pins of the block may be exchanged between nets to enhance the overall chip wireability. Although the syntax defined for the section enables quite an intricate relationship among equivalent groups of pins to be defined, it is unlikely that a physical design algorithm would utilize such a detailed specification.

When pin swapping has indeed been used during physical design, it is necessary to indicate the nature of the transformation back to the design database. A block parameter can be defined to record this information (Figure 4.21).

```
SYSTEM:

   TYPE

      PIN_NAME: PACKED ARRAY [1..8] OF CHAR;

      PIN_LIST_PTR: ↑ LOG_PHYS ;

      LOG_PHYS : RECORD

         LOG_PIN : PIN_NAME ;
         PHYS_PIN : PIN_NAME ;
         NEXT : PIN_LIST_PTR ;
       END ;

   VAR
                  •
                  •
                  •
      PARAMETER  HEAD_PIN_LIST : PIN_LIST_PTR ;
                  •
                  •
                  •
```

Figure 4.21. A collection of database structures can be defined to allow each block description to contain revised logical-to-physical pin associations if pin swapping was used during chip physical design.

4.7 PIN DROPPING

In previous chapters it was mentioned that master image and master slice VLSI design system chip images are commonly developed with a specific *granularity* of circuit pins in each cell. Advantages and disadvantages were given for both narrow and wide cell definitions. Whatever the cell size, design productivity can indeed be enhanced if the technology library database permits the designer to implicitly *pin drop* (i.e., simply leave pins of a book unspecified).

A PIN_DROP section of a design system library block description is depicted in Figure 4.22; also included is a corresponding "BOOK = library_identifier" pointer as it would appear in a design description. The PIN_DROP information includes a list of which pins may indeed remain unspecified and the logic value to subsequently use in the simulation model. In the example given, pins A2 and A3 of the technology book are not identified; these *dropped* pins are to be assigned a fixed 0 value during simulation. If the design description attempts to drop a pin that is not included in this list, an error is generated when the description model is compiled.

The design system decision to support wider use of individual book designs through pin dropping affects other design system tools, in addition to logic simulation. The pin drop list for each instance of the book is also used in assembling the chip shapes file; specific shapes will need to be added and/or deleted from

```
BLOCK CMOS2UM.NOR (OUT:O,A0:I,A1:I,A2:I,A3:I);

  /* THE DESIGN SYSTEM IMAGE SUPPORTS A 2-TO-4 WAY
       NOR WITH ONE BOOK (IN A SINGLE CELL LOCATION) */
  PIN_DROP (A2:0, A3:0);
  PIN_EQUIVALENCE [A0,A1,A2,A3];

  LOGIC   OUT = NOT (A0 OR A1 OR A2 OR A3);


  /* SAMPLE USAGE                */

  PARTS
            °
            °
            °
      G1: NOR ( DECODE, S0, S1 ) BOOK=CMOS2UM.NOR;
            °
            °
            °
```

Figure 4.22.   A special PIN—DROP section is included in the block descriptions of technology library books to indicate which input pins may be left unspecified by the designer and which logic simulation value to assign to a dropped pin.

the circuit layout for the book to correspond to the pins that are unused. (Section 5.2 discusses a technique for assigning a *conditional* use to a collection of shapes in a layout; the evaluation of this condition is based on the set of pins used and dropped.) Likewise, the design system testability analysis and test pattern generation tools must incorporate the pin drop list for each book into the test model, modifying the overall set of possible physical faults accordingly.

The logic description language introduced in Section 4.3 indicates the capability of specifying that block input pins were to be connected to 0 or 1 and that outputs could be denoted as Unconnected. When the design description is ultimately translated to a technology-dependent form, this information is also pertinent to pin dropping. Droppable pins tied to a logic value are candidates for modifying the chip physical design and fault model; book input pins tied to a logic value that are not in the pin drop list will require special consideration during wiring so that an interconnection to a chip supply voltage rail can be produced.

4.8 FUNCTIONAL VERIFICATION AND TIMING ANALYSIS

Two primary features are integral to any logic simulation tool: the interpretation of block delay information and the set of simulation values assignable to a net. This section briefly discusses some of the alternative approaches that have been used in supporting these features.

The set of assignable simulation values may include some or all of the following: 0, 1, U, R, F, C, and S. The Unknown value is usually assigned by default to *all* nets in the model prior to evaluating initial (sim—time = 0) stimulus conditions; it may also be assigned during execution of a simulation run to high-impedance nets that have *decayed* and to nets that did not satisfy one of the timing check functions in the model (pulse width, setup and hold time, or minimum delay). When coupled with the assignment of a technology driving strength (to resolve contention on multiple-source nets, as discussed earlier), the three values 0, 1, and U are commonly used for functional design verification tasks.[6] The values Rising, Falling, Changing, and Stable will be discussed shortly.

The number of delay values assigned to a block in the simulation model can range from one to six, as follows:

$$delay_1: \quad \text{used for all block transitions, rising and falling}$$
$$delay_1, delay_2: \quad \text{rising transition delay, falling transition delay}$$
$$delay_1, \ldots, delay_6: \quad \text{min\_rising, nom\_rising, max\_rising, min\_falling,}$$
$$\text{nom\_falling, max\_falling delays}$$

A *three-valued* functional simulation (0, 1, and U) would also be designated as MIN, NOM, or MAX, depending on the rising and falling delay numbers to be selected from a full set of values. Thus, the design verification test cases can be exercised at both *best-case* (minimum delay) and *worst-case* (maximum delay) conditions to ensure overall functionality over the anticipated range of operating conditions.

One of the shortcomings of the MIN, NOM, or MAX delay simulation approach, however, is the presumption that all block delays will *track* perfectly, that all block delays simultaneously exhibit minimum, nominal, or maximum values. A number of system timing problems can indeed be uncovered if a variation (or spread) in each assigned block delay is permitted throughout the simulation model, an uncertainty introduced in the actual simulation time when a net transition is observed. In this case, a block is assigned two delay values for each transition; using the delay definitions, it would be possible to specify MIN:NOM, NOM:MAX, or MIN:MAX delay modeling. This design verification approach is commonly referred to as *timing analysis*, to distinguish it from the aforementioned *functional* delay simulation technique. A typical system design flow would initially involve functional verification, followed by timing analysis. In this manner, coarse design problems would be detected and corrected as early as possible; more intricate timing skew problems are then uncovered and evaluated.

Timing analysis can be performed in two different ways: *simulation with transition uncertainty* and *network delay path evaluation* (for synchronous systems). Simulating for the purpose of performing timing analysis is similar to functional simulation; a set of test case pattern stimuli is applied against the model, which now contains a range of delays for each possible block transition. Figure 4.23 depicts how the simulation output would represent the uncertainty in the

---

[6] A U value would also be assigned if forcing 1 and forcing 0 values were in contention simultaneously on a net with multiple sources.

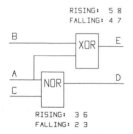

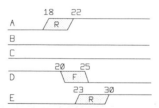

Figure 4.23. Simulation with transition uncertainty is similar to functional verification, except that a range of possible delay values is used for each transition; the simulation values Rising and Falling are added to indicate the interval of uncertainty.

actual block output transition time; in the figure, an R represents a Rising transition in progress, while an F describes a Falling transition. The truth table for the simulation primitive is modified to include R and F input values; an indeterminate or Changing value is also required for the case where more than one transition is pending (Figure 4.24). The interpretation of signal values also requires modification when performing one of the pulse width, setup and hold time, or transition-to-transition minimum delay checks (Problem 4.9). Simulating with transition uncertainty may be used to expose such potential timing problems as glitch pulses due to network hazards and sequencing problems due to races in asynchronous state machine transitions.

For synchronous systems, timing analysis can use the network delay path analysis approach (Figure 4.25). Of primary concern to the designer is that the signal flow through the combinational network results in stable input values to the target register within a system clock cycle (minus the setup time); this design constraint should be observed *for all possible* signal paths through the network.

AND

	Ø	1	R	F	C
Ø	Ø	Ø	Ø	Ø	Ø
1	Ø	1	R	F	C
R	Ø	R	R	C	C
F	Ø	F	C	F	F
C	Ø	C	C	F	C

Figure 4.24. Truth table for an AND primitive for simulation with transition uncertainty; Rising, Falling, and Changing values are added to the set of possible signal values.

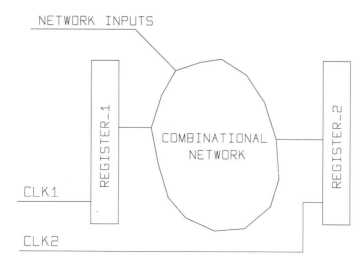

Figure 4.25. The network path delay evaluation approach to timing analysis involves path tracing to determine the longest delay between clocked registers; this delay must be less than the clock separation time between CLK1 and CLK2.

The goal of ensuring that all possible paths will meet system timing requirements separates this approach to timing analysis from conventional simulation. A special timing analysis design system tool is commonly provided to exhaustively search all possible logic paths between registers, rather than depending on the specific set of simulation test cases developed by the designer.

The timing analysis evaluation of a combinational network is depicted in Figure 4.26. The primary inputs to the network are assigned a range of time values when they are potentially Changing; for the remainder of the clock period, they are assumed to be Stable. The Changing output values propagate through the network until the outputs are reached; the outputs should be stable a sufficient time period before the *active* edge of the clock arrives. An analysis of the delays through the system clock powering tree is required to be able to properly specify the network propagate and reference compare clock times; the skews present between the clocks to different registers should be included in the timing analysis comparisons. The method of propagating Changing network values illustrated in Figure 4.26 is relatively simple. No logic functionality is associated with any of the blocks, and a single range of delay values is used. A more detailed alternative would again use separate rising and falling delay times; in this case, it is necessary to represent the block as either an inverting or noninverting logic function. Thus, the range in time of a rising input transition to an inverting function would be used to calculate the block's falling transition time window; the cases of the network inputs rising and falling would be treated independently.

Network path analysis tools are developed to pinpoint timing problems in an efficient and thorough manner, obviating the need for extensive simulation (be-

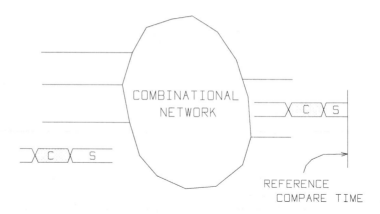

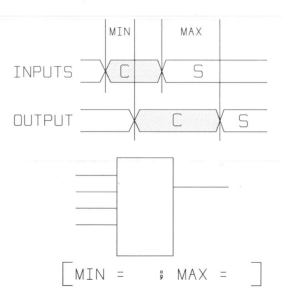

Figure 4.26. Network path delay timing analysis. (a) During the network propagate interval, network inputs are assigned Changing values; subsequently, their simulation value goes to Stable. The network outputs must be S before the reference compare time (derived from the arrival time of the active clock edge to the register storing the output values). (b) The calculation of Changing and Stable block outputs can simply use a range of delay values, and need not be concerned with the particular function of the block.

yond functional verification). Part of that efficiency arises from the coarse modeling of the combinational network, neglecting a block's specific logic function; an exhaustive path analysis is used to calculate transition time uncertainties and ensure thoroughness. This approach has its drawbacks, however; there may be paths identified as timing errors that can never be sensitized by the network logic, paths that were included without regard to function. There may also be paths designed to be stable after *two* clock cycles, rather than one; two machine cycles may indeed be available before the storage register is to be sampled. These special cases require review by the designer and potentially a modification to the network model strictly for the purposes of timing analysis. These modifications might include the addition of delay modifiers to the model and/or the assertion of stable internal network values for a specific timing analysis test case (references 4.1 and 4.2).

An informative means of highlighting network timing problems back to the designer is described in reference 4.1. After the forward calculation of transition times from network inputs, this timing analysis tool also *back-traces* from network outputs toward the inputs (Figure 4.27). Each output is assigned an *arrival time* just sufficient to meet signal stability constraints. From the network outputs, block delays are subtracted to calculate the required arrival time at the block inputs. These arrival times are then assigned to the driving blocks, and the process is repeated until the network inputs have been reached. For a block with several fan-outs, the most aggressive (earliest) arrival time is assigned. Each block output now has an associated delay time (from forward propagation) and a required arrival time (from back-tracing). The required arrival time minus actual delay time difference is denoted as *slack* time; a positive slack implies that the block output is stable sufficiently early, while a negative slack is indicative of a timing problem. Rather than being faced with a report of network outputs at which timing problems were detected, the designer can examine internal blocks for the source(s) of negative slack values; a repowering of the book(s) causing the negative slacks or the local addition of parallel function to reduce fan-out may be sufficient to resolve the timing problem.

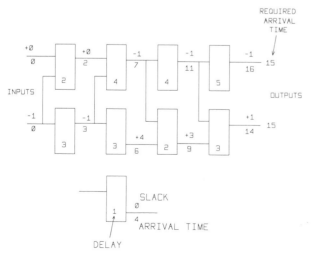

Figure 4.27. After forward-tracing to network outputs, the timing analysis algorithm described in reference 4.1 back-traces toward network inputs, calculating the differences between actual and required transition arrival times.

4.9 LOGIC SIMULATION APPROACHES

There are two steps to performing logic simulation for functional verification once the system functional model has been built. First, the designer's test case stimulus must be developed, followed by execution of the simulation algorithm itself. This section briefly discusses these two aspects of design verification.

Logic Simulation Stimulus

Just as there exist two main approaches to simulation modeling, LOGIC and BEHAVIORAL, so there are also two levels of sophistication in specification of the test case stimuli to apply to a model, *time-based* and *procedural.*

The *time-based* stimulus description consists simply of a set of sequential simulation patterns (Figure 4.28). The format of these patterns is "time:vector"; the simulation time at which the pattern becomes active is followed by a vector of Boolean signal values to be applied (0, 1, Undefined, and − for unchanged). A header in the file of patterns indicates the specific order of model primary inputs to associate with the columns in the vector. An equivalent means of specifying the input pattern would be to define for each signal the simulation times where a change in value is to be asserted (Figure 4.29). This format permits a collection of signals to be specified more efficiently using decimal, octal, or hexadecimal values. A + preceding a numeric time value in this list indicates a relative adder

```
INPUTS   NET₁, NET₂, ..... NET N ;
VECTORS
/*  TIME    SIGNAL VALUES 1 THROUGH N   */
        0:   101UU01U      ...     1
       10:   -10--10-      ...     -
       50:   ---10-1-      ...     0
                   ○
                   ○
                   ○
/*  '-' INDICATES NO CHANGE FROM PREVIOUS VECTOR   */
```

Figure 4.28. A time-based simulation stimulus file consists of a list of records specifying the input pattern vector to be applied at a particular simulation time; a header provides the ordering between model primary inputs and their position in the stimulus pattern.

```
DATA_IN(7:0)   =   0:'FF'H +20:'7E'H 100:'A8'H +40:'FE'H  ....
   /* THE "H" SUFFIX IMPLIES HEXADECIMAL BUS VALUES  */
```

Figure 4.29. An alternative means of specifying input stimulus transitions is to list changes by signal (or bus), rather than by simulation time.

to the preceding absolute simulation time declaration. Problem 4.10 requests that an enhancement to this format be developed to efficiently describe periodic signals.

A procedural stimulus description is analogous to a behavioral logic simulation model; the constructs available to the designer are a subset of those provided in structured programming languages. For example, a procedural description may include loops, conditional constructs, and output statements (for text messages). Logical and arithmetic operators (given the appropriate definitions and scope for different data types) are incorporated into conditional expressions and signal assignment statements. A function to provide direct signal transitions (as in Figure 4.29) would also typically be provided. Figure 4.30 illustrates how these constructs could be used in a stimulus description. Both absolute and relative simulation time definitions are given. When a new construct is entered, the time at which it becomes active would become the reference for all relative time calculations within the construct. Also, for convenience, identifiers can be defined in the header of the description to signify simulation time intervals and short signal names given to represent simulation model primary input net names.

```
PROCEDURE SIM_EXAMPLE        ;
/*  TEST CASE FOR ILLUSTRATING STIMULUS CONSTRUCTS    */
/*  ALL SIGNAL STIMULUS ASSUMED TO BE OF FORCING STRENGTH */
SIGNALS
   /*  SHORT_NAME := SIMULATION_MODEL_PATH_NAME:NET_NAME */
     DATABUS  :=  /SYSTEM/DMA:ASYNC_PORT(7:0);
     OSC   :=  /SYSTEM/DMA:SYS_CLK ;
     STATUS   :=  /SYSTEM/DMA:S1,S0 ;
CONSTANTS
     CLK = 100 ;
STIMULUS
   BEGIN
     0:  DATABUS := '8F'H  ;
     20:  WAVE  OSC,('1'/+50,'0'/+50)  ;
  +2*CLK:  STATUS  := '00'B ;
     500:  FOR I := 1 TO 3 BY 1 ;
             +0:  STATUS  :=  STATUS + 1 ;
            +40:  DATABUS := NOT DATABUS ;
             +0:  WRITELN("THE VALUE OF DATABUS AT TIME: ";
                                     SIMTIME; " IS "; DATABUS );
            +20:  END ;
   +CLK:  IF(OSC = '1') THEN DATABUS := 'FF'H ;
                     o
                     o
                     o
```

Figure 4.30.   Procedural simulation stimulus description. Some of the constructs illustrated include loops, data operators, conditional statements, and message reports.

An important aspect of procedural descriptions is the ability to define a number of *independent* stimulus events and have them execute in parallel; in other words, separate constructs can be *active* simultaneously. A time-based description is effectively a list of patterns. If several different events are to be applied concurrently, they must all be merged into an equivalent binary vector. A procedural description would offer little advantage over a time-based approach if only a single construct were in *control* at any one time; all unrelated events would then have to be contained within that group of statements. Rather, the separation of distinct, parallel events into discrete sections is encouraged; this division will likely enhance the manageability of coding a complex test case. A set of statements can be grouped into a unit to execute in parallel with other units as depicted in Figure 4.31. A *START* statement group is used to define a new unit. The START groupings can be located inside other groups and can therefore be temporarily in control of stimulus events. A START group may terminate its active execution prematurely based on the evaluation of a conditional statement (Figure 4.32). An enhancement to this approach would be to permit the execution of a START group to be active on multiple occurrences within a single test case description based on a conditional expression. As illustrated in Figure 4.33, a START-WHEN group would change from inactive to active at the point in simulation time when the associated conditional expression became true. Like a nonretriggerable "one-shot" monostable circuit, this group would execute and expire. When the group is inactive, the *triggering* conditional expression is continuously being monitored.

Both the time-based and procedural stimulus descriptions should include the means for initializing the values at specific internal nodes of the model and in

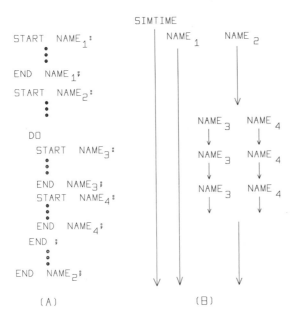

Figure 4.31. Use of START groups in the stimulus description to introduce multiple active events into the simulation. (a) Stimulus description. (b) Sketch of groups that are active for description in part (a).

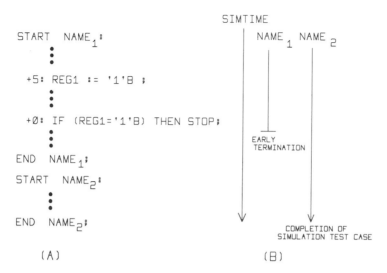

Figure 4.32.   A START group may terminate before the complete execution of all its associated events. (a) Stimulus description, with conditional STOP statement. (b) Sketch of stimulus event activity versus simulation time for part (a).

array locations. At the beginning of a simulation run, all internal nets are commonly assigned an Undefined value; a stabilization analysis is performed (at sim_time = 0) to determine the correct logic value to give to internal nets based on the sim_time = 0 input stimulus conditions. For the purposes of the test case, however, it may be more efficient to directly initialize register contents and RAM array locations, rather than beginning with the sequence of patterns necessary to write the intended signal values.

Each of the two approaches to writing simulation test cases has its own inherent advantages and disadvantages. In many regards, the time-based description is like a programming language that is *interpreted* rather than *compiled*. The sequential list of simulation patterns is read by the simulator as it progresses in time. Changes to this description require no recompilation process, just as changes to interpreted program code can be executed directly. The procedural description does require a compilation process. The various START (and START-WHEN) groups, conditional expressions, looping constructs, and output statements require complex data management; the simulator must be closely tied to the resultant data structures in order to correctly produce the intended stimulus.

The time-based approach is perhaps best suited to evaluating smaller simulation models, while the procedural approach is optimum for system-level networks. The development of large, complex test cases with a time-based approach would be cumbersome. However, when verifying a number of smaller models independently, considerable test case coding effort can be saved by using the simulation outputs from one model to directly apply stimulus to other models.

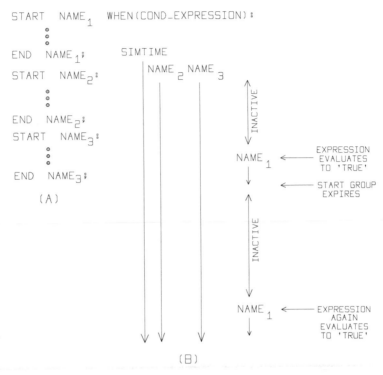

START   NAME$_1$ WHEN(COND_EXPRESSION):

END   NAME$_1$:

START   NAME$_2$:

END   NAME$_2$:

START   NAME$_3$:

END   NAME$_3$:

(A)

SIMTIME

NAME$_2$ NAME$_3$

INACTIVE

NAME$_1$ ← EXPRESSION
                EVALUATES
                TO 'TRUE'

← START GROUP
      EXPIRES

INACTIVE

NAME$_1$ ← EXPRESSION
                AGAIN
                EVALUATES
                TO 'TRUE'

(B)

Figure 4.33. A START-WHEN event group can be active on multiple occurrences within a simulation test case; it changes from inactive to active whenever the associated conditional expression evaluates to TRUE.

This is easily facilitated by a straightforward time-based pattern listing (Figure 4.34). As might be inferred, the relative incompatibility of the two approaches presents considerable difficulty to providing a coherent design system methodology over a range of model sizes. On the one hand, the time-based test cases will effectively be discarded when a procedural description is coded for a larger model; on the other hand, writing individual procedural test cases for different models is inefficient (and error prone) when actual simulation output results from other interrelated networks may be available. A major challenge currently facing design system developers is that of providing a smooth transition between the two approaches to design verification.

A similar challenge exists in the utilization of interactive and batch simulation tools. An interactive simulator allows the execution of the test case to be interrupted; internal model signal values can be examined and potentially altered before continuing. If an undesired network response is observed, that value can be overwritten and the simulation allowed to proceed to isolate additional problems. A batch environment does not facilitate this "on-the-fly" investigation, leading to a more deliberate evaluation of design errors as they are uncovered.

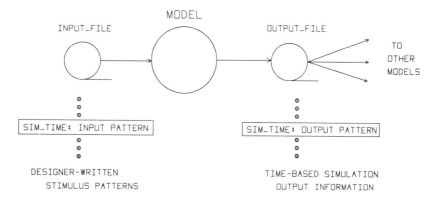

Figure 4.34.   The simulation output pattern file from one model and test case can easily be adapted to provide input patterns to other models, if a time-based stimulus format is used.

Yet, as the system model under evaluation grows, the number of design problems will decrease, while their intricacy will undoubtedly increase; both characteristics tend to diminish the effectiveness of quick network response fixes afforded by interactive tools. A smooth migration from an interactive environment (with time-based test cases, typically) to a batch mode (often with procedural stimulus descriptions) is crucial to maintaining a high level of productivity. Additional complications contribute to this already vexing problem when the computing resource used for these two simulation environments differs. Often, the interactive simulator resides on an engineering workstation, while the batch tool is targeted to run on a mainframe (consistent with the resource required to efficiently manage the size of the network model being analyzed). The designer is thus faced with alternatives involving test case strategy, simulation environment, and targeted computing resource; help in making these decisions is a relatively new aspect of design system support.

The remainder of this section briefly discusses the characteristics of an *event-driven* logic simulator, as well as some of the features commonly provided to the designer to aid with the subsequent analysis of the results. In some cases, a particular feature will be unique to interactive simulation.

Event-driven Logic Simulation

Logic simulation is the task of analyzing the system model response to an input stimulus test case. This entails the evaluation of each block's logic or behavioral description to calculate the values to assign to block outputs, given the current and previous input conditions. If this calculation results in a change in output value, a signal transition is recorded; this transition occurs in *future* simulation time based on the associated block delay. Typically, only a small percentage of the signal values in a network model are affected by a change in the input stimulus pattern or an internal block output transition. An efficient approach for logic simulation is to examine only those blocks that are *active* (i.e., blocks whose inputs have just changed). When an input stimulus pattern and/or a block output transition goes into effect, only those blocks in the fan-out of the changed net values are considered. There are commonly simulation time intervals when no network activity requires evaluation; the simulation time can be advanced directly to the occurrence of the next pending transition. The changes in signal values are denoted as *events*. This approach to network simulation, where inactive blocks are not evaluated and the simulation time advances somewhat sporadically, is referred to as *event-driven* simulation.

The overall structure of an event-driven logic simulator is depicted in Figure 4.35. The algorithm consists of three primary procedures: the evaluation of network responses, the queueing of future scheduled signal transitions, and the updating (and reporting) of network values.

At each new increment of simulation time, the update list provides a list of all primitives whose input values have just changed. The evaluation procedure then goes through this list, exercising the model for each primitive block against the modified input condition. If the block is modeled by a simulation behavioral, that particular routine is called. If the function is combinational, the effect of the input change on the block output is calculated. If no change of the block output results, no further action is necessary. If a change in output value is indeed required, the evaluation procedure *schedules* a future output transition for the current simulation time plus the associated block delay. This future transition is placed in a list, ultimately to be processed by the queueing procedure. (Recall that behaviorals can add to this list directly by a function call from within the routine.) The evaluation procedure completes when the update list is empty (i.e.,

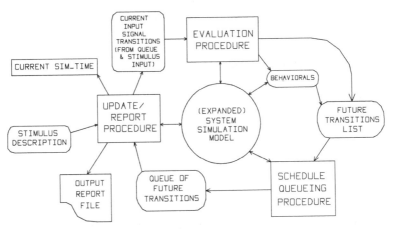

Figure 4.35.   Structure of an event-driven logic simulator, illustrating the three main procedures: evaluation, scheduling, and update and report.

when the responses to all current signal value changes have been determined). Control then passes to the queue management procedure.

The evaluation procedure has produced a future transition list; the queue management portion of the simulation algorithm determines if those transitions are to be propagated, ultimately to reside on the event queue. The event queue can be regarded as a two-dimensional linked list (Figure 4.36). Individual (horizontal) lists represent all the signal transitions pending at a given simulation time; the individual lists are linked in ascending future time. Normally, the queue management procedure inserts items off the future transition list at the appropriate position. If other events are pending at the same time, the horizontal list is extended; if no events are scheduled for the same time, the vertical ordering of the event queue is suitably altered to incorporate a new horizontal row.

The exception to this process is the main feature of the queue management step. To illustrate the exception case, consider the function, block delays, and input transitions in Figure 4.37. At simulation time $t = 10$, the input signal change at A resulted in an output rising transition scheduled for $t = 17$; at $t = 10$, this event is moved from the future transition list to the event queue. At time $t = 12$, the input change at B results in a falling transition at C being placed on the future transition list. However, there is already an event queued for this block output, and, furthermore, the event on the future transition list *precedes* the pending queued event in simulation time. When a future signal transition is presented for which an event is already pending, the queue management procedure must decide which, if any, of these will end up on the event queue. For the particular case illustrated, the simulator could adopt the approach that net C should remain at 0 and that the rising transition scheduled for this output should be *canceled*. This requires deleting a transition from the event queue. When the processing of the future transitions list is complete, control passes to the update and report procedure.

The top of the event queue represents the transitions pending at the next

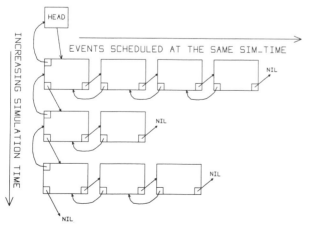

Figure 4.36. The event queue is effectively a two-dimensional linked list of signal transitions, arranged in order of increasing future simulation time.

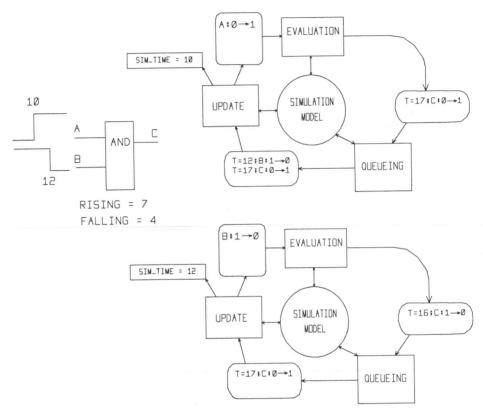

Figure 4.37. The requirement to delete scheduled events (pending transitions) off the event queue; the rising transition scheduled for net *C* at sim_time = 17 is *canceled* due to the subsequent falling transition projected for sim_time = 16.

significant simulation time. The update and report procedure extracts the top of the queue, establishes the new simulation time, and prepares the collection of events for evaluation. This procedure must merge the signal transitions of the stimulus input description with the internal transitions of the event queue. In particular, this routine is also responsible for producing the simulation output files, as requested by the designer. These requests include the specific signals and array location values in the overall model that are of interest, as well as the criteria to be used in choosing whether or not to output those values. Options typically provided to the designer for output of the selected network signals include write on any change, write every periodic time interval, or write only if a given Boolean condition (of signal values and simulation time) is satisfied. This last option permits the simulation output to be tailored to highlight specific details of the test case, minimizing the amount of data to be analyzed and reducing the I/O execution overhead of the overall program.

Not shown in Figure 4.35, but certainly a major part of the simulation procedure, is the stabilization of the network at time $t = 0$ to the initial stimulus.

In addition to the output report features listed, design system simulation tools offer additional command options to aid with analysis of the results. These features include the following:

(1) SAVE/RESTORE (sim_time)

The entire status of the network model is saved when the specified simulation time is reached. Subsequently, this information can be restored to the network model to permit the test case to recommence.

(2) BREAK (Boolean_condition)

Breakpoints can be established in an interactive environment to alert the designer to the presence of a specific network condition. Once interrupted, the designer may need to interrogate the internals of the model (e.g., signal or array values, block delays, a net's source or sink interconnection list). In an unusual circumstance, the transitions pending on the event queue may also be of interest. For example, the event queue could assist with finding network model problems if the stabilization at time $t = 0$ requires an inordinate amount of start-up *pseudosimulation* time.

(3) CONDITIONS (delay_type = MIN, NOM, MAX, MIN:NOM, MIN:MAX, or NOM:MAX)

The most important designer specification is the type of delay analysis to be used for the design. As discussed in Section 4.8, the delay_type parameter could select minimum, maximum, or nominal block delays in functional evaluation or a variation in block delays in timing analysis simulation with transition uncertainty.

This section has only briefly looked at simulation strategies for design verification. Newer design system tools are offering a much more diverse range of modeling options; for example, *mixed-mode* simulators add techniques of transistor-level modeling to more conventional logic and behavioral descriptions. These include both *continuous-waveform* circuit simulation and *switch-level* transistor modeling (with capacitive node charge redistribution). Also, design system simulation support is being offered for system models that incorporate physical hardware implementations and interfaces.

In all its various forms—functional verification, timing analysis with transition uncertainty, exhaustive network path-tracing timing analysis—the task of design verification is the most computational resource intensive. The design system development team is continually faced with the challenge of keeping this resource in check. This task is becoming increasingly difficult in VLSI system environments due to substantial increases in network size and wider diversity of functional modeling alternatives.

PROBLEMS

4.1. How do different automata in a register-transfer level description communicate asynchronously? What measure of *hand-shaking* is required?

4.2. Develop an enhancement to the logic description language of Section 4.3 that enables nested block descriptions located *in separate text files* to be included in a design description.

4.3. Develop a syntax and overall file format for describing the function of a combinational network as a *truth table*. Features to consider include the following:

Representation of network inputs and/or outputs using decimal, octal, or hexadecimal values

Inclusion of *don't cares* in the input vector

What action will be taken (or assumptions made) when an incomplete list of input patterns is provided in the file

4.4. Develop a technique whereby the system description language introduced in this chapter could define a system consisting of several chips, potentially implemented in different technologies (Figure 4.38).

4.5. Consider the use of a two-valued block parameter, which indicates whether the function of the network is purely combinational, or contains sequential elements:

```
type  network_type = [combinational, sequential];
var   parameter  f : network_type;
```

How can the partitioning of the top-down design description into combinational and sequential networks be used in design system applications such as synthesis, timing analysis, and test pattern generation and fault coverage analysis? (Refer to Sections

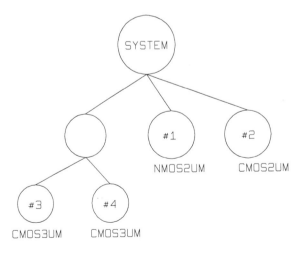

Figure 4.38. System hierarchy consisting of multiple VLSI chips implemented in different technologies.

12.1 through 12.3 for array alternatives in the synthesis of combinational networks; Chapter 13 describes test pattern generation and fault coverage analysis in detail.)

4.6. Research the techniques currently in use by the computer-aided engineering hardware and software industry for incorporating physical electronics with a software simulation model. Specifically, compare the different implementations with respect to the following:

Exercising dynamic hardware

Application of system clocks to the hardware

Definition of the electrical interface (input pins, output pins, bidirectionals)

Measurement of signal values from the outputs of the physical hardware and the logic simulation values assigned in the software model

Use of a single physical unit to represent multiple instances of the hardware in the system design

4.7. Develop an extension to the logic description language presented in this chapter to accommodate a physical hardware replacement of the simulation model of a block. In particular, define the syntax for describing the block pin-to-electrical interface correspondence, the voltage interface levels, and the model delays to use for measured output transitions.

4.8. What design system application tools will need to reference the logical-to-physical pin data provided after physical design, as defined in Figure 4.21?

4.9. Describe how the minimum pulse width, setup and hold time, and signal transition-to-signal transition simulation model checks are to be interpreted when simulating with transition uncertainty, using the set of signal values 0, 1, U, R, F, and C.

4.10. Develop an enhancement to the syntax of Figure 4.29 for the specification of simulation input stimulus to efficiently represent periodic signals.

4.11. The cancellation of a scheduled event off the simulation event queue was illustrated in Figure 4.37. Describe what alternatives could be implemented to report to the designer that a possible output signal *glitch* may indeed have occurred.

4.12. As the verification of a system design nears completion, the number and extent of succeeding design changes diminish. Using the hierarchical database structures alluded to in this chapter, describe how the system design description can be *incrementally updated*; that is, if changes were confined to a single block definition, describe how that new information could be incorporated into the database incrementally, leaving the remainder of the database intact.

REFERENCES

4.1 Hitchcock, R. B., Smith, G. L., and Cheng, D. D., "Timing Analysis of Computer Hardware," *IBM Journal of Research and Development*, 26:1 (January 1982), 100–105.

4.2 McWilliams, T. M., "Verification of Timing Constraints on Large Digital Systems," Ph.D. thesis, Lawrence Livermore Laboratory (May 1980), UCRL-52995.

Additional References

Breuer, M. (ed.), *Digital System Design Automation: Languages, Simulation, and Data Base*. Potomac, Md.: Computer Science Press, Inc., 1975.

Chu, Y., "Why Do We Need Computer Hardware Description Languages?" *IEEE Computer* (December 1974), 18–22.

Hill, F., and others, "Introducing Computer Hardware Description Languages," *IEEE Computer* (December 1974), 27–44.

Hong, S., Cain, R., and Ostapko, D., "MINI: A Heuristic Approach for Logic Minimization," *IBM Journal of Research and Development* (September 1974), 443–458.

Jephson, J., McQuarrie, R., and Vogelsberg, R., "A Three-value Computer Design Verification System," *IBM Systems Journal*, 8:3 (1969), 178–188.

Maissel, L., and Ostapko, D., "Interactive Design Language: A Unified Approach to Hardware Simulation, Synthesis, and Documentation," *19th Design Automation Conference* (1982), 193–201.

Marsh, Robert, "LOGSIM: A General Logic Simulation Program," *Software Age* (November 1969), 28–35.

Sakai, T., and others, "An Interactive Simulation System for Structured Logic Design—ISS," *19th Design Automation Conference* (1982), 747–754.

Su, Stephen, "A Survey of Computer Hardware Description Languages in the U.S.A.," *IEEE Computer* (December 1974), 45–51.

Thomas, D., and Nestor, J., "Defining and Implementing a Multilevel Design Representation with Simulation Applications," *IEEE Transactions on Computer-Aided Design of Integrated Circuits and Systems*, CAD-2:3 (July 1983), 135–145.

Vaidya, A., Dietmeyer, D., and Engh, M., "*WISLAN—Technology Transformation and Optimization,*" *Proceedings of the 6th International Symposium on Computer Hardware Description Languages and Their Applications* (1983).

vanCleemput, W., "An Hierarchical Language for the Structural Description of Digital Systems," *14th Design Automation Conference* (1977), 377–385.

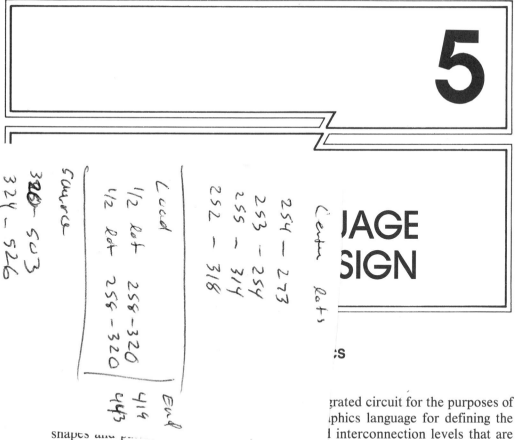

5

...JAGE
...SIGN

...:s

Handwritten notes:

```
Center Lots

254 — 273
253 — 254
255 — 314
252 — 318

Load
1/2 lot 258 — 320
1/2 lot 256 — 320      End  419
                            443

Source
326 — 503
324 — 526
```

...grated circuit for the purposes of ...phics language for defining the shapes and p... l interconnection levels that are part of the final circuit implementation. These shape and pattern descriptions for a particular level are to be transferred photolithographically into an identical pattern of a thin-film coating on a glass plate called a *mask*. During fabrication, each mask pattern is transferred into a photosensitive material called *photoresist,* which has been coated onto the surface of the IC wafer substrate. The presence or absence of this photomasking material on the surface during fabrication, for example, subsequently defines the presence or absence of an insulating dielectric layer between interconnection levels (Figure 5.1). The shape on the mask, subsequently transferred to the photoresist coating, defines the location of a *via*; the mask is referred to as the *VIA* mask. The fabrication of an integrated circuit in current technologies requires defining shapes on ten or more mask levels. Typically, several photomasking steps are required initially to define the devices; the final steps define the local and global circuit interconnections between devices. This chapter presents some ideas for describing the shapes that are used to determine the final IC function.

The physical design description is also used extensively by a number of design system software tools prior to committing the design for manufacture.

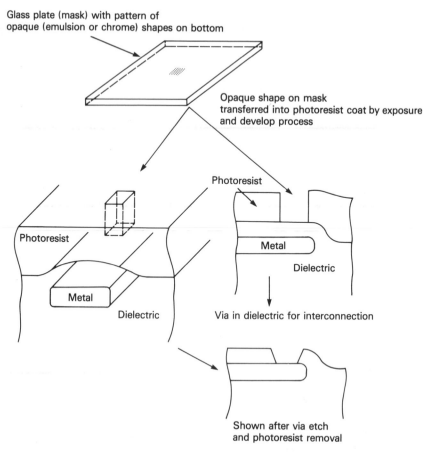

Glass plate (mask) with pattern of
opaque (emulsion or chrome) shapes on bottom

Opaque shape on mask
transferred into photoresist coat by exposure
and develop process

Photoresist

Photoresist

Metal

Photoresist

Dielectric

Metal

Dielectric

Via in dielectric for interconnection

Shown after via etch
and photoresist removal

Figure 5.1. Pattern transfer from the shape on the glass mask plate into the photoresist coat and eventually into a dielectric layer on the wafer surface.

Some of the software applications are plotting, design rule checking, logical-to-physical checking, and postprocessing for mask patterning.

Plotting

The physical design description is converted to the plotter's syntax, and the resulting file is transferred to the plotter to generate a hardcopy plot of all or part of the physical design. In previous IC technologies, plots were often used for visually (i.e., manually) checking layouts for manufacturing design rule violations (e.g., minimum width of a shape, minimum spacing between shapes on the same level, minimum overlap of contact areas by interconnection levels, minimum device dimensions). A high magnification plot was attached to a wall or light table and searched by eye and with ruler for design rule violations. The increase in design rule complexity and the sheer volume of physical design data for current IC technologies have made manual checking very tedious and error prone. The development of design rule checking software tools and the availability of interactive engineering graphics workstations have replaced much of the need for hardcopy plots.

One application for which the graphics design language should provide some support and where plotted (as opposed to drafting) hardcopy is in increasing demand is the generation of engineering diagrams and circuit schematics. Too often postponed until long after the design is completed, the engineering documentation to support the release of a part must be available as early in the design cycle as possible, particularly as VLSI integrated circuits encompass system-related applications (and thus have system-related impact). To as large an extent as possible, the graphics language should enable the engineering diagrams to be an integral part of the total collection of design data.

Design Rule Checking

As previously mentioned, it is necessary to apply the physical design description against the set of manufacturing design rules to check for violations prior to committing the design for manufacture. A design rule checker (DRC) is a design system tool that accepts as input two files, the physical (chip) design description and the *technology-specific* file of checks, measures, and manipulations to be applied against the design to investigate for rule violations. For output, the DRC provides an error file describing the nature of the violation and the location of the specific shape(s) in error. In addition, diagnostic information (such as the number of shapes on each level or the minimum and maximum coordinates of the design) can be provided. The function of the DRC is illustrated in Figure 5.2.

Note that the design rule checking program itself is intended to be technology independent. The specific nature of the shape manipulations, checks, and measures for a particular technology are input to the program. The DRC should support a range of technologies, each potentially with its own device and interconnection levels and design rules.

Just as the physical design of a part is developed initially at the local (circuit) level and subsequently at the (global) interconnect level, so should the design rule

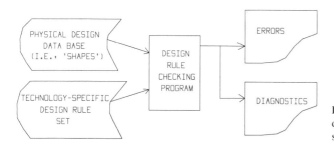

Figure 5.2. Flowchart of the function of the design rule checking (DRC) software tool.

Figure 5.3. Macro *shadow* and pin shapes for global interconnection.

checking be applied. To significantly reduce the data volume and execution time costs, the DRC program should first be applied against individual macros to verify that they are indeed "clean" and, subsequently, against the global interconnection of those macros in the final chip description. Ideally, only a small amount of design information about each macro need be examined for the global design rule checking step; the outline or *shadow* of each macro is checked to verify that the macros are sufficiently separated from each other, and the pin location information for the macro inputs and outputs (*pin shapes*) is checked to verify that the global signal and power supply wires are properly connected to the circuit inputs and outputs (Figure 5.3). The physical design description of the internal macro circuits need not be examined at the global design rule checking step. In addition, the total number of manufacturing design rules that must be checked can be separated into those that pertain to the macro design level and those pertaining to the global interconnect design level, streamlining the checking operation. Design rule checking operations are further discussed in Section 5.4.

Logical-to-physical Checking

The physical design description is also used as an input file to a logical-to-physical (L/P) checking software tool. Logical-to-physical checking, like design rule checking, is most effectively performed at two levels, macro and global. Basically, macro L/P checking involves the following:

1. Using the physical design description of the macro to:
   - Identify all the devices in the layout.
   - Identify the hierarchical output node names of each circuit in the macro.
   - For each circuit, trace from the output node to identify how the devices in each circuit are connected from the output node to the supply voltage.

- For each circuit, identify the device inputs of other circuits to which the output is connected.
- Identify the overall macro input and output nodes, as terminated by pin shapes.
- For each circuit, compare the identified device interconnections with the set of *standard* circuit configurations to find a matching logic gate description.
- Assign the appropriate node name to each logic gate input device.

2. Using the LOGIC model section of the macro's block description to:
   - Compare the logic gates identified in the layout (and their interconnections) with the logical description of the macro.
   - Output the appropriate error messages and coordinates (e.g., missing or incorrect connections, incorrect or conflicting node names, unrecognizable circuit configuration).

The function of the macro L/P checking software is illustrated in Figure 5.4. An example of the circuit tracing and comparison against logical description is given in Figure 5.5 for a simple, one-circuit physical macro.

The macro L/P checker can be developed to be technology independent; one of the input files to the L/P checker contains the technology-dependent information, including the following:

1. The shape operations required to identify and isolate individual devices (and different device types).
2. The level-to-level association information that determines a valid interconnection (Figure 5.6).

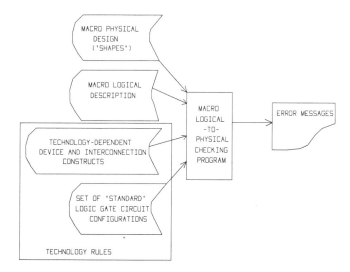

Figure 5.4.   Flowchart of the logical-to-physical checking process (macro level).

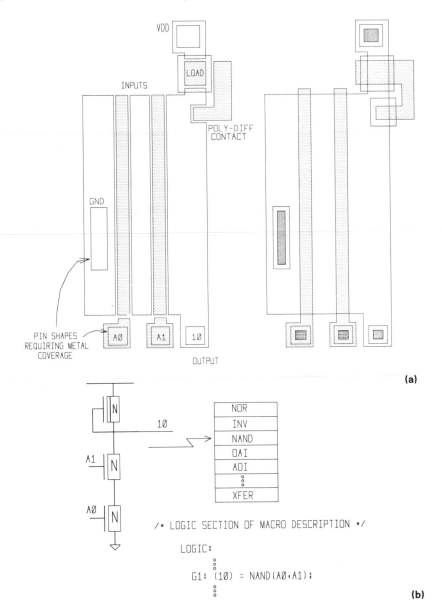

Figure 5.5. Macro logical-to-physical checking. (a) Dissection of a circuit layout into devices and valid interconnections. The dissection and determination of different devices requires identification of the shape geometries that determine different device types. For example, the presence of a DEPL shape determines the gate area as a depletion (load) device. Valid interconnection associations must be made specific to the technology (e.g., DIFF-CON-M1PIN defines a diffusion node contact). (b) After dissection, the schematic must be developed and compared against the library of schematics that the L/P checker is to use for verification.

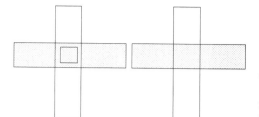

Figure 5.6. The layout on the left is a valid interconnection (which must be traced in tree fashion to the various pin shapes); the layout on the right should not be interpreted as an interconnection.

Note also that the physical designer is restricted to use only those device configurations that constitute the circuits checked for by the L/P checker. The set of possible logic circuit configurations should agree with the set of available simulation primitives used in building the LOGIC section of the macro's technology library block description. The physical designer must work in conjunction with the design system tools developer to determine an acceptable set of circuit constraints.

There must be a provision in the graphics language for assigning a pin or node name to a particular level and location in the physical design. This name is affixed to the circuit node or macro pin shape which resides at that location on that level.

The most complex algorithm in the implementation of a macro L/P checker is the one that must trace out, through the valid interconnection topography, to determine how the individual devices are interconnected within each circuit and, subsequently, how the circuits are interconnected in the macro. The circuit fan-out of some of the current IC technologies can be quite large, and the resulting tree of interconnections that must be extracted can be quite elaborate. In addition to the schematic description, it is extremely advantageous if this path-tracing algorithm has the capability to provide a *physical* description of the interconnection tree as well, specifically, the lengths, widths, and contact sizes of the connections between the devices and the circuits. This facilitates the calculation of the interconnection resistance and capacitance values that (along with the device dimensions) will be used to determine the circuit performance (block delay). This physical design database can become very large. Again, the physical designer must work closely with the software tools developer to extract and keep only the physical design information that will be useful in subsequent circuit analysis.

The *global* L/P checking program performs much the same function as the macro-level program, yet it is simplified somewhat in the sense that no devices are present, only shadows of macros, pin shapes, and shapes on interconnection levels. The tree of global interconnections (between pin shapes) is compared against the (hierarchical) system logic description of the macros, and, as before, appropriate error messages are generated.

In summary, L/P checking is somewhat of a back-end step in the design process. A completed physical macro is checked against its full logical description using the node names, valid interconnection paths, and the set of standard circuit configurations as references. Logical-to-physical checking is not applied typically to array macros (e.g., ROMs, RAMs) nor automatically generated macros (e.g., PLAs), which typically do not use a LOGIC section of the block description for simulation. Global L/P checking is used to verify that the completed chip physical design agrees with its logical description and is especially important if any manual wiring (or rewiring of automatically routed nets) was used.

Postprocessing for Mask Patterning

With current mask-making technology, the final pattern prepared in the coating on the glass mask plate is typically composed of sets of disjoint (or possibly overlapping) rectangles. A machine called a *pattern generator* accepts data in the form of rectangles to "flash" onto the photographic emulsion (or photoresist over a chrome film) coating on the mask. After developing (and etching of the chrome film), the pattern has been transferred to the mask, typically at a $10\times$ multiple of the final IC design dimension. (The techniques for realizing the $1\times$ pattern in a photoresist coating on the IC wafer substrate for fabrication are discussed in Chapter 15.) These rectangular flashes are described by a center coordinate, an *x*- and *y*-dimension, and an angle with respect to the horizontal. The *postprocessing* software tool accepts the physical description of the integrated circuit and, for each level, dissects the description into the appropriate set of rectangles to transmit to the pattern generator to flash on the mask plate. This procedure is illustrated in Figure 5.7.

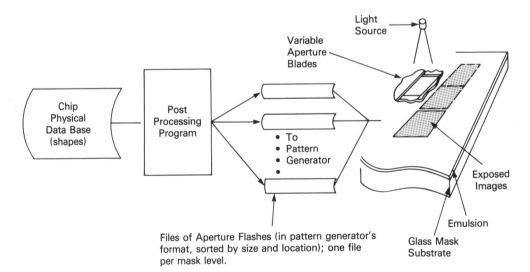

Figure 5.7. Postprocessing procedure for the dissection of design information on each masking level into rectangular flashes for mask generation.

The quantity of data sent to the pattern generator for each mask level is reduced by the following means:

1. Restricting the physical design to a limited number of permissible angles of shapes (typically, orthogonal and $\pm 45°$, or just orthogonal).
2. Restricting the amount of alphabetic character information to be included for identification purposes on the integrated circuit [usually limited to level information used during fabrication and part number (P/N) information used for test and packaging reference].
3. Providing a variable-sized iris for flashing.
4. Allowing for a number of overlapping flashes to produce a more complicated shape (Figure 5.8)

However, the multiple flashes described in item 4 may cause the final developed image on the mask to *bloom* where flashes have overlapped; if the blooming is severe, adjacent shapes may bleed together and not be individually resolvable during wafer fabrication (Figure 5.9).

To transfer the description of shapes in the graphics design language to a data format suitable for the pattern generator, the postprocessing program must not only dissect the shape, but also sort the resulting rectangles to be flashed. The dissection of a complex shape should provide coverage with a minimum number of rectangles. The sorting of the data should take advantage of a minimum

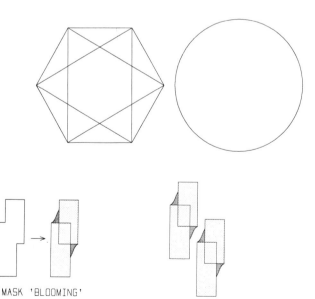

Figure 5.8. Complex shapes may be realized by the overlapping of flashes, such as the overlapping of three rectangles at 60° to realize the six-point circle illustrated.

MASK 'BLOOMING'

UNRESOLVABLE SPACING BETWEEN SHAPES

Figure 5.9. *Blooming* of the resulting mask shape may occur where flashes have overlapped. Severe blooming between adjacent shapes may introduce considerable difficulty in resolving the individual shapes in the photoresist coating on the wafer during fabrication.

number of iris-size changes and a minimum amount of movement of the *xy* stage on which the mask plate is placed for exposure. The specific nature of the post-processing algorithm is strongly related to the requirements, specifications, and tolerances of the pattern generator.

Before leaving this section, one final comment should be made. The features of a graphics language to be described in the remainder of this chapter pertain to a text file of character records (i.e., an ASCII character file). If an interactive engineering graphics workstation is used for graphics entry and editing, the workstation will use an internal representation consisting primarily of binary data, shape *opcodes*, and the like. This internal representation saves on storage requirements and permits faster arithmetic calculations and/or manipulations of the shape information. Yet an ASCII text file representation of the physical design is not without its advantages as well. A text file is machine independent, can be easily transmitted, and can easily be created and edited using a standard text-editing program. Software to convert back and forth from the internal workstation representation to the text file format is nevertheless required. The next section describes the features of a textual graphics language that could be used for IC design.

5.2 LANGUAGE FOR HIERARCHICAL PHYSICAL DESIGN

Prior to describing any language, some definition of the syntax of the language should be provided. For a graphics language in particular, some of the questions to be considered are as follows:

1. What types and formats of numeric coordinate information are supported (e.g., fixed real, floating point real, negative values)? What is the maximum number allowed?
2. Is a global *scaling* (dimensional) multiplier to be applied to all the shapes descriptions? How is such a dimensional constant included?
3. Is the numeric information used to describe the shapes to be given in absolute coordinates or in relative (relocatable) coordinates, relative to the origin of a *cell* or group of shapes?
4. Are the fields of each text record fixed in length and location or simply separated by some delimiter?
5. If the records are fixed in length, how is continuation provided if it is necessary to indicate a longer statement? How long can a shape statement description be?
6. What is the necessary order or grouping of records that describe the entire chip? For more efficient data management, how can these records be grouped into a hierarchical structure of cells?
7. What are the limits on the number of level names, cell names, and shape descriptions in the definition of a cell or chip?

8. How is alphanumeric information included with the shape descriptions (e.g., level names, chip part number, company name or logo)? How is the alphanumeric information to be postprocessed into rectangular flashes?

9. Are comments in the text file of shape descriptions permitted?

10. What angular orientations are permitted for the shape descriptions?

11. What shape types (rectangles, polygons, circles) are allowed?

Some of these questions and others are addressed in the remainder of this section.
The general block structure of a physical design file is illustrated in Figure 5.10.

Header

The first few records in the physical design file constitute the file header and contain information about the chip in general, information pertinent to the physical designers, process engineers, and test and packaging engineers. Such information would include the following:

Part number (P/N)

Unit of measure for all numeric shapes information, that is, global multiplier or *step size* for all shape descriptions (e.g., 0.00025 mm or 0.25 μm)

Overall chip size (e.g., 6.25 mm × 6.25 mm)

Periodicity of chip size plus kerf area

Type of package for the finished chip

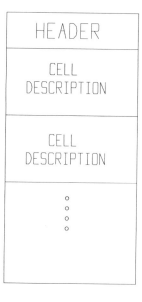

Figure 5.10. General block structure of the physical design file.

Chip pad probing description

Definition of the level names in the physical design, both manufacturable and nonmanufacturable (examples of nonmanufacturable levels are the shadows and pin shapes discussed earlier)

Definition of the cell names to be described in the remainder of the physical design file

The word *cell* has to this point taken on several contexts. In Chapter 3, when describing the master image and master slice design system approaches, the term *cell* was used to describe the smallest area for placing a physical circuit design. In the context of a graphics language, the term *cell* has a slightly different meaning. A cell in this sense refers to a collection of records in the physical design file describing a group of shapes that are to be treated collectively as a unit. This group of shapes is given a cell name. This name is used to nest cells hierarchically, to construct a higher-level layout containing within it a number of other cells. In Figure 5.11, cells A, B, C, and D have been nested inside a physically larger cell. Note that the cell descriptions used to define a higher-level cell can be replicated (as with cell A) and *mirrored* (as with cell D, about the vertical axis). Also note that the physical extent of the shapes contained in a cell description can overlap that of another cell description, as cell C overlaps cells A and D. This nesting process is used repetitively to define larger and larger physical descriptions in constructing circuit and macro layouts and, ultimately, the complete chip description.

The shape/header records could easily be of either fixed or free format; consistent with modern programming language approaches, free format will be assumed. The statements used in the remainder of this section will use a relatively simple statement syntax:

OPCODE : body of statement;

A record appearing in the file header could be

HEADER : parameter$_1$ = . . . , parameter$_2$ = . . . , . . . ,
parameter$_n$ = . . . ;

Example

HEADER : PART_NUMBER = 0123456789, REVISION = C, STEPSIZE = 0.25,
UNITS = MICRON, PERIODICITY = (6500,6500) ;

The colon and semicolon will be used consistently henceforth as delimiters in the graphics design language syntax, the colon for the opcode and the semicolon as the terminator for the statement. Comments may be embedded between state-

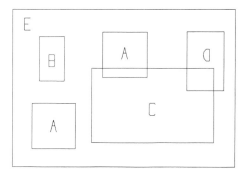

Figure 5.11. Nesting of cells to form a hierarchically higher-level cell definition.

ments with appropriate delimiters, such as braces ({ and }), in concurrence with some of the free-format programming languages.

In particular, the header information will be used by the postprocessing software to develop the absolute physical dimensions of the rectangular flashes on the final mask plate. With reference to the step size and step-size units given in the header, all coordinates and distances contained in subsequent records will be interpreted as integers. All physical dimensions will be treated as integral multiples of the step-size dimension given in the header.

More than one HEADER statement at the beginning of the file is certainly permissible; conflicting parameter definitions should be disallowed.

Cell Description

The collection of records used to describe a cell is illustrated in Figure 5.12a. A cell description starts with the cell name definition and contains any number (including zero) of shapes and other cells nested within. In addition, as mentioned in Section 5.1, it is desirable to assign a name to specific locations (nodes) in the physical macro design.

The first record in the cell description provides the cell name and could be given as

DEFINE : cellname ;

with an appropriate limit on the length of the cell name. There may be a potential use for other opcodes and additional information about the cell as well. For example, to place a physical macro on a macro image chip requires knowledge of the size of the macro in units of macro image cells:

SIZE : (width, height) ;

After the cell definition statement(s), the shape, node name, and cell transform statements follow. Each new cell definition implies that there is a new origin

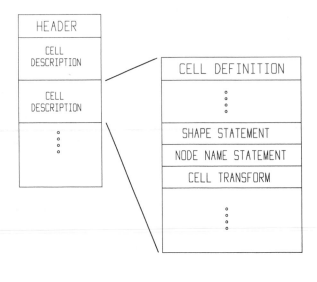

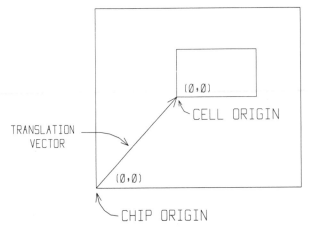

Figure 5.12. (a) The collection of records that comprise a cell description: the cell definition, shape statements, node name statements, and other cell references (transforms). (b) Relationship between the chip origin and cell origin; shapes defined in the nested cell description are relative to the cell origin.

and thus a new coordinate system for the cell being defined. The coordinates given in the statements which follow the cell definition are all relative to the cell origin (Figure 5.12(b)).

Shape Statements

The most general set of shapes that could be present in a graphics language include circles, lines, polygons, rectangles, and alphanumeric characters. As described earlier, these shapes (or, in the case of polygons, individual segments) could be positioned at some angle (other than 0° and ±90°) to the horizontal, or could be restricted to having only x- and y-directed segments (thereby eliminating circles). It is also feasible that the desired pattern on the mask plate could be of the opposite

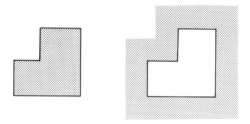

Figure 5.13. Shape descriptions may conceivably refer to the enclosed inside area or to the intersection of the outside area of all shapes on that level.

polarity to the shapes described; that is, a shape description that encloses some finite area may refer to the inside area, or to all the outside area, deleting the inside area (Figure 5.13). The final pattern to be flashed on the mask by the pattern generator may be either the union of all the inside area of the shape descriptions or its complement. For simplicity, both at the design level and with regard to the checking applications software, all shape statements discussed subsequently are inside shapes and include only orthogonal orientations. (The assumption of only orthogonal topology shapes is by no means universal. A large number of custom chip designs and printed circuit boards contain a myriad of shapes and segments that are not horizontally or vertically directed. However, the checking and automatic placement and routing programs are simplified considerably if such an assumption is made.)

Lines.   The first shape statement opcode to be discussed describes a line. Lines (of zero width) are truly not fabricated on the chip itself; nevertheless, such a description may be convenient for the physical designer's use in describing long, straight connections, such as power supply lines or global signal interconnections. A field in the statement gives the width of the line as it would be expanded by the postprocessing program, prior to sending the data to the pattern generator.

Lines are particularly useful, for example, when viewing and editing global interconnections between macros on a graphics workstation. (Imagine the difficulty in examining a large number of parallel, closely spaced rectangles as opposed to line segments.) Also, lines of zero width are useful in generating engineering diagrams and schematics.

Besides the coordinates of the end points and the expanded width measure, the only other necessary data field required to convert the line into a rectangular shape is whether or not the expansion is to be performed at the end points of the line (Figure 5.14). A LINE statement looks like

$$\text{LINE : level, } (x_1, y_1) \text{ , } (x_2, y_2) \text{ , width , ends ;}$$

where (x_1, y_1) and (x_2, y_2) are the coordinates of the end points of the line (relative to the cell origin), level is the mask level, width is the expansion width, and ends is a two-valued variable indicating whether or not the expansion is to be added at the ends.

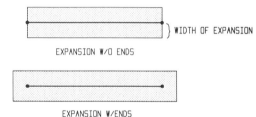

Figure 5.14. LINE statement, including the option of expansion at the end points of the line.

Another possibility would be to replace the second coordinate pair with the increments in the *x*- and *y*-directions to determine the other end point:

$$\text{LINE : level, } (x_1 , y_1) , (\Delta x , \Delta y) , \text{ width , ends ;}$$

(If only orthogonal shapes are allowed, one of the pair Δx or Δy will be zero.) The advantages of the second representation are twofold:

1. Some data storage may be saved since Δx and Δy may require less storage for their representation than the coordinate pair (x_2, y_2).

   Example: $\Delta x, \Delta y$: full word (16-bit) integer
   x_2, y_2: double word (32-bit) integer

2. Applying a transform (e.g., translation, replication, mirroring) to the shape description is simplified (fewer calculations required) if as many coordinates as possible are specified relative to a reference coordinate.

As depicted, the width parameter applies to both sides of the line; the resulting width is twice the parameter specified and is therefore always an even number of design step-size increments. The same dimension is used to determine the extent of the ends added.

Rectangles.   The rectangular shape is probably the most frequently occurring shape of a chip. (As mentioned earlier, all shapes eventually get converted in some manner to rectangular shapes for optical mask manufacture. The other types of shapes included in the graphics language are to provide additional convenience to the physical designer; the wider the variety of graphics types, the more sophisticated the postprocessing program must be to perform the final conversion, however.) The description of a rectangle is quite straightforward (Figure 5.15):

$$\text{RECT : level , } (x , y) , (\Delta x , \Delta y) ;$$

This rectangle statement does not include a parameter for an angular orientation. As with the LINE statement, the shape dimensions are specified in terms of one

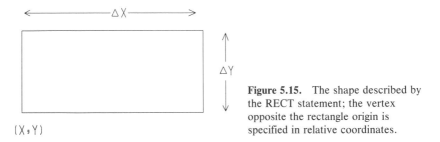

Figure 5.15. The shape described by the RECT statement; the vertex opposite the rectangle origin is specified in relative coordinates.

coordinate pair and increments in both directions, instead of two coordinate pairs. Although both Δx and Δy are positive, as illustrated in Figure 5.15, Δx, Δy, or both may be negative.

Polygons.   Including a polygonal shape description in the graphics language is an enhancement over simply specifying a group of adjacent rectangles. The postprocessing program dissects the polygon into an appropriate set of rectangles for mask pattern generation. A polygon shape description would be as follows:

$$\text{POLYGON : level , (x , y) , } \Delta x_1 , \Delta y_2 , \Delta x_3 , \ldots , \Delta y_n ;$$

The coordinate pair (x, y) is the starting point for the polygon. The Δx's and Δy's represent the change in the x or y coordinate of each vertex of the polygon relative to the previous vertex as the segments of the shape are traversed. By convention, the first segment specified relative to the starting point is a horizontal segment whose length is given by Δx_1. Note that the delta values may be positive or negative and that the last two values need not be included, but rather may be inferred from the enclosing of the area of the shape. An example of the POLYGON statement is shown in Figure 5.16a. If necessary, a limit may be imposed on the maximum number of vertices in any polygonal shape. The polygon shape description must be checked at some point to verify that the shape does not intersect itself; otherwise, some ambiguity may arise about the exact nature of what is inside versus outside area (Figure 5.16b).

Alphanumeric information.   Several examples were given earlier in the chapter in which the presence of alphanumeric information on the chip itself during and after manufacture is quite useful (the company name, the part number and copyright date, each level name, etc.). (An unorthodox yet frequent use is for the designers to try to find the chip area to include their own initials.) A possible format for an alphanumeric character string statement is

$$\text{ALPHA : level , (x , y) , width , height , space , ``} c_1 c_2 \ldots c_n \text{'' ;}$$

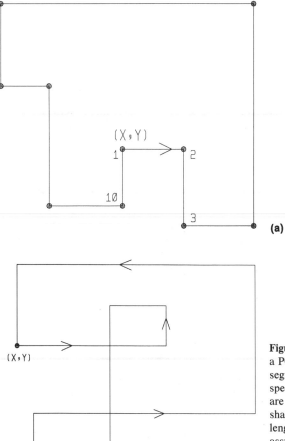

Figure 5.16. (a) A shape described by a POLYGON statement; only eight segment lengths are necessary to specify the shape, since the last two are uniquely determined to close the shape. By convention, the first segment length from the origin of the polygon is assumed to be in the *x*-direction. (b) An inside–outside polygonal shape.

As illustrated in Figure 5.17, each valid character (A–Z, 0–9, +, −, . . .) corresponds to a set of rectangles that represents the character when the graphics language file is expanded. The character size is scaled using the width and height parameters. Another format that could be used is

$$\text{ALPHA : level , (x , y) , (} \Delta x \text{ , } \Delta y \text{) , "} c_1 c_2 \ldots c_n \text{" ;}$$

In this case, the coordinate pair (x, y) is one corner of a *bounds box* and $(\Delta x, \Delta y)$ is the displacement of the opposite corner of the box. The bounds box is a rectangular area that encompasses the alphanumeric string; the width, height, and spacing of the characters is scaled so that the entire string fills (or slightly underfills) the bounds box. If the box is of insufficient area to hold the string of (minimum size) characters, the design rule checking program provides an appropriate error message.

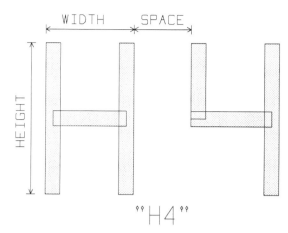

Figure 5.17. Rectangles used to realize the alphanumeric string specified by the ALPHA statement.

Node Name Statements

To perform macro and global logical-to-physical checking, it is necessary to assign a node name to a location and level in the physical design. Likewise, for an automated routing program to be able to determine where a global interconnection wire must be routed, the points where the connections are to be made must be defined. A node name statement looks like

$$\text{NODE} : \text{level} , (x , y) , \text{``} c_1 c_2 \ldots c_n \text{''} ;$$

Node name statements pertain to the description of a physical macro to be placed on the chip. The character string "$c_1 c_2 \ldots c_n$" must agree with the hierarchical node name in the expanded logical description of the macro to satisfy L/P checking. The data for an automated routing program are generated as follows:

1. The results of macro placement give the absolute chip coordinates where the origin of the physical macro design is to be translated.
2. Relative to that origin, the coordinates of each of the macro's inputs and outputs are calculated by adding the (x, y) coordinate pair in the node name statement to the overall translation coordinate pair.
3. These calculated coordinates are written to a file to be used as input to the routing program.

Cell Transform Statements

For graphics language purposes, a cell is defined as a group of shapes that is to be treated as a unit, referred to by its cell name. This group of shapes may *collectively* be used in the definition of another, larger cell with a single cell transform statement, called the USE statement. When a cell description contains other cell transforms (USE statements), this is referred to as a nesting of cells, in accord with the terminology of hierarchical system design; the chip shapes file structure illustrated in Figure 5.10 implies that *all* cell definitions use nested references, as opposed to design partitions. Obviously, a cell may not USE itself, nor may any cyclic cell transforms exist in the shapes file (e.g., cell A USEs cell B in its description, which in turn USEs cell A). Transformations used when nesting cells are as follows:

1. TRANSLATION (of the nested cell relative to the origin of the cell being described)
2. MIRRORING (of the nested cell about one or both of the *x*- and *y*-axes)
3. REPLICATION (use of the nested cell more than once in the cell description)
4. CONDITIONAL USE (use the nested cell in the description of a physical macro only if a given conditional expression about pin usage is satisfied)
5. PIN DROPPING (when using a physical macro, delete one or more pins)

The general form of the USE statement is

$$\text{USE : cellname , T(x,y) M(mirror\_code) R(\#x, \#y, \Delta x, \Delta y)}$$
$$\text{C(logical\_expression)/D(pin}_1, \text{pin}_2, \ldots) ;$$

All the fields in the USE statement except the cell name and Translation field are optional and may appear in any order. (For ease of compilation, all cell names appearing in USE statements should have been previously defined; that is, the cells being USEd should be described in the design file header.)

The *Translation* field specifies the location of the origin of the cell being USEd to the cell being defined. The *Mirroring* field causes the translated cell to (optionally) be mirrored about its *x*-axis, *y*-axis, or its origin (both axes). The valid mirror_codes are

N: no mirroring (default)
X: about its *x*-axis
Y: about its *y*-axis
O: about its origin (both axes)

The *Replication* field results in the translated cell being replicated $\#x$ number of times in the *x*-direction and $\#y$ number of times in the *y*-direction, enabling an entire array of an identical group of shapes to be easily specified. Δx and Δy are the displacements between each instance of the replicated cell, measured from origin to origin of the replicated cell (Figure 5.18). If the cell transform statement contains both a mirroring and a replication field, the cell should be mirrored appropriately and then replicated as described previously.

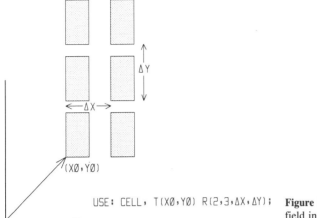

USE: CELL, T(XØ,YØ) R(2,3,ΔX,ΔY);   **Figure 5.18.**   Use of the Replication field in the cell usage statement.

Either a *Conditional* field or a *Drop* field may appear in the cell transform statement, but not both. These fields are used to facilitate assembling a chip description from a design system technology macro library, as introduced in Chapter 3. Predefined physical macro descriptions are copied from the macro library to assemble a chip description. A single macro description may suffice to describe a number of different circuits, which may vary in fan-in or in output pin configuration. The macro description may contain cell transforms that are to be used only if some combination of pins is included; the cell transforms in the macro description would therefore contain the appropriate Conditional fields. In the

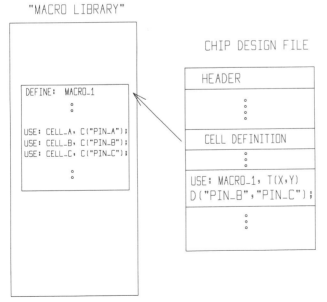

Figure 5.19.   Use of the Conditional and Drop fields in building a chip physical design. The Conditional field is coded in the definition of a macro design, where a cell transform in the macro design is to be used only if a particular pin (or combination of pins) is indeed used; in other words, some pins are *droppable*. The USE statement in the chip design indicates which pins are to be dropped in a particular instance of that macro.

absence of a Conditional field, the default is that the cell transform is to be included.

When a macro from the macro library is USEd in the chip assembly process, a Drop field may be included to specify the desired circuit configuration for a particular instance. In other words, the macro may be used a number of times in assembling the chip, each time subject to a distinct Drop field. When unnesting

```
DEFINE :     AOl2-2 ;
          /*  2-wide, 2-high AOl cell description;   the B1 pin is droppable to implement
              a 2-1 AOl     */
RECT:  DEPL, ( 35, 101), ( 25, 25);
RECT:  BURCON, ( 45, 86), ( 13, 13 );
POLYGON:  DIFF, (0,14), 40, −14, 17, 14, 42, 61, −42, 69, −17, −69 ;
POLYGON:  POLY, (47, 95), 15, 10, −30, 16, 36, −32 ;
NODE:  M1, (19, 4), "A0";
NODE:  M1, (33, 4), "A1";
NODE:  M1, (63, 4), "B1";
NODE:  M1, (77, 4), "B0";
NODE:  M1, (49, 4), "OUT";
NODE:  M1, (9, 35), "GND";
NODE:  M1, (90,35), "GND";
NODE:  M1, (48, 134), "VDD";
USE:  VDDPIN, T(42, 128);
USE:  GNDPIN, T(5, 25) R(2, 1, 81, 0);
USE:  OUTPIN, T(46, 1);
USE:  ENHIN, T(15,0) R(2, 1, 14, 0);
USE:  ENHIN, T(73,0);
USE:  ENHIN, T(59,0) C("B1");
```

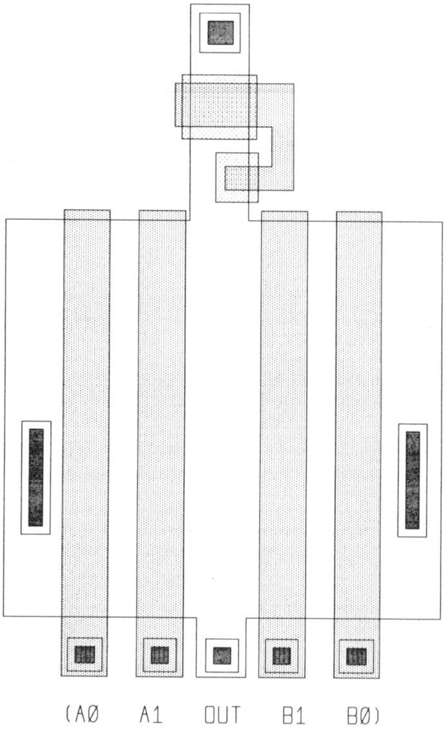

(A0   A1   OUT   B1   B0)

Figure 5.20.   Description of the AOI2—2 cell and its constituent shapes and cell transforms.

the physical design data, the postprocessing program evaluates each macro use to determine which cell transforms in the macro description to include in the final chip design (Figure 5.19). The Drop field is just a list of pin names to be deleted from the macro for a particular use: D (pin_name$_1$, . . .). The default is that all pin names should be used. The Conditional field, if it appears in a cell transform in a macro description, implies that the transform will be ignored if the conditional expression in parentheses is false. The conditional expression becomes considerably more powerful (and likewise more difficult to parse) if the expression contains the basic Boolean operators (OR, AND, NOT) in addition to pin names:

Example

C (NOT ["A0" AND "CLK"])

The parser must assign a priority to the Boolean operators; typically, NOT is the highest, while OR and AND are of equal priority, evaluated left to right in a

Physical Design Language
```
OPCODE  :   body of statement ;
HEADER :   parameter =   ,  parameter =     , . . . . ;
   /*       comments       */
```

Cell Description

```
DEFINE :   cellname ;
     LINE :   level,   ( x₁ , y₁ ) ,   ( Δx , Δy ), width, ends ;
     RECT :   level,   ( x , y ),   ( Δx , Δy );
     POLYGON :   level,   ( x, y ),   Δx₁, Δy₂, Δx₃, . . . ., Δyₙ ;
                            /*     (n+2)-sided polygon   */
     ALPHA :   level,   ( x, y ),   ( Δx, Δy ), "c₁c₂c₃...cₙ";
                            /*   defines size of bounds box   */
```

{ OR

```
     ALPHA :   level,   ( x, y ), width, height, space, "c₁c₂c₃...cₙ";

     NODE:   level,   ( x, y ),  "c₁c₂c₃...cₙ";
```

Cell Transform Statement
```
     USE:   cellname,  T( x,y )  M(mirror_code)  R( #x, #y, Δx, Δy )
                      C(logical_expression)/D(pin₁, . . . , pinₙ) ;
          T  = Translation
          M  = Mirroring
          R  = Replication
          C  = Conditional Use
          D  = Pin Dropping
```

Figure 5.21.   Reference list of graphics language statements.

(sub)expression. Figure 5.20 illustrates the macro library description for the example of a two-high, two-wide AOI circuit. In a more general graphics design language that includes nonorthogonal shapes, the cell transform statement may include an optional *ROtation* field.

Figure 5.21 contains a reference list of graphics language statements.

5.3 VECTOR-BASED VERSUS PIXEL-BASED DESIGN REPRESENTATIONS

In Section 5.2, when describing the fields of the HEADER statement, the *step size* was introduced to signify the physical dimension multiplier to be applied against all shape descriptions. The determination of what the step size should be is by no means a simple task; the manufacturing process development engineer, the design system applications software programmer, and the information services (computer resource and data management) engineer must agree on an aggressive, yet manageable technology definition.

To elaborate, the goal of the process development engineer is to provide a set of technology design rules where a high circuit (layout) density is possible. Consider the design rule describing the minimum overlap of a contact shape by a metal line (Figure 5.22). Due to manufacturing alignment tolerances from level to level and due to the etch bias in transferring the photoresist images into final chip dimensions, the process engineer will likely conclude that a particular overlap design rule is required (a dimension that is commonly much less than the minimum width of any shape or the minimum space between distinct shapes on the same level). This implies that a very fine design increment (step size) should be used, allowing precise level-to-level shape layout to increase the overall circuit density. As another example, consider the design rule governing the minimum space between nondevice polysilicon and diffusion shapes (Figure 5.23). After fabrication,

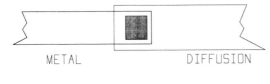

METAL   DIFFUSION

Figure 5.22. A design rule must be specified for the minimum overlap of a contact shape (to diffusion) by a metal shape.

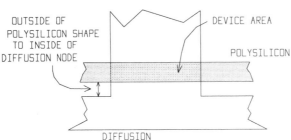

OUTSIDE OF POLYSILICON SHAPE TO INSIDE OF DIFFUSION NODE

DEVICE AREA

POLYSILICON

DIFFUSION

Figure 5.23. A design rule must be specified for the polysilicon-to-diffusion space where the shapes do not intersect. This geometry should merit aggressive design rule development as it is a very common layout topology, especially key to dynamic RAM design.

the resulting polysilicon and diffused lines should intersect in the intended device areas only and should not intersect where their use is for interconnection; no other restrictions exist. As with the contact overlap design rule, a small design rule (step size) would be beneficial in maximizing circuit density. (This second example in particular bears an aggressive design and process philosophy since it is a very repetitive structure in the design of RAM and ROM array macros; a small savings multiplied by a large array factor can result in a significant macro area savings.)

The other alternative is to use a relatively coarse step size that corresponds roughly to the minimum feature size of any shape. In other words, if the minimum process linewidth of any shape is 2.0 microns (μm), then the step size would be 2.0 μm. This approach has commonly been referred to as a *lambda-based* approach to IC physical design, where lambda refers to the coarse step-size measure. The main advantage of this latter approach is its simplicity; the variability among different design rules is reduced, making them easier to remember, interpret, and use. However, using a coarse step size means that the level-to-level shape design rules must also become coarse and may result in a significant loss in circuit layout density. Whereas an overlap or level-to-level space design rule may permit a fine step size (of 0.25 or even 0.1 μm), the design rule minimum would default to the 2.0-μm coarse step size (Figure 5.24).

In addition to choosing the design rule and process step size, another key decision facing the development engineers and programmers is the choice of how shape information is to be efficiently represented in an internal database for the application programs to use. This database results from the evaluation of conditional transforms and cell unnesting. The two options available for this database are to represent the shape information internally in either *vector* or *pixel* form. These classifications are the same as is used in describing the two types of engineering graphics workstations: vector based and pixel based (or raster scan).

Vector form for the shape descriptions implies that, for each mask level, the shapes on the level are broken down into a list of (horizontal and vertical) segments. Alternatively, pixel form means that the entire chip layout area is re-

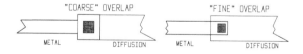

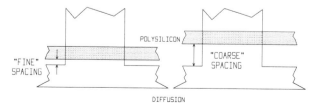

Figure 5.24. Layouts using fine and coarse step sizes for the two design rules of Figures 5.22 and 5.23.

garded as a large array of memory storage. Each storage location typically represents an area one step size square; the presence (or absence) of a shape covering the chip area corresponding to each storage location is indicated by writing a 1 (or 0) into that location. The name *pixel* is short for "picture element"; it is the term used to describe the smallest area on the screen of an engineering graphics workstation that can be individually described. Just as with the internal shape representations, a typical raster-scan workstation uses an array of memory storage to describe the pattern of pixels to be displayed on the screen (a *memory-mapped* implementation; see Figure 5.25). Indeed, the analogy between the means for internally representing shapes and the types of graphics workstations is quite strong. A brief digression into the design and operation of workstations will help in illustrating the advantages and disadvantages of the two forms.

Vector-based workstations represent shapes as the collection of individual vectors or segments; the cathode-ray tube (CRT) of the workstation is illuminated by an electron beam that is steered to provide a segment on the screen, much as a signal trace is illuminated on an oscilloscope used in *XY* (nonsweep triggered) mode with intensity modulation. The electron beam excites the phosphor coating on the screen, which locally emits a visible trace. To represent a typical circuit layout, a large number of vectors must be drawn, each shape being composed of its individual line segments. Workstations employing *storage tubes* use stored charge near the surface of the screen to continuously excite the screen's phosphor

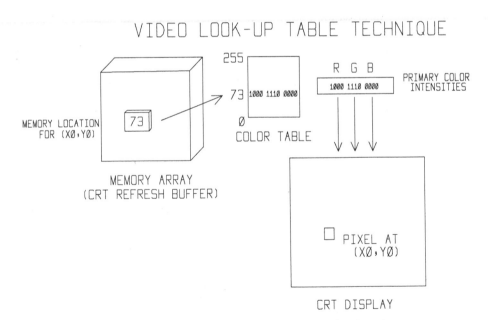

Figure 5.25. Memory-mapped storage for a raster screen. The values in each memory array location are encoded to represent specific pixel attributes, either directly or indirectly (through a look-up table).

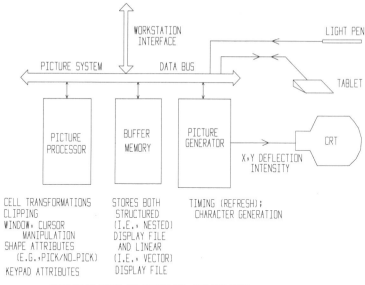

Figure 5.26.   General block diagram of the architecture of a vector-based workstation.

coating and provide a static display of the design. As such, no *refreshing* of the segments is required to avoid a flickering display, which is due to the decay of the excited phosphors. For *interactive* graphics design and editing, a refresh display CRT is more typically used. To prevent a flickering display, each visible segment must be redrawn by the electron beam at least 30 times a second (30 hertz).[1] A refresh display commonly uses a fast access time refresh buffer memory to store the segments to be displayed. This entire buffer memory is read, the segment information processed, and the beam driven and blanked appropriately at least once every thirtieth of a second. A general block diagram of the architecture of a vector-based workstation is given in Figure 5.26. When editing the shape information (or changing the screen's window coordinates to change the size or location of the layout being viewed), the central processing unit of the workstation must interrupt the refreshing of the screen to make the appropriate changes to the storage locations in the refresh buffer memory.

Interaction with the vector workstation usually takes one of three forms:

1. Keyboard (alphanumeric) entry of edit or design commands or file management tasks.
2. Cursor control for defining shape location information (using a keyboard, joystick, mouse, data tablet, etc.).
3. Screen *pick* operations to either pick a particular segment or shape for iden-

---

[1] Long-persistance phosphors may also be used to reduce flickering.

tification and modification or to choose a particular *command menu* option (using a light pen).

A sophisticated graphics system software supervisor may even provide a combination of all three of these. (For more information on interactive graphics system software and hardware, see references 5.4 and 5.5.) The vector-based workstation has several limitations:

1. The screen is monochrome.
2. No *fill* of the inside area of shapes is provided.
3. A *display buffer full* condition results if an attempt is made to display more vectors than can be refreshed in the allotted time.

Nevertheless, the vector-based workstation has several advantages (several of which are just different interpretations of the disadvantages listed):

1. Although the screen is monochrome, different intensity levels (typically OFF, LOW, HIGH) can be used; the software supervisor can use the low and high intensities to provide shapes in *background* and *foreground* modes, where only shapes in the foreground mode would be detectable using a light pen picking device. Controlling the visibility of shapes is considerably more straightforward on a vector display than a raster display.
2. The analog nature of the vector display system allows the viewing of the layout to be very flexible; the physical chip coordinate area displayed can be easily scaled or translated.
3. A potentially large number of shapes (vectors, really) can be displayed, even more so if some flicker in the display can be tolerated. Physical distances can be readily measured by picking on different vectors and using the supervising software to compare their coordinates.
4. Alphanumeric characters can be easily included for text and menu options provided by the supervising software (Figure 5.27).
5. Nonorthogonal shapes and vectors can easily be drawn. (Again, the analogy with an oscilloscope in *XY* mode is appropriate.)

Vector-based workstations have been and continue to be commonly used for interactive engineering graphics applications; references 5.4 and 5.5 include descriptions of several physical design hardware and software systems that utilize vector-based workstations.

The other option for a workstation CRT is to use a raster graphics tube, similar to a standard television tube. Just as with a standard television set, the image is provided by a set of horizontal scan (raster) lines that span the viewing area on the screen from top to bottom. Each horizontal scan line consists of a finite number of picture elements (or pixels), which are illuminated by the scanning

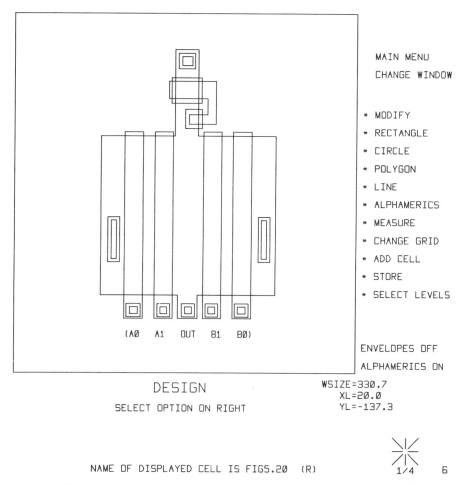

MAIN MENU

CHANGE WINDOW

* MODIFY
* RECTANGLE
* CIRCLE
* POLYGON
* LINE
* ALPHAMERICS
* MEASURE
* CHANGE GRID
* ADD CELL
* STORE
* SELECT LEVELS

ENVELOPES OFF

ALPHAMERICS ON

DESIGN

SELECT OPTION ON RIGHT

WSIZE=330.7
XL=20.0
YL=-137.3

(A0   A1   OUT   B1   B0)

NAME OF DISPLAYED CELL IS FIG5.20   (R)

1/4   6

Figure 5.27.   Typical menu on a vector-based workstation; the menu options are the alphanumeric commands on the right side that are picked with a light pen (after reference 5.1).

electron beam. The pixel area is quantized by the periodicity of holes in a grid screen located behind the surface of the tube (Figure 5.28). Raster screens, just as with vector displays, must be refreshed at least every thirtieth of a second. Whereas nonilluminated areas of a vector display are not scanned at all by the electron beam, every part of the raster screen is scanned; the beam intensity is modulated to vary the screen intensity. In a vector display, the beam is blanked while being driven between end points of disjoint vectors; for a raster display, the beam is blanked during the retrace intervals of the scan (a horizontal retrace between lines and a vertical retrace between screen refreshes).

Unlike vector-based workstations, raster-scan systems provide the options

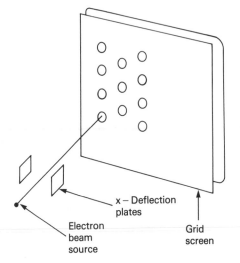

x – Deflection
plates

Electron
beam
source

Grid
screen

Three electron guns are used for a color raster monitor.

Figure 5.28. The pixel area is defined by the periodicity of the grid screen located behind the phosphor-coated tube surface.

of color and of filling internal shape areas. A color raster system consists of primary color pixels of red, green, and blue (commonly abbreviated as RGB). To effectively use color, the different mask levels of a physical layout are assigned different colors; on a particular level, the area of a shape is illuminated with the corresponding color. Using an *additive* representation scheme for each pixel, an interesting and useful feature of color raster displays is that the area of intersection of shapes on different levels (filled with different colors) can itself be viewed as a third color (Figure 5.29). In the example in Figure 5.29, the device area (the intersection of polysilicon and diffusion shapes) is distinguished by a third color for that pixel area.[2] However, note that the presence of a third color in the area of intersection may not be desirable. To avoid confusion, the contact shapes in the diffusion and polysilicon contact areas probably should not be of different colors; in this case, a *paint* subroutine of the software supervisor should be used to override the mixing of colors.

Raster-scan workstations share many of the advantages (and few of the disadvantages) of vector workstations:

1. Many of the same means of interactive data entry, cursor manipulation, and picking can be used with raster displays as with vector displays (joysticks, a mouse, light pens, etc.).

2. Text and menu alphanumeric information can be easily displayed.

3. Nonorthogonal shapes can be displayed, albeit with some *staircasing* due to the quantization of the pixels.

---

[2] A conscious attempt is being made to consistently refer to a pixel as representing an area, whereas a vector drawn on a vector CRT is regarded as a line segment of zero width. This is consistent with the raster screen's ability to fill the area of a shape by illuminating an array of pixels.

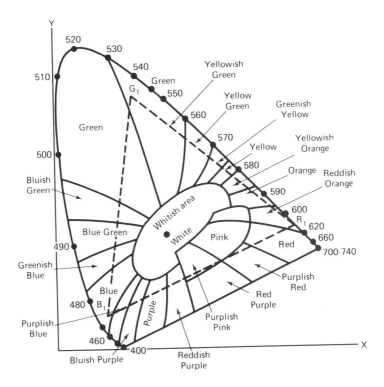

X, Y, Z: Magnitudes of the three "standard" spectral energy distributions, as defined by the Commission Internationale L'Eclairage (CIE)—*NOT* a pure (saturated) primary color.

x, y, z: "Chromaticity values"; X, Y, Z components normalized by the total amount of light—e.g.,

$$x = \frac{X}{X + Y + Z}$$

$$x + y + z = 1$$

Saturated Primaries:

$$\text{Red} \Rightarrow x = 0.67, y = 0.33$$

$$\text{Green} \Rightarrow x = 0.21, y = 0.71$$

$$\text{Blue} \Rightarrow x = 0.14, y = 0.08$$

The area inside this triangle represents the different "hues" that can be provided; the closer the hue approaches the whitish area, the less the "color saturation."   (*continues*)

Figure 5.29.   (a) Chromaticity diagram, illustrating the blending of the three primary color monitor colors: red, green, and blue (RGB). [Diagram from C. N. Herrick, *Color Television: Theory and Servicing*, Reston, VA: Reston Publishing Co., 1977, p. 3.]. Reprinted with permission. (b) Using a color fill (additive) representation of shapes, the blending of colors in areas of overlap highlights specific chip geometries.

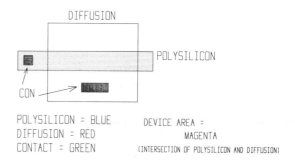

POLYSILICON = BLUE   DEVICE AREA =
DIFFUSION = RED   MAGENTA
CONTACT = GREEN   (INTERSECTION OF POLYSILICON AND DIFFUSION)

Figure 5.29. (*Continued*)

A main advantage of the raster systems over the vector-based workstations is their significantly reduced cost, sharing many of the same enhancements as standard television technology. Another main advantage over vector displays is the option of color and the solid filling of shapes.

The main disadvantage of raster-scan displays is that presently their resolution does not match that of the vector-based displays. A high-quality raster workstation may have a pixel array matrix of 1024 × 1024; current vector-based displays have a larger number of resolvable (vertical or horizontal) segments than 1024.

Resolution is also the reason for ultimately choosing a vector form for the internal shape representations for the graphic design language of this chapter. Although a design rule set based on a coarse increment is undeniably simpler to use, a fine step size permits the maximum layout density to be achieved. Attempting to represent a chip description in memory-mapped fashion using a fine step size would require a prohibitive amount of memory storage. Henceforth, a fine step-size increment will be assumed when discussing technology design rules; the internal representation of shape descriptions will be assumed to be given in vector form.

Creating the internal shape representations requires a procedure comprised of the following steps:

1. Reading in the graphics language description.
2. Performing preliminary syntax and error checking (e.g., invalid opcodes, self-intersecting polygons, invalid alphanumeric characters).
3. Unnesting the cell transforms, evaluating the conditional use fields.
4. Replacing chip alphanumeric information and lines with the equivalent polygons and rectangles.
5. Dissecting the primitive shapes descriptions (RECT, POLYGON) into their constituent horizontal and vertical vectors, storing with each vector a value to indicate the side of the vector that points to the inside area of the shape (Figure 5.30).

```
type
  horv = ( h,v );    /*  vector is horizontal or vertical   */
  lorr  = ( l,r );
  ptrvec  =  ↑ vector;
  vector = record
    x:  integer;    /*  x-origin  */
    y:  integer;    /*  y-origin  */
    l:  integer;    /*  length   */
    d:  horv:    /*  horizontally or vertically directed   */
    s:  lorr;    /*  inside shape area is to the left or right as
                          traversed from the (x,y) origin   */

    next:  ptrvec ;
  end;
```

Figure 5.30.   Pascal record definition of the dissection of rectangular and polygonal shapes into their constituent vectors, including a field for the orientation of the inside area of the shape relative to the traversal of the vector.

Note that the dissection of rectangles into constituent vectors and area orientation is relatively straightforward; however, the dissection of polygons is not. In parsing the textual description of a polygon, it is not possible to determine a priori whether the traversal from the polygon origin starts with the inside shape area on the right or the left of the first vector (see Problem 5.1).

5.4 DESIGN RULE CHECKING OPERATIONS

The discussion in Section 5.3 led to the conclusion that a fine step size would be assumed by the process engineer and the physical designer as the design rules and circuit layouts for the technology were developed. The design system applications programmer is now faced with developing an internal data storage representation that is the most efficient for the various software applications that are exercised against the physical layout; an argument for a vector-based form was given in Section 5.3. This section deals specifically with design rule checking of layouts; Section 5.5 addresses shape postprocessing for generating flashes for a pattern generator.

During the development of a circuit layout by the physical designer, the shape information in that layout must be checked against the design rules (as provided by the process engineer) to ensure that no design rule violations are present. Once a circuit description is clean, shadows and pin shapes for that layout may be used to check the placement and connection of that circuit with other circuits or macros in a higher-level physical description. Adopting this two-part local and global checking approach reduces the number of shapes or vectors to be compared, reducing the execution time considerably. In addition, this approach allows division of the design rule set into rules that pertain to detailed circuit

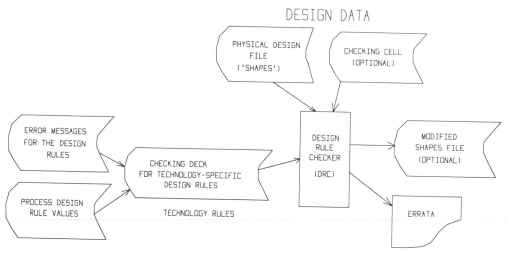

Figure 5.31. Design rule checking process.

design and rules that pertain to the placement and global interconnection of circuit layouts.

The general flow of the design rule checking process is diagrammed in Figure 5.31. A program, called a design rule checker (DRC) accepts three files as input:

1. The text file containing the shape descriptions to be checked (i.e., the circuit layout).

2. The text file containing shape descriptions that are used, in conjunction with the layout, to verify that the design rules around the perimeter of the layout are satisfied (Figure 5.32). This *checking cell* should be included when building a macro library, that is, when it is necessary to check that the circuit layout fits in the appropriate area allocated for that macro on a structured chip image and does not extend into adjacent areas where other circuits will ultimately be placed.

3. The file containing the *checking deck*:[3] a text file that describes the technology-dependent process design rules as well as the various intermediate operations that must be performed on the shape descriptions in order to be able to more easily verify those rules.

The design rule checking program must then read in the shapes file(s), unnest the cell information, and convert the shape descriptions into an internal representation, as described in Section 5.3. The steps in the checking deck are then

---

[3] The term *deck* is synonymous with *file,* a holdover from the days (not so long ago) of punched card input/output.

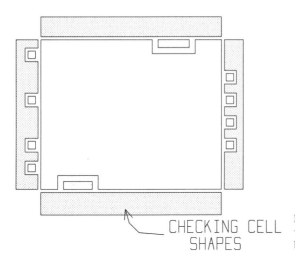

CHECKING CELL
SHAPES

Figure 5.32. *Checking cell* used for verifying design rules around the border of a macro design.

executed; the operations and checks indicated in the deck are applied against the shape descriptions. Note that the design rule checking program can be developed so that its procedures and algorithms are technology (process) independent; all the technology-specific rules and operations are contained within the checking deck.

It should be noted that the first part of the design rule checker, the conversion of shapes into their internal representation, is common to all the physical design system application programs. The chosen (vector) form for the internal representation serves as the common database for those programs. The question of vector versus pixel form for internal shape representation should probably be addressed one last time. Although a vector form is assumed, it will be evident that some of the operations to be described would be just as well if not more suited to a pixel (area) form. Most process design rule checks do indeed require interrogating for the presence or absence of other shapes within a very localized area of a particular shape (e.g., width, spaces, and overlaps). A pixel form would permit locally scanning the layout area against these design rules looking for particular bit patterns in an array of memory storage; otherwise, a significant amount of execution time may be spent scanning the list of shape vectors that are far removed from the one under investigation. Conversely, when scanning a large area that may potentially contain relatively few shapes or shapes with little detail, a raster scanning approach would be extremely wasteful of execution time and storage resource. Problems are included at the end of the chapter that suggest means of making each approach more efficient: reducing the storage inefficiencies of the pixel form by hierarchical representation or run-length encoding, and reducing the execution time problems of vector representation by including pertinent area information.

The remainder of this section will describe some of the common operations included in the checking deck, which must then be supported by the DRC program. The checking deck format to be presented is much like the format of a high-level language program; some of the same programming constructs typically available in a high-level language are used. Adopting some of these constructs provides considerable flexibility in the coding of the checking deck and therefore in the development of the process design rules as well. The process engineer is encouraged to *fine-tune* the design rules for maximum layout density, confident that the power and flexibility exist in the checking deck to adequately verify an intricate rule. As an example, consider a *conditional* design rule, using an IF . . . THEN . . . construct. Suppose that the process engineer measures that the etch bias for a contact hole, that is, the difference in final wafer dimension from the original image size, is a function of the original size of the contact. A minimum-sized signal contact may have a worst-case etch bias that is less than that of a larger power contact (Figure 5.33). The design rules must be developed to ensure that, at the final wafer dimension, the contact area is sufficiently covered by both metal and the diffusion layers. If a single, conservative overlap rule for M1 and DIFF around the contact based on power contact measurements were used, a significant loss in density around signal contacts would result. A conditional design rule (one overlap value if the contact shape is minimum size, or else another for greater than minimum contacts) may enable a significant density improvement to be realized.

In addition to process engineering flexibility, another advantage of coding the checking deck in a high-level form is that it is easier to verify that the deck correctly and sufficiently represents the intended meaning of the full set of design rules; in other words, the checking deck is easier to debug.

The DRC library of procedures provides the ability to output to a text file the error messages for the shape(s) that violates any of the design rules. Such messages should include the coordinates of the vector(s) in error, the associated mask level(s), the design rule number and value, and a descriptive error message of the nature of the check. Note that an interactive graphics workstation is almost

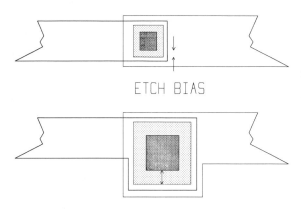

ETCH BIAS

Figure 5.33. The need for *conditional* design rules; in this example, the etch bias between artwork and the wafer-level dimension for the contact is determined to be a function of the original contact size. A conditional design rule may be written requiring a larger metal overlap of the contact shape for contacts larger than a specific size.

a necessity for investigating and correcting any design rule errors for the following reasons:

1. The shapes have been unnested.
2. The coordinates given are therefore the unnested coordinates.
3. The output is typically in the form of vectors, rather than shapes.

Attempting to pinpoint design rule errors in a graphics text file description can be a tedious task.

Another useful procedure to have in the DRC library is the ability to output a modified shapes file in internal representation form for use by other application programs. The modifications to the list of shapes may include selecting particular levels for output or using some of the built-in DRC procedures to create modified shapes on temporary (or *pseudo*) levels.

An example of a checking deck format is given in Figure 5.34; the design rule values and their associated error messages are assumed to be present in separate files, rather than embedded in the sequence of checking operations. This enables these values and messages to be changed easily without affecting the checking deck. A brief flow chart of steps to initiate design rule checking is presented in Figure 5.35. The *unioning* procedure indicated in Figure 5.35 is used to create a single shape from shapes that intersect or abut; this procedure may introduce *inside–outside* shapes (nonsimply connected regions), as illustrated in

```
FILES:
    MESSAGE__FILE:  filename ;  /*  80-character record text file   */
    DESIGN__RULE__FILE:  filename ;  /*  file of real numbers containing design rule
                                        values  */
    OUTPUT__FILES:  filename₁, filename₂, . . .  ;
                       /*  e.g., text file for outputting shapes records from DRC   */


CONSTANTS:
    UNITS = measure ;  /*  e.g., 'microns' –– used with the design rule values to convert
                           these values into the step-size increments of the shapes file   */
    MAX__ERROR__MESSAGES = n ;  /*  integer  */


LEVELS:
    DESIGN; diff, con, m1, shadow, poly, . . . ;
    TEMP: gate, n + dif, p + dif, polywire, . . . . ;  /*  created during execution of the DRC   */


BEGIN
    /*  body of checking operations/procedures   */


END.
```

Figure 5.34.   Sample design rule checking deck format.

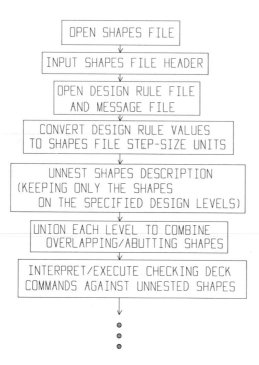

Figure 5.35. Flowchart of the initial steps in the design rule checking procedure.

Figure 5.36. Application programs should ensure that inside–outside shapes are handled correctly (or an appropriate measure taken to ensure that they are not unioned).

The library of design rule checking procedures to be used within the checking deck can be divided into three groups: shapes file input/output routines, shape manipulation routines, and shape checking and diagnostic routines. Examples of

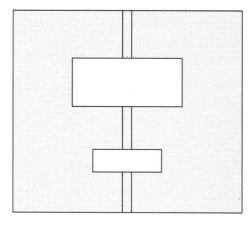

Figure 5.36. The union of overlapping shapes may result in *inside–outside* shapes (nonsimply connected regions).

typical procedures for each of the three groups are presented, followed by some additional constructs that may be useful.

Shape Input/Output Routines

Several application programs may be designed to use the same internal shapes representation. As a result, it may be advantageous to be able to output the shapes to an external file at some intermediate point in the checking deck. Likewise, some means of inputting the shapes records directly from an external file would be necessary. In addition, it may be desirable to use the output file as a *dump* of the internal shapes during the checking procedure; this provides the capability to *restart* the execution of the program at an intermediate point in the sequence of operations and checks if there would be a (nonrecoverable) system error during the execution of a long computer job. Note that an internal shapes representation that is dynamic (e.g., a linked-list structure) must be modified appropriately when writing to a sequential file; similarly, the internal structure must be re-created when reading in from the sequential file. The procedures to be used could be called out by

$$\text{save (filename , 'level}_1\text{' , 'level}_2\text{' , } \ldots \text{);}$$

or

$$\text{save (filename);}$$

where the first example would save only the shapes on the selected levels, while the second example indicates that all shapes are to be saved. The counterpart to the save operation would be

$$\text{restore (filename);}$$

which re-creates an internal shapes representation from an external sequential file.

Shape Manipulation Routines

The following set of shape manipulation routines is by no means exhaustive; only the more common operations are described.

(1) UNION

The designer doing a physical layout of a large macro usually divides the layout into smaller, more manageable cells. To provide the necessary interconnections between these cells internal to the layout, it is common for electrical continuity to be made by abutting or overlapping shapes on the same mask level in the different cells. An example of this technique was given in Figure 5.36. Prior to performing design rule checks (or dissecting shapes for pattern generator flashes), it will be desirable to union these shapes:

union ('oldlevel' , 'newlevel');

where 'oldlevel' is the shape mask level to be unioned and 'newlevel' is the level that is to contain the new shapes created by the union as well as all isolated shapes on 'oldlevel'. Another format for the union procedure would include the specification of three level names, two different levels to be unioned and a third to store the result:

union ('level$_1$' , 'level$_2$' , 'newlevel');

As mentioned earlier, unioning can create inside–outside shapes, which may require special handling by the DRC procedures.

(2) INTERSECT
Determining the intersection of shapes of different mask levels is key to analyzing device-related design rules, verifying the correct usage of pin shapes, and a variety of other applications.

intersect ('level$_1$' , 'level$_2$' , 'newlevel');

or

intersect ('level$_1$', 'level$_2$' , 'newlevel' , 'complementary');

Shapes added to the new level represent the intersection of the area of shapes on the first level with that of shapes on the second. Shapes that only abut (area of intersection is zero) do not create any shapes on the new level. A second format for the intersect operation is provided so that the inside shape area of level$_1$, but outside the area of intersection, can be added to yet a different level, as depicted in Figure 5.37.

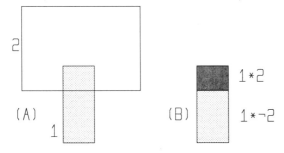

Figure 5.37. The result of performing an intersection of shapes on different levels. (a) Shapes prior to intersection; (b) shapes after intersection; the intersection command permits two (or more) shapes to be produced on different levels, the area of intersection and the complementary area inside the first shape but outside the area of intersection.

(3) MOVE and COPY

These operations provide the flexibility to move or copy shapes from one level to another; their usefulness is enhanced considerably by using them in conjunction with a conditional construct.

move ('oldlevel' , 'newlevel');
copy ('oldlevel' , 'newlevel');

(4) DELETE

Deleting shapes on a level, whether conditionally or not, is used to reduce the dynamic storage requirements and reduce the execution time of many of the algorithms in the various DRC operations:

delete ('level');

(5) EXPAND and ETCH

expand ('oldlevel' , 'newlevel' , design__rule__number);

The expand operation requests that the inside area of a shape on 'oldlevel' is to be *out-diffused* by a particular value (a design rule number that refers to an integral multiple of the step size), and the resulting shape is to be added to 'newlevel'. Note that in the case of an inside–outside shape, the enclosed outside area is reduced, potentially to zero. Likewise, *notches* in a shapes may disappear (Figure 5.38).

etch ('oldlevel' , 'newlevel' , design__rule__number);

Etching a shape is the opposite of out-diffusing it; the integer value referred to by the design rule number specifies the number of step-size increments the

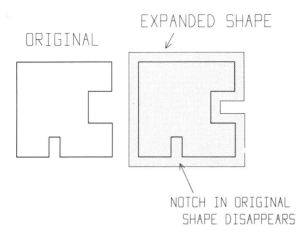

NOTCH IN ORIGINAL
SHAPE DISAPPEARS

Figure 5.38.   The expand operation. Notches in a shape may disappear if the width of the notch is less than twice the expand value.

shape is to be 'shrunk'. Areas of narrow width and all or part of the inside area of an inside–outside shape may disappear (Figure 5.39). An inside–outside shape could dissolve into multiple shapes. The method for etching a shape is illustrated in Figure 5.40; only the inside area not spanned by any of the expanded shape vectors is retained.

(6) CREATE

create ('level' , x—origin , y—origin , Δx , Δy);

Executing the create procedure adds a rectangle to the shapes file on the specified level. This may be useful (in conjunction with intersect) to isolate a particular area in the unnested physical design, performing design rule checks specific to that area. The parameters x—origin, y—origin, Δx, and Δy are all integers.

Shape-checking and Diagnostic Routines

The technology design rules developed by the process engineer may be expressed in a variety of ways (spaces, widths, lengths, coverages, etc.). As a result, a variety of checking routines must be available to ensure that each design rule is correctly

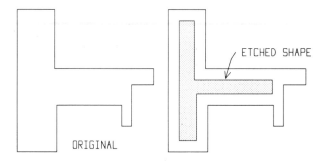

ETCHED SHAPE

ORIGINAL

Figure 5.39. *Etching* a shape.

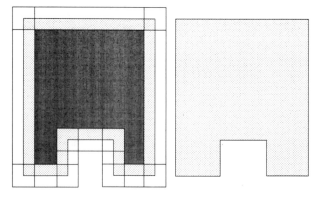

Figure 5.40. Method for etching a shape. All the shape vectors are expanded (with ends) by the etch value; only the shape inside area not spanned by any of the expanded vectors is retained.

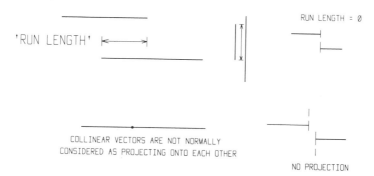

Figure 5.41. *Projection* of one vector onto another. Note the collinear vectors are normally not considered as projecting onto one another; also note that vectors with a *run length* of zero do indeed project onto one another.

and sufficiently verified against the collection of shapes. The checking routines must output to the error message file any shapes or vectors in error, along with the design rule value and any descriptive error message provided. The output procedures of the DRC program should limit the printed errata for any single check to a reasonable number of lines; a total count of the number of errors should also be provided.

Although most design rules specify minimum values for the relationships between shapes, maximum values (and exact values) are used as well. The list of input parameters to each checking routine includes a parameter to specify the relation desired; shapes or vectors *not* satisfying that relation are to be regarded as in error. Some of the checking routines to be described refer to the relationship between vectors; others refer to shapes. Also included in the list of checking routines are those that are used for diagnostic purposes.

Vector-based Checking Routines

(1) PROJection checks

If two parallel vectors project onto each other, as defined in Figure 5.41, the projection check *proj* can be used to compare the distance between those vectors against a particular design rule. (Figure 5.41 also illustrates the definition of the term *run length* between two projecting vectors. Note that collinear vectors are *not* included as projecting onto one another.) To implement a particular design rule check, such as a required spacing or overlap, it is also necessary to specify the orientation of the internal area of the shape relative to each of the vectors in question. The two keywords used to indicate the orientation for a particular design rule check are *inside* and *outside*. The format of the *proj* check is

$$\text{proj ('level}_1\text{', } \begin{bmatrix} \text{inside} \\ \text{outside} \end{bmatrix} \text{, 'level}_2\text{', } \begin{bmatrix} \text{inside} \\ \text{outside} \end{bmatrix} \text{,}$$

$$\text{relational\_operator, design\_rule\_number);}$$

Figure 5.42 illustrates the four cases of inside and outside checking between parallel vectors on two different mask levels. For example, the statement

proj ('diff', inside, 'con', outside, $>=$, 32);

should be interpreted as

> The distance between an inside edge of a 'diff' shape and the outside edge of a 'con' shape must be greater than or equal to the value specified for design rule number 32. Any parallel vectors on these two levels that project onto one another that do not satisfy this requirement are in error.

The proj check is deceptively powerful. Consider the cases illustrated in Figure 5.43 where a variety of physical contacts are to be implemented, all using shapes on the same contact mask level. These physical contacts are the following:

1. Contacts between metal and n^+ or p^+ diffusion nodes
2. Contacts between metal and polysilicon
3. Contacts between metal and abutting n^+ and p^+ nodes (i.e., a butted contact)

By including the particular vector-to-shape orientation, the proj check permits the application of a number of different design rules to appropriately match the particular type of contact being made. Another example of design rule proj checks is given in Figure 5.44. To check that a device, whose area is given by the intersection of 'poly' and 'diff' levels, meets process design rule requirements for

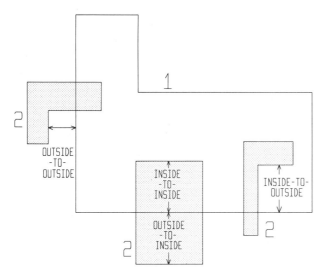

Figure 5.42. Various combinations of inside–outside vector-checking operations.

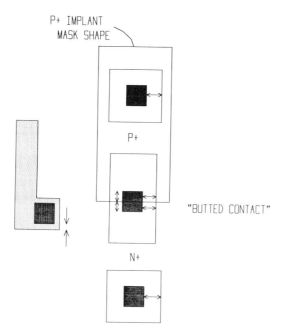

P+ IMPLANT
MASK SHAPE

P+

"BUTTED CONTACT"

N+

Figure 5.43.   A number of different contact types to be made; the arrows indicate the different design rules that may be applied using the projection-checking operation.

minimum width and minimum length, a pair of proj checking operations is executed; prior to these operations, two intersect routines are used to create device-related shapes on new nondesign levels.

In addition, it should be noted that the two mask level names specified in the proj procedure may indeed refer to the same level.

Example

$$\text{proj ('diff' , inside , 'diff' , inside, } >= \text{ , 4);}$$

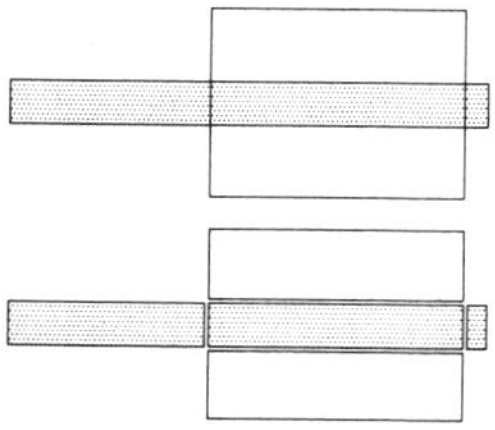

```
/* EXAMPLE CODING:
    CREATE NODES AND DEVICE SHAPES */
        INTERSECT ('DIFF','POLY','GATE','NODE');
/* ISOLATE POLYSILICON WIRING EXTENSIONS PAST GATES */
        INTERSECT ('POLY','GATE','JUNK','POLYWIRE');
/* CHECK FOR MINIMUM DEVICE LENGTH, DEVICE WIDTH,
    AND POLYWIRE EXTENSION */
        PROJ ('NODE','OUTSIDE','GATE','INSIDE',>=,8);
        PROJ ('GATE','INSIDE','POLYWIRE','OUTSIDE',>=,9);
        PROJ ('GATE','OUTSIDE','POLYWIRE','INSIDE',>=,10);
```

Figure 5.44.   Deck for the checking of design rules on minimum device width and length.

(2) COLLINEAR check

It was illustrated in Figure 5.41 that collinear vectors are not normally included in the definition of vectors that project onto one another. To check for collinear vectors that are invalid, the following operations may be performed:

collinear ('level₁' , 'level₂' , design—rule—number);

or

collinear ('level₁' , 'level₂' , design—rule—number , $\begin{bmatrix} \text{inside} \\ \text{outside} \end{bmatrix}$);

The first command indicates that any collinear vectors on levels 1 and 2 are to be flagged as in error. The design rule number simply points to a message; no relational operator is required. The second example narrows the scope of the error to just those collinear vectors where the inside areas of the shapes in question are located on the same or opposite sides of the collinear vectors (Figure 5.45).

Shape-checking Operations

The vector operations are relatively simple to implement, and typically quite efficient to execute since only parallel, projecting vectors are investigated. Because of their efficiency, they should be used whenever the wording and interpretation of a process design rule permits. However, they are certainly lacking in a number of ways and must be accompanied by a number of shape-oriented checks. These shape-oriented checks use some of the same internal procedures as the vector routines, yet usually contain additional or supplementary meaning.

(1) DISTance check

To check the distance between two disjoint (nontouching) shapes against a spacing design rule, the *dist* check should be used. The dist check is different from an outside-to-outside proj check for the following reasons:

1. Only nontouching shapes are checked.
2. Pythagorean distances between vectors that do not project onto one another are also included (Figure 5.46).

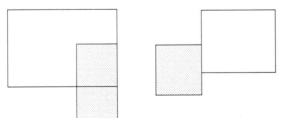

Figure 5.45.   Collinear vectors: in one case, the collinearity is *inside*, while in the other, the collinearity is *outside*.

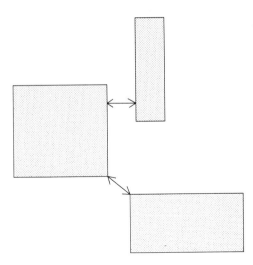

Figure 5.46. DISTance check: the distance between vectors on nonintersecting shapes is checked, including the Pythagorean distance between the end points of vectors that do not project onto each other.

The format of the dist check is

dist ('level$_1$' , 'level$_2$' , relational—operator , design—rule—number);

where 'level$_1$' and 'level$_2$' may be the same. Two shapes on the levels specified in the dist check that *do* intersect are not regarded as in error; rather, they are just disregarded. It may still be necessary to include a vector-based projection routine to check for design errors on the shapes that intersect (Figure 5.47).

(2) WIDTH check

width ('level', relational—operator , design—rule—number);

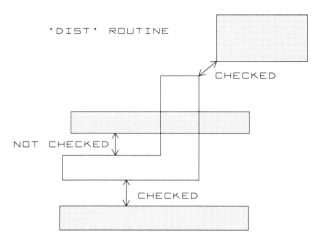

Figure 5.47. Only the distance between nonintersecting shapes is checked by the DIST check. In this particular example, polysilicon intersects diffusion where a device is to be formed; The DIST check only verifies the distance between adjacent poly and diff shapes. Note that it will still be necessary to include a projection check to verify the vector-to-vector distance for the intersecting shapes.

The check for the width of a shape is fundamental to all mask levels. The width check is more comprehensive than an inside-to-inside proj check since, just as with the dist check, Pythagorean distances (if there are any) must also be calculated and compared against the relational operator–design rule combination. This ensures that any *necks* in the shape satisfy the design rule as well (Figure 5.48). It may even be desirable to enhance the width-checking operation to include a separate design rule value for the minimum neck distance. Using Figure 5.48 as an example, it is possible that the changes in surface topography at the neck of the polysilicon shape could be severe for the metal line crossing over; the dielectric layer separating the two levels may conform to the topography underneath (i.e., it may only partially *planarize*). The metal line may have difficulty covering the steps in the topography at the neck of the polysilicon shape, introducing a mechanism for a loss in yield or long-term reliability. As a result, a separate (typically more conservative) *neck* design rule may be required above and beyond the width check.

(3) SURROUND check

The syntax of the surround check is

surround ('level₁' , 'level₂' , relational_operator, design_rule_number);

This operation checks that a shape on level 1 surrounds *each* shape on level 2 by an amount determined by the relational operator and design rule value. In addition, any shape or part of any shape on level 2 that is not covered by a shape on level 1 is also in error (Figure 5.49). The surround check is commonly used to verify that contact openings are properly covered by shapes on the levels being interconnected.

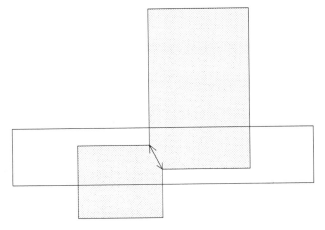

Figure 5.48. Just as with the DIST check, the WIDTH check also includes the Pythagorean (inside area) distance between the end points of vectors that do not project onto one another. It is also possible to implement a separate *NECK* checking routine to apply against a different (typically more conservative) ground rule than the WIDTH check.

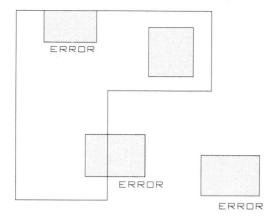

Figure 5.49. The SURROUND checking operation. Shapes that are only partially covered or not covered at all are in error.

(4) AREA and PERIMeter checks

area ('level' , relational_operator , design_rule_number);

perimeter ('level' , relational_operator, design_rule_number) ;

For the *area* check, the inside area of each individual shape is to be calculated and compared against the relational operator/design rule specification. The enclosed outside area for an inside-outside shape is not included. The *perim* check calculates the perimeter of each shape and checks it against the given specifications. These two routines are probably of limited usefulness in implementing design rule checks; however, the algorithm to calculate the area of each shape is used in an important diagnostic routine to be discussed shortly, *sum*.

(5) ERROR on level output
The command

error ('level' , design_rule_number);

generates a line of output for each shape remaining on the specified level; the message provided is the message corresponding to the given design rule number. Thus, any shapes created on a new level by one of the manipulation routines may be easily flagged as in error.

Diagnostic Routines

The following diagnostic routines generate some form of output, but do not really detect errors. A possible application for each routine is also given.

(1) COUNT

count ('level');

This routine counts the number of shapes on the specified level and outputs an appropriate message. This information may be useful to the circuit designer (the number of devices, the number of contacts, etc.). If the shapes file consists of the data after postprocessing into rectangular flashes, this count may be useful to the mask manufacturer, both as an estimate of the time required to generate the mask and as a check that the appropriate number of flashes were indeed made during mask plate exposure.

(2) RECT check

This routine checks that all shapes on the specified level are rectangular; any polygonal shape is output with an appropriate message:

rect ('level');

One example of where this may be useful is for the contact and/or via mask levels where the pattern generator flash counts should be kept to a minimum. Another potential use is for checking device gate shapes; the algorithms for logical-to-physical checking, circuit parameter extraction, and device modeling for circuit analysis are simplified considerably if the assumption of rectangular device dimensions can be made.

(3) SUM of area output

sum ('level');

The sum of the areas of all the shapes on a particular level can be a very useful diagnostic tool. For example, the process engineer will have measured defect density data from process monitor and evaluation runs; the total device area for a particular design, when combined with the pertinent (thin oxide) defect density, will give a measure of the expected yield loss due to this particular failure mechanism.

Additional Constructs

(1) Comments

Comments can be included in a design rule checking deck by surrounding the comment with suitable delimiters (e.g., braces).

(2) Conditional execution and flagging

To maximize the circuit density and/or process yields, it is useful to provide the process engineer with flexibility in the development of the process design rules, while still maintaining a high level of design checking and auditability. A major addition to that flexibility is the inclusion of conditional design rules, that is, design rules applied to a subset of the shapes on a particular level (or levels). The specific subset of shapes used in a checking routine is indicated by the presence or absence of a *flag,* which is assignable to each shape in the shapes file. To *flag* a shape (or its opposite, to *clear* a shape) simply means to set (reset) a Boolean value that adds (deletes) the shape to (from) the active subset.

Before elaborating on the subsetting and conditional checking operations, a brief mention of some simplifying assumptions is in order. To make the interpretation of the checking deck as simple as possible, the following assumptions are made:

1. No nesting of IF . . . THEN . . . statements is used (instead, use a series of flagging and clearing operations to refine the subset of shapes).
2. No ELSE clause is used.
3. It is shapes that are flagged, not the individual vectors of a shape; subsetting uses only the shape-checking routines and not any of the vector-based routines.

To specify that only a subset of shapes on a level (or levels) is to be used for the subsequent routine, the following syntax can be used:

(a) if flag ('level$_1$', 'level$_2$' , . . . , 'level$_n$') then "routine" ;
(b) if clear ('level$_1$', 'level$_2$' , . . . , 'level$_n$') then "routine" ;
(c) if flag ('level$_1$', . . . , 'level$_m$') and clear ('level$_n$' , . . . , 'level$_z$') then "routine" ;

The routine being executed may be any manipulation, diagnostic, or checking routine described earlier. The routine may refer to one or more levels; only the levels listed in the conditional clause use a subset of shapes on that level, while all shapes on any other levels in the routine should be considered.

Example

```
{check overlap of contact by metal to two different design
rules based on a subset of shapes; note that any contact
not completely covered is still caught as being in error}
if flag ( 'con' ) then surround ( 'metal', 'con',  > =, 38 ) ;
if clear ( 'con' ) then surround ( 'metal', 'con',  > =, 39 ) ;
```

To conditionally set or clear the flag for a shape, the following statement syntax is used:

One-level conditions

$$
\text{if} \begin{bmatrix} \text{rect ('level')} \\ \text{area ('level' , relational\_operator , design\_rule\_number)} \\ \text{perim ('level' , relational\_operator , design\_rule\_number)} \\ \text{width ('level' , relational\_operator , design\_rule\_number)} \end{bmatrix}
$$

$$
\text{then} \begin{bmatrix} \text{flag ('level') ;} \\ \text{clear ('level') ;} \end{bmatrix}
$$

If the condition is satisfied, then set (if flag is used) or reset (if clear is used) the flag for that shape; the previous condition of the flag for that shape is overwritten. For shapes that do not satisfy the condition, their flag is unaffected; thus, a cumulative subset of shapes can be built using a series of conditional statements. Note that the 'level' specification in both clauses must agree. Also note that a slight change in interpretation is included; when used as checking operations, only the shapes *not* satisfying the check are involved in the subsequent (error message) operation, while in this case it is the shapes that *do* satisfy the conditional check that are used (in the operation of modifying their flag value).

Two-level conditions

$$
\text{if} \begin{bmatrix} \text{dist ('level}_1\text{' , 'level}_2\text{' , relational\_operator , design\_rule\_number)} \\ \text{surround ('level}_1\text{', 'level}_2\text{' , relational\_operator, design\_rule\_number)} \end{bmatrix}
$$

$$
\text{then} \begin{bmatrix} \text{flag ('level');} \\ \text{flag ('level}_1\text{' , 'level}_2\text{');} \\ \text{clear ('level') ;} \\ \text{clear ('level}_1\text{' , 'level}_2\text{'):} \end{bmatrix}
$$

where 'level' may be either 'level$_1$' or 'level$_2$'. For shape-checking routines that involve two levels, it is possible to flag (or clear) the shapes that satisfy the condition for one or both of the levels used in the routine. As before, any level(s) used in the "then" clause should be referred to in the conditional clause.

In addition to the checking routines shown previously, some additional two-level constructs can be used in the conditional clause; these constructs do not imply a design rule check but rather simply an orientation between shapes on different levels. These constructs are *covers, borders, matches, crosses,* and *touches.* Examples of each are illustrated in Figure 5.50. Briefly, *covers* checks that a shape on level 1 covers a shape on level 2; covering includes shapes that share an inside edge. *Borders* implies that the shapes share all or part of an outside edge (including just a vertex), but with no area of intersection. Two shapes *match* if they occupy exactly the same area, while two shapes *cross* if their area of intersection is greater than zero and not equal to the area of either shape. The construct *touches* includes all the above.

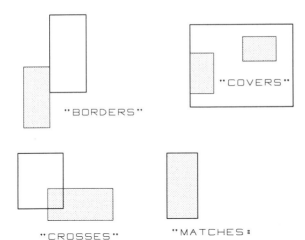

Figure 5.50. Additional two-level constructs for conditional checking; no design rule checking is involved, simply an orientation between shapes.

Finally, to unconditionally set or clear the flag for shapes on a particular level, the *flag* and *clear* statements can be used stand-alone (not in any conditional statement) in the checking deck:

flag ('level') ;
clear ('level') :

Before leaving this section on design rule checking, it is worth stating again that the overhead of using pin shapes, node names, and checking cells for a macro design should allow the design verification task to be managed more effectively. Dividing the checking (and therefore the design) into macro and global requirements reduces the necessary checking resource, detects errors earlier, and encourages the repetitive use of standard, prechecked functions and layouts.

The list of design rule checking statements and a summary of their syntax is included in an appendix at the end of this chapter.

5.5 CHIP ASSEMBLY AND POSTPROCESSING FOR DESIGN SYSTEM APPLICATIONS

Postprocessing refers to taking the chip design information, adding to that the necessary shapes for alignment and process evaluation, and dissecting the resulting collection of shapes into a data structure (consisting of rectangles, typically) suitable for mask manufacture.

The chip design is only a part of the shape information that is patterned onto each mask plate. In addition to the collection of design data, the *mask house* includes some additional shapes specific to the technology's fabrication process.

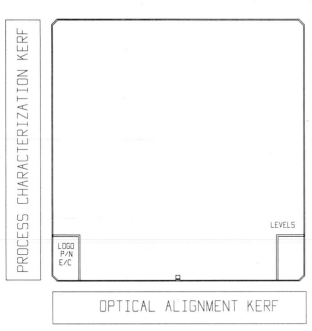

OPTICAL ALIGNMENT KERF

(INCLUDES MEASUREMENT VERNIERS)

Figure 5.51. Chip image to which has been added the optical alignment and process and defect monitor structures in the kerf area. The wafer sawing operation after wafer test cuts through the *streets* containing these kerf structures.

As indicated in Figure 5.51, the additional information lies in the *optical alignment kerf* and the *process characterization kerf*. This additional information can be placed just outside each chip site on the wafer in the *dicing channel* area. The sawing operation that separates the individual chips (or *die*) after fabrication and testing cuts through these structures.[4]

The optical kerf contains the shapes on particular mask levels to which subsequent levels are aligned. Examples of typical mask shape alignment *targets* are shown in Figure 5.52. These sets of shapes are designed to provide the best x, y, and θ alignment aids to reference shapes previously patterned in the wafer surface topography. As discussed in greater detail in Chapter 15, the choice of a prior level to which a new level should be aligned is not trivial. The process engineer must determine which levels will provide a suitable high-contrast image (under a photoresist coat) when aligning a subsequent level. In addition, the level-to-level alignment tolerance statistically increases between two levels that were not directly aligned to one another. Critical device or contact-related levels should be directly aligned to the previous related level. In addition to the alignment targets, the optical kerf also typically contains alignment *verniers* for a relatively accurate measure of the actual alignment value, or *overlay,* after photoresist ex-

---

[4] A number of different die-separation techniques are discussed in Section 15.17; separation by diamond wheel sawing allows these structures to be located in the dicing area, with associated separation design rule restrictions.

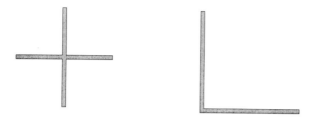

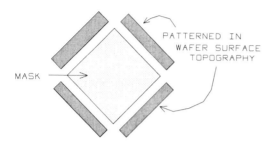

MASK

PATTERNED IN
WAFER SURFACE
TOPOGRAPHY

Figure 5.52. Alignment targets: The bottom example is commonly used with mask alignment equipment with *autoalign* capability; the surrounding bars are present in the topography of the preceding etched level, while the diamond shape on the mask plate is automatically centered within this window.

posure and develop. These vernier measurements are useful for in-line process evaluation for possible rework (if the photoresist image-to-existing vernier alignment is outside suitable limits) and for subsequent correlation with process yield and characterization data.

The process characterization kerf typically contains a variety of structures intended to provide useful feedback data about the particular processing run in a very limited amount of real estate. Ideally, to maximize the number of die per wafer, the spacing between the die should be no wider than what is necessary to ensure a good chip edge dielectric overcoat seal and a reliable sawing operation. The design and layout of the process characterization structures, and even the choice of which structures to include, is not always straightforward. They vary considerably from technology to technology and will vary in scope as a particular technology matures. The two types of structures usually included in this part of the chip kerf are defect density monitors and device and thin-film evaluation circuits. Both sets of structures require the process to complete a number of levels through metallization, whereupon large-area pads can be probed to make suitable electrical measurements.

The defect density monitors include structures for measuring the density of *pinholes* in dielectric layers, open circuits in interconnections, or node-to-node shorts and leakage currents. Typically, to gather statistically significant defect density data, these must be large-area structures, which is contradictory to the goal of minimizing kerf area and maximizing the number of chip sites on the wafer. These structures bear no relation to the design rules; they are developed to accentuate potential process concerns about interconnection-, dielectric-, and topography-related failure mechanisms. Often, these designs include shapes whose layout is *more* aggressive than the normal manufacturing design rules, to gain a

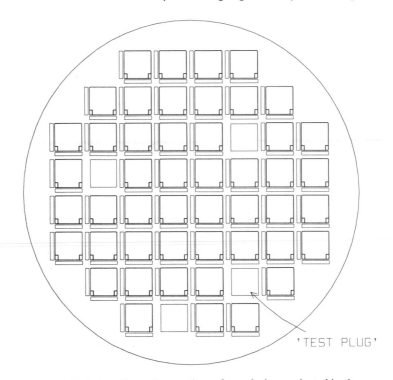

'TEST PLUG'

Figure 5.53. Inclusion of test sites on the wafer replacing product chip sites.

better understanding of process sensitivities. The characterization kerf will likely evolve from a preproduction to production version. Initially, the kerf contains a significant percentage of defect density monitors; as the process matures, some of the structures will be deleted to increase the number of die per wafer. An appropriate chip seal and dicing channel must still be maintained, however.

Another approach to including process characterization structures is to replace one or more full chip sites with a much larger, more exhaustive set of designs (Figure 5.53). These sites are commonly called *test plugs*.[5] The particular approach chosen for including characterization monitors depends on the particular mask-making and mask-alignment strategy used by the process line (and the die-separation technique).

A word about mask strategy: the technology options for transferring the collection of shapes into the mask coating parallel those of the engineering workstation discussed earlier, that is, vector and raster. The vector option refers to the exposure of rectangles of varying size and orientation over the mask area. These rectangles are produced by a strobe flash through a variable rectangular

---

[5] The terms *drop-ins* and *process control monitors* (PCMs) are also used to describe these chip sites containing characterization structures.

aperture. Conversely, it is also possible to produce the desired set of shapes by scanning an electron beam (of fixed or potentially variable size) in raster fashion over the mask surface, turning the beam on and off to delineate the desired pattern. Different photoresist coatings are used on the mask plate for the two exposure types, each tailored for maximum sensitivity to the wavelength or energy of the exposing radiation. A more advanced technique for wafer patterning does not use a mask plate at all, but direct electron beam writing on the photoresist-coated wafer. These pattern-transfer technology alternatives are discussed in more detail in Chapter 15.

The remainder of this section assumes the means for making masks is by optical exposure, necessitating the development of an algorithm for dissecting shapes into a suitable set of rectangles. (Problem 5.3 asks for the development of an algorithm to rasterize the shapes data for electron beam mask manufacture.)

The first question to be addressed when developing an algorithm for post-processing the total collection of shapes data is whether or not the hierarchically structured data (i.e., the cell transforms used) should be unnested before dissecting the shapes. Recall from Section 5.4 that the first step in design rule checking was indeed to unnest the cell transform description and then union any overlapping or abutting shapes resident on the same mask level. To correctly perform the design rule checking, a complete expansion of the cells was required; the resulting number of shapes is potentially very large and therefore cumbersome to manage. The situation with postprocessing the shapes data is different; the option exists of dissecting the elemental shapes into rectangles in each cell description first and then proceeding with the unnesting of the cell transforms. The data volume on which the dissecting algorithm must operate is therefore reduced considerably (assuming the design has been well structured using basic cells repetitively). The disadvantages of postponing the unnesting process until after dissection are twofold, both having to do with overlapping or abutting shapes:

1. Shapes that overlap result in overlapping rectangular flashes and a *blooming* of the final developed mask pattern.

2. The rectangle sizes used for dissecting the shapes in each cell description may be smaller than would otherwise be used for unioned shapes in the unnested description, creating either more flashes or more different flash sizes than necessary.

The first problem can potentially be minimized by enforcing in the design rule checking deck that the extent of overlap among shapes on the same level is relatively small and is not located in close proximity to any critical device or contact structures. (This checking would be performed prior to unioning the levels.) An interesting case is also presented by shapes on the same level that abut; ideally, without unnesting and unioning, the two abutting shapes would be patterned by two abutting flashes that form a continuous interconnection when developed in the photoresist coat or emulsion on the mask plate. However, the design dimension used by the circuit designer does not always directly correspond to the dimension on the mask (Figure 5.54). To be able to more closely hit the targeted physical dimension at the wafer level after fabrication, the process engineer may request that a positive or negative *mask bias* be applied to all rectangular flashes on that level prior to exposure. The goal is to keep the sum of the process bias and mask bias on target, consistent with the circuit designer's models and analysis. (The circuit designer is often unaware that such a trade-off between process and mask bias is being made.) If a negative mask bias is applied to abutting rectangular flashes, the desired connection is no longer continuous. As a result, if shape dissection is to be done before unnesting and unioning, abutting shapes should be regarded as being in error; shapes should therefore overlap by a minimum of twice the largest conceivable mask bias.

Another consideration for the dissection algorithm is the difference in dimension between the minimum design image size and design step size; for example, a particular technology may use a 2μm minimum design image and a 0.25-μm step size. (Figure 5.55 illustrates a pattern generator system used to make $10\times$ reticles for subsequent use in a $10\times$ reduction step-and-repeat mask aligner. The required rectangle dimensions and stage addressability for the technology are indeed provided; note that the stage motion orthogonality specification also influences the dissection algorithm.) A step size smaller than minimum flash size can introduce overlapping flashes; the dissection algorithm should try to minimize the extent of this overlap. Figure 5.56 gives an example of a staircase shape that will necessarily result in overlapping flashes, additional complexity for the dissection algorithm. (The circuit designer should be aware of what the resulting dissection will look like and to what extent the resulting mask image may bloom.)

A number of different algorithms have been developed for dissection; several have been reported in the technical literature (see the additional references at the end of the chapter). The one described here is based on the concept of a *maximum empty rectangle* (MER), as described in references 5.2 and 5.3. Consider a rectangular area from which a number of smaller rectangles have been removed (Figure 5.57). A maximum empty rectangle is a rectangle that is bounded on all four sides by either the boundary of the entire area or an edge of a removed object; in other words, all four sides of the MER abut the edge of the area or of a removed object. The MERs in the complete set for the rectangular area can overlap; each MER has at least two edges that are unique to the set of MERs. The six MERs corresponding to the area in Figure 5.57 are shown in Figure 5.58. Note in particular that not all six MERs are necessary to completely cover the open area; all six are included because each represents a potential rectangle that may be

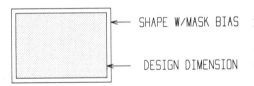

SHAPE W/MASK BIAS

DESIGN DIMENSION

Figure 5.54. *Mask bias* added to all shapes on a particular level. If the mask bias is positive, the mask dimension will be larger than the design dimension.

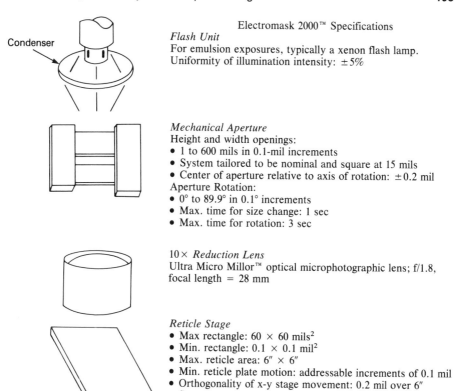

Electromask 2000™ Specifications

Flash Unit
For emulsion exposures, typically a xenon flash lamp.
Uniformity of illumination intensity: ±5%

Mechanical Aperture
Height and width openings:
- 1 to 600 mils in 0.1-mil increments
- System tailored to be nominal and square at 15 mils
- Center of aperture relative to axis of rotation: ±0.2 mil
Aperture Rotation:
- 0° to 89.9° in 0.1° increments
- Max. time for size change: 1 sec
- Max. time for rotation: 3 sec

10× Reduction Lens
Ultra Micro Millor™ optical microphotographic lens; f/1.8,
focal length = 28 mm

Reticle Stage
- Max rectangle: 60 × 60 mils²
- Min. rectangle: 0.1 × 0.1 mil²
- Max. reticle area: 6″ × 6″
- Min. reticle plate motion: addressable increments of 0.1 mil
- Orthogonality of x-y stage movement: 0.2 mil over 6″
- Stage positioning repeatability: ±0.03 mil (over 5 hours)

Figure 5.55.   Pattern generator block diagram.

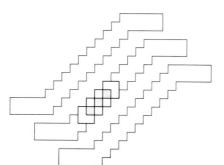

Figure 5.56.   A *staircase* shape
(designed using an aggressive step size
in comparison with the minimum flash
size), which will result in overlapping
flashes (shaded area).

used to create a flash for the pattern generator. A particular MER may be more
desirable than others because of its large size or perhaps because of a common
size and orientation with other flashes already chosen for the mask level. Once
a particular rectangle is selected, it is *removed* from the active area, a new set
of MERs is determined, and the selection process repeats until no more active
area remains. The result of the algorithm is a set of abutting rectangles that attempt

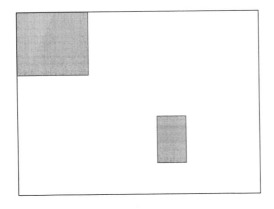

Figure 5.57. Rectangular area to be dissected into rectangular flashes; the shaded area is not to be included.

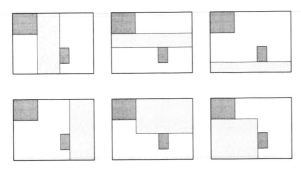

Figure 5.58. The six *maximum empty rectangles* (MERs) for the rectangular area of Figure 5.57. Note that not all six MERs are necessary to completely span the desired area.

to completely span the active area; if the area cannot by fully covered with abutting rectangles of at least minimum dimension, like the staircase shape shown earlier, an extension to the MER selection procedure is required, one that determines the best overlap situation for minimum blooming (Problem 5.4). Briefly, the post-processing algorithm for a polygonal shape would proceed as follows:

1. Assuming the shape is initially described as a list of vectors (similarly to the data structure for design rule checking), discard the horizontal segments and sort the remaining vertical segments by increasing *x*-coordinate (Figure 5.59). The inside–outside orientation of the shape area relative to the vector is now determined by the left-to-right ordering of this sorting procedure.

2. From this list, determine the bounds of the enclosing rectangle (Figure 5.60). This is the initial maximum empty rectangle, which becomes the head (and tail) of a (dynamic) linked list of MERs.

3. Determine the segments along the left edge of the bounding rectangle that are complementary to the polygon segments at this edge (Figure 5.61). Replace the polygon segments for the leftmost edge of the polygon (at the top of the sorted list) with the segments from the bounding rectangle.

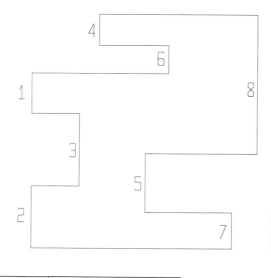

Figure 5.59. From the vector-based description, sort the shape vectors into an increasing *x*-coordinate list.

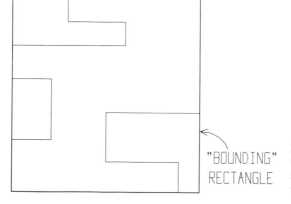

"BOUNDING"
RECTANGLE

Figure 5.60. From the sorted list of shape vectors, determine the bounding rectangle for the overall shape. This rectangle is the initial MER.

4. Repeat step 3 for the right side of the shape; replace the polygon segments for the rightmost edge of the polygon (at the bottom of the sorted list) with the complementary segments from the right side of the bounding rectangle.

5. For the top vector in the list, find the corresponding vector(s) farther down in the list opposite to it that describes a rectangular *slice* to be removed from the bounding rectangle (Figure 5.62). Write the resulting rectangle into a separate list and delete the segments used to generate this rectangle from the sorted list of segments (Figure 5.63).

6. If the sorted list of segments is empty, go to step 7. If the list of segments contains only one element, an error has occurred. Otherwise, return to step 5.

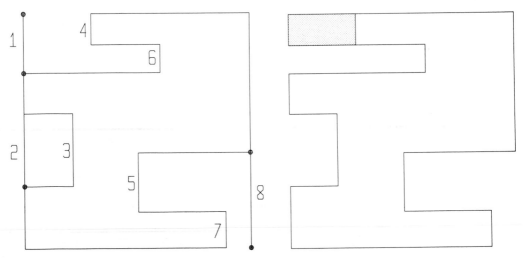

Figure 5.61. Determine the segments of the initial MER on the left and right sides that are complementary to the leftmost and rightmost segments of the original shape. Replace the segments at the top of the sorted list with their complements from the MER. Replace the segments at the bottom of the sorted list with the complementary segments from the right side of the MER.

Figure 5.62. Starting with the top vector in the list, find the outside shape vector opposite it that defines a rectangular area to be removed from the bounding rectangle. Save this rectangular area in another list and delete the vertical segments used to define this rectangle from the list of vectors. This step will correctly identify any enclosed outside area (if the original shape were an inside–outside shape) and write the area to the list of rectangles to be deleted.

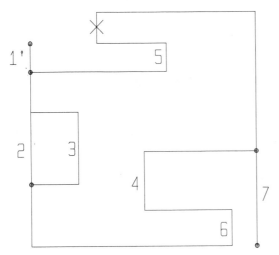

Figure 5.63. Deletion of vectors after a rectangular area to be removed from the bounding rectangle has been found.

7. Each rectangle in the list created represents an area to be deleted from the original bounding rectangle (Figure 5.64). More precisely, each rectangle in the list represents an area to be deleted from the dynamic list of MERs; after the first deletion, the list of MERs no longer contains the bounding rectangle as its only element. For each rectangle in the area deletion list:
   a. Check each MER in the list of MERs for intersection with the rectangle.
   b. If there is no intersection with a particular MER, continue on to the next MER in the list.
   c. If the two intersect, several steps follow:
      (1) Delete the intersected MER from the list of MERs.
      (2) If the rectangle and MER match, go on to check the rectangle against the next MER in the list (return to step 7a).
      (3) Any area that is inside the MER, but outside the area of intersection, may result in the creation of new MERs (Figure 5.65); check each rectangle inside the MER and outside the area of intersection (up to four may be required) against the complete list of MERs; if the new rectangle is not contained entirely within any other MER, then add the rectangle to the list as a new MER that resulted from the area deletion procedure.
      (4) Go on to check the rectangle against the next MER in the list (return to step 7a).

   After comparing the rectangle against the entire list of MERs, the area spanned by the rectangular slice has been removed from the active area; any new MERs that resulted have been added to the list. Repeat this area-deletion procedure for all rectangular slices in the array, effectively deleting the area outside the original shape from its bounding rectangle (Figure 5.66).

Figure 5.64. Once the list of rectangles describing the area to be removed has been completed, return to the original bounding rectangle and begin the process of deleting the rectangular areas from the dynamic, linked list of MERs.

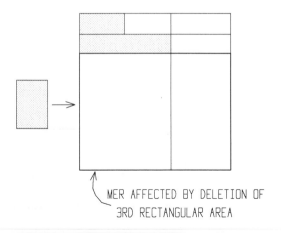

MER AFFECTED BY DELETION OF
3RD RECTANGULAR AREA

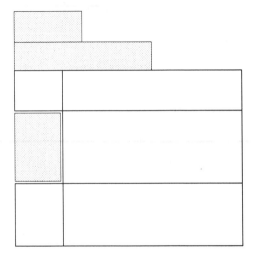

Figure 5.65. Creation of new MERs from the removal of rectangular area from the existing list of MERs. The deletion of the next rectangular area indicated in (a) affects MER_3 and results in the replacement of MER_3 with three smaller MERs (b). There are now a total of five MERs for the remaining area.

At this point, the linked list of MERs represents the largest-area potential rectangles for flashing to cover the original polygon. The procedure for selecting a particular flash from the MER list will likely have some characteristics specific to the pattern-generator tool (e.g., maximum aperture size), as well as possibly some process engineering input (e.g., preferred aperture sizes and/or orientations). Whatever rectangular area is selected from the list of MERs, the procedure for deleting the flash area and updating the list of MERs is identical to that just described. Note that the flash selected need not necessarily match any MER; it may certainly be smaller. Also note that the algorithm works, albeit rather clumsily, for the situation where the original shape is a rectangle (a four-sided polygon); in this case, the initial step of the algorithm, the deletion of outside area from the bounding rectangle, is skipped.

The resulting array of rectangular flashes produced by dissecting all the

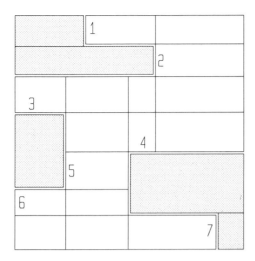

Figure 5.66. The entire list of MERs after deleting all the rectangular area outside the original shape but inside the bounding rectangle. Each of these MERs represents a possible flash for mask exposure. Once a rectangular flash area is subsequently selected (and written to the file of postprocessor flashes for the level in question), that area is deleted from the list of MERs (until all area has been covered and the list of MERs has no members). *Note:* This algorithm as described produces no overlapping flashes, and therefore will not suffice for staircase shapes.

polygonal and rectangular shapes for a particular mask level needs to be "massaged" prior to pattern generation. The pattern-generator tool has restrictions regarding maximum flash frequency, *xy* mask stage velocity and settling time, and settling time for changes in the position of the aperture blades (which varies with the extent of the change). The array of flashes should be sorted so as to minimize the pattern-generator time, thereby increasing the reliability of the mask-making process.

One final comment: as described in more detail in Chapter 15, the pattern generator has a data format for describing each flash that differs slightly from the format used here. As illustrated in Figure 5.67, the pattern generator uses the

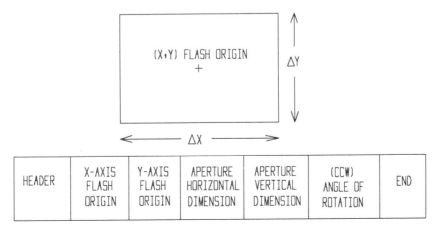

Figure 5.67. Pattern generator data record format; the flash origin is the center of the rectangular area, rather than a vertex.

center of the rectangle as its reference, rather than a vertex. Each element in the sorted array of flashes needs to be translated accordingly.

5.6 SUMMARY

This chapter has presented some basic concepts about shape descriptions and checking tools to support integrated-circuit layout. For purposes of generality, the graphics language developed uses textual descriptions; the more common approach is to use the internal data representation of an engineering graphics system. A vector-based system for shapes representation (using only orthogonal vectors) was used for illustration; an alternative technique, based on a raster-scan approach, is also widely used and has the attractive feature of reduced graphics hardware cost. The discussion on logical-to-physical checking was brief; the tracing and circuit-matching algorithms are extremely complex, as are the techniques for circuit parameter extraction. A chip postprocessing algorithm was presented for determining the flashes to be used for optical pattern generation of mask data; however, the need for dimensional accuracy at VLSI densities is driving the mask-making industry toward employing electron-beam mask manufacture. This approach requires significant modifications to the postprocessing algorithm given here.

This chapter also reinforces the importance of hierarchy in the design process. The hierarchy described in this vein implies the division of the task of VLSI macro and chip layout into more manageable and repetitive structures, which will enable enhanced checking capability and improved designer productivity.

PROBLEMS

5.1. The design rule checking operations and postprocessing algorithms described in this chapter are both based on the dissection of the shapes into their constituent vectors, with a field in each vector record indicating the orientation of the inside shape area relative to the vector. This problem deals with the construction of that vector database.

(a) Develop a Pascal-like record format to represent the physical design database (for the statements of Figure 5.21). Comments within the textual physical design description need not be stored internally. Choose only one of the two types of ALPHA statements to support. The length of the list of parameters in the chip header is unbounded, but each parameter name may be assumed to be limited to a reasonable number of characters.

(b) Write and code an algorithm to fully unnest the cell transform records in the database, evaluating the *Conditional* and *Drop* pin usage fields for macro designs and macro usage. Note that nested cell transforms may contain a number of

Translation, Mirroring, and Replication operations that must be evaluated in appropriate sequence.

(c) Write and code an algorithm to convert the LINE and ALPHA database records into RECTangles. For the ALPHA records, assume there exists a library file of records (one record for each character in the available character set) that describes the shape(s) comprising each character. Develop a record format for the character records in this library file; in particular, this record format should include the appropriate information to be able to *scale* the character to the desired size. Evaluating the ALPHA records requires determining the height, width, and origin of each individual character, reading in and scaling the shape(s) for that character from the library file, and then outputting the appropriate records.

(d) Recall that NODE statements are used primarily for logical-to-physical checking during macro design and for the global wiring procedure to calculate the coordinates for interconnection. However, it is also necessary during macro design to check that the NODE statements are assigned to the correct mask level. As such, it is necessary to include the coordinate and level information about the NODE records in the design database for macro design rule checking. To use the DRC procedures, a shape is created from the NODE record (e.g., a rectangle on the NODE record level about the coordinate one step long on each side). Since a UNIONing operation is the first procedure in design rule checking, the rectangles created from the NODE records on the correct level are absorbed; those assigned to an incorrect level can be flagged later in the checking deck as being too small. Develop a technique for generating these rectangles from the NODE records.

(e) Develop a record format for storing the vectors that comprise each shape, including the orientation of the inside area of the shape relative to the vector and the shape's mask level. Also, retain the chip design header information. (See Problems 5.5a and 5.6a.)

(f) Develop a technique for determining the inside area orientation of segments in a polygon; note that it is not possible to determine from the initial direction of traversal from the polygon origin how the inside area of the shape is oriented. *Suggestion:* What is unique about the vertical vector(s) in the shape with the smallest x-coordinate?

(g) Develop and code an algorithm to take POLYGON and RECTangle records from the shapes database and create the vector database for design rule checking and mask postprocessing, using the structure developed in part (e).

(h) Choose a suitable character set and, using the record format design in part (c), create the library file of records and shapes for ALPHA character look-up and decomposition.

5.2. Write a parser to accept the physical design text file description and build the database developed in Problem 5.1a.

5.3. As lithography dimensions shrink and the quantity of shapes data increases in evolving VLSI technologies, the momentum behind raster-scan electron-beam mask manufacture will grow. This chapter assumed that mask manufacture would be by optical pattern generation, necessitating the postprocessing of shapes into rectangles. In anticipation of the growing use of e-beam mask-making equipment, write and code an algorithm to *rasterize* the vector data base of Problem 5.1e. The e-beam database should consist of a field indicating the starting exposure coordinate and a field that encodes the run length of the beam on that particular scan. All coordinates (as with rectangular flashes) are to be referenced to the center of the beam. Assume that the beam size is constant throughout the mask manufacture and is an input variable to the program. For reasonable edge acuity, the size of the beam (with a Gaussian intensity profile away from the center) must be less than or equal to one-fifth the minimum linewidth dimension. Any (horizontal or vertical) shape that may therefore be insufficiently resolved should be output to an error message file.

5.4. **(a)** Code the postprocessing algorithm given in Section 5.5 (or an algorithm of your own design). The minimum aperture size (at $100\times$) and stage increment (at $10\times$) are input variables to the program; ensure that the measure units in the database header information agree with the units of the aperture. The rectangular flash data file should consist of records indicating the origin (the center of the flash) and its x- and y-opening size. Use the vector database of Problem 5.1e as the format of the input file; assume the input file to the postprocessor contains shapes that have been extracted from the entire vector database and that all reside on a single mask level (along with a copy of the design header). If a shape cannot be completely covered with nonoverlapping rectangles (down to the minimum aperture size), output the entire shape record to an output file and do not use the rectangular flashes found for that shape.

(b) Modify the algorithm in part (a) to create a suitable dissection for shapes that require overlapping flashes.

5.5. This problem addresses concerns about execution time efficiency and data storage volumes of the design rule checking procedure.

(a) Design rule checking involves a comparison between vector descriptions; the majority of the comparisons required are between vectors that are in close proximity. For a large vector database, much execution time would be wasted making comparisons between vectors that are well separated from each other. Develop a means to enhance the execution speed of the DRC to avoid many unnecessary comparisons. One suggestion would be to modify the record format of Problem 5.1e to include a field to store the bounding rectangle of any polygon. Two shapes whose bounding rectangles are separated by a distance larger than the dimension of interest could be skipped by the DRC. (Also, finding the bounding rectangle is a key step to the postprocessing algorithm of Section 5.5.)

(b) The data representation to this point has all been vector based, assuming a fine step size relative to the minimum shape dimension. Assume instead that a coarse step size were used and that a pixel representation were appropriate. However, to reduce the necessary data storage volume, it is necessary to develop a more efficient means of representation than a straightforward memory-mapped implementation; several mask levels could be quite sparse of shapes and/or may only contain shapes of large geometries. Develop a scheme to reduce the pixel representation data storage volume; two suggestions are (1) a hierarchical representation and (2) a run-length encoded representation.

5.6. This problem deals with the development of a DRC tool.

(a) Among the DRC operations are those that set, clear, and interrogate the value of a Boolean *flag* that is associated with each shape. Modify the data record format of Problems 5.1e and 5.5a to include this capability.

(b) Using the vector database format given in part (a), write and code the individual algorithms to implement the various DRC operations listed in the appendix to this chapter. Of particular interest are the possibility of inside–outside shapes being formed when UNIONing and the Pythagorean checks of the DISTance and WIDTH operations.

(c) Write a DRC program compiler that will accept the input files (shapes and DRC text) and execute the desired sequence of operations, producing the appropriate output files.

APPENDIX: DESIGN RULE CHECKING OPERATIONS AND STATEMENTS

Shape Input/Output Routines

save (filename , 'level$_1$' , 'level$_2$' , ...);
save (filename);   { all levels }
restore (filename);

Shape Manipulation Routines

union ('oldlevel' , 'newlevel');
union ('level$_1$' , 'level$_2$' , 'newlevel');
intersect ('level$_1$' , 'level$_2$' , 'newlevel');
intersect ('level$_1$' , 'level$_2$' , 'newlevel' , 'complementary');
move ('oldlevel' , 'newlevel');
copy ('oldlevel' , 'newlevel');
delete ('level');
expand ('oldlevel' , 'newlevel' , design__rule__number);
etch ('oldlevel' , 'newlevel' , design__rule__number);
create ('level' , x__origin , y__origin , Δx , Δy);

Shape Checking and Diagnostic Routines

$$\text{proj ('level}_1\text{' , } \begin{bmatrix} \text{inside} \\ \text{outside} \end{bmatrix} \text{ , 'level}_2\text{' , } \begin{bmatrix} \text{inside} \\ \text{outside} \end{bmatrix} \text{ , relational\_operator,}$$

design__rule__number);
{ relational__operator : $<, <=, =, >, >=, <>$ }
collinear ('level$_1$' , 'level$_2$' , design__rule__number);

collinear ('level$_1$' , 'level$_2$' , design_rule_number , $\begin{bmatrix} \text{inside} \\ \text{outside} \end{bmatrix}$);

dist ('level$_1$' , 'level$_2$' , relational_operator , design_rule_number);

width ('level' , relational_operator , design_rule_number);

surround ('level$_1$' , 'level$_2$' , relational_operator , design_rule_number);

area ('level' , relational_operator , design_rule_number);

perim ('level' , relational_operator , design_rule_number);

error ('level' , design_rule_number);

Diagnostic Routines

count ('level');

rect ('level');

sum ('level');

Additional Constructs

{ comments }

Conditional Execution and Flagging

if flag ('level$_1$' , 'level$_2$' , . . . , 'level$_n$') then "routine" ;

if clear ('level$_1$' , 'level$_2$' , . . . , 'level$_n$') then "routine" ;

if flag ('level$_1$' , 'level$_2$' , . . .) and clear ('level$_a$' , . . . , 'level$_m$') then "routine" ;

One-level conditions

$$
\text{if} \begin{bmatrix} \text{rect ('level')} \\ \text{area ('level', relational\_operator, design\_rule\_number)} \\ \text{perim ('level' , relational\_operator, design\_rule\_number)} \\ \text{width ('level' , relational\_operator, design\_rule\_number)} \end{bmatrix}
$$

$$
\text{then} \begin{bmatrix} \text{flag ('level') ;} \\ \text{clear ('level') ;} \end{bmatrix}
$$

Two-level conditions

$$
\text{if} \begin{bmatrix} \text{dist ('level}_1\text{', 'level}_2\text{', relational\_operator, design\_rule\_number)} \\ \text{surround ('level}_1\text{', 'level}_2\text{', relational\_operator, design\_rule\_number)} \end{bmatrix}
$$

$$
\text{then} \begin{bmatrix} \text{flag ('level');} \\ \text{flag ('level}_1\text{', 'level}_2\text{');} \\ \text{clear ('level');} \\ \text{clear ('level}_1\text{', 'level}_2\text{');} \end{bmatrix}
$$

$$
\text{if } \begin{bmatrix} \text{covers ('level}_1\text{', 'level}_2\text{')} \\ \text{borders ('level}_1\text{', 'level}_2\text{')} \\ \text{matches ('level}_1\text{', 'level}_2\text{')} \\ \text{crosses ('level}_1\text{', 'level}_2\text{')} \\ \text{touches ('level}_1\text{', 'level}_2\text{')} \end{bmatrix} \text{ then } \begin{bmatrix} \text{flag ('level');} \\ \text{flag ('level}_1\text{', 'level}_2\text{');} \\ \text{clear('level');} \\ \text{clear('level}_1\text{', 'level}_2\text{');} \end{bmatrix}
$$

Unconditional

flag ('level');

clear ('level');

REFERENCES

5.1 Carmody, P., and others, "An Interactive Graphics System for Custom Design," *17th Design Automation Conference* (June 1980), 430–439.

5.2 Jayakumar, V., "A Data Structure for Interactive Placement of Rectangular Objects," *17th Design Automation Conference* (June 1980), 237–242.

5.3 Jayakumar, V., and Chatterjee, A., "Description of Rectilinear Areas in Terms of a Minimum Number of Disjoint Rectangles," *Proceedings of the IEEE International Custom Circuits Conference* (1980), 777–779.

5.4 Foley, J. D., and Van Dam, A., *Fundamentals of Interactive Computer Graphics,* Reading, Mass.: Addison-Wesley Publishing Co., 1982.

5.5 Newman, W. M., and Sproull, R. F., *Principles of Interactive Computer Graphics.* New York: McGraw-Hill Book Co., 1979.

Additional References

Avenier, J. P., "Digitizing, Layout, Rule Checking—The Everyday Tasks of Chip Designers," *Proceedings of the IEEE,* 71:1 (January 1983), 49–56.

Baird, Henry, "Fast Algorithms for LSI Artwork Analysis," *14th Design Automation Conference* (June 1977), 303–311.

Lambert, David, "Graphics Language/One: IBM Corporate-Wide Physical Design Data Format," *18th Design Automation Conference* (1981), 713–719.

Lauther, Ulrich, "An Order (N log N) Algorithm for Boolean Mask Operations," *18th Design Automation Conference* (June 1981), 555–561.

Ludwig, John, and others, "A Hierarchical Approach to VLSI Chip Design and Verification," *IEEE International Symposium on Circuits and Systems* (1983), 16–19.

Newell, M., and Fitzpatrick, D., "Exploitation of Hierarchy in Analyses of Integrated Circuit Artwork," *IEEE Transactions on Computer-aided Design of Integrated Circuits and Systems*, CAD-1:4 (October 1982), 192–200.

Rung, R. D., "Determining IC Layout Rules for Cost Minimization," *IEEE Journal of Solid-State Circuits*, SC-16:1 (February 1981), 31–35.

Scheffer, L., and Apte, R., "LSI Design Verification Using Topology Extraction," *Asilomar Conference on Circuits, Systems, and Computers* (1979), 149–153.

Skinner, Frank, "Interactive Wiring System," *IEEE Computer Graphics and Applications,* 1:2 (April 1981), 38–51.

Tucker, Michael, and Haydamack, William, "A System for Modifying Integrated Circuit Artwork through Geometric Operations," *Asilomar Conference on Circuits, Systems, and Computers* (1979), 154–158.

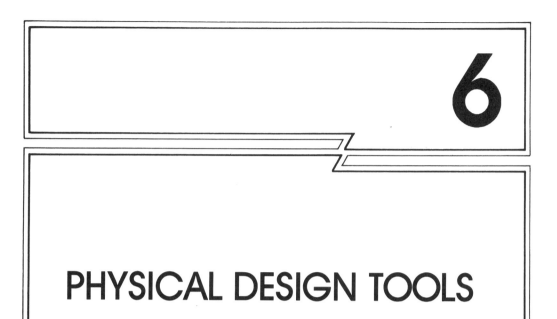

PHYSICAL DESIGN TOOLS

MANHATTAN: Excellent drink; terrible place to find a cab. Also: Geometry in which all edges are parallel to the grid lines.

—Preliminary Electronic Design Interchange
Format (EDIF) Specification, V. 0.08

6.1 PLACEMENT ALGORITHMS

The VLSI design system utilizes a set of programs and image and book rules to initiate the physical design of the chip, based on the logic interconnection description. This section describes some algorithms that have been developed to assist with the task of finding image placement locations for all books in the chip design; subsequent sections discuss the task of determining interconnection wiring routes. Although the tasks of global placement and wiring are closely interrelated, they are commonly approached separately (in large part due to the complexity of each of the tasks). The output from these physical design tools must eventually be translated into a shapes file for fabrication and checking.

Before describing placement techniques, it is necessary to discuss *image rules*, that is, the fixed information extracted from the master image or master slice chip image that is required by the design system program to be able to assign valid placement locations. These rules include the following:

1. Coordinate origins for each internal and chip I/O cell.
2. Restrictions on the book *type* that may be assigned to a particular internal

cell (e.g., a three-cell-wide book cannot be placed within one or two cells from the end of a row of cells).

3. Book orientation restrictions (especially for chip I/O circuit placement).

Each book in the design system technology library will contain the related rule information required by the placement procedure (e.g., book cell size, mirroring and orientation restrictions, and pin locations relative to the book origin).

The extent of the rule information provided in the design system library is directly related to the features of the placement algorithm. For example, the placement program will likely estimate the additional chip interconnection wiring length resulting from the assignment of a book to a particular cell location; the optimum location will be that which tends to minimize this value. The option exists of using a single cell coordinate as the reference for all book pins or calculating the resulting location of each book pin individually. Certainly, the latter approach would be more accurate (especially for larger macros); it would also be more time consuming. The need to provide pin location rules to the placement program varies with the particular approach selected for coordinate calculation.

Cell mirroring is another feature that can be contemplated; significant enhancements in overall wireability might be achieved for a master image chip design if the cell layout can be reflected about one or both axes (Figure 6.1). If this placement feature is indeed incorporated, individual restrictions on mirroring due to an asymmetry in the image or book design need to be provided. Input/output cell locations around the chip periphery present a unique placement consideration, as both mirroring and 90° rotation are involved between the horizontal and vertical sides of the image. As before, appropriate cell placement rules need to be provided. Figure 6.2 illustrates the situation where the first and second metal wiring grids are on different pitches; thus, a horizontal I/O cell is distinct from a vertical I/O cell.

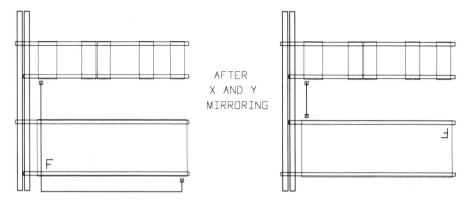

AFTER
X AND Y
MIRRORING

Figure 6.1. An option for the design system placement program is the *mirroring* of the macro layout to potentially improve the total chip interconnection length (master image design systems).

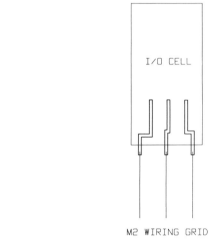

M2 WIRING GRID

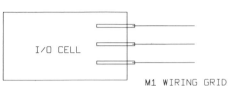

M1 WIRING GRID

Figure 6.2. If the orthogonal wiring grids are of different pitch, the same I/O cell layout cannot be used on vertical and horizontal sides of the chip; alternative physical design cells within the layout must be selected to match the circuit connections to the wiring grid.

Some data present in image and book rules can potentially be extracted automatically from the shapes layout; the remainder is entered manually. Prior to releasing the technology design system to its users, it is necessary to verify that these rules have been coded correctly; usually, several chip test cases are run to determine the validity of these additions to the technology library database.

A number of global chip physical design goals may be the objectives of the placement program (performance, wireability, accommodating some number of simultaneously switching off-chip drivers, etc.). The most frequently used metric, which is part of decision-making during execution of the placement algorithm, is the overall chip wireability.

Wireability can be gauged in terms of the total estimated chip wire length to complete all nets. For example, the total estimated chip wire length could be calculated as follows:

Consider the smallest rectangle that encloses the currently placed pins in a particular net; the routing length of the net is taken to be the length of the horizontal side plus the length of the vertical side of this encompassing rectangle. The total chip routing length is the sum over all nets. The added cost of placing a book at a particular location is the increase in rectangle half-perimeter lengths for all nets connected to the book.

A different wire length measure can be developed that attempts to assign a weight to each interconnection segment in a net based on the number of pins in the net;

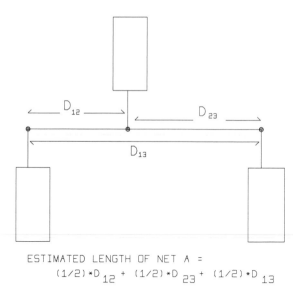

ESTIMATED LENGTH OF NET A =

$(1/2)*D_{12} + (1/2)*D_{23} + (1/2)*D_{13}$

Figure 6.3. The calculation of estimated net wiring length is depicted, using a formula that scales each pin-to-pin distance by a factor $1/(n-1)$, where n is the number of pins in the net.

the greater the number of pins in a net, the less each point-to-point segment should contribute:

> Consider each net in terms of all combinations of pairs of pins in the net; for each pair of pins, the wireability measure is taken to be $d_{ij}/(n-1)$, where d_{ij} is the (Manhattan) distance between pins i and j, and n is the total number of pins in the net.[1] The total net length is the sum of individual measures for each pair of pins; the total chip interconnection length is the sum over all nets. Figure 6.3 illustrates the calculation for a single net.

The most accurate wire estimate is typically based on a calculation of the *Steiner tree* length. A Steiner tree is an interconnection pattern in which the pins of the net are only a subset of the vertices used to complete all Manhattan wire segments. This is in contrast to a *chain* or a *spanning tree* in which only pins in the net are used as end points (Figure 6.4). A Steiner tree estimation algorithm can be used to provide a measure for each net.[2]

Alternatively, wireability measures can be selected to minimize congested areas by comparing estimated wire crossings with available wiring channels. Cells on the chip image are now regarded as incorporating some fixed number of wiring channels in both horizontal and vertical directions (Figure 6.5). With the place-

---

[1] The *Manhattan* distance is the rectilinear distance between two points: $d = |x_1 - x_2| + |y_1 - y_2|$. The origin of the term apparently is related to means of specifying distance measures in that borough of New York City (e.g., six blocks west plus nine blocks north). It is more representative and faster to calculate for chip physical design than Euclidean distance.

[2] Hanan, M., "On Steiner's Problem with Rectilinear Distance," *Journal of SIAM Applied Math*, 14 (1966), 255–265.

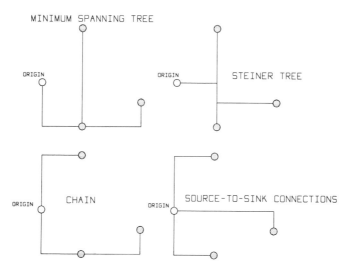

Figure 6.4. Steiner tree, chain, and spanning tree approaches to net wiring. The Steiner tree is unique in that the book pins in the net may be a subset of the vertices in the route.

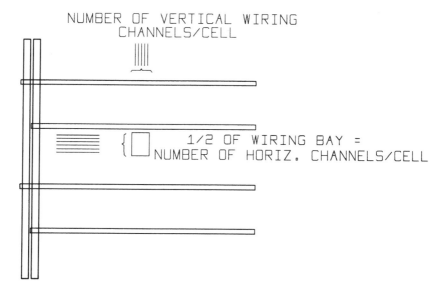

Figure 6.5. Another method of wireability estimation during placement is to compare the available channels at each cell boundary against the number of channel crossings required by an initial, coarse routing of the net. Specific placement locations are unfavorable if the number of channel crossings exceeds the number of available channels for many cells.

ment of each book, a rough wire route is completed. A treelike route can be used to complete each net, without the need for assigning specific channels to individual segments. Updates to the number of wires crossing the wiring boundaries of each cell are calculated, and a comparison is made against the number of available channels. The total number of excess channel overflows can be used (in conjunction with a wire length measure, typically) to direct placement decisions.

Specific performance, simultaneous switching, or other chip physical design requirements can be addressed in a variety of ways. For example, performance-critical nets can be specified in the input description to placement; the algorithm will weigh the length of this net heavily in calculating the chip measure. The books with pins in a critical net will therefore be clustered together. However, the benefits of this approach degrade rapidly if the number of nets so specified is large. In addition, the wire length of other nets associated with these books may suffer considerably. (An alternative to critical net weighting is to provide multiple power/performance levels of individual book designs, levels that can be selected to adjust path delays.) Other physical requirements usually require preplacing specific books, macros, and/or I/O cells. A file of book-to-cell location assignments is initially provided; typically, these assignments are not to be modified during execution of the placement procedure. To ensure that specific signals are present on specific package pins, an initial assignment of I/O cell locations is provided. Likewise, if minor modifications to an existing chip design are required (with little change to the existing performance), it will be desirable to hold the remaining logic at its current location. In both cases, the preplaced books serve as *seeds* for the assignment of unplaced logic.

Placement algorithms typically consist of two major steps: constructive placement and iterative placement improvement. The former involves the assignment of an initial placement location to a book, while the latter may select to modify a placement location in order to investigate if the metric that describes the overall wireability of the result can be improved. This improvement approach involves moving a single book to a vacant location or, more commonly, swapping the placement locations of two identically sized books.[3] If the results of the swap or move are regarded as favorable after comparing against the previous measure, the new assignments are accepted; if not, the original locations are retained. This improvement technique is iterative in nature and can consume significant CPU time. The remainder of this section will discuss some classes of algorithms that have been developed to perform initial book assignment and iterative placement improvement.

---

[3] Algorithms that investigate permutations of the placement locations of more than two books at a time have also been developed; for example, refer to Goto, S., "A Two-Dimensional Placement Algorithm for the Master Slice LSI Layout Problem," *16th Design Automation Conference* (1979), 11–17. Also, a swap may be attempted between books of different size; an adder to the metric would be included, based upon the resulting amount of cell overlap area.

Constructive Placement

Rather than spending CPU time trying to construct a favorable initial placement, it is certainly feasible to generate a *random* initial assignment, without any consideration of resulting chip wire length. This random assignment can then be directly input to the iterative step of the overall procedure. However, experimental results indicate that (for large assignment problems, at least) a constructive placement algorithm followed by iterative improvement provides optimum results.[4] As will be discussed shortly, new iterative placement improvement techniques are being developed; the debate over the value of constructive placement algorithms continues.

Constructive placement consists of two tasks: (1) the selection of the next book among the unplaced set to be assigned, and (2) the positioning of that book in a vacant location on the chip image. Selection is usually based on finding the unplaced book with the maximum *connectivity* to the currently placed set. The connectivity between two books can be calculated as follows:[5]

$$c_{ij} = \sum_{k \in N(i) \cap N(j)} \left(\frac{n_k + \lambda}{n_k} \right)$$

where $N(i)$ and $N(j)$ are the nets associated with the pins of books i and j, n_k is the number of pins in net k, and λ ($=1$) is a weighting parameter. Essentially, $c_{ij} = 0$ if books i and j have no nets in common; otherwise, the connectivity is the sum of the terms $(n_k + \lambda/n_k)$ for each net k the two books have in common. The next book for selection from the set of unplaced books is the one with maximum connectivity to the set of placed books. Two possible interpretations of *maximal connectivity* can be taken: (1) the selected gate i is the one with the highest c_{ij} to some gate j in the set of placed books, or (2) the selected gate i is the one with the greatest *total* connectivity, $\sum_j c_{ij}$, to the entire set of placed books. The parameter $\lambda = 1$ is used in the preceding expression to assign a lesser weight to a net shared by two books if the number of pins in the net is large (and therefore is part of many different books). Other factors have also been utilized in assessing the connectivity between two books; for example, $2/n_k$, $1/(n_k - 1)$, $(n_k + \lambda)/n_k$ with $\lambda = -1$. (Note that the last factor in the previous list assigns a *greater* selection weight to shared nets with a high overall pin count.) In the case of a tie in the connectivity value between two or more books to be selected, some secondary measure can be calculated and compared or an arbitrary decision can be made. If the input to chip physical design contained preplaced functions, these

---

[4] For example, refer to Hanan, M., and others, "Some Experimental Results on Placement Techniques," *13th Design Automation Conference* (1973), 214–224.

[5] This form of the connectivity expression comes from Khokhani, K. and Patel, A. M., "The Chip Layout Problem: A Placement Procedure for LSI," *14th Design Automation Conference* (1977), 291–297.

will serve as the initial set of placed books; otherwise, the *pair* of gates among all unplaced books with the highest c_{ij} value can be selected together as an initial *nucleus* for placement.

The positioning of the selected book commonly involves a calculation of the *center-of-gravity* coordinates relative to the previously placed books with nets in common. For example, from the list of all placed books with nets in common with the selected book, a median *x*- and *y*-coordinate value can be determined. The vacant cell closest to this target value is used. Alternatively, the chip wireability measures described earlier can be used to select the optimum position among the set of vacant cells.

The selection and positioning techniques just described evolved from the initial design automation efforts targeted at the placement of IC packages on printed circuit boards. Subsequently, the same techniques were applied to LSI master slice images. The advent of VLSI has resulted in considerable increases in CPU time and resource required to implement these algorithms. A different strategy to the placement problem would be to approach the task of constructive placement in a *top-down* manner. The sequence of steps used could be described as follows:[6]

I. Partitioning

The chip logic is partitioned into clusters or *supernodes*. A number of techniques can be used to develop chip partitions:

(a) Sequential Method. Individual partitions can be *grown* sequentially, adding one element at a time to a single partition. When either an area or external wiring connection limit is reached (preventing the addition of more elements), the partition being grown is closed and a new one is created. Logic books are added to a partition based on maximum connectivity to function already present in the partition (and/or minimal external connectivity to the chip logic outside the partition being grown).

(b) Simultaneous Method. Begin with an assumed number of empty partitions; to each of these partitions, one at a time, add a *seed* component (based on a size or connectivity measure). Before selecting a seed component for the next partition, add the closest *neighborhoods* of logic to the current selection.[7] Thus, the next seed component will not be closely connected to any of the previous elements. Once all partitions have been seeded, it is necessary to divide up all the remaining elements. The simultaneous approach to the task entails assigning a candidate book to each partition in one step; that is, in every iteration of assignment, each partition receives exactly one additional member. The contents

---

[6] Refer to Feuer, M., and others, "The Layout and Wiring of a VLSI Microprocessor," *IEEE Conference on Circuits and Computers* (ICCC '80) (October 1980), 674–677.

[7] The *neighborhood* of a logic gate consists of the set of all other gates with nonzero connectivity to the gate. The *second* neighborhood of a gate is comprised of those gates with nonzero connectivity to any of the elements in the first neighborhood, and so on.

of each partition are then updated and the assigned books are removed from the list of remaining elements. To select the members to be added for each iteration, a *connectivity matrix* is built; the number of rows of the matrix is equal to the number of partitions, while the number of columns corresponds to the number of unassigned elements. Each entry in the matrix is the connectivity of that element to the existing partition; if adding an element to a partition would exceed area or external wiring connection limits, a large negative number is inserted instead. The solution at each iteration is the particular combination of array values (one per row and from different columns) that leads to a maximum sum. This class of problem is known as a *linear assignment* problem, for which a number of algorithms to find optimum solutions have been developed.

(c) Hierarchical Description.   The criteria used in methods (a) and (b) have distinct disadvantages. The sequential method finds tightly interconnected clusters first, yet will then produce relatively disconnected final partitions. Method (b) attempts to distribute seed elements throughout the logic, and, with some experimentation on the number of partitions and the inclusion of logic neighborhoods, will yield well-connected clusters. However, by investigating the addition of only one element to a partition at a time, external wiring connection measures could be adversely affected. It may be the case where a group of books should be added to absorb some of the external connection count. Neither method utilizes a hierarchical representation of the chip logic, as would be provided by the designer using the description language presented in Chapter 4. Effectively, the partitioning task would be performed by the designer as part of developing the chip description. The difficulties this imposes, however, are primarily due to the wide disparity that may be present in the size and external interconnectivity of upper-level nodes in the description hierarchy. Considerable research effort is currently directed toward implementing top-down constructive placement tasks with a wider variety of partition characteristics.

II. Placement of Supernodes

The partitions are then assigned to regions (*supercells*) of the chip based on interconnectivity to other partitions. This step must be preceded by an estimate of the area and aspect ratio required to accommodate each partition. However, it is extremely difficult to develop an algorithm that will develop an optimal chip *floor plan* in a direct-release design system environment, given the diversity that may be present in the nature of the chip partitions.

III. Supernode Decomposition

The individual books or macros of the design are then assigned a placement location within (or near to) the region of the chip to which its supernode was assigned in the previous step. The placement of individual books within a region would likely utilize the same constructive techniques described earlier for LSI designs, selection and positioning. The decomposition of these partitions will typically require some mixing between neighboring partitions in order to share common power supply routing among books or macros of common width.

Once the constructive assignment has been completed, an iterative improvement procedure can be invoked. However, if the hierarchical chip description provided by the designer is accurately reflected by the constructive placement, the added value of this CPU-intensive step may be slight. The variation in macro cell sizes (and associated macro placement restrictions) characteristic of VLSI designs impose additional considerations for iterative improvement algorithms.

Iterative Improvement

All the iterative improvement algorithms to be described involve the investigation of interchange of the placement locations of two elements on the chip image; it is implicity assumed that the books selected will indeed fit in their new potential position. The algorithms differ primarily in the means for selecting the *secondary* element for interchange with the chosen *primary* book. (If the secondary cell location is vacant, the interchange is treated as if the location contained a *dummy* book design with no connections.) As a large fraction of the CPU time required for implementation of these algorithms is expended in providing estimated routes (and segment lengths) for the affected nets, the particular net representation (e.g., Steiner tree, spanning tree, chain, or bounding half-perimeter) is a major consideration. The choice of whether or not to accept the interchange is based on the change in the wireability measure that would result. A novel method for making this decision, denoted as *simulated annealing*, will be discussed at the end of the algorithm descriptions.

I. Pairwise Interchange

The choice of a pair of books to interchange can be made randomly (i.e., a Monte Carlo selection technique can be repeatedly used). When some consecutive number of possibilities has failed to produce substantial improvements (or sufficient CPU time has elapsed), the process can be halted.

More systematically, if there are n books initially placed on a chip image consisting of m cells, there are $(n*(n-1)/2)+(n*(m-n))$ possible interchanges that can be evaluated. The selection sequence may simply follow some prearranged order of pairs of locations of the chip image, without regard for specific nets or books contributing to the wireability measure. A *cycle* would consist of investigating this complete sequence of pairs for improvement; vacant cells containing dummy books would not be used as primary elements. This interchange cycle would repeat until insufficient improvement results were being obtained.

Alternatively, interchange results for all possible swaps with a selected primary element can be compared; the best *single* interchange is then executed after all have been considered. The primary element is then secured, and a new primary is selected from the remaining population. A cycle consists of sequencing through all books as primaries. The order in which the books are selected can potentially be based on connectivity or contribution to total chip wire length.

II. Neighborhood Interchange

The number of possible interchanges per cycle can be reduced if the potential secondary locations for each primary are limited to within some local surrounding neighborhood of the primary. Assuming the constructive placement provided a near-optimum solution, only localized improvements may need to be considered. Likewise, as an optimum solution is being approached after a few cycles, the size of the neighborhood may be reduced.

III. Steinberg's Algorithm[8]

Consider extracting two books that have no nets in common from the chip image; in terms of the total chip wire length measure, the optimum placement location for one of these books has no impact on the optimum location for the other. In other words, the two books are independent. This algorithm uses a set of independent books that have been extracted. For each book in the set, the change in wire length is calculated for placing that book in any of the locations vacated by the independent set. A matrix of wire length values is produced, where the rows are the books in the independent set and the columns are the vacant locations. The association of each row with a single column to minimize the total wire length measure is another example of a linear assignment problem. A number of disjoint independent sets may be generated and evaluated within each iterative cycle.

IV. Force-directed Techniques

Given a particular placed book, the collection of *vectors* between that book and others with nets in common can be computed. If that book were now released from its location and these vectors were regarded as *forces* acting on that book, the motion of the book to its zero vector sum would result.

A number of *force-directed* techniques have been developed that utilize this concept; they differ primarily in the way the primary book is assigned a new force-directed location. If the zero-vector-sum location is occupied, the nearest vacant cell location can be used. Alternatively, the book at the target zero-vector-sum location can be extracted to make room for the primary; the extracted book now becomes the primary for another force-directed evaluation. Thus, an entire sequence of displacements would result until the current primary element were placed in a vacant location; the overall change in wireability measure as a consequence of the *entire* sequence would be evaluated before accepting all the displacements. A third technique would swap the primary element with a secondary element *in the general direction* of the force vector. Like the pairwise techniques of (I) and (II), the swap would be accepted if the wireability measure were suitably

[8] Steinberg, L., "The Backboard Wiring Problem: A Placement Algorithm," *SIAM Review*, 3:1 (January 1961), 37–50.

improved. A number of incremental swaps, all in the same general direction, could be tried with the same primary until the interchange was no longer favorable. A slightly different interchange approach would be to try to find a secondary element at or near the target location of the primary whose resulting force vector happens to be in the general direction of the primary; thus, both books would potentially benefit from the interchange.

As mentioned earlier, these constructive and iterative placement algorithm descriptions do not include considerations of varying sizes among books being interchanged, the potential for swapping logically equivalent book or macro pins among nets, or reflecting book or macro layouts about some axis to improve the overall wireability measure. Likewise, considerable development effort is required to improve placement techniques that can utilize the chip's hierarchical design representation (and to determine if an accurate reflection of that description is indeed ultimately wireable). All these features become increasingly important considerations for VLSI designs. Future placement algorithm development must assess the trade-offs between additional degrees of freedom in decision alternatives versus the need to constrain the magnitude of the problem being attacked to best manage CPU resources.

To add more consternation to future algorithm development, the common interpretation of the *optimum* placement solution based on a reduction in the total chip wire length measure has recently been regarded as suspect.[9] Figure 6.6 illustrates that an iterative improvement technique may end in a *local* minimum, if only the changes that reduce the chip wire length metric are accepted. To achieve a truly optimum measure, some degree of *unfavorable* changes must be accepted during execution of the algorithm. The technique that was developed in the footnoted references is denoted as *simulated annealing*, an analogy to the very slow cooling process used to provide metals with large crystalline order (as opposed to *quenched* cooling). At each point in the execution of the improvement algorithm, an unfavorable change in the resulting metric has a *finite* probability of acceptance:

$$p = \begin{cases} 1, & \Delta m \leq 0 \\ e^{-(\Delta m/f(t))} & \Delta m > 0 \end{cases}$$

where Δm is the potential change in the metric, and $f(t)$ is a decreasing function of CPU time. The function $f(t)$ is really more accurately regarded as a conditional *schedule* for accepting unfavorable interchanges as the algorithm progresses. All favorable or neutral changes are accepted, while the likelihood of accepting an unfavorable change decreases with the magnitude of the change and with increasing execution time. The acceptance probability for initial CPU times will be high, effectively randomizing the initial placement. Note that for such an optimization

[9] Kirkpatrick, S., and others, "Optimization by Simulated Annealing," *Science* 220:4598 (May 13, 1983), 671–680; and Minot, T., "Workstation Moves in on Mainframes, Tackling Hefty Gate Arrays," *Electronic Design* (March 7, 1985), 99–108.

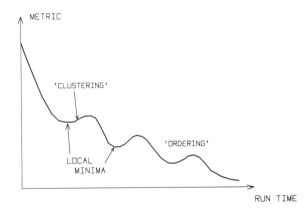

Figure 6.6. Wireability metric versus execution run time for an iterative improvement technique using finite acceptance probabilities of unfavorable changes. This *simulated annealing* approach can be divided into *clustering* and *ordering* phases, in analogy with the crystallization of metals. Note that with this technique the metric does not remain in its first local minimum, which would otherwise be the case.

approach the characteristics of a constructive initial placement are quickly erased; a random initial placement, with a suitable annealing schedule, is perhaps more appropriate. The schedule $f(t)$ should be slowed when the derivative with respect to $f(t)$ of the average value of the metric at each $f(t)$ is large; such will be the case in the initial formation of placement clusters, where additional execution time is required (Figure 6.6). (The annealing of metals involves a temperature schedule, which varies slowly in the vicinity of the freezing point.) A decision is based on comparing the parameter p with a randomly generated number between 0 and 1; if p is greater than the random number, the change is accepted.

This section has briefly described some of the techniques commonly used for design system placement of technology library layouts onto a structured chip image. Constructive and iterative placement procedures were described, along with different measures that are calculated and compared to make decisions regarding location assignments. Although placement and wiring are regarded as distinct tasks (due in large part to the extreme complexity of each task by itself), the metric used to direct placement decisions is typically based on an estimate of subsequent wireability (using calculations of estimated interconnection length and/or wiring channel congestion). Considerable design system development will be required to adapt placement algorithms to more flexible chip images, like the macro image discussed in Chapter 3.

6.2 POWER SUPPLY DISTRIBUTION AND ROUTING

A number of chip images were presented in Chapter 3; most of the master image designs (and all the master slice designs) utilized a predefined power supply distribution network. Some of the more recent master image (custom macro) systems permit more variation in power supply routing for greater macro flexibility. This section does not attempt to offer any solutions to the multitude of problems faced when adopting a more flexible image; rather, some of the more prevalent trade-offs are presented for consideration.

The simplest modification to the power supply distribution of a structured image to accommodate a large macro layout would be to truncate the supply rails at the edge of the macro (Figure 6.7). The supply distribution to internal macro circuits is provided within the layout, with connection to the overall image maintained at the truncating edge. The possible placement locations for such a macro are restricted to the ends of supply rails, as continuity of supply connections through the layout would not be a requirement. If the layout spans one or more wiring bays, the associated channel blockage information is input to the wiring procedure. The advantages of this approach are the ease of supply rail modification on the chip image, the ability to easily analyze the adequacy of the supply distribution for macro current requirements, and the lack of major modifications to the placement procedure. Since the supply distribution to the macro is well defined, the chip image and circuit design development groups can readily model the global supply resistance and inductance for worst-case dc and transient noise-margin analysis. The positioning of the macro layout can easily be incorporated

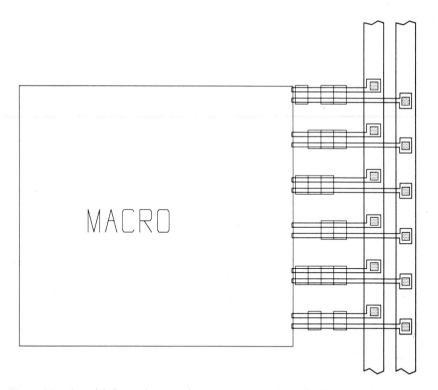

Figure 6.7. A straightforward means of macro power supply routing is to truncate the image supply rails at the macro edge; power supply routing to internal macro circuits is contained within the layout. This approach will typically restrict the macro placement locations to the ends of the structured image rails, as continuity to the other end of the layout would not be required.

into the placement algorithm, given the straightforward restrictions on possible locations; also, a preplacement file for these layouts could be provided by the designer. The structured chip image cell description remains unchanged, while the calculation of available wiring channels and net interconnection distances requires only slight modification.

An enhancement to this approach for structured images would remove the restriction that macros be placed at the ends of the supply rails; instead, a continuous supply interconnection would be provided through the macro layout (Figure 6.8). This macro layout and supply routing strategy does add some complexity to the tasks of automated macro placement and image noise analysis. The chip image cell description and wireability measures would need little or no modification, as before. If the macro layout can be suitably designed, the power supply distribution need not even be segmented, as depicted in Figure 6.8b.

The most flexible chip custom macro images remove the rigid distinction between internal circuit area, wiring bays, and supply rail positions. The intent of such an approach is to make maximum utilization of chip area, permitting wiring bays to expand and contract in order to accept the required number of intercon-

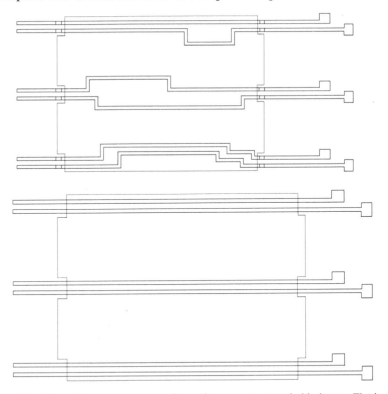

Figure 6.8.   Through-macro power supply routing on a structured chip image. The image rails are segmented by the macro in (a), while the layout in (b) maintains the image supply routing.

nections. In *coursed* macro images, the power supply rails are defined by the final position of the macro tier; in a gridless image, the power supply rails would be routed globally between macros. In the former case, the routing of the supply rails is not an integral part of the wiring procedure, whereas the more general case requires simultaneous consideration of global signal and supply interconnections. The noise-margin analysis for the unconstrained image is considerably more difficult. The definition of wire length distance in calculating a wireability measure will also require a modified interpretation.

As is the case with most design system features, a number of trade-offs must be addressed in providing the requisite tools; flexible images with variable power supply routing are no exception. The need for a more flexible design system chip image is largely dictated by the anticipated extent of macro utilization. The ability to accommodate a more flexible image (with variable supply routing) is constrained by the additional placement and wiring algorithm complexity required and the additional resource required to analyze noise margins. These same constraints apply to the possibility of variable supply and signal pad locations, with related impacts in the areas of chip testing and packaging.

6.3 WIRING ALGORITHMS

This section discusses some of the different approaches to the task of finding routes for the global interconnections between macro and book pins, as specified in the logic design description. It is assumed that internal macro circuit connections are contained within the macro layout and that pin shape and pin name coordinate information is provided by the output of the placement procedure. The task of combining global signal and supply interconnections, as discussed in Section 6.2, is not explicitly discussed, but rather is assumed to be an adjunct to the overall procedure. Before discussing some of the better-known wiring algorithms, a few definitions specific to the wiring task are in order.[10]

Definitions

Directed Graph.   A directed graph is a general means of representing the requirement that one signal lies *above* or *below* another signal in a particular wiring bay. For example, Figure 6.9b illustrates the directed graph for the nets depicted in Figure 6.9a; in this instance, net a must be above nets c and d, while net b must be below net c. Figure 6.9c illustrates the nature of the terms *above* and *below*.

---

[10] A number of wiring-related terms have already been defined in Section 6.1 (e.g., Manhattan distance, Steiner tree, chain, and minimum spanning tree).

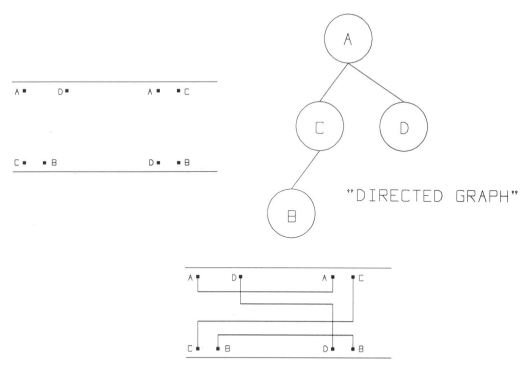

Figure 6.9. A directed graph can be used to represent how nets are to be assigned to horizontal wiring channels in a wiring bay; nets are said to lie *above* or *below* other nets, based on their relative ordering.

Cycle or Cyclic Conflict. A cyclic conflict arises when two or more signals must be both above *and* below one another; this situation manifests itself as a loop in the directed graph (Figure 6.10b). A local wiring bay algorithm typically must work around the cyclic conflict in order to find a solution. This may involve either pin swapping among logically equivalent pins to eliminate the conflict, assigning segments in a net to more than one wiring channel, or, for master image designs, using an intrabay segment on a device-related interconnection level. (Local wiring bay algorithms are also denoted as *channel routers*.)

Dogleg. Doglegging refers to the use of multiple horizontal segments per net in a wiring bay, in order to reduce the number of wiring channels required and/or to resolve a cyclic conflict (Figure 6.10c).

Wrong-way Segment. For a two-level metal interconnection technology, one level will be primarily dedicated to horizontal segments and the other to vertical segments. In the implementation of a dogleg wiring bay route, the option exists of using a single interconnection level with a wrong-way segment to con-

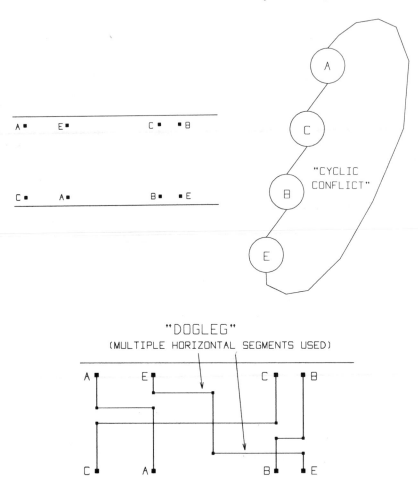

Figure 6.10. Cyclic conflict in the directed graph, derived from the pin locations along the edges of the wiring bay. The conflict can potentially be resolved by using wiring *doglegs*, where a net is assigned to multiple wiring channels in the wiring bay.

tinue the interconnection. This approach has the disadvantage that a number of adjacent channels in the primary direction are now blocked by the wrong-way segment; however, it provides the distinct advantage that two vias between wiring levels are no longer required, a potential manufacturing yield enhancement. As will be discussed shortly, wiring algorithms can *discourage* the use of vias by assigning an additional wire length weighting factor each time a route changes interconnection planes. Wrong-way segments can also be introduced as part of a *clean-up* phase after the initial wiring is completed. A dogleg segment can be made into a wrong-way segment on a single level (and the vias deleted) if the resulting wrong-way segment does not intersect any existing interconnections.

Wavefront.   Instead of attacking the overall wiring task one wiring bay at a time, *global* routing algorithms often attempt to find the interconnection for an entire net in one sequence of steps. To find a route from a starting vertex to target vertices, a class of global algorithms known as *maze-runners* initiates an outward wavefront from the starting vertex. This wavefront continues to expand around blockages until target points are reached. Maze-runners have the advantage that a solution for the route will ultimately be found, if indeed any solution exists (for the current channel blockage characteristics of previously wired nets). Their primary disadvantage is the extensive memory resource and CPU time required to propagate the wavefront throughout the chip wiring grid. This resource problem is aggravated by the possibility of the wavefront expanding on multiple wiring planes. This technique will be illustrated shortly in discussing the Lee connection algorithm.

Escape Line, Escape Point.   Another class of global wiring algorithms attempts to reduce the extreme resource requirements of maze-runners by replacing the wavefront with a list of potential interconnection line segments between a starting and a target vertex. A *line-search* algorithm utilizes the concept of *escape lines* and *escape points* around existing wire segments and macro blockages (Figure 6.11a); only line segment and line orientation information are used in the calculation so as to efficiently utilize memory storage. The escape points are investigated in alternate fashion between origin and target until an intersection between the associated origin and target escape lines is detected. Figure 6.11b gives an example of the line-search technique for a route between points A and B consisting of the segment end points (A, p_4, p_3, p_2, B); escape point p_1 is found by the algorithm, but is not used in the final route. A single wiring plane configuration is depicted in the figure, although the extension of the algorithm (and the definitions of escape lines and points) to multiple planes is straightforward. (For a more detailed description of line-search techniques, see reference 6.1.)

The remainder of this section briefly describes algorithms that have been used in the maze-runner, channel-router, and *hierarchical* approaches to the wiring task. Often, the chip physical design flow incorporates multiple approaches. A global hierarchical approach provides input to a channel router, which could be followed by a maze-runner to clean up remaining unwired nets.

Maze-runner Algorithms

The most general maze-runner algorithm is known as *Lee's algorithm* (reference 6.2). (A number of variants have been developed to try to enhance execution time.) The algorithm assumes that the chip wiring space can be represented as a two-dimensional rectangular array (for each wiring plane) and that each grid location can be assigned a numeric value during wavefront expansion of each net. For the purpose of this discussion, the value assigned to a grid location will be

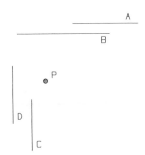

SEGMENTS B AND D ARE 'COVERS' OF P.

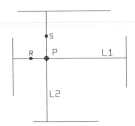

LINE SEGMENT L1 IS THE HORIZONTAL ESCAPE LINE.
LINE SEGMENT L2 IS THE VERTICAL ESCAPE LINE.
POINT R IS A VERTICAL ESCAPE POINT.
POINT S IS A HORIZONTAL ESCAPE POINT.

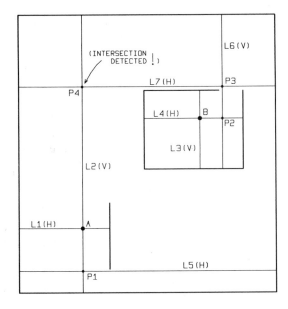

Figure 6.11. *Line-search* algorithm for automated wiring; escape line segments (and associated escape points) are determined between starting and target vertices (in alternate fashion). Part (b) depicts the procedure for routing between points A and B; the escape lines and escape points used are numbered in ascending order. Note that $l_2(v)$ and $l_5(h)$ extend from boundary to boundary, and therefore constitute a sufficient set of escape lines from point A.

the lowest-cost wavefront number that has reached the location. (Techniques for encoding the *incremental* path to a grid location in a more efficient memory representation have been proposed, saving storage at the expense of additional CPU data manipulation during wavefront expansion; see references 6.3 and 6.4.)

The first task in implementation of the Lee algorithm is the *ordering* of the nets to be processed. As nets are wired, their routes serve as blockages for succeeding nets; thus, the ordering criteria are important factors in maintaining overall wireability. Characteristics to consider in developing these criteria include the estimated wire length, the size of a net's minimum bounding rectangle, the number of *other* nets with pins located inside the bounding rectangle, and performance-critical nets. For optimum results, shorter runs should be completed first; nets with large bounding rectangles and/or a high number of other nets intersecting that rectangle should receive lower priority.

The next step is to initialize the wiring grid array for each interconnection plane; grid points that are part of channel blockages should be properly designated so as to be removed from consideration. As wiring segments are completed for the ordered list of nets, the occupied channels will likewise be indicated as unavailable. The remainder of the array should be initialized to the *available* value.

For the first and every succeeding net to be evaluated, a source and an initial target pin are chosen. For a multiple-pin net, the two pins that are closest together are commonly selected, one as source and the other as target. (The procedure for adding additional target points after partial segments in the net have been completed will be discussed shortly.) The wavefront is now initiated from the source pin. Each neighboring grid location is investigated to determine if that location is available. For each available neighbor, a *cost* of expanding the wavefront into that neighboring location is calculated and assigned to the matrix location for that grid point. A sample cost function is illustrated in Figure 6.12. From the starting vertex on the first metal plane, three cost values are indicated: $c_{h,M1}$ is the adder for expanding into a horizontal neighbor, $c_{v,M1}$ is associated with expanding into a vertical neighbor on the same interconnection plane, and c_{via} is the cost of changing interconnection planes. (The definition of an available location includes the location on a different plane directly above or below the point being expanded.) The cost values are selected to discourage wrong-way segments and level-to-level vias (e.g., $c_{h,M1} \ll c_{v,M1}$, $c_{h,M1} < c_{\text{via}}$).

The rectilinear neighbors that have just been assigned a value by the expansion algorithm are used to construct an initial *frontier* list; that is, these locations will each serve as expansion points to calculate minimum cost values for propagating the wavefront further. A simple illustration of the Lee algorithm is depicted in Figure 6.13a. A single wiring plane is shown, and no wrong-way penalty is assigned in that both horizontal and vertical costs are the same. Figure 6.13b denotes the sequence in which grid locations would be assigned costs by the expansion algorithm. In the general case, there will be frontier locations on all interconnection planes. The expansion of the wavefront into available neighboring locations continues to iterate until the target location is reached or the

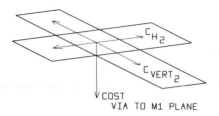

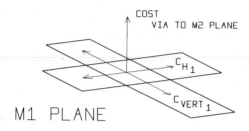

Figure 6.12. The Lee algorithm uses a set of costs to calculate values to assign to each grid location neighboring the current wavefront.

frontier list becomes empty. In the former case, a solution has been found, whereas the latter condition indicates that *no* solution is possible for the current blockage information.

Once a solution has been found, it is necessary to back-trace through the wavefront cost values toward the starting location to define the net interconnection segments. Beginning with the cost associated with the target location, search for the neighboring cell that led to that value. In other words, find the cell whose own value plus associated expansion cost toward the target cell adds up to the target value. That cell now serves as the reference point for the next retrace evaluation. The retrace procedure iterates in this manner using the current reference to determine the path that produced the target value. Once the starting location is reencountered, the segments and vias in the net have been defined. These segments and vias are then written to a file for generating shapes data; their matrix locations are now designated as unavailable, and the remainder of the available wiring grid matrix locations are cleared of any wavefront values in preparation for processing the next net.

Some aspects of the expansion and back-trace steps require unique consideration. For example, there can be several possible paths to a target location with the same minimal cost (Figure 6.14). The back-trace procedure is faced with a decision alternative when more than one neighboring cell results in the same cost for the reference cell. A systematic means for adopting a particular selection is required.

The expansion cost values depicted in Figure 6.12 were assumed to be dependent only on interconnection plane and direction; these values were design independent. In the general case, however, an expansion cost value toward an

3	2	1	2	3	4	5	6	7	8
2	1	A	1	2	3	4	5	6	7
3	2	1	2	3	4	5	6	7	8
4				5	6	7	8	9	
5	6	7	8	7	6	7	8	9	10
6	7	8	9	8				11	
7	8	9	10	9	10	11	12		12
8	9	10	11	10	11	B 12			
9	10	11	12	11					
10	11								

(a)

(b)

14	8	2	6	12	17	21	26	33	41
9	3	A	1	5	11	16	20	25	32
15	10	4	7	13	18	22	27	34	42
19				23	28	35	43	50	
24	30	38	45	37	29	36	44	51	57
31	39	47	52	46				64	
40	48	54	58	53	59	66	73		71
49	55	61	65	60	67	B			
56	62	69	72	68					
63	70								

Figure 6.13. The Lee algorithm: (a) the matrix costs assigned by the expanding wavefront; (b) the sequence of grid location assignments.

available location could be a function of the current status of the design. For example, if the minimum via pitch for the technology were greater than the interconnection wiring pitch, adjacent vias would be disallowed. As a result, after a net is completed it would be necessary to locally alter the via cost (to a prohibitive value) for locations neighboring a via in the net (Figure 6.15). Design-

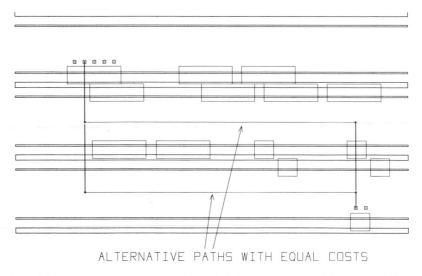

ALTERNATIVE PATHS WITH EQUAL COSTS

Figure 6.14. There may exist alternative paths between vertices with the same minimum cost in the Lee algorithm.

(DYNAMIC) VIA COST MATRIX

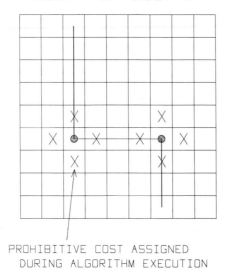

PROHIBITIVE COST ASSIGNED
DURING ALGORITHM EXECUTION

Figure 6.15. If the technology design rules and image definition discourage or prohibit adjacent wiring plane vias, a cost matrix corresponding to the chip grid can be used to locally modify the via costs during execution of the algorithm.

dependent expansion costs could also be used if adjacent parallel interconnection segments were to be discouraged to minimize signal coupling noise. The expansion costs for these general cases would need to be represented in a manner that would record their positional dependence during execution of the algorithm.

Again considering the case of more general cost values, it must be a feature of the algorithm that the wavefront expansion step should not terminate when the target is initially reached, but rather it ceases the expansion calculations only when all frontier values exceed that assigned to the target location. In other words, the initial wavefront that reaches the target by expanding all frontier points into adjacent neighbors with each iteration may be superseded by a *longer yet lower cost* path. The expansion value assigned to a grid location may be subsequently replaced by a smaller one as the frontier continues to expand.

A number of techniques have been developed to reduce the time-consuming wavefront expansion step. Figure 6.16 illustrates four possibilities for constraining the wavefront to reduce the number of expansion cost calculations. The first technique merely opts to select as the starting point the vertex that is closest to a blockage region (Figure 6.16a), rather than the one that may be more centrally located. If a blocked area is not conveniently located, an artificial bounding area can be established around the pair of vertices (Figure 6.16c); if no route exists in the bounding region, the unexpanded frontier points at the edge of the region would be considered. Figure 6.16b depicts a technique that initiates wavefronts simultaneously from both vertices, expanding each alternately until a point of intersection is reached (similarly to the line-search algorithm illustrated earlier). The last technique (Figure 6.16d) is one that selects to expand only those elements of the frontier list that are immediately closer to the target vertex. When applied in conjunction with expansion costs that discourage changes in course, this selection criterion will tend to provide a very narrow initial search path. If the most direct route is blocked, locations along the route nearer the target are selected first among the list of frontier points to be further expanded. The line-search class

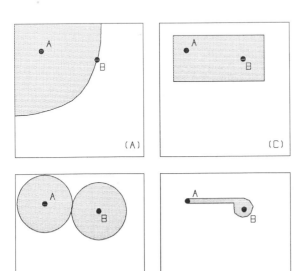

Figure 6.16. The Lee algorithm can potentially be enhanced by one of a number of speed-up techniques. These techniques include (a) starting point selection, (b) double wavefronts, (c) initial framing of starting and target vertices, and (d) selective wavefront expansion of only those frontier points closer to the target.

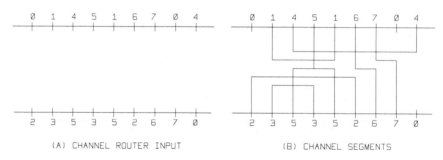

(A) CHANNEL ROUTER INPUT   (B) CHANNEL SEGMENTS

Figure 6.17. A channel router utilizes the description of a single wiring bay at a time.

of algorithms is to some extent an enhancement of this directed expansion technique, where the escape line accelerates the search process by considering distant (nonneighboring) locations.

A modification to the Lee algorithm to determine a route for a net with greater than two pins is relatively straightforward. Initially, a centrally located pin in the net is selected as the starting vertex. The wavefront expansion from this point proceeds as before until the initial encounter with another target pin in the net. The route between the two pins is again determined by back-tracing. Subsequently, *all* points on this existing route serve as starting points for another wavefront expansion. This expansion proceeds until another pin is encountered. As each new pin is reached, the resulting grid points in the net each serve as a starting point for expansion toward remaining pins. This enables the total net interconnection to be of Steiner-tree form, where segment end points need not coincide with pins in the net.

Channel-router Algorithms

A class of wiring algorithms has been developed that is well suited to the use of wiring bays in design system chip images; specifically the global routing task is broken down into finding a solution for one wiring bay at a time. A channel-router algorithm begins with a description of the nets that must be interconnected on opposite sides of the wiring bay (Figure 6.17). The algorithm proceeds to determine intrabay vertical and horizontal segments, assigning segments to specific wiring channels. The goal of the algorithm is to minimize the number of horizontal wiring channels required (custom macro image) or to stay within the maximum number of wiring channels available (fixed macro image or master slice). As discussed in Section 6.1, the placement algorithm may incorporate wireability measures when locating books or macros on a structured image to ensure that the number of channels available would likely be sufficient.

A number of special considerations are required in developing a suitable channel-routing algorithm:

1. A two-layer metal master image technology will typically offer the polysilicon level as an additional option for intrabay segments, to eliminate conflicts with global second metal routes and/or reduce the number of yield-detracting contacts or vias. However, the interconnection resistance of these connections is unattractive. Polysilicon segments should *not* be used in series with a high fan-out circuit output pin, and only sparingly with circuit (gate) inputs. The channel router would need to be equipped with suitable decision-making criteria in order to assign the appropriate interconnection level to a segment and to add or delete the requisite contact or vias.

2. Channel routers may select to incorporate doglegs in the wiring bay solution, using multiple horizontal segments in different channels to complete a net. As discussed earlier, assigning a single horizontal segment or channel to a net is not feasible in the case of a cyclic conflict. Conversely, the polysilicon interconnection segments could be used to eliminate the conflict.

3. Wrong-way horizontal segments on the vertical second metal interconnection layer could be used to reduce the required number of wiring channels (and/or to eliminate cyclic conflicts), as illustrated in Figure 6.18.

4. As all segments in the bay are determined in one iteration of the channel router, the concept of *dynamic weighting* of grid points (as discussed with the Lee algorithm) is difficult to incorporate. For example, the undesirability of parallel adjacent routes would not be particularly straightforward to describe to the algorithm; the same difficulty applies to restrictions on adjacent vias.

The failure of the channel router to complete a set of connections in the wiring bay (due to cyclic conflicts, lack of available wiring channels, or whatever) requires the creation of a *net overflow* file. These nets may subsequently be able to be routed manually, as discussed in the next section. Segments that have been defined for part of the net may also be output to assist with the manual processing.

A brief description of a channel-router algorithm follows based on that given in reference 6.5. For simplicity, it is assumed that no cyclic conflicts are present, and no dogleg routes will be produced. Furthermore, no limit on the number of available channels is imposed. All pins are at the edge of the wiring bay, and it is assumed that all connections from these locations into the bay utilize vertical

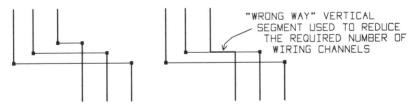

Figure 6.18. A wrong-way segment on the second metal interconnection level can be used to reduce the number of wiring channels required to implement a wiring bay solution.

segments. Descriptions of channel-router algorithms with additional features can be found in references 6.5, 6.6, and 6.7.

A list describing pin locations at the edges of the wiring bay is provided (Figure 6.17a). Net numbers in common represent pins to be connected with intrabay segments, while a zero value indicates an unused pin location. From this representation, a directed graph can be constructed, as defined earlier. Under the assumption that all connections in a net are to be completed with a single horizontal segment (and that only two metal interconnection planes are available), the directed graph represents the order that must be maintained in assigning the horizontal segments in the bay from top to bottom. In other words, if net a is above net b in the directed graph, the horizontal segment used to connect all pins in net a must be above that segment for net b. Having constructed the directed graph, it is necessary to determine the extent of the horizontal segment for each net that reaches all pins in the net. The largest number of horizontal segments that overlap at any point along the bay is a lower bound on the number of channels required. The goal of the algorithm is to attempt to *merge* two nonoverlapping segments into the same wiring channel, while continuing to satisfy all the constraints of the directed graph (Figure 6.19).

As the wiring bay is scanned from left to right, a set of net numbers for nets whose horizontal segment is present at that position is maintained. The length of this list expands and contracts as new segments arise and old segments terminate. To be able to merge the segments of two nets, it is necessary that the first terminate before the next is encountered; this condition is evident when neither of the two lists at horizontal positions i and $(i + 1)$ is a subset of the other; that is $L(i) \not\subseteq L(i + 1)$ *and* $L(i + 1) \not\subseteq L(i)$. Using the terminology of reference 6.5, each instance when this condition occurs represents a new *zone* in the wiring bay; each zone is characterized by the nets present in the longest list between zone boundaries

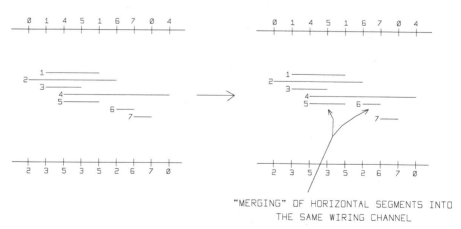

"MERGING" OF HORIZONTAL SEGMENTS INTO
THE SAME WIRING CHANNEL

Figure 6.19. Two nonoverlapping horizontal segments can be merged into the same channel if all the constraints of the directed graph remain satisfied.

(Figure 6.20). Two nets can be merged into the same horizontal channel if (1) they are never present together in the same zone, and (2) there exists no constraints between the two nets in the directed graph. When two nets are merged, the separate nodes of the directed graph are combined, maintaining the set of constraints regarding channel assignments above and below (Figure 6.21a). The zone lists are also updated to contain the new merged net (Figure 6.21b).

The referenced algorithm works at each new zone boundary, attempting to merge nets that originate in the zone with horizontal segments that have previously terminated. In the general case, there will be a number of nets in the new zone to consider. For each new zone, merging proceeds in the following manner:

(0) Prepare two lists of nets, those that originate in the current zone and those that have terminated in some previous zone.

(1) For each of the nets originating in the current zone, determine the longest possible path from top to bottom of the directed graph through the net. Compare the values for all originating nets and sort them in descending order; that is, the net on the longest path will be considered first.

(2) From the list of segments that have terminated in a previous zone, disregard any that lie on a path with the net selected in step 1 through the directed graph and thus cannot be merged.

(3) If any terminating segments remain, select the one that will result in little or no increase in the longest path in the directed graph for the merged net. Update the directed graph and zone representations to reflect the combined net.

(4) Remove the individual terminating and originating nets that have been merged from consideration, and if any originating nets remain, return to step 1. If all originating nets have been considered, move to the next zone and return to step 0.

COLUMN	S(I)	ZONE
1	2	
2	123	
3	12345	1
4	12345	
5	1245	
6	246	2
7	467	
8	47	3
9	4	

Figure 6.20. A *zone* is defined as a range of vertical channels in the wiring bay; the list of nets requiring crossing horizontal segments is used to characterize the zone. The boundaries of the zone occur where the lists of channels just outside are neither subsets or supersets of the longest list in the zone.

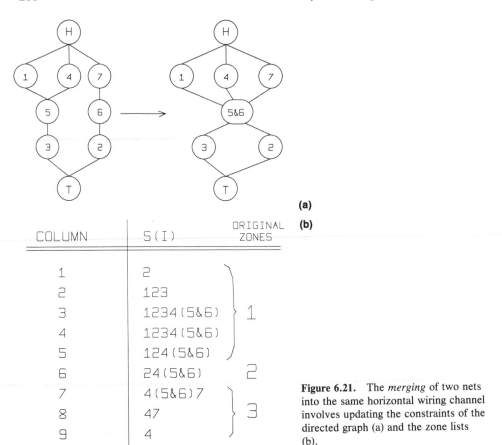

COLUMN	S(I)	ORIGINAL ZONES
1	2	
2	123	
3	1234(5&6)	⎫ 1
4	1234(5&6)	⎬
5	124(5&6)	⎭
6	24(5&6)	2
7	4(5&6)7	⎫
8	47	⎬ 3
9	4	⎭

Figure 6.21. The *merging* of two nets into the same horizontal wiring channel involves updating the constraints of the directed graph (a) and the zone lists (b).

(b)

In discussing the zone representation of the wiring bay, it was mentioned that the size of the largest zone is a minimum bound on the number of wiring channels required; that limit does not apply if the longest path through the directed graph is of greater number. The actual number of wiring tracks used is equal to the number of distinct nodes in the directed graph that remain after merging. The merging process attempts to keep this number to a minimum by avoiding increases in path length, which would otherwise worsen the best achievable channel count and tend to inhibit subsequent merges.

Once the merging process has been completed, it is necessary to assign horizontal tracks to merged nodes of the final directed graph. The paths through the graph must be consistent with the top-to-bottom ordering of segments in the bay. The relative positioning of nodes that are not part of a directed path is not constrained (Figure 6.22).

Channel-routing algorithms have been developed to specifically address the wiring bay configuration that is common to many chip images. The global routing

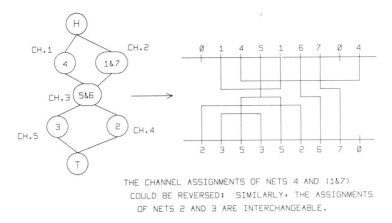

THE CHANNEL ASSIGNMENTS OF NETS 4 AND (1&7)
COULD BE REVERSED; SIMILARLY, THE ASSIGNMENTS
OF NETS 2 AND 3 ARE INTERCHANGEABLE.

Figure 6.22.   Assignment of merged nets in the directed graph to horizontal channel positions in the wiring bay.

step for a VLSI part is broken down into iterations of the channel-routing task for computational efficiency. To be able to achieve this efficiency, however, it is necessary to first consider the entire chip routing task and prepare the interbay connections for subsequent decomposition. This initial process is the function of a global hierarchical router.

Hierarchical Global Wiring

To prepare a VLSI chip design for wiring bay routing, it is first necessary to provide vertical channel assignments for interbay segments. A net that enters and/ or exits a wiring bay needs a target vertical channel location at the edge of the bay for the local channel router to reach with a connection. This vertical assignment is the function of a hierarchical global router, such as described in reference 6.8.

    To initiate the global routing process, the total chip image is divided into abutting rectangular regions, each of which encompasses many horizontal and vertical wiring channels and pin connections. These regions share edges with their neighbors, and each edge has a global wiring channel capacity. This parameter is equal to the number of available wiring channels that intersect this edge (minus a factor related to the number of nets that are totally contained within the region and a factor related to the presence of any local channel blockages resulting from macro placements).

    An initial global route for each net is determined through *regions* on the chip, using pin locations defined by region; a maze-runner (or line-search) algorithm may be used to determine this preliminary route. (As the number of matrix locations or regions is small, these algorithms will execute quite efficiently. Any one of a number of different net configurations, Steiner tree, spanning tree, chain, could be used.) Regarding the collective sum of *all* preliminary routes, the results of this global process will likely contain edges where the global wiring is severely congested (i.e., where channel demand exceeds channel capacity). A rerouting of congested nets is required, starting with the most severe edge and with the net(s) contributing to the greatest number of channel demand excesses over all edges. A new global route can now be determined, using maze-runner expansion costs reflective of the degree of congestion in crossing region boundaries, thereby moving toward an ultimately wireable solution.

This region definition and edge capacity wiring analysis can continue in a top-down fashion, using results from the previous definition to determine more detailed (yet still global) routes. Once a global routing solution of sufficient region detail is complete, each net is assigned vertical pin locations for the wiring bay edges intersected by the global route. These assignments are combined with the pin locations in the wiring bay to proceed with the iterations of the channel-routing procedure.

Wiring Space Requirements and Wireability Estimation

In developing a design system structured chip image with fixed wiring bays, it is crucial to provide sufficient wiring channel capacity in the image to ensure that chip physical design will commonly require little manual intervention. If too little wiring space is provided, considerable time and effort (i.e., money) will be spent in attempting to embed overflow nets or to *depopulate* the image by removing logic function. On the other hand, providing an excess of horizontal and vertical wiring tracks (or keeping the circuit count purposefully low) will indeed make the physical design a *turnkey* step, yet chip costs will be higher. A great deal of data has been collected in order to be able to provide accurate estimates of required wiring space for future chip images; an example of such a prediction model is to be found in reference 6.9.

Briefly, the input parameters to the model include the number of cells to be wired, the average number of connections per cell, and the average length of an interconnection between cells; this last parameter is a function of many variables, including cell count, placement algorithms, and general logic interconnectivity. Assuming that a representative value for this parameter can be determined, the model provides an estimate for the total number of wiring tracks (horizontal and vertical) required per cell. This information can then be used in developing the design system chip image.

The evolution of VLSI design systems toward higher macro utilization offers considerable avenues for the development of enhanced wiring tools. The management of more flexible chip images will require greater interrelation between placement and wiring procedures.

6.4 MANUAL INTERVENTION

Despite efforts to automate the physical design task for a VLSI chip design, a number of steps may require designer intervention:

Preplaced Macros/Books/Chip Inputs and Outputs.   The designer may select (or may be required) to assign fixed placement locations to a subset of the chip logic design, to act as a seed for the selection of unplaced books for assignment. In addition, the chip I/Os may need to be suitably distributed to satisfy simultaneous switching or fan-out current sourcing and sinking requirements of the chip image power supply distribution.

Embedding of Overflows.   Nets whose global interconnections were not completed by the wiring procedure will be described in an overflow file; the designer may choose to try to complete the wiring task manually (if the number of overflows is not large). Embedding overflows requires rip-up and reroute of other nets in order to make wiring channels locally available. (Assuming a maze-runner algorithm was exercised at the end of the wiring procedure, it can be assumed that an interconnection for the net does not exist. It should be emphasized that wiring algorithms prioritize nets for global routing, vertical channel assignment, and horizontal channel merging *using heuristic criteria*; the particular sequence of alternative decision paths chosen during wiring algorithm execution may ultimately lead to unwireable conditions for specific nets.) During the embedding process, the integrity of the design is in a state of flux as existing connection segments are removed and new routes defined; an interactive graphics application tool is indispensable for viewing the design, highlighting incomplete interconnections, and managing and enforcing the overall design integrity.

Engineering Change Mode.   A second pass of a given design involving minor logic modifications typically requires that as few changes as possible be made to the bulk of the physical design in order that performance measures remain relatively constant. The designer will likely choose to execute the physical design step in *engineering change* mode, where the placement locations for unchanged portions of the logic are maintained (and as many of the wiring interconnections kept as is feasible). The physical design tools accept the new logic description and provide the necessary deletions and additions to book placements and interconnection segment routes.

A difficult situation is encountered by the designer when estimated performance values for a logic path or net are not achieved as a result of the chip physical design. Alternatives to contemplate will vary considerably in required resources:

1. Net rip-up and reroute to try to find a shorter route and less wiring load capacitance.
2. Logic revisions to buffer heavily loaded nets and/or provide parallel logic function to reduce fan-out loading.
3. Reexecute the placement (and subsequently wiring) procedure with different coefficients assigned in the calculation of the wireability metric.

Unfortunately, the characteristics of these alternatives are sufficiently diverse so as to make a direct comparison between them rather difficult; little design system support is commonly available to aid the designer in selecting the particular option that will provide the highest measure of confidence that the desired performance improvements can indeed be achieved.

PROBLEMS

6.1. Develop an enhancement to the channel-router algorithm described in Section 6.3 to best utilize an additional polysilicon interconnection layer for vertical intrabay segments. This layer should be used for circuit input pins only; modify the channel-router pin information data description appropriately.

6.2. Describe the features of an interactive graphics design system application tool that should be provided to assist with the task of embedding overflow nets. Aspects of the tool to consider include both user interaction and maintenance of design integrity; also, describe how the tool might integrate a maze-runner algorithm to investigate individual overflow nets during editing.

REFERENCES

6.1 Hightower, David, "A Solution to Line-Routing Problems on the Continuous Plane," *ACM–IEEE Design Automation Conference* (1969), 6.75–6.93.

6.2 Lee, C. Y., "An Algorithm for Path Connections and Its Applications," *IRE Transactions on Electronic Computers*, EC-10:3 (September 1961), 346–365.

6.3 Rubin, Frank, "The Lee Path Connection Algorithm," *IEEE Transactions on Computers*, C-23:9 (September 1974), 907–914.

6.4 Hightower, David, "The Lee Router Revisited," *IEEE International Conference on Computer Design: VLSI in Computers* (1983), 136–139.

6.5 Yoshimura, T., and Kuh, E., "Efficient Algorithms for Channel Routing," *IEEE Transactions on Computer-Aided Design of Integrated Circuits and Systems*, CAD-1:1 (January 1982), 25–35.

6.6 Deutsch, D., "A Dogleg Channel Router," *13th Design Automation Conference* (1976), 425–433.

6.7 Hassett, James, "Automated Layout in ASHLAR: An Approach to the Problems of 'General Cell' Layout for VLSI," *19th Design Automation Conference* (1982), 298–304.

6.8 Nan, N., and Feuer, M., "A Method for the Automatic Wiring of LSI Chips," *IEEE International Conference on Computer Design: VLSI in Computers* (1978), 11–15.

6.9 Heller, W. R., Donath, W. E., and Mikhail, W. F., "Prediction of Wiring Space Requirements for LSI," *14th Design Automation Conference* (1977), 32–43.

Additional References

Albano, A., and Sapuppo, G., "Optimal Allocation of Two-Dimensional Irregular Shapes Using Heuristic Search Methods," *IEEE Transactions on Systems, Man, and Cybernetics*, SMC-10:5 (May 1980), 242–248.

Chen, K., and others, "The Chip Layout Problem: An Automatic Wiring Procedure," *14th Design Automation Conference* (1977), 298–302.

Donath, W. E., "Placement and Average Interconnection Lengths of Computer Logic," *IEEE Transactions on Circuits and Systems*, CAS-26:4 (April 1979), 272–277.

———, "Wire Length Distribution for Placements of Computer Logic," *IBM Journal of Research and Development*, 25:2 (May 1981), 152–155.

Haims, M. J., "On the Optimum Two-Dimensional Allocation Problem," *15th Design Automation Conference* (1978), 298–304.

Heyns, W., "A Line-Expansion Algorithm for the General Routing Problem with a Guaranteed Solution," *17th Design Automation Conference* (1980), 237–242.

IEEE Transactions on Computer-Aided Design of Integrated Circuits and Systems, Special Issue on Routing in Microelectronics, CAD-2:4 (October 1983).

Lee, D. T., Hong, S. J., and Wong, C. K., "Number of Vias: A Control Parameter for Global Wiring of High-Density Chips," *IBM Journal of Research and Development*, 25:4 (July 1981), 261–271.

REVIEW OF SOLID-STATE DEVICE CONCEPTS

7.1 SILICON CRYSTALLOGRAPHY

The majority of chemical elements are crystalline solids at room temperature. The condition of lowest system potential energy that maintains the regular order of the spatial configuration of the atoms in the crystal is provided by the chemical bonds present between adjacent atoms.

Definitions

Chemical Bond. The spatial (probabilistic) configuration of the two electrons shared between adjacent atoms, which results in an overall lower potential energy

Bonding or Valence Electrons. The electron(s) in the outermost orbitals of an atom that participate in the formation of a chemical bond between atoms

Covalent Bond. A chemical bond in which the two bonding electrons are shared *equally* (in a symmetric spatial probability distribution) between/around the bonded atoms (Figure 7.1)

Ionic Bond. A bond in which one or more electrons are effectively *transferred* from atom A to atom B (a very asymmetrical spatial distribution); each atom has a configuration of completely filled (or empty, from the opposite per-

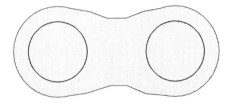

Figure 7.1. Symmetric spatial probability distribution for electron(s) in a covalent bond.

spective) outermost electron orbitals, that is, an inert gas configuration (Figure 7.2)

Metallic Bond.   An electrostatic attractive force between positively charged metal ions and a *gas* of free electrons that are not bound to any atom(s) but rather are free to move throughout the crystal

An *ideal* crystal is constructed by the infinite repetition in space of identical structural units. Each of these structural units, or *unit cells,* has associated with it a number of lattice points, repeated in space by the dimensions of the unit cell. Each lattice point, although strictly just a mathematical point in space, corresponds physically to the location of one or a group of atoms in the crystal; in the case of a lattice point group, one atom of the group is assigned to the lattice point, while the remaining atom(s) are chemically bonded to the atom at the lattice point (and to other atoms associated with other lattice sites). No crystal is in actuality ideal; at the very least, all crystals have *surfaces*. In addition, point imperfections are common; some of these localized irregularities are chemical or elemental impurities at atomic sites (*substitutionally*), impurities or host atoms *not* in regular atomic sites (*interstitially*), and vacant atomic sites, called *voids*. At any finite temperature, there will be a volume density of *broken* (missing) chemical bonds and crystalline voids associated with the available thermal energy; using Maxwell–Boltzmann statistics (for the energy distribution of independent particles with discrete energies), both of these concentrations are proportional to the factor $\exp(-E_D/kT)$, where k is Boltzmann's constant and E_D is an energy of dissociation.

There are four common crystalline lattices or unit cells (Figure 7.3). The cubic cell types have a characteristic dimension that uniquely describes the pe-

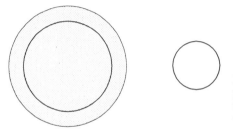

Figure 7.2. Asymmetric spatial probability distribution for electron(s) in an ionic bond; a complete charge transfer is indicated.

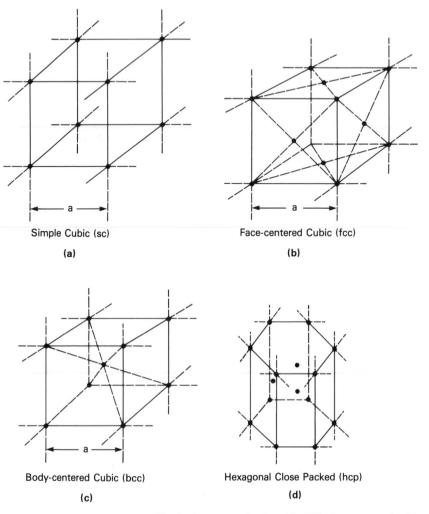

Figure 7.3. The common crystalline lattice types: simple cubic (SC), face-centered cubic (FCC), body-centered cubic (BCC), and hexagonal close-packed (HCP). Note the three internal lattice sites of the HCP lattice type.

riodicity of the crystal, while the hexagonal close-packed unit cell has two characteristic measures.

Silicon, the crystalline substrate for the vast majority of integrated-circuit fabrication, is an element in column IV of the Periodic Table. It has four valence electrons (outermost shell electrons). To achieve a *complete* outermost shell of eight electrons (and reduce its overall energy), each atom shares (covalent bonding) its four valence electrons with one electron from each of its four nearest neighbors. This is known as the *tetrahedral* bond arrangement (Figure 7.4). The

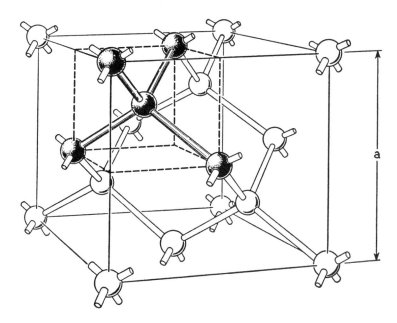

Figure 7.4. Tetrahedral bonding configuration of silicon (the diamond structure): face-centered cubic with two atoms per lattice point. The two atoms are located at (0 0 0) and ($\frac{1}{4}$ $\frac{1}{4}$ $\frac{1}{4}$). (From W. Schockley, *Electrons and Holes in Semiconductors*. New York: Van Nostrand Reinhold, 1950, p. 6. Copyright © 1950 by Van Nostrand Reinhold Co., Inc. Reprinted with permission)

Figure 7.5. Crystalline structure of the semiconductors silicon and germanium.

two major semiconductor substrate materials used for IC manufacture, silicon
and germanium, crystallize in the tetrahedral or *diamond* structure (Figure 7.5).
The configuration is face-centered cubic, with two atoms per lattice point such
that all atoms are equally spaced.

Since all crystals have surfaces and, in particular, since the characteristics
of the MOS device are very dependent on the properties of the active device
surface of the wafer substrate, it is essential to know the surface orientation of
the crystalline wafer used for fabrication. Crystal surfaces are specified in the
following manner:

1. Find the intercepts of the surface plane on the coordinate axes (in multiples
of the lattice constant).

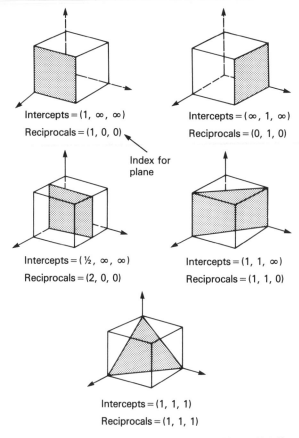

Figure 7.6. Crystalline indexing scheme for surface planes. Planes (1 0 0) and (0 1 0) are
equivalent and are generally grouped into the {1 0 0} *family* of surfaces; the braces signify
any of the equivalent surfaces.

2. Take the reciprocals of these intercepts and express all three coordinates as integers (keeping the same ratios as the reciprocals).

Figure 7.6 gives some examples; note that in a cubic crystal system there are many equivalent groups of planes due to the symmetry of the lattice points. The different families of surfaces have different areal densities, which results in a number of different physical and electronic properties (e.g., different oxidation growth rates and different free carrier transport properties). For MOS integrated-circuit technologies, the {100} surface is the most common.

7.2 SEMICONDUCTOR PROPERTIES

The following empirical rules are characteristic of materials that have come to be classified as *semiconductors*:

1. Semiconductors usually have resistivities in the range of 10^{-3} to 10^9 ohm $*$ cm at 300 kelvins (Figure 7.7).

$$\text{resistance } R = \rho * \text{length/area}, \quad \rho = \text{resistivity}$$

2. The resistivity of a pure semiconductor decreases approximately exponentially as temperature increases.

3. The resistivity of a semiconductor depends critically on the purity of the semiconductor sample and on the species of impurity present.

4. The charge carriers responsible for current flow in semiconductors can be positive, negative, or both, depending on the semiconducting material and its purity.

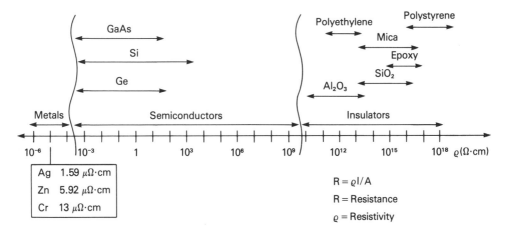

Figure 7.7.   Room temperature resistivities of various solids, including an indication of the three types of materials: metals, insulators, and semiconductors.

5. The resistivity of semiconductors is reduced by illumination of the sample with light of some wavelength in the infrared, visible, or near ultraviolet range.

6. Reasonably pure semiconductors are transparent to radiation in at least some portion of the infrared or visible spectrum.

Energy States, Energy Bands, and Free Carrier Densities in Semiconductors

The electrical conductivity (σ) of a material is given by

$$\sigma = q * (\mu_n * n + \mu_p * p)$$

where q is the electronic charge in units of coulombs ($q = 1.6 \times 10^{-19}$ C), and μ_n, μ_p are electron and hole carrier mobilities in units of cm^2/volt $*$ sec. The units of the electrical conductivity σ are (ohm $*$ cm)$^{-1}$. n refers to the density of electrons in a semiconductor that are free to conduct current in the material. Electrons that remain bound to their host atom do not conduct current and are not included in the calculation. Likewise, p refers to the density of *holes* that are free to conduct current. The current conduction of holes is a phenomenon particular to semiconductor materials; it is not present in metals. [The equation for the conductivity of a metal reduces to $\sigma = (q * \mu_n * n)$.] The origin and nature of holes and the means of determining the free carrier densities n and p are elaborated on in the remainder of this section. (The discussion that follows is by no means rigorous; it is much more of a plausibility argument.)

In the early 1900s, a theory for the electronic structure of atoms was proposed. The theory did predict and concur with experimental data for the wavelengths of the spectral lines of the hydrogen atom. Although this theory has been refined to a considerable extent by quantum mechanical considerations, it still remains an illustrative model for the behavior of electrons in a solid.

This model, proposed by Niels Bohr in 1913, states that instead of the infinite number of orbits (and energies) allowable in classical Newtonian mechanics, it is only possible for an electron to orbit the nucleus of an atom in *discrete* orbits with *discrete* values of energy. These results were based on an assumption that the angular momentum of the electron orbit is *quantized*. The bound electron energies are depicted by discrete energy lines or *levels*, referenced to a free electron at 0 electron volts (eV). The terminology used is that electrons occupy electron *states* at discrete energy levels. The number of electrons that can occupy any state is exactly two, one for each *spin* orientation of the quantum mechanical particle. The theory also indicates that electromagnetic radiation (light) is emitted if an electron, initially moving in an orbit of energy $E_{initial}$, changes its motion so that it moves in an orbit of energy E_{final}. The frequency of the emitted radiation (ν) is equal to the difference in energy divided by Planck's constant, h:

$$\nu_{photon} = \frac{E_{initial} - E_{final}}{h}$$

As stated earlier, this postulated model was used to accurately describe the wavelengths of light emitted by hydrogen atoms when they decay from a higher to a lower total energy (Figure 7.8). Since the energies are given as negative, the difference ($E_{\text{initial}} - E_{\text{final}}$) is positive.

The major modification to this theory, which results after applying quantum mechanical principles, is that the electrons do not occupy fixed circular orbits around the nucleus of the atom with radius proportional to the square of the quantum number (Figure 7.8c), but instead are *spread out* in three dimensions. The electron's temporal and spatial position is no longer well defined, but is described by a *wave function*. The magnitude of the wave function for an electron

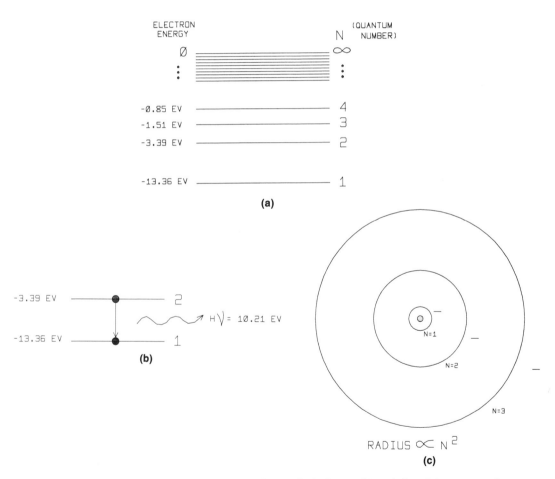

Figure 7.8.   (a) Electron energy level diagram for hydrogen. (b) Emission of electromagnetic radiation from the decay of an electron from a higher to a lower energy level. (c) In the circular electron orbits around the nucleus of the atom postulated by Bohr, the electron radius is proportional to the square of the quantum number.

in a potential energy system defines the probability density for the spatial distribution of the electron; that is, the electron will likely be found where the magnitude of its wave function is large and will never be found where the wave function is zero. For a one-dimensional atomic coulomb potential well with a bound electron, the magnitude of the resultant wave function for the lowest energy level is sketched in Figure 7.9.

As *two* potential wells are (conceptually) brought close to one another, there will be an increasing interaction between the bound electrons and the combined potential energy function. Figure 7.10 illustrates the linear combination of two *ground state* electron wave functions as the distance between the two wells is reduced. There are four possible ways the linear combination of the electron wave functions can occur (Figure 7.11). As a result of bringing two wells in interaction with one another, two distinct resultant wave functions were produced. In one of the resultant wave functions, a zero occurs between the two wells. This means that neither electron will be found between the two wells. In the other resultant wave function, the electrons will very likely be found between the two charges. The resultant wave function without the zero crossing is *lower* in total energy; each electron reduces its potential energy by residing in a region where it is under the attractive force of both positive charges. As the two wells were brought together, two electrons (both initially in the ground state) can reside in either of two energy levels. In other words, one energy level has *split* into two energy levels, one of lower energy (corresponding to the wave function without the zero crossing) and one of higher energy (the resultant wave function with the zero crossing). The splitting of energy levels is illustrated in Figure 7.12. The example

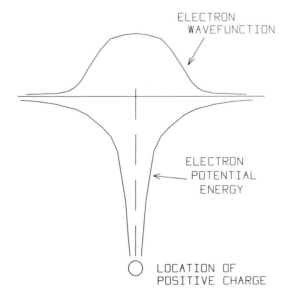

ELECTRON
WAVEFUNCTION

ELECTRON
POTENTIAL
ENERGY

LOCATION OF
POSITIVE CHARGE

Figure 7.9. Electron potential energy and quantum-mechanical wavefunction for a one-dimensional potential well.

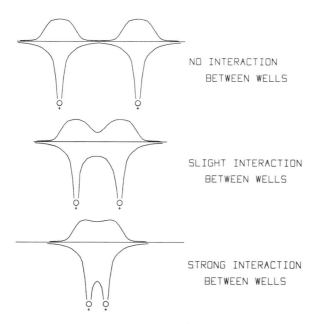

NO INTERACTION
BETWEEN WELLS

SLIGHT INTERACTION
BETWEEN WELLS

STRONG INTERACTION
BETWEEN WELLS

Figure 7.10.   Linear combination of two ground-state electron wave functions under slight and strong interaction.

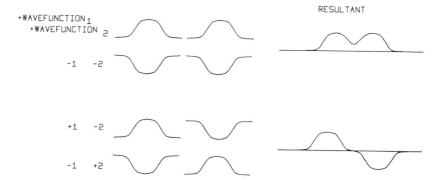

Figure 7.11.   Possible linear combinations of two electron ground state wave functions.

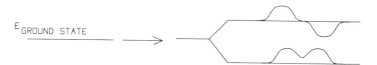

Figure 7.12. *Splitting* of an energy level into two energy levels due to the strong interaction of two electron wave functions.

described is analogous to the case of two pendulums:

> Consider two pendulums of the same length and mass. If they are noninteracting, they both oscillate with the same frequency (both are in their ground state). Now assume the two pendulums are coupled by a spring between them; this coupled system now has two (total) energy *states* (Figure 7.13). The situation for three coupled pendulums is illustrated in Figure 7.14; there exists for this example three distinct modes.

The extension of this analogy to *n* coupled systems implies that there are *n* different energy levels for the total system energy.

For the case of the atoms in a crystalline solid, the number of coupled systems *n* becomes very large. The energy levels that result from the electron interactions essentially become indistinguishable and are regarded as continuous; Figure 7.15 illustrates the splitting of a single energy level into a *band* of levels, or allowed states, as a result of the (artificial) experiment of reducing the spacing between atoms in the crystal. The splitting of *several* discrete electron levels for a particular element is illustrated in Figure 7.16. Two features are of particular note:

1. Higher electron energy levels correspond to larger orbits; as a result, the splitting of levels begins at a larger separation. (The electrons in the larger orbits begin to interact sooner as the separation between atoms *r* is decreased.)

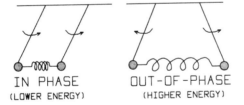

IN PHASE   OUT-OF-PHASE
(LOWER ENERGY)   (HIGHER ENERGY)

Figure 7.13. Analogy between two coupled pendulums and two strongly interacting ground-state electron wave functions.

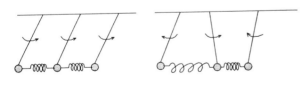

Figure 7.14. The analogy between interacting wave functions and a coupled pendulum system is extended to three: three distinct energy levels for the coupled system are produced.

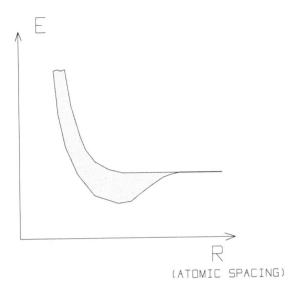

Figure 7.15. The splitting of a single energy level into a *band* of energy levels upon strong interaction between individual atoms.

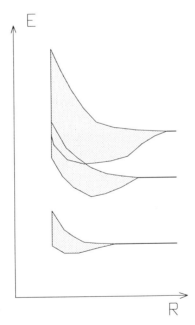

Figure 7.16. The splitting of three energy levels into bands; note that the bands may overlap and that the splitting is strongly dependent on the quantum number.

2. The extent of the splitting of energy levels into an energy band increases with increasing energy.

The *density of states* function (units are states per cm^3 * eV) describes the number of allowed electron states per unit volume in the material as a function

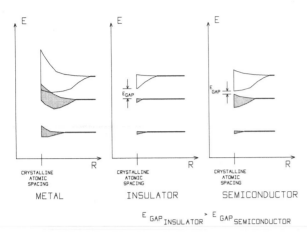

Figure 7.17. Occupied energy states in each of the various energy bands of a metal, insulator, and semiconductor. In a metal, there is no gap between the occupied and lowest unoccupied energy state. No electron excitation out of the ground state is shown. For the cases of an insulator and a semiconductor, there is a gap in the energies of the highest occupied energy state and the lowest available unoccupied electron energy state. The magnitude of the gap is considerably less for the semiconductor as compared to the insulator.

of the electron energy. This density function, when multiplied by the probability that a state at a particular energy is occupied, can then be used to determine the number of electrons occupying states in a band of energies:

$D(E)$: density of states function

$D(E) * dE$: volume density of electron states in the energy range E to $(E + dE)$

$p(E)$: probability density function

$$n = \int_{E_1}^{E_2} D(E) * p(E) \, dE = \text{number of electrons present in} \atop \text{band of energies from } E_1 \text{ to } E_2$$

The sketches in Figure 7.17 are similar to Figure 7.16, except that only the energy states that are *actually occupied* are shaded in. The highest-energy occupied states are filled with the valence electrons of the atoms in the crystal. The three different cases illustrated in Figure 7.17 represent the situation for the three different types of materials: metals, insulators, and semiconductors. The difference between the three types of materials is determined by the difference in energy between the highest-energy occupied level and the lowest-energy unoccupied level. In the case of a metal, the bands overlap and the assumption of a continuum of energy states implies that the difference in energy between an occupied and unoccupied state is infinitesimal. For the cases of a semiconductor and an insulator, a *gap* in available electron energy levels exists between the occupied *valence* band of electron energies and the next highest unoccupied level; the energy gap (E_g) is defined as the energy difference between the top of the valence band and the bottom of the empty (or *conduction*) band.[1]

---

[1] It should *not* be inferred from Figures 7.16 and 7.17 that the atomic spacing in the crystal is strongly correlated to whether the element is a metal, semiconductor, or insulator. Rather, these figures are meant to illustrate that, *whatever* the particular equilibrium spacing in the crystal, the overlap or gap present in the splitting of energy levels into bands determines the type of material.

An electron conducts current by moving (under the influence of an electric field) into empty energy states spatially distributed throughout the crystal. The electron can absorb the energy to initially jump from a valence band state into an empty conduction band state from a variety of sources (light, thermal energy, crystal strain and stress, etc.); different device types can be used to detect these various sources of energy and from these sources produce free carriers available for the conduction of current.

The distinction between a metal, semiconductor, and insulator involves the minimum amount of energy required to move electrons into available energy states for conduction (Figure 7.18), where the horizontal axis in Figure 7.18 now corresponds to the one-dimensional distance in the crystal in the direction of current flow. In the absence of a potential difference across the ends of the crystal (and in the absence of the local addition of impurities), the potential energy and electron density are uniform. For the semiconductor material at finite temperatures, sufficient absorbed thermal energy will promote an electron from the valence band to the conduction band (Figure 7.19). When an electron in a semiconductor is elevated in energy to the conduction band, it leaves a *hole*, an empty electron state in the crystal. In other words, in a semiconductor, electron–hole *pairs* are thermally generated. Holes also contribute to current conduction (and thus are included in the expression for the conductivity of the material).

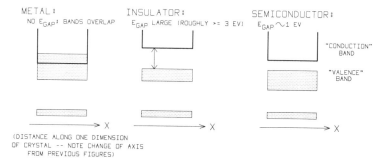

Figure 7.18. Occupied and unoccupied energy states for a metal, insulator, and a semiconductor, similar to Figure 7.17. Note that the horizontal axis is now the distance along the crystal in the *x*-direction.

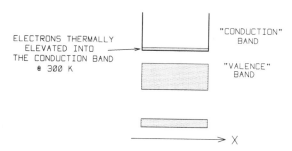

Figure 7.19. Partial occupancy of the conduction band states by electrons thermally elevated from the previously occupied states in the valence band.

Conceptually, the hole *moves* as bound electrons in the valence band move in the direction opposite to an applied electric field to fill the empty state created in the valence band by the thermal generation of a hole–electron pair. The hole moves in the direction of the electric field as a positive charge would traverse through the crystal. The hole will be regarded (like the electron) as a classical mechanics "particle" of unit positive charge, which is free to carry current *independently* of conduction band electrons; it is a concise means of representing the quantum mechanical motion of bound valence-band electrons. The hole is not a localizable particle like the electron, and its *effective mass* (the ratio of an applied electric field force to the *acceleration* of a hole at the top of the valence band) differs from the effective mass of the electron in the conduction band.

The density of electron states function and the free electron concentration in the conduction band as a function of energy for an *intrinsic* material are plotted in Figures 7.20 and 7.21 (a refinement of Figure 7.19). The probability density function (pdf) $p(E)$ for the electron occupation of conduction band states is the pdf appropriate for the limitation of only two electrons per state, although no distinction is made between discrete states in the empty conduction band.[2] The pdf used is the *Fermi–Dirac distribution* function:

$$p_e(E) = \frac{1}{1 + \exp[(E - E_F)/kT]}$$

The probability function appropriate for holes occupying hole states at the top of the valence band is the probability that the associated electron state is *unoccupied*:

$$p_h(E) = 1 - p_e(E) = \frac{1}{1 + \exp[(E_F - E)/kT]}$$

The energy value E_F in the Fermi–Dirac distribution is called the Fermi energy; it is the electron energy where the probability of state occupation is exactly equal to one-half. The Fermi energy may be regarded as the unknown parameter to be determined in a particular equilibrium free carrier statistics analysis; knowing the Fermi energy is equivalent to knowing the equilibrium carrier concentrations. Due to the rapid increase in the density of states in the conduction band and due to the exponential decrease in the probability function with increasing energy, a slight change in the Fermi energy *reference* level (e.g., 100 meV) results in a tremendous change in the free carrier concentration and the material conductivity. For the situation illustrated in Figure 7.21, the free hole and electron carrier concentrations are exactly equal and the one-half probability of occupied states (the Fermi energy) lies in the middle of the energy gap. Indeed, for all device types to be discussed, the free carrier concentrations will be much less than the volume density of states; the probability of occupation of a conduction

---

[2] The limitation on the number of electrons per state is known as the Pauli exclusion principle. Two electrons of opposite *spin* are allowed per electron energy state.

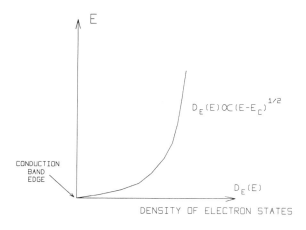

Figure 7.20. Plot of the density of states function for electrons in the conduction band as a function of electron energy. A similar diagram can be drawn for the density of hole states: zero at the top of the valence band and increasing with positive hole energy, the downward direction on the electron energy axis.

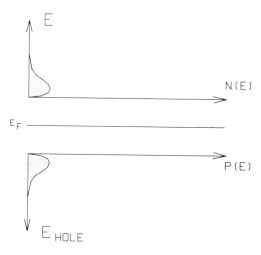

Figure 7.21. Density of states occupied by electrons in the conduction band and holes in the valence band as a function of energy. The occupation probability function is called the Fermi–Dirac distribution function and leads to the definition of a parameter on the energy band diagram known as the Fermi energy, E_F.

band state by an electron (or a valence band state by a hole) is much less than one-half, and E_F is well distant from either the valence or conduction band edges. In other words, the number of occupied conduction band states will be small, allowing two simplifications to be made:

(1) The limitation of two electrons of opposite spin per state is not of great concern; the number of electrons in the conduction band combined with the number of infinitesimally spaced energy states in the band implies that the spin interaction is weak. As a result, the conduction band electrons are essentially independent and the Maxwell–Boltzmann statistical distribution for the pdf of noninteracting particles should be valid:

$$p_e(E) \propto \exp\left(\frac{-E}{kT}\right)$$

This distribution can be derived from the Fermi–Dirac distribution as a direct consequence of the assumption the $(E - E_F)$ is much greater than kT:

$$p_e(E) = \frac{1}{1 + \exp[(E - E_F)/kT]} \approx e^{E_F/kT} * e^{-E/kT}, \qquad E - E_F \gg kT$$

Note the interpretation of the Fermi energy as part of the normalization of the probability function. The exponential term $\exp -[(E - E_F)/kT]$ is common to all (equilibrium and slight perturbations from equilibrium) carrier concentration problems.

(2) The total number of free electrons in the conduction band is the integral of the product of the density of states function times the Maxwell–Boltzmann probability function over the conduction band energies:

$$n = \int_{\text{conduction band}} D_e(E) * p_e(E) \, dE$$

However, again making the assumption of limited free electron concentrations, most of the free electrons will occupy states very close to the lower edge of the conduction band. (Likewise, free holes will occupy only those states very close to the top of the valence band.) As a result, the density of states function $D_e(E)$ can be simplified to an interpretation where an *effective* density of states is concentrated *specifically* at a *single* energy value, the lower edge of the conduction band (the upper edge of the valence band for holes):

$$D_e(E)_{\text{conduction band}} \to D_{\text{eff}C} * \delta(E - E_{C(\text{lower})})$$

The determination of the *effective density of states* is driven by the fact that the free carrier concentration n must remain the same:

$$n = \int_{\text{conduction band}} D_e(E) * p_e(E) \, dE$$

$$= \int D_{\text{eff}C} * \delta(E - E_C) * p_e(E) \, dE$$

$$n = N_{\text{eff}C} * p_e(E_C) = N_C * \exp\left[\frac{-(E_C - E_F)}{kT}\right]$$

$$p = N_{\text{eff}V} * p_h(E_V) = N_V * \exp\left[\frac{-(E_F - E_V)}{kT}\right]$$

The reference to the conduction and valence band of energies has been collapsed to simply refer to the edges of the bands E_C and E_V separated by an energy gap E_g. [The change in notation from $D(E)$ to N_C and N_V is used as a reminder that the density of states function has units of states/cm^3 $*$ eV; multiplying $D(E)$ by dE in an integral expression gives the density of states between energies E and $E + dE$. The N_C and N_V notation is used to signify that these densities are in units of states/cm^3 and that the energy dependence has been eliminated by referring strictly to the energies $E_{C(lower)}$ and $E_{V(upper)}$. In short, N_C and N_V are *not* a function of electron or hole energy.]

Quantitatively, the number of holes and electrons that are thermally generated (and therefore contribute to the semiconductor conductivity) is derived using the relationships

$$n = N_C * \exp\left[\frac{-(E_C - E_F)}{kT}\right], \quad p = N_V \exp\left[\frac{-(E_F - E_V)}{kT}\right]$$

and the knowledge that for thermal generation in an intrinsic material (no impurities present), free electrons and holes are equal in number: $n = p = n_i$ (intrinsic carrier density). The product of the two carrier concentrations is given by

$$n * p = n_i^2 = N_V * N_C * \exp\left[\frac{-(E_C - E_V - E_F + E_F)}{kT}\right]$$

$$= N_V * N_C * \exp\left[\frac{-(E_C - E_V)}{kT}\right] = N_V * N_C \exp\left[\frac{-E_g}{kT}\right]$$

Example Calculation (silicon)

$$n = p = n_i = \sqrt{N_C N_V} \cdot e^{-E_g/kT}$$

At 300 K: $kT = 0.026$ eV, $E_g = 1.12$ eV,

$$N_C = 2.8 \times 10^{19} \text{ cm}^{-3}$$

$$N_V = 1.1 \times 10^{19} \text{ cm}^{-3}$$

$$n_i \approx 1 \times 10^{10} \text{ cm}^{-3}$$

Intuitively, the expression for the intrinsic carrier density n_i indicates that the greater the density of unoccupied states available (N_C) or occupied states from which to make a transition in energy (N_V), the greater the likelihood of the transition of an electron from an occupied state to an unoccupied one, and therefore the greater the free carrier density; however, by Maxwell–Boltzmann statistics, the greater the amount of energy involved in making the transition (relative to the energy kT), the less likely such a transition will occur.

The mechanism that exactly balances the rate of thermal generation of free electron–hole pairs in equilibrium (in the absence of an applied external stimulus) is the rate at which free holes and electrons are lost due to *recombination*, that is, the capture of a free electron and free hole in the crystal (the transition of a free electron down to the valence band and the transfer of energy to another free carrier or the emission of a photon).

The free carrier densities are (locally or uniformly) modified in the semiconductor substrate by the introduction of impurities in the crystal with a volume density much larger than the intrinsic carrier density n_i. The extent to which the intrinsic free carrier densities are modified by the addition of impurities is deter-

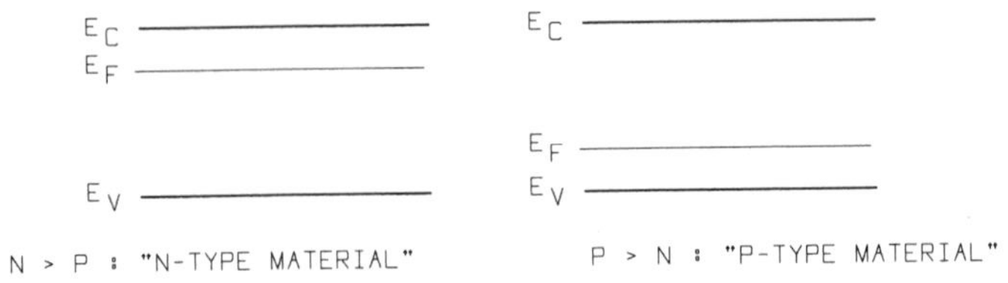

E_C ———————— E_C ————————
E_F ————————

E_V ———————— E_F ————————
                 E_V ————————

N > P : "N-TYPE MATERIAL"   P > N : "P-TYPE MATERIAL"

Figure 7.22. Energy band diagram for an *n*-type material (electrons are in the majority) and a *p*-type material (holes are the majority carriers). Note the position of the Fermi energy relative to the conduction and valence band edges for the *n*- and *p*-type materials.

mined by the impurity density (per cm³) or, equivalently, by the position of the Fermi energy (the probability function) relative to the conduction and valence band edges. The positioning of the Fermi energy closer to the conduction band edge than the valence band edge correlates to the case where the density of free electrons (occupying states in the conduction band) is greater than the density of unoccupied states (holes) in the valence band; conversely, a Fermi energy value closer to the valence band implies a greater density of free holes than electrons (Figure 7.22). This imbalance is the result of impurity introduction into the crystal; the effects of different types of impurities on the magnitudes of the free carrier densities are discussed in greater detail in Section 7.4.

7.3 CARRIER MOBILITY

Using the results of the previous section, the intrinsic conductivity of a semiconductor material is given by

$$\sigma_{\text{intrinsic}} = q * (\mu_n * n_i + \mu_p * n_i)$$

μ_n and μ_p are the carrier *mobilities*, to be discussed shortly. For silicon at room temperature, the following conductivity is calculated:

$$n_i \approx 1 \times 10^{10} \text{ cm}^{-3}$$

$$\mu_{n_{\text{intrinsic}}} \cong 1350 \text{ cm}^2/\text{V·sec};$$

$$\mu_{p_{\text{intrinsic}}} \cong 480 \text{ cm}^2/\text{V·sec}$$

$$\sigma_i \cong 3 \times 10^{-6} \ (\Omega\text{·cm})^{-1}$$

$$\rho_{\text{intrinsic}} \cong 3 \times 10^5 \ (\Omega\text{·cm}): \quad \text{intrinsic resistivity}$$

The carrier mobility μ is defined by the relation

$$v = \mu * \mathcal{E}, \quad v = \text{carrier velocity (cm/sec)}$$

$$v_n = \mu_n * \mathcal{E}, \quad v_p = \mu_p * \mathcal{E}, \quad \mathcal{E} = \text{applied electric field (V/cm)}$$

This expression assumes that the mobility may be regarded as a proportionality constant over the range of device operation and terminal voltages; that is, the carrier velocity is proportional to the applied electric field for the range of electric fields corresponding to the applied device voltages. MOS devices depend on the carrier velocities at the silicon–silicon dioxide surface; as a result, the mobility for MOS devices is typically denoted as μ_{eff} and called the *effective channel* (or surface) mobility to distinguish it from the *bulk* substrate carrier mobility. (In Section 8.6, the surface carrier mobility will no longer be treated as a constant but will be assumed to be dependent on the magnitude and direction of the applied electric field when deriving the most detailed MOS device current equation models. The device *V–I* characteristics derived in this chapter will use μ_{eff}, treating it as a constant.) A discussion of the nature of the bulk and surface carrier mobilities occupies the remainder of this section.

For electric fields small in magnitude, the carrier drift velocity v_{drift} is proportional to the electric field; the equation for the carrier velocity with the mobility as a proportionality constant is accurate. In a bulk semiconductor material, there are two primary *scattering* mechanisms that significantly affect the carrier velocity: (1) collisions with ionized impurities in the crystal (the free electrons and holes are charged particles that interact with atoms with a net charge in the crystal structure), and (2) free carrier–phonon collisions. (A phonon is a particlelike entity that represents or "carries" the energy of an elastic wave in the crystal due to thermally generated atomic vibrations.)

At low temperatures, the phonon mechanism is weak, and the scattering is due primarily to the ionized impurities. At high temperatures, the phonon mechanism is strong and dominates the scattering of mobile electrons. These two mechanisms are independent, and the functional form for the bulk carrier mobility can be written as

$$\frac{1}{\mu_{bulk}} = \frac{1}{\mu_{phonon}} + \frac{1}{\mu_{impurity}}$$

where the terms in the expression are understood to be strong functions of temperature. If a scattering mechanism dominates and has a low mobility value, it effectively determines the carrier mobility. This functional form will be used again in describing the effective surface mobility.[3] For electrons in silicon at room temperature with a background impurity concentration of $10^{16}/cm^3$, the bulk mobility is roughly 1000 $cm^2/V * sec$, down from the value of 1350 $cm^2/V * sec$ given earlier for intrinsic silicon (no impurities) at room temperature.

---

[3] It is not uncommon to include another independent scattering mechanism in the expression for the bulk mobility, that is, the scattering due to other charged free carriers. The mobility expression then becomes

$$\frac{1}{\mu_{bulk}} = \frac{1}{\mu_{phonon}} + \frac{1}{\mu_{impurity}} + \frac{1}{\mu_{free\ carriers}}$$

The mechanisms described for the bulk carrier mobility in the semiconductor are *isotropic*. For an MOS device, however, the presence of the surface introduces a scattering mechanism that is definitely anisotropic. Due to the large surface-to-volume ratio of the conducting channel (inversion) layer of an MOS device, surface boundary scattering (neglected in the bulk scattering mechanisms) will also reduce the effective carrier mobility in the device channel. Again making the assumption that all the previously described scattering mechanisms are independent, the carrier channel mobility for the device can be written as

$$\frac{1}{\mu_{eff}} = \frac{1}{\mu_{bulk}} + \frac{1}{\mu_{surface}} = \frac{1}{\mu_{phonon}} + \frac{1}{\mu_{impurity}} + \frac{1}{\mu_{free\ carrier}} + \frac{1}{\mu_{surface}}$$

Again note how the functional form includes all the scattering mechanisms as independent and as restricting the overall carrier channel mobility, some potentially dominating over others.

7.4 IMPURITY INCORPORATION IN SILICON

As discussed in Section 7.2, free carrier concentrations are altered locally by the introduction of specific impurities into the crystal structure. The introduction of impurities is often referred to as *doping* the crystal; the impurities are referred to as *dopants*.

The two major processing techniques for introducing impurities into the crystal are (1) high-temperature diffusion (from an external source), and (2) ion implantation (followed by a high-temperature anneal). Impurities in the crystal may either reside at atomic sites (substitutionally) or between atomic sites (interstitially). Interstitial impurities are more common at very high impurity concentrations (greater than 10^{19} impurities per cubic centimeter). Interstitial impurities result in a local volume where the crystal structure is quite distorted. As a consequence, they produce irregular electron potentials and the band structure is difficult to characterize. An integrated-circuit fabrication process sequence will commonly include sufficient high-temperature steps to reduce the density of interstitial impurities by thermal annealing.

There are two general classes of impurities in silicon; the resulting regions of the semiconductor material where one (or more) impurity species has been added are referred to as *n*-type or *p*-type, depending on which class is locally in the majority. The two classes of impurities are *donors* (resulting in an *n*-type material) and *acceptors* (resulting in a *p*-type material). Introduction of a donor impurity into the silicon crystal changes the conductivity of the material locally by adding (*donating*) a free electron to the electron concentration (at temperatures sufficient to promote the electron to an unoccupied state in the conduction band). Silicon has a valence of four; that is, four outer electrons are shared (bonded) with the four nearest crystalline neighbors. A donor impurity has a valence of five; four valence electrons are involved with the covalent crystal bonding and the remaining outer electron is donated to the free electron concentration in the crystal. A donor impurity for silicon, therefore, typically resides in group V of the chemical periodic table.

For the donor atom to be effective in releasing a free electron to the crystal, there are two major requirements:

1. The donor atom must reside "comfortably" in the crystal (i.e., not induce a substantial strain); group V atoms that are much greater in atomic number than silicon are also larger in size and do not make good substitutional impurities.

2. Once in the crystal structure, the donor atom must have a low *ionization energy*. The ionization energy is the amount of energy required to free a valence electron from its host atom in the silicon crystal (with the surrounding crystal structure dielectric constant). Figure 7.23 illustrates the electron energy level for the outermost electrons of the donor atom relative to the energy band diagram for the silicon crystal. Figure 7.24 lists the ionization energies for the common donor impurities in silicon.

After donating an electron to the crystal, the resulting host (donor) atom is ionized, with a *net charge* of $+1q$. If the free electrons are distributed uniformly

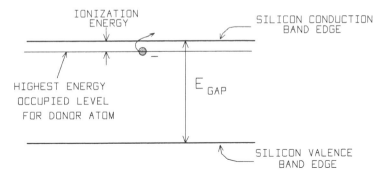

Figure 7.23.   Location of donor impurity occupied energy level relative to the conduction band edge of silicon.

ELEMENT	E_I (EV)
PHOSPHORUS (P)	0.044 EV
ARSENIC (AS)	0.049 EV
ANTIMONY (SB)	0.039 EV
	<< 1.1 EV = E_{GAP} IN SI @ 300 K

Figure 7.24.   Common donor impurities in silicon and their ionization energies (i.e., the binding energy of the fifth valence electron to the group V impurity atom resident in the silicon crystalline structure).

in the presence of the donor atoms, the entire region can be regarded as charge neutral. If the free electron concentration is *depleted* in this region, the remaining $+1q$ charge on each *fixed* donor atom means that the region has an overall positive charge. [*Depletion regions* are present at every *p*-type to *n*-type junction region and, as a result, there are local noncharge neutral regions throughout each integrated circuit (Section 7.5).]

There are also impurity donor atoms that are not as effective in changing the free electron concentration; the ionization energy for these impurities is significantly larger than that for the impurities listed previously, in the neighborhood of one-half the total energy gap. These impurities are called *deep donors* (Figure 7.25). Although deep donors do not ionize as readily as *shallow* donors, the presence of deep donor impurity atoms in the silicon crystal has another effect. When the concept of the intrinsic carrier density (n_i) was introduced, it was indicated that the characteristic amount of energy associated with the recombination of an electron–hole pair was equal to the energy gap, E_g. The presence of deep donor impurities with electron states at midband energies increases the probability that recombination will occur; the transition probability increases as the transition energy is reduced. Figure 7.25 illustrates a two-step recombination process: (1) the capture of a hole by a deep donor electron, and (2) the transition of an electron from the conduction band to the now unoccupied electron state at midband. Due to the resulting increase in the recombination rate, deep donors are also denoted as *recombination centers*.

Introduction of an acceptor impurity into the crystal changes the material's conductivity (locally) by accepting an electron and adding it to the fixed impurity atom. In the process, a hole is created. An acceptor impurity has a valence of three in order that the addition of an electron from a nearby silicon atom provides the fourth electron to allow the impurity to form a covalent bond with its four nearest neighbors in the crystal. An acceptor impurity therefore is typically found in column III of the chemical periodic table.

As with donor atoms, for an acceptor impurity to be effective, it is important

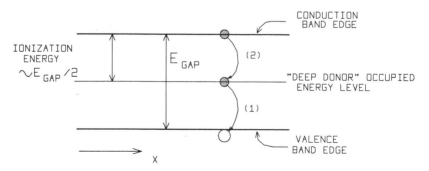

Figure 7.25. Deep donor impurity occupied energy level and the enhanced recombination transition probabilities.

that (1) the impurity reside substitutionally in the crystal structure, and (2) once in the crystal, the amount of energy required to release an electron from its silicon atom and transfer it to the acceptor impurity be low, much less than E_g (Figure 7.26).

Figure 7.27 gives a table of common acceptor impurities in silicon and the hole ionization energies. After accepting an electron, the acceptor atom is no longer charge neutral, but carries a *fixed* charge of $-1q$. If the free holes are distributed uniformly in the presence of the acceptor atoms, the entire region can be regarded as charge neutral. If the free hole concentration is depleted in this region, the remaining $-1q$ charge on each fixed acceptor atom implies that the region has an overall negative charge density.

The notation to be used in subsequent discussions of impurity and free carrier charge densities is as follows:

n: free electron concentration

p: free hole concentration

N_D^{+1}:   fixed donor atom concentration

N_A^{-1}:   fixed acceptor atom concentration

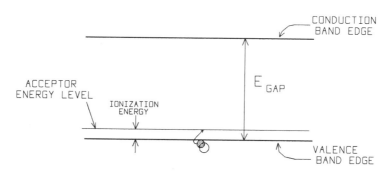

Figure 7.26.   Location of an acceptor impurity energy level in silicon relative to the valence band edge.

ELEMENT	E_I (EV)
BORON (B)	0.045 EV
ALUMINUM (AL)	0.057 EV
GALLIUM (GA)	0.065 EV
INDIUM (IN)	0.16 EV

Figure 7.27.   Common acceptor impurities in silicon and their ionization energies.

It will always be implicitly assumed that the operating temperature range of interest is such that whatever shallow impurities are present in the crystal (donors or acceptors) they are always ionized: N_D^{+1} for donors, N_A^{-1} for acceptors. This assumption leads to the result that, in some region of the crystal, if $N_D > N_A$, then $n > p$ and the region is n-type, whereas if $N_A > N_D$, then $p > n$ and the region is p-type. The term *compensation* refers to the technique of changing a region in the crystal from one type to another. For example, the *net* donor impurity concentration, $N_{D,eff}$, in an n-type region with an acceptor impurity density is given by the difference of the impurity concentrations: $N_{D,eff} = (N_D - N_A)$ (a p-type region subsequently converted to n-type by the addition of a greater donor impurity concentration).

The introduction of impurity atoms alters the free electron (n) and free hole (p) carrier concentrations significantly. It was stated earlier that electron–hole pairs are generated thermally. Let that rate of generation of pairs be denoted by g. It was also stated earlier that when a free electron and hole recombine, an electron–hole pair is lost. Neglecting recombination centers, the probability of collision increases as either the number of free holes or free electrons increases. The equilibrium rate of recombination, R, is therefore proportional to both p and n:

$$R = A * p * n, \quad \text{units of } R: (\text{cm}^3 * \text{sec})^{-1}$$

where A is a proportionality constant; the parameter A is a function of the recombination energy (E_g), temperature, and the mass and *capture cross-sectional area* of the free holes and electrons. At equilibrium, the carrier densities are time invariant, identically balanced by equal rates of generation and recombination:

$$\frac{dn}{dt} = \frac{dp}{dt} = g - A * p * n = 0$$

$$p * n = \frac{g}{A} = \text{constant} \quad \left(\begin{array}{c}\text{strong function}\\\text{of temperature}\end{array}\right)$$

It was stated earlier that without impurities present electrons and holes are thermally generated in pairs and $p = n = n_i$ and $p * n = (n_i)^2$. This second relationship also holds in the presence of impurity atoms, where p and n differ from their intrinsic values. The expression $p * n = (n_i)^2$ is known as the *law of mass action*; it merely indicates that in equilibrium the product of the free carrier concentrations is a constant for a particular recombination rate. The individual concentrations, p and n, may vary over a wide range of values while the law of mass action remains valid.

The net charge density in a region of the crystal is given by

$$\rho = q * (N_D^{+1} - N_A^{-1} + p - n)$$

IF	$N_0 \approx$	$P_0 \approx$
$(N_D - N_A) \gg N_I$	$(N_D - N_A)$	$N_I^2 / (N_D - N_A)$
$(N_A - N_D) \gg N_I$	$\dfrac{N_I^2}{(N_A - N_D)}$	$(N_A - N_D)$

'O' SUBSCRIPT IMPLIES AN EQUILIBRIUM VALUE

Figure 7.28. Summary of the relations for the free carrier concentrations in equilibrium for *n*- and *p*-type materials. The *o* subscript implies an equilibrium value.

The charge density is equal to zero in equilibrium in a region of uniform net impurity concentration:

$$N_D^+ - N_A^- + p_0 - n_0 = 0$$

where the zero subscripts refer to equilibrium conditions. In an *n*-type (net donor concentration) material

$$N_D - N_A + \frac{n_i^2}{n_0} - n_0 = 0$$

where the law of mass action was used to replace p_0 with a term dependent on n_0. Solving the resulting quadratic for n_0,

$$n_0 = \tfrac{1}{2} * \{(N_D - N_A) + [(N_D - N_A)^2 + 4n_i^2]^{1/2}\}$$

Similarly, the free hole concentration in a *p*-type material of uniform (net) acceptor concentration is given by

$$p_0 = \tfrac{1}{2} * \{(N_A - N_D) + [(N_A - N_D)^2 + 4n_i^2]^{1/2}\}$$

Figure 7.28 lists the resulting free carrier concentrations in equilibrium for *n*- and *p*-type materials, if the assumption can be made that the net impurity concentration is much greater than the intrinsic value.

7.5 *PN* JUNCTIONS

A *p*-type to *n*-type junction is formed in the semiconductor material wherever adjacent regions exist in the semiconductor of opposite type; in one region, donor impurities are in the majority, while in the other acceptor impurities are prevalent. The two common means for introducing impurities locally in the crystal are diffusion and ion implantation; the substrate semiconductor wafer also has a (uniform) background concentration of dopants prior to beginning fabrication. The wafer type, background impurity concentration, and crystalline surface orienta-

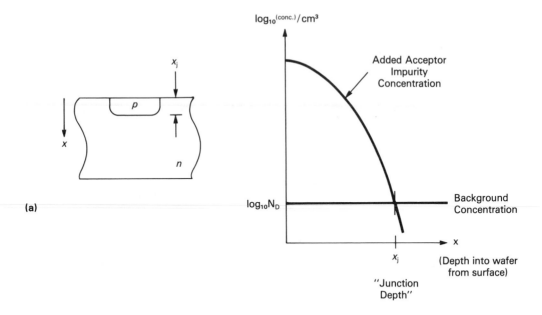

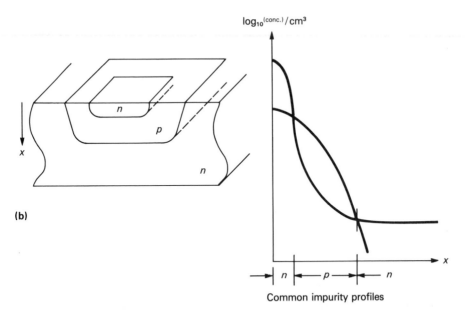

Figure 7.29. Formation of *pn* junctions by the addition of impurities to the background substrate (a) or by the *compensation* locally of the opposite type material (b). Associated with each cross section is a sketch of the impurity concentration *profile* as a function of the depth into the wafer substrate from the surface (not to scale).

tion differ from fabrication process to process, depending on the desired final device set to be provided and the desired device characteristics. *PN* junctions can therefore be formed in two ways (Figure 7.29): (1) the local addition of a dopant material into the background substrate, or (2) the compensation of a region in the crystal by the addition of a larger concentration of dopant of the opposite type (to a shallower depth in the substrate).

Figure 7.29 also illustrates a common doping profile, that is, the concentration of impurities as a function of the distance into the crystal from the device surface. The metallurgical junction depth is indicated as the point where the net impurity concentrations of donor and acceptor impurities are equal. (Another means of junction formation involves the addition of *silicon* to the top surface of the wafer in the same crystalline orientation as the surface, or *epitaxially*. In the epitaxial growth process, in the gaseous environment containing the silicon compound, another gas containing the desired dopant as a constituent of the gas molecule is added at a particular partial pressure. The dissociation of these gases results in the growth of doped crystalline silicon onto the wafer surface; using the appropriate dopant can provide a *pn* junction at the epitaxial layer–substrate interface. Epitaxial growth is discussed further in Chapter 15.)

Although the impurity concentration profiles of Figure 7.29 clearly indicate that the concentration in both of the doped regions is not uniform, it is nevertheless worthwhile (and in many cases a good approximation) to analyze the following conceptual experiment assuming uniformly doped regions. Consider two artificially disjoint *n*- and *p*-type semiconductor regions, which are then placed in contact (Figure 7.30). The principles of diffusion state that free mobile electrons

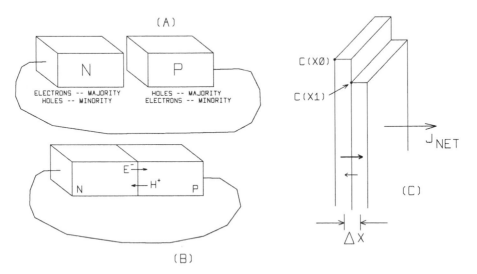

Figure 7.30.   A conceptual experiment: the adjoining of two uniformly doped materials of opposite type to form a *pn* junction. As a result, there will be a net diffusion flow of mobile carriers due to the concentration gradient at the junction (c).

and holes tend to move from regions of high concentration to regions of low concentration. Diffusion of a nonuniform concentration of mobile carriers is simply a consequence of the assumed random thermal motion of the carriers. As illustrated in Figure 7.30c, there will be a *net* flow of carriers in the direction of lower concentrations, since, at any time, *more* carriers will have a velocity component in this direction than in the counter direction. The diffusion current density (per unit area) is proportional to the concentration gradient (the derivative of the concentration in the one-dimensional case):

$$J \propto - \frac{dC}{dx}$$

where J is the carrier current density (units are carriers per square centimeter per second) and the minus sign is required since the direction of particle current is positive when the derivative is negative. The diffusion current for holes and electrons is given by

$$J_{\text{diffusion, holes}} = +q \left(-D_p * \frac{dp}{dx} \right)$$

$$J_{\text{diffusion, electrons}} = -q \left(-D_n * \frac{dn}{dx} \right)$$

where D_p and D_n are the hole and electron *diffusion coefficients*. These coefficients are functions of temperature and carrier mobility (in cm²/sec).

As a result, with the two materials in contact, electrons will diffuse from the *n*-type material and holes will diffuse from the *p*-type. Electrons that diffuse from the *n*-type region leave behind positively charged donor impurity ions N_D^+; holes that diffuse from the *p*-type region leave behind negatively charged acceptor impurity ions N_A^- (Figure 7.31). Recall that electric field lines emanate from fixed positive charges and terminate on fixed negative charges. Mobile positive charges move in the direction of the electric field. Note in Figure 7.31 that the direction of the electric field developed *opposes* the further diffusion of holes from *p*-type to *n*-type regions and likewise opposes the further diffusion of electrons from *n*-type to *p*-type regions. Therefore, every *pn* junction acquires a *built-in field* and *built-in contact potential* ($V = - \int \vec{\mathscr{E}} \cdot \vec{dl}$), which opposes the further equilibrium diffusion of holes and electrons. The fixed charge density for the *abrupt* junction example is plotted in Figure 7.32; the extent of the exposed fixed ionized impurities is denoted by x_n and x_p into the *n*- and *p*-type regions, respectively. Since the *n*- and *p*-type regions in this experiment were each initially charge neutral, the resulting system of contacted regions is charge neutral *overall*, despite the local nonzero charge densities in the vicinity of the junction and the counteracting drift and diffusion currents. Overall charge neutrality implies that (per unit area) $N_D * x_n = N_A * x_p$. The region between $-x_n$ and x_p is assumed to be completely *depleted* of mobile holes and electrons due to the presence of the built-in field; this volume is typically denoted as the *depletion region*.

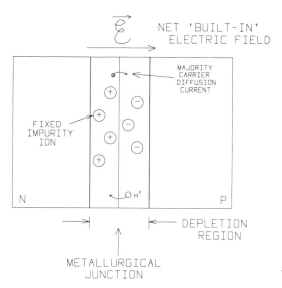

Figure 7.31. *Net* fixed impurity charge density in the vicinity of the *pn* junction due to the depletion of mobile (majority) carriers.

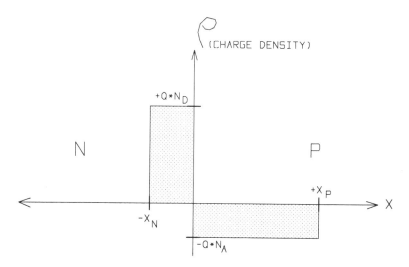

Figure 7.32. Plot of the charge density (ρ) for the abrupt junction approximation with uniformly doped *p*- and *n*-type regions.

The built-in field and contact potential will reach a magnitude such that the depletion region current produced by the diffusion flow of carriers from their majority to minority regions is exactly balanced by the depletion region drift (electric-field driven) current of carriers from their minority into majority regions; the net junction current is exactly zero. The drift current in equilibrium includes the current of thermally generated free electron–hole pairs in the nonzero field regions that are *swept apart* by the built-in field.

Although not strictly valid in this artificial example (dn_0/dx and dp_0/dx are undefined at the junction), the equilibrium expressions for the depletion region net hole and electron currents for a general nonuniform impurity concentration profile are

$$\text{Diffusion current + drift current} = 0$$

$$J_{\text{holes}} = q\left(-D_p * \frac{dp_0}{dx} + \mu_p * \mathcal{E}_0(x) * p_0(x)\right) = 0$$

$$J_{\text{electrons}} = q\left(D_n * \frac{dn_0}{dx} + \mu_n * \mathcal{E}_0(x) * n_0(x)\right) = 0$$

$$\text{Drift current} = q * \text{concentration} * \text{carrier velocity} = \begin{bmatrix} q * \mu_n * \mathcal{E} * n \\ q * \mu_p * \mathcal{E} * p \end{bmatrix}$$

If the concentration gradient is large, significant diffusion current results; a large built-in electric field is therefore also present in equilibrium.

Using the Poisson equation describing the emanation and termination of electric field lines on charges,

$$\epsilon_{\text{Si}} * \frac{d\mathcal{E}}{dx} = \rho(x)$$

where ρ is charge density in coulombs/cm^3 and ϵ_{Si} is the dielectric permittivity of silicon, the charge density plot of Figure 7.32 can be used to generate the plot of the built-in electric field (Figure 7.33). To solve for x_n, x_p, and the magnitude of the electric field, it is necessary to determine the total built-in voltage. The key principle used to proceed with the calculation (and the analysis of any multiconductor carrier analysis problem) is *in equilibrium, the Fermi energy level is*

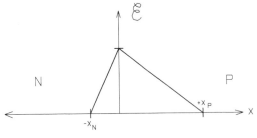

$$\mathcal{E}(X) = \begin{cases} (1/\mathcal{E}_{SI}) * (X + X_N) * Q * N_D, & -X_N <= X <= 0 \\ (1/\mathcal{E}_{SI}) * (X - X_P) * -Q * N_A, & 0 <= X <= X_P \end{cases}$$

Figure 7.33. Plot of the built-in electric field in the vicinity of the junction resulting from the nonzero charge density in the junction depletion region.

constant throughout the entire (multiconductor) system.[4] Using this principle and the relationships for the carrier concentrations in the *bulk n-* and *p*-type regions, the total built-in junction potential voltage can now be calculated (Figures 7.34 and 7.35):

$$n_0 \bigg|_{n\text{-type bulk}} = N_C * \exp\left[-\left(\frac{E_{C_n} - E_F}{kT}\right)\right] = N_D$$

$$p_0 \bigg|_{p\text{-type bulk}} = N_V * \exp\left[-\left(\frac{E_F - E_{V_p}}{kT}\right)\right] = N_A$$

$$n_0 * p_0 = N_D * N_A = N_C * N_V \exp\left[-\frac{(E_{C_n} - E_{V_p})}{kT}\right]$$

$$E_{C_n} - E_{V_p} = E_g - [(-q) * (V_{\text{built-in}})]$$

$$N_D * N_A = \underbrace{N_C * N_V\, e^{-E_g/kT}}_{n_i^2} * \exp\left[-\frac{(q * V_{\text{built-in}})}{kT}\right] \Rightarrow$$

$$V_{\text{built-in}} = \frac{-kT}{q} * \ln\left(\frac{N_D * N_A}{n_i^2}\right)$$

The built-in voltage function is related to the *electron* potential energy by $\Delta E_{\text{electron}} = -q * V_{\text{built-in}}$; a more positive voltage corresponds to *less* electron potential energy and lower values on the electron energy band diagram of Figure 7.35b. The *n*-type region in the *pn* junction is positive with respect to the *p*-type region. Note that the total built-in voltage is dependent only on the carrier concentrations in the bulk (zero electric field) regions and not on the specifics of the impurity profile. If the assumptions of the abrupt depletion region approximation were relaxed and the nonzero field regions were not totally depleted of mobile carriers, the electric field peak magnitude would be reduced, yet the built-in potential would be unaltered (Figure 7.36).

---

[4] Although the introduction of the Fermi energy level in Section 7.2 referred to its value as a *normalization* parameter for the occupation probability distribution function, its physical interpretation for a semiconductor in equilibrium is that the Fermi energy is equivalent to the *electrochemical potential energy* of the system (which is constant throughout a system in equilibrium). The electrochemical potential is a thermodynamics parameter used to describe the total free energy of a system:

$$\text{Electrochemical potential, } \overline{\mu} = \frac{\partial(\text{free energy})}{\partial(\text{number of particles})}\bigg|_{\text{constant } T,V}$$

See, for example, the discussion in A. van der Ziel, *Solid State Physical Electronics*, Prentice-Hall, Inc., Englewood Cliffs, N.J., 1976, pp. 55–57.

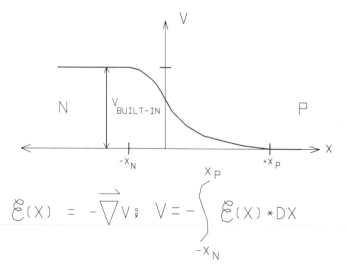

$$\vec{\mathcal{E}}(X) = -\vec{\nabla}V; \quad V = -\int_{-X_N}^{X_P} \vec{\mathcal{E}}(X) * DX$$

Figure 7.34. Plot of the junction built-in potential function. Note that a more positive potential corresponds to lower electron energy on the electron energy band diagram.

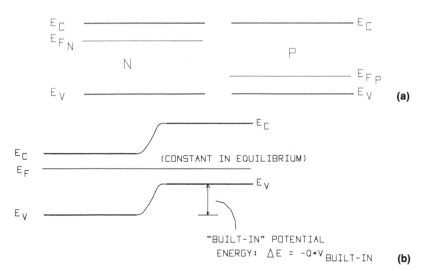

Figure 7.35. The equilibrium energy band diagrams for *n*- and *p*-type materials (a) and for the *pn* junction (b).

The energy band diagram of Figure 7.35b is a useful tool for illustrating the behavior of a *pn* junction under nonequilibrium conditions with an applied stimulus, to be discussed shortly. The following are several qualitative features of note about the band diagram:

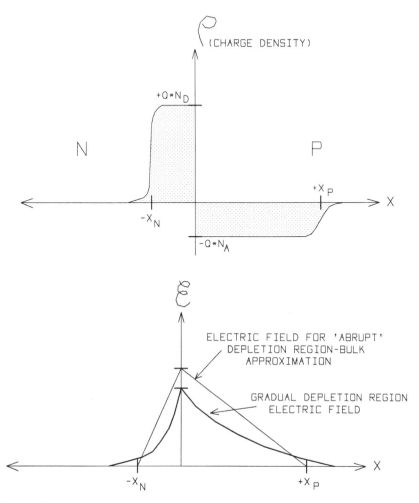

Figure 7.36. Charge density and electric field in the vicinity of the *pn* junction if the abrupt depletion region approximations were relaxed. Since the built-in potential depends on the impurity concentrations in the bulk (nonzero field) regions, it remains unchanged.

1. Electron potential energy increases in an upward direction; hole potential energy increases in a downward direction.

2. Regions where the energy band diagram has no slope are regions of constant potential; this implies that the electric field in these regions is zero:

$$-(dV/dx) = \mathscr{E} = 0.$$

3. The energy gap is constant throughout the material; only the relative potential energy changes.

4. If a voltage is applied to the *pn* junction through appropriate metal contacts to the bulk *p* and *n* regions of the same polarity as the built-in voltage, the peak magnitude of the electric field, x_n, and x_p will all increase; in other words, the *band-bending* will increase.

5. If a voltage is applied to the *pn* junction of the opposite polarity as the built-in voltage, the peak electric field, x_n, and x_p will all decrease.

The situation described in item 4 is referred to as *reverse bias*; the situation in item 5 is known as *forward bias*. The circuit element formed by the *pn* junction is a *diode* or *rectifier*.

Reverse Bias

If an additional applied reverse voltage V_R is added to the built-in contact potential, the extent of the depletion region increases and an additional fixed charge ΔQ is exposed (Figure 7.37). The relationship between an additional applied two-terminal potential difference and the resulting charge that each terminal must provide implies the junction has an associated capacitance:

$$C(V) = \frac{dQ}{dV} \approx \frac{\Delta Q}{\Delta V}$$

Note in this expression that it is implied that the junction capacitance is not a constant value, but rather is a function of the reverse applied junction potential difference. This capacitance function can be derived as follows:

1. $\dfrac{d\mathscr{E}}{dx} = \dfrac{\rho}{\epsilon_{Si}}$

2. Find the field in the *n*-type region:

$$\rho = q * N_D, \qquad -x_n \le x < 0$$

$$\frac{d\mathscr{E}_n}{dx} = \frac{qN_D}{\epsilon_{Si}}, \qquad \mathscr{E}_n(x = -x_n) = 0 \Rightarrow \mathscr{E}_n = q\frac{N_D}{\epsilon_{Si}}(x + x_n)$$

3. Find the field in the *p*-type region:

$$\rho = -q * N_A, \qquad 0 < x \le x_p$$

$$\frac{d\mathscr{E}_p}{dx} = \frac{-qN_A}{\epsilon_{Si}}, \qquad \mathscr{E}_p(x = x_p) = 0$$

$$\mathscr{E}_p = \frac{qN_A}{\epsilon_{Si}}(x_p - x)$$

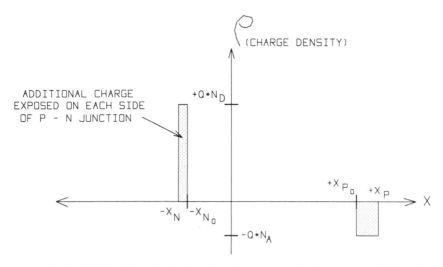

Figure 7.37. Additional fixed ionic impurity charge exposed in the vicinity of the junction depletion region if a voltage is applied across the junction to increase the total junction potential difference (*reverse bias*).

4. Check:

$$\mathcal{E}_n(x = 0) = \mathcal{E}_p(x = 0)$$

$$\frac{qN_D}{\epsilon_{Si}} * x_n = \frac{qN_A}{\epsilon_{Si}} * x_p \Rightarrow N_D * x_n = N_A * x_p$$

Checks with the assumption of overall charge neutrality.

5. $\quad -\dfrac{dV}{dx} = \mathcal{E}, \qquad V_{\text{TOT}} = -\displaystyle\int_{-x_n}^{x_p} \mathcal{E}(x) * dx$

$$V_{\text{TOT}} = -\left\{ \frac{qN_D}{2\epsilon_{Si}} * x_n^2 - \frac{qN_A}{2\epsilon_{Si}} * x_p^2 \right\}$$

6. Using the result in step 5 and the substitution $N_D * x_n = N_A * x_p$,

$$x_n = \left(\frac{2\epsilon_{Si}}{qN_D} * \frac{N_A}{N_A + N_D} * |V_{\text{TOT}}| \right)^{1/2}$$

$$x_p = \left(\frac{2\epsilon_{Si}}{qN_A} * \frac{N_D}{N_A + N_D} * |V_{\text{TOT}}| \right)^{1/2}$$

7. $\dfrac{Q_{\text{TOTAL}}}{\text{unit area}} = q * N_D * x_n = q * N_A * x_n \quad \left(\begin{array}{l} \text{on either side of} \\ \text{abrupt junction} \end{array} \right)$

$$= q * N_D * \left(\frac{2\epsilon_{Si}}{q * N_D} * \frac{N_A}{N_A + N_D} * |V_{\text{built-in}} + V_{\text{reverse}}| \right)^{1/2}$$

Either of two interpretations is suitable for determining the $C(V)$ expression:

$$\text{I.} \quad \frac{C}{\text{unit area}} = \frac{dQ_{\text{TOT}}}{dV_{\text{reverse}}}$$

$$= \left(\frac{\epsilon_{Si} * q}{2} * \frac{N_A * N_D}{N_A + N_D} \right)^{1/2} * (\mid V_{\text{built-in}} + V_{\text{reverse}} \mid)^{-1/2}$$

or

$$\text{II.} \quad \frac{C}{\text{unit area}} = \frac{\epsilon_{Si}}{d}$$

a *parallel plate* expression, where d is the incremental *separation* between charges on the plates (i.e., a *dipole*):

$$d = (x_n + x_p)$$

$$\frac{C}{\text{unit area}} = \frac{\epsilon_{Si}}{(x_n + x_p)}$$

This yields the same result as in equation I.

The derivation in method II uses the analogy between the incremental depletion region charge of a reverse-biased *pn* junction and the fixed plates of a parallel plate capacitor; in the case of the junction, however, the differential separation between the plates is a function of the applied voltage.

The incremental junction capacitance *decreases* as the applied reverse bias increases. The junction capacitance (of *n*-type circuit nodes in a *p*-type substrate, for example) is part of the load capacitance that a circuit on a chip must charge or discharge to vary the node voltage; to minimize this capacitance and enable faster switching circuit performance, (1) the layout area and perimeter of the node are kept to a minimum, and (2) a constant reverse-bias voltage can be applied to the chip substrate to increase the total reverse bias on *all* circuit nodes for all node voltages (Figure 7.38).

In addition to the junction capacitance, the reverse-biased *pn* junction also conducts current between the two terminals of the diode. The equilibrium balance between depletion region drift and diffusion currents that resulted in zero net current is disrupted by the applied reverse-bias voltage. There are two components of the reverse-biased current to consider:

1. The current due to the free electron–hole pairs thermally generated in the *increased* reverse-bias depletion region that are swept out by the reverse-bias electric field.

2. The current resulting from the *injection* of carriers from the region in which the carrier is in the minority to the region where the carrier becomes part of the majority carrier nonequilibrium concentration, that is, electrons from

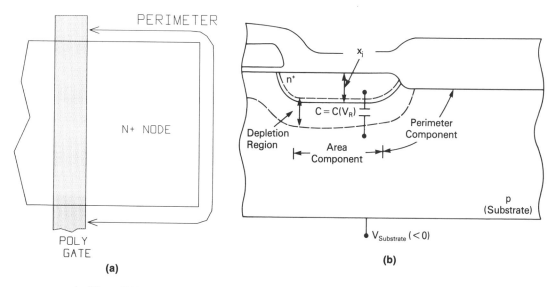

Figure 7.38. Reverse-bias voltage applied to the chip substrate relative to *all* junction (device) nodes: (a) layout; (b) cross section.

the *p*-type into the *n*-type region and holes from the *n*-type into the *p*-type region (Figure 7.39); due to the applied reverse-bias voltage, the depletion region drift current now exceeds the diffusion current. The term *injection* refers to the change in free carrier concentrations at the *bulk* edges of the depletion region.

The current component in item 1 is difficult to determine analytically; qualitatively, it exhibits the following dependencies:

a. The net generation minus recombination rate in the depletion region is greatest where both the carrier concentrations are small, a distance that is some fraction of the total depletion width; recall from Section 7.4 that the recombination rate is proportional to the product of the carrier concentrations.

b. The magnitude of the generation current should vary with the extent of the depletion region, $x_n + x_d$ (nonequilibrium values), roughly increasing as the square root of the sum of the built-in and applied reverse-bias voltages.

c. This generated current should exhibit a strong temperature dependence, similar to that of the intrinsic carrier concentration, n_i.

The current component in item 2 can be analyzed more thoroughly, if several key assumptions are made:

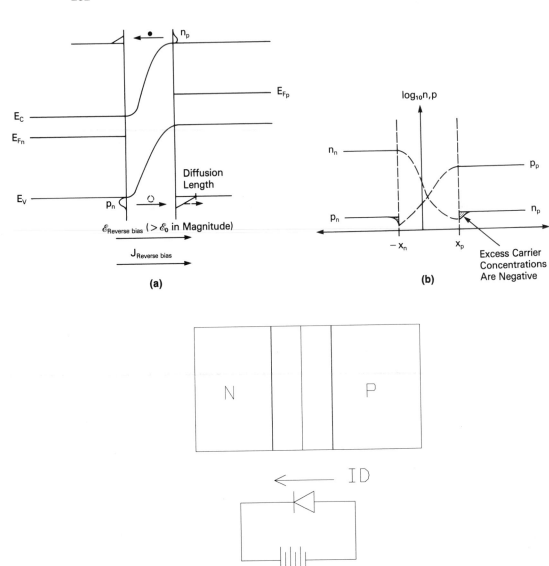

Figure 7.39. Reverse-bias current mechanism of the injection of mobile carriers on either side of the depletion region from the bulk region in which the carrier is in the minority into the region where the carrier is in the majority. In (b), the *excess* carrier concentrations at the edge of the depletion region are shaded; in reverse bias, these excess carrier concentrations are *negative*. (c) Circuit configuration for reverse bias.

a. The concentration of free carriers in the depletion region is small relative to the fixed charge densities and can be neglected; this *abrupt depletion region* assumption has indeed been used up to this point in the discussion.

b. There is no recombination of injected carriers in the depletion region; carriers injected from one bulk region are assumed to eventually recombine in the opposite bulk region. The carriers that recombine with the injected carriers in the bulk regions are replenished by the flow (drift-current component) of carriers in from the contacts at the ends of the diode. The carrier concentrations at the contact ends of the bulk regions (assumed to be well distant from the depletion region) do not vary from their equilibrium values; the drift current results from the steady-state (time-invariant) flow of carriers to support the necessary recombination. With the assumption of no recombination in the depletion region and with thermal generation reverse bias currents accounted for in component 1, the current continuity equations in the depletion region for steady state conditions become (time derivatives are zero):

$$\vec{\nabla} \cdot \vec{J}_n + \cancel{\frac{\partial n}{\partial t}}^{0} = \cancel{G}^{0} - \cancel{R}^{0}, \quad \vec{\nabla} \cdot \vec{J}_p + \cancel{\frac{\partial p}{\partial t}}^{0} = \cancel{G}^{0} - \cancel{R}^{0} \Rightarrow \frac{dJ_n}{dx} = 0, \quad \frac{dJ_p}{dx} = 0$$

$dJ_n/dx = 0$ and $dJ_p/dx = 0$ imply that the hole and electron currents in the depletion region are constant.

c. Although the drift current of carriers in from the contacts implies a field present (and a resistive voltage drop) across the bulk regions, for the range of diode operation of interest (i.e., *low injection*), it will be assumed that the magnitude of this field and voltage drop is negligible. In other words, the applied voltage is assumed to modify only the built-in potential and the field in the depletion region. (This assumption should be verified after finding the solution of a diode problem.)

The preceding assumptions have some significant ramifications:

1. The bulk regions are therefore still assumed to be charge neutral. However, since the injection of carriers results in an excess carrier concentration at the edge of the bulk region, it must be assumed that *both* carrier concentrations increase by an equal amount to keep the bulk region charge neutral:

$$\rho = q(p - n + N_D - N_A)$$

$$= q[(\underbrace{p_0 + \Delta p}) - (n_0 + \Delta n) + N_D - N_A] \cong 0$$

equilibrium value

excess carrier concentration

Equilibrium values were derived from $p_0 - n_0 + N_D - N_A = 0$, implying $\Delta p = \Delta n$ (in either bulk region). Note that the product $p * n = (p_0 + \Delta p) * (n_0 + \Delta n)$ is no longer equal to n_i^2 in this nonequilibrium, steady-state condition. This assumption regarding the bulk charge density at the edges of the depletion region is often called the assumption of *quasi-neutrality*.

2. The definition of low injection permits the change in minority carrier concentration at the edge of the bulk region to be many orders of magnitude, while the corresponding change in the majority carrier concentration ($\Delta n = \Delta p$) remains negligible. Over the length of the bulk, therefore, the minority carrier concentration varies by orders of magnitude, which results in a minority carrier diffusion current.

3. The carrier current densities, J_e and J_h, both have drift and diffusion components that are perturbed from their equilibrium values by excess carrier injection. It will be assumed, in the case of low injection, that the majority carrier concentration remains uniform throughout the bulk (at its equilibrium value) and therefore has no diffusion current component, only a drift current component. The product of the field in the bulk (assumed to be small) and the minority carrier density results in negligible minority carrier drift current in the bulk; the large change in concentration over the extent of the bulk regions implies the minority carrier diffusion component *is* significant.

4. The steady-state application of an external bias is a nonequilibrium situation; hence, the Fermi energy level for the system is not constant. For calculating the carrier concentrations and currents in the depletion region and at the edges of the bulk regions in nonequilibrium, it is common to define distinct (quasi-) Fermi energy levels for electrons and holes (functions of x), which collapse together in each of the neutral bulk regions. These two neutral bulk region levels are separated in electron energy by $\Delta E = q * V_{\text{applied}}$.

Figure 7.39 illustrates the carrier distributions on both sides of the depletion region in reverse bias. As mentioned previously, minority carriers are injected across the depletion region into their majority carrier bulk regions. For the case of reverse bias, the *excess* minority carrier densities on both sides of the junction are *negative* (Figure 7.39b), as is the diode current defined in Figure 7.39c. Summarizing the pertinent assumptions and their consequences:

(i) The applied voltage is dropped across the depletion region.

(ii) The current at each edge of the depletion region consists of the minority carrier diffusion current into the bulk from a gradient in excess minority carrier density.

(iii) The carrier currents are constant within the depletion region (and thus are equal to the minority carrier bulk diffusion currents at the bulk–depletion region edges, that is, at $-x_n$ and x_p).

The analysis of the reverse-biased diode problem proceeds by calculating the current density, $J = J_e + J_h$, at any point along the diode, since, in steady state, the *total* current density is constant throughout, although the individual carrier current densities will vary. The point along the diode chosen to evaluate the total current density will be a depletion region–bulk edge; at that point, the minority carrier diffusion current into the bulk is calculated. Using the result in item iii, that the carrier currents are constant in the depletion region, the majority carrier current at that point is equal to its minority carrier diffusion current at the opposite depletion region edge. Thus, the total diode current (nonthermal generation component) can be expressed as the sum of the minority carrier diffusion currents at their respective depletion region–bulk edges (Figure 7.40).

The nonequilibrium diode current under low injection is assumed to be a small fraction of the equilibrium oppositely directed drift and diffusion currents. Thus, the nonequilibrium analysis will be approximately valid if, in the expressions involving the potential difference between the two bulk regions, the term $(V_{\text{built-in}} + V_{\text{applied}})$ is used rather than $V_{\text{built-in}}$. Referring to the equilibrium band diagram of Figure 7.41a, the equilibrium *minority* electron carrier density can be determined as follows:

$$p_{p0} = N_A = N_V * \exp\left[-\left(\frac{E_F - E_{V_p}}{kT}\right)\right]$$

$$n_{p0} = N_C * \exp\left[-\left(\frac{E_{C_p} - E_F}{kT}\right)\right]$$

$$n_{n0} = N_C * \exp\left[-\left(\frac{E_{C_n} - E_F}{kT}\right)\right] = N_D$$

$$E_{C_n} - q * V_{\text{built-in}} = E_{C_p} \qquad (V_{\text{built-in}} < 0)$$

$$n_{p0} = N_C * \exp\left[-\left(\frac{E_{C_n} - E_F}{kT}\right)\right] * \exp\left[\frac{qV_{\text{built-in}}}{kT}\right]$$

$$= n_{n0} * \exp\left[\frac{qV_{\text{built-in}}}{kT}\right]$$

The nonequilibrium minority carrier density is therefore expressed as

$$n_p(+x_p) = n_n * \exp\left[\frac{q(V_{\text{built-in}} + V_{\text{applied}})}{kT}\right]$$

The *excess* minority electron carrier concentration at the depletion region–p type bulk edge is therefore given by

$$\Delta n_p \big|_{+x_p} = n_p(+x_p)\big|_{\text{nonequilibrium}} - n_{p0}$$

$$= N_D * \exp\left[\frac{qV_{\text{built-in}}}{kT}\right] * \left(\exp\left[\frac{qV_{\text{applied}}}{kT}\right] - 1\right)$$

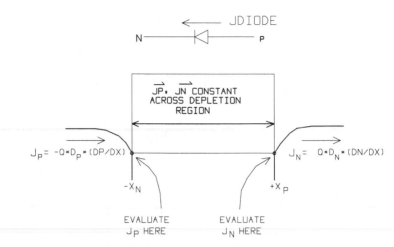

Figure 7.40. The reverse-bias diode current is evaluated by calculating the excess carrier diffusion currents at the edges of the junction depletion region.

A similar derivation for the *excess* hole concentration at the depletion region–*n* type bulk edge ($-x_n$) gives

$$\Delta p_n \big|_{-x_n} = N_A * \exp\left[\frac{qV_{\text{built-in}}}{kT}\right] * \left(\exp\left[\frac{qV_{\text{applied}}}{kT}\right] - 1\right)$$

In the case of reverse bias, V_{applied} is negative, and therefore so are Δn_p and Δp_n (Figure 7.39). The product of $p * n$ at either edge of the depletion region can be evaluated using the expressions for $p_p = N_A$, $n_p(+x_p)$, $n_n = N_D$, and $p_n(-x_n)$:

$$p * n \big|_{+x_p, -x_n} = n_i^2 * \exp\left[\frac{qV_{\text{applied}}}{kT}\right] \neq n_i^2$$

Note also that, for large reverse bias, the excess minority carrier concentrations at the bulk edges Δn_p and Δp_n approach the negative of their equilibrium values n_{po} and p_{no}; in other words, the *net* minority carrier concentration approaches zero, in concurrence with the assumption of no mobile carriers present in the depletion region.

The minority carrier diffusion currents into the bulk at the coordinates $-x_n$ (holes) and x_p (electrons) are given by

$$J_p = -q * D_p * \frac{dp}{dx}\bigg|_{-x_n}, \qquad D\text{: carrier diffusion coefficients}$$

$$J_n = q * D_n * \frac{dn}{dx}\bigg|_{+x_p}$$

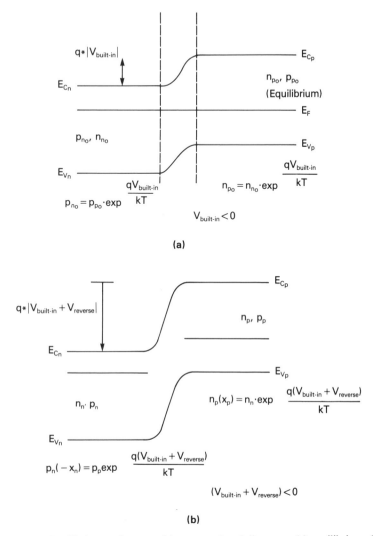

Figure 7.41. Equilibrium and reverse-bias energy band diagrams: (a) equilibrium. (b) reverse bias.

To be able to determine the slopes of the excess carrier concentrations at the depletion region–bulk edges, it is necessary to describe the recombination process in the bulk region. In Section 7.4, two competing processes, generation and recombination, were described; in particular, it was stated that the recombination rate is proportional to the carrier concentration. For example, for holes

$$\frac{dp}{dt} = g_p - R_p$$

where R_p is proportional to the concentration p. Of particular interest in this steady-state, reverse-bias analysis is the recombination process of *excess* carriers at points in the bulk region. Again, for holes

$$\frac{\partial(\Delta p)}{\partial t} = g_{\Delta p} - R_{\Delta p}, \qquad \text{where } R_{\Delta p} \propto \Delta p$$

As illustrated in Figure 7.42, the "generation" of excess carriers at a point in the bulk is not due to the equilibrium thermal generation discussed earlier, but rather the rate of change of carriers flowing into and out of the point:

$$g_{\Delta p} = -\frac{1}{q} * \frac{dJ_{\Delta p}}{dx}$$

The excess carrier recombination is characterized by a parameter called the excess carrier *lifetime* in the bulk region:

$$R_{\Delta p} \propto \Delta p, \qquad R_{\Delta p} \equiv \frac{\Delta p}{\tau_p}, \qquad \tau_p: \text{ excess carrier lifetime}$$

A physical interpretation of this proportionality constant is that the lifetime is the average duration of an excess minority carrier (under low injection) in a bulk

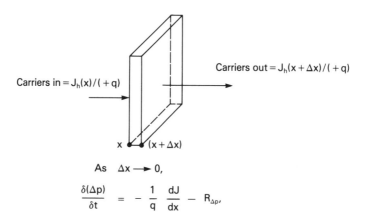

Carriers in $= J_h(x)/(+q)$

Carriers out $= J_h(x+\Delta x)/(+q)$

x   (x + Δx)

As Δx → 0,

$$\frac{\delta(\Delta p)}{\delta t} = -\frac{1}{q}\frac{dJ}{dx} - R_{\Delta p'}$$

where J refers to the electric current and J/(+q) refers to the particle current (per unit area).

Figure 7.42.   Continuity equation in the bulk; the particle current into the differential region is given by $J/+q$ (for holes).

region before recombining with a majority carrier. Its value depends on a number of physical parameters: the majority carrier density, temperature, the concentration and energy level of recombination centers in the bulk, and others; it typically is on the order of 0.5 μsec (holes) to 1.0 μsec (electrons) for a relatively lightly doped bulk region free of recombination centers at room temperature.

For the steady-state example under analysis, at any point in the bulk, the time derivative of the excess carrier concentration is zero. Using the forms derived for generation and recombination in the bulk, for holes:

$$0 = -\frac{1}{q}\frac{dJ_{\Delta p}}{dx} - \frac{\Delta p}{\tau_p} = D_p * \frac{d}{dx}\left(\frac{d(\Delta p)}{dx}\right) - \frac{\Delta p}{\tau_p}$$

$$\frac{d^2(\Delta p)}{dx^2} = \frac{\Delta p}{D_p * \tau_p}$$

For electrons,

$$0 = \frac{1}{q} * \frac{dJ_{\Delta n}}{dx} - \frac{\Delta n}{\tau_n} = D_n * \frac{d^2(\Delta n)}{dx^2} - \frac{\Delta n}{\tau_n}$$

The solutions to the preceding equations for the boundary conditions in this case of interest are sketched in Figure 7.43 and are given by

$$\Delta p(x) = \Delta p(-x_n) * \exp\left[\frac{x - (-x_n)}{\sqrt{D_p \cdot \tau_p}}\right], \qquad x \leqq -x_n$$

$$\Delta n(x) = \Delta n(x_p) * \exp\left[\frac{-(x - x_p)}{\sqrt{D_n \cdot \tau_n}}\right], \qquad x \geqq x_p$$

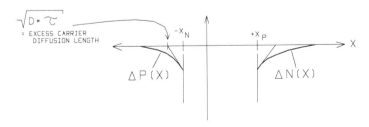

Figure 7.43.   Plot of the excess carrier concentrations as a function of the distance into the bulk; the term $(D * \tau)^{1/2}$ is denoted as the excess minority carrier diffusion length.

These relationships and the ones given earlier for the excess carrier concentrations at the edges of the depletion region finally permit the calculation of the diode current:

$$J_{\text{diode}} = -J_{\Delta n}|_{x_p} - J_{\Delta p}|_{-x_n} \quad \text{(from Figure 7.40)}$$

$$J_{\Delta p}|_{-x_n} = -q * D_p * \left.\frac{d(\Delta p)}{dx}\right|_{-x_n} = -q * D_p * \frac{\Delta p(-x_n)}{\sqrt{D_p * \tau_p}}$$

$$= -N_A * \exp\left[\frac{qV_{\text{built-in}}}{kT}\right]$$

$$* \left(\exp\left[\frac{qV_{\text{applied}}}{kT}\right] - 1\right) * \frac{qD_p}{\sqrt{D_p \tau_p}}$$

$$J_{\Delta n}|_{x_p} = q * D_n * \left.\frac{d(\Delta n)}{dx}\right|_{x_p} = -q * D_n * \frac{\Delta n(x_p)}{\sqrt{D_n * \tau_n}}$$

$$= \frac{-qD_n}{\sqrt{D_n \tau_n}} * N_D * \exp\left[\frac{qV_{\text{built-in}}}{kT}\right] * \left(\exp\left[\frac{qV_{\text{applied}}}{kT}\right] - 1\right)$$

$$\frac{J_{\text{diode}}}{\text{unit cross-sectional area}} = q * \left(\frac{D_n * N_D}{\sqrt{D_n \tau_n}} + \frac{D_p * N_A}{\sqrt{D_p \tau_p}}\right)$$

$$* \exp\left[\frac{qV_{\text{built-in}}}{kT}\right] * \left(\exp\left[\frac{qV_{\text{applied}}}{kT}\right] - 1\right)$$

$$= q * N_A * N_D * \exp\left[\frac{qV_{\text{built-in}}}{kT}\right] * \left(\frac{D_n}{N_A\sqrt{D_n * \tau_n}}\right.$$

$$\left. + \frac{D_p}{N_D * \sqrt{D_p \cdot \tau_p}}\right) * \left(\exp\left[\frac{qV_{\text{applied}}}{kT}\right] - 1\right)$$

This expression can be simplified in two ways:

1. Using the expression for the built-in voltage in terms of the doping densities:
$n_i^2 = N_A * N_D * \exp\left(\dfrac{qV_{\text{built-in}}}{kT}\right)$.

2. The terms $(D_p * \tau_p)^{1/2}$ and $(D_n * \tau_n)^{1/2}$ have the dimensions of length; it is common to replace these terms with the parameters L_p and L_n, respectively, that is, the *diffusion lengths* of the carriers.

Using substitutions 1 and 2, the diode current can be expressed as

$$\underbrace{\frac{J_{\text{diode}}}{\text{unit area}} = q * n_i^2 * \left(\frac{D_n}{N_A L_n} + \frac{D_p}{N_D L_p}\right)}_{J_{\text{sat}}} * \left(\exp\left[\frac{qV_{\text{applied}}}{kT}\right] - 1\right)$$

The product of the coefficients that multiply the voltage-dependent term is commonly referred to as the *saturation current density*; its value represents the negative diode current under reverse bias (negative V_{applied}) conditions where the exponential voltage term can be neglected in magnitude with respect to unity.

Summary of reverse bias.   The reverse bias *pn* junction exhibits a junction capacitance related by $C = \Delta Q/\Delta V$ and a reverse current that has two major components: depletion region current and excess injected minority carrier diffusion current. The former current component was qualitatively expected to increase with the square root of the magnitude of the built-in and applied reverse-bias voltage, while the latter saturates quickly for a reverse bias much larger in magnitude that a few kT/q.

Both reverse-bias current components are very small in magnitude; if the node-to-substrate junctions on the chip are suitably reverse biased, the operation of MOS devices near the surface can typically be modeled without much concern as to the diverting of logic circuit currents due to reverse-bias junctions (Figure 7.44). There is still a potentially significant capacitive displacement current into the node-to-substrate junction capacitance as the node voltage changes value, however. The operation of dynamic circuits that rely on the temporal charge stored on a node capacitance *is* dependent on the junction reverse current; the permitted change of voltage on the nodal capacitance (due to reverse currents), while maintaining reliable sensing by support circuitry, dictates the maximum time duration before sensing (or refreshing) the stored value. Particularly important is the large percentage increase in reverse current with increases in operating temperature, due primarily to the n_i^2 factor in J_{sat}.

As illustrated in Figure 7.29, in an IC application, the circuit nodes are present to a junction depth of x_j into the substrate. The node-to-substrate junction is therefore modeled as two circuit elements, a voltage-dependent capacitor and a reverse-bias current source; both models must include an accurate measure of the total surface area of the junction, which then multiplies the per unit area equations given earlier. Using the shape on an IC layout that defines a device node as a reference, it is necessary to include terms dependent on both the area and perimeter of the shape to determine the junction surface area (Figure 7.44). The reverse current source is small enough in magnitude for most nodal areas and static logic circuits to be neglected for circuit analysis and simulation.

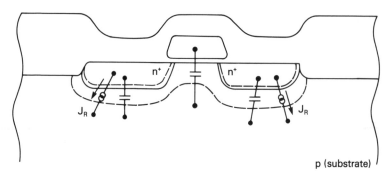

For most static circuits, the reverse current
($J_R * A$) can be neglected in modeling.

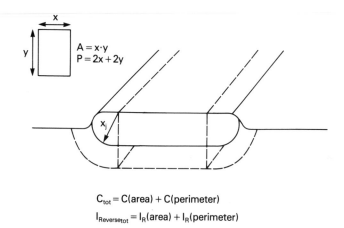

$$C_{tot} = C(area) + C(perimeter)$$

$$I_{Reversetot} = I_R(area) + I_R(perimeter)$$

Figure 7.44. Reverse-bias current source model for each junction node on a chip; usually neglected in all but dynamic or precharged circuit designs that temporally store charge on junction nodes.

Forward Bias

The operating environment of the MOS circuits on a chip is intended by design to preclude the forward biasing of any junction on the chip. However, there is a special circuit type (the substrate voltage generator) whose operation can involve the periodic forward biasing of a junction; in addition, it is possible that capacitive coupling or inductive current effects will produce sufficient noise on an interconnect line contacted to a circuit node to potentially forward bias the node-to-substrate junction. Therefore, although the need to include a forward-bias model of a junction in a circuit analysis is rare, its behavior is included for any pertinent special cases.

The applied voltage conditions for the forward biasing of a junction (and the resulting energy band diagram) are sketched in Figure 7.45. The excess minority

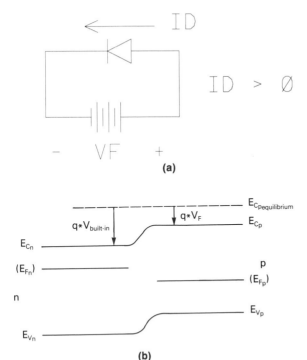

(a)

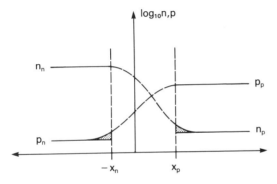

(b)

Figure 7.45. Applied bias polarity for
pn junction forward bias (a) and the
resulting (nonequilibrium) electron
energy band diagram (b).

carrier concentrations on both sides of the depletion region are plotted in Figure
7.46. Note that, as before, forward bias represents a nonequilibrium condition;
as a result, the Fermi energy level is not constant but rather is offset (comparing
the two bulk region values) by $q * V_{applied}$. The analysis of the forward-bias junction typically proceeds by dividing the operating range into four intervals, based
on the magnitude of the applied voltage:

1. Depletion region recombination
2. Minority carrier low injection diffusion current

Figure 7.46. Excess carrier
concentrations for the minority carriers
at the depletion region edges are
depicted, in the case of majority carrier
injection in forward bias.

3. High injection

4. Bulk resistive effects

High injection refers to the condition where the magnitude of the carriers injected across the reduced depletion region and into the opposite bulk becomes comparable to the majority carrier concentration in that region. The assumption used previously that the percentage change in majority carrier concentration in quasi-neutrality is small is truly no longer valid, and a different technique of analysis is required. At even higher currents, the voltage drop across the series resistance in the bulk regions becomes significant, violating another low-injection assumption.

In the range of very low applied forward-bias voltage, there exists two current components: (1) currents due to recombination of injected carriers in the depletion region, and (2) the low-injection minority carrier diffusion currents. Component 1 is the dual of the generation current component considered for reverse bias earlier. Under reverse bias, the depletion region electric field increases, which enhances generation currents and reduces the probability of recombination currents in the depletion region; the opposite conditions are true for forward bias. For low forward bias, component 1 is larger than 2; as the forward-bias voltage increases, component 2 increases at a rate greater than 1, until in the low-injection range it is the dominant component.

To analyze the low-injection forward-bias current of the diode, the same assumptions apply as were used in the reverse-bias analysis, changing recombination assumptions to pertain to generation. As a result, the same Maxwell–Boltzmann carrier statistics for the excess minority carrier concentration at the depletion region–bulk edges are valid:

$$\Delta p(-x_n) = N_A * \exp\left[\frac{qV_{\text{built-in}}}{kT}\right] * \left(\exp\left[\frac{qV_{\text{applied}}}{kT}\right] - 1\right)$$

$$\Delta n(+x_p) = N_D * \exp\left[\frac{qV_{\text{built-in}}}{kT}\right] * \left(\exp\left[\frac{qV_{\text{applied}}}{kT}\right] - 1\right)$$

In this case, however, V_{applied} is positive and the excess minority carrier concentrations increase exponentially. (This is in marked contrast to the saturation of the excess carrier concentrations at the negative of the equilibrium value, resulting in a net minority carrier concentration approaching zero at the depletion region–bulk edge for the reverse-bias case.)

The analysis used for determining the forward-bias excess minority carrier diffusion current is identical to that used for the condition of reverse bias, yielding the same current density expression:

$$J_{\text{diode}} = q * n_i^2 * \left(\frac{D_n}{N_A L_n} + \frac{D_p}{N_D L_p}\right) * \left(\exp\left[\frac{qV_{\text{applied}}}{kT}\right] - 1\right)$$

Due to its applicability in both forward- and reverse-bias conditions, this equation is commonly used as a universal or *ideal* diode expression; however, it should be emphasized that its range of applicability is limited. A plot of J_{diode} versus V_{applied} using this expression is given in Figure 7.47.

The forward-bias junction also exhibits a capacitive behavior for varying node-to-substrate voltage differences. The capacitance has two components, corresponding to the two types of charge that must be supplied to change the depletion region width and alter the exponentially varying magnitude of injected charge in the opposite bulk region. The expression for the depletion layer capacitance is the same as was used for reverse-bias operation; in forward bias, however, the depletion region width is reduced and the differential capacitance increases with increasing bias. (Note also in the abrupt depletion region approximation used that the capacitance expression has a pole at $V_{\text{applied}} = V_{\text{built-in}}$; this can present significant convergence problems to a circuit simulation program if this expression were to be used for the capacitive element at high forward biases. An alternative expression should therefore be substituted at high forward bias.)

The excess minority carrier capacitance can be evaluated by determining the change in total excess minority charge (hole and electron) with a change in applied voltage, maintaining quasi-neutrality assumptions. For example, for the excess hole concentration in the *n*-type region (Figure 7.48),

$$\frac{Q}{\text{unit area}} = q \int_{-\infty}^{-x_n} \Delta p(x) * dx$$

$$= q \int_{-\infty}^{-x_n} \exp\left[\frac{x - (-x_n)}{L_p}\right] * N_A$$

$$* \exp\left[\frac{qV_{\text{built-in}}}{kT}\right] * \left(\exp\left[\frac{qV_{\text{applied}}}{kT}\right] - 1\right) dx$$

$$\frac{C_{\Delta p}}{\text{unit area}} = \frac{dQ}{dV_{\text{applied}}} = \frac{q^2}{kT} * L_p * p_{n0} * \exp\left[\frac{qV_{\text{applied}}}{kT}\right] \qquad \text{(incremental)}$$

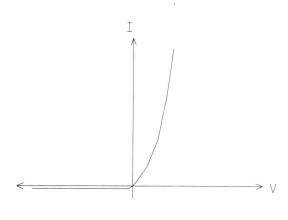

Figure 7.47. Plot of the ideal diode equation.

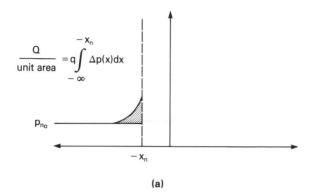

(a)

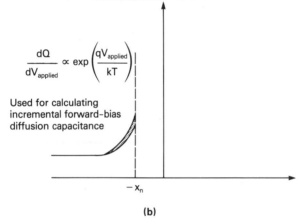

(b)

Figure 7.48. Excess carrier concentration at the edge of the depletion region (a) and the resulting (forward bias) *incremental* junction capacitance (b).

A similar expression can be derived for the excess electron concentration in the *p*-type region. Note from this expression that for reverse-bias conditions ($V_{applied}$ negative) the excess minority carrier capacitance is negligible compared to the depletion region capacitance since the excess minority charge saturates for large reverse bias.

Summary.   The important aspect of this discussion to remember for conventional MOS circuit design and analysis is the origin of the circuit node-to-substrate reverse-bias depletion region capacitance and the voltage dependence of that element's value. The node-to-substrate capacitance will vary as the node switches between logic signal voltages, at its maximum value when the reverse bias is a minimum. The circuit designer and process engineer must work closely to determine the node and substrate impurity concentration values for optimum circuit performance; the extent to which the junction node capacitance enters as a consideration in this judgment involves an assessment of its magnitude in com-

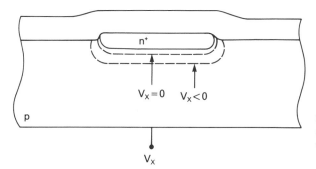

Figure 7.49.   Reduction of junction capacitance with the use of a substrate voltage (similar to Figure 7.38).

parison to other performance-limiting factors, especially the typical interconnection wiring capacitance between circuits. In addition, its value (and the value of the reverse saturation current) plays a significant role in the design of any dynamic circuitry that relies on node charge storage. One technology design feature used to reduce the junction capacitance (and modify other circuit element parameters as well) is a negative substrate voltage for *p*-type substrates (Figure 7.49).[5] The maximum node capacitance value is therefore reduced, since the reverse bias across all junctions has increased. The circuit simulation model for the junction capacitance element should include both area and perimeter geometrical coefficients.

7.6 THE MOS DEVICE (*n*-CHANNEL AND *p*-CHANNEL)

This section on MOS device characteristics and modeling is intended to be illustrative, as opposed to rigorous. It is essentially the same derivation for long, wide devices as first presented by Sah (reference 7.1), and the resulting equations describing the device behavior have become known as the Sah equations. As device geometries have been reduced with continuing technological enhancements, these expressions for MOS device behavior have become less and less accurate. Some of the additional effects observed with minimum geometry devices will be described in Chapter 8.

Two types of MOS devices are discussed in this section, *n*-channel and *p*-channel; a third device type, the *depletion mode n*-channel MOS device, is discussed briefly in Section 7.8. Cross sections of the *n*-channel and *p*-channel devices are given in Figure 7.50 with the corresponding circuit schematic symbol, which will be used throughout this text.[6] The two nodes of the device between

---

[5] As VLSI lithography device dimensions continue to be reduced, the suitability of a large node-to-substrate junction reverse bias is diminished; device *punchthrough* is a breakdown mechanism that is accelerated at higher node-to-substrate reverse bias and at smaller device dimensions (Section 8.8).

[6] These symbols are by no means standard; an appendix to this chapter presents some alternative representations to be found in the literature.

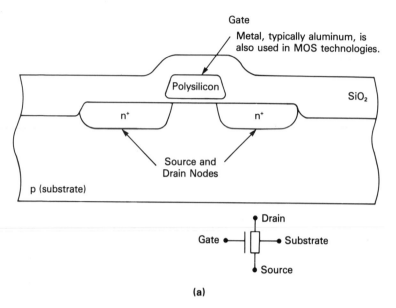

(a)

Figure 7.50. Cross sections (and circuit schematic symbols) for *n*- and *p*-channel MOS devices. (c) The fabrication of a *p*-channel device by the local addition of a *well* of donor impurities to form a substrate in which a *p*-channel device may be fabricated. The *n*-well to *p*-type substrate junction depth is much larger than the *p*-channel device node junction depth into the well. The additional processing steps necessary to fabricate the well permit both *n*- and *p*-channel devices to be incorporated into circuits on the chip. A *p*-well (in an *n*-type substrate) may likewise be used to fabricate *n*-channel devices.

which the device current is to be analytically described are denoted as the *drain* and the *source*. There is no physical difference between the two nodes; the impurities added to the substrate to form all device nodes are added in a single process step and may all be assumed to have an identical junction depth and impurity profile. The device input used to modulate the current between drain and source nodes (for an applied drain-to-source voltage difference) is denoted as the *gate*. (The term gate is also used to refer collectively to all the devices comprising a logic circuit, i.e., a logic *gate*; the particular meaning implied should be determined from the context of its use.)

In the schematic symbol of Figure 7.50, the substrate connection is explicitly included; the effects of the substrate voltage on device behavior will be evident in the upcoming discussion. Since the substrate voltage value is globally the same for all devices of the same type, it is common to drop the substrate connection from the schematic and only indicate its overall value.[7] However, it should be emphasized that the MOS device is indeed a *four-terminal* device.

---

[7] There may be local, transient perturbations for a device from the global value due to substrate currents; however, these transients are typically analyzed only for unique circuit types and transient conditions.

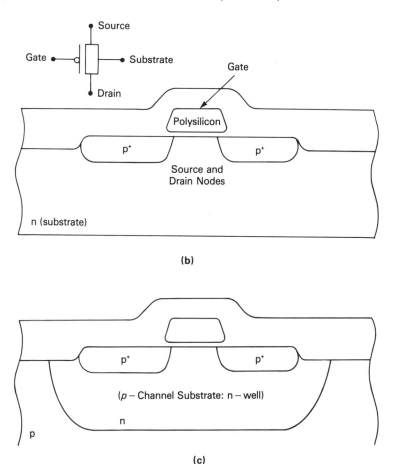

(b)

(c)

Figure 7.50.   (*Continued*)

Since there is no physical distinction between drain and source, the following circuit convention is used for the subsequent analysis:

> The source node of an *n*-channel device is the node at the lower voltage of the two nodes; device current I_{DS} as measured from drain to source is positive. The source node of a *p*-channel device is the node at the *higher* voltage of the two nodes; in this case, the device current I_{DS} as measured from drain to source is negative.

Device current between drain and source consists (in this simplistic analysis, at least) of the electric-field-driven (drift) velocity of majority carriers between drain and source due to an applied voltage difference V_{DS}. That is, the flow of electrons produces the current in an *n*-channel device, while the flow of holes provides the *p*-channel device current. As fabricated, there exists no continuous region of the majority carrier type to support the carrier flow between source and drain; the source and drain nodes are effectively isolated from one another by the back-to-back reverse-biased node-to-substrate junctions. The function of the gate device input is to *induce* a region at the silicon surface of the opposite type as the substrate, a region that thereby provides a continuous channel of majority carriers to enable nonzero current flow between source and drain. The gate input is isolated from the channel surface by a dielectric (silicon dioxide, relative dielectric permittivity = 3.9) produced during fabrication prior to the addition of the gate interconnect material. The thickness, dielectric strength, and resistivity of this dielectric are key parameters of the technology and of the device behavior; the fabrication of this dielectric is a critical processing step. The vertical electric field that emanates from the gate at the gate-to-oxide interface terminates on mobile and fixed charges in the channel region close to the oxide-substrate surface. The channel region is *inverted* in carrier type (enabling device current to flow) by the application of a gate voltage sufficient in magnitude and polarity to repel the mobile substrate majority carriers and attract a sufficient concentration of minority carriers to the region. The gate–dielectric–substrate *sandwich* is commonly analyzed as a parallel-plate capacitor structure. For the purposes of illustration, an *n*-channel MOS device will initially be used to analyze the device behavior; modifications appropriate for a *p*-channel device will then be discussed.

Capacitance of the MOS Structure

As illustrated in Figure 7.51, three regions characterize the MOS input capacitance for the *n*-channel device.

I. Accumulation

If a negative voltage (with respect to the substrate, source, and drain nodes) is applied to the gate, the resulting electric field attracts *holes* (majority carriers in the *p*-type substrate) to the surface. No device channel for current conduction between drain and source is formed. The differential gate input capacitance to the substrate node is the parallel-plate structure formed by the gate and substrate surface.

II. Depletion

As the gate voltage becomes more positive, the resulting electric field repels holes from the surface, leaving a region depleted of mobile carriers (i.e., a depletion region similar to that of the *p-n* junction node capacitance). The field lines are terminated on the fixed ionized acceptor impurity charge, of concentration N_A^-. When the gate voltage is sufficiently positive to form a depletion layer, two capacitors in series are effectively formed, the parallel plate oxide capacitance in series with the depletion region capacitance (a strong function of the applied

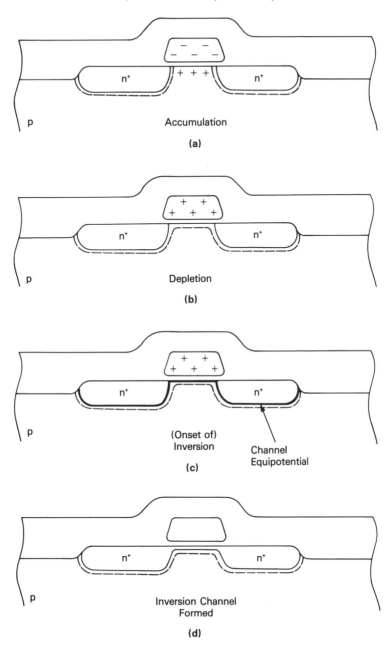

Figure 7.51. Three regions of the MOS device input capacitance: (a) accumulation, (b) depletion, and inversion [onset: (c), channel: (d)]. The sequence of cross sections from (a) to (d) reflects a more positive gate voltage with respect to source and drain nodes of the device.

gate voltage). The resulting gate input capacitance is given by

$$\frac{1}{C_{tot}} = \frac{1}{C_{oxide}} + \frac{1}{C_{depletion\ region}}$$

As the gate voltage is increased, the depletion layer extends further into the substrate. As a result, $C_{depletion\ region}$ and C_{tot} decrease with increasing gate voltage.

III. Inversion

By increasing the gate voltage, the extent of the depletion region grows until it reaches the width of the $n +$ node-to-substrate junction depletion regions (Figure 7.51c). Assuming the drain and source node voltages are equal, it is illustrative at this point to consider the equipotential line comprised of the source node, the channel surface, and the drain node. If the gate voltage is increased further, electrons from the source and drain regions are attracted (under the gate) to the channel region; the surface has *inverted* in majority carrier type, and a path for current flow now exists between source and drain (Figure 7.51d). The gate-to-source (or, equivalently, gate-to-drain) voltage difference at which this occurs is called the device *threshold voltage*. When the inversion layer forms, the capacitance of the MOS gate input once again consists of a parallel-plate capacitor where the distance between the plates is the oxide thickness, t_{ox}. The differential change in the charge is provided to the device channel from the source and drain nodes.

Of specific interest in analyzing the circuit performance of the interconnection of MOS logic circuits is the modeling of the MOS device input capacitance, specifically the voltage dependent capacitive elements present between the terminals of the device. Figure 7.52 illustrates the capacitive elements used to model the input capacitance; for a complete device capacitance model, two additional elements are added, a capacitance between drain–substrate and between source–substrate nodes.

A circuit simulation program analyzes lumped capacitors between circuit nodes; no representation of distributed element values is typically possible. As a

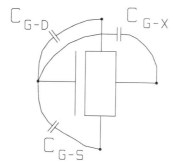

Figure 7.52. The capacitive elements used in device modeling for the MOS transistor input capacitance.

result, it is often necessary to make an assignment of the magnitude of charge (or displacement current) provided from each distinct network node; such is the case for the input capacitance of the MOS device.

When the device is off, the channel region is in accumulation or depletion. The gate input capacitance is divided among the source, drain, and substrate nodes (Figure 7.53). In the figure, C_{GS} and C_{GD} are given by the fixed geometrical *overlap* capacitances between the gate input surface and the extent of the overlap of the *n*-type nodes below (denoted as L_0 in the figure). As the gate voltage changes, these overlap capacitances remain constant. Any change in accumulation or depletion region charge at the channel surface due to a change in gate voltage must be provided by substrate currents; in accumulation or depletion, the gate-to-substrate capacitance, C_{GX}, is equal to the voltage-dependent series capacitance described just previously.

As the gate input voltage to the MOS device increases beyond the threshold voltage (hereafter denoted as V_T), the inversion layer forms in the device channel area and charge is supplied to the channel region (to terminate the vertical gate oxide electric field) from the source and drain nodes. Effectively, the substrate node is isolated from the gate node and $C_{GX}(V)$ has dropped to zero. Changes in the channel charge require a capacitive current from channel to substrate through the channel depletion region capacitance; this current flows through the source-to-substrate and drain-to-substrate capacitances (Figure 7.54). To describe the device capacitance currents in inversion, it is necessary to assign some fraction of the total distributed channel capacitive current to each of the drain and source node lumped capacitances. The value of this fraction is strongly dependent on the voltages V_T, V_{GS}, V_{GD}, and therefore V_{DS}. Two limiting cases will be dis-

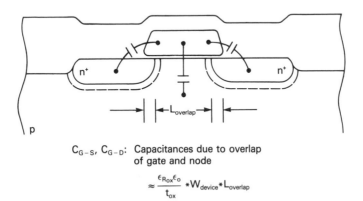

C_{G-S}, C_{G-D}: Capacitances due to overlap
of gate and node

$$\approx \frac{\epsilon_{Rox}\epsilon_0}{t_{ox}} * W_{device} * L_{overlap}$$

C_{G-X}: Gate-to-channel capacitance; change in channel
charge provided by substrate node

Figure 7.53.   Division of the MOS gate input capacitance into lumped elements C_{GS}, C_{GD}, and C_{GX}. For the case where the device is in either accumulation or depletion, C_{GS} and C_{GD} are equal to the overlap capacitances between gate and device nodes.

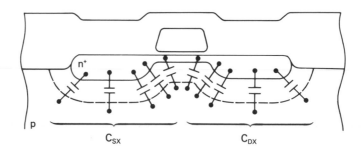

Figure 7.54. Channel-to-substrate capacitance divided into the two lumped elements, C_{SX} and C_{DX}.

cussed, corresponding to what is denoted as the *linear* and *saturated* regions of device operation.

Recall that in defining the device threshold voltage it was assumed that the source and drain nodes were at the same potential. The extent of the depletion region into the substrate surrounding the two nodes was the same, and the threshold equipotential line was horizontal. As the input voltage increases, the device channel forms and the source and drain nodes both provide charge to the channel region; due to the symmetry afforded by the assumption $V_{DS} = 0$, it can be assumed that both nodes provide equal capacitive current to the channel region: $C_{GS} = C_{GD} = \frac{1}{2} * C_{G\text{-channel}} = \frac{1}{2} * (\epsilon_{ox} * \text{area}/t_{ox})$.

As V_{DS} increases from 0 so that device (drift) current now flows in the channel region from drain to source, the variation in channel potential means that the magnitude of the vertical field and therefore the channel charge decreases toward the drain node. In the limiting case of very large V_{DS} (relative to the voltage difference $V_{GS} - V_T$), the vertical field no longer sustains the inverted channel charge at the drain end of the channel; the common terminology is that the channel has *pinched off* at the drain end. The drain end of the channel once again is only a depletion region, rather than an inversion layer.

The first case, where V_{DS} is small compared to $(V_{GS} - V_T)$, is denoted as the *linear* or *triode* range of device operation; past pinch-off, the device is described as *saturated*.[8] First the linear and subsequently the saturated device capacitance variations are described.

When the device is on and operating in the linear region (i.e., when $V_{GD} \cong V_{GS} > V_T$), the total channel capacitance is given by $C_{G\text{-channel}} = (\epsilon_{ox} * \text{area}/t_{ox})$. The channel capacitance is split into the two capacitances C_{GD} and C_{GS}, which are then combined in parallel with the fixed overlap capacitances (Figure 7.55). As V_{DS} increases, the fraction of the channel capacitance assigned to C_{GD} decreases from $\frac{1}{2}$ to zero; in saturation, after the channel has pinched off, only the source node is assumed to provide the change in channel charge corresponding

---

[8] This terminology should not be confused with the two primary regions of *bipolar* device operation, that is, *active* for large V_{CE} and *saturated* for small V_{CE}.

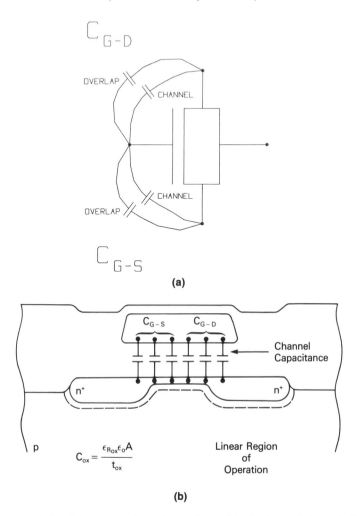

Figure 7.55. Split of the gate-to-channel capacitance into the lumped elements C_{GS} and C_{GD} when the device is on and operating in the linear region (b) and the saturated region (c). *(continues)*

to a change in gate voltage (Figures 7.55b and c). As a result, the C_{GS} fraction of the total channel capacitance must increase with increasing V_{DS}. Section 8.6 discusses how this fraction varies with V_{DS} and V_{GS}; for the necessary level of accuracy for a transient circuit simulation (and for increased speed of execution of the element model), it is common to assign the C_{GS} saturation fraction a constant value, that is, two-thirds of the total channel capacitance ($\epsilon_{ox} * A/t_{ox}$).

Figure 7.56 illustrates a slightly different perspective on the input capacitance elements as a function of the input voltage referenced to the source node (i.e., varying V_{GS} while holding V_{DS} constant). The device is off for V_{GS} less than

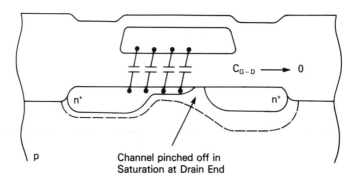

Channel pinched off in
Saturation at Drain End

Saturation Region
of
Operation

(c)

Figure 7.55. (*Continued*)

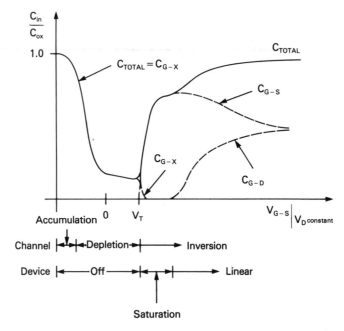

Figure 7.56. Plot of the input capacitance and its association with the lumped model elements: C_{GX}, C_{GD}, and C_{GS} over a range of gate input voltages (with the drain-to-source voltage constant).

V_T, initially enters the saturation region, and finally ends up operating in the linear region as the *overdrive* ($V_{GS} - V_T$) becomes much greater than V_{DS}. Note that the capacitive load presented by the MOS device input to a driving circuit is initially quite small due to the depletion region capacitance in series with C_{ox}. As the input charging current from the driving circuit causes the device to turn on, the capacitive load increases considerably, reducing the input signal transition rate. This results in a slower increase in device current and therefore reduced circuit performance.

Of particular importance is the effect of the substrate voltage on the preceding analysis of device input capacitance. The distinction between the *depleted* and *inverted* channel layer occurs when the width of the channel depletion region is equal to the width of the depletion region between the source node and the substrate. The gate-to-source voltage at this point is called the threshold voltage.[9] For an *n*-channel device situated in a *p*-type substrate, a negative voltage can be applied to the substrate, with two major consequences:

1. This voltage reverse biases all *pn* junctions on the chip. Part of the total substrate current is comprised of the reverse saturation leakage current of the reverse-biased junctions. This small current requirement allows a negative substrate voltage generator to be included on chip, as opposed to providing a separate supply.

2. The reverse-bias voltage also affects all the junction capacitances:

$$C_{\mathrm{junction}} \propto (\mid V_{\mathrm{built\text{-}in}} + V_{\mathrm{reverse}} \mid)^{-1/2}, \qquad V_{\mathrm{reverse}} = V_{\mathrm{node}} - V_{\mathrm{sub}}$$

Increasing the magnitude of the substrate voltage (i.e., making V_{sub} more negative) decreases the diffusion junction capacitance since it widens the extent of the junction depletion region. In addition, this substrate voltage increases the gate voltage required to create a *channel* depletion region of the same width. The capacitances given in Figure 7.56 shift to the right with increasing $\mid V_{\mathrm{sub}} \mid$. In other words, applying a negative substrate voltage increases (makes more positive) the threshold voltage of the *n*-channel device.

The next discussion describes a working expression for the *n*-channel device threshold voltage as a function of the pertinent processing and operating voltage parameters.

---

[9] The definition of the threshold voltage is taken to apply to the condition where the drain-to-source voltage is close to zero; for large drain-to-source voltages, a different interpretation is required. For large V_{DS}, a common experimental measure is to define the threshold voltage as the gate input voltage where a minimum current density (current normalized by the device width-to-length ratio) flows from drain to source. For V_{GS} very close to this threshold value, the MOS device also exhibits small *subthreshold* current, as discussed in Chapter 8.

Threshold Voltage

This discussion will employ the following notation:

N_A: Acceptor impurity concentration in the p-type substrate

t_{ox}: Gate oxide thickness

ϵ_{ox}: Dielectric constant of silicon dioxide, SiO_2

   A: Area of channel surface (device width $*$ device length measured at the wafer level)

Q_{total}: Total charge at the silicon–silicon dioxide interface due to the applied gate voltage

   The method used for developing a working expression for the device threshold is to first determine the effective gate oxide voltage that produces the vertical electric field lines to induce the channel and then determine the magnitude of the channel charge so induced. Referring to Figure 7.57, the two quantities, the effective gate oxide voltage and the channel charge, are related by the expressions

$$\frac{V_{\text{gate oxide}}}{t_{ox}} = \mathscr{E}_{ox}$$

$$\epsilon_{ox}\vec{\mathscr{E}}_{ox}\Big|_{\substack{\text{Si–SiO}_2 \\ \text{interface}}} = \epsilon_{Si}\vec{\mathscr{E}}_{Si}\Big|_{\substack{\text{Si–SiO}_2 \\ \text{interface}}} = \vec{D}$$

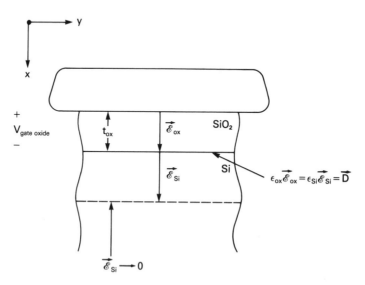

Figure 7.57. Vertical gate oxide field and its relation to the electric field at the silicon dioxide–silicon surface.

$\vec{D}$ is the electric flux density, which is constant at the interface of two dielectric media (in the absence of any charge at that interface).

$$\int_{\text{surface}}^{\text{bulk}} \epsilon_{\text{Si}} \, d\mathscr{E} = \int_{\text{surface}}^{\text{bulk}} \rho \, dx \quad \left(\text{from } \epsilon_{\text{Si}} * \frac{d\mathscr{E}}{dx} = \rho \right)$$

$$| \vec{\mathscr{E}}_{\text{bulk}} | = 0, \qquad \int_{\text{surface}}^{\text{bulk}} \rho \, dx = Q_{\text{total}}$$

Q_{total} is negative, since the channel electrons and the fixed acceptor charge density (N_A^-) are both negative.

$$| \epsilon_{\text{Si}} \mathscr{E}_{\text{Si}} | = | Q_{\text{total}} | = \text{total channel charge (per unit area)}$$

$$\frac{V_{\text{ox}}}{t_{\text{ox}}} = \frac{| Q_{\text{total}} |}{\epsilon_{\text{ox}}} \Rightarrow | Q_{\text{total}} | = \frac{\epsilon_{\text{ox}}}{t_{\text{ox}}} * V_{\text{ox}}$$

$$\frac{\epsilon_{\text{ox}}}{t_{\text{ox}}} = C_{\text{ox}}, \text{ gate oxide capacitance per unit area}$$

Figure 7.58 illustrates how the *effective* gate oxide voltage is established. There are several points to illustrate about the figure:

(1) Between the gate material and the gate oxide and between the oxide and the silicon substrate, contact potentials are introduced. (A contact potential is also commonly referred to as a *work-function* difference or the difference in electron affinities between the two materials.) Some measure of control and/or variation is afforded by the choice of gate material, specifically, metal (aluminum) or polysilicon; for the case of polysilicon, the choice of impurity concentration and type alters the contact potential (as it varies the Fermi energy in the material); for example, an *n*-type polysilicon gate has a contact potential smaller in magnitude than a *p*-type polysilicon gate. The contact potential of the gate (metal or polysilicon) to oxide is positive; the contact potential of the silicon substrate to the oxide is also positive. An equivalent statement is to say that the electron energy at the conduction band edge of the *oxide insulator* is higher than the conduction band edges of either the gate or the substrate, corresponding to a more negative electrical potential in the oxide.

(2) Ionic contaminants incorporated into the gate oxide during fabrication alter the vertical gate field. The most common ionic contaminant is sodium (Na^+), which *adds* to the vertical gate field; considerable contamination in the oxide will provide a gate oxide field sufficient to invert the surface even with $V_{GS} = 0$. Eliminating this source of contamination is a necessity for fabricating reproducible, stable switching devices; particularly significant is the fact that sodium ions in silicon dioxide are quite mobile in the presence of an applied electric field (especially at elevated temperatures), thereby causing the device threshold voltage to drift over time.

(3) The preceding derivation equating the electric flux density $\vec{D}$ at the oxide–

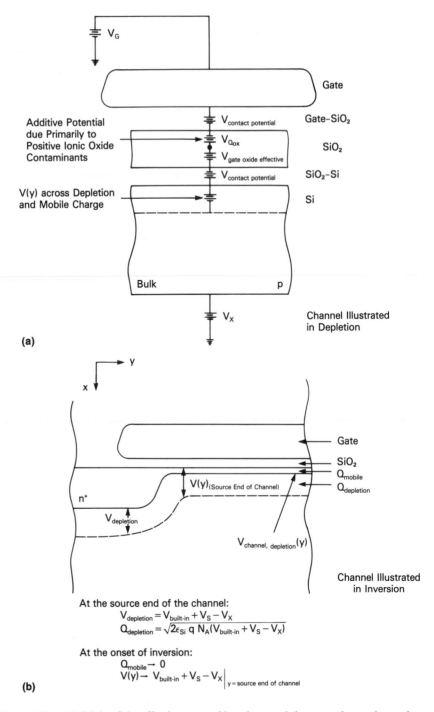

(a)

(b)

At the source end of the channel:
$$V_{depletion} = V_{built-in} + V_S - V_X$$
$$Q_{depletion} = \sqrt{2\epsilon_{Si} \, q \, N_A(V_{built-in} + V_S - V_X)}$$

At the onset of inversion:
$$Q_{mobile} \rightarrow 0$$
$$V(y) \rightarrow V_{built-in} + V_S - V_X \Big|_{y=source\ end\ of\ channel}$$

Figure 7.58. (a) Origin of the effective gate oxide voltage and the terms due to the work function differences (contact potentials) of the gate–oxide and oxide–substrate interfaces. The channel is illustrated in depletion. (b) Relationship between the depletion channel charge density and the voltage across the depletion region in the effective gate voltage loop equation. The channel is depicted in inversion.

310

silicon surface assumed that there was no fixed charge present at this interface. A positive fixed charge density is assigned to the surface, based on experimental measures; it is commonly attributed to ionized silicon complexes present in the transition region between amorphous oxide and silicon crystal. Rather than modifying the derivation given earlier, for the purposes of this discussion, this charge density will be lumped into a single effective term with the ionic contamination as in paragraph (2).

Referring again to Figure 7.58, the effective gate oxide voltage can be derived from the loop equation:

$$V_G - V_{\substack{\text{contact potential,} \\ \text{gate–SiO}_2}} + V_{Q\text{ox}} - V_{\text{ox}} + V_{\substack{\text{contact potential,} \\ \text{SiO}_2\text{–Si}}} - V(y) - V_x = 0$$

$$(N_a^+, \text{ surface states})$$

giving

$$V_{\text{ox}} = t_{\text{ox}} * \mathscr{E}_{\text{ox}} = V_G - V_{\text{gate–SiO}_2} + V_{Q\text{ox}} + V_{\text{SiO}_2\text{–Si}} - V(y) - V_x$$

The voltage difference $V(y)$ in this expression is dropped across the mobile and depletion region charge present at the silicon dioxide–silicon interface:

$$|Q_{\text{total}}| = |Q_{\text{mobile}} + Q_{\text{depletion (bound)}}|$$

The depletion region charge can be determined (referring to Figure 7.58b) using the results of the *pn* junction reverse-bias analysis:

$$Q_{\text{depletion}} = \sqrt{2 * \epsilon_{\text{Si}} * q N_A * V_{\text{depletion}}(y)}$$

where the voltage difference across the reverse-biased depletion region is given by

$$V_{\text{depletion}}(y) = V_{\text{built-in}} + V_{\text{channel, depletion}}(y) - V_X$$

Using the relationship derived for V_{ox} and Q_{total},

$$V_{\text{ox}} = \frac{t_{\text{ox}}}{\epsilon_{\text{ox}}} * |Q_{\text{total}}|$$

the following relationship between the gate voltage and the channel charge can be written:

$$V_G - V_{\substack{\text{gate–SiO}_2, \\ \text{contact potential}}} + V_{Q\text{ox}} + V_{\substack{\text{SiO}_2\text{–Si,} \\ \text{contact potential}}} - V(y) - V_x$$

$$= \frac{t_{\text{ox}}}{\epsilon_{\text{ox}}} * (Q_{\text{mobile}} + \sqrt{2\epsilon_{\text{Si}}qN_A * (V_{\text{built-in}} + V_{\text{channel, depletion}}(y) - V_x)})$$

At the *onset* of inversion (i.e., when the mobile channel charge is just equal to zero), the voltage $V(y)$ across the channel (depletion region) charge is equal to $V_{\text{built-in}} + V_S - V_X$, where $V_S \approx V_D \approx V_{\text{channel, depletion}}(y)$. The oxide voltage–

channel charge relationship becomes

$$\{V(y) = V_{\text{built-in}} + V_S - V_X, \qquad Q_{\text{mobile}} \to 0\}$$

$$V_G - \underset{\text{contact potential}}{V_{\text{gate-SiO}_2,}} + V_{Q\text{ox}} + \underset{\text{contact potential}}{V_{\text{SiO}_2\text{-Si,}}} - V_S - V_{\text{built-in}}$$

$$= \frac{t_{\text{ox}}}{\epsilon_{\text{ox}}} \sqrt{2\epsilon_{\text{Si}} q N_A (V_{\text{built-in}} + V_S - V_X)}$$

The threshold voltage is the gate-to-source voltage at the onset of inversion:

$$V_T = V_{GS}\big|_{Q_{\text{mobile}} \to 0} = \underset{\text{contact potential}}{V_{\text{gate-SiO}_2,}} - V_{Q\text{ox}} - \underset{\text{contact potential}}{V_{\text{SiO}_2\text{-Si,}}}$$

$$+ V_{\text{built-in}} + \frac{t_{\text{ox}}}{\epsilon_{\text{ox}}} \sqrt{2\epsilon_{\text{Si}} q N_A (V_{\text{built-in}} + V_{SX})}$$

Note that the device threshold voltage is subject to process and operating voltage tolerances, due to the statistical distribution of process parameters and the specified range of operating conditions, specifically the following

$V_{Q\text{ox}}$: related to the presence of any oxide contamination

$V_{\text{built-in}}$: temperature dependent, weakly process dependent

t_{ox}, N_A: $\pm 5\%$ to 10%

Power supply tolerances: (e.g., V_X) $\pm 10\%$

If some of the oxide charges are mobile, the threshold voltage is also subject to drift. A working threshold voltage expression is given by

$$V_T = \underbrace{\underset{\text{contact potential}}{V_{\text{gate-SiO}_2,}} - V_{Q\text{ox}} - \underset{\text{contact potential}}{V_{\text{SiO}_2\text{-Si,}}} + V_{\text{built-in}}}_{\substack{C_0 \text{ (only a function} \\ \text{of temperature)}}} + \underbrace{\frac{t_{\text{ox}}}{\epsilon_{\text{ox}}} \sqrt{2q\epsilon_{\text{Si}} N_A} * \sqrt{V_{\text{built-in}} + V_{SX}}}_{C_1}$$

$$V_T = C_0 + C_1 * \sqrt{V_{\text{built-in}} + V_{SX}}$$

In the n-channel device model described to this point, the substrate impurity concentration N_A was used for the channel depletion region charge density. The N_A impurity concentration is present in the device threshold voltage expression in two terms, the built-in voltage and in the multiplier for the dependence on the source-to-substrate voltage depletion region. To minimize the variations in V_T with respect to V_{SX} (and to minimize the pn junction nodal capacitance), it would be preferred that this impurity concentration be small. However, reducing N_A also reduces $V_{\text{built-in}}$ and therefore V_T. The desired process threshold voltage is chosen by the process and circuit design engineering team to provide the optimum performance and noise margin of switching circuits using the devices; specifically, if a particular input voltage is to be used to hold the device off ($V_{in} < V_T$), the device threshold voltage should be large enough to provide sufficient margin for circuit and supply noise present on that input. As a result, reducing N_A substantially may provide too low a threshold voltage. At the expense of additional process complexity, both process goals, low dV_T/dV_{SX} (and $C_{junction}$) and a particular design $V_{T,min}$, can be achieved. In a very shallow layer at the channel surface, the impurity concentration can be increased by the use of an ion implantation processing step (Figure 7.59).

Essentially, this additional impurity concentration leads to a threshold voltage *shift* from the original value, an additive term much like V_{Qox}, only opposite in sign. For an *n*-channel device, the implanted impurity to raise the threshold voltage should be an acceptor, such as boron. It will therefore require a more positive gate voltage to provide the necessary vertical field that terminates on the additional ionized acceptors in the channel depletion region. (It is also common to *lower* the *n*-channel threshold voltage for a subset of the devices on the chip to a *negative* value by implanting donor impurities, such as arsenic or phosphorus. Such a device is no longer used as a normal switching device in a logic circuit, but, suitably connected, is very useful as a circuit load (pullup) device; the *depletion mode* device, with a threshold voltage less than zero, is discussed in Section 7.8.) It is assumed that the implanted layer is sufficiently shallow so that the extent of the source/channel depletion region is well below, into the uniformly lightly doped substrate. The implanted layer then does not affect the term dV_T/dV_{SX}.

Using the device threshold voltage and mobile channel charge equations of this section, the next section develops equations that describe the *n*-channel device current. Before leaving this section, it is appropriate to develop the dual threshold voltage expression for the *p*-channel device (Figure 7.60).

For the *p*-channel device, the *n*-type substrate is connected to the most positive voltage available on chip, again to assure that all node-to-substrate junctions are at reverse (or zero) bias. For the channel depletion region to form, a gate oxide electric field directed *toward the gate* is required; since the polarity

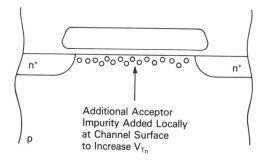

Additional Acceptor
Impurity Added Locally
at Channel Surface
to Increase V_{T_n}

Figure 7.59. Addition of impurities (by ion implantation) at the channel surface to locally modify the device threshold voltage; for a shallow surface layer, the resulting model for the threshold voltage is an adder to the V_T expression that is fixed in value; the body effect term in the threshold voltage expression is unaltered.

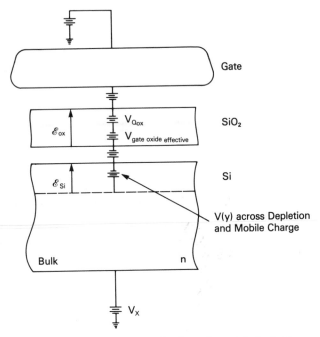

Note: $V_{\text{gate oxide}}$, $V(y)$, $\mathscr{E}_{\text{ox}}$, and $\mathscr{E}_{\text{Si}}$ change polarity for the p-channel device from the n-channel description.

Figure 7.60. Terms in the effective gate oxide loop equation for the *p*-channel device; note the change in polarity of the voltage terms of the effective gate oxide voltage and the voltage drop across the channel depletion region: $V_{\text{gate oxide}}$, $V(y)$, $\vec{\mathscr{E}}_{ox}$, and $\vec{\mathscr{E}}_{si}$ change polarity for the *p*-channel device from the *n*-channel description.

of the depletion region charge has changed, it is necessary that the direction of the electric field at the oxide–silicon interface also be reversed. The inversion layer forms in the channel area, again defined with $V_{\text{source}} = V_{\text{drain}}$, for gate-to-source voltages *more negative* than the threshold voltage, $V_{T, p\text{-channel}}$. The effective gate oxide voltage, V_{ox}, is given by

$$V_{\text{ox}} = t_{\text{ox}} * \mathscr{E}_{\text{ox}} = -(V_G - \underset{\text{contact potential}}{V_{\text{gate–SiO}_2}}$$

$$+ V_{Q\text{ox}} + \underset{\text{contact potential}}{V_{\text{SiO}_2\text{–Si,}}} + V(y) - V_X)$$

Note that the polarity of the V_{ox} and $V(y)$ terms has been modified from Figure 7.58 for Figure 7.60 to account for the change in the direction of the oxide field. Using the result (same as for an *n*-channel device) that

$$\frac{\epsilon_{\text{ox}}}{t_{\text{ox}}} * V_{\text{ox}} = (|\, Q_{\text{mobile}} + Q_{\text{depletion}}\, |)_{\text{channel}}$$

the relationship between the gate oxide voltage and the channel charge is

$$\frac{\epsilon_{ox}}{t_{ox}} * V_{ox} = Q_{mobile} + \sqrt{2\epsilon_{Si}qN_D(V_{built-in} + V_X - V_{channel, depletion}(y))}$$

Again, at the onset of inversion Q_{mobile} goes to zero and the voltage across the channel depletion region goes to $(V_X - V_S + V_{built-in})$. The oxide voltage–channel charge equation at the onset of inversion for the p-channel device is

$$V_{ox}\big|_{Q_{mobile}\to 0} = -(V_G - \underset{\text{contact potential}}{V_{gate-SiO_2}} + V_{Qox} + \underset{\text{contact potential}}{V_{SiO_2-Si,}}$$

$$-V_S + V_{built-in}) = \frac{t_{ox}}{\epsilon_{ox}}\sqrt{2\epsilon_{Si}qN_D(V_{built-in} + V_X - V_S)}$$

giving a threshold voltage expression of:

$$V_{GS} = V_T = \underset{\text{contact potential}}{V_{gate-SiO_2,}} - V_{Qox} - \underset{\text{contact potential}}{V_{SiO_2-Si,}}$$

$$-V_{built-in} - \frac{t_{ox}}{\epsilon_{ox}}\sqrt{2\epsilon_{Si}qN_D(V_{built-in} + V_X - V_S)}$$

	Threshold Voltage Expression
V_{T_n}	$\underset{\text{Gate-SiO}_2}{V_{\text{contact potential,}}} - V_{Qox} - \underset{\text{SiO}_2-Si}{V_{\text{contact potential,}}} + V_{built-in} + \frac{t_{ox}}{\epsilon_{ox}}\sqrt{2q\epsilon_{Si}N_A(V_{built-in} + V_S - V_X)}$
V_{T_p}	$\underset{\text{Gate-SiO}_2}{V_{\text{contact potential,}}} - V_{Qox} - \underset{\text{SiO}_2-Si}{V_{\text{contact potential,}}} - V_{built-in} - \frac{t_{ox}}{\epsilon_{ox}}\sqrt{2\epsilon_{Si}qN_D(V_{built-in} + V_X - V_S)}$

		1	2	Contact Potential Magnitude*
Gate to Oxide		Al	SiO$_2$	$3.2V$
		p-Type Polysilicon	SiO$_2$	$\left(3.25 + \frac{E_{gap}}{2q} + \frac{kT}{q}\ln\frac{N_A}{n_i}\right) V$
		n-Type Polysilicon	SiO$_2$	$\left(3.25 + \frac{E_{gap}}{2q} - \frac{kT}{q}\ln\frac{N_D}{n_i}\right) V$
Oxide to Substrate		SiO$_2$	n-Type Si Bulk (p-channel)	$\left(3.25 + \frac{E_{gap}}{2q} - \frac{kT}{q}\ln\frac{N_D}{n_i}\right) V$
		SiO$_2$	p-Type Si Bulk (n channel)	$\left(3.25 + \frac{E_{gap}}{2q} + \frac{kT}{q}\ln\frac{N_A}{n_i}\right) V$

* Reference: W. Carr and J. Mize, *MOS/LSI Design and Application*. New York: McGraw-Hill, 1972.

Figure 7.61. Table of n-channel and p-channel threshold voltage expressions and a table of contact potentials for a variety of gate and substrate materials. (Adapted from W. Carr and J. Mize, *MOS/LSI Design and Application*. New York: McGraw-Hill, 1972. pp. 33–38.)

This relationship is similar to the relation for the *n*-channel device; note in particular which terms change in sign between the two expressions.

The *p*-channel switching device in a logic circuit on a chip has a threshold voltage, $V_{T,p}$, which is negative; that is, with the source node at the higher of the two channel node voltages, the device is off for gate input voltages equal to or not more than a threshold voltage lower than the source node voltage. The device is on for gate-to-source voltage differences more negative than $V_{T,p}$. The gate input capacitances versus gate-to-source voltage relationships for the *p*-channel device are mirrored about the threshold voltage from the curves plotted for the *n*-channel device in Figure 7.56.

The pertinent device threshold voltage equations for *n*-channel and *p*-channel devices are summarized in Figure 7.61.

7.7 MOS CURRENT EQUATIONS (*n*-CHANNEL AND *p*-CHANNEL)

This section develops an expression for the device current as a function of the node voltage differences, using the results of Section 7.6. As in the previous section on device input capacitance and threshold voltage, this section will first analyze the *n*-channel device and then extend the results to the *p*-channel device.

Recall from the previous section that the charge density in the channel is given by (Figure 7.62)

$$Q_{\text{total}} = Q_{\text{depletion}} + Q_{\text{mobile}} = \frac{\epsilon_{\text{ox}}}{t_{\text{ox}}} * V_{\text{ox}}$$

or

$$Q_{\text{mobile}} \text{ (per unit area)} = \frac{\epsilon_{\text{ox}}}{t_{\text{ox}}} * V_{\text{ox}} - Q_{\text{depletion}}$$

where

$$V_{\text{ox}} = V_G - \underset{\text{contact potential}}{V_{\text{gate-SiO}_2,}} + V_{Q\text{ox}} + \underset{\text{contact potential}}{V_{\text{SiO}_2\text{-Si,}}} - V(y) - V_X$$

At the onset of inversion, with $V_S = V_D$, the voltage $V(y)$ is equal to $(V_S - V_X + V_{\text{built-in}})$ everywhere along the channel (i.e., for all y); the depletion region charge is equal to

$$Q_{\text{depletion}} \big|_{V_{GS} = V_T} = \sqrt{2q\epsilon_{\text{Si}} N_A (V_S - V_X + V_{\text{built-in}})}$$

Consider now a drain-to-source voltage difference V_{DS}; before proceeding with an evaluation of the mobile charge density at a point in the channel with nonzero V_{DS}, a key assumption will be made. The term $V(y)$ in the preceding vertical gate oxide field loop equation is the voltage drop across the entire induced charge region (inversion and depletion channel charge). For gate-to-source voltages larger than the threshold voltage, $V(y)$ and the channel *surface* potential both increase

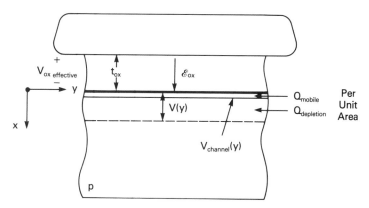

Figure 7.62. Voltage term V(y) in the voltage loop equation (similar to Figure 7.58a); note that for the case where the device is on, the channel charge density includes the mobile carriers.

to support a voltage drop across the resulting inversion charge. The inversion channel charge increases exponentially (Maxwell–Boltzmann approximation) with an increase in surface potential; a change in surface potential of only $(\ln 10) * (kT/q)$ results in an order of magnitude difference in the channel charge density. The approximation to be used neglects this increase in surface potential with increasing V_G. In other words, the potential difference across the *inversion* charge at any point in the channel will be assumed to be zero; the potential at this point will be denoted simply as $V_{\text{channel}}(y)$. The voltage across the induced channel depletion region charge therefore becomes

$$V(y) = V_{\text{channel}}(y) - V_X + V_{\text{built-in}}$$

The mobile charge density therefore varies as

$$\frac{Q_{\text{mobile}}}{\text{unit area}} = \frac{\epsilon_{\text{ox}}}{t_{\text{ox}}} * [V_G - V_{\text{channel}}(y) - \underset{\substack{\text{contact potential}}}{V_{\text{gate–SiO}_2}} + V_{Q\text{ox}} + \underset{\substack{\text{contact potential}}}{V_{\text{SiO}_2\text{–Si,}}}$$

$$- V_{\text{built-in}} - \frac{t_{\text{ox}}}{\epsilon_{\text{ox}}} \sqrt{2\epsilon_{\text{Si}}qN_A * (V_{\text{channel}}(y) - V_X + V_{\text{built-in}})}]$$

An equivalent way of regarding the device in inversion (with the aforementioned approximation) is to consider the derivative

$$\left. \frac{dQ_{\text{mobile}}}{dV_G} \right|_y = \frac{\epsilon_{\text{ox}}}{t_{\text{ox}}} = C_{\text{ox}}$$

the gate oxide capacitance per unit area. The inversion layer in a small slice of the device channel acts as the lower plate of the gate oxide capacitance, a plate isolated from the substrate by a depletion region with reverse bias: $V_{\text{channel}}(y) - V_X + V_{\text{built-in}}$. The variation in $V_{\text{channel}}(y)$ is due strictly to the applied voltage difference from drain-to-source nodes, V_{DS}.

The device current is the drift flow (velocity) of mobile charges, driven by the drain-to-source electric field (Figure 7.63):

$$I = \frac{Q_{mobile}}{A} * W * | \vec{v}_y |$$

where W is the width of the device and $\vec{v}_y$ is the velocity of the electrons along the length of the channel. Using the definition of the effective carrier mobility (Section 7.3), the carrier velocity can be related to the channel electric field (in the y-direction):

$$\vec{v}_y = \mu_{effective} * \vec{\mathscr{E}}_y$$

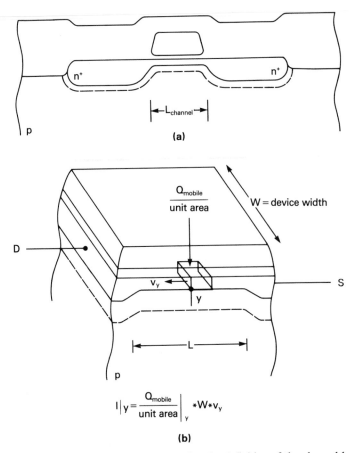

(a)

(b)

Figure 7.63. (a) Device cross section illustrating the definition of the channel length. (b) Definition of the device current in terms of the flow of mobile charge past a point in the channel. (The device width, W, is also illustrated.)

The electric field at a point along the channel is related to the one-dimensional derivative of the channel voltage:

$$| \vec{\mathcal{E}}_y | = \frac{dV_{\text{channel}}(y)}{dy}$$

The device current can now be written as

$$I = \frac{Q_{\text{mobile}}}{A} * W * \mu_{\text{effective}} * \frac{dV_{\text{channel}}(y)}{dy}$$

$$= \frac{\epsilon_{\text{ox}}}{t_{\text{ox}}} [V_G - V_{\text{channel}}(y) - V_{\substack{\text{gate–SiO}_2, \\ \text{contact potential}}} + V_{Q\text{ox}} + V_{\substack{\text{SiO}_2-\text{Si}, \\ \text{contact potential}}} - V_{\text{built-in}}$$

$$- \frac{t_{\text{ox}}}{\epsilon_{\text{ox}}} \sqrt{2q\epsilon_{\text{Si}}N_A(V_{\text{channel}}(y) - V_X + V_{\text{built-in}})}] * W * \mu_{\text{eff}} * \frac{dV_{\text{channel}}}{dy}$$

Multiplying both sides of the equation by dy and integrating from the source end of the channel to the drain end, over the device channel length, L, yields

$$\int_0^L I\, dy = \int_{V_S}^{V_D} \frac{\epsilon_{\text{ox}}}{t_{\text{ox}}} * W * \mu_{\text{eff}} * [V_G - V_{\text{channel}}(y)$$

$$- V_{\substack{\text{gate–SiO}_2, \\ \text{contact potential}}} + V_{Q\text{ox}} + V_{\substack{\text{SiO}_2-\text{Si}, \\ \text{contact potential}}} - V_{\text{built-in}}$$

$$- \frac{t_{\text{ox}}}{\epsilon_{\text{ox}}} \sqrt{2q\epsilon_{\text{Si}}N_A(V_{\text{channel}}(y) - V_X + V_{\text{built-in}})}]\, dV_{\text{channel}}$$

A simplifying assumption is commonly made to develop a working expression for the device current. The term $\sqrt{2q\epsilon_{\text{Si}}N_A} * (V_{\text{channel}}(y) - V_X + V_{\text{built-in}})$ appears in the integrand of the current equation. If the assumption is made that $(V_D - V_S) \ll (V_S - V_X + V_{\text{built-in}})$, or, in other words, $(V_D - V_X + V_{\text{built-in}}) \approx (V_S - V_X + V_{\text{built-in}})$, the square-root term in the integrand of the current equation can be assumed to be essentially constant over the range of integration:

$$\int_0^L I\, dy = \int_{V_S}^{V_D} \frac{\epsilon_{\text{ox}}}{t_{\text{ox}}} * \mu_{\text{eff}} * W * [V_G - V_{\text{channel}}(y) - \{ V_{\substack{\text{gate–SiO}_2, \\ \text{contact potential}}} - V_{Q\text{ox}}$$

$$- V_{\substack{\text{SiO}_2-\text{Si}, \\ \text{contact potential}}} + V_{\text{built-in}} + \frac{t_{\text{ox}}}{\epsilon_{\text{ox}}} \sqrt{2q\epsilon_{\text{Si}}N_A(V_S - V_X + V_{\text{built-in}})}\}]\, dV_{\text{channel}}$$

replace $V_{\text{channel}}(y)$ by V_S

The term in the braces is therefore regarded as a constant in the integration; it is also the same expression as was developed for the device threshold voltage V_T

(thus the reason for this *gradual channel approximation*). The integrand therefore becomes (also treating the effective mobility as a constant)

$$\int_0^L I \, dy = I * L = \frac{\epsilon_{ox}}{t_{ox}} * \mu_{eff} * W * \int_{V_S}^{V_D} (V_G - V_T - V_{channel}(y)) * dV_{channel}$$

$$I = \left(\frac{\epsilon_{ox}}{t_{ox}} \mu_{eff}\right) \left(\frac{W}{L}\right) \left(V_{GS} - V_T - \frac{V_{DS}}{2}\right) (V_{DS}),$$

$$\text{where } V_{GS} = V_G - V_S, \quad V_{DS} = V_D - V_S$$

This expression describes the *linear* (or *triode*) operating region of the *n*-channel MOS device characteristic. It is valid for the range of node voltages where $V_{GS} > V_T$, $(V_{GS} - V_T) > V_{DS}$. Note that the restriction that V_{DS} be small is consistent with the assumption made earlier in replacing the channel voltage $V_{channel}(y)$ inside a square root in the integrand with V_S. A family of linear region device curves is sketched in Figure 7.64.

The important features to note about the linear device current equation are as follows:

1. I_{DS} is independent of the individual values of W or L; it is only a function of the ratio W/L.

2. The device *transconductance* is defined by dI/dV_{GS} and is proportional to $(\epsilon_{ox} * \mu_{eff}/t_{ox})$. The factor W/L is usually dropped from the expression for the *process* transconductance, which is denoted by γ:

$$\gamma = \frac{\epsilon_{ox} * \mu_{eff}}{t_{ox}}$$

3. In silicon, μ_{eff} for electrons (*n*-channel device) is greater than μ_{eff} for holes (*p*-channel device), implying better circuit switching performance. This is

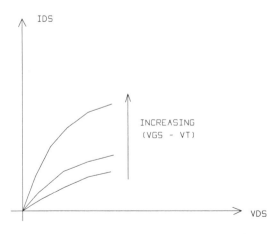

Figure 7.64. Linear region device curves I_{DS} versus V_{DS}, with the device overdrive $(V_{GS} - V_T)$ as a parameter.

one of the driving reasons behind the early evolution from *p*-channel to *n*-channel MOS device large-scale integrated circuits and also behind the growing interest in the semiconductor Ga–As as a substrate, in which the effective carrier mobility is considerably increased.

Rather than making a simplifying assumption, if the voltage integral along the channel from source to drain is evaluated directly, the resulting current expression is a bit more difficult to work with manually, but nevertheless is relatively straightforward to incorporate into a device model for computer-aided circuit simulation:

$$I_{n\text{-channel}} = \frac{\epsilon_{ox}}{t_{ox}} \mu_{eff} \frac{W}{L} \Bigg[(V_G - \underset{\text{contact potential}}{V_{\text{gate}-SiO_2}} + V_{Q_{ox}} + \underset{\text{contact potential}}{V_{SiO_2-Si,}}$$

$$- V_{\text{built-in}})(V_D - V_S) - \tfrac{1}{2}(V_D^2 - V_S^2) - \frac{t_{ox}}{\epsilon_{ox}} \sqrt{2q\epsilon_{Si}N_A}$$

$$* (2/3) \{(V_D - V_X + V_{\text{built-in}})^{3/2} - (V_S - V_X + V_{\text{built-in}})^{3/2}\} \Bigg]$$

Channel Pinch-off

If $V_G - V_D$ is less than V_T, the effective gate oxide voltage and vertical oxide field are no longer sufficient to produce an inversion layer at the drain end of the channel (Figure 7.65). As V_{DS} increases, the channel is described as *pinching off* at the drain node. In Figure 7.65, note the larger depletion region extent into the substrate at the drain node. Since the drain of the device is at a higher potential than the source, the reverse-bias voltage to substrate, $V_R = V_D - V_X$, is greater and therefore the depletion region wider.

In the simplified linear current expression, the pinch-off current can be derived by setting $V_G - V_D = V_T$ or, equivalently

$$V_{DS} = V_{GS} - V_T = V_{\text{pinch-off}}$$

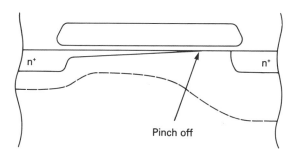

Pinch off

Figure 7.65. Channel cross section when the channel is pinched off by a high drain-to-source voltage.

Substituting the pinch-off voltage V_p into the linear current expression yields the device current at pinch-off:

$$I_{DS} = \frac{\epsilon_{ox}}{t_{ox}} * \mu_{eff} * \left(\frac{W}{L}\right) \left(V_{GS} - V_T - \frac{V_{DS}}{2}\right) (V_{DS}) \,|_{V_{DS} = V_{GS} - V_T}$$

or

$$I_{DS} = \frac{\epsilon_{ox}}{t_{ox}} * \mu_{eff} * \left(\frac{W}{L}\right) * \frac{(V_{GS} - V_T)^2}{2}$$

Note also that the pinch-off voltage can be found by determining the maximum of the linear current expression, that is, by solving for the equation $dI_{DS}/dV_{DS} = 0$.

For drain-to-source voltages larger than V_p, the device current *saturates* (i.e., remains constant). The inverted channel layer remains under the gate from the source node to a point where the channel voltage is such that $V_{channel}(y) - V_S = V_p$; the additional applied voltage $V_{DS} - V_p$ is dropped across the depletion region between drain node and inversion layer. The high electric field in this depletion region sweeps the electrons injected at the end of the inverted channel to the drain. (This depletion region field acts much like the reverse-bias base-to-collector *pn* junction in a bipolar transistor, with the drain node acting as the collector.) The device current is nevertheless dictated by the voltage along the channel and the resulting mobile channel charge; despite increases in V_{DS}, the device current remains approximately constant. This region of device operation is denoted as the *saturation* region (Figure 7.66).

Referring to Figure 7.66, two characteristics of the device current are of particular note:

1. In the ideal saturated current expression, I_{DS} is not a function of V_{DS}; the device behaves as a current source.

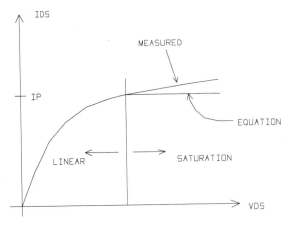

Figure 7.66. Plot of the linear and saturated device current for a particular value of gate-to-source voltage. The boundary between the linear and saturated operating regions is illustrated; note that the measured device current continues to increase with drain-to-source voltage in saturation due to the shortening of the effective inversion channel length.

2. In the pinch-off region, the additional voltage $V_{DS} - V_p$ is dropped across the channel depletion region. With increasing V_{DS}, therefore, the channel depletion region grows wider; the remaining inversion channel length decreases. The I_{DS} current increases slightly since the ratio W/L_{inv} increases. The magnitude of this increase in device current with increasing V_{DS} (shorter pinch-off channel length) is discussed further in Section 8.6 when discussing computer simulation models for second-order device effects.

The pinch-off voltage to use in the current expression resulting from a direct integration with respect to channel voltage may be determined from the expression for the channel mobile charge as a function of $V_{channel}(y)$:

$$\frac{Q_{mobile}}{A} = \frac{\epsilon_{ox}}{t_{ox}} * \left[V_G - V_{channel}(y) - \underset{\text{contact potential}}{V_{gate-SiO_2,}} + V_{Qox} + \underset{\text{contact potential}}{V_{SiO_2-Si,}} \right.$$

$$\left. - V_{built-in} - \frac{t_{ox}}{\epsilon_{ox}} \sqrt{2q\epsilon_{Si}N_A(V_{channel}(y) - V_X + V_{built-in})} \right]$$

At the onset of pinch-off, the mobile charge at the drain end of the channel, $V_{channel}(y) = V_D$, goes to zero:

$$V_G - V_D - \underset{\text{contact potential}}{V_{gate-SiO_2,}} + V_{Qox} + \underset{\text{contact potential}}{V_{SiO_2-Si,}} - V_{built-in}$$

$$- \frac{t_{ox}}{\epsilon_{ox}} \sqrt{2q\epsilon_{Si}N_A(V_D - V_X + V_{built-in})} = 0$$

Solving the quadratic for V_D yields

$$V_{Dsat} = V_G - \underset{\text{contact potential}}{V_{gate-SiO_2,}} + V_{Qox} + \underset{\text{contact potential}}{V_{SiO_2-Si,}} - V_{built-in} + \frac{t_{ox}^2}{\epsilon_{ox}^2}\epsilon_{Si}qN_A$$

$$* \left[1 - \sqrt{1 + \frac{2\epsilon_{ox}^2}{t_{ox}^2\epsilon_{Si}qN_A}(V_G - V_{gate-SiO_2} + V_{Qox} + V_{SiO_2-Si} - V_X)} \right]$$

An important point to make is the indirect influence of the source-to-substrate voltage on the device current in various switching circuit applications (known as the *body effect*). Consider the charging current provided by the load device in the circuit in Figure 7.67. The *n*-channel device is connected such that $V_G = V_D$ or $V_{GS} = V_{DS}$, so the device always operates in the saturated mode: $(V_{GS} - V_T) < V_{DS}$. In the simplified current expression,

$$I_{DS} = \gamma * \left(\frac{W}{L}\right) * \frac{(V_{GS} - V_T)^2}{2}$$

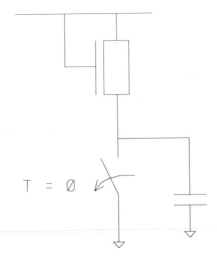

Figure 7.67. Use of a saturated *n*-channel enhancement-mode device as a load element in a logic circuit application.

the source-to-substrate voltage enters as an adder to the device threshold voltage as the output (source) node starts to rise. The decrease in $V_G - V_S = V_G - v_{out}$ and the increase in V_T both combine to cause a decrease in device current, and thus poorer performance, on the rising transition. The influence of the body effect on the device current will be further discussed in Section 9.2, when discussing switching circuit load device options.

The device current relationships for the *p*-channel device are derived in a similar manner to those of the *n*-channel device; however, the choice of the *p*-channel source node to be at a more positive potential results in an expression for the drain-to-source current I_{DS} that is negative. Whereas the *n*-channel device characteristics are plotted in the first quadrant of the I_{DS} versus V_{DS} graph, the *p*-channel device characteristics are commonly plotted in the third quadrant: $V_{DS} < 0, I_{DS} < 0$.

The mobile charge density for an arbitrary point y in the channel is given by

$$\frac{Q_{mobile}}{A} = -\frac{\epsilon_{ox}}{t_{ox}}\left[V_G - V_{channel}(y) - \underset{\substack{\text{contact potential}}}{V_{gate-SiO_2,}} + V_{Q_{ox}} + \underset{\substack{\text{contact potential}}}{V_{SiO_2-Si,}} \right.$$

$$\left. + V_{built-in} + \frac{t_{ox}}{\epsilon_{ox}}\sqrt{2q\epsilon_{Si}N_D(V_X - V_{channel}(y) + V_{built-in})}\right]$$

The device current is again written as

$$I = \left(\frac{Q_{mobile}}{A}\right) * W * \mu_{eff_p} * \frac{dV_{channel}(y)}{dy}$$

which gives

$$
I = -\frac{\epsilon_{ox}}{t_{ox}} * W * \mu_{eff_p} \left[V_G - V_{channel}(y) - \underset{\text{contact potential}}{V_{gate-SiO_2,}} \right.
$$

$$
+ V_{Q_{ox}} + \underset{\text{contact potential}}{V_{SiO_2-Si,}} + V_{built-in}
$$

$$
\left. + \frac{t_{ox}}{\epsilon_{ox}} \sqrt{2q\epsilon_{Si}N_D(V_X - V_{channel}(y) + V_{built-in})} \right] * \frac{dV_{channel}}{dy}
$$

Multiplying both sides of the equation by dy and integrating from the source end over the channel length L gives

$$
I * L = \int_{V_S}^{V_D} -\frac{\epsilon_{ox}}{t_{ox}} * W * \mu_{eff_p} \left[V_G - V_{channel}(y) - \underset{\text{contact potential}}{V_{gate-SiO_2,}} + V_{Q_{ox}} \right.
$$

$$
+ \underset{\text{contact potential}}{V_{SiO_2-Si,}} + V_{built-in}
$$

$$
\left. + \frac{t_{ox}}{\epsilon_{ox}} \sqrt{2q\epsilon_{Si}N_D(V_X - V_{channel}(y) + V_{built-in})} \right] dV_{channel}
$$

Replacing $V_{channel}(y)$ by V_S in the square-root term in the integrand as before permits the definition of the device threshold voltage to be used:

$$
I_{DS} = -\frac{\epsilon_{ox}}{t_{ox}} * \mu_{eff_p} * \frac{W}{L} * \int_{V_S}^{V_D} [V_G - V_T - V_{channel}(y)] \, dV_{channel}
$$

Evaluating this integral yields

$$
I_{DS} = -\frac{\epsilon_{ox}}{t_{ox}} * \mu_{eff_p} * \frac{W}{L} \left(V_{GS} - V_T - \frac{V_{DS}}{2} \right) (V_{DS})
$$

where

$$
V_T < 0, \quad (V_{GS} - V_T) < 0, \quad V_{DS} < 0
$$

The range of validity of this linear current expression is

$$
|V_{GS} - V_T| \geq |V_{DS}|
$$

For drain-to-source voltages larger in absolute value than $|V_{GS} - V_T|$, the channel is pinched off at the drain end and the saturation current is given by

$$
I_{DS} = -\frac{\epsilon_{ox}}{t_{ox}} * \mu_{eff_p} * \frac{W}{L} * \frac{(V_{GS} - V_T)^2}{2}
$$

A plot of the I_{DS} versus V_{GS}, V_{DS} characteristics of the *p*-channel device (in the third quadrant) is given in Figure 7.68. The derivation of the *p*-channel device current expression without the simplifying assumption (which yields the 3/2 power

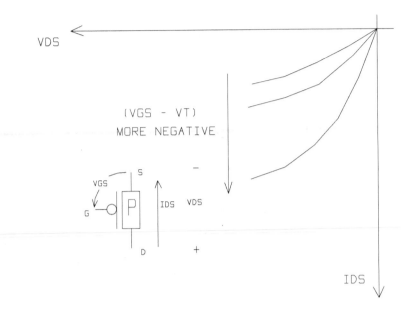

Figure 7.68. Plot of the p-channel device characteristics I_{DS} versus V_{DS} with the overdrive $(V_{GS} - V_T)$ as a parameter.

terms) is requested in Problem 7.2. As with the n-channel device, the influence of the source-to-substrate junction bias on the device operation is an indirect one, through the expression for the p-channel threshold voltage (the *body effect*). A larger substrate-to-source reverse bias makes $V_{T,p}$ more negative, reduces the junction capacitance, and reduces the magnitude of the derivative $dV_{T,p}/dV_{XS}$.

Due to the change in node voltage and current polarities (V_{GS}, V_T, $V_{GS} - V_T$, V_{DS}, I_{DS}) for the p-channel device, as compared to the n-channel device, the p-channel device can conceptually be more difficult to use in design. However, the identical nature of the simplified linear and saturated current equations made possible by the adoption of these polarities is strong motivation for their use. When discussing the p-channel device as the load device in logic circuit applications (Section 9.2), it will be necessary to plot the drain-to-source current load line curve in the first quadrant to match with the $I–V$ curves of the n-channel active devices.

7.8 MODIFICATIONS FOR THE *N*-CHANNEL DEPLETION-MODE DEVICE

In the previous sections it was assumed that the sign of the n-channel device threshold voltage was positive; the difference $V_{GS} - V_T$ is commonly referred to as the *overdrive* on the device. The n-channel device (with its positive threshold) is commonly referred to as an *enhancement-mode* device. (Likewise, the *p*-channel device with a negative $V_{T,p}$ would also be called an enhancement-mode device, with $V_{GS} - V_T$ less than zero as the overdrive.) *N*-channel enhancement-mode devices are commonly used as the switching devices in logic circuit applications. The logic voltage levels of the technology are chosen so as to ensure that, when input to a switching device, the device is either on or off.

As just described in Section 7.7, it is also possible to use the *n*-channel enhancement-mode device as a load device in logic switching circuits if the gate input is connected to an appropriate supply voltage. However, as that analysis indicated, the load charging current decreases sharply as the output node rises, due to the decrease in overdrive; V_{GS} decreases as the source node voltage increases, and V_T increases due to the influence of the substrate body effect. In addition, with the device gate connected to *VDD*, the maximum output voltage reached by the enhancement-mode load device is $VDD - V_T(V_{SX})$, at which point the load device current goes to zero.

In Section 7.6, it was mentioned that the device threshold voltage may be shifted (without changing the dV_T/dV_{SX} dependence) by the addition of a shallow layer of impurity atoms at the oxide–silicon interface. The addition of acceptor impurities to the *p*-type substrate makes the *n*-channel threshold voltage more positive, while the addition of donor impurities makes $V_{T,n}$ less positive. Sufficient additional donor impurities will result in a $V_{T,n}$ that is negative; that is, the inverted channel layer is present even for the hard-wired connection: $V_{GS} = 0$. An *n*-channel device with a negative threshold voltage is referred to as a *depletion-mode* device. For a 5-V logic technology, a typical depletion device threshold may be $V_{T,n(\text{depletion})} = -2\text{V}$, with an applied source-to-substrate reverse bias of 3V.

The depletion mode *n*-channel device is used as the load device in switching circuits if the process technology includes the additional steps required to locally add the necessary donor impurities (for the depletion-mode device areas only). The use of the depletion-mode device as a load device is illustrated in Figure 7.69; note that even though the gate-to-source voltage is identically zero the device channel for current conduction is nevertheless still present. With a negative $V_{T,n}$, the circuit output voltage now reaches the supply voltage value, not just $VDD - V_T$. In addition, the loss in overdrive voltage is confined to the change in $V_{T,n(\text{depletion})}$ due to the body effect as V_{SX} increases. Section 9.2 discusses the advantages and disadvantages of the various load device types in greater detail.

The depletion device threshold voltage is assumed to be realized by a simple subtractor from the enhancement device threshold. The process tolerances that affect the enhancement device threshold (e.g., the gate oxide thickness, t_{ox}) therefore (ideally) have an identical effect on the depletion device threshold. If the desired enhancement device threshold voltage is attained in a particular process by the addition of acceptor impurities, it is preferable that the depletion mode device areas receive the identical added dose during the same process step; the subsequent donor dose is then appropriately increased to achieve the desired

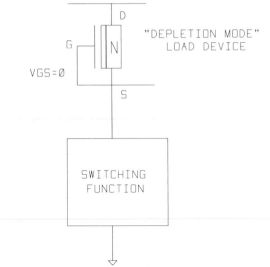

Figure 7.69. Use of an *n*-channel depletion-mode device as the load element in a logic circuit application.

depletion-mode device threshold voltage. Thus, an attempt is made to ensure that the depletion mode and enhancement mode devices *track* as closely as possible; the only independent process step is the local addition of donor impurities in the depletion-mode device areas. If the enhancement-mode device V_T is on the positive end of its process tolerance (say, due to a higher than nominal acceptor impurity dose), the depletion-mode device V_T will be *less* negative. Conversely, if $V_{T,n(enhancement)}$ is slightly less positive than nominal, so $V_{T,n(depletion)}$ will be more negative. As will be analyzed in Section 9.4 on dc circuit design constraints, this is indeed the desired tracking direction.

The same form for the device current equations apply for the *n*-channel depletion-mode devices as for its enhancement-mode counterpart; the depletion mode V_T should be substituted for the enhancement mode V_T. The same criteria apply as to the definition of the linear and saturated (pinch-off) regions of operation for the device. In Section 7.3, the effects of the impurity and mobile carrier concentrations in the crystalline material on the carrier mobility were briefly discussed; it should be expected that the measured *effective* electron mobility, μ_{eff}, differs for enhancement- and depletion-mode devices under the same applied terminal voltage conditions.

7.9 SUMMARY

This chapter has only briefly discussed some of the terminology and basic relationships of solid-state electronics and, in particular, *pn* junction theory. The goal of this discussion was to develop a working knowledge of the origin of depletion

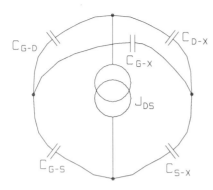

Figure 7.70. Lumped element circuit model for the MOS device for circuit simulation: one (voltage-dependent) current source and five (voltage-dependent) capacitors.

regions in semiconductor materials and how equilibrium and applied electric fields emanate and terminate within a solid-state device. From this discussion, an analysis of the MOS (field effect) transistor was presented, again directed toward the derivation of working device current relationships.

However, there are many inadequacies with this approach. The six-element device model (the current source and five node-to-node capacitances of Figure 7.70 have values that are strongly voltage dependent (for given process and temperature assumptions); any circuit analysis of transient performance requires the use of a circuit simulation software tool to iteratively solve the necessary network equations. In addition, as the cross-sectional and lateral device dimensions have been reduced in the evolution of VLSI processes, significant second-order effects have been measured and require substantial modifications to the working model device current and threshold voltage expressions. Again, an efficient software simulation tool is called for. Chapter 8 discusses circuit simulation and analysis techniques in some detail; included in the discussion are some of the device effects that an accurate analysis requires and only a circuit simulation tool can provide. As a result, the derivations and discussions presented in this chapter should be applied only where the required accuracy of the analysis permits them to suffice.

PROBLEMS

7.1. The derivation of the reverse-bias junction capacitance expression in the chapter used the assumption of an abrupt junction profile between p- and n-type regions, yielding the $(V_{\text{built-in}} + V_R)^{-1/2}$ dependence. Repeat the calculation of the reverse-bias junction capacitance for the assumption of the linearly graded junction, whose profile is illustrated in Figure 7.71.

7.2. Derive the p-channel device current equation without the simplifying assumption that $V_{\text{channel}}(y) - V_X$ can be approximated by $V_S - V_X$, resulting in the 3/2 power terms.

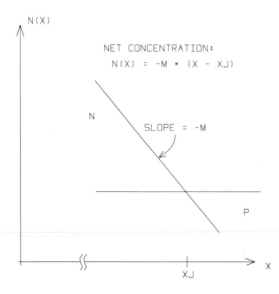

Figure 7.71. Expanded view of a linearly graded junction. The net concentration in the vicinity of the metallurgical junction is given by $N(x) = -m * (x - x_j)$.

7.3. Plot a set of nominal, low, and high $V_{T,n}$ versus V_{SX} curves for the following process assumptions (25°C):

n-type polysilicon gate ($N_D = 2 \times 10^{19}/\text{cm}^3$)

$t_{ox} = 400\text{Å} \pm 5\%$

$N_A = 1.5 \times 10^{15} \pm 20\%/\text{cm}^3$

$Q_f = 2 \times 10^{10}$ charges/cm^2 (fixed charges at the Si–SiO$_2$ interface; assume no mobile sodium in the oxide)

$V_{Qox} = Q_f/(\epsilon_{ox}/t_{ox})$ oxide voltage drop equivalent to interface charge Q_f

$\epsilon_{ox} = 3.9 * \epsilon_0$

$\epsilon_{Si} = 11.8 * \epsilon_0$

$V_{built-in} = 0.65$ V

$n_i = 1.5 \times 10^{10}/\text{cm}^3$

7.4. Repeat Problem 7.3 with the addition of (a) an enhancement threshold shifting implant of $5 \times 10^{11}/\text{cm}^2$ (boron); assume the entire implant dose is active in shifting the threshold. (b) The enhancement device threshold implant in part (a) followed by a depletion device threshold shifting implant of $2.0 \times 10^{12}/\text{cm}^2$ (arsenic).

7.5. For the enhancement device of Problem 7.4a, what is the nominal and worst case minimum logic 1 output voltage achieved when using a saturated enhancement-mode load device (Figure 7.72)? (Be sure to include the body effect dependence of the source-to-substrate voltage on $V_{T,n}$.) For the power supply values in the circuit analysis, use

$$VDD = 5 \text{ V} \pm 10\%$$

$$V_X = -3 \text{ V} \pm 15\%$$

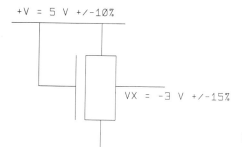

Figure 7.72.   Saturated enhancement-mode load device.

APPENDIX: ALTERNATIVE MOS DEVICE SYMBOLS

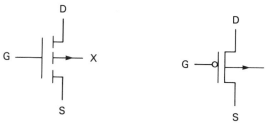

n-channel enhancement mode device

p-channel enhancement mode device

n-channel depletion mode device

REFERENCES

7.1 Sah, C. T., "Characteristics of the Metal-Oxide-Semiconductor Transistors," *IEEE Transactions on Electron Devices*, ED-11:7 (July 1964), 324–345.

Additional References

Chern, J., and others, "A New Method to Determine MOSFET Channel Length," *IEEE Electron Device Letters*, EDL-1:9 (September 1980), 170–173.

Critchlow, D., Dennard, R., and Schuster, S., "Design and Characteristics of n-Channel Insulated-gate Field-effect Transistors," *IBM Journal of Research and Development*, 17:5 (September 1973), 430–442.

Edwards, J. R., and Marr, G., "Depletion-Mode IGFET Made by Deep Ion Implantation," *IEEE Transactions on Electron Devices*, ED-20:3 (March 1973), 283–290.

Hofstein, S. R., and Heiman, F. P., "The Silicon Insulated-Gate Field-Effect Transistor," *Proceedings of the IEEE*, 51:9 (September 1963), 1190–1202.

———, and Warfield, G., "Carrier Mobility and Current Saturation in the MOS Transistor," *IEEE Transactions on Electron Devices*, ED-12:3 (March 1965), 129–138.

Lawrence, H., and Warner, R. M., "Diffused Junction Depletion Layer Calculations," *Bell System Technical Journal*, 39:2 (March 1960), 389–403.

Shichman, H., and Hodges, D., "Modeling and Simulation of Insulated-Gate Field-Effect Transistor Switching Circuits," *IEEE Journal of Solid-State Circuits*, SC-3:3 (September 1968), 285–289.

Wordeman, M. R., "Characterization of Depletion Mode MOSFETs," *IEDM Technical Digest* (December 1979), 26–29.

———, and Dennard, R. H., "Threshold Voltage Characteristics of Depletion-Mode MOSFET's," *IEEE Transactions on Electron Devices*, ED-28:9 (September 1981), 1025–1031.

8

DEVICE STRUCTURES, DEVICE MODELING, AND CIRCUIT SIMULATION

8.1 INTRODUCTION TO DEVICE MODELING FOR CIRCUIT SIMULATION

The circuit designer's fundamental tool for the analysis of the noise margin and transient characteristics of a design and/or physical layout is the circuit simulation program. This program calculates a *network solution* (i.e., node voltages and branch currents) at incremental points of simulation time, in approximation of a continuous (waveform) value; this behavior is a result of a specified time-varying stimulus (or possibly unstable initial conditions). Estimates of the derivatives of network node voltages and branch currents are determined initially at each new simulation time by linear extrapolation of previous time-step values. A *dc solution* is a single calculation of the network values with stable, time-invariant conditions specified.

To keep the circuit simulation program applicable over a wide range of technologies, a limited number of primitive circuit elements are supported; this set of primitives typically includes resistors, capacitors, inductors, dependent voltage and current sources, and independent voltage and current sources. The passive circuit elements may in general be linear or nonlinear; that is, their specified values may be fixed or may be dependent on the voltage across or current through that element. For example, a nonlinear, voltage-dependent capacitance results in a branch current of $i_C = C * (dv/dt) + v * (dC/dt)$. The dependent sources may also incorporate complex functional dependencies on a combination of element voltages and/or currents. The independent sources are commonly used to provide

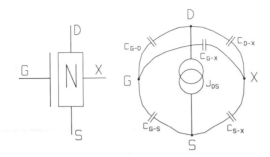

Figure 8.1. Equivalent primitive element model for an *n*-channel FET; all six primitives are nonlinear.

supply voltages and time-varying network stimuli.[1] Circuit simulators are described in more detail in Section 8.7; this section continues with a discussion of the device modeling techniques used to apply a specific technology to a general simulator.

Note that the previous list of primitives accepted by a general simulator did not include any device-level components; rather, it is common to utilize a library of *device models,* where each device type is comprised of the interconnection of primitives that collectively describe the device behavior. The circuit schematic input to the simulator provided by the designer refers to this library of models (and other primitives). An input postprocessing step will unnest these models into an expanded network of primitives for the simulation. An example of a circuit simulation model used to describe an *n*-channel FET is illustrated in Figure 8.1. The five capacitors are all nonlinear (i.e., voltage dependent), while the current source is a function of the voltage differences across all device nodes. More complex device models can certainly be used in an attempt to better correlate calculated response to measured device characteristics.

Figure 8.2 depicts part of an input text file that could be used to describe a circuit design in a specific technology. For example, the library model NFET is referenced; the node interconnection names in the schematic are assigned to the device terminals, and device width and length parameters are provided. Using the library description of Figure 8.1, this model would be unnested and the width and length parameters incorporated into the primitive descriptions. The unnested element names for the NFET will include the $Q1$ qualifier for identification and uniqueness.

A sophisticated simulation input description processor will allow hierarchical models to be defined in the text file network description; that is, the circuit design schematic may be most efficiently described by incorporating *in-line* model definitions. As depicted in Figure 8.3, devices may be grouped into a model definition that is subsequently used in multiple occurrences in defining the overall network. The main model definition is fully unnested down to the device primitive

---

[1] In the general case, *any* element value could include the simulation time variable into its definition. An example where this general capability is particularly useful is given in Section 8.7.

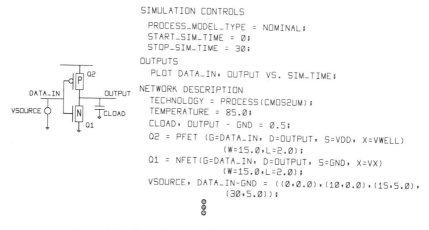

```
                              SIMULATION CONTROLS
                                PROCESS_MODEL_TYPE = NOMINAL;
                                START_SIM_TIME = 0;
                                STOP_SIM_TIME = 30;
                              OUTPUTS
                                PLOT DATA_IN, OUTPUT VS. SIM_TIME;
                              NETWORK DESCRIPTION
                                TECHNOLOGY = PROCESS(CMOS2UM);
                                TEMPERATURE = 85.0;
                                CLOAD, OUTPUT - GND = 0.5;
                                Q2 = PFET (G=DATA_IN, D=OUTPUT, S=VDD, X=VWELL)
                                            (W=15.0,L=2.0);
                                Q1 = NFET(G=DATA_IN, D=OUTPUT, S=GND, X=VX)
                                            (W=15.0,L=2.0);
                                VSOURCE, DATA_IN-GND = ((0,0.0),(10,0.0),(15,5.0),
                                            (30,5.0));
```

Figure 8.2. Text file model description to a circuit simulator.

```
SIMULATION CONTROLS
  PROCESS_MODEL_TYPE = NOMINAL;
  START_SIM_TIME = 0;
  STOP_SIM_TIME = 30;

NETWORK DESCRIPTION
  TECHNOLOGY = PROCESS(CMOS2UM);
  TEMPERATURE = 85.0;

  G1 =  NAND2(A0=S0, A1=S1, OUT=CMD);
  G2 =  NAND2(A0=RDY,A1,=SEL,OUT=PARITY_ENABLE);

/* IN-LINE MODEL DEFINITION NESTED IN MAIN MODEL*/
MODEL NAND2 (A0, A1, OUT);

    Q1 = NFET(G=A0, D=OUT, S=N1, X=VX)(W=15.0,L=2.0);
    Q2 = NFET(G=A1, D=N1, S=GND, X=VX)(W=15.0,L=2.0);
    Q3 = PFET(G=A0, D=OUT, S=VDD, X=VWELL)(W=10,L=2);
    Q4 = PFET(G=A1, D=OUT, S=VDD, X=VWELL)(W=10,L=2);
```

Figure 8.3. Hierarchical input model description to the circuit simulator. MODEL NAND2 is an in-line model definition used in building a higher-level circuit function.

element level for the network simulation. Multiple qualifiers are concatenated to the device primitive names for reference.

To enhance the execution efficiency of the simulation program, the calculation of an element value for all devices of the same type (e.g., J_{DS} for all NFETs) should be provided by a *single* routine (function or procedure). The calculation of this value for different devices in the circuit would call this routine with parameter list values specific to the device. For example, as depicted in Figure 8.4, a routine to calculate the nonlinear junction capacitance for a specific technology

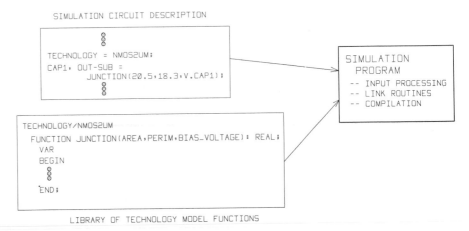

```
        SIMULATION CIRCUIT DESCRIPTION

      ┌─────────────────────────────────────┐
      │              ⊗                       │
      │              ⊗                       │
      │  TECHNOLOGY = NMOS2UM;               │          ┌─────────────────────┐
      │  CAP1, OUT-SUB =                     │          │  SIMULATION         │
      │          JUNCTION(20.5,18.3,V.CAP1); │ ───────▶ │  PROGRAM            │
      │              ⊗                       │          │                     │
      │              ⊗                       │          │  -- INPUT PROCESSING│
      └─────────────────────────────────────┘          │  -- LINK ROUTINES   │
                                                        │  -- COMPILATION     │
 ┌──────────────────────────────────────────┐          │                     │
 │ TECHNOLOGY/NMOS2UM                        │          └─────────────────────┘
 │                                           │            ▲
 │  FUNCTION JUNCTION(AREA,PERIM,BIAS_VOLTAGE): REAL;     │
 │    VAR                                    │           │
 │    BEGIN                                  │───────────┘
 │              ⊗                            │
 │              ⊗                            │
 │    END;                                   │
 └──────────────────────────────────────────┘

        LIBRARY OF TECHNOLOGY MODEL FUNCTIONS
```

Figure 8.4. A set of functions is provided to calculate nonlinear element values. These routines are linked to the simulation program from a technology library.

would typically be provided; that routine would be called with a parameter list specifying the area and perimeter of the junction, as well as the voltage across the element. Ideally, these routines to calculate element values should include the ability to evaluate worst-case, best-case, nominal, and statistically based circuit characteristics. The technology-specific information must therefore also be provided to select the appropriate fabrication process variables to use for these cases. A common *global* variable list contains the process-related (as opposed to the physical design-related) coefficients to use in the routines, for example, t_{ox}, μ_{no}, μ_{po}, x_{jn}, x_{jp}, ΔL_n(design − wafer), ΔW_n(design − wafer), ΔL_p, and ΔW_p. Different initial global variable assignments can provide nominal, best-case, and worst-case analyses. A statistical analysis requires the selection of process variable coefficients from their statistical distributions using suitably weighted random number seeds. (As several subsets of the global process variables are closely correlated, defining this statistical selection process can be tricky. In addition, it is desirable to allow some narrow device-to-device variations (or *mismatch*) around each global process parameter value; refer to the discussion of parameter tracking in Sections 8.9 and 9.4. This further complicates the development of a statistical modeling capability; see Problem 8.1.)

In summary, the database for simulation of a circuit design in a particular technology consists of library models, library routines, and a global process model for initializing the global process variable list with different values for different simulation cases.

Detailed circuit simulation of chip-to-chip performance and the characteristics of the printed circuit card interconnection will often require a mixed technology analysis capability. Problem 8.2 asks for a discussion of the simulator and technology model library enhancements that would be required to support a multiple-technology simulation.

The development of a model library and global variable list for a technology requires close interaction between circuit designer, process engineer, and device characterization engineer. The goals of this development effort are to provide efficient routines for the calculation of process parameters and model element values, and ultimately simulation results that accurately represent the measured hardware. Addressing the first objective may lead to the utilization of table-lookup or multidimensional (Newtonian) interpolation techniques to determine a parameter value from an array of values, as opposed to evaluating an intricate scientific expression. In addition, it may be feasible to bypass certain calculations if the intended simulation application warrants. For example, it may be satisfactory in some applications to neglect the voltage-dependent nature of the device gate input and junction capacitances; the maximum value of these functions over the range of device operation may be assigned initially and used throughout the simulation. This worst-case assumption will likely not be overly pessimistic, if the interconnection wiring capacitance load is the dominant component of the total output load; it will, however, result in a significant execution time reduction. The calculation of device *subthreshold* currents (Section 8.5) may also be omitted if the purpose of the simulation is of a nature such that these currents are not of significant magnitude. Once the J_{DS} current-calculating routine determines that the device is indeed in this region of operation, the calculation may be skipped and the result assigned a value of zero. For applications where a higher level of accuracy is required, this calculation would be performed.

Providing accurate models that reflect measured hardware characteristics is a very difficult task, particularly when the fabrication process is in the early stages of development. The factors complicating this effort include the following:

1. The small amount of hardware available to collect a statistically significant database.
2. Existing device modeling theories may need refinement to predict device behavior at reduced dimensions.
3. The "process" is likely to be evolving to enhance the resulting chip yields and reliability.

Item 3 can have a major impact on the circuit design engineering group. In an effort to synchronize design system availability with the emergence of the fabrication process to *production-level status,* the library of circuits must be designed and characterized early in the development cycle. Modifications to the process that occur late in this cycle may necessitate major circuit revisions based on the analysis of existing designs with updated models. This loop—process changes, model updates, and circuit analysis and redesign—must be suitably constrained to minimize the resource requirements of the design system development.

8.2 JUNCTION CAPACITANCE AND JUNCTION LEAKAGE

Expressions for junction capacitance and junction leakage were given in Section 7.5; the voltage and temperature dependencies of these element values were also indicated. This brief section will discuss the circuit simulation of these structures and introduce the concept of *area* and *perimeter* components.

Section 8.1 discussed measures that may be taken to reduce the simulation execution time. For most applications, it will *not* be necessary to include a junction leakage current source between each device node and the substrate. The magnitude of this current is *not* significant for most circuit delay simulations, even at a worst-case (maximum) temperature. (Junction leakages *should* be included, however, when characterizing the minimum clocking frequency or refresh rate of dynamic circuits.)

The junction capacitance should be incorporated into a circuit simulation model when it constitutes a significant portion of the total capacitive load on a switching node; Section 8.1 discussed the initialization of this element to a maximum, voltage-independent value for efficiency. For some circuit designs, the capacitive charge transfer ratio between source and drain nodes of a switching device is a key parameter (Figure 8.5). In these circuit simulations, an accurate description of the related junction capacitance is crucial.

In reference to Figure 8.5, each junction node has an associated *area* and (nondevice) *perimeter*. Figure 8.6 depicts a wafer cross section of a junction node;

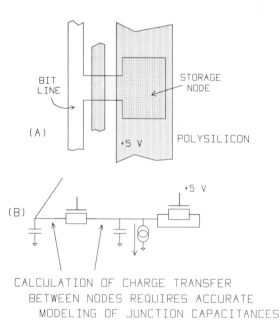

Figure 8.5. Some circuit design applications require extremely accurate modeling of junction capacitances, where the capacitive charge transfer between device nodes is used to sense a (stored charge) logic value. The particular circuit design in the figure is a dynamic RAM memory cell.

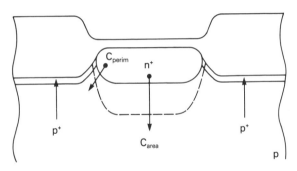

Figure 8.6. The introduction of impurities under the field oxide regions to raise the field oxide (parasitic device) threshold voltage introduces the need for distinct junction node area and perimeter coefficients for capacitance and leakage current.

the technology illustrated incorporates a *recessed* field oxide region between nodes. Also, the field oxide (parasitic) device threshold voltage is raised by the addition of impurities under the field oxide regions. (The field implant impurity type is the same as the background substrate type.) As a result, the perimeter of junction nodes represents a different dielectric/depletion region structure than does the bottom area of the junction. This distinction introduces the need for different area and perimeter models of junction capacitance and leakage current. A device cross section is given in Figure 8.7, again illustrating the node area and perimeter; note that the node perimeter that is equivalent to the device channel edge is *not* included in the perimeter calculations. This region does not abut the field oxide structure, and the associated *gate overlap* capacitance is commonly added to the device capacitance models.

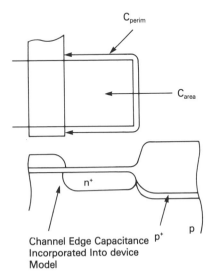

Channel Edge Capacitance
Incorporated Into device
Model

Figure 8.7. The node-to-substrate capacitance for a device channel edge is not assigned to the node, but rather is incorporated into the device model.

8.3 SHEET RESISTIVITY AND CONTACT RESISTANCE

The resistance of the fabrication process interconnection layers and the layer-to-layer contact/via resistance is important to include in simulations of dc circuit noise margins, power supply distribution $I * R$ voltage drops, and circuit performance.

Resistances are calculated from the physical layout using a characteristic parameter of each interconnection layer known as the *sheet resistivity*. The resistance of a segment of material of uniform resistivity ρ is given by $R = \rho * l/A$, where l is the length of the segment and A its cross-sectional area ($A = w * t$). The design-specific features in the resistance of an interconnection segment are its length l and its width w; the material resistivity and its thickness are defined by the fabrication process. The sheet resistivity, ρ_s, of a material is calculated such that the resistance of a segment is given by $R = \rho_s * l/w$. In short, the resistance is equal to the sheet resistivity times the number of *squares* of the segment (Figure 8.8). The units of the sheet resistivity parameter are ohms per square of the interconnection material; the value of the parameter (and its tolerance) for each different layer is determined by the process and applies to all segments on that layer.

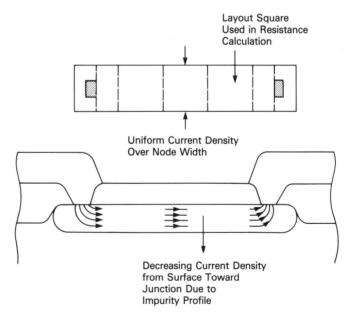

Figure 8.8.   Assuming a uniform current density over the *width* of a node, the resistance of a junction stripe is calculated by multiplying the sheet resistivity of the junction profile times the number of wafer dimension layout *squares* (l/w).

For a material of uniform resistivity ρ and thickness t, the sheet resistivity is given by $\rho_s = \rho/t$. For junction nodes, the calculation of the sheet resistivity is complicated by the nonuniform distribution of impurities over the junction cross section. In this case, the sheet resistivity is calculated from the integral expression

$$\rho_s \cong \int_0^{x_j} q * \mu(n) * n(x)\ dx$$

where $n(x)$ is the net carrier concentration, $\mu(n)$ is the carrier mobility, and x_j is the junction depth into the substrate. For detailed calculations of the resistance of a diffused segment, the following node characteristics should also be included:

1. The *lateral out-diffusion* of impurities during processing, effectively increasing the wafer-level junction width.
2. The pinching of the junction at high node-to-substrate reverse-bias voltages, effectively increasing the resistance.
3. The nonuniform current-distribution effects (i.e., current crowding) in the neighborhood of contacts and bends in the resistor geometry.

Note that the length of the segment used in the resistance calculation is the wafer-level distance between contacts; the border around the contacts is provided to accommodate some misalignment and develop and etch bias tolerance between photolithography levels and effectively does not contribute to the current distribution in the segment. As the carrier mobility μ is a strong function of temperature, so is the sheet resistivity of the interconnection material.

The models for the *contact* resistance between different pairs of interconnection layers vary in functional complexity, depending on the difference in resistivity of the two materials. Between materials of near-identical resistivity (e.g., between first- and second-level metal wires) a uniform current distribution through the contact may be assumed; the contact resistance could then be modeled as inversely proportional to the contact area. For dissimilar materials (e.g., between a metal wire and a junction node) the current distribution in the vicinity of the contact is nonuniform, crowding toward the *leading edge* of the lower-resistivity material (Figure 8.9). A model for the contact resistance in this case is given in Figure 8.10, which depicts a network consisting of differential resistive elements; the junction node is replaced by sheet resistance elements of value $dR = \rho_s * dx/w$, and the contact resistance is modeled with elements $dR' = \rho_c/(w * dx)$. In the previous expressions, ρ_s is the junction sheet resistance (ohms per square), ρ_c is the *specific contact resistance* [ohms $*$ (microns)^2], and w is the leading edge width of the contact area. The specific contact resistance is dependent on the metal, the semiconductor surface impurity concentration, and the subsequent temperature and time processing after metal deposition and patterning. The so-

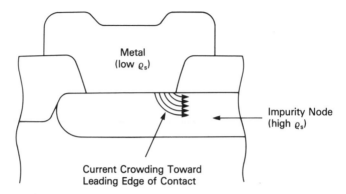

Figure 8.9. For a contact between interconnection layers of different resistivities, a non-uniform current distribution through the contact results, crowding toward the *leading edge* of the contact geometry.

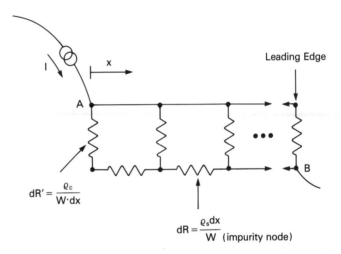

Figure 8.10. Differential transmission line model for calculating contact resistance between metal and junction node; ρ_c is the *specific contact resistance*; units are ohms $*$ (micrometers)^2. (For more details on the contact resistance derivation from this model, consult reference 8.11.)

lution for the resistance between points a and b in Figure 8.10 is

$$R_{a-b} = \frac{\sqrt{\rho_s * \rho_c}}{w} * \coth\left(l * \sqrt{\frac{\rho_s}{\rho_c}}\right), \qquad w \gg l$$

Throughout this discussion, it has been implicitly assumed that the impurity concentration at the semiconductor surface is sufficiently high to form an ohmic (nonrectifying) contact.

8.4 INTERCONNECTION CAPACITANCE

The global wiring between circuits represents a component of the capacitive load on the driving circuit output. An accurate calculation of this element is essential for the signal transient simulations used to characterize the circuit delay. Typical cross-sectional structures used to calculate the capacitive coefficients of interconnection wiring are illustrated in Figure 8.11. These structures include the following:

- **(a)**   An isolated wire embedded in a dielectric above the substrate plane
- **(b)**   An isolated wire sandwiched between reference planes
- **(c), (d)**   The preceding structures above with adjacent parallel wires on the same plane.

In a VLSI design system environment, the most common structure is that of Figures 8.11c and d, a high density of minimum width wires at minimum pitch on each wiring plane. Also included in Figure 8.11 are approximate analytical expressions to calculate line capacitance for some of the given structures; for more detailed two- or three-dimensional geometries (e.g., nonrectangular wiring cross sections), a finite-element analysis tool is commonly used to determine the line-to-reference and line-to-line coupling capacitance values. When modeling an interconnection segment on a particular wiring level, it is common to assume that the other wiring level(s) constitute a reference plane for capacitive currents. Likewise, adjacent wires on the same wiring level are typically assumed to be ''quiet'' for transient simulation of circuit performance.

Since the line widths for interconnections on all wiring planes are fixed as part of the design system chip image development, it is possible to calculate a *per unit length* capacitive multiplier for segments on each level. (The dense signal wiring configurations of Figures 8.11c and d are commonly assumed; end effects are usually neglected.) The output of the automatic wiring program provides the global segments in each net. These routed signal lengths are then used to calculate the net interconnection capacitance and eventually the driving block delay and signal transition times.

When developing a chip wiring image, the choice of the wiring pitch on each level is a key decision. To maximize the number of available wiring channels, the minimum fabrication process pitch is selected; however, the line-to-line coupling capacitance introduces the following considerations:

1. This capacitive component increases as the spacing between adjacent lines is reduced; the resulting per unit length capacitive coefficient will not be at a minimum.

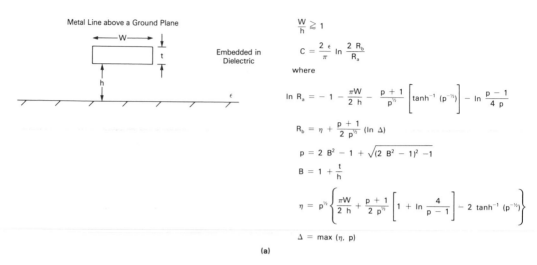

Metal Line above a Ground Plane

Embedded in Dielectric

$$\frac{W}{h} \gtrsim 1$$

$$C = \frac{2 \epsilon}{\pi} \ln \frac{2 R_b}{R_a}$$

where

$$\ln R_a = -1 - \frac{\pi W}{2 h} - \frac{p+1}{p^{\frac{1}{2}}} \left[\tanh^{-1} (p^{-\frac{1}{2}}) \right] - \ln \frac{p-1}{4 p}$$

$$R_b = \eta + \frac{p+1}{2 p^{\frac{1}{2}}} (\ln \Delta)$$

$$p = 2 B^2 - 1 + \sqrt{(2 B^2 - 1)^2 - 1}$$

$$B = 1 + \frac{t}{h}$$

$$\eta = p^{\frac{1}{2}} \left\{ \frac{\pi W}{2 h} + \frac{p+1}{2 p^{\frac{1}{2}}} \left[1 + \ln \frac{4}{p-1} \right] - 2 \tanh^{-1} (p^{-\frac{1}{2}}) \right\}$$

$$\Delta = \max (\eta, p)$$

(a)

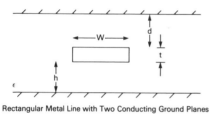

Rectangular Metal Line with Two Conducting Ground Planes

$$\frac{W}{h} \gtrsim 0.5, \frac{d}{h} \gtrsim 0.5$$

$$C = \frac{2 \epsilon}{\pi} \ln \frac{R_B}{R_A}$$

where

$$\ln R_A = -\frac{\pi W}{2 h} - 2 \alpha \tanh^{-1} \sqrt{\frac{p+q}{p(1+q)}} + 2 \gamma \tanh^{-1} (p^{-\frac{1}{2}}) + \ln \frac{4 p}{p-1}$$

$$\ln R_B = \gamma^{-1} \left\{ \frac{\pi W}{2 h} + 2 \alpha \tanh^{-1} \sqrt{\frac{1+q}{p+q}} + \gamma \ln \frac{p-1}{4} - 2 \tanh^{-1} (p^{-\frac{1}{2}}) \right\}$$

$$\alpha = \frac{h+d+t}{h}$$

$$\gamma = \frac{d}{h}$$

$$p = \frac{q^2}{\gamma^2}$$

$$q = \frac{1}{2} \left[\alpha^2 - \gamma^2 - 1 + \sqrt{(\alpha^2 - \gamma^2 - 1)^2 - 4\gamma^2} \right]$$

(b)

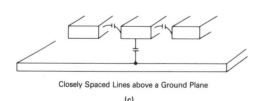

Closely Spaced Lines above a Ground Plane

(c)

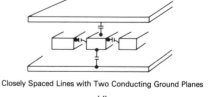

Closely Spaced Lines with Two Conducting Ground Planes

(d)

Figure 8.11. Integrated-circuit geometries used for calculation of per length wiring capacitance coefficients (reference 8.12).

DIELECTRIC	DIELECTRIC CONSTANT
SiO_2 (THERMAL)	3.9
$SiO_2 - P_2O_5$	4.1
Si_3N_4	7.0
POLYIMIDE	3.4
Si	11.7

Figure 8.12. The dielectric properties of some of the insulating materials used in integrated circuit manufacture. (See, for example, H. Wolf, *Silicon Semiconductor Data*. Elmsford, N.Y.: Pergamon Press, 1969.)

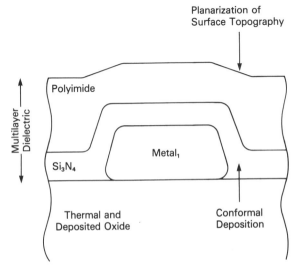

Figure 8.13. A multiple-layer dielectric may be used by the fabrication process to utilize the optimum features of different materials; for example, nitride provides a good contamination barrier, while polyimide can planarize the surface topography to some degree.

2. If a quiet line is bordered by two adjacent simultaneously switching nets, the coupled signal transient onto the quiet line may result in a logic noise margin error. It may be necessary to add a constraint to the wiring algorithm restricting the length of adjacent parallel segments.

The dielectrics used in IC fabrication vary in composition, means and conformality of deposition, diffusivity of contaminants, and, in particular, dielectric constant. Some common dielectrics are listed in Figure 8.12. Multidielectric structures are often used between interconnection layers to utilize the optimum features of individual materials. For example, as illustrated in Figure 8.13, one dielectric may be selected for its low permeability to contaminants and another for its ability to provide some measure of planarization over existing topography. Finite-element analysis tools are indispensable with such intricate geometries in calculating accurate capacitance measures.

8.5 SECOND-ORDER MOS DEVICE EFFECTS

As device dimensions have been reduced in evolving VLSI technologies, the model for MOS device characteristics developed in Chapter 7 has proved insufficient for accurate correlation to measured data and therefore accurate circuit simulation. A number of second-order effects have been isolated, properties that require enhancements and/or modifications to the existing device model. This section discusses some of these anomalous characteristics and how they manifest themselves at reduced vertical and photolithographic dimensions.

Short-channel Effect

As the device length is reduced, it has been discovered that the magnitude of the gate electric field required to invert the semiconductor surface and form a conducting channel is also reduced. The device threshold voltage is correspondingly altered. For n-channel enhancement-mode devices, the threshold voltage is less positive, whereas for n-channel depletion-mode devices, the threshold voltage is more negative. For p-channel (enhancement) devices, the threshold voltage is *less* negative. As a result, a parameter $\Delta V_T(L)$ is added to the expression for the threshold voltage.

A model has been proposed to predict the variation of the device threshold voltage as a function of the channel length. The proposed model is based on calculating modifications to the total (depletion) charge in the channel region. The modifications are due to a geometrical approximation (reduction) to the depletion region charge in the channel that terminates the electric field lines emanating from the gate charge. This trapezoidal approximation is sketched in Figure 8.14. A fraction of the depletion charge in the channel has been assigned to the source/drain junction depletion regions. As a result, to induce the necessary depletion region charge to invert the surface, less gate charge is required. Since the extent of the junction depletion regions under the channel dictates the amount of charge reduction, the short channel effect by itself is a function of the device length, the

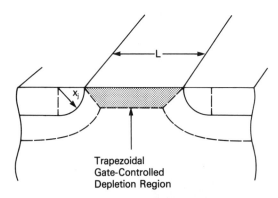

Trapezoidal
Gate-Controlled
Depletion Region

Figure 8.14. Trapezoidal approximation used in calculating the magnitude of the short-channel effect.

source/drain junction depth, the source-to-substrate voltage, and the *drain-to-source* voltage. It should be noted that modeling the device threshold voltage as a function of the device's drain-to-source voltage requires an *iterative* solution for the analysis or simulation of the device's operation in a circuit design. The iterative nature of the algorithm used for solving the network equation matrix in a computer simulation program is well suited to such modeling intricacies.

As the device length is reduced, the magnitude of the short channel effect $|\Delta V_T(L)|$ increases, as an increasing fraction of the depletion region channel charge is attributable to the junction depletion regions; similarly, the magnitude of the short channel effect increases with increasing source-to-substrate and drain-to-substrate reverse bias.

Using *minimum length* enhancement devices for the inputs of a logic data circuit reduces the layout area and input capacitance, but also requires that the off voltage driving this input be likewise reduced to match the reduction in input threshold voltage. Proper (worst-case) dc circuit design techniques, as discussed in Chapter 9, should be observed to maintain sufficient noise margin between circuits.

Narrow-width Effect

In contrast to the short channel effect, where the correction term to the device threshold voltage corresponds to a reduction in gate-controlled charge, narrow-width devices demonstrate that an *increase* in gate field is necessary to invert the surface and form a channel. For narrow-width n-channel enhancement-mode devices, the threshold voltage becomes more positive, whereas for n-channel depletion-mode devices, the threshold voltage is less negative; for p-channel (enhancement) devices, $V_{T,p}$ becomes more negative.

Referring to Figure 8.15, the extent of the depletion region into the substrate under the field (thick) oxide is less than that under the gate (thin) oxide. As the magnitude of the gate charge increases, the depletion region extends further into the substrate, much more rapidly under the thin oxide. The depletion region profile will not be rectangular at the field oxide–gate oxide edge, but will be roughly

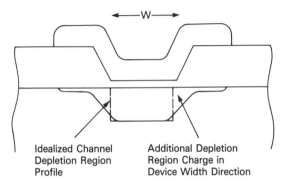

Idealized Channel
Depletion Region
Profile

Additional Depletion
Region Charge in
Device Width Direction

Figure 8.15. Additional channel depletion region charge that must be induced by the gate electric field for narrow width devices.

parabolic in the vicinity of this boundary. (The gradient of the depletion region profile must be continuous.) As a result, the shallow junction depletion region under the field oxide on both sides of the gate oxide tends to inhibit inversion near this interface; an increased gate field is necessary to invert the surface across the entire channel width. With increasing source-to-substrate reverse bias, the depletion region depth under the gate area must be greater to result in an inversion layer; in addition, the deviation in the depletion region profile from ideal at the thick–thin oxide interface will also be greater. Therefore, the magnitude of the narrow-width effect, $\Delta V_T(W)$, increases with increasing source-to-substrate reverse bias.

An alternative interpretation of the narrow-width effect introduces the model definition of an effective, electrical channel width, or W_E ($W_E < W_{\text{wafer}}$) in the device current equations.

In an attempt to write an analytic expression for the device threshold voltage, two modeling approaches have been used:

1. Treat the two effects as independent of each other:

$$V_T = V_{T(\text{large geometry})} + \Delta V_T(W) + \Delta V_T(L)$$

   (where V_T is positive or negative, depending upon the device type).

2. Include a coupling term for minimum-sized devices:

$$V_T = V_{T(\text{large geometry})} + \Delta V_T(W) + \Delta V_T(L) + \Delta V_T(W, L)$$

Some of the interpretation of the short-channel and narrow-width effects depends on the criteria used for measuring the threshold voltage (refer to the discussion in Section 8.9 on electrical device parameter measurement techniques). The threshold voltage definition also applies directly to the next effect to be discussed.

Device Subthreshold Current

In analyzing dc circuit behavior and logic interfacing noise margins, an important concern is the subthreshold current; an MOS device does *not* conduct "zero" current for gate voltages below threshold. Rather, the drain-to-source current I_{DS} varies exponentially with gate voltage in the vicinity of the threshold voltage; the relationship $\ln(I_{DS})$ versus V_{GS} is linear in the subthreshold region (Figure 8.16). The significance of the subthreshold current on dc circuit analysis is illustrated in Figure 8.17. The logic 0 input voltage at the *n*-channel enhancement device gate must yield a logic 1 output voltage. This input voltage is *not* $V_{T,n(\text{enhancement})}$; it is equal to the voltage required to reduce the device subthreshold current to a value such that the voltage drop across the load device is less than $VDD - V_1$. Another very important consequence of device subthreshold currents pertains to dynamic circuit design, where the leakage currents from a capacitive node tem-

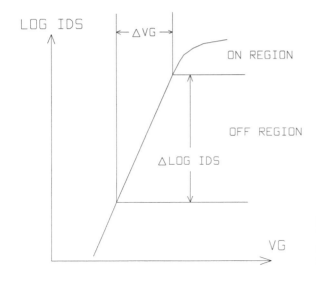

Figure 8.16. Plot of the ln I_{DS} versus V_{GS} relationship for the device subthreshold operating region (adapted from reference 8.2).

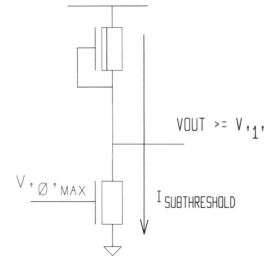

Figure 8.17. The device subthreshold current is included in dc circuit analysis simulations to determine the maximum input 0 voltage that still maintains a valid output 1 level.

porally storing a specific initial charge should be minimized. As illustrated in Figure 8.18, the device subthreshold current of the off charge transfer device adds to the junction leakage current, resulting in an increased rate of charge loss. The device threshold voltage has truly become an experimental measurement subject to different interpretations; the expression $V_T = V_{GS}$ at zero drain-to-source current is not well defined.

For relatively low values of drain-to-source voltage V_{DS}, the current in the subthreshold region is already *saturated,* independent of further increases in V_{DS}.

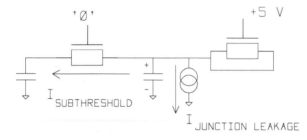

Figure 8.18. Modeling the device subthreshold current is necessary in analyzing the losses in the stored charge on isolated nodes.

A useful parameter to characterize the subthreshold effect is the reciprocal of the slope of the semilogarithmic plot of the device current versus gate voltage:

$$\alpha = \frac{dV_{GS}}{d(\log_{10} I_{DS})}, \quad \text{in mV/decade}$$

In other words, α is the change in gate voltage required to alter the drain-to-source current by a factor of 10.

Although a derviation of an analytic expression for the subthreshold parameter α is beyond the scope of this discussion, some of the functional dependencies can be understood:[2]

$$\frac{dV_{GS}}{d(\log_{10} I_{DS})} \quad \text{is proportional to} \quad \frac{1}{C_{ox}} = \frac{t_{ox}}{\epsilon_{ox}}$$

$$\frac{dV_{GS}}{d(\log_{10} I_{DS})} \quad \text{is proportional to} \quad C_{depletion} = \frac{\epsilon_{Si}}{t_{\,channel\ depletion\ region}}$$

As the source-to-substrate reverse bias is increased, the channel depletion thickness is increased; α is reduced, implying a steeper slope of the device turn-on characteristics. Lighter background impurity concentrations also result in an increased channel depletion region thickness, and again a steeper turn-on slope. Given the dc circuit design constraints described earlier, a sharper turn-on characteristic (smaller α) is desirable. The theoretical minimum value for α is

$$\alpha_{min} = \left(\frac{kT}{q}\right) * (\ln 10)$$

Figure 8.19 plots the drain-to-source current versus gate-to-source voltage, with the substrate voltage as a parameter (*n*-channel device, $t_{ox} = 350$ Å, $V_{DS} = 4$ V); an enclosed annular FET structure was used to eliminate any surface (leakage) currents between source and drain under the field oxide, a concern when measuring extremely low current levels. To maximize the steepness of the characteristic (minimize α) for logic and dynamic circuit applications, a substrate bias

---

[2] For a more detailed discussion, consult reference 8.1.

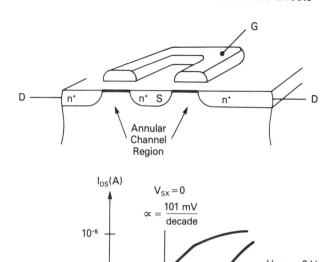

Figure 8.19.   Subthreshold current behavior of an annular FET device (adapted from reference 8.2).

is preferable. The background concentration should be low, and any threshold-shifting surface implant dosages should be shallow.

For logic circuit applications, the subthreshold device current is a characteristic that can be detrimental to circuit behavior. The exponential transconductance of the device characteristics in this region may have application in the amplification of small-dynamic-range analog signals.

Field-dependent Carrier Mobility

In the discussion on MOS device characteristics in Chapter 7, the carrier channel mobility was assumed to be a constant over the range of terminal voltages and device operation. To indicate this assumption, the mobility was written as μ_{eff} and was designated the *effective channel* mobility to distinguish it from the *bulk carrier* mobility. In this section, a heuristic model is proposed for the channel mobility; μ is no longer regarded as a constant, but will be regarded as a function of the channel electric field. In order to apply this modification to the device

V-I characteristics, the expression for $\mu = \mu(\mathscr{E}_{channel})$ will be translated into terms related to the applied terminal voltages. This discussion closely follows that of reference 8.3.

With the reduction in device physical dimensions provided by VLSI technologies, the magnitudes of the lateral (transverse) and vertical electric fields are increasing. At these higher magnitudes a dependence of the channel carrier mobility on the field components is measured. For high electric fields, it is experimentally observed that the *bulk* carrier mobility also does not remain constant but decreases with increasing electric field. Several functional forms have been proposed to model these *velocity saturation* effects, as sketched in the *v* versus $\mathscr{E}$ relationship of Figure 8.20. The two forms that have commonly appeared in the technical literature are as follows:

1. The hyperbolic relation: $v(\mathscr{E}) = \mu_{eff} * [\mathscr{E}/(1 + \mathscr{E}/\mathscr{E}_c)]$, where $\mathscr{E}_c$ is the critical field.
2. The square-root relation: $v(\mathscr{E}) = \mu_{eff} * \mathscr{E}/\sqrt{1 + (\mathscr{E}/\mathscr{E}_c)^2}$, where $\mathscr{E}_c$ has the same meaning as in form 1.

For the purposes of the following discussion, the hyperbolic relation will be used, only written in a slightly different form:

$$\frac{1}{\mu} = \frac{1 + (\mathscr{E}/\mathscr{E}_c)}{\mu_{eff}} = \frac{1}{\mu_{eff}} + \frac{\vec{\alpha} \cdot \vec{\mathscr{E}}}{\mu_{eff}}$$

where $1/\mu_{eff} = 1/\mu_{impurity} + 1/\mu_{phonon} + 1/\mu_{surface}$. Several points to be made about this expression are the following:

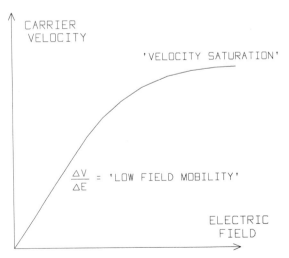

Figure 8.20. *Velocity saturation* of carriers at high electric fields in a bulk material.

1. The low-field mobility is indeed μ_{eff}, as before.
2. The expression for the total mobility resembles those given earlier for the independent scattering mechanisms; however, it differs in that the denominator of the second term in the preceding equation is μ_{eff}, the effective channel mobility. The high-field reduction in carrier mobility is assumed to affect *all* the scattering mechanisms.
3. The notion of a critical electric field strength, $\mathscr{E}_c$, has been replaced by a vector *dot product* in order to individually treat each field component and relate it to the applied terminal voltages:

$$\vec{\alpha} = \beta\hat{x} + \gamma\hat{y} + \delta\hat{z}$$
$$\vec{\mathscr{E}} = \mathscr{E}_x\hat{x} + \mathscr{E}_y\hat{y} + \mathscr{E}_z\hat{z}$$
$$\vec{\alpha} \cdot \vec{\mathscr{E}} = \beta\mathscr{E}_x + \gamma\mathscr{E}_y + \delta\mathscr{E}_z$$

where β, γ, and δ are proportionality constants.

The components of the electric field are assumed to have the following functional dependencies:

1. Vertical direction, $\mathscr{E}_z$:

$$\mathscr{E}_z \propto \frac{V_{GS} - V_T - (V_{DS}/2)}{t_{ox}} \qquad \text{(linear)}$$

This is a linear region approximation for the average vertical field between gate and channel; thus, the voltage $V_{DS}/2$ is subtracted. For the device operating in the saturation region, to approximate the average vertical field on the remaining surface inversion layer, replace $V_{DS}/2$ in this expression with $V_{\text{pinch-off}}/2$:

$$\mathscr{E}_z \propto \frac{V_{GS} - V_T - (V_{\text{pinch-off}}/2)}{t_{ox}} \qquad \text{(saturation)}$$

2. Device length direction, $\mathscr{E}_y$:

$$\mathscr{E}_y \propto \frac{V_{DS}}{L} \qquad \text{(linear)}$$

$$\mathscr{E}_y \propto \frac{V_{\text{pinch-off}}}{L - l_{\text{dep}}} \qquad \text{(saturation)}$$

where l_{dep} is the extent of the pinch-off depletion region between drain and channel in saturation.

3. Device width direction, $\mathscr{E}_x$: In the width direction, the direction perpendicular to current flow, the field $\mathscr{E}_x$ is small; however, recalling the discussion of the narrow-width effect on the device threshold voltage, an x-directed field, $\mathscr{E}_x = -\partial V/\partial x$, is present at the edges of the channel. For a wide device,

the gradient of the channel potential in the vicinity of the thin–thick oxide interface influences only a small fraction of the total channel width. At reduced dimensions, the gradient of the potential influences an increasing fraction of the channel width and hence the carrier mobility. The x-directed field strength is assumed to be inversely proportional to the device width: $\mathscr{E}_x \propto (W)^{-1}$.

The resultant expression for the carrier mobility for both low- and high-field conditions is

$$\frac{1}{\mu} = \frac{1}{\mu_{\text{eff}}} + \frac{1}{\mu_{\text{eff}}} \left\{ \frac{\beta}{W} + \gamma * \frac{V_{DS}}{L} + \theta * \left(V_{GS} - V_T - \frac{V_{DS}}{2} \right) \right\}$$

where $\theta = \delta/t_{\text{ox}}$ and V_{DS} is replaced by $V_{\text{pinch-off}}$ and L goes to $L - l_{\text{dep}}$ in saturation.

The proportionality constants β, γ, and θ in the expression for μ can be determined for a particular technology from experimental measures of device current:[3]

1. Measure the current and calculate the mobility for a long, wide device, varying $V_{GS} - V_T$:

$$\mu = \frac{I_{DS}}{[(\epsilon_{\text{ox}}/t_{\text{ox}})](W/L)[V_{GS} - V_T - (V_{DS}/2)]V_{DS}} \qquad \text{(linear)}$$

$$\frac{1}{\mu} \cong \frac{1}{\mu_{\text{eff}}} + \frac{1}{\mu_{\text{eff}}} * \theta * \left(V_{GS} - V_T - \frac{V_{DS}}{2} \right), \qquad \frac{\partial(\mu_{\text{eff}}/\mu)}{\partial(V_{GS} - V_T)} = \theta$$

Figure 8.21 illustrates an experimental measure of μ_{eff}/μ versus $V_{GS} - V_T$; the vertical axis of the graph is normalized to $\mu_{\text{eff}}/\mu = 1$ as the curve is extrapolated to $V_{GS} - V_T = 0$ (providing the value of the low-field mobility μ_{eff}). The slope of the curve is the proportionality constant θ.

2. To determine the effects of the transverse field along the device length from drain to source on the carrier mobility, measure the device current for several wide devices with varying channel lengths. Calculate μ as before, neglecting the vertical field term for small gate voltage overdrive:

$$\frac{1}{\mu} = \frac{1}{\mu_{\text{eff}}} + \frac{1}{\mu_{\text{eff}}} \left\{ \gamma * \frac{V_{DS}}{L} + \overbrace{\theta * \left(V_{GS} - V_T - \frac{V_{DS}}{2} \right)}^{\text{small}} \right\}$$

$$\frac{\partial(\mu_{\text{eff}}/\mu)}{\partial(1/L)} = \gamma * V_{DS}$$

---

[3] The discussion and experimental data illustrating the field-component mobility model assume that the wafer-level device width and length are known; refer to Section 8.9 for a description of the techniques used to determine these dimensions.

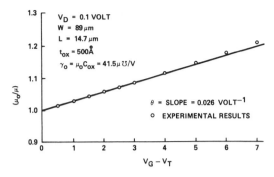

Figure 8.21. The coefficient θ is determined from a measure of μ_{eff}/μ versus $V_{GS} - V_T$ for a long, wide device operating in the linear region. (From P. Wang, "Device Characteristics of Short-Channel and Narrow-Width MOSFET's." *IEEE Transactions on Electron Devices*, Vol. ED-25, No. 7 (July 1978), p. 783. Copyright © 1978 by IEEE. Reprinted with permission)

Figure 8.22 depicts an experimental measure of μ_{eff}/μ versus $1/L$ and the transverse proportionality constant γ. In the experiment, the gate-to-source voltage must be adjusted for different device lengths to keep $V_{GS} - V_T$ constant, eliminating any variation due to the short-channel effect.

3. To determine the effects of the field in the width direction, current measurements and mobility calculations are repeated, in this case for long devices with varying widths. The input voltage V_{GS} is adjusted to compensate for narrow-width effects on the device threshold voltage:

$$\frac{\partial(\mu_{\text{eff}}/\mu)}{\partial(1/W)} = \beta$$

Figure 8.23 illustrates an experimental result of the determination of the proportionality constant β; as was expected, the decrease in μ (or conversely, the increase in μ_{eff}/μ) is not as pronounced as for the other two field components. The validity of the previous assumptions,

independence of bulk and surface scattering mechanisms,

reduction of μ by the three field components, and

proportionality constants used to describe each high-field mobility reduction mechanism,

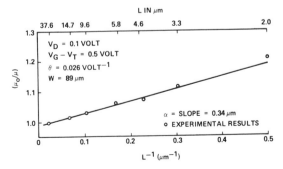

Figure 8.22. The transverse proportionality constant is determined from a measure of μ_{eff}/μ for devices of varying length. (From P. Wang, "Device Characteristics of Short-Channel and Narrow-Width MOSFET's." *IEEE Transactions on Electron Devices*, Vol. ED-25, No. 7 (July 1978), p. 783. Copyright © 1978 by IEEE. Reprinted with permission)

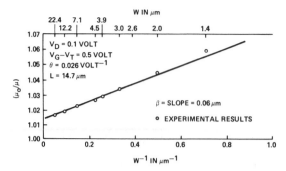

Figure 8.23. Experimental data used to determine the weak effect of the device width on the carrier channel mobility. (From P. Wang, "Device Characteristics of Short-Channel and Narrow-Width MOSFET's." *IEEE Transactions on Electron Devices,* Vol. ED-25, No. 7 (July 1978), p. 783. Copyright © 1978 by IEEE. Reprinted with permission)

can be veriried to some degree by using the independently determined constants (β, γ, and θ from Figures 8.21 through 8.23) and checking the accuracy of the device model when all three mobility-reduction terms are significant. Figure 8.24 illustrates the data measured for a narrow, short device; the experimentally measured data and model predictions for the channel mobility μ are plotted versus $V_{GS} - V_T$, with V_T adjusted for short-channel and narrow-width effects.

Device Saturation Characteristics

The discussion in Chapter 7 on the MOS device *I-V* characteristics defined *saturation* as the region of operation where the device current becomes independent of the drain-to-source voltage. This condition was described by introducing the concept of *pinch-off,* where the surface inversion layer is no longer present in the vicinity of the drain node. The drain-to-source voltage that defines the boundary between saturation and linear regions was called the *pinch-off voltage* and was described by the relation $V_{\text{pinch-off}} = V_{GS} - V_T$. In actuality, as depicted in Figure 8.25, the device current increases with increasing V_{DS}; the output conductance of the MOS device operating in the saturation region, $(\partial I_{DS}/\partial V_{DS})\,|_{V_{GS}}$, is nonzero. This finite output conductance is commonly attributed to the spreading or widening of the pinch-off depletion region along the channel, resulting in a reduction of the effective channel length (Figure 8.26). A *channel shortening model* for the

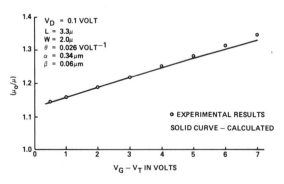

Figure 8.24. A test of the validity of the assumptions used for the field-dependent mobility model; the channel mobility for a narrow, short device is plotted versus the predictions of the composite mobility model. (From P. Wang, "Device Characteristics of Short-Channel and Narrow-Width MOSFET's." *IEEE Transactions on Electron Devices,* Vol. ED-25, No. 7 (July 1978), p. 783. Copyright © 1978 by IEEE. Reprinted with permission)

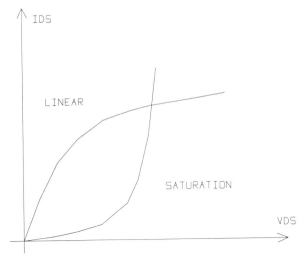

IDS

LINEAR

SATURATION

VDS

Figure 8.25.   Increase in I_{DS} in the saturation region with increasing V_{DS} past pinch-off.

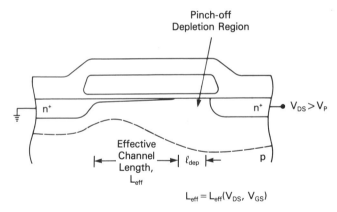

Pinch-off
Depletion Region

n^+ n^+ ● $V_{DS} > V_P$

Effective
Channel
Length,
L_{eff} ℓ_{dep} p

$L_{eff} = L_{eff}(V_{DS}, V_{GS})$

Figure 8.26.   Device cross section depicting *channel shortening* for increasing V_{DS} past pinch-off.

device saturation characteristics will be discussed, as first presented in reference 8.4.

An increase in drain-to-source voltage beyond the pinch-off voltage results in a depletion region of length l_{dep} being formed between drain and channel. The model assumes that the increase in I_{DS} is strictly due to the reduction in inversion layer channel length for increasing drain-to-source voltage above $V_{pinch-off}$; the voltage difference between the channel pinch-off point and the source node remains constant. The saturation device current increases past the drain-to-source pinch-off value by the following proportion:

$$I_{DS}\,|_{V_{DS}>V_{pinch-off}} = I_{DS_{sat}} * \left(\frac{L}{L - l_{dep}}\right)$$

where L is the nominal device channel length. For a long-channel device where $L \gg l_{\text{dep}}$ over the range of operation in the saturation region, the device current remains constant. For short-channel devices, l_{dep} may be a significant fraction of the device channel length, causing the saturation current to increase. The remainder of this discussion will describe how l_{dep} may be evaluated in terms of the device's physical parameters and the applied terminal voltages.

The length of the depletion region in saturation, l_{dep}, will be taken to be the ratio of the voltage difference between drain and channel ($V_{DS} - V_{\text{pinch-off}}$) and the average *transverse* electric field in the depletion region, near the Si–SiO$_2$ surface:

$$l_{\text{dep}} = \frac{V_{DS} - V_{\text{pinch-off}}}{\mathscr{E}_{\text{transverse, average}}}$$

The average transverse field is a composite of the electric field lines present due to the applied terminal voltages. A sketch of the field distribution near the drain end of the device in saturation is given in Figure 8.27. Three components identified in the figure contribute to the transverse electric field:

1. The field in the depletion region of a *pn* junction consisting of n^+ drain contact and the *p*-type substrate; this component is denoted by $\mathscr{E}_1$ in Figure 8.27.
2. The fringing field of the drain-to-(effective) gate potential difference, $V_D - V_G'$, denoted in the figure as component $\mathscr{E}_2$.
3. The fringing electric field of the potential difference between the (effective) gate voltage and the potential at the end of the channel inversion layer, $V_G' - V_{\text{pinch-off}}$, denoted by $\mathscr{E}_3$.

For the remainder of the discussion, it will be assumed that the source is at ground potential so that the S subscript may be dropped, and the electric field due to the gate input voltage is altered by the presence of charge in the oxide or at the Si–SiO$_2$ surface:

$$V_G' = V_G + \frac{Q_f}{C_{\text{ox}}}$$

or

$$Q_{\text{Gate}}' = V_G' * C_{\text{ox}} = V_G C_{\text{ox}} + Q_f$$

The field contribution in the transverse direction for component $\mathscr{E}_1$ can be estimated from reverse-bias *pn* junction theory. The calculation of the transverse contributions of $\mathscr{E}_2$ and $\mathscr{E}_3$ would require a two-dimensional solution of Poisson's

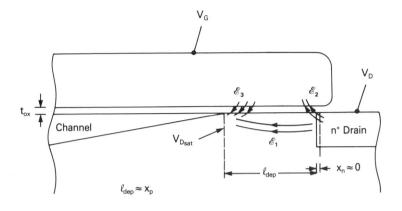

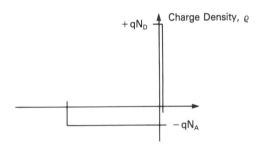

Figure 8.27. Electric field distribution for the MOS device operating in saturation. Components of the transverse field $\mathcal{E}_T$ are shown, indicated as $\mathcal{E}_1$, $\mathcal{E}_2$, and $\mathcal{E}_3$.

equation in this region and will only be approximated instead. Recall that the goal is to calculate

$$\mathcal{E}_{\text{transverse average}} = \frac{V_D - V_{\text{pinch-off}}}{l_{\text{dep}}}$$

where

$$\mathcal{E}_{\text{transverse average}} = \mathcal{E}_{1\text{average}} + \mathcal{E}_{2\text{transverse average}} + \mathcal{E}_{3\text{transverse average}}$$

Referring to Figure 8.28 for the calculation of $\mathcal{E}_{1\text{average}}$,

$$\mathcal{E}_{1\text{average}} = \frac{1}{2}\,|\,\mathcal{E}_{1\text{max}}\,| = \left(\frac{qN_A}{2\epsilon_{\text{Si}}}\right)^{1/2} * (V_D - V_{\text{pinch-off}})^{1/2}$$

The transverse components of $\mathcal{E}_2$ and $\mathcal{E}_3$ are approximated by assuming that they

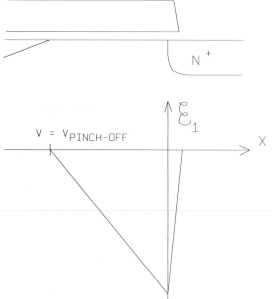

Figure 8.28. Pinch-off depletion region field plot for calculating $\mathscr{E}_{1\text{ transverse, average}}$.

are proportional to the *normal* field components of $\mathscr{E}_2$ and $\mathscr{E}_3$ across the oxide:

$$\mathscr{E}_{2\text{normal}} = \frac{V_D - V'_G}{t_{\text{ox}}}$$

$$\mathscr{E}_{3\text{normal}} = \frac{V'_G - V_{\text{pinch-off}}}{t_{\text{ox}}}$$

The normal field components of $\mathscr{E}_2$ and $\mathscr{E}_3$ will be related to the transverse components by two factors:

1. $(\epsilon_{\text{ox}}/\epsilon_{\text{Si}})$; the higher dielectric constant of Si relative to SiO_2 *reduces* the transverse field component.
2. The assumed proportionality constants between the normal and transverse fields.

The proportionality constants (α and β) are intended to represent geometrical, field-fringing factors and are to be regarded as independent of applied voltages and the device's physical parameters, such as channel length or gate oxide thickness. As a result, the average transverse gate oxide field components are written as

$$\mathscr{E}_{2\text{transverse}} = \alpha \left(\frac{\epsilon_{\text{ox}}}{\epsilon_{\text{Si}}}\right) \left(\frac{V_D - V'_G}{t_{\text{ox}}}\right) \quad \text{and} \quad \mathscr{E}_{3\text{transverse}} = \beta \left(\frac{\epsilon_{\text{ox}}}{\epsilon_{\text{Si}}}\right) \left(\frac{V'_G - V_{\text{pinch-off}}}{t_{\text{ox}}}\right)$$

The total average field is the sum of the three components:

$$\mathcal{E}_{\text{total average}} = \underbrace{\left(\frac{qN_A}{2\epsilon_{\text{Si}}}\right)^{1/2} (V_D - V_{\text{pinch-off}})^{1/2}}_{\substack{\text{Due to substrate} \\ \text{impurity concentration}}}$$

$$\underbrace{+ \; \alpha * \frac{\epsilon_{\text{ox}}}{\epsilon_{\text{Si}}} * \frac{V_D - V_G'}{t_{\text{ox}}} + \beta * \frac{\epsilon_{\text{ox}}}{\epsilon_{\text{Si}}} * \frac{V_G' - V_{\text{pinch-off}}}{t_{\text{ox}}}}_{\substack{\text{Due to gate} \\ \text{terminal contribution}}}$$

The extent of the channel depletion region, l_{dep}, is written as

$$l_{\text{dep}} = \frac{V_D - V_{\text{pinch-off}}}{\mathcal{E}_{\text{total average}}}$$

$$\frac{1}{l_{\text{dep}}} = \left(\frac{qN_A}{2\epsilon_{\text{Si}}}\right)^{1/2} \frac{1}{(V_D - V_{\text{pinch-off}})^{1/2}} + \left(\frac{\epsilon_{\text{ox}}}{\epsilon_{\text{Si}}}\right)\left(\frac{1}{t_{\text{ox}}}\right)$$

$$\times \left[\frac{\alpha \cdot (V_D - V_G') + \beta \cdot (V_G' - V_{\text{pinch-off}})}{V_D - V_{\text{pinch-off}}}\right]$$

The output conductance of the device operating in the saturation region is given by

$$g_{DS\text{sat}} = \left. \frac{\partial I_{DS}}{\partial V_{DS}} \right|_{\substack{\text{saturation,} \\ V_{GS} = \text{constant}}} = \frac{\partial I_{DS}}{\partial l_{\text{dep}}} * \frac{\partial l_{\text{dep}}}{\partial V_{DS}}$$

From the expression

$$I_{DS} = I_{DS\text{sat}} * \left(\frac{L}{L - l_{\text{dep}}}\right)$$

$$\frac{\partial I_{DS}}{\partial l_{\text{dep}}} = \frac{I_{DS\text{sat}}}{L[1 - (l_{\text{dep}}/L)]^2}$$

$$g_{DS\text{sat}} = \frac{I_{DS\text{sat}}}{L[1 - (l_{\text{dep}}/L)]^2} * \frac{\partial l_{\text{dep}}}{\partial V_D}$$

If the assumption is made that

$$t_{\text{ox}} \ll (V_D - V_{\text{pinch-off}})^{1/2} \left(\frac{2\epsilon_{\text{Si}}}{qN_A}\right)^{1/2}$$

corresponding to the process parameters of thin gate oxide and low substrate impurity concentration, then the expression for l_{dep} reduces to

$$l_{dep} \approx \frac{\epsilon_{Si}}{\epsilon_{ox}} \frac{t_{ox}(V_D - V_{pinch\text{-}off})}{\alpha(V_D - V'_G) + \beta(V'_G - V_{pinch\text{-}off})}$$

$$g_{DS\,sat} = \frac{I_{DS\,sat} * t_{ox}}{L[1 - (l_{dep}/L)]^2} * \frac{\epsilon_{Si}}{\epsilon_{ox}} * \frac{(\beta - \alpha)(V'_G - V_{pinch\text{-}off})}{[\alpha(V_D - V'_G) + \beta(V'_G - V_{pinch\text{-}off})]^2}$$

Experimental data to test the adequacy of this channel-shortening model are presented in Figure 8.29 (from reference 8.4); the constants used for the transverse field-fringing coefficients were $\alpha = 0.2$ and $\beta = 0.6$.

Other sets of assumptions have also been used to calculate the inversion channel length in saturation; for example, reference 8.5 describes a model where the critical electric field, $\mathcal{E}_c$, and the drain node junction depth, x_j, are parameters in the inversion length expression. In this analysis, the transverse field at the edge of the inversion layer is assumed to be equal to $\mathcal{E}_c$, and the field lines from drain to channel are distributed over the junction depth of the drain node.

The development of accurate MOS device models for computer simulation

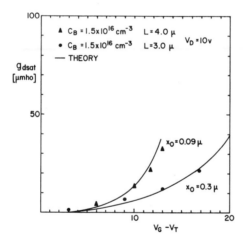

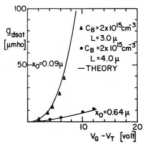

Figure 8.29. Experimental data to assess the applicability of the channel-shortening model; the transverse field proportionality constants used were $\alpha = 0.2$ and $\beta = 0.6$. (From D. Frohman-Bentchkowsky and A. S. Grove, "Conductance of MOS Transistors in Saturation," *IEEE Transactions on Electron Devices*, Vol. ED-16, No. 1 (January 1969), p. 112. Copyright © 1969 by IEEE. Reprinted with permission)

is an ongoing effort. As VLSI technology device dimensions continue to be reduced, new device effects will emerge as significant, requiring modifications to
existing models. For example, enhanced models incorporating both inversion
layer and subsurface channel depletion region current components along the two-
dimensional device cross section are becoming of extreme importance. Yet the
underlying criteria used to evaluate new device theories will continue to be (1)
their relative accuracy to measured characteristics, and (2) their efficiency of
execution in a computer simulation environment. Trade-offs between these criteria, based on the targeted application, will continually have to be addressed.

8.6 MOS DEVICE CAPACITANCE MODELS

This section provides working expressions for the voltage-dependent capacitances
of the MOS device model: C_{GX}, C_{GS}, C_{GD}, C_{SX}, and C_{DX}. These equations describe the *intrinsic* device capacitances; the *extrinsic* gate overlap and nodal junction capacitances should be added to complete the total device model. The extrinsic element values are functions only of the voltage across the capacitance
terminals.

Modeling of each of the intrinsic capacitances is more complex due to the
possible dependence on all four node voltages: V_G, V_D, V_S, and V_X. Examining
the voltage dependence on V_G, for example, the intrinsic capacitance expressions
should concur with the curves sketched in Figure 7.56 for different regions of
device operation. The expressions to be given are per unit channel area; $C_{ox} =
\epsilon_{ox}/t_{ox}$ is used to indicate the maximum inversion gate oxide capacitance. The
notation V_{FB} is used to replace the terms

$$V_{FB} = \left(\underset{\substack{\text{contact potential}}}{V_{\text{gate}-\text{SiO}_2,}} - V_{Q_{ox}} - \underset{\substack{\text{contact potential}}}{V_{\text{SiO}_2-\text{Si},}} \right)^\dagger$$

An *n*-channel device (*p*-substrate, of uniform impurity concentration N_A) will be
used for illustration.

The device analysis proceeds initially by writing charge equations for the
device in its different operating modes. These equations describe the gate charge,
Q_G, the bulk depletion region charge, Q_B, and the inversion channel charge, Q_I,
per unit area. The mobile inversion channel charge will be divided between source
and drain device nodes in a voltage-dependent manner, indicative of the charge
contribution (and therefore capacitive current) of each node to the mobile inversion charge density. The overall charge neutrality of the static MOS device dic-
tates that the charge densities should satisfy the equation $|Q_G| = |Q_B + Q_I|$,
where the polarities of the gate and surface charges are opposite. The total gate
charge Q_G is the sum of the applied charge and any oxide charges; the oxide

---

† This expression describes the *flatband voltage* of the MOS device energy band structure.

charge density is assumed to be voltage independent. The effect of any oxide charge is incorporated into the flatband voltage term.

It should be emphasized that the given charge equations reflect an assumption of steady-state conditions, rather than small-signal high-frequency behavior. In other words, the charge densities are the solutions to the electrostatic charge neutrality MOS condition.

Region of operation (accumulation): $V_G - V_{FB} - V_X < 0$

$$Q_G = C_{ox} * (V_G - V_{FB} - V_X)$$

$$Q_B = -Q_G$$

$$Q_I = 0$$

No mobile inversion channel charge is present in accumulation.

Region of operation (depletion): $V_{FB} + V_X - V_S < V_{GS} < V_T$

$$Q_G = C_{ox} * (V_G - V_{FB} - V_{depl}(y) - V_X)$$

$$Q_B = -qN_A x_{depl} = -Q_G$$

$$Q_I = 0$$

As with the device in accumulation, no mobile inversion channel charge is present in depletion. x_{depl} is the extent of the surface depletion region into the substrate. The term $V_{depl}(y)$ represents the voltage across the surface depletion region:

$$V_{depl} = \frac{qN_A}{2\epsilon_{Si}} * x_{depl}^2$$

(from the one-dimensional Poisson equation $\nabla^2 V = -\rho/\epsilon$).

Equating the bulk and gate charge expressions yields a quadratic in x_{depl}; solving this equation for x_{depl} and hence Q_G gives

$$Q_G = \frac{\epsilon_{Si} q N_A}{C_{ox}} \left[-1 + \sqrt{1 + \frac{4C_{ox}^2}{2\epsilon_{Si} q N_A} (V_G - V_{FB} - V_X)} \right]$$

Region of operation (linear): $V_T < V_{GS_{sat}} < V_{GS}$,   where $V_{GS_{sat}} = V_T + V_{DS}$

With the presence of the inversion channel charge, the charge density calculations become more involved than the straightforward (two-terminal) expressions of the accumulation and depletion regions. The general method is to write expressions for the *total* (gate, inversion, or bulk depletion) charge, integrating over the channel length; the surface potential at a point y in the channel is denoted as $V_{\text{channel}}(y)$. For example, the magnitude of the total inversion charge is given by

$$Q_{I_{\text{total}}} = -W \int_0^L |Q_I(y)| \, dy$$

$$= -W * \int_0^L [C_{\text{ox}}(V_G - V_{FB} - V_{\text{built-in}} - V_{\text{channel}}(y))$$

$$- \sqrt{2q\epsilon_{\text{Si}}N_A(V_{\text{channel}}(y) - V_x + V_{\text{built-in}})}] \, dy$$

To facilitate the evaluation of the integral (and those for the bulk and gate charges), the following general current expression can be used to translate the variable of integration from the channel position y to the channel potential V_{channel}:

$$I = W * \mu_{\text{eff}} * \frac{Q_I(y)}{\text{unit area}} * \frac{dV_{\text{channel}}}{dy}$$

The integral for the total mobile inversion charge can now be written as

$$Q_{I_{\text{total}}} = \int_{V_S}^{V_D} \frac{-W^2 \mu_{\text{eff}}}{I} * Q_I^2(V_{\text{channel}}) * dV_{\text{channel}}$$

Similarly, the *total* bulk depletion region charge is given by

$$Q_{B_{\text{total}}} = -W \int_0^L \sqrt{2\epsilon_{\text{Si}}qN_A} * \sqrt{V_{\text{channel}}(y) - V_x + V_{\text{built-in}}} * dy$$

or, using the current equation to again make the translation in integration variable,

$$Q_{B_{\text{total}}} = \int_{V_S}^{V_D} \frac{-W^2 \mu_{\text{eff}}}{I} \sqrt{2\epsilon_{\text{Si}}qN_A} * \sqrt{V_{\text{channel}} - V_x + V_{\text{built-in}}}$$

$$* Q_I(V_{\text{channel}}) * dV_{\text{channel}}$$

The total gate charge can be expressed in magnitude as the sum of the previous two results.

To develop a more meaningful expression for the inversion charge density (and ultimately the capacitive elements C_{GS} and C_{GD}), the *gradual channel approximation* will be used for the inversion channel charge over the range of integration V_S to V_D; that is,

$$Q_I(y) \cong -C_{\text{ox}}(V_G - V_T - V_{\text{channel}}(y))$$

where

$$V_T = V_{FB} + V_{\text{built-in}} + \frac{1}{C_{\text{ox}}} \sqrt{2q\epsilon_{\text{Si}}N_A(V_S - V_x + V_{\text{built-in}})}$$

Likewise, the current value may be replaced by

$$I = \mu_{\text{eff}} * \frac{W}{L} * C_{\text{ox}} * \left(V_{GS} - V_T - \frac{V_{DS}}{2} \right) V_{DS}$$

The inversion charge density integral reduces to

$$\frac{Q_I}{\text{unit area}} = \frac{Q_{I\text{total}}}{W * L} = \frac{-C_{\text{ox}}}{[V_{GS} - V_T - (V_{DS}/2)]V_{DS}}$$

$$* \int_{V_S}^{V_D} (V_G - V_T - V_{\text{channel}})^2 * dV_{\text{channel}}$$

Evaluating the inversion charge density integral (followed by some algebraic manipulation) gives

$$\frac{Q_I}{\text{unit area}} = -C_{\text{ox}} \left\{ \left(V_{GS} - V_T - \frac{V_{DS}}{2} \right) + \frac{V_{DS}^2}{12[V_{GS} - V_T - (V_{DS}/2)]} \right\}$$

The bulk depletion region charge density may be evaluated directly from its integral expression or may likewise be simplified by an appropriate assumption. In this case, the assumption used will be

$$\frac{Q_B}{\text{unit area}} = -\sqrt{2q\epsilon_{\text{Si}}N_A} * \left(\alpha * \sqrt{V_S - V_X + V_{\text{built-in}}} \right.$$

$$\left. + \beta\sqrt{V_D - V_X + V_{\text{built-in}}} \right)$$

where α and β are the fractional coefficients of the channel inversion charge density assigned to C_{GS} and C_{GD}, respectively. These coefficients will be derived shortly. Note that the gate voltage V_G is absent from this expression; this implies that the capacitance value C_{GX} goes to zero in the linear region of operation. The reasoning behind this assumption is twofold:

1. The gate is effectively isolated from the bulk by the presence of the inversion layer charge; changes in the gate voltage in the linear region result in changes in the inversion charge as provided by the source and drain terminals of the device.
2. The coefficients for the contribution of C_{GS} and C_{GD} to changes in the inversion charge are also to be used for the currents through C_{SX} and C_{DX}; using these coefficients twice will avoid another analytic calculation and save simulation execution time.

The gate charge in the linear region is given by

$$Q_G = -(Q_I + Q_B).$$

Region of operation (saturation): $V_T < V_{GS} < V_{GS,\text{sat}} = V_{DS} + V_T$

$$V_{\text{pinch-off}} = V_{GS} - V_T$$

In saturation, the device channel is pinched off in the vicinity of the drain node; the inversion charge density in the pinch-off region is assumed to be zero (i.e., the drain node no longer contributes to changes in the induced inversion charge).

The expression for the inversion charge density in the saturation region can be derived by evaluating the linear region expression at $V_{DS} = V_{\text{pinch-off}} = V_{GS} - V_T$:

$$Q_{I_{\text{saturation}}} = Q_{I_{\text{linear}}}\bigg|_{V_{DS}=(V_{GS}-V_T)} = -\tfrac{2}{3} * C_{\text{ox}} * (V_{GS} - V_T)$$

An expression for the bulk charge density in saturation can likewise be derived by evaluating the integral equation of the linear region at saturation conditions; as was the case for the linear mode, however, some simplifying assumptions will be made regarding the functional dependencies:

1. Evaluating the bulk charge density at saturation conditions implies that $Q_{B,\text{sat}}$ is independent of V_D; that is, changes to the bulk charge in the intrinsic pinch-off region due to changes in the drain node voltage are neglected. As discussed with the channel-shortening model, the electric field concentration near the drain pinch-off region is primarily in the transverse and upward directions; the influence of the lateral edge of the drain node on the bulk field in saturation is small in comparison to that of the planar drain node area.

2. An accurate approximation for the gate electric field in the channel past the pinch-off region is difficult to develop. The change in the saturation bulk charge density due to variations in gate voltage will be neglected (i.e., $Q_{B,\text{sat}}$ is also independent of V_G).

3. The same fractional coefficient used for the inversion channel charge density ($\tfrac{2}{3}$) will be used for the source-to-substrate bulk charge density.

With these assumptions, the bulk charge density in saturation can be written as

$$Q_{\text{bulk}_{\text{sat}}} = -\tfrac{2}{3}\sqrt{2\epsilon_{\text{Si}}qN_A(V_S - V_x + V_{\text{built-in}})}$$

The gate charge density is equal to $Q_{G,\text{sat}} = -(Q_I + Q_B)_{\text{sat}}$.

Given the charge density equations, the capacitance element values can now be calculated. The capacitive elements of the MOS device simulation model can be derived from the charge derivatives:

$$i_G = \frac{dQ_G}{dt} = \frac{\partial Q_G}{\partial V_{GS}} * \frac{dV_{GS}}{dt} + \frac{\partial Q_G}{\partial V_{GD}} * \frac{dV_{GD}}{dt} + \frac{\partial Q_G}{\partial V_{GX}} * \frac{dV_{GX}}{dt}$$

$$i_G = C_{GS} * \frac{dV_{GS}}{dt} + C_{GD} * \frac{dV_{GD}}{dt} + C_{GX} * \frac{dV_{GX}}{dt}$$

Region of Operation	Charge Density
Accumulation	$Q_G = C_{ox}(V_{GX} - V_{FB})$
Depletion	$Q_G = \dfrac{\epsilon_{Si}qN_A}{C_{ox}}\left[-1 + \sqrt{1 + \dfrac{4C_{ox}^2}{2q\epsilon_{Si}N_A}(V_{GX} - V_{FB})}\right]$
Saturation	$Q_I = -\frac{2}{3}C_{ox}(V_{GS} - V_T)$ $Q_B = -\frac{2}{3}\sqrt{2\epsilon_{Si}qN_A}(V_S - V_X + V_{built\text{-}in})$
Linear	$Q_I = -C_{ox}\left[\left(V_{GS} - V_T - \dfrac{V_{DS}}{2}\right) + \dfrac{V_{DS}^2}{12\left(V_{GS} - V_T - \dfrac{V_{DS}}{2}\right)}\right]$ $Q_B = -\sqrt{2q\epsilon_{Si}N_A}\left(\alpha \cdot \sqrt{V_S - V_X + V_{built\text{-}in}} + \beta \cdot \sqrt{V_D - V_X + V_{built\text{-}in}}\right)$ $\alpha = \left[1 - \dfrac{(V_{GD} - V_T)^2}{((V_{GD} - V_T) + (V_{GS} - V_T))^2}\right] = \dfrac{\partial Q_i|_{linear}}{\partial V_{GS}}$ $\beta = \left[1 - \dfrac{(V_{GS} - V_T)^2}{((V_{GD} - V_T) + (V_{GS} - V_T))^2}\right] = \dfrac{\partial Q_i|_{linear}}{\partial V_{GD}}$

Element	Accumulation	Depletion	Saturation	Linear
C_{GX}	C_{ox}	$\dfrac{C_{ox}}{\left(1 + \dfrac{4C_{ox}^2}{2q\epsilon_{Si}N_A}(V_G - V_{FB} - V_X)\right)^{1/2}}$	0	0
C_{GS}	0	0	$\frac{2}{3} \cdot C_{ox}$	$\frac{2}{3}C_{ox}[\alpha]$
C_{GD}	0	0	0	$\frac{2}{3}C_{ox}[\beta]$
C_{SX}	0	0	$\frac{1}{2} \cdot \frac{2}{3} \dfrac{\sqrt{2q\epsilon_{Si}N_A}}{\sqrt{V_S - V_X + V_{built\text{-}in}}}$	$\alpha \cdot \frac{1}{2} \dfrac{\sqrt{2q\epsilon_{Si}N_A}}{\sqrt{V_S - V_X + V_{built\text{-}in}}}$
C_{DX}	0	0	0	$\beta \cdot \frac{1}{2} \dfrac{\sqrt{2q\epsilon_{Si}N_A}}{\sqrt{V_D - V_X + V_{built\text{-}in}}}$

Figure 8.30. Charge densities and equations used to calculate the per unit area capacitance element values for the MOS device model.

where

$$C_{GS} = \frac{\partial Q_G}{\partial V_{GS}}\bigg|_{\substack{V_{GD}, V_{GX},\\ \text{constant}}} \qquad C_{GD} = \frac{\partial Q_G}{\partial V_{GD}}\bigg|_{V_{GS}, V_{GX}}, \qquad C_{GX} = \frac{\partial Q_G}{\partial V_{GX}}\bigg|_{V_{GS}, V_{GD}}$$

Similarly, the other model capacitances are given by

$$C_{SX} = \frac{\partial Q_S}{\partial V_{SX}}\bigg|_{V_{GS}, V_{DS}} \quad \text{and} \quad C_{DX} = \frac{\partial Q_D}{\partial V_{DX}}\bigg|_{V_{DS}, V_{GD}}$$

where the charge densities Q_S and Q_D are the partitions of the inversion and bulk charge densities assigned to the source and drain nodes, respectively. Taking the derivatives indicated yields the table of capacitance equations in Figure 8.30. As described previously, the expressions of Figure 8.30 represent the intrinsic device capacitances; to these equations, the extrinsic values should be added. In particular, to avoid simulation convergence difficulties, the voltage-dependent capacitance looking into a device node should *not* be permitted to go to zero in the model during a network analysis.

To reduce the run time of a transient circuit simulation (at the sacrifice of accuracy), it is possible to consider deleting capacitive elements from the network or device input description whose magnitudes are not significant and/or removing the voltage dependence of the capacitance expressions. The suitability of these approaches depends on the nature of the network and the intent of the analysis.

8.7 ELEMENTS OF CIRCUIT SIMULATION

The circuit simulation program compiles a modeling language description of the network under analysis, applies stimuli to the network as described by the time-dependent current and voltage sources, and determines a network solution at each specific time step. The values of the requested network parameters are recorded for each solution and output to the designer. The extrapolation of previous and current network branch currents and node voltages is used to develop initial estimates for the next step time and network matrix. Given these values, the central algorithm of the network analyzer iterates toward an acceptable solution. The increments in simulation time between successive solutions will be adjusted by the program, based on relative value limits specified by the designer. When network derivatives are small, the time step can be dynamically increased to reduce the total number of time steps and therefore the overall execution time.[4] This section describes some of the features commonly available with circuit simulation programs; some of the techniques and options available for network analysis are

---

[4] It may be necessary to provide the capability to specify solution time(s) exactly and override this variable time step feature. For example, if a stable network is to receive a very narrow pulse as input, specific times corresponding to particular points on the pulse may be requested as solution times. This will avoid loss of detail if the pulse characteristics were to be skipped.

also highlighted. The simulator is applicable to a wide variety of problems outside the realm of circuit analysis, as it is indeed a general numerical analysis tool for a set of nonlinear differential equations. In many cases, an electrical analogue of a nonelectrical or a mixed electromechanical system can be developed. The transient or stability behavior of the system can be analyzed from the simulation of its electrical equivalent. Problems in the areas of heat transfer, fluid flow, and vibration analysis can be solved by means of their electrical equivalents.

The modes of analysis that are afforded to the designer by the simulator include transient, dc, and frequency analysis. The transient analysis begins with initial network branch current, node voltage, and model element values and proceeds in simulation time by finding successive solutions to the set of nonlinear differential equations that describe the circuit. In general, any source or element value could be specified as time varying, as could any network parameter (e.g., temperature). The initial values represent a *dc solution* of the network behavior at the starting simulation time. This initial solution is commonly the result of a *pseudotransient charge-up* analysis, performed prior to advancing the simulation time clock. Source values are applied, while capacitive voltages and inductive currents are held at *zero* and then released to permit iteration toward an overall stable dc operating point. If multiple stable operating points exist at time zero, it may be necessary to initially specify nonzero charge-up voltages and currents to direct the pseudotransient analysis toward a *particular* stable point. Alternatively, the circuit designer may include dummy elements in the overall network description, whose time-zero value forces a particular network condition, yet whose time-dependence effectively eliminates the element from the details of the simulation (Figure 8.31). It may also be possible to specify that the simulation begin *without* the pseudotransient step of determining a dc solution (i.e., to observe the behavior of the network to the application of the power supplies). In any case, whether starting from a stable dc condition or not, the transient simulation proceeds until the specified *stop time*.

The dc mode of analysis is the same as the pseudotransient analysis just described. In addition to being the starting point for a transient simulation, a dc

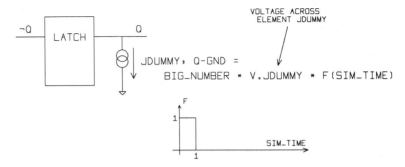

Figure 8.31. Additional dummy elements may be added to the network model to assist the pseudotransient charge-up analysis toward one of a number of stable operating points.

analysis is useful for logic circuit noise margin analysis and for determining the operating point device parameters for a small-signal analysis, to be discussed next.

The ac or frequency mode of analysis is applicable for characterizing the magnitude and phase of network voltages and currents in a small-signal, linear circuit analysis. The specification of the input stimuli now consists of the super-position of dc and small-signal sources; the variable frequency is now used (rather than time) and is varied between designer-specified limits. The output transient waveform analysis is likewise replaced by Bode plots of network transfer function. Any element value may be frequency dependent. The dc operating point is used to determine the partial derivatives of nonlinear element values at that point, effectively linearizing the circuit prior to analysis. A complex parameter with frequency-dependent magnitude and phase would be evaluated through the an-alysis; its dc value uses the real value of the parameter. A nonfrequency-depen-dent parameter would be evaluated only initially, during the linearization process.

All three modes of analysis benefit considerably from the ability to select element values (or the coefficients used to calculate those values) from a specified statistical distribution. Prior to beginning any network analysis, parameters as-sociated with statistical distributions would be identified and a value selected for each parameter from its probability density function. Subsequently, the network simulation would proceed to completion using these values.

A common probability density function to be specified is the Gaussian or normal distribution, given by

$$p(x) = \frac{1}{\sqrt{2\pi} * \sigma} \exp\left[-\frac{1}{2}\left(\frac{x - \mu}{\sigma}\right)^2 \right]$$

where μ is the mean of the variable and σ its standard deviation. For example, a statistically varying process parameter such as the gate oxide thickness could be depicted as belonging to the distribution TOX $= N(300, 10) * (1E - 8)$. The oxide thickness is nominally 300 Å, with a standard deviation of 10 Å; the mul-tiplicative factor $(1E - 8)$ converts the units to centimeters for use in subsequent device calculations.

The statistical capabilities of the simulator are facilitated by the use of a random-number generator; this random number (between 0 and 1), in conjunction with the cumulative distribution function of the probability density, is used to select the parameter value. To keep the statistically varying parameters indepen-dent, a new random number is generated for each distribution prior to beginning simulation. It may also be possible to permit the designer to specify the random number directly for all distributions, overriding the random-number generator; for example, a specified random number of 0.5 would provide a network analysis with all nominal parameters. It may also be possible to request that the transient, dc, or ac analysis be repeated, while continuing to use the random number gen-erator to calculate varying process values. Repeating a number of cases of the same simulation run will give the designer a measure of the overall statistical variation of a network voltage or current. The results of many cases would be output in histogram (dc) or waveform envelope form (transient, frequency). Execution of a number of identical simulations, with parameters randomly selected from their distributions for each case, is denoted as a Monte Carlo analysis.

Although the individual distributions could be regarded as independent for calculating parameter values, it is nevertheless straightforward to implement some measure of *tracking* between different process parameters that are not truly independent. For example,

$$VTE = N(1.0, 0.05); \text{ enhancement mode } V_T$$

$$VTD = VTE + N(-3.5, 0.08); \text{ depletion mode } V_T$$

Tracking variations between the same parameter of different instances of the device model can also be provided when the input description is expanded into its corresponding network, based on a distance measure between devices (Figure 8.32).

A general structure of the input description to the simulator is illustrated in Figure 8.33. Programming constructs such as functions and subroutines are available to the model definitions to calculate element values. The model and simulation directive statements are preprocessed into high-level source code routines prior to compilation and execution. There is no order of execution of the statements within a model; rather, the model statements are expanded to develop a total equivalent network interconnectivity matrix. Any simulation time-independent routines that result from the input preprocessing are executed first, and only once for each case of the simulation. These routines typically extract values from statistical distributions and set up constants for the run. The tightest loop of routines is that associated with calculating time-varying element values and iterating to find the network solution satisfying the error limits at each time step.

The simulation input begins with the specification of the run controls (e.g., the length of simulation time to execute in a transient run). Of particular impor-

```
/* PROCESS PARAMETER DISTRIBUTIONS  */
      %
DISTANCE = 300.0 ;
      %
TOX = N(300,10) * 1E-8 ; /* GATE OXIDE IN CM */
      %
/* NETWORK DESCRIPTION   */

Q1 = MODEL NFET(G=IN, D=OUT, S=GND, X=VSUB)
        (W=25.0,L=2.0);
      %
/* MODEL DESCRIPTION   */
  MODEL NFET ( G, D, S, X );
      %
    TOXN = TOX + N(0,F(DISTANCE));
  /* UNIQUE OXIDE THICKNESS FOR EACH DEVICE  */
      %
```

Figure 8.32. Implementation of parameter mismatches between different device models. F(distance) is a function provided to describe the increase in the statistical variation between device parameters with increasing distance between the devices.

SIMULATION CONTROLS:
    ⋮
    STOP_TIME =   ; /* for transient simulations */
    RELATIVE_ERROR =   ;
    ABSOLUTE_ERROR =   ;
    CASES =   ; /* for multiple statistical analyses */
    RANDOM_NUMBER_SEED =   ; /* provides the seed to use for generating the first
                                random number for a statistical analysis */
    ⋮
    END;
OUTPUTS:
    /* output requests can specify any network node voltage, branch current, element value, or
        any general network parameter */
    PRINT   , , ... ;
    PLOT (vs TIME)   , , ... ;
    HISTOGRAM (intervals = n)   , , ... ;
    ENVELOPE   , , ... ;
    END;
MAIN_MODEL:
    INCLUDE: /CMOS/DEVICE_MODELS/NFET, /CMOS/DEVICE_MODELS/PFET,
             /CMOS/CAPACITANCE_MODELS/DIFFUSION, /CMOS/PROCESS_MODEL, ... ;
    /* network parameters */
    TEMPERATURE = 25.0 ;
    /* power supplies */
    E_1, GND − VDD = 5.0 ;
    E_2, GND − SUB = −2.0 ;
    E_IN, GND − INPUT = WAVE ((0,0), (10,0), (15,5), (18,5), (20,0), ...);
    /* connections for Gate, Drain, Source, and Substrate nodes */
      Q1 = MODEL_NFET (,   ,   ,   )(W =   , L =);
      Q2 = MODEL_PFET (,   ,   ,   )(W =   , L =);
    /* nested circuit model */
      GATE1 = MODEL_NAND (, ,);
    /* additional elements */
      C_LOAD, OUTPUT − GND = 0.8 ;
    END; /* main model */
REFERENCE_NODE : GND ;
GLOBAL_PARAMETERS : TEMPERATURE, NA, ND, XJ, ... ;
    /* user-defined functions for calculating parameter values */
FUNCTION f1 (, ,): real;
    begin
      ⋮
    end ;
MODEL_NAND (, ,);
    begin
      ⋮
    end;
RE_EXECUTE :
    TEMPERATURE = (10, 85);
END.

Figure 8.33.   General structure of simulation input model and test case description.

tance is the specification of error limits (relative and absolute) that are used to control the accuracy of the accepted network solution at each point in simulation time, as well as the extrapolation of network parameters to succeeding simulation points. Stringent error controls will result in a more accurate solution (and may reduce the likelihood of network convergence problems), yet will increase the number of time steps and therefore the overall run time. The general relation used to determine the adequacy of the network solution and time extrapolation is

$$| p(t_n) - p(t_{n-1}) | < (\text{relative error}) * | p(t_n) | + (\text{absolute error})$$

where t_n is the current simulation time (after n increments) and $p(t)$ refers to any of the network model parameters. Careful attention must be paid to the set of units specified for currents, voltages, and element values to ensure that the entire set is consistent and the error limits are correctly stated to keep percentage errors within acceptable bounds. In most network simulations, node voltages and branch currents will vary over several orders of magnitude. This will affect the *percentage* error of the solution at different time steps. For example, assume that the relative error is 0.008 and the absolute error is 0.0005. For currents that vary over four orders of magnitude, the variation in percentage error will be (units are specified as milliamperes)

Network branch current value	Possible total error	% Error
1 mA	0.0085	0.8%
0.001 mA	0.000508	50%

In developing the simulation input, the designer must decide if a 50% possible error at reduced current (or voltage) levels will seriously affect the expected results of the simulation; the absolute error should be selected accordingly. Problem 8.4 addresses the task of developing a compatible set of units for all models.

After the simulation controls, the next section of the simulation input description specifies the variables to be output from the simulation. These variables could include node voltages, branch currents, element values, element voltage differences, process parameter values, in short, just about any network characteristic. To request a node voltage, it is necessary to identify the equivalent reference (or ground) node from the network node list; this is accomplished by the REFERENCE __NODE statement following the main model description. The output formats include (1) printouts of the requested values (at regular time intervals), (2) waveform plots versus simulation time (or some other independent variable), and (3) histograms of the distribution of values from a statistical simulation. Envelope plots of waveforms from statistical transient runs are also indicated.

The main model description represents the highest level of hierarchical model nesting. Included within this description are (1) individual (linear or nonlinear) elements (R's, L's, C's, . . .), (2) independent or dependent sources, (3) expressions to calculate or assign parameter values, and (4) references to models nested within the main model. The name assigned to the model reference statement is a qualifier used when referring to the elements and parameters of the nested model (e.g., Q1.CGS). These qualifiers must be unique to each reference statement nested with a model definition. The transient stimulus to the network is also listed in the main model. Elements are defined by a qualifier and associated schematic node names; the first letter of the qualifier is commonly used to designate the type of element (e.g., E for voltage source, J for current source). The models nested within the network description include the process parameter definitions, the subcircuits (defined subsequently in the input deck), device models, wiring capacitance models, and the like. The models for the devices, the capacitances, and the process parameters are included into the simulation description from a permanent model library. This library is the central reference from which technology details are incorporated into all related circuit simulations. The development and maintenance of this library is the task of the characterization engineering group and provides the link between the circuit design and process engineering areas.

When a model is incorporated into a higher-level structure, parameters within the nested model may be assigned a new value in the change list that follows the interconnection description. For example, default dimensions given within the device model are replaced with those specified in the parameter change list. Additionally, a list of global variables is provided from the main model to incorporate into all nested model descriptions (e.g., the process variables used in the device model equations).

A *re-execute* section is also shown in Figure 8.33; after the simulation has been completed (for all cases, if a statistical analysis), it is *rerun* with changes in parameter values. A significant fraction of the *total* simulation computer time is spent in the input preprocessing step, that is, developing the network interconnectivity, constructing the high-level source code routines, and compiling those routines into executable code. This time may be saved for subsequent runs of the same network by re-executing the simulation with different parameter values of interest. The network itself cannot be modified; only those parameters whose original values were assigned constants may be assigned new values. (The new values are placed directly into the appropriate storage location in the executable code.)

The simulation program will provide a library of functions and subroutines useful in describing the time variation of input stimuli and recording that of the network response. For example, the WAVE statement is used in Figure 8.33 to describe an input waveform. Other commonly available functions include the following:

INTEGRAL (argument, $time_0$, $time_1$);
DERIVATIVE (argument, $time_0$);

MAXIMUM (argument, $time_0$, $time_1$);
MINIMUM (argument, $time_0$, $time_1$);

To increase the effectiveness and efficiency of the simulation, the following practices are suggested:

1. The units of the elements should be selected to keep the network voltages and currents within a limited range of magnitudes; this will permit the specification of meaningful error limits and enhance the overall convergence of the solution algorithms.
2. Utilize zero-valued voltage sources in the network for ammeters and zero-valued current sources for voltmeters, as opposed to some resistive equivalent of very small or very large value.
3. Use dependent current and voltage sources liberally throughout the network to implement (a) nonlinear behavior, (b) dummy elements to assist the pseudotransient solution toward an initial stable solution, and (c) fan-out currents (Figure 8.34). In Figure 8.34, the current source is used to multiply the time-varying current into the single device model load to effectively represent a large fan-out, *without* a large network description; the voltage source in series with the device gate is used as an ammeter. In general, the function defining the dependent source value and its derivative should be continuous.
4. The simulation program may initially attempt to use "large" time steps in periods of little network variation, prior to iterating toward a solution. It is therefore necessary that dependent element and source values be suitably limited for independent variable values that have been extrapolated beyond their expected domains.

All these practices are applicable in a much broader sense. In executing circuit simulations, the designer should try to anticipate any possible difficulties with successfully solving the problem under analysis and should have a high degree of confidence in the anticipated results. In other words, the intent of circuit simulation is almost to *confirm* the expected behavior of the design; if the de-

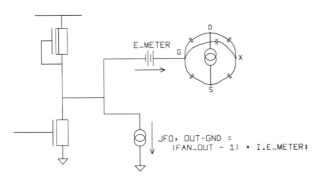

Figure 8.34. Voltage source E—METER (of value zero) is effectively used as an ammeter. The transient current through E—METER is a parameter in a dependent current source specification.

signer's insight into the problem is not substantiated by the results of the simulation, further investigation is warranted. This investigation begins by examining the validity of the network description itself and may lead to examining the implementation of the technology model libraries. The circuit design group should provide feedback to the modeling group on the execution efficiency, convergence, and usability of the device, element, and process models.

8.8 ADDITIONAL DEVICE PARAMETERS

This section briefly discusses some of the IC structures and operating modes that can lead to device parameter drift and/or to circuit or chip failure. The process engineer and circuit designer must work in tandem to ensure that all chip designs are suitably protected from the destructive behavior characteristic of some of these effects. This protection entails additional constraints on the physical layout.

Field Oxide Threshold Voltage

An interconnection line routed over two isolated diffusion nodes provides the structure for a *field oxide* device (Figure 8.35). It is necessary that the field oxide device *never* be conducting, connecting two otherwise isolated nodes. This requirement is satisfied if $V_{T,\text{field oxide}} \gg VDD$, where the thickness of the field oxide is used in the threshold voltage calculation. Normally, the field oxide is sufficiently thick to result in the desired large $V_{T,\text{field oxide}}$; however, the thicker the field oxide, the greater the step in topography for a metal interconnect line when making contact to a diffusion node. Large changes in topography are undesirable from a step coverage standpoint, resulting in thinning of the interconnection line and reduced measures of yield and reliability. As a result, a reduced field oxide thickness is preferred. To keep the field oxide threshold voltage sufficiently high, a field implant process step is performed and/or a *recessed* oxide structure is grown.

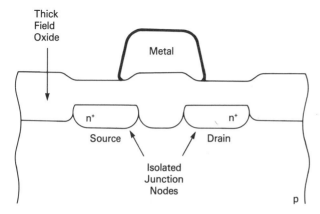

Figure 8.35. Cross section of a field (or thick) oxide FET device.

A field implant raises the impurity concentration at the semiconductor surface, thereby raising the field threshold voltage. As mentioned in Section 8.2, the presence of the field implant changes the junction capacitance and leakage current, introducing the need for both area and perimeter components.

The circuit designer must be particularly cognizant of the field oxide threshold voltage limitation in the design of any circuits that will capacitively couple a node voltage above the supply (refer to the discussion of voltage-doubler circuits in Section 11.4).

Secondary Impact Ionization Substrate Currents

One component of the chip substrate current is the reverse saturation leakage current from the reverse-biased diffusion node-to-substrate junctions. Another component to the substrate current is the leakage current from *gated surfaces* (i.e., device channel area and perimeters). A third and definitely not insignificant component of the substrate current (particularly at low temperatures) is due to the effect called *impact ionization*.

When a device is operating in the saturation region, a high electric field is present in the channel between the drain node and the inverted portion of the channel surface. For an *n*-channel device, electrons moving from source to drain, constituting a positive drain current, are accelerated by the high field at the drain end of the channel. This may result in the generation of an additional free hole–electron pair; mobile carriers are generated from the energy given up by the channel electron due to a collision with the substrate lattice. The behavior of the free electron will be described shortly. The behavior of the free hole is the subject of this discussion and the discussion following on excess secondary electron leakage.

The free holes contribute to the chip substrate current (Figure 8.36). This hole current contribution to the substrate current may result in the following conditions:

1. A *locally* higher (less negative) substrate voltage in the immediate vicinity of the device.

2. A *globally* higher (less negative) substrate voltage; if a substrate voltage generator circuit is included on-chip, the output of the voltage generator will be less negative with increasing substrate current.

The fraction of the drain current I_{DS} that contributes to the impact ionization substrate current is a function of the following:

$V_{DS} - V_{\text{pinch-off}}$:   the impact ionization current increases with increasing V_{DS} $- V_{\text{pinch-off}}$, as this increases the accelerating field in the saturation region of the channel.

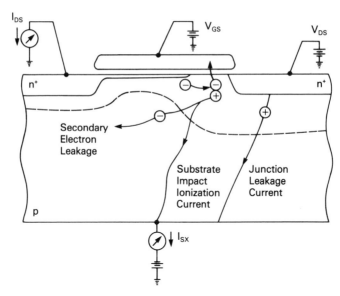

Figure 8.36. Cross section of a device in saturation, illustrating the additional (hole) substrate current due to impact ionization.

V_{GS}: the impact ionization current decreases with increasing V_{GS}.

L: the impact ionization current increases with decreasing device length.

Typical values for the substrate impact ionization current range from 10^{-4} to 10^{-6} times I_{DS} for n-channel depletion-mode load device lengths of 3 to 5 microns, $V_{DS} = 5$ V, $V_{GS} = 0$ V, $V_T = -2.2$ V, and $V_X = -3$ V. Note that this configuration is an example of a device that is operating in the saturation region under dc conditions. Transient switching through the saturation region will result in a transient impact ionization current.

Excess Secondary Electron Leakage

The conditions of n-channel device operation that produce the hole current discussed previously also may produce an excess electron current in the substrate. Holes generated by impact ionization near the drain of the device are accelerated through the drain–substrate depletion region. They may gain sufficient energy to cause a secondary ionization, generating an excess concentration of electrons (minority carriers) in the p-type substrate. The excess electrons may increase the leakage current of adjacent diffusion nodes (if the bulk recombination rate is low) and may, in particular, reduce the storage time of dynamic circuit nodes. This excess electron current will be proportional to the hole impact ionization current, with a proportionality constant on the order of 10^{-4}; that is, $I_{\text{excess electron}} = 10^{-4}$

* $I_{\text{substrate,impact ionization}}$. Plots of the magnitudes of the impact ionization and excess electron leakage currents are given in Figure 8.37.

Hot Electron Effects

The long-term (or end-of-life) drift of MOS device parameters is in large part dependent on *hot electron* effects, specifically, the emission of electrons into the gate oxide of an *n*-channel device. These electrons may emanate from either the substrate or the device channel. The trapping of these electrons in the oxide results in a drift of the device's threshold voltage and/or a degradation in device characteristics.

Two hot electron effects are of importance:

1. Substrate hot electrons due to substrate leakage currents (Figure 8.38): Electrons generated in the channel depletion region or diffusing from the bulk neutral region of the substrate drift toward the Si–SiO₂ surface. They are

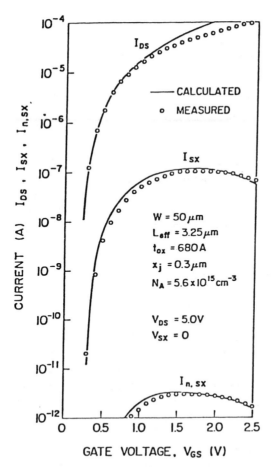

Figure 8.37. Relative magnitudes of secondary device currents due to high drain fields in saturation. I_{sx}: hole impact ionization current; $I_{n,sx}$: secondary electron leakage current. (From P. K. Chatterjee, "Device Modeling for Submicron FET Integrated Circuits." *Proceedings of the Fourth Biennial University/Government/Industry Microelectronics Symposium,* 1981, p. VII-42. Copyright © 1981 by IEEE. Reprinted with permission)

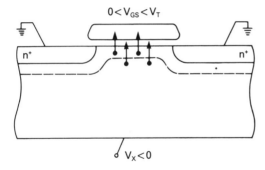

Figure 8.38. Origin of substrate hot electrons, which may gain sufficient energy to enter and be trapped in the gate oxide.

accelerated by the high field in the surface depletion region. Those incident on the surface with enough energy to overcome the Si–SiO$_2$ contact potential energy barrier are emitted into the gate oxide. The substrate hot electron effect increases with increasing temperature and increasing background impurity concentration.

2. Channel hot electrons due to saturation device currents at large drain-to-source voltages (Figure 8.39): Electrons flowing from source to drain gain energy in the high-field (saturation) region near the drain. Those arriving at the substrate–oxide surface with enough energy to surmount the barrier potential are emitted into the gate oxide. Free electrons generated by impact ionization may also contribute to the injected electron current. The channel hot electron emission current is restricted to be in the vicinity of the drain node and is a function of the applied terminal voltages, the drain node junction depth and profile, the device channel length, and temperature.

Hot Electron Effect Contribution to End-of-life Device Parameter Drift

(1) Substrate hot electron effect.   Because the gate oxide contains empty electron states (also known as *traps*), some of the injected hot electrons will not reach the gate interconnection. Those that do reach the gate constitute

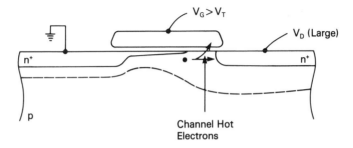

Figure 8.39.   Origin of saturation channel hot electrons.

a positive gate current. Assuming that the trapping species has a cross section σ and a spatial density per unit area of N_T, the trapped charge density as a function of time is given by (reference 8.7)

$$N(t) = N_T \{1 - \exp[-\sigma * N_{\text{injected}}(t)]\}$$

where the density of injected electrons $N_{\text{injected}}(t)$ is given by

$$N_{\text{injected}}(t) = \frac{1}{q} \int_0^t J_{\text{injected}}(\tau) * d\tau$$

In short, the density of electron states in the gate oxide that remain *unoccupied* decreases exponentially with time; eventually, the traps will all be filled.

Assuming that the trapping of the injected electron current density is *uniform* over the thickness of the gate oxide (t_{ox}) *and* over the area of the device (for the *substrate* hot electron effect), the resultant shift in the *n*-channel device threshold voltage over time is

$$\Delta V_T = \frac{q}{2} * \frac{N_T}{C_{\text{ox}}} \{1 - \exp[-\sigma * N_{\text{injected}}(t)]\}$$

The trapped electron charge density acts like a contribution to the fixed oxide charge term in the expression for the device threshold voltage. Due to the polarity of the trapped charge, the resulting shift in the *n*-channel device threshold voltage is *positive* (Figure 8.40). The injected current (and therefore the injected electron density) is determined experimentally from a measurement of the gate current, using the relationship

$$I_{\text{gate}} = I_{\text{injected}} - q * A * \frac{dN}{dt}$$

where N is the trapped charge density per unit area.

The substrate hot electron effect is a function of the device length (recall

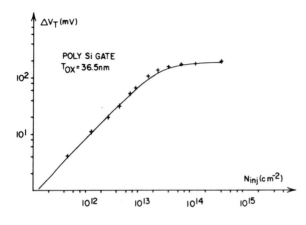

Figure 8.40. Threshold voltage shift versus injected electron density. (From P. Cottrell, R. Troutman, and T. Ning, "Hot Electron Emission in N-Channel IGFET's." *IEEE Transactions on Electron Devices*, Vol. ED-26, No. 4 (April 1979), p. 526. Copyright © 1979 by IEEE. Reprinted with permission)

from the discussion of short-channel effects that, as the device length decreases, the depletion regions between the source/drain diffusions and the substrate play an increasingly important role). As the device length decreases, the substrate hot electron shift in the threshold voltage *decreases* in magnitude; a larger fraction of the injected electron density is swept to the diffusion nodes, as opposed to the device surface (Figure 8.41). A quantitative measure of the effect of channel length on the substrate hot electron effect is given in Figure 8.42. The hot electron emission current is increased at the edges of the device width, due to the increased electric field at the transition from gate to field oxide (Figure 8.43). The edge field is increased (and therefore so is the hot electron effect) by narrower-width devices and increased field oxide surface doping (as provided by a field threshold implant).

(2) Channel hot electron effect.   Again, a fraction of the emitted electrons is trapped in the gate oxide. The trapped charge reduces the gate electric field *locally* (in the vicinity of the drain junction) and results in a decrease in device current. This local distribution of trapped oxide charge is particularly significant if the device is operated bidirectionally. Figure 8.44 gives experimental data plotting the fractional change in device current versus the time duration in

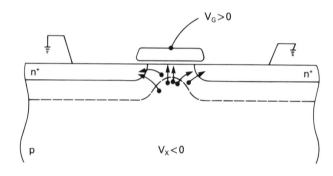

Figure 8.41.   As the device length is reduced, the substrate hot electron injection into the gate oxide is also reduced.

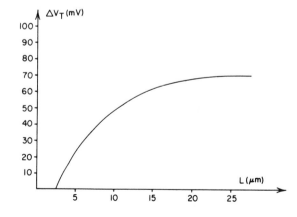

Figure 8.42.   Dependence of the substrate hot electron effect on the device channel length; the plot describes the V_T shift due to oxide trapping after 19 hours at $V_{SX} = 15$ V. (From P. Cottrell, R. Troutman, and T. Ning, "Hot Electron Emission in N-Channel IGFET's." *IEEE Transactions on Electron Devices,* Vol. ED-26, No. 4 (April 1979), p. 526. Copyright © 1979 by IEEE. Reprinted with permission)

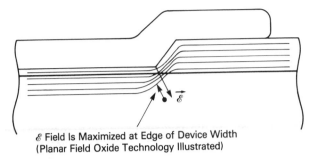

 ℰ Field Is Maximized at Edge of Device Width
(Planar Field Oxide Technology Illustrated)

Figure 8.43. The substrate hot electron emission is increased at the channel width edges due to the higher electric field at the transition from thin to field oxide. (Note the crowding of equipotential lines at the device edge.)

which the device was biased in saturation with the opposite drain-to-source voltage polarity.

  The previous discussions used *n*-channel devices for illustration and presentation of experimental results. Hot hole effects in *p*-channel devices are not as prevalent due to the reduced carrier velocities.

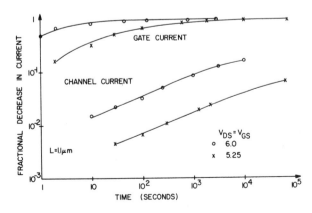

Figure 8.44. Experimental data illustrating the fractional decrease in reverse-mode current after a period of time operating in forward (saturation) mode. (From P. Cottrell, R. Troutman, and T. Ning, "Hot Electron Emission in N-Channel IGFET's." *IEEE Transactions on Electron Devices*, Vol. ED-26, No. 4 (April 1979), p. 526. Copyright © 1979 by IEEE. Reprinted with permission)

Forward-bias Junctions and Guard Rings

When an $n+$ diffusion to p-substrate junction is driven or capacitively coupled into its forward-biased region of operation, electrons are injected into the substrate. This excess minority carrier density either recombines with majority carriers in the substrate or increases the leakage current of adjacent nodes. The amount of excess leakage current collected at an adjacent node depends on the geometries of the injecting and collecting nodes, the distance between them, the location of other neighboring nodes, and the location of the closest substrate contact.

In general, conditions should be avoided that lead to forward biasing of diffusion–substrate junctions. Where forward biasing is a possibility, the physical layout should include a diffusion node collector surrounding the junction, in *guard ring* fashion. Such a condition may arise for chip output pads. A transmission-line reflection due to an impedance mismatch at the far end of a printed circuit card wiring trace may result in an incident pulse edge back at the pad that will forward bias a device node-to-substrate junction. The chip physical layout may incorporate a guard ring surrounding pad driver devices, biased to collect the injected excess minority carriers before they reach sensitive (internal) circuit nodes. This guard ring and collecting node design constraint is particularly important in CMOS logic technologies, where the large transient excess minority carrier current constitutes a trigger current for an SCR (*pnpn*) structure which can cause latch-up.

Gate Oxide and Junction Breakdown

The gate oxide breakdown voltage can be calculated directly from the dielectric breakdown strength of silicon dioxide, which is in the neighborhood of 5×10^6 V/cm.[5] By design, this should provide sufficient margin for logic technology applications. However, the gate breakdown voltage is reduced by the presence of localized oxide imperfections (e.g., *pinholes* and *weak spots*). The gate oxide also demonstrates a cumulative degradation over time when subjected to high electric field stress, due to cumulative electron trap state generation and occupation (and the internal oxide field that results; reference 8.8). In either case, the oxide breakdown is a consequence of localized resistive heating and eventually leads to *destructive* arcing. The key design consideration is that the oxide breakdown is indeed destructive. The energy of an electrostatic transient on a chip package pin must be diverted from the gate oxide input of a receiver circuit by the addition of an input protection structure, which utilizes the *nondestructive* nature of junction avalanche breakdown. (An electrostatic transient may not cause breakdown

---

[5] In actuality, the breakdown strength is a function of oxidation and anneal conditions, oxide charges, substrate impurity concentration, surface crystallographic orientation, surface preparation, and a number of other factors.

at once, but will nevertheless bring the device oxide closer to breakdown. Section 11.9 discusses input pad protection circuits.)

For a high impurity concentration device node located in a lightly doped substrate, the junction breakdown mechanism is denoted as *avalanche multiplication*. In this region, the saturation current carriers create additional free hole–electron pairs in the reverse-bias depletion region by impact ionization, the same phenomenon described earlier. At sufficiently high fields, secondary ionizations will occur; effectively, the carrier density exiting the depletion region will have multiplied by some large factor. The junction breakdown voltage can be calculated from the maximum (or critical) depletion region electric field, which for a diffused node in a lightly doped substrate (e.g., $N_D \gg N_A$) is related to the reverse-bias voltage by

$$V_B = \frac{\epsilon_{Si} * \mathscr{E}_{crit}^2}{2 * q * N_A}$$

For a background doping of $10^{15}/cm^3$, the critical electric field is roughly 2×10^5 V/cm, yielding a breakdown voltage in excess of 100 V.[6] Unfortunately, this will not offer much protection to gate oxides. The incorporation of a field oxide threshold voltage implant region adjacent to the diffused node will reduce the breakdown voltage to some degree; however, the most common structure used for its reduced breakdown voltage is a *gate-field enhanced* drain node (Figure 8.45). With this structure, the presence of the gate terminal (and its associated voltage bias) increases the electric field intensity in the depletion region between the drain node and the device accumulation layer. As a result, the avalanche mechanism in this region is substantially enhanced, which results in a much reduced breakdown voltage (sufficiently low to protect a device input oxide capacitance). The gate terminal can implement either a thick or thin oxide device cross section, depending on the magnitude of the desired avalanche breakdown voltage reduction. Junction breakdown is not a destructive phenomenon if interconnection current density limits and resistive thermal dissipation are controlled. It is indeed well suited to offer chip pad input protection.

Punchthrough

The term *punchthrough* has been applied to a second-order effect occurring in both bipolar and field-effect devices. In both cases, it refers to the condition where the depletion region extending from one terminal of the device (the collector for bipolar junction transistors, the drain for a MOS field-effect transistor) approaches that of another terminal of the device (bipolar emitter or MOS source node) due

---

[6] S. M. Sze, *Physics of Semiconductor Devices*, New York: John Wiley & Sons, Inc., 1969.

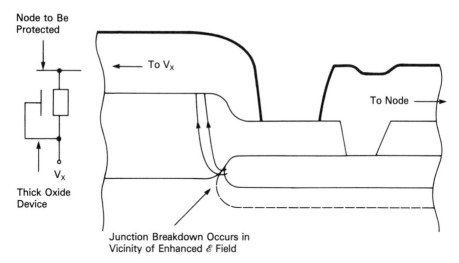

Figure 8.45. Cross section of junction breakdown structure used to protect device gate oxides. The lower breakdown voltage is a result of the enhanced electric field in the vicinity of the drain node due to the thick oxide gate bias.

to a large reverse-bias voltage on the first terminal. For increasing drain-to-source voltage beyond punchthrough, carriers injected by the source node into the channel depletion region are swept by the lateral electric field and collected by the drain node. As a result, a large drain-to-source current may flow in a normally *off* device ($V_{GS} < V_T$). As the device length decreases, the drain depletion region reaches the source node depletion region at a lower drain-to-source voltage. Therefore, the punchthrough voltage is a strong function of channel length: $V_{\text{punchthrough}} \propto L^2$.

Punchthrough need not necessarily apply only to drain-to-source current; it may be extended to node-to-node current between adjacent diffusions, particularly if their separation is less than or comparable to the minimum device length.

As device geometries continue to be reduced, punchthrough will no longer be simply a second-order effect, but may be a severely limiting factor. This is particularly true considering that the supply voltage *VDD* has typically not been reduced in proportion to the channel length, in order to maintain compatibility with older technologies. Modifications to the standard +5-V supply voltage will very shortly have to be addressed.

For a device operating beyond punchthrough, both the gate voltage and the substrate voltage modulate the drain-to-source punchthrough current. For example, increasing the substrate reverse bias will alter the source and drain depletion region field distribution in a manner so as to *reduce* the drain-to-source punchthrough current.

Electromigration of Integrated-circuit Metallurgy
(Current Density Limitations)

A phenomenon observed with the metallurgy used for IC interconnections is that of *electromigration,* that is, the transport of mass within a metal line subjected to high current densities. This transport of material eventually manifests itself as an open circuit in the interconnection, due to the creation of a localized void in the material.

A thermally excited metal atom in its crystalline lattice location (or on the metal surface or at a crystalline grain boundary) may be dislodged from that location due to a collision with a conducting electron. The energy imparted to the metal ion may free that atom from its bound potential energy state. As a result of the momentum transfer to the ion, the ions move *in the direction* of electron flow in the metal (i.e., *in the opposite direction* to the current flow in the wire). Conversely, vacancies in the metal lattice move in the direction of current flow; dislodged metal atoms tend to fill the vacancies as they are in motion. Vacancies in the metal lattice tend to condense, whereupon a void results. The dislodged atoms condense at the grain boundaries of the metallurgy, forming *whiskers* and *hillocks.*

A simple theory for the calculation of the *mean time to fail* for the interconnection is of the form (reference 8.9):

$$\frac{1}{\text{mean time to fail}} \propto A^{-1} * J^2 * \exp[-(E_A/kT)]$$

where A is the cross-sectional area of the interconnection, J is the dc current density, and E_A is a characteristic activation energy, which is a strong function of the crystalline structure and composition of the film. For small grain size structure thin films, the mass transport is enhanced by the presence of the grain boundaries; an activation energy of 0.48 eV was reported for small grain structure aluminum (reference 8.10). Conversely, an activation energy of 0.84 eV was reported for very large grained aluminum films. Integrated-circuit applications utilize a dielectric film over the interconnection layer, which tends to reduce the surface transport of metal atoms and inhibit the formation of hillocks. The predominant IC electromigration failure mechanism is localized melting due to the temperature increase of current flow in the vicinity of a high vacancy concentration.

The electromigration of metal atoms is reversible in the sense that a reversal of the direction of current flow reverses the momentum direction imparted to the atoms. The reversibility of this phenomenon is dependent on the extent to which the vacancies have concentrated into a lower potential energy void in the metal; the larger the void size, the less likely that migrated ions will reverse their action. This can be demonstrated by evaluating the resistance of the interconnection under high current density stress over time. The resistance increases as electromigration progresses, but begins to decrease (reversing the process) with a reversal in the current direction. The impact on the chip designer is that the net zero charging and discharging transient current in global signal interconnections is not an electromigration concern. However, design system image power supply and any internal circuit interconnections with a net time-averaged dc current must be designed to be of sufficient width to ensure a long mean time to failure.

The presence of a silicon solute into an aluminum wire in the immediate vicinity of a contact also presents electromigration concerns. Silicon dissolved in the aluminum also migrates in the direction of electron flow. This introduces another IC failure mechanism, the transport of silicon into the aluminum at high contact current densities, resulting in an *etch pit* of the node surface area. This pit can continue to grow through the shallow junction node, leading to an electrical short to the substrate. In addition to the wiring current density, the designer must also ensure that the contact current density is suitably limited.

Considerable research efforts are being dedicated to metallurgy options for IC interconnections to reduce the electromigration mass transport; presently, aluminum–copper alloys are commonly used (e.g., Al–5% Cu).

This section has described some of the additional device modeling and layout considerations for phenomena associated with the evolution of VLSI technologies. These characteristics are not to be overlooked when developing the process technology, constructing the chip image, or calculating long-term chip reliability.

8.9 ELECTRICAL TEST STRUCTURES FOR PARAMETER EVALUATION

The discussion in Section 8.5 presented some of the experimental procedures used to measure the functional dependencies of second-order MOS device parameters. In this section, additional experimental techniques are introduced to measure device and process electrical parameters.

Threshold Voltage

Recalling the discussion of subthreshold current in Section 8.5, it is evident that the definition of the device threshold voltage as the gate-to-source voltage V_{GS}, where the device current I_{DS} goes to zero, is subject to interpretation. Two possible methods for experimentally measuring the device threshold voltage are by means of (1) device current density and (2) extrapolation of linear region current measurements.

1. *Current density method:* The threshold voltage is defined to be the gate-to-source voltage at the input with a particular saturated current density flowing from drain to source; for example, 50 nA $* (W_{\text{wafer}}/L_{\text{wafer}})$, where $W_{\text{wafer}} = W_{\text{design}} - \Delta W$ and $L_{\text{wafer}} = L_{\text{design}} - \Delta L$. The experimental determination of the device width bias, ΔW, and the device length bias, ΔL, will be discussed shortly.

2. *Linear region measurement method:* For devices with a low width-to-length ratio, the measurement of a small current value required by the current density method may introduce a significant error due to the tolerance limitations of the measurement equipment or of the ΔW and ΔL parameters. The linear method involves plotting V_{GS}, I_{DS} measurements, with V_{DS} fixed at a low value (e.g., 0.1 V); the device is biased to operate in the linear region. As illustrated in Figure 8.46, the linear region measurements are curve-fitted and extrapolated back to the intercept on the V_{GS} axis. This calculated point is equated to $V_T + V_{DS}/2$.

$\Delta L, \Delta W$

In the expressions for device current written previously, the device dimensional factors W and L represented the actual physical (wafer level) device dimensions. These dimensions are related to the design (or artwork) level dimensions by the process biases, ΔL and ΔW.

The change in the device length from the design dimension is due to the source–drain node outdiffusion and the polysilicon linewidth process bias (for

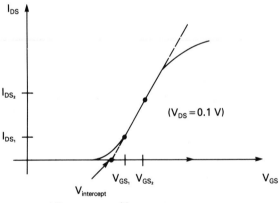

$$I_{DS} = \mu \frac{\epsilon_{ox}}{t_{ox}} \left(\frac{W}{L}\right) \left(V_{GS} - V_T - \frac{V_{DS}}{2}\right) (V_{DS})$$

Slope of Experimental Line: $\dfrac{I_2 - I_1}{V_2 - V_1}$

Solving Equation of Line for $I = 0$:

$$V_{intercept} = \frac{I_2 V_1 - I_1 V_2}{I_2 - I_1}$$

Threshold Voltage: $V_T = V_{intercept} - \dfrac{V_{DS}}{2}$

Figure 8.46. Linear region method for experimentally measuring the device threshold voltage.

polysilicon gate technologies). The extent of the junction outdiffusion is controlled by the following:

Ion implant energy and dose

Implanted (donor/acceptor) species

Time and temperature of subsequent (high-temperature) processing steps.

To experimentally determine ΔL, several wide-width devices with varying channel lengths are used. Wide-width devices are selected in order to minimize any narrow-width effects on device threshold or carrier mobility. The linear region device current equation should incorporate the short-channel effects on the device threshold and carrier mobility:

$$I_{DS} = \mu \frac{\epsilon_{ox}}{t_{ox}} \frac{W}{L} \left(V_{GS} - V_T - \frac{V_{DS}}{2} \right) V_{DS}$$

where $V_T = V_T(L, W)$, $L = L_{design} - \Delta L$, $W = W_{design} - \Delta W$, and

$$\mu = \mu_{eff} * \left[\frac{1}{1 + (\beta/W) + \gamma * (V_{DS}/L) + \theta * [V_{GS} - V_T - (V_{DS}/2)]} \right]$$

For wide-width devices, the following relationships can be used: $W \approx W_{design}$, $1/W \approx 0$, and $V_T = V_T(L)$. The reciprocal of the linear current equation reduces to

$$\frac{1}{I_{DS}} = \frac{L_d - \Delta L}{\mu_{eff} * C_{ox} * V_{DS} * W_d} \left\{ \frac{1 + \gamma * [V_{DS}/(L_d - \Delta L)]}{V_{GS} - V_T - (V_{DS}/2)} + \theta \right\}$$

The experimental method requires measuring I_{DS} versus $V_{GS} - V_T(L) - (V_{DS}/2)$, varying $V_{GS} - V_T$. This measurement is repeated for a number of devices with different design lengths. For each device, plot $1/I_{DS}$ versus $[V_{GS} - V_T(L) - (V_{DS}/2)]^{-1}$. This data can be fitted with a straight line whose slope is

$$\frac{d(1/I_{DS})}{d\{[V_{GS} - V_T - (V_{DS}/2)]^{-1}\}} = \frac{L_d - \Delta L + \gamma * V_{DS}}{\mu_{eff} * C_{ox} * V_{DS} * W_d}$$

and whose intercept point at $[V_{GS} - V_T - (V_{DS}/2)]^{-1} = 0$ is equal to

$$\left. \frac{1}{I_{DS}} \right|_{intercept} = \frac{L_d - \Delta L}{\mu_{eff} * C_{ox} * V_{DS} * W_d} * \theta$$

Plots of the slope and intercept values versus design dimension L_d can again be fitted by straight lines. The derivative of the plot of the slopes versus L_d is $1/(\mu_{eff} * C_{ox} * V_{DS} * W_d)$, which facilitates the calculation of μ_{eff}, since the remaining terms are known. The plot of the intercept points $(1/I_{DS})_{intercept}$ versus L_{design}, when extrapolated to the L_d axis, yields ΔL.

The change in the device width direction, ΔW, is due primarily to the process etch bias of the masking layer that defines the gate oxide area. For a technology with a planar field oxide, this will be the field oxide etch opening lithography layer. For a nonplanar (recessed) field oxide, the bias of the selective masking layer differentiating the field oxide areas (plus the intrusion of the recessed oxide growth into the channel area) determines the device width bias. To experimentally determine ΔW, a technique similar to that used for calculating ΔL can be adopted. Measure $I_{DS,\text{linear}}$ versus V_{GS} for a number of long devices of varying widths, determine the width-related factors of the current equation, and plot the curve-fitted data versus W_{design} to isolate ΔW:

$$\frac{1}{I_{DS}} = \frac{L}{\mu_{\text{eff}} * C_{\text{ox}} * V_{DS} * (W_d - \Delta W)} * \left[\frac{1 + \gamma * \cancel{\frac{V_{DS}}{L}}^{\,0} + \cancel{\frac{\beta}{W}}^{\,0}}{V_{GS} - V_T - (V_{DS}/2)} + \theta \right]$$

Initially, measure and plot $1/I_{DS}$ versus $[V_{GS} - V_T - (V_{DS}/2)]^{-1}$ for each of the devices; for each plot, extract the intercept at the $[V_{GS} - V_T - (V_{DS}/2)]^{-1} = 0$ axis:

$$\left. \frac{1}{I_{DS}} \right|_{\text{intercept}} = \frac{L * \theta}{\mu_{\text{eff}} * C_{\text{ox}} * V_{DS} * (W_d - \Delta W)}$$

Subsequently, plot the value of $I_{DS\text{intercept}}$ versus W_d for the range of devices measured and fit to a straight line; extrapolate the line to the W_{design} axis to determine ΔW.

The technique applicable for both the ΔL and ΔW measurements is depicted in Figure 8.47. In Figure 8.47d, note that the ΔW process bias is depicted as a *negative* quantity (i.e., the wafer-level device width is *greater* than the artwork dimension). This will commonly be the case for planar technologies, where a subtractive etch of the field oxide is performed.

Sheet Resistivity

The sheet resistivity measurement can be made using a *resistive cross* (also known as a *van der Pauw cross*). Figure 8.48 illustrates a *four-point probe* layout for the sheet resistivity measurement. A current is forced between two adjacent pads, with the corresponding voltage difference measured between the remaining two pads. The sheet resistance is determined from the van der Pauw formula:[7]

$$\rho_s = f * \frac{\pi * R_{\text{measured}}}{\ln 2}$$

where $R_{\text{measured}} = V/I$ and f is a correction factor related to the symmetry of the

---

[7] L. J. van der Pauw, "A Method of Measuring Specific Resistivity and Hall Effect of Disc or Arbitrary Shape," *Philips Research Reports,* 13:1 (February 1958) 1–9.

(a) Plot $(I_{DS})^{-1}$ vs. $(V_{GS} - V_T(L) - \frac{V_{DS}}{2})^{-1}$:

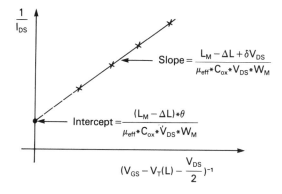

(b) Plot the slope and intercept of each of the graphs in (a) versus the mask channel dimension L_M for the different devices:

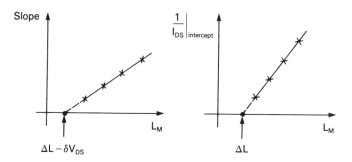

(c) Plot $(I_{DS})^{-1}$ vs. $(V_{GS} - V_T - \frac{V_{DS}}{2})^{-1}$ for each different device width:

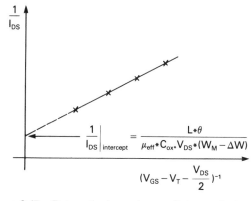

(continues)

Figure 8.47. Data gathering and curve fitting method used to determine ΔL, (a), (b), and ΔW, (c), (d).

(d) Plot $I_{DS}|_{intercept}$ vs. W_M for the different devices plotted in (c); extrapolate to determine ΔW:

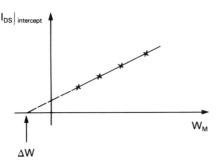

Figure 8.47. (*Continued*)

structure and the ratio of the arm length to width. For the cross of Figure 8.48, it is assumed that the impedance of the voltage taps does not significantly perturb the current flow in the cross; subsequently, $f \approx 1$. To eliminate any measurement-related voltage offset error, the resistance can be measured with both polarities of current and at two different 90° orientations.

The accuracy of the measurement system is indicated by the error term

$$\alpha = \frac{|R_+ - R_-| + |R_+^{90} - R_-^{90}|}{2 * R_{average}}$$

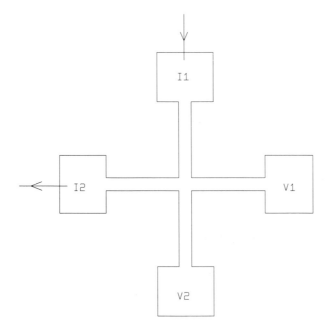

Figure 8.48. Four-point probe measurement structure for determining the sheet resistivity, ρ_s. This structure is known as a van der Pauw cross.

An alternative structure for determining both the sheet resistivity and the electrical width bias is illustrated in Figure 8.49. Unlike the cross structure, which typically requires an accurate measure of relatively low voltages for moderate forcing currents (1 to 10 mV), the stripe structure of Figure 8.49 provides an easier measurement to execute (and potentially automate) with typical tester hardware. The measurements of the voltage drop for both the narrow and wide stripes, given a fixed value of forcing current, can be used to determine the process width bias W_{bias} for the interconnection level:

$$V_{\text{Wide}} = \left(\frac{L_{\text{Wide}}}{W_{\text{Wide}} - W_{\text{bias}}}\right) * I * \rho_s ; \qquad V_{\text{Narrow}} = \left(\frac{L_{\text{Narrow}}}{W_{\text{Narrow}} - W_{\text{bias}}}\right) * I * \rho_s$$

Solving these two relationships for W_{bias} and ρ_s yields

$$W_{\text{bias}} = \frac{V_N W_N L_W - V_W W_W L_N}{V_N L_W - V_W L_N}$$

$$\rho_s = \frac{V_N}{I_{\text{force}}} * \frac{W_N - W_{\text{bias}}}{L_N}$$

Contact Resistance

A measurement of the contact resistance (and specific contact resistivity) is commonly made using a *contact chain*. For a chain of contacts, only two probe pads are required. A current is forced and the resulting voltage difference is measured between the same pads. For an accurate measurement of the average contact resistance, the series resistances of the 'links' in the chain must be subtracted

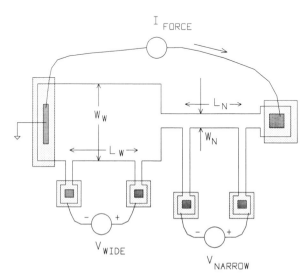

Figure 8.49. Sheet resistivity measurement test structure (note both narrow- and wide-width stripe measurements, used to determine $\Delta W_{\text{electrical}}$).

from the total calculated resistance. Dividing this difference by the number of contacts in the chain results in the average contact resistance:

$$\overline{R_{con}} = \frac{(V_{measured}/I_{forced}) - R_{series}}{\text{no. of contacts}}$$

The series resistance term can be written as

$$R_{series} = \rho_s * \frac{L - L_{bias}}{W - W_{bias}} * \text{no. of links in contact chain}$$

where ρ_s is the sheet resistivity of the series interconnection, L is the design dimension between contacts, and W is the design width of a link in the chain. Additionally, L_{bias} is the process bias due to the contact hole etch (measured optically), while W_{bias} is the process bias for the electrical width of the interconnection, as determined from the sheet resistivity measurement. The specific contact resistivity can then be calculated from the results for the contact resistance, using the expression given in Section 8.3.

Electrical Screen Tests

In addition to process characterization tests, a number of electrical tests are applied to *each* chip site at wafer-level testing, prior to exercising the chip against its Boolean test patterns. In this manner, those sites with catastrophic (electrical) defects are detected and skipped, saving tester time. These tests can also provide valuable diagnostic information about the particular process run and/or the design itself. Some of the common electrical screen tests include the following:

(1) Pad-to-substrate leakage. To ensure that signal I/O (and power) pads are not directly shorted to the substrate, all pads should be tested for (maximum) pad-to-substrate reverse-biased leakage currents; measured values above the anticipated junction leakage currents would be regarded as a failing site. The chip failure modes that lead to a pad current failure include metal interdiffusion through the junction and device input gate oxide dielectric shorts. During this test, power is not applied to the chip; each pad-to-substrate junction is measured independently.

In addition, to verify that the test probe is indeed making connection to the pad, a forward-bias test between pad and substrate could be performed. (Note that all I/O pads should be connected to a protective circuit containing a junction or device node-to-substrate diode. This provides some measure of electrostatic discharge dissipation and should bypass the permanently destructive dielectric breakdown of the input gate oxide.)

(2) High-impedance testing. Another common electrical screen test is to verify the high-impedance condition of disabled off-chip driver circuits. This testing condition is not a regular part of the chip test pattern sequence, yet should be included to ensure chip functionality. An additional chip input is typically dedicated as a Test Enable signal, which may be directly asserted at wafer-level testing to hold off all driver devices. The measurement of the high-impedance condition may take a variety of forms, yet all are designed to ensure that no active pullup or pulldown current is being provided at the chip pad.

Additional electrical screen tests may be developed to give a higher measure of confidence that a site that passes the logic test pattern set does indeed meet performance and power dissipation criteria. Examples of these measurements include the following:

1. Logic paths between input and output pads where the path delay can be easily determined
2. Average power supply current to the chip during application of test patterns
3. Average substrate current

Problems 8.6 through 8.8 request that additional parametric and chip screening tests be developed to try to more fully encompass the acquisition of process characterization data and chip functionality.

PROBLEMS

8.1. Develop a methodology for providing statistical process modeling for circuit simulation. Specifically, describe how the process and device models are to be constructed to allow some *mismatch* in the global process parameters between devices. The potential mismatch should be a function of the *separation* between devices, a parameter specified by the designer in the main circuit model.

8.2. (a) Describe how the process and device model libraries should be constructed to facilitate multiple-technology simulation. What restrictions are imposed on device model and process parameter names?

(b) Enhance the structure developed in part (a) to be able to simulate among different chips in the same technology, where the circuit models of separate chips use different sets of process parameter values. How do you propose the main model be hierarchically structured to best facilitate single-technology, multiple-chip simulation?

8.3. For ρ_s small, show that the expression for the contact resistance between materials of similar resistivities reduces to

$$R_c = \frac{\rho_c}{W * l}$$

8.4. Develop a number of compatible sets of simulation units for the following simulation variables: voltage, current, resistance, inductance, capacitance, and time (frequency). The following relationships should be satisfied by the set of units selected:

$$v = i * R, \qquad i_c = C * \frac{dv}{dt}, \qquad v_L = L * \frac{di}{dt}$$

Which set is the most useful for IC designs? For transmission line analysis?

8.5. Why are aluminum–copper and aluminum–copper–silicon metallurgies used for IC wiring levels? What are the common percentages of Cu and Si that are used? What additional fabrication complexities does a composite metallurgy introduce in terms of deposition, etching, and annealing?

8.6. Develop test measurement techniques and design the appropriate layout structures to facilitate the following process parameter characterization:

Field oxide threshold voltage

V_T versus V_{SX}

V_T tracking/mismatch versus separation

t_{ox}

t_{ox} tracking/mismatch versus separation

8.7. For chip testing, assume that each design has an input pin (or combination of pins) that can be asserted to put all the push–pull off-chip driver circuits into their high-impedance condition. Develop a test measurement technique to determine that an off-chip driver correctly functions as a high-impedance connection. To assess chip functionality, a high-impedance maximum leakage current specification must be imposed. What maximum leakage current was assumed in developing the measurement technique?

8.8. Another desirable electrical screen measurement is the input switching threshold of all chip receiver circuits. Develop a chip design technique to easily facilitate a measurement of individual chip receiver switching thresholds, using a *single* output pad for all receiver measurements and a minimal amount of additional internal circuitry. (*Hint:* Consider a *ladder* circuit, where a string of two-input NAND books are used, one input driven by the previous book in the ladder and the other driven by a receiver output.)

REFERENCES

8.1 Brews, John, "Subthreshold Behavior of Uniformly and Nonuniformly Doped Long-Channel MOSFET," *IEEE Transactions on Electron Devices,* ED-26:9 (September 1979), 1282–1291.

8.2 Rideout, V. L., Gaensslen, F. H., and LeBlanc, A., "Device Design Considerations for Ion Implanted n-Channel MOSFETs," *IBM Journal of Research and Development* (January 1975), 50–58.

8.3 Wang, Paul, "Device Characteristics of Short-Channel and Narrow-Width MOSFET's," *IEEE Transactions on Electron Devices,* ED-25:7 (July 1978), 779–786.

8.4 Frohman-Bentchkowsky, D., and Grove, A. S., "Conductance of MOS Transistors in Saturation," *IEEE Transactions on Electron Devices,* ED-16:1 (January 1969), 108–113.

8.5 Merckel, G., and others, "An Accurate Large-Signal MOS Transistor Model for Use in Computer-Aided Design," *IEEE Transactions on Electron Devices,* ED-19:5 (May 1972), 681–690.

8.6 Chatterjee, P. K., "Device Modeling for Submicron FET Integrated Circuits," *1981 University, Government, and Industry Microelectronics Symposium (UGIM),* pp. VII 32–43.

8.7 Cottrell, P., Troutman, R., and Ning, T., "Hot-Electron Emission in N-Channel IGFET's," *IEEE Transactions on Electron Devices,* ED-26:4 (April 1979), 520–533.

8.8 Cobbold, R., *Theory and Applications of Field-Effect Transistors.* New York: Wiley–Interscience, Sections 6.1–6.4.

8.9 Bobbio, A., and others, "Electromigration Failure in Al Thin Films under Constant and Reverse DC Powering," *IEEE Trasactions on Reliability,* R-23:3 (August 1974), 194–202.

8.10 Black, J. R., "Electromigration—A Brief Summary and Some Recent Results," *IEEE Transactions on Electron Devices,* ED-16:4 (April 1979), 338–347.

8.11 Murrmann, H., and Widmann, D., "Current Crowding on Metal Contacts to Planar Devices," *IEEE Transactions on Electron Devices,* ED-16:12 (December 1969), 1022–1024.

8.12 Chang, W. H., "Analytical IC Metal-Line Capacitance Formulas," *IEEE Transactions on Microwave Theory and Techniques,* MTT 24:9 (September 1976), 608–611.

Additional References

Antognetti, P., and others, "CAD Model for Threshold and Subthreshold Conduction in MOSFET's," *IEEE Journal of Solid-State Circuits,* SC-17:3 (June 1982), 454–459.

Antoniades, D. A., "Calculation of Threshold Voltage in Nonuniformly Doped MOSFET's," *IEEE Transactions on Electron Devices,* ED-31:3 (March 1984), 303–307.

Baccarani, G., and Wordeman, M. R., "Transconductance Degradation in Thin-Oxide MOSFET's," *IEEE Transactions on Electron Devices,* ED-30:10 (October 1983), 1295–1304.

Barker, J. R., and others, "On the Nature of Ballistic Transport in Short-Channel Semiconductor Devices," *IEEE Electron Device Letters,* EDL-1:10 (October 1980), 209–210.

Baum, G., and Beneking, H., "Drift Velocity Saturation in MOS Transistors," *IEEE Transactions on Electron Devices,* Correspondence (June 1970), 481–482.

Brews, J. R., Nicollian, E. H., and Sze, S. M., "Generalized Guide for MOSFET Miniaturization," *IEEE Electron Device Letters,* EDL-1:1 (January 1980), 2–4.

Büget, U., and Wright, G. T., "Space-Charge-Limited Current in Silicon," *Solid-State Electronics,* 10:3 (March, 1967), 199–207.

Chern, J. G., and others, "A New Method to Determine MOSFET Channel Length," *IEEE Electron Device Letters,* EDL-1:9 (September 1980), 170–173.

Ruehli, A. E., and Brennan, P., "Capacitance Models for Integrated Circuit Metallization Wires," *IEEE Journal of Solid-State Circuits,* SC-10:6 (December 1975), 530–536.

Cooper, J. A., and Nelson, D. F., "Measurement of the High-Field Drift Velocity of Electrons in Inversion Layers on Silicon," *IEEE Electron Device Letters*, EDL-2:7 (July 1981), 171–173.

DeLaMoneda, F., Kotecha, H., and Shatzkes, M., "Measurement of MOSFET Constants," *IEEE Electron Device Letters*, EDL-3:1 (January 1982), 10–12.

Eitan, B., and Frohman-Bentchkowsky, D., "Hot-Electron Injection into the Oxide in n-Channel MOS Devices," *IEEE Transactions on Electron Devices*, ED-28:3 (March 1981), 328–340.

——, and ——, "Surface Conduction in Short-Channel MOS Devices as a Limitation to VLSI Scaling," *IEEE Transactions on Electron Devices*, ED-29:2 (February 1982), 254–266.

Fichtner, W., and others, "Semiconductor Device Simulation," *IEEE Transactions on Electron Devices*, ED-30:9 (September 1983), 1018–1030.

Fu, K. Y., "Mobility Degradation due to the Gate Field in the Inversion Layer of MOSFET's," *IEEE Electron Device Letters*, EDL-3:10 (October 1982), 292–293.

Fukuma, M., and Okuto, Y., "Analysis of Short-Channel MOSFET's with Field-Dependent Carrier-Drift Mobility," *IEEE Transactions on Electron Devices*, ED-27:11 (November 1980), 2109–2114.

Gaensslen, F. H., "Geometry Effects of Small MOSFET Devices," *IBM Journal of Research and Development*, 23:6 (November 1979), 682–688.

——, and Aitken, J. M., "Sensitive Technique for Measuring Small MOS Gate Currents," *IEEE Electron Device Letters*, EDL-1:11 (November 1980), 231–233.

Geipel, H. J., and Fortino, A. G., "Process Modeling and Design Procedure for IGFET Thresholds," *IBM Technical Report*, TR.19.0455, June 15, 1978.

Hanafi, H., Camnitz, L., and Dally, A., "An Accurate and Simple MOSFET Model for Computer-Aided Design," *IEEE Journal of Solid-State Circuits*, SC-17:5 (October 1982), 882–891.

Hu, G. J., and others, "Design and Fabrication of P-Channel FET for 1-μm CMOS Technology," *International Electron Devices Meeting* (1982) *Technical Digest*, 710–713.

Jenkins, F. S., and others, "MOS-Device Modeling for Computer Implementation," *IEEE Transactions on Circuit Theory*, CT-20:6 (November 1973), 649–658.

Kim, M. J., "MOS-FET Fabrication Problems," *Solid-State Electronics*, 12:7 (July, 1969), 557–571.

Leburton, J. P., and Dorda, G. E., "V – E Dependence in Small-Sized MOS Transistors," *IEEE Transactions on Electron Devices*, ED-29:8 (August 1982), 1168–1171.

—— and others, "Analytical Approach of Hot Electron Transport in Small Size MOSFET's," *Solid-State Electronics*, 24:8 (1981), 763–771.

Leistko, O., Grove, A. S., and Sah, C. T., "Electron and Hole Mobilities in Inversion Layers on Thermally Oxidized Silicon Surfaces," *IEEE Transactions on Electron Devices* (May 1965), 248–253.

Lombardi, C., and others, "Hot Electrons in MOS Transistors: Lateral Distribution of the Trapped Oxide Charge," *IEEE Electron Device Letters*, EDL-3:7 (July 1982), 215–217.

Mar, J., and others, "Substrate Current Modeling for Circuit Simulation," *IEEE Transactions on Computer-Aided Design of Integrated Circuits and Systems,* CAD-1:4 (October 1982), 183–185.

Masuda, H., Nakai, M., and Kubo, M., "Characteristics and Limitation of Scaled-Down MOSFET's due to Two-Dimensional Field Effect," *IEEE Transactions on Electron Devices,* ED-26:6 (June 1979), 980–986.

Melstrand, O., and others, "A Data Base Driven Automated System for MOS Device Characterization, Parameter Optimization and Modeling," *IEEE Transactions on Computer-Aided Design of Integrated Circuits,* CAD-3:1 (January 1984), 47–51.

Meyer, John, "MOS Models and Circuit Simulation," *RCA Review,* 32:1 (March 1971), 43–63.

Neumark, G. F., and Rittner, E. S., "Transition from Pentode- to Triode-Like Characteristics in Field Effect Transistors," *Solid-State Electronics,* 10:4 (April, 1967), 299–304.

Ogura, S., and others, "Design and Characteristics of the Lightly Doped Drain-Source (LDD) Insulated Gate Field-Effect Transistor," *IEEE Transactions on Electron Devices,* ED-27:8 (August 1980), 1359–1367.

———, "A Half Micron MOSFET Using Double Implanted LDD," *International Electron Devices Meeting* (1982) *Technical Digest,* 718–721.

Oh, S. Y., Ward, D. E., and Dutton, R. W., "Transient Analysis of MOS Transistors," *IEEE Transactions on Electron Devices,* ED-27:8 (August 1980), 1571–1578.

Okumura, K., and Miyoshi, M., "A Novel Method for Measurement of the Channel Length of Short Channel MOSFET's," *Journal of the Electrochemical Society,* 129:6 (June 1982), 1338–1341.

Paulos, J. J., Antoniadis, D. A., and Tsividis, Y. P., "Measurement of Intrinsic Capacitances of MOS Transistors," *IEEE International Solid-State Circuits Conference* (1982), 238–239.

Perloff, David, "A Four-Point Electrical Measurement Technique for Characterizing Mask Superposition Errors on Semiconductor Wafers," *IEEE Journal of Solid-State Circuits,* SC-13:4 (August 1978), 436–444.

Ratnakumar, K. N., and Meindl, J., "Short-Channel MOST Threshold Voltage Model," *IEEE Journal of Solid-State Circuits,* SC-17:5 (October 1982), 937–947.

Ratnam, P., and Salama, C., "A New Approach to the Modeling of Nonuniformly Doped Short-Channel MOSFET's," *IEEE Transactions on Electron Devices,* ED-31:9 (September 1984), 1289–1298.

Reddi, V., and Sah, C. T., "Source to Drain Resistance beyond Pinch-Off in Metal-Oxide-Semiconductor Transistors (MOST)," *IEEE Transactions on Electron Devices,* ED-12:3 (March 1965), 139–141.

Richman, Paul, "Modulation of Space-Charge-Limited Current Flow in Insulated-Gate Field-Effect Tetrodes," *IEEE Transactions on Electron Devices,* ED-16:9 (September 1969), 759–766.

Rideout, V. L., and Silvestri, V., "MOSFET's with Polysilicon Gates Self-Aligned to the Field Isolation and to the Source and Drain Regions," *IEEE Transactions on Electron Devices,* ED-26:7 (July 1979), 1047–1052.

Sabnis, A. G., and Clemens, J. T., "Characterization of the Electron Mobility in the Inverted ⟨100⟩ Si Surface," *International Electron Devices Meeting* (1979), *Technical Digest,* 18–21.

Sansbury, J. D., "MOS Field Threshold Increase by Phosphorus-Implanted Field," *IEEE Transactions on Electron Devices,* ED-20:5 (May 1973), 473–476.

Schrieffer, J. R., "Effective Carrier Mobility in Surface-Space Charge Layers," *Physical Review,* 97:3 (February 1, 1955), 641–646.

Shima, T., and others, "Three-Dimensional Table Look-Up MOSFET Model for Precise Circuit Simulation," *IEEE Journal of Solid-State Circuits,* SC-17:3 (June 1982), 449–454.

Special Issue on "Hot Electrons in Semiconductors," *Solid-State Electronics,* 21:1 (January 1978).

Sun, S. C., and Plummer, J., "Electron Mobility in Inversion and Accumulation Layers on Thermally Oxidized Silicon Surfaces," *IEEE Transactions on Electron Devices,* ED-27:8 (August 1980), 1497–1508.

Taylor, G. W., and others, "A Description of MOS Internodal Capacitances for Transient Simulations," *IEEE Transactions on Computer-Aided Design of Integrated Circuits and Systems,* CAD-1:4 (October 1982), 150–156.

Tewksbury, S. K., "N-Channel Enhancement-Mode MOSFET Characteristics from 10 to 300 °K," *IEEE Transactions on Electron Devices,* ED-28:12 (December 1981), 1519–1529.

Toyabe, T., and others, "A Numerical Model of Avalanche Breakdown in MOSFET's," *IEEE Transactions on Electron Devices,* ED-25:7 (July 1978), 825–832.

Troutman, R. R., "Subthreshold Design Considerations for Insulated Gate Field-Effect Transistors," *IEEE Journal of Solid-State Circuits,* SC-9:2 (April 1974), 55–61.

Tsividis, Y. P., "Relation between Incremental Intrinsic Capacitances and Transconductances in MOS Transistors," *IEEE Transactions on Electron Devices,* ED-27:5 (May 1980), 946–948.

Ward, D. E., and Doganis, K., "Optimized Extraction of MOS Model Parameters," *IEEE Transactions on Computer-Aided Design of Integrated Circuits and Systems,* CAD-1:4 (October 1982), 163–168.

———, and Dutton, R. W., "A Charge-Oriented Model for MOS Transistor Capacitances," *IEEE Journal of Solid-State Circuits,* SC-13:5 (October 1978), 703–707.

Weeks, W., and others, "Algorithms for ASTAP—A Network-Analysis Program," *IEEE Transactions on Circuit Theory,* CT-20:6, (November 1973) 628–634.

Williams, R. A., and Beguwala, M. M., "Reliability Concerns for Small Geometry MOSFET's," *Solid State Technology* 24:3 (March 1981), 65–71.

Wright, G. T., "Current/Voltage Characteristics, Channel Pinchoff and Field Dependence of Carrier Velocity in Silicon Insulated-Gate Field-Effect Transistors," *Electronics Letters,* 6:4 (February 19, 1970), 107–109.

Yamaguchi, K., "Field-Dependent Mobility Model for Two-Dimensional Numerical Analysis of MOSFET's," *IEEE Transactions on Electron Devices,* ED-26:7 (July 1979), 1068–1074.

Yau, L. D., "A Simple Theory to Predict the Threshold Voltage of Short-Channel IGFET's," *Solid-State Electronics,* 17:10 (1974), 1059–1063.

MOS CIRCUIT DESIGN

9.1 NOTATION AND DEFINITIONS

This chapter presents some of the design considerations and techniques for MOS technology logic circuits, both for proper dc circuit behavior and transient (switching) circuit performance. Before commencing, some mention should be made of the design philosophy that will be adopted in developing these circuits for a design system technology library:

> Two fundamentally different design approaches are used to compensate for processing tolerances; some manufacturers use performance screening/sorting at wafer test, while others design their parts to one specification and maximize their yield by using 3σ statistical worst-case (Figure 9.1) process assumptions.[1]

The circuit analyses illustrated in this chapter assume a worst-case design approach; this requires that information be available not only about the nominal process, but also about the (statistical) variation inherent in each of the process steps. In addition, the circuit designer must be aware of the specified tolerances on the applied power supply voltages and the expected range of junction temperatures.

The 3σ worst-case *device current* is the typical measure of worst-case process conditions; an analysis of transient circuit performance, for example, should

---

[1] A Gaussian (or normal) distribution is usually assumed for all statistically varying parameters. The 3σ value is commonly designated as the worst-case value. (σ is the standard deviation of the distribution.)

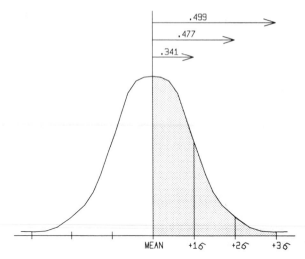

Figure 9.1. Each fabrication parameter distribution is assumed to be Gaussian, where the worst-case value is three standard deviations away from the mean of the distribution.

use the worst-case *minimum* device current to maximize the delay values extracted from simulations. Yet the number of processing parameters that affect the device characteristics are many; setting *each* to its 3σ worst-case value (minimum or maximum, as appropriate) to minimize or maximize the device currents is too pessimistic and will result in a device current calculation far past *its* distribution's 3σ value. (Many of the different process parameter distributions are independent; only a few are correlated, whose values track on their respective distributions.) Determining the necessary combination of device parameter variations to describe the 3σ currents is the initial step in approaching a worst-case circuit design or analysis.

The remainder of this section presents some of the notation and definitions used throughout this chapter.

Device and Schematic Notation

In all cases, the drain and source nodes of MOS devices are interchangeable. The standard convention is that the drain node is the device node at the more positive potential for *n*-channel devices and at the lower potential for *p*-channel devices, in circuits where the device currents are unidirectional. For circuits where the device current is bidirectional, an explicit notation is recommended. If the substrate node connection is omitted, it will be implicitly assumed that the substrate is connected to the appropriate fixed supply voltage (Figure 9.2).

Definitions

Load Devices, Driver Devices, Transfer Gates.   The load devices in a MOS logic circuit are those that provide the capacitive charging current to pull up the circuit output node to a logical 1 value. The logic circuit may use a passive load

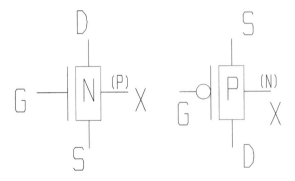

Figure 9.2.   General device and schematic notation for *n*- and *p*-channel devices. The substrate connection is explicitly indicated.

device (with no connection to the set of logic inputs), an active load device(s), or a combination of both (Figure 9.3).

The driver devices in a logic circuit are the active devices that provide the discharge current path for the circuit output node to fall to a logical 0 value (Figure 9.4).

The *transfer gate* is an implementation using one or two MOS devices that acts as a switch to connect or electrically isolate two nodes from different circuits (Figure 9.5). The transfer gate current flows in the direction dictated by the initial voltage and drive capability at the two nodes when the gate is closed; if neither node is asserting a drive current, a transfer of charge between the capacitive loads on the nodes will result in a final equilibrium voltage value. (The input transition to the gate and the resulting capacitive currents $I_{C_{Gs}}$ and $I_{C_{GD}}$ will introduce additional charge into the charge transfer between node capacitances, as depicted in Figure 9.6.)

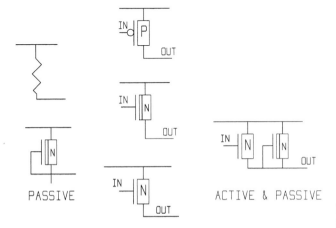

Figure 9.3.   Logic circuits may use a passive load device, an active load device, or a combination of both.

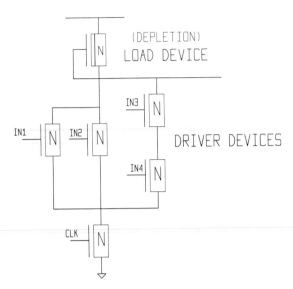

Figure 9.4. The driver devices are those active devices that provide the discharge current path for the circuit output to fall to a logical 0 value.

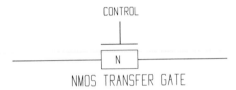

Figure 9.5. Two versions of a *transfer gate*: the first uses a single *n*-channel MOS device, while the second is a CMOS implementation. The CMOS version requires both the control input signal and its complement. The transfer gate acts as a switch to electrically connect or isolate two nodes from different circuits.

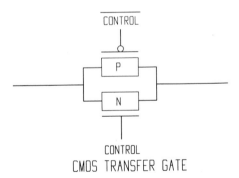

Transfer Characteristic, Unity Gain Voltage. The transfer characteristic for a logic circuit is the plot of the circuit's dc output voltage versus input voltage (Figure 9.7). If the block is an inverting function, the slope of the transition region between logic states is negative; for noninverting functions, the slope is positive. It is also possible that the circuit design provides some *hysteresis* in the transfer

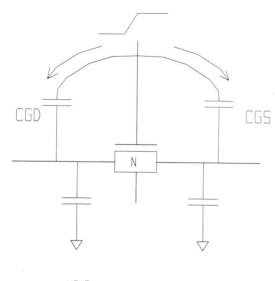

Figure 9.6. A transition on the transfer gate control signal input may feed through to the circuit nodes through the device currents I_{CGS} and I_{CGD}.

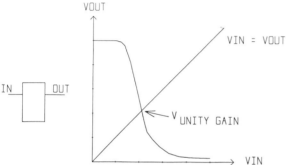

Figure 9.7. The transfer characteristic for a logic block is the plot of the circuit's dc output voltage versus input voltage; the unity-gain voltage is the point of intersection of the transfer characteristic and a line from the origin with slope of $+1$.

characteristic, where the v_{out} versus v_{in} values in the transition region are dependent on the initial logic output value of the block (Figure 9.8).

The unity gain voltage is the point on the (nonhysteresis) transfer characteristic where $v_{in} = v_{out}$ (i.e., the point of intersection of the transfer characteristic and a line from the origin with slope of $+1$).

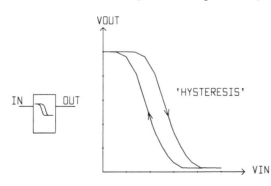

Figure 9.8. Transfer characteristic with hysteresis. The transfer characteristic values in the transition region are dependent on the initial logic output value of the block.

The transient behavior of a logic circuit, $v_{out}(t)$ versus $v_{in}(t)$, is a strong function of the capacitive loading on the circuit's output node, the circuit power, the circuit gain, and the input signal transition rate. The input signal to the logic gate crosses through the circuit's unity gain value prior to the output voltage, due to the *external* discharging or charging capacitive current, which increases the voltage drop across the switching device. The logic circuit's unity gain voltage can therefore serve as a *reference* for measuring the capacitively loaded circuit performance; the block delay (for a particular capacitive load and input signal rise or fall time) is measured as the time between the input signal crossing through the unity gain voltage to the time when the output signal crosses through that point (Figure 9.9). The fraction of the output transition time between its initial logic value and the unity gain point is absorbed into the circuit's delay specification. As will be discussed in Section 9.5, a consistent means of specifying individual block delays permits the accurate summation of those delays when determining an overall path delay (between sequential elements or to chip I/O). If the circuits in a path have different unity gain points, the delay definition is modified to encompass the time between the input signal crossing the circuit's unity gain point and the circuit output crossing the unity gain voltage of the *next* circuit in the path.

Noise Margin.   Figure 9.10a depicts two circuits and a noise voltage source; the circuits are labeled as the *driving* and *driven* logic gates. For this discussion, assume that the two circuits are both inverters, and each has the min/max transfer characteristics given in Figure 9.10b. The transfer characteristics include the variation in the v_{out} versus v_{in} behavior over process and supply tolerances; the particular junction temperature used for the analysis should be chosen so as to *minimize* the resulting noise margin. Also included in this figure are the definitions of the logic 0 and 1 voltage levels: $v_{in,0\text{-MAX}}$, $v_{in,1\text{-MIN}}$, $v_{out,0\text{-MAX}}$, and $v_{out,1\text{-MIN}}$. The input voltage levels are selected by the circuit designer and/or technology developer; the output voltage levels are extracted from the transfer characteristic

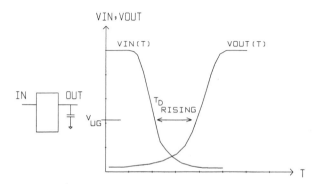

Figure 9.9.   The block delay is measured as the time between the input signal crossing through the unity-gain voltage to the time when the output signal crosses through that point.

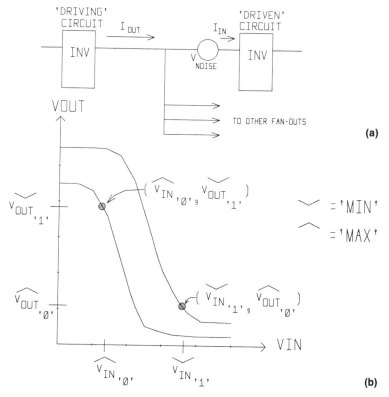

Figure 9.10. Logic schematic (a) and transfer characteristic (b) used in noise margin calculations.

and are temperature dependent. The noise margin for the two logic levels is given by

$$\text{NM}_0 = v_{\text{in,0-MAX}}|_{\text{defined}} - v_{\text{out,0-MAX}}|_{\text{measured from characteristic}}$$

$$\text{NM}_1 = v_{\text{out,1-MIN}}|_{\text{measured}} - v_{\text{in,1-MIN}}|_{\text{defined}}$$

Section 9.4 discusses the techniques for verifying that each *pair* of driving and driven circuit designs satisfies the criteria necessary to ensure an adequate noise margin between circuits, with special consideration for higher fan-in blocks.

Beta Ratio.   The correct dc operation of many of the logic circuit designs to be presented in this chapter depends strongly on the ratio of passive load and driver device dimensions; these designs are appropriately denoted as *ratioed* circuits. (Conversely, other circuit design approaches are *ratioless*.) The parameter

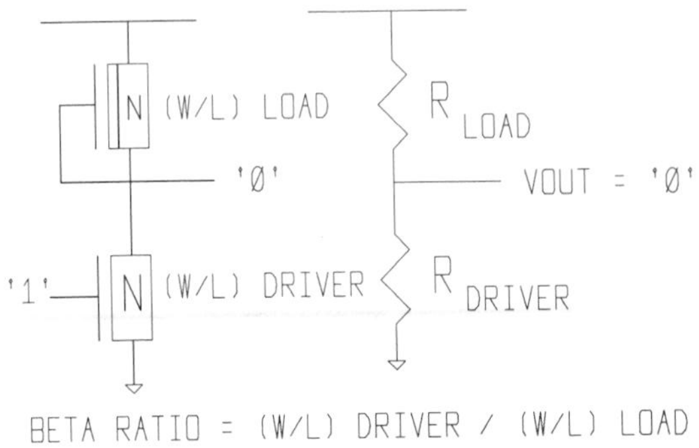

Figure 9.11. For ratioed circuits, the *beta ratio* is the parameter used to give an indication of the logic 0 voltage level.

used to describe the circuit is the *beta ratio*; in equation form, the beta ratio for each driver device in the circuit design is equal to

$$\beta = \frac{(W/L)_{\text{driver}}}{(W/L)_{\text{load}}}$$

The device width and length dimensions used in the previous equation refer to the wafer level dimensions (postfabrication), not the initial layout/lithography measure; the bias between artwork- and wafer-level dimensions should be included in the calculation.

The beta ratio is the device-design equivalent to a resistive divider ratio; as illustrated in Figure 9.11, it is indicative of the logic 0 voltage level to be expected when the driver device(s) is operating in its linear region. As driver devices are placed in series and/or in parallel in a logic circuit implementation, the circuit is now characterized by an *effective* beta ratio; for *n* devices in series, the effective beta ratio is given by

$$\beta_{\text{eff}_{\text{series}}} = \frac{1/[(L_1/W_1) + (L_2/W_2) + \cdots + (L_n/W_n)]}{(W/L)_{\text{load}}} = \frac{1}{\sum_i (1/\beta_i)} = \left(\sum_i (\beta_i)^{-1}\right)^{-1}$$

while for *n* devices in parallel, the effective beta ratio is

$$\beta_{\text{eff}_{\text{parallel}}} = \frac{(W_1/L_1) + (W_2/L_2) + \cdots + (W_n/L_n)}{(W/L)_{\text{load}}} = \sum_i \beta_i$$

For the purposes of noise margin analysis, the beta ratio used for the driving circuit should be the *smallest* ratio of any combination of driver device inputs that is to provide a logical 0 output, thus maximizing the divider output voltage.

9.2 LOAD DEVICE OPTIONS

Five options are available for the choice of load device(s) to implement a particular logic function (MOS technology):

1. Resistive load
2. Enhancement-mode device, saturated region of operation
3. Enhancement-mode device, linear region of operation
4. Depletion-mode device
5. *p*-channel MOS device(s)

If the load device characteristics are plotted on a graph of I_{load} versus v_{out}, the area under the curve is an indication of the performance of the load device in charging the output node capacitance; the greater the area under this curve, the greater the available charging current and the better the rising transition circuit performance.

For a given circuit power dissipation, the ideal load device from a performance standpoint is a current source, whose load current is independent of output node voltage (Figure 9.12). The power dissipated by the circuit is $I_o * VDD$ for a logic 0 output value, while no power is dissipated when the output voltage is 1 (*VDD*). The worst-case power dissipation is therefore given by the product of $(I_{o,max} * VDD_{max}) * $ (logic 0 duty cycle factor). This current source plot will be used as a reference for other load device options; that is, circuits using other load device types will be normalized to the same 0-level power dissipation for comparing the resulting performance measure.

Resistive Load

The characteristics of a resistive load are shown in Figure 9.13, normalized to the same dc power at $v_{out} = 0$. No suitable means for implementing a linear resistor of appropriate value is typically available in most fabrication processes; the interconnection layers are designed to be of a low resistivity and are thus not area efficient for the design of a circuit load. (Some technologies do provide a *very*

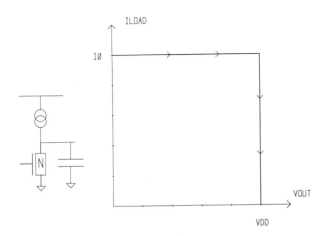

Figure 9.12. *Ideal* load device type, the current source.

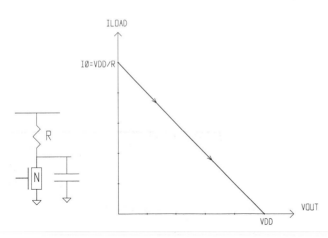

Figure 9.13. Characteristics of a resistive load.

high resistivity interconnect; refer to the discussion of static RAM design and process options in Section 12.4.)

Saturated *n*-Channel Enhancement Load

The saturated enhancement mode *n*-channel device load curve is given in Figure 9.14; the load device always operates in the saturation region, since $V_{GS} - V_T < V_{DS}$, $V_{GS} = V_{DS}$. Of particular note about this option is the fact that $v_{\text{out,max}} \neq VDD$, but rather

$$v_{\text{out,max}} = VDD - V_{T,\text{load}} \big|_{v_{\text{out,max}}}$$

($V_{T,\text{load}}$ is evaluated at $V_S = v_{\text{out,max}}$.) As v_{out} rises, the load device charging current decreases rapidly; the load device turns off when $V_{GS} = VDD - v_{\text{out}}$ reaches the threshold voltage. The threshold voltage increases as the output node rises, since the device source-to-substrate reverse bias voltage is also increasing. Since the output 1 voltage is reduced, the overdrive ($v_{\text{out,1}} - V_S - V_T$) on driven device inputs is also reduced. To maintain a suitable driven circuit output 0 voltage level, the driven circuit's beta ratio must be increased. This results in a significant area and circuit density penalty. The major advantage of this load type is its simplicity, both in processing and circuit physical design. The fabrication process menu need only include a single device type; all load and driver devices are *n*-channel enhancement-mode devices. The physical circuit or chip design requires the routing of a single supply voltage, unlike the two supply voltages required by the next load device type.

Linear *n*-Channel Enhancement Load

The linear enhancement-mode load of Figure 9.15 requires that $V_{GS} - V_T > V_{DS}$ over the entire range of output values. A full output 1 voltage of VDD is achieved by connecting the load device gate input to a separate supply voltage of greater

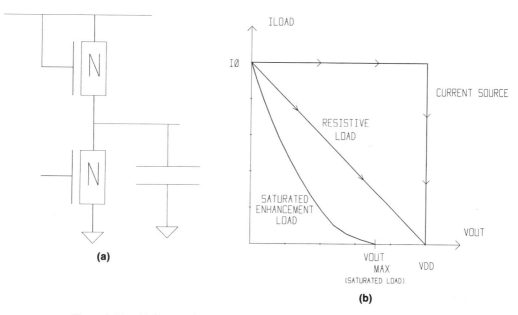

(a)

(b)

Figure 9.14.   (a) Saturated enhancement-mode *n*-channel load device circuit. (b) Comparison of current source, saturated enhancement-mode device, and resistive load for charging current performance.

value than the logic supply. The choice of the additional load supply voltage is governed by the limits on the field oxide threshold voltage, as defined in Section 8.8. Figure 9.16 compares the drive current available from the various load options discussed to this point.

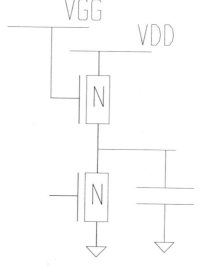

Figure 9.15.   Linear-mode *n*-channel load device. The requirement of the load gate voltage *VGG* is $VGG > (VDD + V_{T,\text{load-MAX}})$.

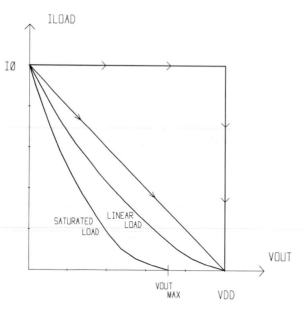

Figure 9.16. Comparison of the current source, resistive load, saturated enhancement-mode load, and linear enhancement-mode load for charging current performance.

The major advantage of a linear enhancement-mode load device is that $v_{\text{out,max}}$ is now equal to VDD, with the resultant decrease in the beta ratio used for driven circuit designs. The disadvantage is that another supply voltage is required (although very little input current from the supply is necessary). Routing this additional supply voltage throughout the chip image to the load devices can be cumbersome.

n-Channel Depletion-mode Load

The *n*-channel depletion-mode load device is illustrated in Figure 9.17; its characteristics eliminate several of the disadvantages of the previous device types; that is, a full output 1 voltage of VDD is provided without requiring an additional supply voltage. Note that the device gate input is connected directly to the source node of the device; for all output voltages, $V_{GS} = 0$. The depletion-mode load device operates in the saturation region when the output voltage is 0, since

$$V_{DS} > V_{GS}^{\;\;0} - V_{T,D}$$

$$VDD - v_{\text{out}} > -V_{T,D}$$

On a rising transition, the device moves from the saturation region to the linear region of operation as the output node charges. Figure 9.18 depicts the $V_{GS} = 0$ curve for the depletion-mode device, both in characteristic curve form and as a load curve (with the value of v_{out} on the horizontal axis). It would appear from the figure that the I_{load} versus v_{out} characteristics for the depletion-mode load

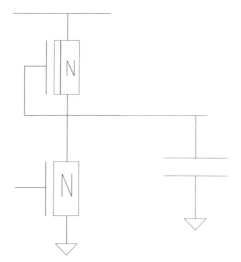

Figure 9.17. Depletion-mode *n*-channel load device.

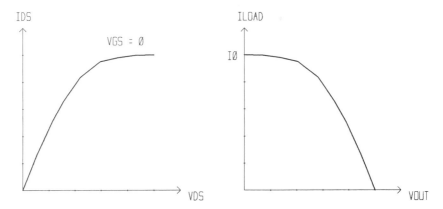

Figure 9.18. $V_{GS} = 0$ V curve for the depletion-mode device, both in characteristic curve form and as a load curve. (*Note:* The body effect on $V_{T,n}$ is not included.)

device most closely resembles that of the ideal current source. Note, however, that the translation of the characteristic curve to a load curve *did not* include the effects of the source-to-substrate voltage (in this case, $v_{out} - V_X$) on the device threshold voltage. The depletion-mode load curve, including the body effect on the device threshold voltage, is plotted in Figure 9.19; the depletion-mode load curve without the body effect is included for comparison. Although the depletion-mode device represents perhaps the best performance of the options investigated, it can be improved (at the expense of additional circuit area) by implementing a push–pull circuit configuration.

A push–pull circuit consists of two stages: the internal stage (with small load

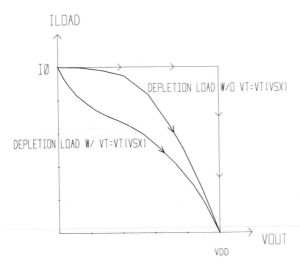

Figure 9.19. Comparison of the current source and depletion-mode device. [The depletion-mode device curve, neglecting the $V_{T,D} = V_{T,D}(V_{SX})$ functional dependence, is also plotted for reference.]

capacitance) and the external stage (which drives a large load capacitance, typically). A push–pull inverter circuit is depicted in Figure 9.20. For the purposes of illustration, the following assumptions will be made:

1. Since the internal node is lightly loaded (as compared to the output node), it will reach *VDD* on a rising transition while v_{out} is still at a low level. Therefore, during the initial portion of a rising transition,

$$V_{GS,\text{external}} = V_{\text{internal node}} - v_{\text{out}} \gg 0$$

This results in a large amount of device current when v_{out} is small. As v_{out} rises to *VDD*, $V_{GS,\text{external}} = VDD - v_{\text{out}}$ goes to zero.

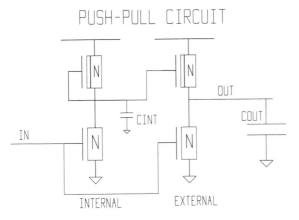

Figure 9.20. Push–pull circuit, consisting of both an internal and an external stage. The voltage $V_{GS,\text{external}}$ is no longer identically equal to 0 V, but rather the gate is connected to the output of the internal stage.

2. The ratio of power as divided between the external and internal stages is 2:1; that is, the total circuit power is split in the following manner:

$$P_{\text{total}} = P_{\text{internal}} + P_{\text{external}} = (\tfrac{1}{3} * P) + (\tfrac{2}{3} * P)$$

As illustrated in Figure 9.21, the push–pull depletion-mode load device provides the best power-performance trade-off of the options considered so far, at the expense of the additional circuit area for the internal stage. For more complex logical functions than the inverter illustrated in Figure 9.20, the internal stage of the push–pull circuit also grows in complexity; the internal stage must provide the gate input of the push–pull load device with the same logic voltage value the output will reach, and therefore must have an identical driver device configuration as the external stage.

On a rising transition, it was assumed v_{internal} would reach VDD very quickly; on a falling transition, v_{internal} will reach a low voltage comparable to the external circuit's 0 voltage level; as a result, $V_{GS,\text{external load}} = v_{\text{internal}} - v_{\text{out}} \approx 0V$, when the output is 0. The push–pull load device therefore acts as a standard depletion-mode lcad device for dc power calculations.

To this point, the discussion of the load device options has primarily dealt with the performance of the rising, or 0 to 1, transition. When considering the opposite (1 to 0) transition, the driver device has two current components, load

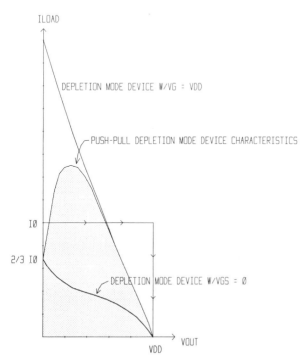

Figure 9.21. Characteristics of a push–pull depletion-mode load device; the shaded area is the area under the i_{load} versus v_{out} curve.

device current and output capacitance discharge current. The driver device must be made sufficiently wide to accommodate the load device current component while providing a particular discharge current component for performance. Making the driver device wider to handle both current components has several consequences:

1. More chip area is consumed $(-)$.
2. The input gate capacitance (C_{in}) of the driver device to preceding circuits increases $(-)$.
3. As C_{out} discharges, the capacitive current component decreases, approaching zero once dc levels have been established; at this point, the only driver device current is from the load device. Increasing the width of the driver device reduces $v_{out}(0)$, improving the logic 0 level noise margin $(+)$.

To improve the 1 to 0 transition performance, the capacitive load discharge current should be maximized and the load device current component minimized. The optimum I_{load} versus v_{out} curve for the 1 to 0 transition is not the ideal current source curve given earlier, but rather one that minimizes the load current in the range of a high output voltage. Figure 9.22 illustrates this revised ideal curve in the neighborhood of VDD; this curve is not consistent with the curves given earlier for the standard and push–pull depletion-mode load devices.

In reference to the logic 0 state, all the load devices previously mentioned dissipate dc power when the circuit output is low (Figure 9.23). The load device type that provides the *best* power–performance trade-off is one where the load current curve is of the form illustrated in Figure 9.24. In addition to the behavior illustrated in Figure 9.22 in the neighborhood of VDD, the load current should

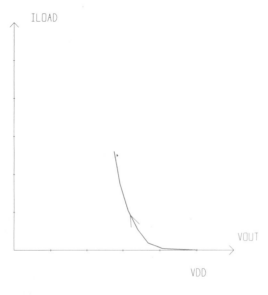

Figure 9.22. Revised ideal load current curve in the neighborhood of VDD; to enhance the capacitive discharge current and therefore the performance, the load device current component should be minimized.

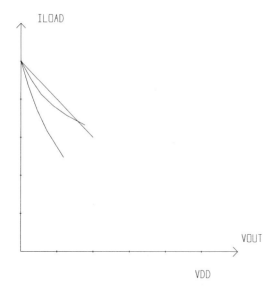

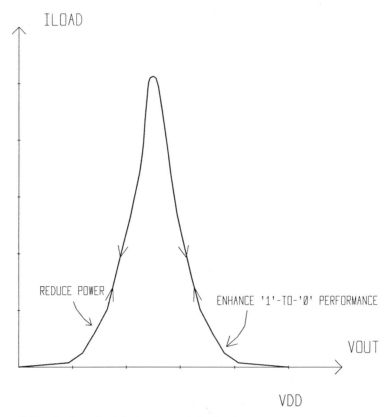

Figure 9.23. The previous load device options all dissipate power when the circuit output is 0: $P = v_{\text{load}} * i_{\text{load}}$; $v_{\text{load}} = (VDD - v_{\text{out}}) \approx VDD$; $P \approx VDD * i_{\text{load}} *$ (logic 0 duty cycle).

Figure 9.24. Optimum load device current behavior that provides the best power performance trade-off.

also go to zero as the output voltage goes to 0 to minimize the power dissipation. To keep the 0 to 1 transition performance, the area under the I_{load} versus v_{out} curve should be maintained. The load device that satisfies these criteria is a *p*-channel device whose gate input is in common with the driver device input (Figure 9.25).

p-Channel Load Device

Unlike the previous load devices, the *p*-channel (or *complementary*) load is not a passive pull-up, but is switched by the circuit input. When the input voltage is 1, the *n*-channel driver device is on, while the *p*-channel load device is off; the gate-to-source voltage of the load device, $(v_{\text{in}}(1) - VDD) \approx 0$, is not sufficiently negative to form the device channel. When the input voltage is 0, the driver device is off, while the load device provides the charging current to raise the output node to *VDD*. The *static* dc power dissipation of the circuit is zero; the dual switching behavior of the driver and load devices reduces the power dissipation to the time-averaged transient device current times device voltage (Figure 9.26).

The logic circuit and fabrication technology that includes both *n*-channel and *p*-channel devices is denoted as *complementary MOS* (CMOS). The key advantage of CMOS is the low static power dissipation of its logic circuits; at VLSI circuit densities, the chip power dissipation is a major packaging and reliability concern. CMOS technology has rapidly evolved as a solution to achieving high circuit counts within package and system power dissipation limits.

The major disadvantages of CMOS technology circuits are as follows:

(1) Fabricating both *p*- and *n*-channel devices together on the same chip introduces considerable additional process complexity (and therefore higher fabrication cost). CMOS technology requires the addition of processing steps to provide (1) both n^+ and p^+ device nodes, (2) *local* substrate regions of *p*- and *n*-type

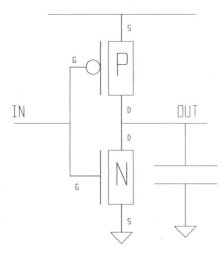

Figure 9.25. *P*-channel load device. In the schematic, the gate node of the load device is connected to the logic input; this will provide the load current behavior diagrammed in Figure 9.24.

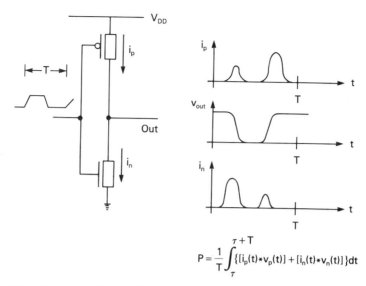

$$P = \frac{1}{T}\int_{\tau}^{\tau+T}\{[i_p(t)*v_p(t)] + [i_n(t)*v_n(t)]\}dt$$

Figure 9.26.   The power dissipated by the *complementary* MOS circuit is the time-averaged device current times device voltage for the load and driver devices.

to fabricate *n*- and *p*-channel devices, respectively, and (3) suitable threshold voltage values (and $\partial V_T/\partial V_{SX}$) for the two device types to provide sufficient overdrive and noise margin.

(2) To implement a more complex logic function than the inverter of Figure 9.25, more than one load device is required; indeed, an equal number of load and driver devices are necessary. As illustrated in Figure 9.27, these additional *p*-

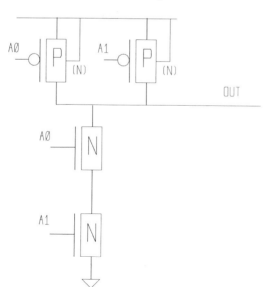

Figure 9.27.   To implement a more complex CMOS logic function than an inverter, additional load devices are required; the load devices are connected in dual fashion to the *n*-channel driver devices. The circuit is a CMOS two-input NAND gate. (The *A*0 and *A*1 input signals are common to both a load and a driver device). A *local n*-well substrate connection for the *p*-channel devices is indicated; low-resistivity contacts and additional supply voltage routing are required to bias the local substrate regions.

channel load devices are connected in *dual* fashion to the driver device configuration; *n*-channel devices in series (parallel) must have corresponding *p*-channel devices in parallel (series) connected to the same set of inputs. This results in a circuit density penalty as compared to the passive load implementations; in particular, the lithography rule governing the space required between a device and the local substrate region of the opposite device type is typically quite large. Additional area is consumed by the need for low-resistance contacts and supply voltage routing to the local substrate regions.

(3) The adjacent complementary device and substrate types used to implement CMOS circuits result in the presence of a four-layer reverse-biased *pnpn*

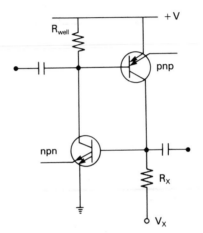

Lumped CMOS SCR Latch-up Model

(a)

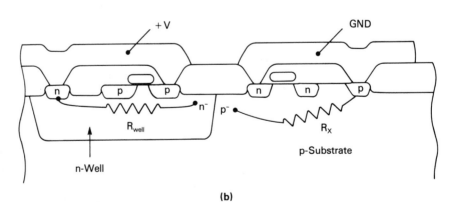

(b)

Figure 9.28. A latch-up schematic model (a) and process cross section (b) illustrating the *pnpn* junction orientation. The schematic model uses lumped elements to approximate distributed *R*s and *C*s.

structure, the same structure as a silicon-controlled rectifier (SCR). In the absence of a triggering current, the SCR acts as back-to-back reverse-biased junctions and does not conduct; this is the normal, electrically isolated condition between CMOS devices of different type. However, a short-duration transient forward-bias base current into one of the *npn* or *pnp* regions may fire the SCR, which results in a very low impedance path between the two terminals of the four-layer structure. The corresponding condition in the adjacent CMOS devices is denoted as *latch-up*; the resulting current represents a failure in the isolation between devices, often a *destructive* failure due to the large current densities in the interconnections between the terminals of the structure. Extreme precautions must be taken in CMOS layouts to ensure that forward-bias transient conditions (caused by inductive $L * di/dt$ noise or capacitive coupling) are limited in magnitude and that the resulting base current can be shunted away from adjacent devices. These transients are a particular problem on pad driver circuits, where an impedance mismatch at the far-end of a transmission line or printed circuit board wiring trace results in a reflection transient at the driver end, which may forward bias a device junction. A lumped-element latch-up schematic model and a corresponding cross section are given in Figure 9.28.

Despite these disadvantages, CMOS technology is the fundamental direction in VLSI process development. Many of the logic circuits illustrated in this chapter depict both CMOS and *n*-channel depletion-mode load device circuits.

9.3 ELEMENTAL CIRCUIT BLOCKS

This section presents FET circuit designs for a number of logic primitives; typically, *n*-channel depletion-mode load device circuits (both standard and push–pull implementations) and CMOS circuits are illustrated. Specific device dimensions are determined by dc power and noise margin limitations and transient performance criteria to be discussed in subsequent sections of this chapter. In many cases, layout examples are included with the description of the circuit; for clarity, some of the device-related lithography levels have been left out of these examples, such as *n*- and *p*-channel threshold voltage tailoring implant mask shapes.

Inverter

The circuit design and layout for an inverter is given in Figure 9.29 for the *n*-channel depletion-mode and *p*-channel load device options of Section 9.2. The gate of the *p*-channel load device is connected to (and switches in complementary fashion to) the gate of the *n*-channel driver device. Also, the gate of the external push–pull load device is driven by the output of the internal stage, which implements the same overall logic function, but at reduced power dissipation. This is

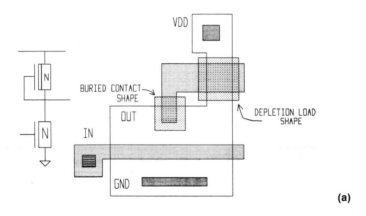

(a)

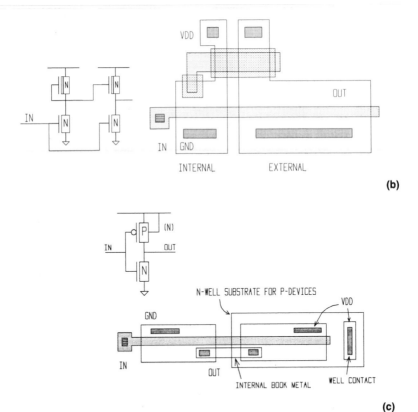

(b)

(c)

Figure 9.29. Schematics and simplified chip layouts for inverter blocks. (a) *n*-channel depletion-mode load circuit. (b) Push–pull depletion-mode load circuit. Note the difference in circuit power between the internal and external stages. (c) CMOS implementation. The *p*-channel device is fabricated in an *n*-well. The implant block mask shapes to separate *n*-channel devices (and n^+ nodes) from *p*-channel devices (and p^+ nodes) are not illustrated.

evidenced by the difference in device *W/L* for the internal and external depletion-mode loads; the driver devices are similarly sized as sufficient driver-to-load device beta ratios must be maintained for both stages.

NOR

NOR gates are illustrated in Figure 9.30; a three-input NOR is used for example in the circuit layouts. The depletion-mode push–pull circuit requires additional devices (and area) to implement the NOR function for the internal stage. The

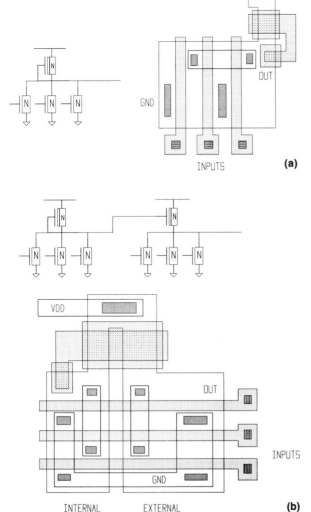

(a)

(b)

(continues)

Figure 9.30.   Schematics and layouts for three-input NOR logic circuits. (a) *n*-channel depletion-mode load circuit. (b) Push–pull depletion-mode load circuit. (c) CMOS implementation.

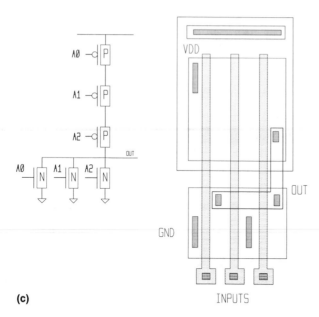

(c) INPUTS **Figure 9.30.** (*Continued*)

CMOS circuit requires the dual of the *n*-channel driver devices implemented as *p*-channel loads; the resulting series interconnection of these lower transconductance, *p*-channel load devices can present a significant pull-up resistance and slow rising transition; as such, high fan-in CMOS NOR gates are not commonly included in a design system library.

NAND

Figure 9.31 illustrates the implementation and layout of NAND logic gates. For the two depletion-mode device types, the driver devices *must* be increased in width to reduce the effective resistance of the series connection, thereby maintaining a valid logic 0 voltage level.

For the depletion-mode load device circuits, the preferred circuit implementation for high fan-in logic gates is the NOR; the area penalty for high fan-in NAND circuits makes this choice unattractive. For CMOS circuits, both NOR and NAND logic functions require a series interconnection of devices; the optimum is that which chooses to put the device type with the higher transconductance (channel carrier mobility) in the series connection.

The series interconnection of devices in a logic circuit implementation can present some concerns with respect to two design factors: (1) glitches on circuit output nodes due to charge transfer to and from the intermediate node(s) of the connection, and (2) the variation in circuit performance as a function of which of the logically symmetric input signals are switching. An example of the first situation is illustrated in Figure 9.32; the sequence of transitions on the inputs of the depletion-mode load NAND gate has discharged the intermediate node, while

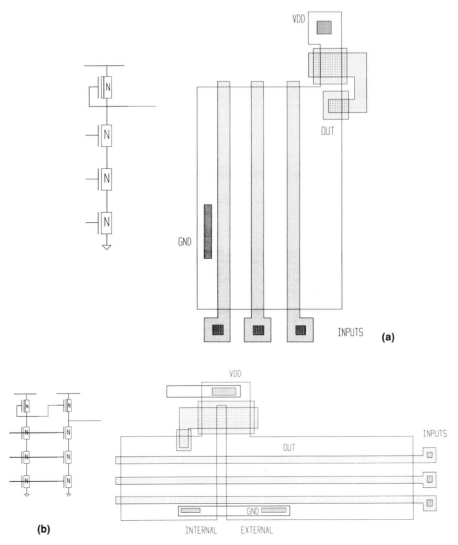

Figure 9.31. Schematics and layouts for a three-input NAND circuit. (a) Depletion-mode load device circuit. (b) Push–pull depletion-mode load circuit. (c) CMOS technology circuit.

the output remains at 1. When the input to the top device of the series connection switches to 1, the intermediate node is charged to 1 by a combination of charge transfer from the output node capacitance and from load device current. The charge transfer ratio $C_{\text{intermediate}}/(C_{\text{out}} + C_{\text{intermediate}})$ and the load device size should be such that the magnitude and duration of the output node glitch is minimized; in particular, the layout should minimize the internal node capacitance. The minimization of the internal node capacitance is also a key design criteria

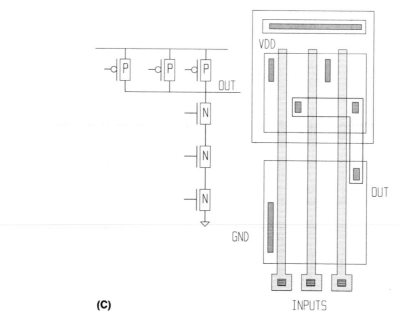

(C)

Figure 9.31. (*Continued*)

when considering the pin-specific performance variations described previously. The output transition delay can be a strong function of which device in the series connection is switching and the initial voltages on the intermediate node(s). The variation in delays across symmetric inputs to a logic circuit introduces the question of how to develop the delay model for logic simulation, whether to include unique delay information for each pin or to simply assign best case or worst case

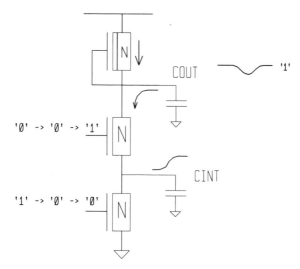

Figure 9.32. A charge transfer between internal node and output capacitances can result in a glitch on an otherwise stable logic output voltage.

delays to the overall block. The answer to this question depends on the trade-offs between the desired accuracy of the simulation model, the level(s) of description the simulator accepts, and the resulting differences in simulator execution time. Section 9.5 discusses in further detail the need for delay assignment to input pins of logic circuits.

AND-OR-INVERT

The AND-OR-INVERT (AOI) is a combination of series and parallel driver devices in a more complex configuration. As illustrated in Figure 9.33, the AOI circuit is physically realized by dot-ANDing multiple NAND circuits together. The AOI function may be several NAND circuits wide; the fan-in (height) of each NAND function is limited by the same area and performance concerns as described earlier, usually two and rarely more than three. The AOI is specified in terms of the width and height of the NAND function, either collectively or explicitly (e.g., three-wide, two high AOI, or 2-2-2 AOI). Note that all NAND functions need not be of the same fan-in (e.g., a 2-2-1 AOI). The AOI function is implemented very efficiently in standard depletion-mode load circuits and is very useful in the realization of multiplexing functions.

OR-AND-INVERT

The OR-AND-INVERT (OAI) function is illustrated in Figure 9.34; additional driver devices are added in parallel to the series NAND interconnection. The OAI is not used as frequently or in as wide a variety as AOI circuits in depletion-

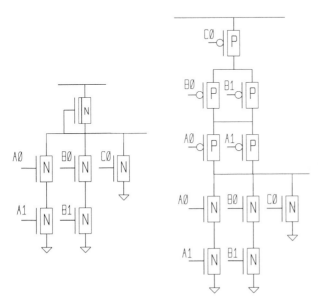

Figure 9.33. Depletion-mode *n*-channel load and CMOS implementations of the And-Or-Invert (AOI) logic function. The particular function illustrated is a 2-2-1 AOI.

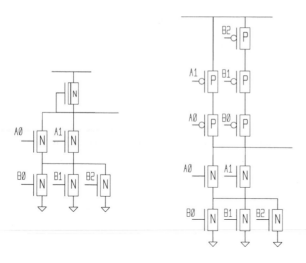

Figure 9.34. Depletion-mode *n*-channel load and CMOS implementations of an Or-And-Invert (OAI) function.

mode load circuit libraries due to the additional intermediate node capacitance and the associated problems mentioned earlier. CMOS circuit libraries typically offer limited AOI and OAI books.

The concept of dotting additional function to the output node can be extended beyond simple AOI and OAI structures to realize a complex logic function, as illustrated in Figure 9.35.

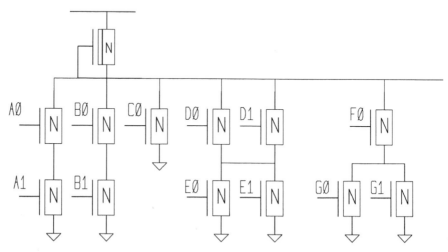

Figure 9.35. Dotting of additional AOI and OAI structures to the output node of a depletion-mode *n*-channel load device circuit.

Exclusive OR

The Exclusive OR (XOR) function is a key function to include and efficiently implement in the design library as it is fundamental to many higher-level functions (e.g., parity generators and checkers). Figure 9.36 illustrates one possible implementation, using two-input NAND blocks with three stages of logic delay. An alternative version is given in Figure 9.37, with two circuits and two stages of logic delay, illustrating that the complex AOI circuit capability can save power and delay over straight NOR and NAND realizations. A third (depletion load) circuit option is depicted in Figure 9.38 (which implements the Exclusive NOR,

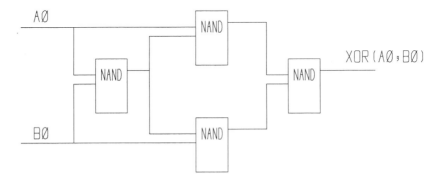

Figure 9.36.   NAND block implementation of the exclusive-OR (XOR) function.

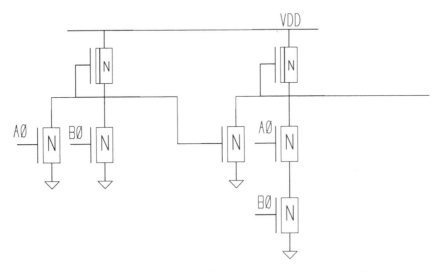

Figure 9.37.   Alternative version of the XOR function, with two levels of block delay.

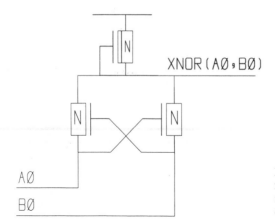

XNOR (AØ , BØ)

Figure 9.38. Alternative circuit option for implementing the XNOR function; a minimal number of devices and block delays are used.

or XNOR, the complement of the XOR). This circuit results in only one block delay, but nevertheless may not be the optimum choice for inclusion in the circuit library, as this third version requires that the driving circuits be able to sink additional dc 0 level current. The proper dc design of the driving circuits is complicated by the fact that this interconnection may be global; the $I * R$ resistive voltage drop in metal, contact, and any higher resistivity interconnect will worsen the circuit's 0 level noise margin (Figure 9.39). The driving circuit also requires special design in order to sink the current from *two* load devices. This implementation may be suitable for macro circuit designs where the driving circuit design and dc noise margin analysis are both under the designer's direct control, but should not be selected for global design system usage.

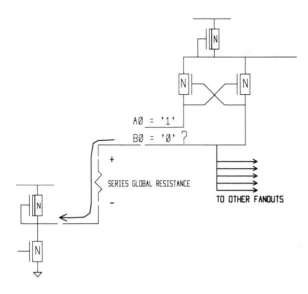

Figure 9.39. Using the XNOR circuit of Figure 9.38, the dc design of driving circuits is complicated by the global interconnection and resulting $I * R$ voltage drop.

A CMOS technology implementation can be efficiently developed in a manner similar to the design of Figure 9.38, utilizing transfer gates, to be discussed next (see Problem 9.12).

Transfer Gates

The transfer gate is not a normal Boolean logic primitive; rather, this term denotes a circuit that acts as an analog, bidirectional switch. Two transfer gate possibilities are illustrated in Figure 9.40; the simpler of the two requires only a single control signal, while the CMOS transfer gate requires the control signal and its complement. The function of the transfer gate is rather straightforward; when the control signal is active, a device channel exists for current to flow from the higher node potential to the lower; when the gating signal is inactive, the input and output nodes are disconnected (the devices are off). Transfer gates are useful in a number of applications, but particularly for controlling the driving source onto a net to which many possible sources need to be connected (Figure 9.41a). In the figure, one and only one of the control signals should be active at any time to avoid contention on the net; when all driving sources are isolated, the net is effectively in a high-impedance state (and may hold its stored capacitive charge for some period of time). Transfer gates may be placed in series to develop a current-steering path for a data signal (Figure 9.41b).

The output voltage of the nMOS transfer gate is limited to a maximum of $v_{\text{out,max}} = V_{\text{control}} - V_T$, with V_T evaluated at $V_{SX} = v_{\text{out,max}} - V_X$, much like the saturated enhancement-mode n-channel load device. The CMOS implementation, driven with complementary controlling voltages on the n- and p-channel device inputs, permits the output voltage to reach a value equal to the full driving voltage. The output of one n-channel transfer gate should not be used as the controlling signal for another, as the loss in maximum output voltage of twice V_T would typically not provide sufficient overdrive to circuits driven by this second output.

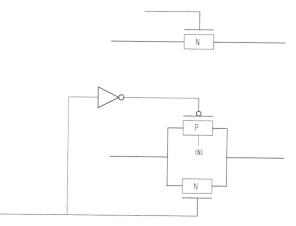

Figure 9.40. NMOS and CMOS implementations of the transfer gate function (similar to Figure 9.5).

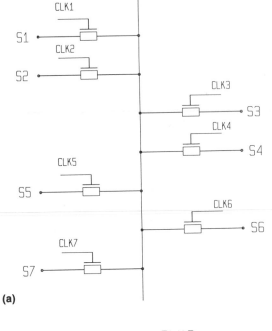

(a)

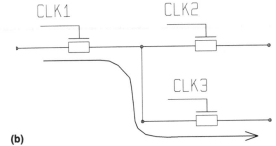

(b)

Figure 9.41. (a) Transfer gates are useful for controlling one of many possible sources onto a net. One and only one of the CLK signals should be active at any time to avoid contention. (b) Transfer gates can be placed in series to form a current-steering path for a signal.

Despite these varied applications, transfer gates can be problematic for general design system usage. Primitive logic simulation models for the transfer gate are difficult to develop, although more recent simulators are designed to incorporate the concept of *driving signal strength* to resolve contention between circuits with dottable outputs, as well as to represent a stored-charge, high-impedance value. Delay parameters for the bidirectional transfer gate are considerably more difficult to incorporate into a design system's delay calculator than for conventional logic circuits; the transfer gate output node transition is a unique function of both the input signal and control signal rise and fall times. For a slow transition on the control signal input, the output will follow the transition; the driving circuit will be able to maintain the input voltage with little perturbation. For faster transitions on the control signal, the charge transfer from input to output nodes will cause a glitch in the input voltage until the driving circuit responds with a restoring current. The delay model for the block driving the transfer gate is also complicated by the variation in output load, depending on whether the transfer gate is on or off; primitive block delay models do not commonly incorporate dependencies on the signal values of other circuits, such as the control signal input to the transfer gate. In addition, if the data input signal is changing as the controlling signal goes *inactive*, the transfer gate output may or may not be well defined; an input hold-time definition relative to the control signal's inactive edge may need to be verified, much as for sequential (latching) circuits.

Additional problems related to transfer gates arise from the situations illustrated in Figure 9.42. Assume that the transfer gates in the figure are very wide devices, in order to minimize the series discharging or charging time constant; they therefore have a significant channel capacitance when the gate is on. A transient on the data input will capacitively couple to the control input. The control signal will likely fan-out to a number of gates (transfer gates and otherwise), and the capacitive coupling on this net may lead to additional circuit delays (Figure 9.42a), or could potentially forward bias a CMOS technology *pn* junction, triggering a latch-up condition (Figure 9.42b).

In a design system environment, transfer gates are typically restricted in use to *within* book or macro circuits, where the pertinent device and capacitive parameters can be extracted from the physical design. As a result, conditions such as those illustrated in Figure 9.42 can be analyzed for their effects on the macro's circuit behavior. One such macro is the latch or register macro, to be discussed next.

Latches

Sequential circuits are undoubtedly the most crucial circuit element in any design system library. Their performance is of the utmost concern in many applications, particularly register-intensive, data-flow designs where the fan-out or interconnection load can be large. Sequential circuits present additional constraints on simulation and checking, constraints as represented by parameters such as data setup and hold times and minimum clock pulse width. In addition, these circuits are unique in that they may have multiple data input ports (with multiple clocks) and that there may be multiple output pins used (e.g., true and complement data outputs). Sequential circuits are prime candidates for developing behavioral-level models for simulation, as discussed in Chapter 4; the behavioral model can verify the necessary circuit design constraints using local variables to record the history of input signal transitions. The behavioral description may also enhance the execution speed of simulation over an equivalent interconnection of logic primitives.

Sequential circuits may be divided into two categories: edge-triggered (flip-flops) or level-sensitive (latches). The difference between the two groups is dependent on the response of the outputs to data input changes at different points in the clock pulse waveform. Flip-flop outputs change only in response to the data input value during the active edge or transition of the clock signal; subsequent

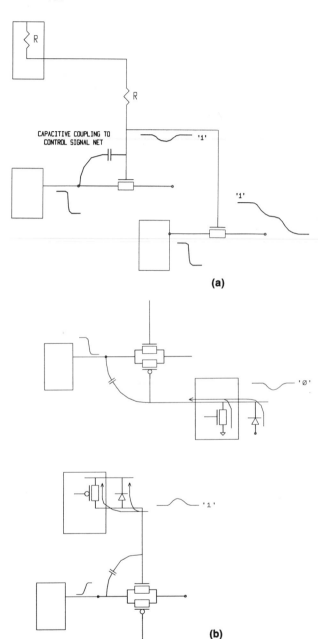

CAPACITIVE COUPLING TO
CONTROL SIGNAL NET

'1'

'1'

(a)

'∅'

'1'

(b)

Figure 9.42. (a) Transients on the transfer gate input signal will couple to the control input through the gate-to-device node capacitance, leading to additional circuit delays. (b) Transients on the transfer gate data input signal that couple to the control signal node may possibly forward bias a CMOS technology *pn* junction, providing a potential latch-up triggering condition.

data input changes when the clock signal is stable are locked out. Latch outputs continue to change in response to changes at the data input as long as the clock signal is at its active level, be it either 1 or 0; as the clock signal returns to its inactive level, the most recent output value is retained. Both latches and flip-flops have usage criteria for data setup and hold times (measured relative to the clock edge that stores the data value) and for minimum clock pulse widths. In Chapter 13, a set of design criteria for enhancing the overall chip testability will be introduced; this level-sensitive scan design technique (LSSD) imposes the restriction that *no* edge-triggered sequential elements be utilized in the chip design (i.e., the design system library contain *no* flip-flop books). Consistent with that *design for testability* philosophy, the remainder of this section will concentrate on various design options for latches. Latch types to be illustrated include single- and multiple-port data latches (also known as *polarity-hold* latches), and set–reset latches.

Set–reset latches.   A schematic for a set–reset latch is given in Figure 9.43; a pair of cross-coupled 2-1 AOIs are used. If $S = R = 0$, the latch outputs do not change in response to an active clock pulse. If either set or reset equals 1, the Q latch output goes to 1 or 0, respectively. For use in a design system library, a buffer circuit is usually added to the book design to facilitate driving a larger global capacitive load with a minimal amount of additional circuit area and power; as illustrated in Figure 9.44, only the external stage of a push–pull inverter is required since both polarities of the latch output are available. In CMOS implementations, a normal inverting buffer may be used. The size and power of the

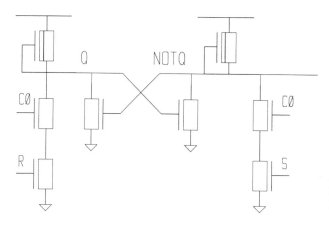

Figure 9.43. Schematic for set–reset latch (*n*-channel depletion-mode load technology).

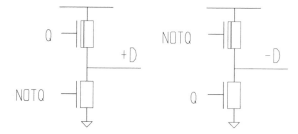

Figure 9.44. Buffer circuit options for set–reset latch; both internal latch polarities are available.

buffer may be available over some range in the design system library for optimizing to the physical design's load.

Regarding the circuit's performance, the slowest transition on either latch output relative to a clock or data input change is the 0 to 1 transition; the opposite AOI of the cross-coupled pair must first change from 1 to 0 in order that the other may rise. As the latch works by discharging one of the AOI nodes to 0, a CMOS implementation need not require the full design of two AOIs; rather, as illustrated in Figure 9.45, two cross-coupled inverters may be used with NAND connections dotted to the output nodes. The *p*-channel load and NAND driver device sizes must be ratioed such that the logic 0 node output is sufficiently low to initiate the latch to change state through the cross-coupled feedback. Since a buffer circuit is typically included to drive a global load capacitance, the *p*-channel load device width-to-length ratio may indeed be relatively small.

The condition where the set and reset inputs are both equal to 1 is normally an invalid condition for this latch type, as it drives both latch AOI outputs to 0 throughout the active clock level. When the clock signal starts to fall, a *race* condition exists between the two outputs, as the node that rises fastest will then turn on the opposite AOI sooner, sending that circuit's output back to 0 (Figure 9.46). The difference in transition time depends on differences in loading and any initial difference in the 0 levels of the two circuit nodes. The latch circuit's stable state is therefore unpredictable when both set and reset are active, unless specific circuit design measures are taken to ensure that one of the set–reset input pair is dominant.

Data latches.   The *data* or *polarity-hold* latch is similar to the set–reset latch, with the reset signal provided by the inverse of the data input (Figure 9.47). Additional input sources (or *ports*) can be provided by increasing the width of the cross-coupled AOIs (Figure 9.48); this essentially incorporates a multiplexing function directly into the latch circuit, if indeed the chip design is developed such that no two clocks are active simultaneously. Note, however, that with this im-

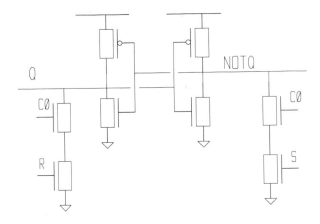

Figure 9.45.   Set–reset latch implementation in a CMOS technology; the full AOI circuit implementation for both halves of the latch is not required, as the latch operates by discharging one of the latch nodes to 0.

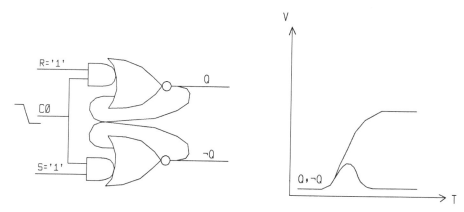

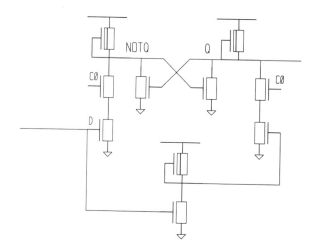

Figure 9.46. Race condition between the two latch outputs for the release of the condition Set = Reset = 1.

Figure 9.47. Schematic for a data or polarity-hold latch circuit (n-channel depletion-mode load device technology). This implementation differs from the S/R latch of previous figures by the addition of a data input signal inverter to connect to the R input.

plementation the addition of input ports increases the number of devices and circuits required with a corresponding area penalty; four active devices (wide devices, as they are in a NAND connection) and an inverter are added with each additional data port. An alternative latch circuit is illustrated in Figure 9.49, which attempts to reduce the additional overhead of extra data input ports. This latch type is driven in single-ended fashion, with the data signal dotted to the latch feedback node through a transfer gate and with the data clock as the controlling input. The feedback inverter in the latch must be sized so that its driving strength is relatively weak compared to that of the data buffers in the input port; when a clock is enabled, the driving strength of the data buffer must be able to override the value held on the net by the feedback inverter.

In addition to the simulation model constraints described earlier, latch cir-

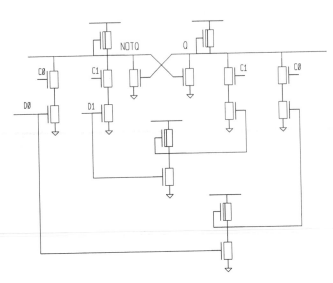

Figure 9.48. Additional ports may be added to the data latch with the addition of AOI branches and data inverter circuits.

cuits present unique modeling characteristics for test pattern generation and circuit fault analysis. Circuit faults may indeed result in errant behavior, which may be readily detected or may result in transient conditions where the presence of the fault may go undetected at the imposed voltage, temperature, and temporal conditions during wafer-level testing. Two such fault conditions are depicted in Figures 9.50 and 9.51. In the first, an output stuck-at-1 fault on the data signal inverter is to be detected by applying a 1 value at the data input during test. If the fault is indeed present, both latch outputs go to 0 during the active clock pulse; when the clock is removed, a race exists between the two output nodes, just as with the Set = Reset = 1 condition described earlier. In this example, the $\overline{Q}$ output must be designed to be dominant to be able to isolate the valid from the

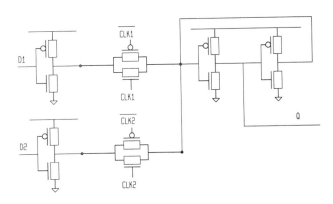

Figure 9.49. Alternative latch implementation, with the cross-coupled AOI circuits replaced by cross-coupled inverters; the latch is driven in single-ended fashion by a data buffer circuit when a clock signal is active.

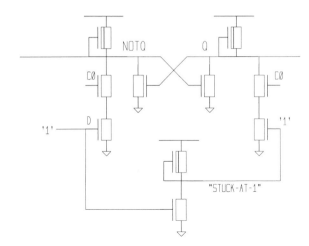

Figure 9.50. Testing for a stuck-at-1 fault on the data inverter of a cross-coupled AOI data latch leads to a possible race condition on the latch outputs if the fault is indeed present. To detect the presence of this fault condition, the latch circuit must be designed to differentiate between the good and faulty behavior.

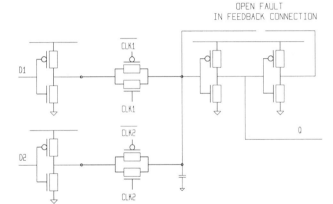

Figure 9.51. Fault condition in the feedback connection of a latch circuit; the stored charge on the feedback node capacitance may continue to provide the correct latch output value for some period of time until the leakage currents result in an indeterminate value.

faulty behavior. The example in Figure 9.51 refers to the latch schematic of Figure 9.49, where the presence of the feedback path in the latch itself is in question. The value provided by the data buffer is retained on the capacitance of the latch input node; the purpose of the feedback inverter is to maintain that value when the clock is inactive. If the feedback path is indeed open, the charge on that node will be modified by junction leakage currents, yet the discharge or charge rate will be sufficiently slow such that the observation of the latch output by some subsequent test pattern will likely capture the correct value.[2] The absence of the feedback inverter (which provides the necessary leakage current) would go un-

---

[2] The circuit of Figure 9.51 with the feedback inverter completely absent is referred to as a *dynamic latch*, with the stored charge used to represent a data value. This latch can only be used in an application where it is clocked at a sufficient rate to ensure the stored value is still interpreted correctly. As such, it is not an attractive candidate for a general usage design system library.

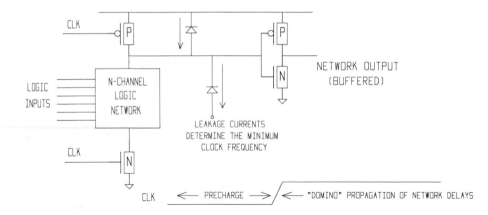

Figure 9.52. Dynamic logic circuit type: domino CMOS. During the precharge portion of the clock cycle, all internal logic nets are set to 1, while all buffered logic outputs are 0. When the CLK signal goes to 1, the internal net will be discharged to 0 if a current path is present through the (complex) *n*-channel logic network. The network output then goes to 1 if the internal net is so discharged, which then propagates to other *n*-channel logic networks. In essence, the internal nets of logic functions at succeeding levels of the network design will "fall like dominoes."

detected and could subsequently lead to an intermittent application-dependent failure.

The circuits presented in this section are representative of the elemental logic circuits used in a variety of technologies. The combinational circuits were *static*; that is, they do not require any input signals to precharge any nodes, nor is their output dependent on any dynamic charge storage. The variations in dynamic logic circuit options are many, like the *domino CMOS* function illustrated in Figure 9.52. These options have a number of advantages in terms of area, power dissipation, and performance, yet are often difficult to incorporate into a design system environment due to their unique timing and clocking constraints.

9.4 DC CIRCUIT DESIGN

This section presents both general and detailed design considerations for proper dc circuit operation of *n*-channel depletion-mode load logic circuits. Reference has been made previously to the circuit's beta ratio, the ratio of the width-to-length dimensions of the active device(s) to that of the load device. This section discusses the criteria for determining a suitable beta ratio. Initially, a general description of the dc design approach for an inverter is given; extensions to NAND and NOR circuits will then be illustrated. Following that discussion, the specific technique used to analyze the physical layout of each circuit design in a design system library for worst-case dc noise margin requirements will be outlined.

Inverter Design Example

The first step in the design of a depletion-mode load logic circuit (inverter or otherwise) is to choose the load device dimensions (W/L) for the circuit's targeted performance and power dissipation; given the power specification (or equivalently, the 0 to 1 transition performance), determine the necessary 0-level device current and then solve for the load device's W/L ratio. Next, the driver device is sized in order to assure an output 0 voltage level with adequate noise margin; the output voltage of the driving stage must be appropriately less than the threshold voltage input of an enhancement-mode device of a driven stage to ensure that the device is off. Assuming that the fan-outs of the driving circuit present strictly capacitive loading, no dc output current from the driving stage is required; the load current and the driver device current are therefore equal. Equating the load current with the driver device current allows for the determination of the driver device dimensions (Figure 9.53).

NAND and NOR Circuit Design

The calculation of the required beta ratio for a NOR circuit is identical to that of the inverter, as the circuit must provide a valid 0 output voltage when any one of the parallel input devices is on. If more than one input to the NOR is 1, the effective beta ratio increases, which reduces the down-level voltage and improves the dc noise margin.

For NAND circuits, the necessary beta ratio for *each* driver device is determined by multiplying the inverter/NOR beta ratio by the number of devices in series. Some additional device width may be necessary in this series connection due to the loss in overdrive voltage ($V_G - V_S - V_T)_{\text{driver}}$; the source node of the upper devices is some fraction of $v_{\text{out}}(0)$.

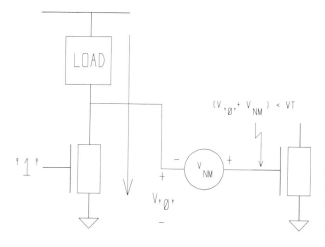

Figure 9.53. The driver device dimensions are selected so as to provide a good 0 logic voltage level; in making this calculation, the dc load and driver device currents are equated.

Worst-case dc Design Analysis

Although the guidelines given previously for selecting a circuit's beta ratio are adequate, it is nevertheless necessary to evaluate *in detail* the physical layout of the books or macros of the design system library to ensure their valid dc behavior in a general usage environment. This involves including such second-order considerations as the following:

1. Temperature, supply, and substrate voltage tolerances
2. Resistances of diffusions and contacts in the layout
3. Power supply and ground shift voltages on the internal image distribution
4. Junction leakage currents
5. Device subthreshold currents

The last item is a key dependency to include; rather than simply using a device threshold voltage as the limit for setting the maximum 0 output (plus noise margin) voltage, it is necessary to examine each *driven* circuit to determine the actual input voltage level at which the driven circuit's 1 output is still maintained (Figure 9.54). Subthreshold currents will result in a voltage drop across the load device and a reduced output voltage. Every driving circuit-driven circuit combination possible in the design system library should therefore be analyzed to verify the noise margin. For macros, this implies each internal macro circuit output should be compared against the other circuit inputs in the net. For books and macro I/Os, each output down level should be calculated and compared against the minimum of all book and macro input off voltages, since any book or macro output could be globally connected to any other book or macro input in the library.

An enhanced circuit model to replace that of Figure 9.53 is given in Figure 9.55, which includes all the circuit design parameters affecting the dc circuit noise margin calculation. The additional circuit elements should be used in conjunction with appropriate worst-case device current parameters. The chip temperature is maximized to reduce the driver device current, thereby increasing the driving circuit's 0 voltage level. Proper dc circuit operation is demonstrated when an

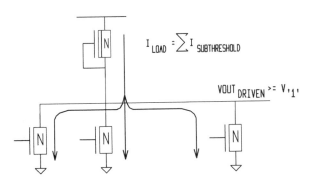

Figure 9.54. In calculating the worst-case maximum input voltage (0) to a driven circuit, it is necessary to model the subthreshold device currents of the input devices. The resulting load device voltage drop should not be so large as to drop the driven circuit's output voltage below a logical 1 value.

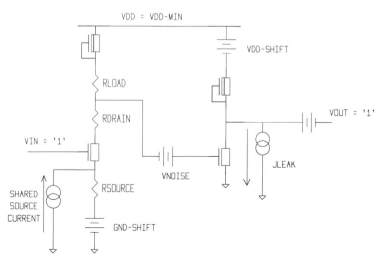

Figure 9.55. Complete model for the driving circuit-to-driven circuit dc noise margin calculation.

input voltage to the *driving* circuit equal to the minimum logic 1 voltage value results in an *output* voltage of the *driven* circuit that is greater than or equal to that value. The two logic inversions have therefore fully restored the minimum 1 voltage assumed for the technology. The verification that these noise margins are met or exceeded can proceed for all circuits in the library, as follows:

1. Regarding the circuit as the driving stage, calculate the on voltage with the device input(s) at the minimum logic 1 voltage. For an inverter or NAND, the v_{on} calculation is straightforward; for a NOR (or AOI or OAI), calculate the value of v_{on} for all of the possible individual current paths through the driver device network, keeping the maximum value.

2. Regarding the circuit as the driven stage, calculate the off voltage at the device input(s) such that the circuit's output is equal to $v_{1,\text{min}} + v_{\text{noise},1}$. The input device model should include the subthreshold current expression for this calculation. For a NAND circuit, use the maximum width-to-length ratio of a single device in the series connection, assuming all other inputs are 1. For a NOR circuit, use the *sum* of all device beta ratios, assuming all are possibly in their subthreshold region of operation.

3. For the circuits in the device library (books and macros), compare the v_{off} voltage of each driven circuit to the v_{on} voltage of its driving circuit. The difference should always be greater than the targeted 0-level noise margin.

The calculation of v_{off} and v_{on} for each library circuit (using worst-case device currents, supply voltages, and temperature conditions) requires a number of circuit simulations, which include all the pertinent additional schematic components in Figure 9.55:

(1) The *VDD* power supply has been reduced from its nominal value by an assumed tolerance of 10%; although this doesn't maximize the output voltage of the driving stage (v_{on}), this particular choice for the supply tolerance minimizes the output voltage of the driven stage (v_{off}).

(2) The effects of any series resistance in the on circuit are included by adding the following lumped elements (as extracted from the physical layout):

R_{load}:   includes the diffusion and contact resistance in the on path from supply to ground that does *not* contribute to the down-level output voltage.

R_{drain}:   includes the diffusion and contact resistance between the load and driver device that does contribute to the down-level output voltage.

R_{source}:   includes the diffusion and contact resistance associated with the terminal of the device connected to ground.

The voltage drop across R_{source} has an adverse effect on the driver device characteristics in addition to its contribution to the down-level output voltage; the device overdrive is reduced, much like the upper devices in a NAND series configuration.

(3) If a multiple number of circuits share all or part of the same diffusion path to a ground contact, the dc current through these adjacent circuits will raise the output level voltage of the circuit being designed; the effects of additional circuit currents on the voltage developed at the source node of the driver device under consideration are modeled by including the current source: $J_{shared\ source}$.

(4) Since it is quite likely that a large number of circuits will share the same metal ground distribution on the chip, the effects of a voltage drop between the circuit's ground connection and the actual chip pad ground are included by adding a fixed voltage source $V_{ground\ shift}$ at the ground connection of the driving circuit; this voltage source will tend to maximize the 0-level output voltage. For general design system usage, the circuit under evaluation may ultimately be placed anywhere on the chip image in relation to a ground pad and with a wide variety of other circuits sharing the same ground distribution; therefore, during design system library development, an assumption must be made as to what the worst-case ground shift allowed will be. This assumption should be verified after the physical design of individual design system chips is complete. The driven circuit will likely be located some distance away from the driving circuit on the chip, so no correlation to the ground shift of the driving circuit is usually assumed for the driven circuit; the worst case is to assume that there is no ground shift for the driven circuit at all, since this will minimize v_{off} and dictate that the 0-level output voltage of the driving circuit (v_{on}) be lower.

(5) Just as the ground distribution shared by a number of circuits can lead to ground shift for a circuit trying to maintain a good down level, a number of

circuits sharing the *VDD* power supply distribution can lead to significant shifts in the actual power supply voltage applied to a circuit trying to maintain an up level; this is included in Figure 9.55 with the voltage source VDD_{shift}. As with the case for ground shift applied to the driving circuit but not the driven circuit, the supply voltage shift is applied only to the driven circuit in order to minimize its up-level voltage. The value of the supply voltage shift must be assumed in this worst-case design approach, and then verified not to have been exceeded after the complete physical design of a design system chip is available.

(6) The reverse-biased *pn* semiconductor junction (between the output node of the driven stage trying to maintain a 1 voltage level and the chip substrate) will have a leakage current to the substrate that contributes to the voltage drop across the load device, further reducing the output up-level voltage. The effects of this current are included in the current source labeled J_{leak}. After the circuit design of each library element has been completed, the actual junction area or perimeter can be determined, and the validity of the leakage current assumption can be checked.

An elaborate dc circuit design noise margin analysis is not usually required of CMOS logic circuits, which dissipate no dc power.

NMOS off-chip driver and chip receiver circuits present unique dc analysis considerations, since external current loads (for driving other technologies) must be satisfied, and the necessary 1 and 0 voltage interface levels that must be provided are likely to differ from the levels present between internal circuits. Off-chip driver and receiver circuit designs are discussed in more detail in Chapter 11.

DC circuit verification is an integral part of developing a design system library of circuits for general use. Chapter 10 discusses in greater detail the need for tool support for checking circuit and macro additions to the design system library to ensure compatibility. In addition, since the circuit level calculations and comparison of v_{on} and v_{off} require assumptions regarding the worst-case global supply voltage and ground distribution drops, a final check should be made for each design to verify that these values are not exceeded. The incorporation of a global voltage-drop analysis tool into a back-end design checking methodology is also discussed in Chapter 10.

Worst-case dc Power Calculation

The worst-case maximum dc circuit power is calculated with the 3σ maximum load device current, $VDD = VDD_{max}$, and the temperature at a maximum. (Although the device current is greater at lower temperatures, the chip power at *maximum* temperature is the cooling constraint that the system environment using that chip must accommodate.) The source node of the load device can roughly be assumed to be at ground potential, neglecting the effect of the 0 level voltage on the load current. The chip power is the summation of the individual circuit power dissipations, with an appropriate assumption about the duty cycle for a logical 0 circuit output value.

Whereas the total NMOS depletion-mode load circuit power dissipation is dominated by the dc logic 0 level current, the CMOS circuit's power dissipation is strictly the time-averaged product of the transient device current times device voltage. It is therefore proportional to the circuit's switching frequency. Problem 9.1 asks for the derivation of an expression for the power dissipation of a CMOS circuit, switching a capacitive load C_{out} at a particular frequency f.

Parameter Tracking

This section introduced some of the worst-case design considerations for dc circuit design and analysis; additional circuit elements were included in an attempt to accurately model the physical layout, but little mention was made of the characteristics of the devices themselves. The characteristics of different devices on the chip are *not* independent of one another; they are coupled by the tracking of each parameter over the area of the chip. As discussed briefly in Section 8.7, each processing parameter has an allowed tolerance or statistical distribution that encompasses the characterization data gathered from many wafers in a number of fabrication runs. Tracking implies that, if a parameter is assumed to have a particular value for a particular device on a chip, the *chip* distribution of that parameter (for modeling purposes) has a much smaller standard deviation than the overall process distribution.

As an example how tracking affects worst-case dc circuit design, consider the circuit of Figure 9.56 (the driving stage from Figure 9.55). Considering the two devices Q_{load} and Q_{driver}, the maximum worst-case voltage v_{out} would occur if Q_{driver} were a 3σ minimum-current enhancement-mode device and Q_{load} were a 3σ maximum-current depletion-mode device; this choice would maximize the voltage drop across Q_{driver} and minimize the voltage across Q_{load}, effectively increasing v_{out}. However, some of the process parameters that determine the enhancement-mode driver device characteristics also pertain to the depletion-mode load device characteristics; it is not valid (far too pessimistic) to assume that the different devices could have values for the same parameter at diametrically opposite ends of that parameter's distribution. Tracking refers to the amount of correlation between two different values of the same parameter at different locations on a chip; the correlation is a function of the distance separating the devices—the less the separation, the closer the values of individual process parameters must be for the two devices. Referring again to Figure 9.56, it is valid to assume that device Q_{driver} may indeed be a 3σ minimum-current enhancement-mode device due to some combination of physical process parameters; the parameters used to calculate the depletion-mode device Q_{load}'s current must track those assumed for device Q_{driver} for the distance separating the two devices. Figure 9.57 illustrates how the values of a particular device parameter, the gate oxide thickness (t_{ox}), may be chosen when modeling the two devices.

WORST CASE V$_{'ON'}$ CALCULATION

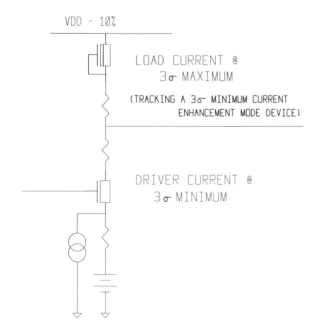

VDD - 10%

LOAD CURRENT @
3σ MAXIMUM

(TRACKING A 3σ MINIMUM CURRENT
ENHANCEMENT MODE DEVICE)

DRIVER CURRENT @
3σ MINIMUM

Figure 9.56. When calculating the worst-case dc logic 0 voltage level, the 3σ minimum driver device current should be used; the load device current should be maximized, yet must *track* the parameters used to calculate the driver current.

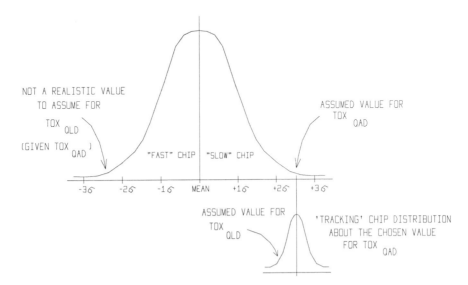

NOT A REALISTIC VALUE
TO ASSUME FOR

TOX$_{QLD}$

(GIVEN TOX$_{QAD}$)

ASSUMED VALUE FOR
TOX$_{QAD}$

"FAST" CHIP   "SLOW" CHIP

-3σ   -2σ   -1σ   MEAN   +1σ   +2σ   +3σ

ASSUMED VALUE FOR
TOX$_{QLD}$

'TRACKING' CHIP DISTRIBUTION
ABOUT THE CHOSEN VALUE
FOR TOX$_{QAD}$

Figure 9.57. Selection of a parameter value for a slow (minimum current) device and the tracking distribution around that global parameter for maximum current device calculations.

In summary, selecting the right values for the process parameters to model a worst-case dc circuit design requires that values from process-wide distributions be selected and possible chip variations subsequently superimposed. The process variations may result in either a "fast" (maximum current) or "slow" (minimum current) chip; individual devices will vary about the chip mean with a smaller deviation. The process and tracking distributions are commonly derived empirically from large amounts of process monitor and characterization data.

9.5 CIRCUIT DESIGN FOR PERFORMANCE CALCULATION

To verify the performance of a network design, the chip designer is faced with the task of choosing a simulation methodology that will provide the necessary circuit delay information so that path delays can be calculated. One option (with the most accurate results) would be to execute a device-level, continuous-time circuit simulation; however, for any substantial number of circuits in the network or any significant length of simulation time, the computational resource required would be prohibitively expensive. An alternative approach, the approach consistent with design system development, is to provide a generalized *delay equation* for the books in the technology library, an equation that incorporates as input variables the design-dependent, circuit-level parameters that dictate the block delay in a logic path. For example, assuming a depletion-mode load NOR circuit, this set of parameters may include the circuit power, the beta ratio, the input signal transition time, and the output (capacitive) loading, consisting of voltage-independent (e.g., interconnection wiring) and voltage-dependent (e.g., junction area or perimeter, device gate input) components. Delay algorithms can also be developed for other load device types (e.g., enhancement mode, CMOS) and other circuit types (e.g., NAND, AOI). Also, if the delay equation requires an input signal transition time value, an equation must be developed for calculating the output signal's transition time (or dv/dt) in addition to its delay.

Delay data taken from a number of small circuit-level simulations run over a reasonable domain of the input-variable set can be curve-fitted to the form of the delay equation to provide the necessary coefficients. Separate delay and transition time relationships may be developed for output rising and falling transitions. Using delay equations, block delays can be calculated based on the actual or estimated chip physical design information and then back-annotated to the logic simulation model for design verification (Figure 9.58). This is a much more efficient (and yet still quite accurate) approach, as compared to an extensive device-level simulation.

In order that individual block delays may be summed to accurately determine path delays using a logic simulator, it is important that the block delay be defined carefully and consistently; the circuit simulations used to extract the data for determining rising or falling delay equation coefficients should all be run under the same set of supply voltage, temperature, and process assumptions. In addition,

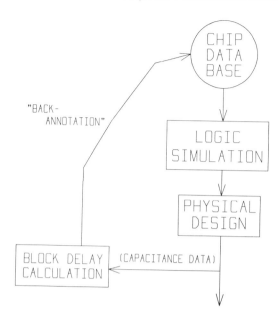

Figure 9.58. Block delays based on the physical interconnection capacitance are calculated after the physical design is complete and back-annotated to the logic simulation model.

the end points used to extract a block's delay value on the circuit simulator's node waveform chart should be chosen so that the block's delay is always positive, and the summation of these delays would be consistent with the total path delay as determined from a larger-model simulation.[3]

One possible delay definition is presented in Figure 9.59; the circuit delay is defined as the time interval between the point where the input voltage crosses the circuit's unity gain voltage to the point where the output crosses the unity gain voltage of the *driven* block. In short, delays are consistently measured between unity gain points of block inputs. If all logic circuits in the technology library have the same unity gain voltage, the reference point for delay measure is fixed. On the other hand, if the unity gain voltages differ among the library elements, it is necessary to refine the delay calculation method in the following manner:

1. In addition to the circuit's input capacitance, the technology data base should also store the circuit's unity gain voltage, potentially on a pin-by-pin basis (Figure 9.60).

2. The delay end points for extracting sample data from circuit simulation results should now be from the input unity gain voltage (as before) to a reference point outside the library's range of unity gain values (Figure 9.61).

3. The additional delay to the driven circuit's unity gain voltage can be calculated from the output signal transition dv/dt equation and the voltage difference ($V_{ref} - V_{unity\,gain}$). This delay can then be incorporated into a trans-

---

[3] The algorithm(s) used in a logic simulator schedules a block's output transition in *future* simulation time; negative block delays are not valid.

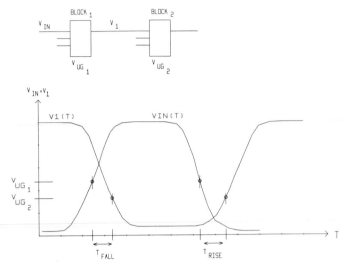

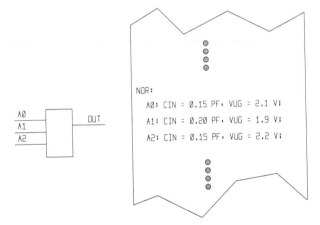

Figure 9.59. The block delay is defined as the time interval between the point where the input voltage crosses the block's unity-gain voltage to the point where the output crosses the unity-gain voltage of the *driven* block.

NOR:

A0: CIN = 0.15 PF, VUG = 2.1 V;

A1: CIN = 0.20 PF, VUG = 1.9 V;

A2: CIN = 0.15 PF, VUG = 2.2 V;

Figure 9.60. If the unity-gain voltage differs among the design system library's circuits, the technology data base should also store the individual circuit's unity-gain voltage, potentially on a pin-by-pin basis.

parent logic block delay in series with the circuit's input pin; as a result, the variation in unity gain voltages among pins in the fan-out of a net can each individually be included in logic simulation of the network.

Off-chip driver and receiver circuits have unique voltage interface requirements and will have unique criteria for the end points used to extract delay values for the delay equations of these block types. In addition, the range of expected loadings (connected to off-chip driver outputs) and transition times (on chip receiver inputs) is considerably wider than what may be present internally; as such,

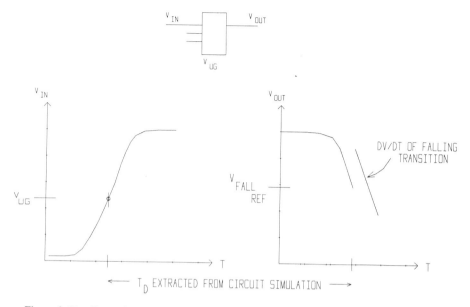

Figure 9.61. From the circuit simulation analysis, circuit delays are now measured from the input unity-gain voltage (as before) to a reference point outside the library's range of unity-gain voltages; the additional delay to a specific driven circuit's unity-gain voltage can then be extrapolated from the *dv/dt* of the signal transition.

the circuit simulations used to provide data values for curve-fitting must span this larger external set of conditions. However, developing accurate delay and transition time expressions over such a wide range of values may be difficult; as an alternative, in lieu of an equation, the design system's delay calculator and back-annotation tool could also accept tabular input (Figure 9.62). A file of records for the particular book may be provided in the technology data base which would be used for the following:

1. To extract a base block delay value.
2. To calculate a delay adder for the input signal transition time (interpolate between the two closest tabular values).
3. To calculate a delay adder and output signal transition time for the capacitive load (interpolate between the two closest tabular values).

The division of delay values into three constituent adders is intended to be as representative as possible of the dependencies observed in the circuit simulations; the goal is to produce a correct block delay summation for any combination of input transition time and output loading.

The use of behavioral-level logic simulation models was discusssed in Chapter 4. To efficiently simulate macro-level functions in the design system library,

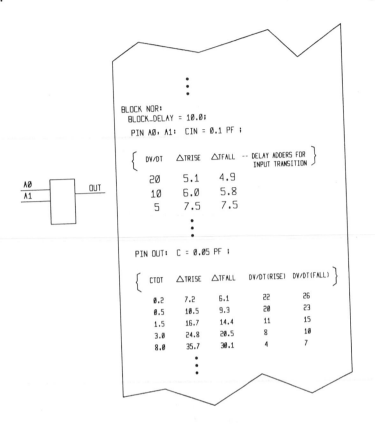

BLOCK NOR:
  BLOCK_DELAY = 10.0;
  PIN A0, A1: CIN = 0.1 PF ;

DV/DT	△TRISE	△TFALL	-- DELAY ADDERS FOR
			INPUT TRANSITION
20	5.1	4.9	
10	6.0	5.8	
5	7.5	7.5	

  PIN OUT: C = 0.05 PF ;

CTOT	△TRISE	△TFALL	DV/DT(RISE)	DV/DT(FALL)
0.2	7.2	6.1	22	26
0.5	10.5	9.3	20	23
1.5	16.7	14.4	11	15
3.0	24.8	20.5	8	10
8.0	35.7	30.1	4	7

A0
A1 ──── OUT

Figure 9.62. Rather than fitting the circuit simulation results to a delay equation, it may also be feasible to provide tabular input of the delay versus capacitive loading and input signal transition time.

these models provide the logic function of the macro to the simulator in a procedural description, eliminating the need for an expanded Boolean representation. The network delays of the macro were incorporated back into the simulator by means of calls to the simulator's scheduling routine from within the behavioral procedure; the delay values were hard-coded into the behavioral. Therefore, the macro delays are somewhat coarse as they cannot include dependencies on input signal transition times and output loading. If the simulation model for the behavioral includes transparent logic blocks at all macro pins, then delay adders surrounding the behavioral description may absorb additional delay dependencies. The delay calculation tool adds transition time dependencies to the inputs and capacitive loading delay adders to the outputs; a separate set of tabular delay records is included for each macro pin.

Note that the design system library complexity has a direct bearing on the features required of the logic simulator for accurate design verification after the chip physical design is complete. For a number of applications, the ability to incorporate delay values directly on pins is crucial to including delay dependencies on transition time and loading parameters. Indeed, it may be feasible to eliminate the use of block delays entirely; instead, the block's simulation model could just be its logical (Boolean) function; all delay information could be assigned to the input and output pins of the block.

The performance modeling of the transfer gate may also require some unique characteristics of the delay calculator tool, since the output rising delay is a very strong function of the control signal transition time; the transfer gate output node "follows" the control signal input for slow transitions. As a result, any inaccuracies in the calculation of the control signal transition time propagate widely as delay inaccuracies in the logic simulation. Other circuit types, such as the dottable open-drain driver discussed in Chapter 11, present similar difficulties to a general delay calculator tool. In this case, the global series interconnection resistance between the net capacitance and its discharge path is a parameter of the physical design that may need to be incorporated into a delay equation for this book type (if indeed this book type is included in the design system library). Circuits employing dynamic (or precharged) nodes also require unique delay algorithms; the precharge (rising) delay calculation is similar to that of the transfer gate, while the discharge (falling) transition is enhanced over that of static logic circuits due to the absence of load device current.

Gate-limited Delay

The ability to trade off additional circuit power to achieve better performance is eventually limited; this region of operation is denoted as *gate-limited* delay. As an example, consider the *n*-channel depletion-mode load inverter circuit illustrated in Figure 9.63; the individual contributions to the total capacitive load on the output of the inverter are also depicted in the figure. For ratioed circuits, the device and junction capacitances *scale* with the circuits' power dissipation; increasing the power dissipation of the circuits in a delay path to gain performance reaches a point of diminishing returns as the gate capacitance becomes the dominant load (Figure 9.64). Assume that a suitable delay equation can be written

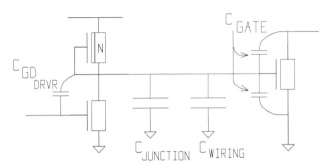

Figure 9.63. Components of the capacitive load on the output of an *n*-channel depletion-mode load circuit. Some of these components will scale with the circuit power in the delay chain; others will not.

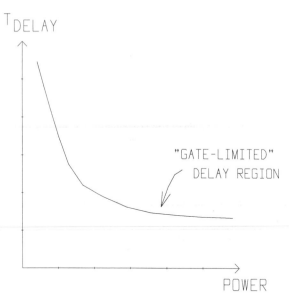

T_{DELAY}

"GATE-LIMITED"
DELAY REGION

POWER

Figure 9.64. Increasing the circuit power in a delay path to improve the performance of the chain reaches a point of diminishing returns as the gate capacitance becomes the dominant load; this range is denoted as the gate-limited delay region.

that is *linear* in the delay components of each of the capacitive terms. A possible form for the expression might be

$$t_{delay} = \alpha * \left(\frac{C_{junction}}{P}\right) + \beta * \left(\frac{C_{wiring}}{P}\right) + \gamma * N_{fan\text{-}out} * \frac{C_{gate}}{P} + \gamma * \frac{C_{GD_{driver}}}{P}$$

where α, β, and γ are constants, P is the power of the circuits in a logic path, and $N_{fan\text{-}out}$ is the fan-out. The device capacitance is proportional to the circuit power, resulting in the delay expression

$$t_{delay} = \alpha \left(\frac{C_{junction}}{P}\right) + \beta \left(\frac{C_{wiring}}{P}\right) + (\delta * N_{fan\text{-}out} + \epsilon)$$

The delay is limited by $(\delta * N_{fan\text{-}out} + \epsilon)$ regardless of the power dissipation of the circuits in the path.

The previous simple delay equation did not incorporate any dependency on input signal rise and fall times; this may be suitable for circuits in a *balanced* logic chain. In a balanced chain, the ratio of the capacitive output load to circuit power is the same for all blocks; all blocks have the same rise times and fall times, so the sensitivity of the block delay to the transition time can be included in the coefficients α and β. Figure 9.65 illustrates the distinction between a balanced and unbalanced logic chain. The linear delay equation written earlier is *not* valid for unbalanced chains; some additional functional dependence on the transition time is necessary. To compensate for unbalanced chains, the linear-type equation

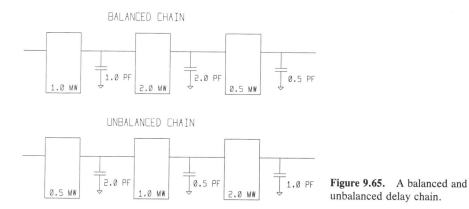

Figure 9.65. A balanced and unbalanced delay chain.

can be modified by an empirically determined multiplier:

$$t_{\text{delay}_{\text{linear,balanced}}} = \alpha * \left(\frac{C_{\text{junction}}}{P}\right) + \beta * \left(\frac{C_{\text{wiring}}}{P}\right) + (\delta * N_{\text{fan-out}} + \epsilon)$$

$$t_{\text{delay}_{\text{unbalanced}}} = M * t_{\text{delay}_{\text{linear,balanced}}}$$

where M is the delay multiplier related to the imbalance between successive circuits in the chain. Figure 9.66 illustrates how this multiplier modifies the block delay as a function of input transition time; the transition time is expressed in terms of the load capacitance-to-circuit power ratios of the two circuits.

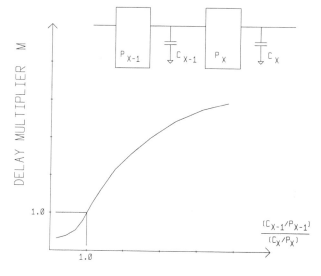

Figure 9.66. Use of a delay equation multiplier to describe the variation between balanced and unbalanced circuit chains.

PROBLEMS

9.1. Referring to the circuit schematic in Figure 9.67, assume that the capacitive load C_{out} is charged and discharged with a frequency f by the CMOS inverter circuit. Develop an expression for the power dissipated by the CMOS circuit, assuming that all the circuit current charges or discharges the load; neglect the power dissipation due to the current in both devices during the input signal transition. (Consider the time-averaged energy added to and extracted from the capacitance C_{out}.)

9.2. What are the channel carrier mobility or transconductance differences between n- and p-channel devices? How does this affect CMOS logic circuit design if balanced rising and falling delays are a design goal for the circuit library? For circuit area considerations, which family of logic circuits, NAND or NOR, is preferable?

9.3. A number of VLSI technologies include a device offering called the *natural threshold voltage* enhancement-mode n-channel device. Investigate and describe the characteristics of the natural V_T device. What processing modifications are required to include this device in a normal NMOS technology? What circuit applications can best utilize the features of this device type?

9.4. FET circuits do not exhibit the saturated region delay time characteristic of bipolar device technology logic circuits. Research and briefly describe the nature of bipolar logic's saturation delay time, as well as the technique(s) used to eliminate this component of delay.

9.5. (a) Research the technical literature and textbooks of early PMOS technology and circuit design. Develop schematics for p-channel enhancement-mode device logic circuits. From your research, determine the years over which PMOS was the dominant IC FET technology, and give some background as to the reasons for its rise and subsequent fade from prominence.

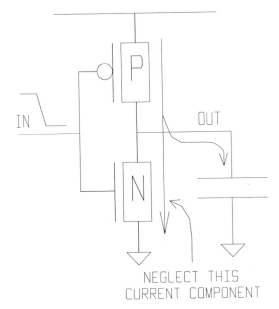

NEGLECT THIS
CURRENT COMPONENT

Figure 9.67. A CMOS inverter circuit used to calculate power dissipation as a function of C_{out} and the switching frequency f.

(b) What is the distinction between positive logic and negative logic?
(A good reference for PMOS technology is W. N. Carr and J. P. Mize, *MOS/LSI Design and Application*, McGraw-Hill Book Co., New York, 1972.)

9.6. Figure 9.3 illustrates a load device option that consists of an enhancement-mode device in parallel with a standard depletion-mode load; this combination can be used as the load for the external stage of a push–pull circuit. What are the advantages and disadvantages of this option? (Be sure to address power dissipation, performance, delay calculation and modeling, and, in particular, testability. Recall that the wafer tester typically verifies overall functionality at slower performance ratings than the chip design specification. How would the circuit behave at the tester if only one of the two load devices were present? How would this affect the circuit's performance?)

9.7. **(a)** Research the causes of latch-up in CMOS circuits, especially pad driver circuits. Determine the conditions under which triggering currents may also be introduced in *internal* circuits; in particular, investigate the behavior of CMOS transfer gates.

(b) Describe the various circuit design, layout, and fabrication process measures developed to reduce the likelihood of latch-up.

9.8. The CMOS circuits illustrated in this chapter all assumed that an *n*-well was locally provided as the substrate for implementing *p*-channel devices; the *p*-type background wafer was the substrate for all *n*-channel devices. Alternatively, a *p*-well or twin-well fabrication process may be used. Research the various process options for CMOS device substrates and briefly describe the differences in layout, supply voltage connection, and device characteristics for the different processes.

9.9. Design a data latch circuit schematic in both NMOS and CMOS technologies that includes an asynchronous reset input signal. Evaluate the performance (in terms of unit block delays), the power dissipation, and the stuck-fault testability of your design. (For more information about testability, refer to Section 13.1.)

9.10. Develop a logic model for an edge-triggered flip-flop, indicating how changes in the data input value are not observed at the output at time separated from the active transition on the clock input. Indicate what data setup and hold time requirements must be satisfied, assuming all logic blocks of the flip-flop have an equal (unit) delay. Compare this logic model against that of the level-sensitive latch. How many feedback paths are required?

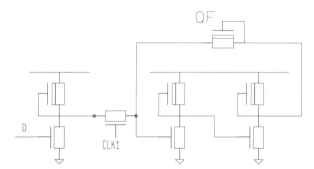

Figure 9.68. NMOS latch circuit with depletion-mode feedback device Q_f.

9.11. **(a)** Describe the operation of the NMOS technology latch circuit depicted in Figure 9.68. What design considerations should be used in sizing the device Q_f?

(b) Design a CMOS technology latch circuit like that of Figure 9.49 but without the requirement that the data buffer must be able to overdrive the feedback inverter; use a CMOS transfer gate in the feedback loop (in a similar manner as device Q_f is utilized in the latch circuit of Figure 9.68).

9.12. Design an XOR logic gate circuit in a CMOS technology; use CMOS transfer gates to provide an efficient implementation.

REFERENCES

Berndlmaier, E., and others, "Delay Regulation—A Circuit Solution to the Power/Performance Tradeoff," *IBM Journal of Research and Development*, 25:2 (May 1981), 135–141.

Chatterjee, P., and others, "The Impact of Scaling Laws on the Choice of n-Channel or p-Channel for MOS VLSI," *IEEE Electron Device Letters*, EDL-1:10 (October 1980), 220–223.

Chawla, B., Gummel, H., and Kozak, P., "MOTIS—An MOS Timing Simulator," *IEEE Transactions on Circuits and Systems*, CAS-22:12 (December 1975), 901–910.

Cooper, James, "Limitations on the Performance of Field-Effect Devices for Logic Applications," *Proceedings of the IEEE*, 69:2 (February 1981), 226–231.

Divekar, D. A., and Dowell, R. I., "A Depletion-Mode MOSFET Model for Circuit Simulation," *IEEE Transactions on Computer-Aided Design of Integrated Circuits and Systems*, CAD-3:1 (January 1984), 80–87.

Dumlugol, D., and others, "Local Relaxation Algorithms for Event Driven Simulation of MOS Networks Including Assignable Delay Modeling," *IEEE Transactions on Computer-Aided Design of Integrated Circuits and Systems*, CAD-2:3 (July 1983), 193–201.

Elmasry, M. I., "Capacitance Calculations in MOSFET VLSI," *Electron Device Letters*, EDL-3:1 (January 1982), 6–7.

Frohman-Bentchkowsky, D., and Vadasz, L., "Computer-Aided Design and Characterization of Digital MOS Integrated Circuits," *IEEE Journal of Solid-State Circuits*, SC-4:2 (April 1969), 57–64.

Hayes, Jim, "MOS Scaling," *IEEE Computer* (January 1980), 8–13.

Kang, S., "A Design of CMOS Polycells for LSI Circuits," *IEEE Transactions on Circuits and Systems*, CAS-28:8 (August 1981), 838–843.

Kokkonen, K., and Pashley, R., "Modular Approach to C-MOS Technology Tailors Process to Application," *Electronics* (May 3, 1984), 129–133.

Kopp, R., and Stevens, D., "Overlay Considerations for the Selection of Integrated-Circuit Pattern-Level Sequences," *Solid State Technology* (July 1980), 79–87.

Masuhara, T., and others, "A High-Performance n-Channel MOS LSI Using Depletion-Type Load Elements," *IEEE Journal of Solid-State Circuits*, SC-7:3 (June 1972), 224–231.

Puri, Yogishwar, "On the Logic Delay in MOS LSI Static NOR Designs," *IEEE Journal of Solid-State Circuits*, SC-14:4 (August 1979), 716–723.

Ratnakumar, K., and Meindl, J., "Performance Limits of E/D NMOS VLSI," *IEEE International Solid-State Circuits Conference* (1980), 72–73, 260.

Rubinstein, J., and others, "Signal Delay in RC Tree Networks," *IEEE Transactions on Computer-Aided Design of Integrated Circuits and Systems*, CAD-2:3 (July 1983), 202–212.

Ruehli, A. E., and Brennan, P., "Accurate Metallization Capacitances for Integrated Circuits and Packages," *IEEE Journal of Solid-State Circuits*, SC-8:4 (August 1973), 289–290.

Rymaszweski, E., and others, "Semiconductor Logic Technology in IBM," *IBM Journal of Research and Development*, 25:5 (September 1981), 603–616.

Saraswat, K., and Mohammadi, F., "Effect of Scaling of Interconnections on the Time Delay of VLSI Circuits," *IEEE Transactions on Electron Devices*, ED-29:4 (April 1982), 645–650.

Silburt, A., and Foss, R., "VLSI Changes the Rules for Coping with Substrates," *Electronics* (November 17, 1982), 155–159.

―――, and others, "An Efficient MOS Transistor Model for Computer-Aided Design," *IEEE Transactions on Computer-Aided Design of Integrated Circuits and Systems*, CAD-3:1 (January 1984), 104–114.

Tokuda, T., and others, "Delay-Time Modeling for ED MOS Logic LSI," *IEEE Transactions on Computer-Aided Design of Integrated Circuits and Systems*, CAD-2:3 (July 1983), 129–135.

Uehara, T., and vanCleemput, W., "Optimal Layout of CMOS Functional Arrays," *IEEE Transactions on Computers*, C-30:5 (May 1981), 305–312.

Wang, C., and others, "Advanced CMOS Process for LSI Circuit Development," *IEEE International Conference on Computer Design: VLSI in Computers* (1983), 375–378.

Wollesen, D., "CMOS LSI—The Computer Component Process of the 80's," *IEEE Computer* (February 1980), 59–67.

10

DESIGN SYSTEM LIBRARY DEVELOPMENT

10.1 BASIC DESIGN SYSTEM TENETS

The developers of the design system library are faced with the trade-off between the need for flexibility in logic design and the physical constraints of area and performance. The decisions of the library development group as to what to include versus what to preclude need not be final; the library can continue to grow with more higher-level functions as the corresponding models and layouts are provided. The major effort in library development is formulating and enforcing the guidelines and requirements for all books, from unit logic to multiple-cell macros. Some of these guidelines are discussed in the remainder of this section.

Fan-in

The NOR and NAND unit logic blocks should be available with a range of fan-in values. The circuit library and chip image development groups must decide the fan-in limit for a single-cell implementation and the maximum fan-in available overall (possibly requiring a multiple-cell layout). The first decision typically relates to the number of pin locations provided per cell and the amount of flexibility desired in accommodating logic changes after the first-pass physical design has been completed. Each circuit input and output occupies one wiring channel on the second metal wiring grid; to enhance wireability, the cell width usually incorporates a number of additional second metal wiring channels to enter or exit the cell at each input (Figure 10.1). The decision regarding the overall fan-in maximum (single- or multiple-cell) is based on the following:

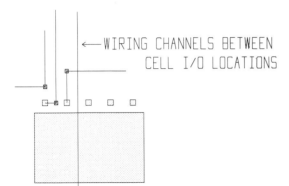

Figure 10.1. The cell design usually includes additional (second metal) wiring channels between pin locations for enhanced overall image wireability.

1. The anticipated width of data fields, address decode lines, and the like, which would typically be used in a chip logical design for the technology.
2. The circuit area and/or performance disadvantages associated with high fan-in books (e.g., device widths for series connections in ratioed NAND designs and CMOS circuits).
3. The incorporation of algorithms in logic synthesis tools to efficiently utilize high fan-in book offerings.

The last item is a key point to consider; if logic synthesis of higher-level network descriptions into technology books is to be incorporated into the design methodology, the technology library should include the book designs for which synthesis transformations can readily be provided.

Cell and Macro Layout

The shapes layout of the circuit library involves several design considerations and the adoption of design restrictions for enhanced checking. Specifically, some of the constraints to consider include the following:

1. All pin shapes should coincide with wiring grid points.
2. All shapes in the vicinity of the cell boundary should be sufficiently distant from the cell edge in order that the placement of books in adjacent cells does not violate any technology lithography rules (Figure 10.2).
3. If circuit layouts contain shapes on global interconnection levels, a means of defining and storing the associated blocked channels must be provided in the technology library database for use by the wiring program. A CAD program for generating the blockage information for the technology database *directly* from the layout is particularly useful to include in the design system tool set.
4. If the macro design is such that the operation of the macro internal circuits

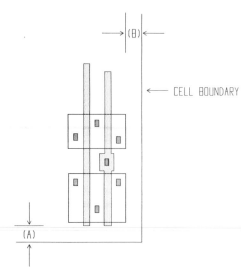

Figure 10.2. Shapes internal to the circuit design within the cell should be sufficiently spaced from the cell boundary so that the placement of designs in adjacent cells does not violate any technology lithography rules. For example, the separation (a) should be at least one-half the technology lithography rule for polysilicon-to-polysilicon space, while separation (b) should be at least one-half the node-to-node spacing.

prohibits *any* external capacitive coupling to internal nodes, the entire layout area is a blockage for global interconnection routing. This information should also be used to influence the placement of that macro to avoid wiring congestion.

5. Some leverage may be gained in the cell layout by permitting shapes on levels not associated with global interconnections to protrude into the wiring bay. An example of a polysilicon interconnection used in the cell design that extends into the wiring bay is given in Figure 10.3.

Global Current Distribution

To reduce supply voltage and ground shifts due to resistive losses, all global supply distribution to the cells in the chip image should be provided on metal interconnect; higher resistivity *cross-unders* (Figure 10.4) should be avoided if at all possible. This will ease the task of analyzing the actual supply shifts from the completed physical design and comparing those results against the values assumed when performing circuit-level dc noise margin checking, as discussed in Section 9.4.

Another major design consideration relates to circuit design options that require global dc *signal* currents. Normally, FET device inputs present a strictly capacitive load to the output of the driving circuit. The propagation of a logic signal is usually regarded as the discharging or charging of the combination of interconnection, wiring, and device input capacitance. A circuit design requiring dc input current (at either or both logic levels) is not a good candidate for a general usage design system library; fan-out checking is now required to ensure that the

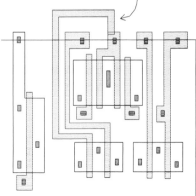

Figure 10.3. Some circuit design leverage may be gained if shapes are permitted to extend into the wiring bay beyond the cell boundary, shapes on nonglobal interconnection levels.

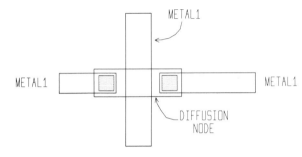

Figure 10.4. *Cross-unders* for continuing power supply distribution throughout the chip image should be avoided if at all possible due to the resistance of the vias and/or cross-under interconnect layer.

additional external driving circuit current(s) do not result in an inadequate output voltage level. This fan-out analysis must necessarily include developing a detailed model of the global signal interconnection resistances.

Dynamic Circuits

Dynamic circuits, utilizing the stored charge (or precharge) of a node capacitance to represent a data bit value, are difficult to incorporate into a design system library due to the unique logic behavioral and test modeling required to represent the minimum clocking frequencies and the loss of data over time due to leakage currents. Of particular concern is the additional complexity in wafer-level testing introduced by the presence of dynamic circuits. VLSI test methodologies typically assume that chip functionality is demonstrated by the absence of dc stuck-faults at all signal nets; chip diagnostic tools used to assist with fault isolation and failure analysis utilize the propagation of logic stuck-at-1 or stuck-at-0 values through a network. The loss of the stored charge at a node results in intermediate logic voltage levels, which could lead to logic failures that are not easily traceable. Therefore, to as large an extent as possible, logic circuits should appear to be static at wafer-level test.

Book Design and Power/Performance Optimization

To enhance the ability to meet overall chip performance goals, a range of power/performance levels of different book designs is commonly provided in the design system library. The assignment of the appropriate performance level of the function to the chip physical database is done based on the delays calculated from either an anticipated fan-out plus interconnection wiring load or from the actual physical routed interconnections after the wiring program has been run. The design system and technology library developers are faced with the questions of when to make or update this assignment and whether all performance levels of a particular function should occupy an identical number of cells (with their circuit I/O pins on identical grid points). If level assignment is done prior to placement based on estimated loading, all levels need not span the same cell count; however, subsequent assignments may require significant manual editing of the physical design to be able to accommodate a higher-performance, higher-cell-count design.

Transfer Gates

Chapter 9 illustrated that transfer gates can be both useful and problematic in circuit designs. Difficulties associated with delay calculation, simulation and/or test modeling, output logic 1 voltages less than the supply, and possible capacitive coupling among circuit nodes tend to discourage the general utilization of transfer gates as a design system library primitive. A more conservative approach would be to adopt the following principle:

> When designing with transfer gates, *all* connections to the nodes of the pass device(s) shall be internal book or macro connections; no device connection should be permitted to be a global connection.

In this manner, circuit simulations (with known driving and driven block physical parameters) may be run to extract delay information for simulation (behavioral) modeling and to verify that no anomalous dc interface levels nor capacitive coupling noise transients are present.

One final design system tenet to be addressed is the logic and circuit design constraints that are consistent with *design for testability* principles. Chapter 9 briefly mentioned that, in accordance with DFT guidelines, flip-flop books would not be available in the design system library. In addition, a number of design criteria must be met (and checked by the design system tools) regarding logic feedback paths, system clock gating, and internal net controllability and observability. The testing methodology incorporated into the design system pervades the logic and circuit design philosophies and must be addressed early in library and tool development. Chapter 13 will describe the traditional stuck-at-fault modeling approach and some of the design and algorithmic techniques developed to aid with fault detection and diagnosis; in addition, alternative design approaches specifically addressing VLSI testing resource constraints will be briefly discussed.

10.2 BOOK COMBINATIONS IN THE DESIGN SYSTEM LIBRARY

A representative set of primitive logic blocks for a design system library is given in Figure 10.5. The OR and AND functions are commonly implemented by an inverting buffer connected to the output of a NOR or NAND circuit, respectively. The range of fan-in values specified for a particular function indicates the amount of flexibility provided within the allocated cell area to accommodate logic changes after the first-pass physical design has been completed.

The higher fan-in functions may be implemented by designing with smaller fan-in circuits, such as the example given in Figure 10.6. In addition, Figure 10.7 illustrates that the technology library may contain unit logic blocks with an input inverted to more fully utilize the cell area and save an inverter cell. Although some savings may indeed result, adding a number of books like the one in Figure

```
INVERTER

NONINVERTING BUFFER

2 - 4 INPUT NOR

2 - 4 INPUT NAND

5 - 9 INPUT NOR (2-CELL)

5 - 9 INPUT NAND (2-CELL)

2 - 4 INPUT AND

2 - 4 INPUT OR

2-2 / 2-1 AOI

2-2 / 2-1 OAI

2 INPUT XOR

2 INPUT XNOR
```

Figure 10.5. Common primitive logic functions to include with the design system library. For this set, a primitive cell width of five pins was assumed.

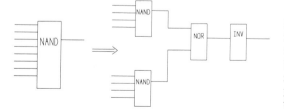

Figure 10.6. Implementation of a high fan-in logic primitive in terms of smaller fan-in circuits; the trade-offs between the two alternatives are with respect to circuit area, circuit performance, and input loading from the driving circuit.

2 - 4 INPUT NOR WITH ONE INPUT INVERTED

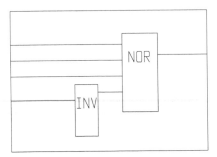

Figure 10.7. Design system technology library book consisting of a primitive gate with one input inverted.

10.7 to the library should be weighed against the disadvantages of larger technology library database size and the difficulty in developing logic synthesis transforms to match this book to the decomposition of the Boolean functions in a network.

Figure 10.8 lists the common sequential elements included in the technology library, while Figure 10.9 gives the typical I/O pad circuit repertoire. A list of typical technology library macro functions is given in Figure 10.10; these designs may be either physical (multiple cell) macro layouts or implemented as logical macros (i.e., the global interconnection of logic primitives).

The sequential functions refer to *L1/L2* and *L1/L2** latch offerings; these are the *LSSD* latch types provided to satisfy the design for testability guidelines, as will be discussed in Chapter 13. (The L1/L2 latch pair is similar to a master–slave flip-flip combination, with the output of the first latch providing the input to the second. However, unlike the master–slave implementation, the L1 and L2 latch clocks are *independent* signals; the L2 latch clock is not to be derived from the

```
DATA LATCH (L1/L2)  (+/- L1, +/- L2 OUTPUTS)

SET/RESET L1/L2  (+/- L1, +/- L2 OUTPUTS)

2-PORT L1/SINGLE PORT L2 (+/- L1, +/- L2 OUTPUTS)

L1 DATA PORT/L2 DATA PORT LATCH (+/- L1, +/- L2)

4-BIT REGISTERS:

      SINGLE-PORT L1/L2

      2-PORT L1/L2

      L1 DATA PORT/L2 DATA PORT

8-BIT REGISTERS:

      SINGLE-PORT L1/L2

      2-PORT L1/L2

      L1 DATA PORT/L2 DATA PORT

SYNCHRONIZING RECEIVER/LATCH
```

Figure 10.8. Common sequential elements included in a design system library. All the sequential elements listed are latches, and all are present in L1/L2 latch pairs.

```
RECEIVERS:
    TTL - INTERFACE --
        INVERTING OR NONINVERTING
        WITH/WITHOUT HYSTERESIS
    FET - INTERFACE --
        INVERTING OR NONINVERTING
        WITH/WITHOUT HYSTERESIS
OFF-CHIP DRIVERS:
    TTL OUTPUT LEVELS OR FET OUTPUT LEVELS
    INVERTING OR NONINVERTING
    2, 4, 8, OR 12 MA CURRENT SINK CAPABILITY (DC)
    PUSH-PULL OR HIGH-IMPEDANCE
        (HIGH IMPEDANCE MAY BE EITHER OPEN-DRAIN W/EXTERNAL PULLUP
        OR PUSH-PULL W/ADDITIONAL OUTPUT ENABLE INPUT)
  BIDIRECTIONAL:  (OFF-CHIP DRIVER/RECEIVER DOTTED)
        SAME OPTIONS AS ABOVE, HOWEVER AN OUTPUT ENABLE
        INPUT IS REQUIRED SO THAT THE RECEIVER MAY BE
        DRIVEN EXTERNALLY;  THE RECEIVER IS ALWAYS ACTIVE
```

Figure 10.9. Common design system library chip input/output books. The high-impedance off-chip drivers may be either an open-drain circuit (with external pull-up) or a push–pull connection with an additional *Output Enable* book input pin. The options for the bidirectional books are the same as listed for the different combinations of receivers and drivers. An Output Enable book input is required to place the driver in the high-impedance condition so that the receiver may be driven at the chip pad; the receiver is always active.

```
    HALF ADDER
    FULL ADDER
    SELECT 1 OF 2 (MUX)
    SELECT 1 OF 4 (MUX)
    2 - TO - 4 LINE DECODE
    3 - TO - 8 LINE DECODE
    4 - BIT MAGNITUDE COMPARATOR (EXPANDABLE)
    9 - BIT PARITY CHECKER/GENERATOR
    4 - BIT LOOK-AHEAD CARRY GENERATOR
    4 - BIT BINARY ADDER
    16 - BIT BINARY ADDER
    4 - BIT ALU
    N - BIT (UP/DOWN) COUNTERS (W/SYNCHRONOUS LOAD,
            SYNCHRONOUS CLEAR, EXPANDABLE)
```

Figure 10.10. Common design system library macro functions. These designs could be implemented as either *physical* macro layouts or possibly as *logical* macros (i.e., the specification of interconnections between primitive books to be placed individually and wired globally).

L1 clock signal.) The individual sequential circuits are available with a number of different input ports and a variety of buffered output pin combinations.

The chip I/O circuit offerings include a number of unique design options:

tri-state, common input/output (*bidirectional*), and circuits requiring an external pull-up. The signal receivers may be available for different voltage-level interfacing (e.g., FET or TTL). The off-chip drivers are available with a number of different output current capabilities for connection to technologies whose receivers have nonzero dc input current. The receiver circuit may be designed with some hysteresis in its dc transfer characteristic for increased noise immunity. No direct wire–input *receiver* book is included in the list in Figure 10.9; all chip input signals must go through a receiver *circuit*. No direct wiring from chip pad to internal circuit is therefore possible. This will help to ensure that external logic voltage levels are fully restored and the driven circuits less susceptible to noise margin problems.

One final circuit design offering to be considered for the technology library is a specific latch design for use as a *synchronizer*, that is, a circuit that captures the value of an asynchronous chip input signal in reference to the internal chip system clock signal. Since latch setup and hold times relative to the system clock cannot be guaranteed for the asynchronous data, a unique design for the latch may be required to reduce the probability of latch *metastability*, which may lead to chip failure. Section 14.7 discusses metastability in further detail.

PROBLEMS

10.1. In an effort to further enhance performance beyond that available from the highest power level of a particular function, some design systems and technology circuit libraries permit the design of *paralleled* logic gates. As illustrated in Figure 10.11, two logic gates are connected in parallel fashion with common inputs and with outputs dotted together. Discuss the advantages and disadvantages of this design approach, particularly with respect to any additional simulation and/or modeling requirements on design system tools. What additional checking capabilities should be provided?

10.2. Figure 10.12 illustrates the design of the L1/L2 latch pair. Two L1 ports are illustrated, with clock signals A0 and C0; no L2 input port is shown, only the input connection to the latch from the output of the L1. The *shift* clock input to the L2 latch, B0, is not derived from the L1 system clock(s), unlike an edge-triggered master–slave flip-flop design. Figure 10.13 illustrates the output of the L2 latch connected to the A0 input port of the L1 latch of another L1/L2 latch pair. Assuming

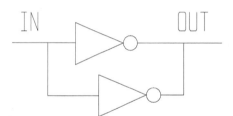

Figure 10.11. Paralleling of identical circuits to improve performance.

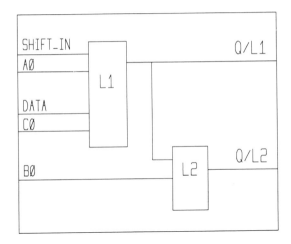

Figure 10.12. Block diagram of the L1/L2 latch pair. Two L1 latch ports are illustrated, one controlled by the A0 clock input and the other gated by the C0 clock input. The L1 latch output feeds the only input port of the L2 latch, which is controlled by the clock input B0.

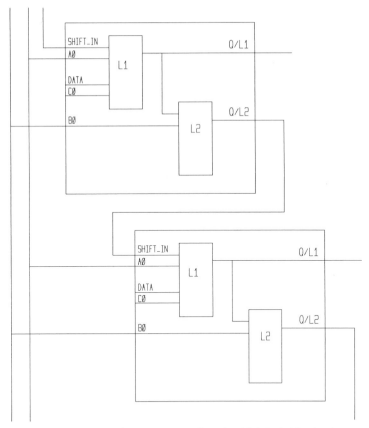

Figure 10.13. Block diagram of the implementation of multiple L1/L2 latch pairs connected in shift-register fashion; the L2 latch output of the preceding pair is connected to the succeeding L1 latch input through the A0 port.

471

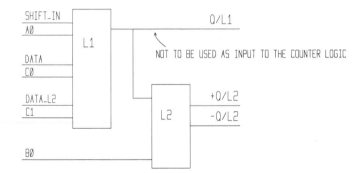

Figure 10.14. Block diagram of the L1/L2 latch pair to be used for implementing a 16-bit counter. The L1 latch contains three input ports: the A0 port for shift-register connection (not to be used for designing the counter), and the C0 and C1 ports for initial parallel loading and the data count input, respectively. Note that the L1 latch output is *not* to be used to feed the count logic. (Why is this limitation imposed? What restrictions apply when using level-sensitive latches for implementing this function?)

clocks A0 and B0 are *nonoverlapping* at their active level, describe how the connection in Figure 10.13 implements a *shift-register* design. Show how the length of this shift register may be arbitrarily extended. How is a parallel load of this shift register accomplished? What happens if shift-register clocks A0 and B0 are both brought active simultaneously for an extended period of time?

10.3. **(a)** Using L1/L2 latch pairs as illustrated in Figure 10.14 and clocks C0, C1, and B0, develop a logic schematic for a 16-bit up-counter; clock C0 is to load the initial count into all 16 bits, while clocks C1 and B0 are to toggle the count. The A0 clock port is dedicated to implementing the L1/L2 latch pairs in shift-register fashion and is not used to construct the counter. Assume that the largest fan-in NOR/NAND circuit available to decode the toggling of higher-order bits is *four* inputs; both L2 latch polarities are available. After the L2 latch output values change, what is the longest path delay (in number of blocks) through which the count decode passes before the C1 clock captures the count into the L1 latches? How is this delay affected by the limitation on the maximum NOR/NAND fan-in? What is the worst-case block delay if the fan-in limit is two? If the fan-in limit is eight?

(b) Repeat the design and analysis for a 16-bit down-counter.

11

SPECIAL CIRCUIT TYPES

11.1 INTRODUCTION

In this chapter, some unique circuit types are introduced that are either required for all chip designs (e.g., off-chip drivers and receivers) or used to satisfy a unique application requirement (e.g., on-chip substrate voltage generation). These circuit types are to some degree the outcasts of a strict design system methodology, as the following examples indicate:

1. For general applicability, the off-chip driver circuit must have a sourcing and sinking current capability for fan-out technologies, specifically for interfacing with standard TTL logic parts. This is the first example (in the FET technologies discussed) of a circuit design required to accommodate an *external* dc current (Figure 11.1).

2. The input to the receiver circuit may or may not be from a compatible logic family; that is, the voltage levels at the receiver input may be shifted and/or the window between 1 and 0 may be reduced relative to the internal circuits.

3. The receiver circuit design may within limits be permitted to have a nonzero dc input current. This relaxes one of the more stringent design system tenets. However, the receiver dc input current should definitely not be greater than an (LS)TTL load.

4. Off-chip driver circuits are required that, in addition to providing valid logic 1 and 0 levels, must also be able to be placed in a high-impedance state so

(LS,ALS) TTL INTERFACE LEVELS

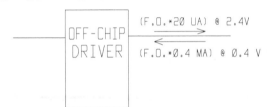

Figure 11.1. Off-chip driver interface requirements.

that other circuits that have been dotted with the off-chip driver can control the value of the signal line. This high-impedance state must not exceed a maximum leakage current specification (Figure 11.2).

5. To reduce the power dissipated on the chip by an off-chip driver circuit, it may be required that the load device of the OCD circuit be located externally and dotted to this special OCD circuit. A pull-up resistor is the most common external element used (Figure 11.3). It should be noted, however, that the use of an external pull-up introduces a question of economics. The benefits of removing the off-chip driver circuit load device are reduced power consumption on-chip and reduced supply voltage noise. However, the disadvantages of this approach are as follows:

   a. The cost of the external pull-up device.

   b. A precise specification of the pull-up device is required to obtain tight control of off-chip driver delays, implying higher cost.

   c. The placement and routing tools must support a variety of components at the printed circuit board level; the additional printed circuit board area

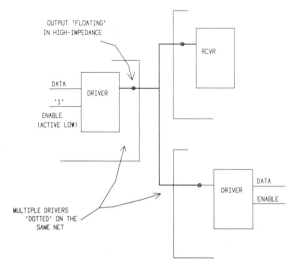

Figure 11.2. Off-chip driver with high-impedance dottable output. The high-impedance state is controlled by the value of the Enable signal.

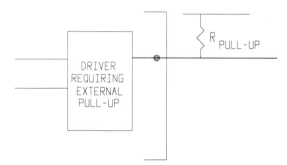

Figure 11.3. External pull-up resistor used as load for an off-chip driver.

is a cost adder (although this is reduced somewhat by the availability of resistor packs: single package, multiple resistor modules).

d. Additional net capacitance results from the printed circuit board wiring to the R-pack.

e. Delay calculation algorithms must be more sophisticated to support the different technologies.

f. Additional documentation for manufacturing and system test must be provided, specifically tracking which output pins require which specific external pull-up devices at the next higher level of system packaging.

6. To increase the effective pin count, it is necessary to utilize the number of chip pads available in the most efficient manner possible. This implies designing bidirectional (or common input/output) circuits, where an off-chip driver circuit is dotted (on-chip) with a receiver circuit and both are connected to the same chip pad. The specific function (driver or receiver) in use is controlled by an internal signal; the system designer must determine which I/Os can be suitably multiplexed in time and are thus suitable for common I/O pads (Figure 11.4). The previously listed requirements for driver

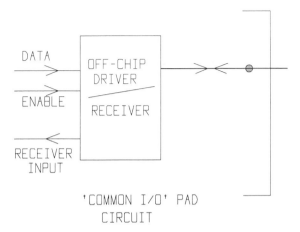

'COMMON I/O' PAD
CIRCUIT

Figure 11.4. Common I/O off-chip driver and receiver circuit.

and receiver design regarding high impedance and input/output current specifications still apply.

7. Since the anticipated capacitances that off-chip driver circuits must charge or discharge are much larger than typical on-chip values (by $10\times$ to $100\times$), the device sizes required are much larger than those typically used for logic circuits. This imposes severe constraints on both the physical designer and the chip image developer:

   a. A *large* device width presents layout difficulties; typically, the device gate is implemented by a number of smaller devices that are connected in parallel (Figure 11.5).

   b. If multiple off-chip driver circuits switch simultaneously, the resultant current spike may cause substantial supply line bounce. The end result is:

   (1) Large positive swings on a *VDD* line increase the susceptibility to device punchthrough and potentially destructive currents; negative swings on the *VDD* line can significantly degrade logic 1 output levels (Figure 11.6).

   (2) Large negative swings on a *GND* line can forward bias a node-to-substrate *pn* junction; this current can locally debias the substrate and inject excess carriers into the substrate in the vicinity of dynamic nodes (Figure 11.7).

   (3) Large swings in either direction on a *GND* line may cause a logic gate or latch to be falsely switched (Figure 11.8).

   These problems can be alleviated to some degree by separating the *VDD* and *GND* metal lines used for pad circuits from those used for internal logic circuits, using distinct pairs of *VDD* and *GND* pads for internal circuits from *VDD* and *GND* pads for chip I/O circuits, and adding decoupling capacitors

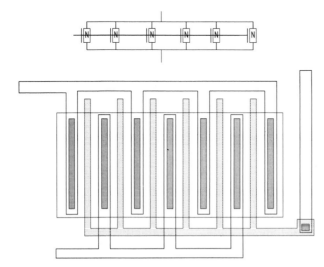

Figure 11.5. Off-chip driver pull-up (or pull-down) transistor implemented as a number of devices connected in parallel.

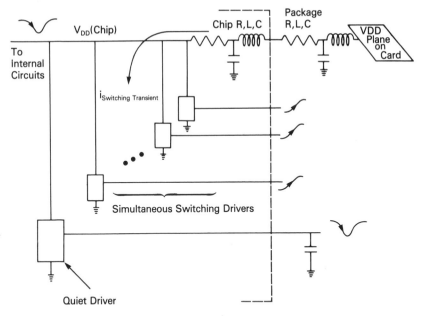

Figure 11.6. Noise model of *VDD* chip supply interconnections due to simultaneously switching drivers. The quiet driver, in providing a logic 1 output, transmits the supply noise off-chip.

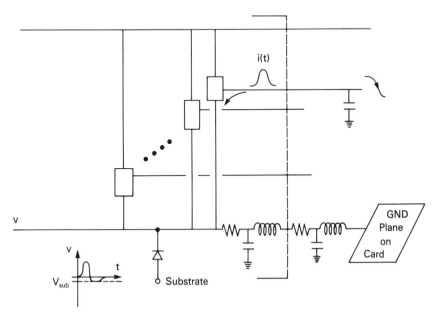

Figure 11.7. Noise model of the *GND* chip supply interconnections due to simultaneously switching drivers.

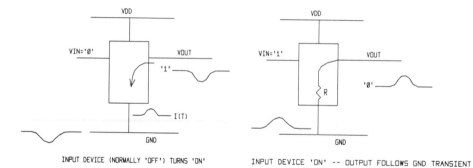

INPUT DEVICE (NORMALLY 'OFF') TURNS 'ON'   INPUT DEVICE 'ON' -- OUTPUT FOLLOWS GND TRANSIENT

Figure 11.8. Transmission of *GND* noise to on-chip circuitry.

to the supply pins near the package on the printed circuit board (or as part of the chip package).

c. Since the off-chip driver circuit devices tend to be large, their inputs present a large load capacitance to the internal circuit connected to the driver. The off-chip driver book should include an intermediate buffer so as to reduce this loading effect (Figure 11.9). Similarly, the receiver should be buffered to drive the internal loading capacitance from a chip input.

d. A general conclusion that can be reached from the previous examples is that calculating the dc validity and the ac performance for OCD/receiver circuits is significantly more complicated than for standard logic blocks.

8. A voltage applied to the substrate of the chip can be used to reverse-bias all *pn* junctions on the chip, reduce junction capacitances, and modify device threshold voltages, that is, (a) make $V_{T,n \text{ Enhancement}}$ and $V_{T,n \text{ Depletion}}$ more positive, and (b) reduce $\delta V_T / \delta V_{SX}$. The current requirement for the substrate supply is quite small, essentially just the junction leakage current and the substrate impact ionization current. This total is typically in the 10- to 100-microampere range (although chip operation could still be acceptable with significantly higher values).

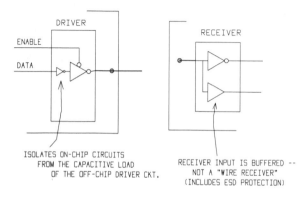

ISOLATES ON-CHIP CIRCUITS
FROM THE CAPACITIVE LOAD
OF THE OFF-CHIP DRIVER CKT.

RECEIVER INPUT IS BUFFERED --
NOT A "WIRE RECEIVER"
(INCLUDES ESD PROTECTION)

Figure 11.9. Buffering of off-chip driver and receiver circuits.

The added expense of providing another power supply for the low-current substrate voltage (in the system product in which the chip will be used) may not be desirable. Special circuit types (which are definitely *not* part of a design system methodology) can be designed to generate a negative substrate voltage for a *p*-type substrate and sink the necessary leakage current. However, prior to including a substrate voltage generator, some of the following design system questions should be addressed:

1. Where can it be placed on-chip so as not to interfere with other circuitry and the overall chip image?

2. How is the substrate voltage to be distributed to different areas of the chip? (Recall that the substrate is a relatively high resistivity material; the background impurity concentration is quite low. The *local* substrate voltage, that is, the substrate voltage in the vicinity of an individual device, may differ from the global substrate voltage due to the $I * R_{sub}$ voltage drop from the device to the substrate contact (Figure 11.10).

3. If a substrate voltage is applied externally to the top surface, a contact to the substrate is obviously required. If the substrate voltage is generated on-chip, can the substrate contact process masking step be eliminated (or is it still required to better distribute the substrate voltage on metal to all corners of the chip)? What additional processing complexity is required to provide an *ohmic* contact to the lightly doped substrate?

4. As stated earlier, the substrate voltage supply must sink the chip leakage and impact ionization current. Junction nodes may be leaky while the related circuits still function correctly. To keep chip yields as high as possible, a large generator current-handling capability is required. How large a leakage current capability should an on-chip substrate voltage generator have? What is the relationship for the substrate voltage output of the generator versus

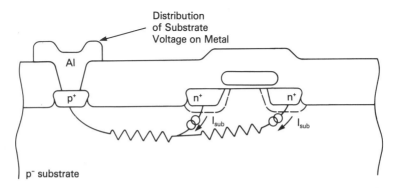

Figure 11.10.   Top-side substrate connection and resulting (distributed) substrate resistance model.

the leakage current? To increase the current-handling capability, can more than one substrate voltage generator be placed on the chip?

5. After applying power to the chip (*VDD*), how long should the voltage generator be allowed to discharge the substrate capacitance to its dc value before functional operation can safely proceed? How can the start-up of the generator be guaranteed when power is applied? How large is the substrate capacitance that must be discharged?

6. What range of substrate voltages can be generated on-chip? What control is available to keep the generated voltage within process design limits?

The special circuit designs that address these unique design questions and requirements are the subject of this chapter. Many of the circuit types to be discussed possess a variety of different implementations in different technologies. For example, in the cases where an *n*-channel enhancement-mode FET is used in a unique configuration, a *p*-channel device, connected in dual fashion, could equivalently be used. The collection of special circuits given in this chapter is not a complete one, but nevertheless is illustrative of some of the unique considerations that must be addressed by a general chip design system.

11.2 ENHANCEMENT-MODE DIODE

Every diffusion node on a chip represents a *pn* junction diode; however, this *pn* diode is *not* useful in the realization of a rectifying circuit element in a chip design. Although its *I-V* characteristics are not as attractive as that for a junction diode, an *n*-channel enhancement-mode FET device can be connected so as to provide a rectifying function (Figure 11.11). The connection is identical to that of a saturated *n*-channel load device; however, in several of the following applications it will be used in a distinctly different manner.

Using the device equation presented in Chapter 7 for the MOS device op-

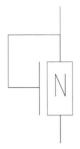

'ENHANCEMENT MODE DIODE'

Figure 11.11. *n*-Channel FET connected as rectifying element (diode).

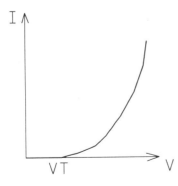

Figure 11.12. *n*-Channel FET enhancement-mode diode *I-V* characteristics.

erating in the saturated mode,

$$I_{DS} = \mu * \frac{\epsilon_{ox}}{t_{ox}} * \frac{W}{L} * \frac{(V_{GS} - V_T)^2}{2}$$

the *I-V* characteristic for the *n*-channel enhancement-mode FET connected as a diode is plotted in Figure 11.12. Recall from the discussion of device characteristics that $V_T = V_T(V_{SX})$; if the device is used in a circuit where the source (i.e., the cathode of the diode) is not connected to ground potential, the threshold voltage where the device cuts off will be increased. The first illustration of the use of an enhancement-mode diode device is in a bootstrapped load device circuit.

11.3 BOOTSTRAPPED CIRCUITS

In the discussion on load device options in Chapter 9 for FET logic circuits, two *n*-channel enhancement-mode load types were presented: saturated and linear. The disadvantage with the saturated load device is that the maximum output voltage achievable is $VDD - V_T$ [$V_T = V_T(V_{SX})$], rather than VDD. In addition, the charging current decreases rapidly as the output voltage approaches $v_{out,max}$. When the logic circuit output voltage is 0, the device dissipates power with the load device gate-to-source voltage equal to $V_{GS,load} = VDD - v_{out}$, or roughly VDD. The linear load device had its disadvantages also; although $v_{out,max} = VDD$ for this load device option, it required the generation and distribution of another power supply voltage to the gate inputs of the load devices. In addition, when $v_{out} = 0$, the device dissipates power with the load device gate-to-source voltage equal to $V_{GS,load} = VGG - v_{out}$, or roughly VGG ($>VDD$).

A *bootstrapped* load device (Figure 11.13) alleviates some of the difficulties of these two types of load devices at the expense of area and the introduction of

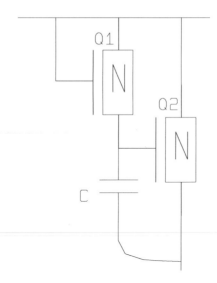

Figure 11.13. Bootstrapped load device.

a dynamic node into the circuit design. Referring to the bootstrapped inverter circuit design in Figure 11.14, when the input is at a logical 1 level, $v_{\text{out}} = 0$ as with a standard inverter. During this time, the capacitance C is charged through enhancement device diode $Q1$ to a voltage difference of $VDD - V_{T,Q1}$. Once the storage node of device $Q1$ reaches $VDD - V_{T,Q1}$, device $Q1$ turns off. The dc load current is roughly given by

$$I_{Q2} = \gamma * \left(\frac{W}{L}\right)_{Q2} * \frac{[(V_{DD} - V_{T_{Q1}}) - V_{T_{Q2}}]^2}{2}$$

The gate-to-source voltage for load device $Q2$ is roughly $VDD - V_{T,Q1}$, a value less than that for the load device in either the linear or saturated option. Assume now that the input to the circuit goes to 0 and thus that the input device turns off. The voltage developed across capacitance C remains constant as there is no discharge path for this capacitor; the device $Q1$ connected as a diode does not conduct in the opposite direction. The gate-to-source voltage across load device $Q2$ remains at $VDD - V_{T,Q1}$ as the output rises. (Recall that, with the gate at a fixed potential, be it either VDD or VGG, the gate-to-source voltage overdrive on the load device is reduced as the output voltage rises.) The node voltage at the gate of device $Q2$ rises to $VDD - V_{T,Q1} + VDD$ and v_{out} charges fully toward VDD, as with the linear mode load device. This circuit is used where (1) large capacitive loads must be driven with a minimum of power (e.g., clock drivers), or the technology only provides an enhancement-mode load and it is necessary or desired to output a value of VDD, rather than $VDD - V_{T,Q\text{load}}$.

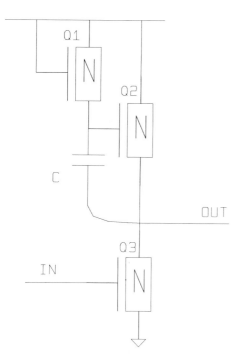

Figure 11.14. Bootstrapped inverter circuit design.

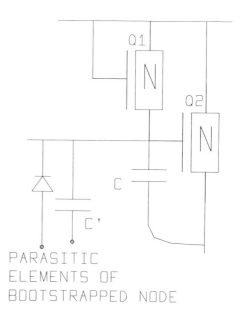

Figure 11.15. Enhanced model of bootstrapped node, including parasitic elements.

It should be mentioned that this circuit is not static in operation since the leakage at the source node of the enhancement-mode diode has been neglected (Figure 11.15). Therefore, there is a minimum toggling frequency for the circuit to guarantee its proper operation. (Another parasitic element that has been ignored is the capacitance from the source node to ground, or V_X.)

Bootstrap Protection

During the normal operation of the bootstrap inverter, the coupled node voltage at the gate input of the load device approaches $2 * VDD - V_{T,Q1}$. This voltage may approach the process thick oxide threshold voltage; this could turn on two otherwise isolated diffusion nodes if a first metal line carrying the bootstrapped voltage was routed between. To ensure that this condition is never encountered, a bootstrap *clamping* device can be added (Figure 11.16). Device Q_{clamp} is used to limit the bootstrapped voltage to $VDD + V_{T,Qclamp}$, instead of $2VDD - V_{T,Q1}$. The output voltage of the bootstrapped inverter still reaches VDD.

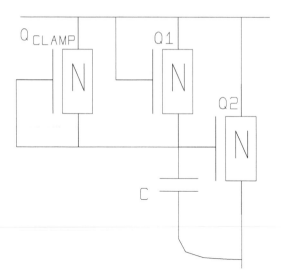

Figure 11.16. Bootstrapped node clamp device.

11.4 VOLTAGE DOUBLER

The next special circuit type to be discussed is a *voltage-doubler* circuit, one that uses bootstrapped devices and capacitive coupling to charge a chip capacitance to a voltage higher than the supply voltage *VDD*. The schematic for this circuit is given in Figure 11.17.

The first stage of the circuit design is a bootstrapped *push–pull* inverter circuit; the second stage is a simple bootstrapped load inverter, driven out of phase with the first stage. Referring to Figure 11.17, recall from the previous section that when

$$v_{in} = 1, \qquad v_1 = VDD - V_{T,Q1}, \quad v_2 = 0, \quad v_3 = 0$$

$$v_{in} = 0, \qquad v_1 = 2VDD - V_{T,Q1}, \quad v_2 = VDD, \quad v_3 = VDD - V_{T,Q3}$$

Note that the bootstrapped node voltage v_1 is applied directly to the gate input of device $Q4$. When $v_{in} = 0$, the output voltage of the first stage v_3 turns device $Q7$ on, discharging node v_5 to near ground. At the same time, the bootstrapped voltage $v_1 = (2VDD - V_{T,Q1})$ is applied to device $Q4$. This higher gate voltage on $Q4$ allows node v_4 to reach *VDD*. This input condition charges capacitor C_B to approximately *VDD*. When v_{in} goes to 1, node v_3 goes to ground, turning off device $Q7$. Node v_1 falls to $VDD - V_{T,Q1}$. Device $Q4$ acts as the desired rectifying element, turning off and preventing C_B from discharging $V_{GS,Q4} = [(VDD - V_{T,Q1}) - VDD]$. As node v_5 rises toward *VDD*, capacitor C_B couples node voltage v_4 to 2 * *VDD*. Device $Q6$ is also connected as an enhancement-mode diode; as node voltage v_4 rises to 2 * *VDD*, the voltage on load capacitance C_L will reach $2 * VDD - V_{T,Q6}$. When v_{in} returns to 0, node voltage v_4 returns to *VDD*; device

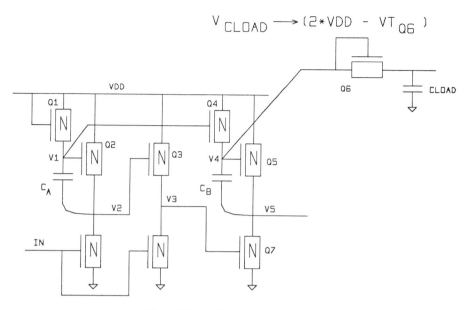

Figure 11.17. Voltage-doubler circuit.

$Q6$ is off, and the voltage across C_L remains at $(2 * VDD - V_{T,Q6})$. The pertinent node voltage waveforms are diagrammed in Figure 11.18. Again, the parasitic leakages and capacitances have been neglected.

The combination of device $Q6$ and capacitance C_L acts as a *peak detector*.

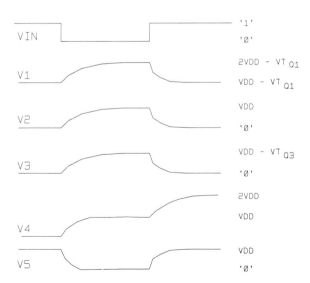

Figure 11.18. Voltage-doubler circuit node waveforms.

This output voltage could be used to drive a low-current load, such as the gate supply for enhancement-mode load devices.

11.5 OSCILLATOR DESIGN

There is little need for a free-running on-chip oscillator circuit in a technology design system. This type of circuit design does find two useful (nonsystem related) applications, however: (1) a parametric test structure, and (2) the input signal for a charge coupling (pumping) circuit (e.g., a voltage-doubler, V_X generator). The general block diagram for a free-running on-chip oscillator is given in Figure 11.19. The oscillator minimally consists of an *odd* number of inverter blocks connected in *ring* fashion; the output of the last stage is used as the input to the first inverter in the chain. A slight modification to the design of Figure 11.19 produces a *gated* oscillator. While the gate input signal is low, the output of the oscillator remains stable; when the gate input is high, the oscillator is free running.

During the initial stages of a technology process bring-up, a typical circuit to implement on a test chip is an unloaded (or lightly loaded) oscillator. The experimental data are used to verify the circuit modeling parameters for simulation. The test structure is physically laid out as a macro so that it may be assumed that the individual device parameters track as closely as possible. The oscillator output is commonly connected to a buffer circuit to drive the chip pad and test probe capacitance.

An occasionally overlooked constraint on the design of an oscillator is its start-up behavior. When power is first applied to the chip, as the power supply line charges toward *VDD*, it is desired that the oscillator kick in and begin operating in the astable mode. If all the inverting stages are identical (i.e., if they all have roughly the same *unity-gain* voltage), it is possible that for very slowly rising *VDD* the oscillator would stall and the output voltage would level out at or near $V_{\text{unity gain}}$. To reduce the likelihood that the oscillator will stall on power-

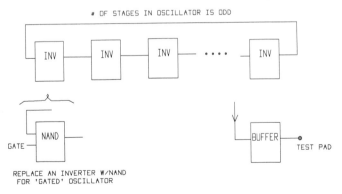

Figure 11.19. Block diagram of gated on-chip oscillator.

up, one of the blocks in the chain should be designed such that its unity-gain voltage is well separated from the unity gain voltages of the remaining stages and/ or a transient pulse can be generated after the supply has reached a sufficient level to perturb one of the circuits in the chain.

11.6 VOLTAGE DIVIDER

A potentially useful circuit is one that provides a reference voltage (between VDD and GND) to other circuits. This function is typically performed in analog circuit designs by a resistive voltage divider and/or a series of forward-biased diode voltage drops. Since a suitable high resistivity material is typically unavailable for MOS (digital) integrated-circuit design, an alternative method of implementation must be used. In addition, to be able to design for a reference voltage with a relatively tight tolerance, the circuit implementation should take advantage of device and process tracking as much as possible. To be able to assume that the devices used in the circuit design are closely matched, the design and physical layout should observe the following constraints:

1. Use devices with large W and L values so as to minimize design tolerances with respect to ΔW, ΔL, $\Delta V_T(W)$, $\Delta V_T(L)$.
2. Use the same type of device (either enhancement or depletion) to reduce the sensitivity to a threshold-shifting ion implant.
3. Try to develop the design so that it is relatively independent of any single device parameter, but rather is a stronger function of a *ratio* of device parameters.
4. In the physical layout, orient the devices that must track each other all in the same direction in order to reduce the sensitivity to x- or y-directed alignment errors.
5. Since the tracking spread of a device parameter increases as the distances between devices increases, matched devices should be placed as close to each other as possible.

The first example of a voltage divider is a unity-gain reference voltage circuit (Figure 11.20). This circuit suffers from several design drawbacks:

1. Two different types of devices are used; this increases the sensitivity to the process parameters: $\Delta V_{T,\text{implant}}$ and $\mu_{\text{depletion}}/\mu_{\text{enhancement}}$.
2. Both devices are operating in their saturated region: $V_{GS} - V_T < V_{DS}$. The effects of this aspect of the design are illustrated in Figure 11.21.

An improved version of the voltage divider is shown in Figure 11.22. Both devices are enhancement-mode devices, so the sensitivity to the threshold-shifting

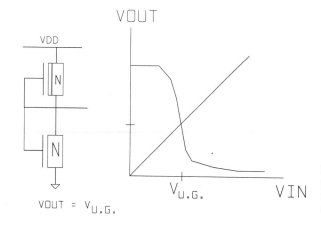

VOUT = V$_{U.G.}$

Figure 11.20. Voltage-divider circuit providing unity-gain output voltage of enhancement and depletion inverter stage.

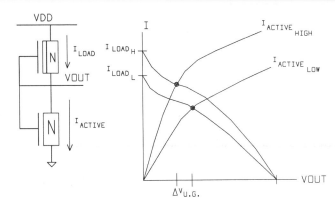

Figure 11.21. Voltage-divider output voltage variation for circuit of Figure 11.20; the high and low curves reflect device current tolerances. The resulting tolerance on the output reference voltage is also indicated.

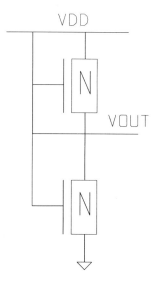

Figure 11.22. Improved voltage-divider circuit using only enhancement-mode devices.

ion implant and the enhancement-to-depletion mode device transconductance ratio is removed. The load device is a saturated enhancement-mode device; the driver is a linear-mode device. The resultant reduction in the tolerance on the targeted reference voltage (as compared with the previous circuit) is illustrated in Figure 11.23. This circuit will be used again in Section 11.7.

The last voltage-divider circuit presented uses two depletion-mode devices, both biased so as to operate in the linear mode (Figure 11.24). The design of the reference voltage and the resulting tolerance are depicted in Figure 11.25; again note how the tolerance on v_{out} is reduced due to device tracking.

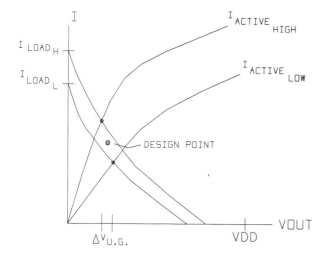

Figure 11.23. Voltage-divider output variation for the circuit of Figure 11.22.

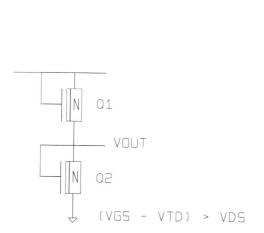

Figure 11.24. Voltage-divider circuit using only depletion-mode devices.

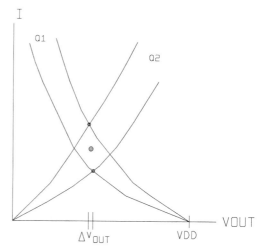

Figure 11.25. Voltage-divider output voltage variation for the circuit of Figure 11.24.

11.7 SUBSTRATE VOLTAGE GENERATION

The motivations for using a negative V_X (for a p-type substrate) are as follows:

Reverse-bias all diffusion junctions

Reduce junction capacitance

Modify $V_{T,n \text{ Enhancement}}$ and $V_{T,n \text{ Depletion}}$

Decrease $\delta V_T / \delta V_{SX}$

Reduce the subthreshold parameter α to improve device turn-off:

$$\left(\alpha = \left. \frac{\Delta V_{GS}}{\Delta \log_{10} I_{DS}} \right|_{V_{GS} \approx V_T} \right)$$

These motivations exist whether V_X is applied externally or generated on-chip.

The equilibrium substrate voltage produced by a substrate voltage generator is such that the time-averaged integral of the substrate discharge current is equal to the total substrate leakage current. Since the time integral of a current is equivalent to a quantity of charge, this can be stated another way: the amount of positive charge removed from the substrate capacitance periodically is equal to the charge that accumulates on that capacitance due to leakage current during the same period (Figure 11.26). A primitive circuit model for the substrate discharge is given in Figure 11.27. For the portion of the generator's period when the switch is open, the voltage V_D is applied across capacitor $C_{\text{discharge}}$ by other circuitry

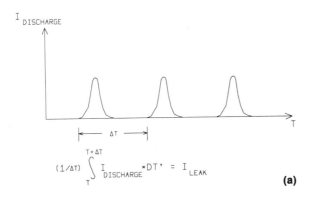

(a)

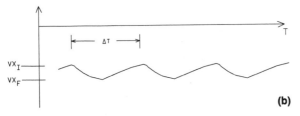

(b)

Figure 11.26. (a) Periodic discharge of substrate to balance leakage current. (b) Substrate voltage variation (ripple) over the period of the generator's charge pump circuitry.

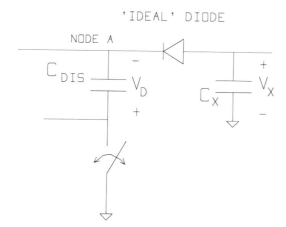

Figure 11.27.   Idealized illustration of substrate discharge action.

(not shown in the figure). When the switch closes, node voltage v_A is coupled by the voltage across $C_{\text{discharge}}$. If voltage $(-V_D)$ is more negative than the substrate voltage V_X, the ideal diode conducts and the substrate capacitance is further discharged.[1] If the diode were truly ideal, the following charge transfer relationship would hold:

$$C_D * (-V_{D,\text{initial}}) + C_X * V_{X,\text{initial}} = (C_D + C_X) * V_{X,\text{final}}$$

$$V_{X,\text{final}} = \frac{-V_{D,\text{initial}} * C_D + V_{X,\text{initial}} * C_X}{C_D + C_X}$$

$$\Delta Q_X = C_X * (V_{X,\text{final}} - V_{X,\text{initial}})$$

$$\Delta Q_X = \frac{C_D C_X}{C_D + C_X} * (-V_{D,\text{initial}} - V_{X,\text{initial}})$$

ΔQ is negative since $(-V_{D,\text{initial}})$ was more negative than $V_{X,\text{initial}}$. The equilibrium assumption states that $\Delta Q = I_{\text{leak}} * T$, assuming I_{leak} is a constant:

$$I_{\text{leak}} = \frac{1}{T} * \frac{C_D C_X}{C_D + C_X} * |V_{D,\text{initial}} + V_{X,\text{ initial}}|$$

To increase the leakage current-handling capability of the generator, the period of the discharge cycle should be minimized (subject to the condition that the capacitance C_D be able to charge and discharge fully between $V_{D,\text{initial}}$ and $-V_{X,\text{final}}$). Since the magnitude of the chip leakage current is a function of the chip operating conditions (e.g., impact ionization currents and temperature), the equilibrium substrate voltage is subject to considerable variation over the chip's operating range. As a result, additional design constraints are involved.

---

[1] The ideal diode would be a rectifying element with no reverse leakage current and no forward voltage drop.

A block diagram and circuit schematic for a general substrate generator is given in Figure 11.28. The combination of the push–pull buffer circuit, device Q_D, and the capacitor C_{DIS} is typically referred to as a *charge pump* circuit. Basically, the substrate generator works as follows on the 0 and 1 values of the oscillator output.

When the oscillator output is 0, the charge pump capacitor is part of the current path illustrated in Figure 11.29. Device Q_D conducts and the capacitor C_{DIS} charges to a voltage difference of $VDD - V_{T,Q_D}$. During this portion of the cycle, the substrate capacitance is not being discharged; it is effectively isolated from the generator.

As illustrated in Figure 11.30, when the oscillator output is 1, the voltage across the charge pump capacitor couples node voltage v_A below ground. Device Q_D, connected as an enhancement-mode diode, does not conduct. If node voltage v_A couples below V_X, the ideal diode conducts, discharging the substrate capacitance.

The substrate voltage will be reduced by the on voltage of the push–pull buffer circuit and by the forward voltage drop across the actual physical implementation of the ideal diode. Speaking of physical implementation, the layout of the charge pump capacitor and the design of the ideal diode should be highlighted. To fabricate a capacitor of moderate value in a minimum area, a parallel-plate structure is typically used with the gate oxide as the dielectric. In a polysilicon-gate MOS technology, the lower capacitor plate can be realized by the addition of the depletion-mode load device implant in the plate area (Figure 11.31). If one of the plates of the capacitor is tied to a fixed voltage, the designer has the option of choosing either the implanted plate or the gate plate for the fixed voltage (Figure

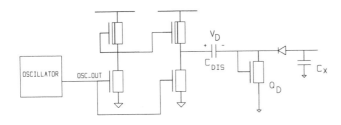

Figure 11.28. Schematic of substrate generator circuitry.

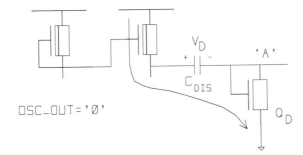

OSC_OUT='Ø'

Figure 11.29. Equivalent charging path for charge pump capacitor on the half-cycle when the oscillator output is 0.

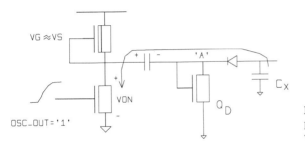

Figure 11.30. Equivalent discharge path for the substrate on the half-cycle when the oscillator output is 1.

11.32). The second option can provide additional node capacitance and is the implementation typically used in dynamic RAM design. If neither node is tied to a fixed supply, the connections used will typically be dictated by the circuit design, as is the case for the design of the charge pump capacitor for a substrate voltage generator.

Since the implanted surface layer for the lower plate is relatively resistive, the relaxation time constant for the plate to come to a constant voltage may be

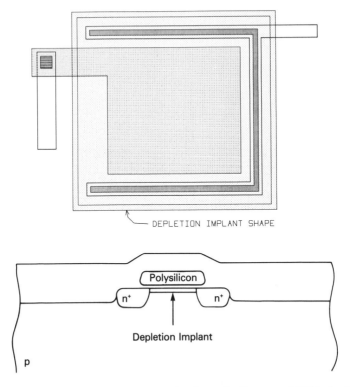

Figure 11.31. Layout of a gate oxide capacitor in a polysilicon gate MOS technology; the lower plate is realized by a depletion-mode implant through the gate oxide prior to polysilicon deposition and patterning.

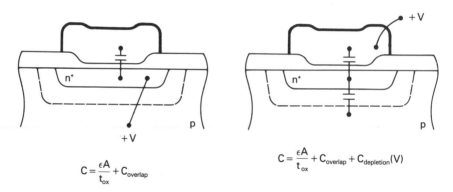

$$C = \frac{\epsilon A}{t_{ox}} + C_{overlap}$$

$$C = \frac{\epsilon A}{t_{ox}} + C_{overlap} + C_{depletion}(V)$$

Figure 11.32. Capacitor biasing options for the case where one plate is at a fixed voltage.

significant; in other words, this two-dimensional *RC* network may be a perform-ance-limiting factor. The best design procedure is to avoid problematic geome-tries: keep the distance between a metal-to-n^+ node contact and the extreme corners of the capacitor area to a minimum. For the polysilicon gate technology, this can be most easily realized by connecting a number of depletion-mode devices in parallel, each with drain connected to source (Figure 11.33).

The physical implementation of the ideal diode with anode connected to the substrate and cathode connected to one of the plates of the charge pump capacitor can be accomplished by one of two options. The first to be considered is to utilize the *pn* junction between the substrate and the implanted plate of the capacitor. The plate of the capacitor corresponding to the n^+ diffusion/depletion implant area is also used as the cathode of the substrate-to-capacitor diode. Since the area of the plate is typically quite large (to realize a charge pump capacitor of significant magnitude), the forward-bias voltage drop across the junction when the substrate is being discharged will be reduced. It was mentioned earlier that the v_{on} voltage drop of the push–pull buffer and the forward-bias voltage drop reduce the mag-nitude of the equilibrium substrate voltage; reducing the forward-bias voltage drop allows for a more negative substrate voltage to be achieved.

However, recall that permitting a diffused junction to become forward biased introduces other potential problems. In forward bias, the pn^+ junction injects a very large number of electrons into the substrate. These electrons recombine with holes in the substrate, constituting the largest fraction of the substrate discharge current. Since a high-resistivity substrate is typically used, the concentration of holes in the substrate is small; the recombination time for the injected electrons is large. For example, for a lightly doped substrate, typical numbers may be

$$\tau \approx 0.2 \text{ msec}, \qquad D \approx 18 \text{ cm}^2/\text{sec}$$

$$L = \sqrt{D * \tau} = 600 \text{ } \mu\text{m}$$

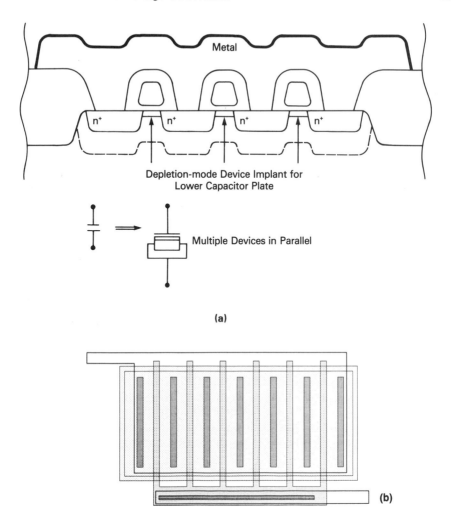

Figure 11.33. (a) Implementation of a parallel-plate capacitor as a number of depletion-mode devices with a common gate connection (upper capacitor plate) and all source and drain nodes connected together (resulting in a relatively low resistance lower plate). (b) Layout of a parallel-plate capacitor implemented as a number of depletion-mode devices with common gate and shorted drain and source nodes.

where τ is the carrier lifetime, D is the excess minority carrier diffusion coefficient, and L is the diffusion length. This excess electron density could have disastrous effects on adjacent dynamic circuit nodes, where the amount of charge stored would be upset by the collection of any excess electrons. To reduce the likelihood that the forward biasing of this junction will perturb any nearby circuit nodes, two design (layout) procedures should be adopted:

1. Locate the V_X generator as distant as possible from any active circuitry, particularly any dynamic storage nodes (typically, a corner of the chip is used).

2. Surround the injecting n^+ node with a collecting n^+ node biased to a positive potential with respect to the substrate.

The purpose of this additional node is to attempt to collect the excess electron density before it has an opportunity to diffuse to a circuit node. However, this additional collecting node reduces the efficiency of the generator by reducing the substrate recombination current. This increases the length of time required to initially discharge the substrate capacitance as well as give the V_X versus I_{leak} relationship a steeper slope.

The other implementation of the ideal diode that will be presented recognizes that the pn^+ junction diode described just previously is inevitably present in the design. This implementation attempts to reduce the loss due to the forward-bias diode voltage drop by adding in parallel with the junction another rectifying element with a smaller forward voltage drop when discharging the substrate. This rectifying element is the enhancement-mode diode; its use is illustrated in Figure 11.34. At first glance, using a device threshold voltage of 0.8 V (typical, with V_{SX} = 0), it would appear that the junction diode (with a typical forward voltage drop around 0.6 V) would dominate the parallel combination of the two circuit elements; the enhancement-mode diode would not conduct at all. Although the enhancement device threshold may indeed be 0.8 V at $V_{SX} = 0$, the mode of operation for discharging the substrate is with the source-to-substrate voltage a few hundred millivolts *below* zero; the graph of the curve $V_{T,n \text{ Enhancement}}$ versus V_{SX} needs to be extended to the left of the $V_{SX} = 0$ axis to determine if the enhancement-mode diode will be effective in shunting the junction diode. This situation is illustrated in Figure 11.35.

Assuming that the enhancement-mode diode *is* effective in reducing the forward voltage drop loss, its use provides one major benefit and one small disadvantage. The disadvantage is that an additional processing step is required to include the ability to make an *ohmic* contact to the substrate to connect the gate and drain of the enhancement-mode diode to the substrate. This may be required in any case; a substrate contact typically surrounds the internal chip area to help

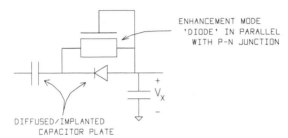

ENHANCEMENT MODE
'DIODE' IN PARALLEL
WITH P-N JUNCTION

$+$
V_X
$-$

DIFFUSED/IMPLANTED
CAPACITOR PLATE

Figure 11.34. Enhancement-mode diode in parallel with lower capacitor plate junction used as a rectifying element for the substrate discharge current.

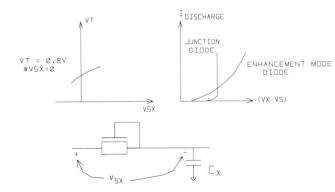

Figure 11.35. Threshold voltage and resulting *I-V* curves for $V_T(V_{SX} < 0)$, illustrating the effectiveness of the enhancement-mode diode in shunting the lower capacitor plate junction.

provide a more uniform voltage distribution. This same contact is the basis for the major advantage of including the enhancement-mode diode to shunt the junction diode (Figure 11.36). The substrate discharge current no longer consists of excess electron recombination current, but rather is primarily hole current. The probability of the excess injected electrons affecting a nearby circuit node has been greatly reduced. Again, it should be emphasized that the effectiveness of the enhancement-mode diode device depends a great deal on the V_T versus V_{SX} relationship for V_{SX} values slightly less than zero.

The circuit design for the substrate voltage generator presented in Figure 11.28 is able to provide a maximum negative substrate voltage of $-(VDD - V_{T,Q_D})$, assuming no leakage current and neglecting the losses due to the additional voltage drops when discharging the substrate. If this voltage is insufficient, or the losses are too large, a modification to the circuit design of Figure 11.28 can be made in order to increase the magnitude of the targeted substrate voltage. This

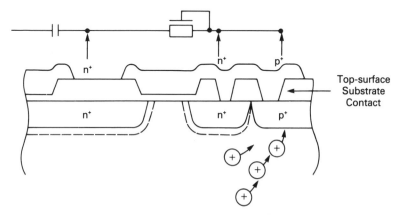

Figure 11.36. The enhancement-mode diode structure for a rectifying connection to the substrate offers the advantage that the substrate current is primarily hole current, rather than the excess minority carrier current of the forward-biased *pn* junction.

double charge pump design is presented in Figure 11.37. When the oscillator output is 1, charge pump capacitor C_{D2} charges to $VDD - V_{T,Q_{D2}}$. The enhancement pull-down transistor connected to C_{D2} is held off by the output of the internal stage of the inverting push–pull buffer connected to C_{D1}. When the oscillator output goes to 0, the charge pump capacitor C_{D2} couples node B in Figure 11.37 below ground, to $-(VDD - V_T)$, as was the case with the single charge pump generator. As a result, the source node of device Q_{D1} is not at ground, but has been coupled down to $-(VDD - V_T)$. The output of the push–pull buffer is now rising toward 1; the charging path for the charge pump capacitor C_{D1} is shown in Figure 11.38. The voltage across capacitor C_{D1} charges toward the value $2*(VDD - V_T)$. When the oscillator output returns to 1, the charging circuit for charge pump capacitor C_{D2} is isolated from the rest of the generator by enhancement-mode diode Q_{D1}, and the voltage at node A is coupled by capacitor C_{D1} and the

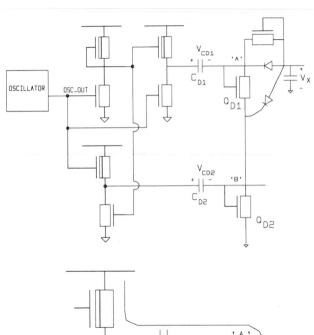

Figure 11.37. Double-charge pump circuit for increased substrate discharge capability.

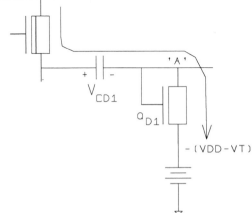

Figure 11.38. Equivalent charging circuit for charge pump capacitor C_{D1} when the oscillator output is 0.

falling output voltage of the push–pull buffer to approximately $-2 * (VDD - V_T)$. If the substrate is less negative than this value, a discharge current will flow through the enhancement-mode diode (or the pn^+ junction diode). Ideally, therefore, the substrate voltage would reach $-2 * (VDD - V_T)$. In actuality, this value will be reduced in magnitude by substrate leakage current, v_{on} for both charge pump circuits, and the forward voltage drop across the discharging diode. Note that an additional diode has been included in the circuit schematic of Figure 11.37 between node B and the substrate. During start-up, when the substrate capacitance is initially being discharged, node voltage B will indeed be coupled below the transient substrate voltage and will contribute to the discharge current.

Up to this point in the discussion, it has been assumed that the equilibrium substrate voltage provided by the generator and the desired substrate voltage design point (for a particular device threshold and junction capacitance) have been one and the same. However, these designs provide a relatively large range of final substrate values with variations in chip leakage current, and therefore an undesirably large tolerance on V_T.

To reduce these variations, a *compensating* circuit is often included in the design. The purpose of the compensating circuit is to inhibit the charge pump action as the desired design point, say, a particular enhancement device threshold, is approached. (It is assumed that the uncompensated generator would otherwise reach an equilibrium substrate voltage well below the design point; using a double charge pump generator, for example, makes this a reasonable assumption.)

An example of a compensated generator is shown in Figure 11.39. Essen-

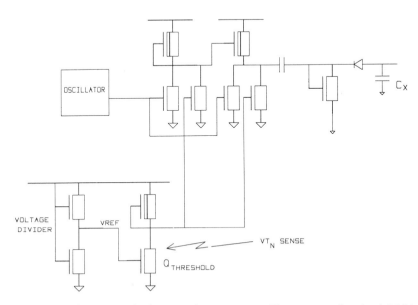

Figure 11.39. Compensated substrate voltage generator. The compensation circuit inhibits charge pumping action when a particular threshold voltage is reached.

tially, the only changes to the design of Figure 11.28 are the addition of a voltage divider with a high-gain inverting stage and the modification of the push–pull inverting buffer to a push–pull NOR. As the substrate capacitance is discharged, the n-channel device enhancement threshold increases. As long as the threshold voltage of device $Q_{\text{threshold}}$ is below the reference voltage V_{ref}, the inverting high-gain stage is on and its output voltage does not affect the operation of the generator. As the substrate continues to discharge, device $Q_{\text{threshold}}$ starts to turn off; the output of the high-gain stage inhibits (reduces) the charge pumping action of the push–pull NOR circuit. The resulting design point is illustrated in Figure 11.40, where it should be noted that the V_T versus V_{SX} curve for the enhancement device threshold is relatively flat, as is typical for lightly doped substrates. Figure 11.41 illustrates that the resulting substrate voltage for the compensated generator may have a large range when the effects of the tolerance on the device threshold are taken into account; although the n-channel device threshold voltage is still relatively tightly controlled, the resulting V_X has a large possible range. The wide range of substrate voltage possible from the compensating circuit introduces the following additional considerations:

1. The range of the thick oxide threshold voltage increases; the greatest sensitivity is to low values of V_X where the thick oxide threshold voltage is reduced.

2. The range of the diffusion node-to-node punch through voltage increases; the greatest sensitivity is to high values of V_X where the node-to-node punchthrough is reduced.

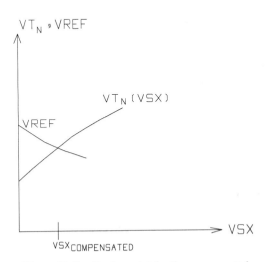

Figure 11.40. Design point for the compensated substrate voltage generator.

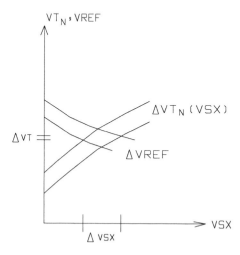

Figure 11.41. Tolerance on the substrate voltage of the compensated substrate voltage generator.

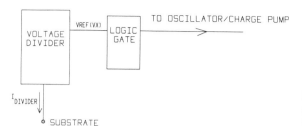

Figure 11.42. Block diagram of a regulator for a substrate voltage generator.

An alternative approach is to accept the tolerance on the device threshold voltage and attempt to regulate the substrate voltage output of the generator instead. The circuit design presented in block diagram form in Figure 11.42 attempts to regulate the generator output to a particular substrate voltage, as opposed to a specific device threshold. The unique feature of this design approach is that one end of the voltage divider is connected directly to the substrate; as the substrate discharges, the output of the voltage divider can be designed to cross the switching threshold of the logic gate. Note that the divider current *adds* to the substrate leakage current and should therefore be minimized. Problem 11.5 requests that a more detailed design be developed.

As smaller device dimensions are provided by VLSI technology evolution, the advantages of a negative substrate voltage for *p*-type substrates in terms of junction and *n*-channel device characteristics must be weighed against the increasing risk of device punchthrough.

11.8 OFF-CHIP DRIVER CIRCUITS

The output of an off-chip driver circuit is connected to a chip pad; the circuit buffers the value of an internal signal to drive the large capacitive load typically connected to the pad. There are several design considerations, unique to off-chip driver circuits, that apply to most of the OCD circuit types to be presented. Briefly, they are as follows:

1. The interface (dc) voltage levels to be provided depend on the particular logic technology (or technologies) to which the package pin may conceivably be connected. If the chip design is intended to be of general use, TTL logic voltage levels are usually assumed: 1 = 2.4 V, 0 = 0.4 V. If the particular application is known, an OCD circuit type may be required that provides *n*MOS logic levels, say, 1 = 4.0 V and 0 = 0.4 V (or CMOS full-rail levels).

2. The interface specifications for an OCD circuit should include (in addition to the voltage level specifications described previously) the sourcing and sinking current at which those dc voltage logic levels are guaranteed. Recall that, for TTL technology logic inputs, $I_{in,0} = -1.6$ mA/gate and $I_{in,1} = +40$

μA/gate; other TTL family versions differ. The power distribution to the OCD circuit should be such that nonzero values of sourcing and sinking current should not introduce a supply or ground voltage drop that would significantly degrade the interface voltage level. Depending on the maximum pad current density, multiple GND pads specifically for the OCD circuits may be required to handle the resulting rms currents when a number of OCDs are each to sink the currents from a high TTL fan-out and/or to reduce simultaneous switching noise. Multiple *VDD* pads may likewise be required.

3. The output load capacitance on a package pin typically varies from application to application; when designing an OCD circuit for a particular performance, a worst-case value of capacitance (say, 85 pF) is chosen as the design point.[2] In applications where these assumptions are insufficient, additional line driver interface integrated circuits are added to the printed circuit card.

TTL-compatible Off-chip Driver

A common off-chip driver configuration is the totem-pole connection of two *n*-channel enhancement-mode devices (Figure 11.43). The maximum output high 1 voltage level is $VDD - V_{T,Q1}$; if a positive output current is specified, the actual output voltage will be less. For example, assuming a TTL fan-out of 5, the output high specification may state $v_{out,1} = 2.4$ V at $I_{out} = 200$ μA. The minimum output low 0 voltage level is ground; if a negative (sinking) output current is specified, the actual output voltage will be higher. Again assuming a fan-out of 5, the output low 0 specification would typically state $v_{out,0} = 0.4$ V at $I_{out} = -8$ mA. From these dc considerations alone, it is evident that very close attention must be paid to supply and ground distribution and contact resistance to minimize voltage drops, as well as line and contact current density limits for maximum chip reliability.

To satisfy the dc interface specifications and the switching performance requirements, very high device width-to-length ratios will be required. A large device width presents layout difficulties. As was depicted in Figure 11.5, a number of smaller devices are commonly connected in parallel. Some consideration should be given to the design of the circuit(s) providing the input signals D and $\overline{D}$ to the two enhancement-mode devices. In particular, two design constraints should be imposed:

1. The input capacitance of the two enhancement devices is quite large and should be isolated from the internal circuits; this will reduce the capacitive load on the internal circuits and reduce the sensitivity of the output transition time to the internal signal's rising and falling transition.

---

[2] The value of 85 pF is a common specification for the load of a tester probe card.

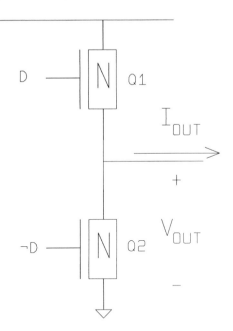

Figure 11.43. Totem-pole n-channel enhancement-mode device output stage for off-chip driver (TTL compatible).

2. The time duration when both devices $Q1$ and $Q2$ may be conducting should be minimized to reduce the power dissipated during the switching transition.

An OCD circuit design that partially satisfies these constraints is illustrated in Figure 11.44. The inputs to the totem-pole devices $Q1$ and $Q2$ are driven by circuits incorporated into the OCD layout.

A common approach to controlling the current spike when both devices are conducting is to add another input signal to the OCD circuit, usually denoted as

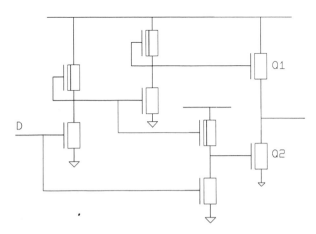

Figure 11.44. Buffered off-chip driver circuit design.

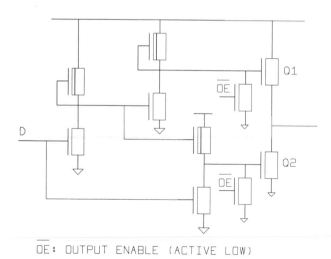

Figure 11.45. Off-chip driver circuit design with an Output Enable signal, providing reduced driver current spikes and an implementation of a high-impedance output node.

$\overline{OE}$: OUTPUT ENABLE (ACTIVE LOW)

Output Enable. A modified circuit design including this input is illustrated in Figure 11.45. While the data input signal to the OCD is changing, the Output Enable ($\overline{OE}$) signal is held high (inactive) so that both devices $Q1$ and $Q2$ are off. After the signal has stabilized, the Output Enable signal is driven low; one and only one of the gate inputs to $Q1$ or $Q2$ will rise, either charging or discharging the output load.

Although introduced as a measure to conserve power during the switching transition, the Output Enable signal also performs another function; when the Output Enable signal is inactive, both devices $Q1$ and $Q2$ are off. This provides a high-impedance condition at the chip output pad. Another circuit, dotted with the output node, can drive the node to either logic value if enabled.

Open-drain Driver

If a logical 1 chip output value is required so as to provide FET voltage levels, a common approach is to delete the enhancement-mode load device and its driving circuitry from the TTL-compatible driver and provide a pull-up to *VDD* (or possibly some other supply voltage) off-chip. The remaining circuit is shown in Figure 11.46. The configuration is called an *open-drain* driver.[3] As with the totem-pole configuration, an Output Enable signal can be included to implement a multiway dotted connection to other OCD circuits tied to the same node. However, unlike the previous example, the output node does not float if all drivers are inactive, but will be pulled up to $+V$ through the external pull-up.

The design of the open-drain device requires sinking the input current of the

[3] In bipolar technologies, the term *open collector* is used.

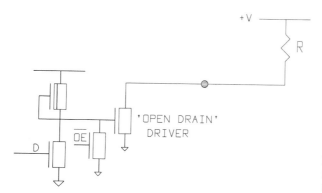

Figure 11.46. Open-drain driver circuit design (external pull-up resistor shown).

technologies connected to the chip output as well as the current through the pull-up (for a logical 0 value). As a result, the minimum value of the external pull-up resistance is an integral part of the specification and design of the open-drain OCD.

CMOS technology OCDs are similar to the circuit designs shown, with the additional option of a p-channel driver device; the p-channel load device provides a logic 1 chip output voltage equal to VDD. The same concerns regarding transient power dissipation mentioned earlier apply to the case where both the p-channel pull-up and n-channel pull-down are conducting; an Output Enable signal is again used to try to minimize this current spike. A concern with utilizing p-channel OCD pull-up devices is the potential of forward biasing a node-to-well junction if a positive reflection were incident back at the chip pad; to avoid injection currents that may lead to latch-up, careful layout practices should be observed (e.g., adjacent nondevice collecting nodes).

11.9 RECEIVER CIRCUITS AND SCHMITT TRIGGER; INPUT PAD PROTECTION

It is *not* common procedure to connect a chip input signal directly to internal circuit inputs; although a direct connection may result in the least delay, it would nevertheless increase an internal circuit's sensitivity to card-level noise voltages. It is also likely that some level shifting (or level restoration) between the logic voltage levels of the driving technology and the chip technology may be required. As a result, special receiver circuits are typically connected to chip input pads. These circuits include design considerations for electrostatic discharge transient protection, level shifting, and buffering of the input signal. The design of the buffer circuit will not be discussed in this section; a conventional buffering circuit can be used.

TTL Receiver Circuit

A level-shifting receiver circuit to translate the TTL logic voltages to voltages suitable for internal FET circuits is shown in Figure 11.47 (nMOS technology). Devices $Q2$ and $Q4$ are connected as a unity-gain stage. The output voltage is treated as a reference voltage; it is designed to be approximately equal to 1.4 V $+ V_{T,Q2}$. Device $Q1$ conducts when the input voltage v_{in} is less than the reference voltage minus the threshold voltage of device $Q1$; in other words, device $Q1$ conducts when $v_{in} < (V_{ref} - V_{T,Q1})$. If the reference voltage is approximately 1.4 V $+ V_{T,Q2}$, and assuming that $V_{T,Q2} = V_{T,Q1}$, then device $Q1$ is on when

$$v_{in} < 1.4 \text{ V} + V_{T,Q2} - V_{T,Q1} \cong 1.4 \text{ V}$$

When $v_{in} = 0.4$ V ($= 0$, TTL levels), the current through $Q3$ and $Q1$ should provide a sufficient voltage drop across $Q3$ to provide a valid FET logic 0 level. Note that for $v_{in} = 0$ this receiver circuit requires the sinking of a dc current (through $Q3$ and $Q1$) by the driving technology's off-chip driver circuitry. By making device $Q3$ a very high impedance load, this current is minimized (and a good FET down level is provided), yet the switching performance from $v_{in} = 0$ to $v_{in} = 1$ correspondingly suffers.

Another circuit commonly used as a receiver (although certainly not restricted to this application) is the Schmitt trigger (Figure 11.48). The feature of note for this circuit design is the *hysteresis* in the switching characteristics (Figure 11.49). When $v_{in} < V_T$, the output voltage is high since both devices $Q2$ and $Q3$ are off. As a result, the gate input to device $Q4$ is high. As v_{in} rises through V_T, device $Q3$ begins to conduct; current flows through $Q4$ and $Q3$, causing a voltage drop across $Q3$. Device $Q2$ is still off since $(V_G - V_S)_{Q2} = (v_{in} - V_S)_{Q2} < V_T$; the voltage drop across $Q3$ reduces the difference: $(V_G - V_S)_{Q2} = v_{in} - V_{DS,Q3}$. Voltage v_{in} must increase further before device $Q2$ conducts, pushing the transfer curve in Figure 11.49 to the right. The slope of the dc transfer curve is relatively steep due to the feedback provided by $Q4$; as v_{in} increases and voltage v_{out} begins to fall, the current through device $Q4$ diminishes, the voltage drop across $Q3$

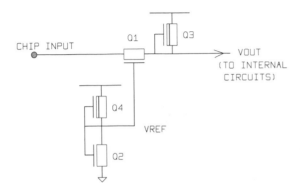

Figure 11.47. nMOS level-shifting receiver circuit. This circuit has a nonzero input current requirement.

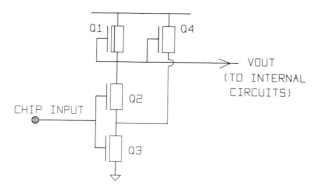

Figure 11.48. *n*MOS Schmitt trigger circuit, used as a receiver.

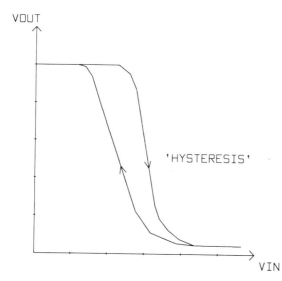

Figure 11.49. Hysteresis in Schmitt trigger transfer characteristics.

decreases, the overdrive $V_{GS} - V_T$ for $Q2$ increases, the current through load device $Q1$ increases, causing the output voltage to fall further, and so on. When v_{in} reaches 1, the output voltage is low and device $Q4$ is completely off.

As v_{in} returns from 1 to 0, the rising transition on the output transfer curve does not retrace the curve of the previous transition, introducing the hysteresis in the transfer characteristic. Device $Q4$ does not contribute to the current through device $Q3$ for a large portion of the curve since $(V_G - V_S)_{Q4} = (v_{out} - V_{DS})_{Q3} < V_{T,Q4}$. The transfer curve more closely resembles that of the equivalent NAND circuit, treating devices $Q2$ and $Q3$ in series and with device $Q4$ absent.

The hysteresis in the transfer characteristic provides for an improved noise margin. The Schmitt trigger, when used as a receiver, also provides the advantage that no sinking current is required of the driving technology.

The primary disadvantage of the Schmitt trigger is that it is best suited for FET interface voltage levels; the design of the Schmitt trigger for TTL interfacing with *n*-channel device enhancement threshold voltages of approximately 1.0 V is considerably trickier, as a window around 1.4 V is desired.

Input Pad Protection

All chip input pads should be connected to a protective structure to reduce the likelihood of destructive electrostatic discharge (ESD), causing gate oxide dielectric breakdown. (ESD often arises from improper package-handling procedures.) For example, assume that the dielectric breakdown strength for silicon dioxide is 5×10^6 cm V/cm, the gate oxide thickness is 300 Å, and the gate input capacitance is on the order of 0.5 picofarad (pF):

$$V_{\text{breakdown}} = (5 \times 10^6 \text{ V/cm}) * (300 \times 10^{-8} \text{ cm}) = 15 \text{ V}$$

$$q_{\text{breakdown}} = C * V = 0.5 \text{ pF} * 15 \text{ V} = (7.5 \times 10^{-12} C) * (6.2 \times 10^{18} e^-/C)$$

$$= 50 \times 10^6 \text{ electrons}$$

As a reference, the charge stored on the cell capacitance of a typical dynamic RAM chip is on the order of 10^6 electrons.

There are three commonly used means of input protection: diode breakdown, node-to-node punchthrough, and gate-field induced breakdown.

Diode breakdown.   A diffused region is placed in series with the input pad and gate input of the receiver circuit (Figure 11.50). Assuming the junction breakdown (plus distributed *RC* time constant to the gate input) can absorb a steep discharge transient, the gate of the device will be protected from an excessive destructive voltage. It should also be noted that this connection will clamp a negative-going transition at the chip input to a diode drop below the substrate

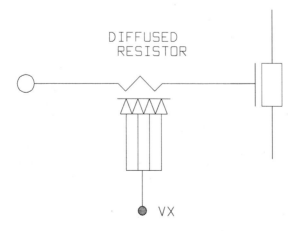

Figure 11.50.   Diode breakdown input pad protection.

voltage. In CMOS technologies, an additional protection diode can be added, utilizing the junction between a p^+ node connected to the pad and an n-well substrate tied to the positive supply.

Node-to-node punchthrough.   The term punchthrough was defined in Chapter 8 and referred to the situation where the depletion region surrounding the n^+ drain node extended along the channel region and contacted the n^+ source node depletion region. Carriers injected at the source node are driven by the high field between drain and source and a large current flows. It also refers to the equivalent situation where the isolated note-to-node voltage difference is high enough to result in a large current flow between the two nodes. Assuming the node-to-node spacing is adjusted appropriately, the punchthrough voltage can be designed to be less than the gate oxide dielectric breakdown voltage. Figure 11.51 illustrates the layout of the punchthrough structure. Typically, a specific protective layout is evaluated to meet an ESD transient reliability measure and then connected to every chip signal I/O pad.

Gate-controlled breakdown structure.   Recall from the discussion of device saturation characteristics in Chapter 8 that the field in the vicinity of the drain node, at the saturated end of the channel, was modeled as consisting of three components that contributed to the total field; the stronger electric field at the surface due to a drain-to-gate field is utilized in the gate-controlled breakdown structure.

An illustration of the structure is given in Figure 11.52; a corresponding layout is shown in Figure 11.53. The increased magnitude of the electric field in the corner region reduces the drain-to-substrate breakdown voltage. The actual input voltage at which the gate field-induced surface junction breakdown occurs

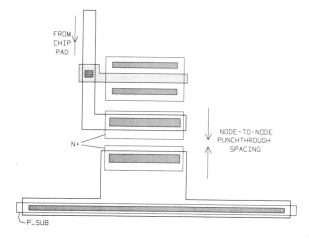

Figure 11.51. Punchthrough input pad protection layout.

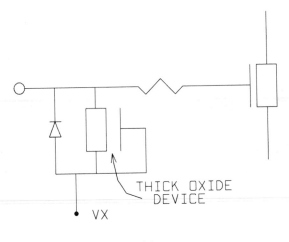

(a)

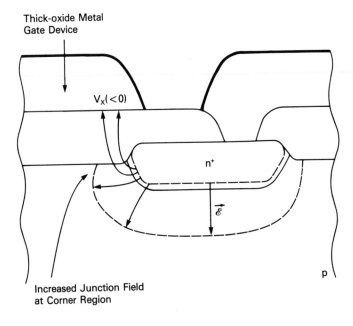

(b)

Figure 11.52. Circuit schematic (a) and cross section (b) of a thick-oxide metal-gate device for node-to-substrate breakdown and input pad protection.

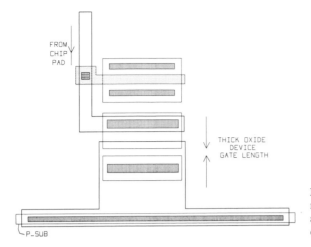

FROM
CHIP
PAD

THICK OXIDE
DEVICE
GATE LENGTH

P_SUB

Figure 11.53. Layout of thick-oxide
metal-gate protective device; the gate
and source node of the protective
device are connected to the substrate.

is controlled by the gate voltage and the oxide thickness. A typical structure is
to lay out a thick oxide metal gate device and connect the protective device's
gate and source nodes to the substrate.

PROBLEMS

11.1. The term *reverse engineering* refers to the development of a circuit schematic for a
chip from a microphotograph taken at high magnification through a microscope. This
procedure is commonly used for competitive analysis of a technology or part developed by another manufacturer (i.e., to be able to assess and compare the lithography and circuit densities of a competitor's technology against your own). Chip
microphotographs are also used for failure analysis of a part that has failed under
stress testing. The chip package is commonly etched or mechanically lapped to expose the chip cavity. To facilitate analyzing the circuitry, it is often necessary to
etch off the protective chip overcoat (and possibly the top global metal interconnection layer). This problem (and the next three) request the reverse engineering of
the special circuit types discussed in this chapter and incorporated in commercial
integrated circuits.

  Reverse-engineer the off-chip driver circuits of a commercial (VLSI) integrated
circuit such as a microprocessor. Evaluate pins that are high-impedance outputs as
well as pins with common I/O function. (To ease the reverse engineering procedure,
choose those parts whose technology specifics have been well documented in the
technical literature.) Specifically, compare the techniques used for obtaining large
width-to-length ratios (e.g., serpentine devices, parallel devices) and the metallization and contact area (which must sink or source the dc fan-out currents while
maintaining a valid logic level). If possible, from the microscope or a microphotograph, measure the output device dimensions; using estimates for the carrier mobility
(and the oxide thickness from the literature), calculate the logic voltage levels at the

chip pin when sinking or sourcing the specified dc fan-out current. (Estimate the device threshold voltage and its body effect coefficient.)

11.2. Reverse-engineer several receiver circuits from commercial integrated circuits; include both receivers and common I/O circuits. Are Schmitt trigger circuits used? From the chip microphotograph, measure the device dimensions, and using the pertinent (and assumed) technology parameters, analyze or simulate the receiver output voltage for the specified maximum 0 and minimum 1 input voltage levels.

11.3. Research the technical literature on dynamic RAM designs to find a commercial part that contains a substrate generator circuit to bias the chip substrate. Reverse engineer this circuit macro on the part to develop a substrate generator schematic. What kind of oscillator and charge pump circuitry is used? Is additional circuitry included that attempts to compensate the resulting device threshold or regulate the substrate voltage, or does the generator operate "wide open"?

11.4. A circuit equivalent for the "human" electrostatic discharge (ESD) source is shown in Figure 11.54. Develop a laboratory bench setup using this model to test the ESD protection for the input pins of commercial integrated circuits. (*Take the necessary precautions for high-voltage circuits and wiring!*) Evaluate several commercial MOS integrated circuits and measure their ESD breakdown voltage; choose MSI or LSI parts that are easily testable. Expose the chips in their packages and describe the failure mechanism of a chip input circuit that has been destroyed by ESD. Reverse-engineer the input protection circuitry for the chips used and compare the circuitry with the experimentally measured results.

11.5. Using the block diagram of Figure 11.42 as a guide, design a voltage-divider circuit and a logic gate that will act to inhibit the substrate generator oscillator when the desired substrate voltage is reached; simulate the design at several substrate voltages in close proximity to the targeted value to determine the logic gate output voltage as a function of V_X. For technology parameters, the following list may be used for calculations:

$$+V = 5 \text{ V} \pm 10\%$$

$$V_X = -3.0 \text{ V} \pm 10\% \quad \text{(design target)}$$

$$V_{T,n\text{Enhancement}} = 0.7 + 0.25 * (0.65 + V_{SX})^{1/2}$$

$$V_{T,n\text{Depletion}} = -2.8 + 0.25 * (0.65 + V_{SX})^{1/2}$$

Assume values for the device transconductances. Keep the divider current into the substrate at a minimum, say below 5 μA.

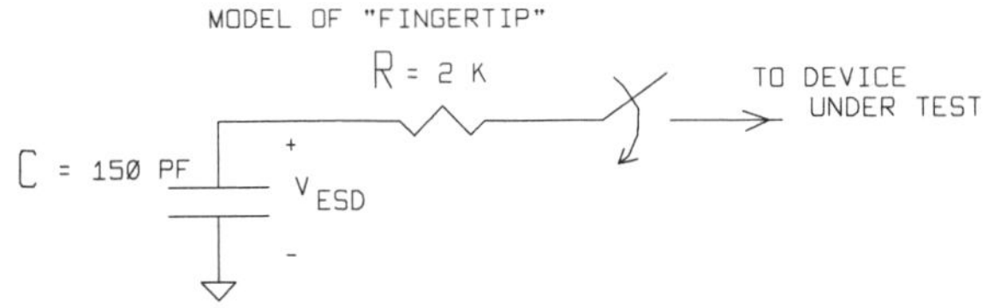

Figure 11.54. Circuit used for evaluating ESD input protection circuitry; the 150-pF human capacitance is precharged to the ESD test voltage.

REFERENCES

Iyer, R., "Protective Device for MOS Integrated Circuits," *Proceedings of the IEEE (Letters)* (July 1968), 1223–1224.

Martino, W., and others, "An On-Chip Back-Bias Generator for MOS Dynamic Memory," *IEEE Journal of Solid-State Circuits*, SC-15:5 (October 1980), 820–825.

Puri, Yogishwar, "Substrate Voltage Bounce in NMOS Self-Biased Substrates," *IEEE Journal of Solid-State Circuits*, SC-13:4 (August 1978), 515–519.

ARRAY CIRCUIT DESIGN

12.1 PROGRAMMED LOGIC ARRAY DESIGN

In a programmed logic array, or PLA, design techniques are used to implement logic functions using simple (memorylike) array structures. The physical design, or personalization, of a PLA can be generated *automatically*, maximizing designer productivity; the quick turnaround provided by the PLA generation tool enables design and engineering changes to be implemented very quickly and with potentially little impact on the remainder of the chip physical design. This is in contrast to standard macro design, where the physical layout of the circuits and their interconnections is very difficult to automate (compile), and where engineering changes may have a significant impact on the surrounding physical design.

Recall that *any* logic function can be expressed in a sum-of-products (or product-of-sums) form of the input variable set. A PLA implements the sum-of-products logic function for a potentially large number of inputs and can provide a potentially large number of different logical functions (outputs) of the input set.

A block diagram of a general PLA structure (including both combinational and sequential logic functions) is presented in Figure 12.1. The first array structure is used to generate the necessary product terms (AND array), and the second array generates the selected sum of the available product terms (OR array). The latches are used in feedback loops to allow sequential logic functions to be implemented. A simplified block diagram of a PLA, omitting the feedback latches and the front-end decode, is given in Figure 12.2.

Since it would be anticipated that a significant number of the product terms in the AND array will require the complement of an input variable, inverters are

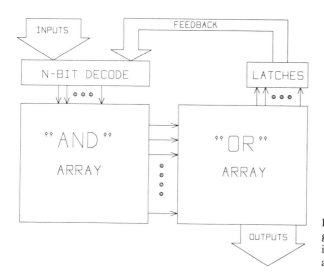

Figure 12.1. Block diagram of a general PLA. The implementation illustrated includes both combinational and sequential logic functions.

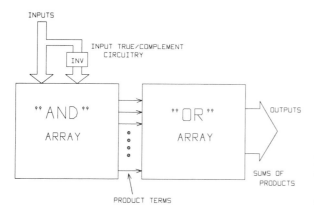

Figure 12.2. Block diagram of a combinational PLA, omitting the sequential functions and the *n*-bit input decode.

included on the front end of the PLA input set in order to make the complement of the input variables available to any product term. The inverters also act as buffers to drive the large input capacitance of the array; the PLA design typically includes additional buffers providing the true input values (not shown in Figure 12.2).

To be more consistent with the limitations of (*n*MOS) FET technology, that is, that high fan-in NAND (or AND) circuits are not advisable, an equivalent NOR representation using complemented inputs is used for the product terms. For the sum terms, a NOR implementation is again used, allowing a large number of product terms to be included in a sum-of-products logical function. The actual final output is inverted (buffered) to provide the true sum-of-products function. An example of a simple PLA used to generate two logical functions of two input variables (with two separate product terms) is given in Figure 12.3, both in logic

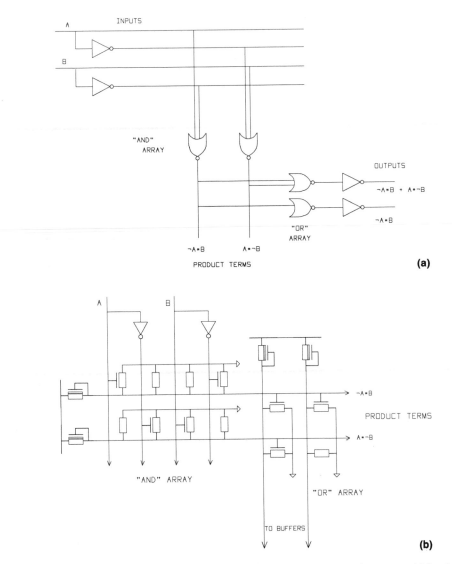

Figure 12.3. PLA with two inputs, two output functions, and two product terms. (a) Logic schematic. (b) Circuit schematic. Note the personalization of the PLA by the presence of gate connections to devices.

and circuit schematic form. The drawing of the circuit schematic in such an unusual form is intentional; it is intended to give an indication of the regularity of the physical layout used to implement both the AND array and the OR array. Note from the circuit schematic how the array is personalized; for a polysilicon gate FET technology, the presence of a thin oxide shape and metal contact that crosses a polysilicon input line adds that input to the NOR logic circuit (Figure 12.4). Note that PLAs are inherently *not* high-performance (single block delay) circuits due to the nature of their physical layout:

1. Long, high capacitance (polysilicon) lines for signal inputs, long product term metal lines for array outputs.
2. Potentially high gate input capacitance on signal inputs (in the AND array) and on product terms (inputs to the OR array).
3. Significant diffusion area (and therefore high junction capacitance) for product term and OR array outputs.
4. Potentially large ground shifts from long diffusion runs for ground to all devices (in both arrays); this can be alleviated to some degree with additional metal line ground connections interspersed between other array metal lines.

   The straightforward physical implementation of a PLA from a logical description results in a large area, yet relatively sparse layout with several performance-limiting factors. The notion of a relatively sparse PLA is borrowed from matrix algebra; a sparse matrix is one where a large number of elements are zero. Similarly, a sparse PLA is one in which a large number of product terms do *not* refer to the majority of the variables (or their complements) in the input set. In other words, a large number of product terms have "don't cares" for the value of a subset of the input variables. In the physical implementation, a number of the gate oxide personalization array locations are vacant for each of the product terms. Enhancements to PLA generation from a logical description are required to reduce some of these performance-limiting factors (and the overall macro layout area). *Folding* and *sharing* techniques will be discussed for reducing the number

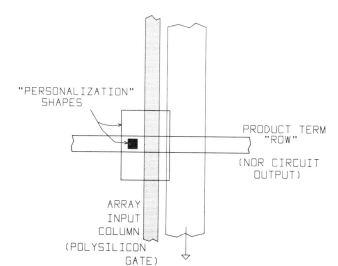

"PERSONALIZATION" SHAPES

PRODUCT TERM "ROW"

(NOR CIRCUIT OUTPUT)

ARRAY INPUT COLUMN
(POLYSILICON GATE)

Figure 12.4. Layout illustrating the personalization shapes added at an array location to implement a device or logic NOR circuit input at that location.

of unused array locations and improving the performance of the AND and OR array circuits.

Another particular case of interest can arise in the physical implementation of the PLA, one that can be alleviated by altering the logical representation. Consider the following common situation:

> A physical implementation of a PLA with input set $A, B, \ldots, N$ requires four input (column) lines to provide the signals $A, \overline{A}, B, \overline{B}$. Each product term that includes one of these signals requires a gate oxide shape (device) present in the row corresponding to that product term. Assume now that, in every product term in which either A or $\overline{A}$ is present, B or $\overline{B}$ is also found. Also, for every product term in which B or $\overline{B}$ is present, A or $\overline{A}$ is also required. The two signals A and B (or their complements) never appear independently. At the expense of a small amount of additional area at the input side of the AND array, a number of devices (and their associated input capacitances) may be eliminated by modifying the logical description of the PLA, eliminating all references to inputs A and B and replacing them with all the possible combinations of these inputs.

This approach to improving PLA performance, called *partitioning*, and how it may be implemented will be discussed as soon as a convenient scheme for representing the logical model of a PLA is presented.

PLA Logic Specification

The regularity of the physical design of a PLA suggests a matrix or table format for specifying the logical function in sum-of-products form. The format typically used is a matrix (corresponding to the actual physical orientation), where the columns in the matrix correspond to the PLA input signals to the AND array and to the OR array output signals, while the rows of the matrix correspond to the signals present in each product term. The implementation of the set of logical functions into a PLA (without any folding or partitioning) requires the following:

1. Two columns in the AND array for each input signal (the signal plus its complement).
2. One row in the AND array for each unique product term.
3. One column in the OR array for each output signal of the PLA.

A sketch of a general PLA matrix logic representation is given in Figure 12.5. Rather than including a matrix column for both the input signal and its complement, the notation used is the following:

AND array
> 1: In an element of the AND array matrix implies the *true* input signal is present in the product term

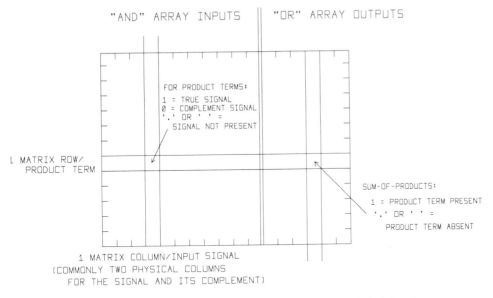

Figure 12.5.   Matrix representation of PLA logical functions.

0:   in an element of the AND array implies that the *complemented* input
signal is present in the product term

" " or . :   the input is not required in the realization of the product term and
a device should not be present at the intersection of the product
term row and the input signal column

OR array

1:   in an element of the OR array implies that this product term is
present in the sum-of-product terms for the logical function

" " or . :   do not include this product term in the sum-of-products for the logical function

Note that this notation is independent of the actual physical realization of the
PLA in (*n*MOS) FET technology using NOR circuitry in both the AND and OR
arrays. A 1 in an element of the AND array, implying the true signal in the product
term, requires the complemented signal as the gate input to the device in the NOR
circuit. Likewise, a 0 in an element of the AND array logical representation requires the true signal as the physical device input.

A value of 0 in an element in the logical OR array matrix does not make
sense in a sum-of-products realization. The output of the OR array when implemented using NOR–NOR circuitry is actually an *inverted* sum of products. As
illustrated earlier, inverting buffer circuits are typically included and connected

to the output of the OR array to provide the proper output polarity and significantly improve the overall PLA performance.

Bit Partitioning and Decoding

The approach referred to earlier for reducing the AND array delay is called *bit partitioning* (also input *decoding*). This reduces the number of personalized signals in a product term and also potentially reduces the required number of columns in the AND array. An example of two-bit partitioning is given in Figure 12.6. The four input signals $(A, \overline{A}, B, \overline{B})$ are decoded and replaced by the output of the four NAND gates: $(A * B, A * \overline{B}, \overline{A} * B, \overline{A} * \overline{B})$. An example of the use of the two-bit decoder is

$$\text{Product term:} \quad A * \overline{B} * D * \overline{F}$$
$$\text{Actual implementation:} \quad \overline{A} + B + \overline{D} + F \quad \text{(without decoding)}$$
$$A * \overline{\overline{B}} + \overline{D} + F \quad \text{(with decoding)}$$

Decoding could be used to reduce the AND array number of columns if, for example, only three out of the four possible combinations of the signals $(A * B, A * \overline{B}, \overline{A} * B, \overline{A} * \overline{B})$ are required to implement the necessary product terms. Note that implementation of two-bit partitioning deletes all references to the input signals (and their complements): A, $\overline{A}$, B, and $\overline{B}$. It was stated previously that this partitioning could only be performed if the input signals A or $\overline{A}$ never appeared independently of signals B or $\overline{B}$ so that deleting the columns $(A, \overline{A}, B,$ and $\overline{B})$ in

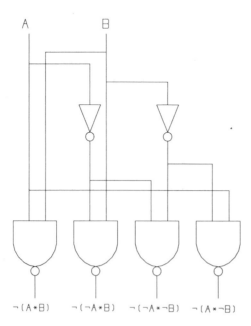

$\lnot(A*B)$ $\lnot(\lnot A*B)$ $\lnot(\lnot A*\lnot B)$ $\lnot(A*\lnot B)$

Figure 12.6. Logic diagram of a 2-bit partitioning network.

favor of ($\overline{A*B}$, $\overline{A*\overline{B}}$, $\overline{\overline{A}*B}$, and $\overline{\overline{A}*\overline{B}}$) was possible. In actuality, this restriction can be relaxed so that partitioning of a pair of inputs can still be regarded as favorable. For example, if input signals A and B appear to be ideally suited for decoding, but the product term $\overline{A}*C*\overline{D}$ is required (A appearing independently of B), an additional product term (and AND array row) could be added:

$$\overline{A}*C*\overline{D} = \overline{A}*(B+\overline{B})*C*\overline{D}$$

$$= \overline{A}*B*C*\overline{D} + \overline{A}*\overline{B}*C*\overline{D}$$

$$= \overline{\overline{A}*B} + \overline{C} + D + \overline{\overline{A}*\overline{B}} + \overline{C} + D$$

The single product term has been expanded to two product terms to facilitate the partitioning of signals A and B. *n-bit* partitioning and decoding is usually limited to either $n = 1$ (the default) or $n = 2$.

Sharing

The physical layout area of the PLA can be reduced by *sharing* the device personalization area between adjacent array inputs which are never both present in the same logic term. In a polysilicon-gate technology, the AND array output node and contact area at each personalization location is shared between adjacent columns (Figure 12.7a). (The two adjacent AND array column lines correspond to an input signal and its complement.) Since it will never occur that both polarities of a signal will be present in the same product term, the area allocated for personalizing the array between adjacent columns need only be large enough for a single output node connection. The personalization shapes are mirrored about the output node location to include either of the two adjacent columns (Figure 12.7b). The ground rail is also shared between adjacent input columns.

Folding[1]

The task of implementing a set of Boolean functions using a PLA macro is threefold:

1. Logical design
2. Topological design (i.e., the construction of the logical matrix)
3. Physical design: the generation of the circuits and shapes to construct and personalize the two arrays, partition and decode the inputs, and buffer the OR array outputs

The logical design phase consists of reducing a set of Boolean expressions into "minimal," two-level, sum-of-products form, simultaneously trying to reduce the

---

[1] This section is drawn from references 12.1 and 12.2.

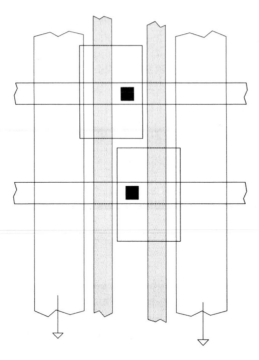

(a)

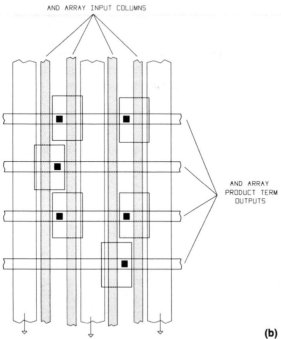

AND ARRAY INPUT COLUMNS

AND ARRAY
PRODUCT TERM
OUTPUTS

(b)

Figure 12.7. (a) Sharing of product term device locations between adjacent AND array input columns. (b) Mirroring of personalization layouts (with sharing) throughout the AND array.

necessary layout area and improve the overall PLA performance. Two-level logic minimization techniques and logic partitioning and decoding are used to attempt to optimize the logical description. The term "minimal" is included in quotes since the absolute, minimal logical description for each Boolean function may preclude the ability to partition a pair of inputs, as discussed earlier.

Once the functional form for the network has been optimized for implementing a PLA, the logic must be translated into the logical matrix, the topological representation scheme used for the personalization of the PLA. This translation involves the following:

1. Ordering the AND array inputs (from left to right).
2. Ordering the product terms (the AND array rows) from top to bottom in the matrix.
3. *Folding* pairs of columns in the AND (OR) arrays to allow signals to be input (output) at both the top and bottom of the physical arrays.
4. *Folding* pairs of rows in the OR array to allow product term output signals from a split AND array to enter from both the left and right sides of the OR array.

Once the logic matrix has been suitably manipulated, the physical design phase, particularly the IC layout, is relatively straightforward. Indeed, a computer-aided design tool can be used to generate the IC layout directly from the final logical representation (Section 12.3).

Row and column folding of a PLA are techniques for reducing the sparsity of the initial logical matrix in order to reduce the chip area occupied by the PLA. Typical densities for personalized array locations using conventional logic translation techniques are 10% for the AND array and 4% for the OR array. The technique of column folding attempts to enhance this factor by putting AND array inputs and OR array outputs on both the top and bottom of the matrix. An example of column folding in a portion of the AND array is shown in Figure 12.8. The folded sections of the AND array correspond to pairs of inputs (located opposite one another at the top and bottom of the matrix) that can be separated from each other. The through inputs in the figure are those that connect to product term rows in both folded sections. The most flexible or variable form of folding would be done on a column-by-column basis. Personalization cuts in the column line allow AND array inputs to face each other on top and bottom. A through input column would extend to the edge of the array, no cut being made. A cut is simply a break in the line (shape) for that column.

Column folding can also be implemented in the OR array; that is, OR array outputs may also be directed to the top or bottom of the logical matrix. This is illustrated in Figure 12.9. Column folding in the OR array is enabled by the judicious ordering of product term rows. The column folding techniques for the

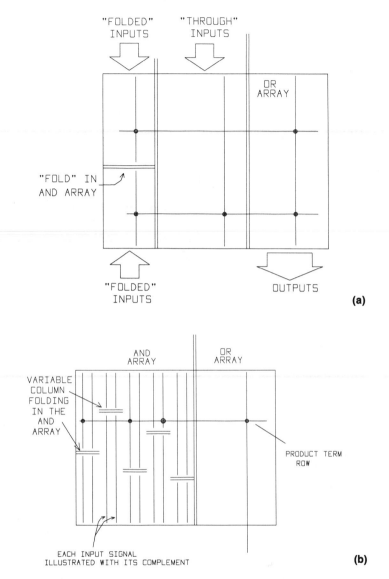

Figure 12.8. (a) Column folding in the AND array. Only a subset of the columns are folded; the remainder are "through" columns. (b) Column folding on a column-by-column basis.

AND and OR arrays are obviously *coupled* very strongly; the two arrays are treated simultaneously. The wiring channels between the AND and OR arrays in Figure 12.9 allow for dotting product term outputs.

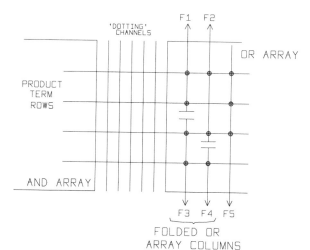

Figure 12.9. Column folding in the OR array. Additional dotting channel area is illustrated between the AND and OR arrays.

Digression: dotting of AND array product terms and dynamic PLAs

Figure 12.9 illustrates the technique of OR array column folding and also shows undedicated wiring channels for dotting product terms together to increase the effective width of the product term. For example,

$$\overline{(A + \overline{B} + C)} * \overline{(\overline{D} + F + \overline{G})} = \overline{A} * B * \overline{C} * D * \overline{F} * G$$

Recall from the discussions of FET circuit technology that dotting two outputs together is a "dot AND" connection; if the output of either circuit is 0, the output of the dotted connection is also discharged to "0". However, the implementation of dotting is ill advised if standard n-channel depletion-mode load devices are used to create the NOR circuits in the AND array. Making the dotted connection connects two load devices in parallel, effectively creating a load device with dimensions $2 * (W/L)_{load}$. To provide a valid dc down level, each enhancement device must likewise be doubled in width (to maintain the same device ratio of ratios), introducing a severe penalty in the overall PLA macro area.

To facilitate dotting and simultaneously reduce the block delay, a dynamic PLA circuit design approach can be adopted. During a portion of the PLA execution timing when the data inputs to the PLA are invalid (changing), a clock signal can be used to perform two functions (Figure 12.10):

1. Inhibit the input decoder circuits from the devices in the AND array (i.e., ensuring that all column line inputs to the AND array carry a logical 0).

2. Precharge the product term rows and the outputs of the PLA to a logical 1.

At the end of the precharge clock pulse, the data are assumed to be valid and the appropriate product terms and PLA outputs will be discharged. An illustration of the use of dotting for dynamic PLAs is shown in Figure 12.11.

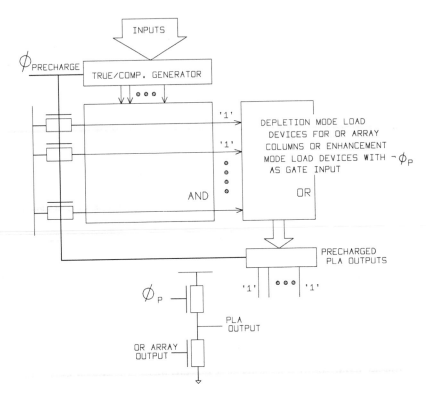

Figure 12.10. Use of a precharge clock signal to implement a dynamic PLA.

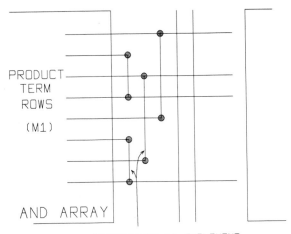

CONNECTIONS TO IMPLEMENT
'DOT-AND' FUNCTION
BETWEEN PRODUCT TERMS

Figure 12.11. Product row dotting in a dynamic PLA.

End of digression

As illustrated earlier, column folding in both AND and OR arrays reduces the number of array columns by putting independent inputs and outputs on the top and bottom of the logical matrix. Row folding, on the other hand, reduces the number of product term rows by typically splitting the AND array and placing the two subarrays on either side of the OR array (Figure 12.12). The criterion for being able to pair and fold two product term rows is that the two product terms are not *both* required by any sum-of-products logical function output. If two columns are to be folded, it is likewise required that they be disjoint; the entries defined in one column of the logical matrix (the 1s and 0s, not the don't cares) must all be in different positions than any of the entries in the other column. Folding column i with column j requires that *all* product terms containing i (or its decodes) be on top of or on the bottom of *all* the product terms containing j (or its decodes). Therefore, folding two columns of a PLA introduces constraints on the folding of other columns as well. A PLA folding example illustrating the constraints involved is given in Figure 12.13. In Figure 12.13b, the particular ordering of the product term rows chosen allowed three sets of column pairs to be folded: (A, D), (B, C) and (F_1, F_2). It is assumed that the input signal and its complement were both cut at the same array location to allow for sharing in the AND array. Figure 12.13c illustrates row folding; the two rows are disjoint in that they do not *both* have personalization entries in the same column of the logical matrix.

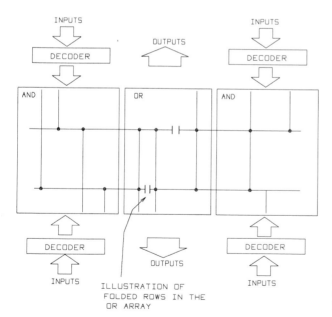

ILLUSTRATION OF
FOLDED ROWS IN THE
OR ARRAY

Figure 12.12.   Row folding in the OR array, resulting in an AND-OR-AND implementation.

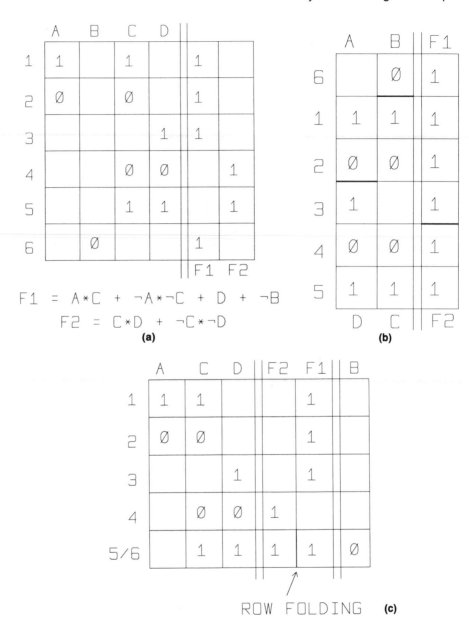

$$F1 = A*C + \neg A*\neg C + D + \neg B$$
$$F2 = C*D + \neg C*\neg D$$

(a)

(b)

ROW FOLDING   **(c)**

Figure 12.13. (a) Column folding example. (b) Folded implementation of the matrix in (a). (c) Row folding of the matrix in (a).

As described just previously, a necessary constraint on the ability to fold a pair of columns (rows) is that the two signals should be disjoint; that is, throughout the logical matrix they should not both have a personalization entry in the same row (column). However, this is not the only constraint. An *ordered* folded pair of columns, (C_i, C_j), implies that the set of all rows where C_i is personalized, $R(C_i)$, must all be placed above the set of rows where C_j is personalized, $R(C_j)$. [An *ordered* folded pair of rows, (R_i, R_j), implies that the set of all columns where R_i is personalized, $C(R_i)$, be placed to the left of the set of all columns where R_j is personalized, $C(R_j)$. The following description pertains to column folding; it could just as easily pertain to row folding, by swapping rows with columns, and vice versa.] For more than one ordered folded pair of signals to be implementable, the relative position of the columns to be folded must be so that no *cyclic* constraint on the required relative positioning of the rows results. If there is no cyclic constraint, then there exists an ordering (not necessarily unique) of the rows so that *all* the required relative positions of the rows may be satisfied. Referring to Figure 12.14, folding columns A and B forces rows (1, 2) to be above (or below) rows (3, 4). This ordering prohibits the folding of pairs (C, D) and (F_1, F_2). In this example, only one pair of columns and no rows may be folded.

In general, it is desirable to find the largest set of ordered folded pairs without any cyclic constraints to minimize the overall PLA area. The general case of the implementation of a PLA with n-bit decoding on the front end of the AND array does not significantly alter the folding criteria; the input signals that have been partitioned together remain together. The general case of the implementation of a PLA with sequential feedback latches again does not significantly alter the folding criteria. The final placement of the location of the OR array columns dictates the placement of the latches on the outputs of those columns. To implement sequential feedback, additional wiring channels are allocated to allow for a connection from latch output to AND array input (Figure 12.15).

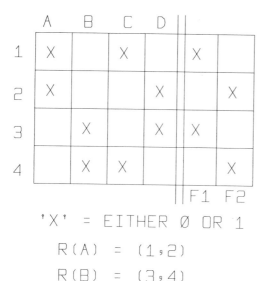

'X' = EITHER 0 OR 1

R(A) = (1, 2)

R(B) = (3, 4)

Figure 12.14. Cyclic conflict between the pairs of foldable columns (A, B), (C, D), and (f_1, f_2).

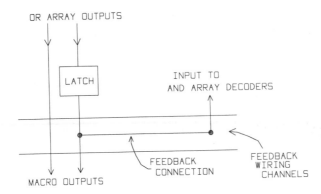

OR ARRAY OUTPUTS

LATCH

INPUT TO
AND ARRAY DECODERS

MACRO OUTPUTS

FEEDBACK
CONNECTION

FEEDBACK
WIRING
CHANNELS

Figure 12.15. Additional macro wiring channels to implement output feedback connection for a sequential PLA network.

12.2 READ-ONLY MEMORY DESIGN

A read-only memory (ROM) is another type of array circuit design, similiar to a PLA, where the logical network function provided at the outputs of the ROM is fixed by the design and fabrication process. Conceptually, the distinction between a ROM and a PLA design is twofold: interpretation and completeness. A ROM is used to store fixed data; a PLA provides logical sum-of-products outputs. The inputs to a ROM are interpreted as *addresses* and are decoded initially; the AND array inputs to a PLA are regarded as logical signals, which provide product term outputs. Array locations are *accessed* in the ROM array by activating *word lines*; in the OR array of a PLA, product term row signals are active. When accessing a ROM array, only one decoded word line is active in the array; in the case of a PLA, many product term outputs may be active in the OR array.

The main distinction, conceptual and physical, between a ROM and a PLA is as follows:

> In the implementation of a ROM, *all* logical combinations of the full input set are decoded; that is, all 2^n possible product terms are included, where n is the number of address bit inputs. In a PLA, only the required (unique, minimized) product term rows are generated. The design of a PLA incorporates the particular network functions to be implemented into the minimization, folding, and sharing procedures. The design of a ROM (both the logical and physical design) is done independently of the data stored in the array, except for the single personalization design and process step.

A general block diagram of a ROM is given in Figure 12.16. If the block labeled Address Decode is replaced by one labeled AND array, the diagram of Figure 12.16 would be identical to that of a PLA. As with PLA design, precharge circuit techniques may be used to improve the data access time (performance) and/or reduce the dc power dissipation. The actual physical implementation of the input decode differs from that of the AND array of the PLA. Since the number

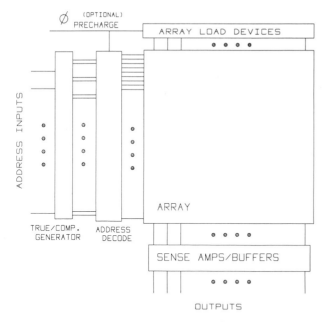

Figure 12.16. General block diagram of a ROM (read-only memory).

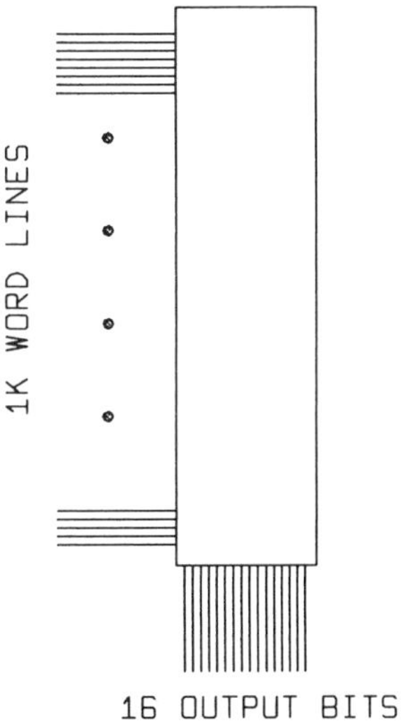

Figure 12.17. Awkward array dimension (1K × 16 example) without array column decoding using a subset of the address inputs.

of decoded array addresses may be large, it is *not* feasible to implement each word line (PLA product term) as a row input to the memory array; the resulting array configuration would be unmanageable (Figure 12.17). It is therefore necessary to use other than a simple gate decode to produce a reasonable array shape. An alternative array configuration uses a *drain decode* scheme. In the example given in Figure 12.18(a), four columns of the array are collectively used to represent a single output bit for the full address set; a subset of the address inputs is decoded to enable the array value to be read from the appropriate column.

Personalization of the ROM is accomplished in a similar manner to a PLA; the presence or absence of a device (intersection of shapes) dictates whether the particular array location stores a 0 or a 1 (Figure 12.18(b)). The disadvantage of the drain decode scheme is that each column of devices effectively becomes a NAND circuit, with two devices in series to ground providing the discharge path for the output. At the expense of additional load devices to drive each column, an alternative scheme eliminates the drain decode devices and (using the same subset of address inputs) selects one of the columns to provide the output bit. The column decode technique is illustrated in Figure 12.19.

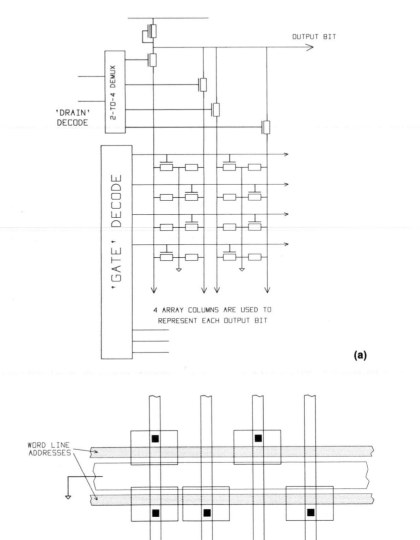

Figure 12.18. *Drain decode* scheme for ROM array implementation. The device personalization layout shapes are identical to that given earlier for PLA array locations.

Decoding a set of row address inputs is commonly implemented with a high fan-in NOR (or NAND) circuit (Figure 12.20). The fan-in required of the logic gate is equal to the number of address inputs being decoded. The physical layout of the decoder could resemble that of the NOR functions in the AND array of a

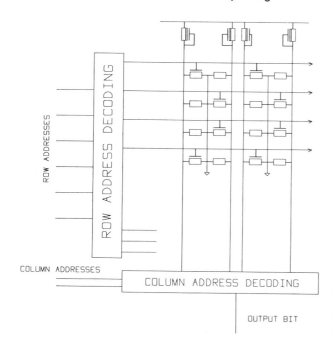

Figure 12.19. Block diagram of *column decode* technique, illustrating the use of column addresses (a subset of the input address set).

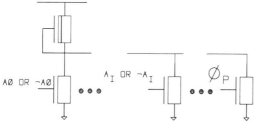

Figure 12.20. High fan-in NOR circuit, used for input address decoding. *Example:* Word line 28 = 11100_2, implemented as $\text{NOR}(\overline{A4}, \overline{A3}, \overline{A2}, A1, A0)$.

PLA or could be implemented with more conventional logic circuit layouts (preferably buffered circuits to drive the address line capacitance).

A more representative block diagram for a ROM is given in Figure 12.21.

Applications of ROMs

For each of the following applications, a ROM design will be conceptually and physically easier to implement than attempting the minimization and folding techniques associated with a PLA design.

1. Look-up table for mathematical calculations, such as the evaluation of trigonometric functions, logarithmic and exponential functions, and square

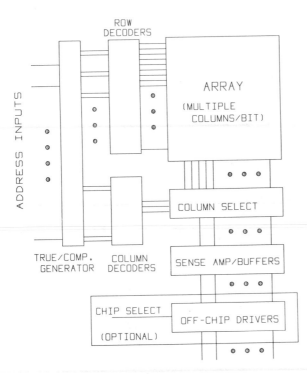

Figure 12.21. Detailed block diagram of a ROM.

roots. These calculations, using a series expansion for the particular function, are typically very time consuming when programmed using a software subroutine. If these calculations are performed frequently, it may be more efficient to include a ROM look-up table [i.e., a code-conversion system between the binary argument of the function (the input address) and the output value (in binary, to some predefined accuracy) of that function evaluated at the argument].

2. Microprogram control store: In a microprogrammed digital system, the execution of an encoded macro-instruction involves generating a sequence of signal vectors used for gating or control purposes. A ROM may be used to generate the appropriate binary sequence if the address input to the ROM is altered by means of a counter (during execution of the macro-instruction). The number of control signals generated would be equal to the number of outputs of the ROM.

3. Character generator: All digital systems rely on the display of alphanumeric characters for input/output: CRTs, seven-segment displays, dot-matrix printers, and so on. In most of these applications, the pattern of pixels, segments, or dots used to display the set of characters is stored in a ROM. A ROM offers a relatively versatile way to allow a designer to individualize the format of the set of characters displayed in a system. For example, a typical im-

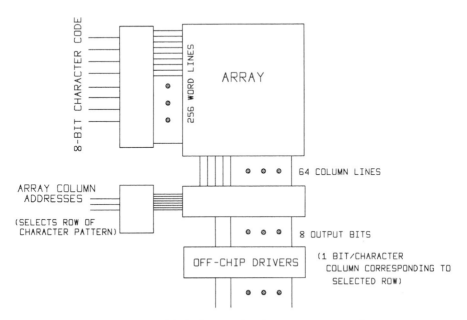

Figure 12.22.   Block diagram of a character-generator ROM.

plementation of a character generator ROM for a printer application would be

$$256 \quad \times \quad 8 \times 8 \quad = 16\text{K bits}$$

number of	character or
alphanumeric	font format is
characters and	provided in an
graphic fonts	8×8 array

A possible organization for a character generator ROM is shown in Figure 12.22.

12.3 DESIGN SYSTEM TOOLS FOR ROM AND PLA GENERATION

The PLA and ROM array designs discussed in the previous two sections are examples of macro designs where a considerable portion of the physical layout of the array, particularly the personalization of the array, can be performed with the assistance of design system program tools. These design system tools can be used to construct the cell transforms necessary to locate and replicate the basic array building blocks for a particular array configuration and/or place the specific shapes necessary to personalize each array location. Once personalized, the resulting collection of shapes (background and personalization) can be added to the chip physical design file. This section discusses briefly some of the considerations of physical design and design system tool development for automated macro generation.

The number and variety of configurations of ROM macros typically provided in a design system library is relatively few. As a result, the majority of the ROM physical design is commonly handcrafted—the array circuits, the row address decoder (word line driver) circuits, and the output driver circuits. Only the personalization gate shapes internal to the array are usually added by a design system tool from a file describing the desired internal data storage pattern. The handcrafting of the ROM is not as monumental a task as may first appear since the number of basic building blocks is not large. Once a word line and column output pitch for the array configuration are determined, the necessary replication of the basic building blocks can be implemented quite easily. However, the development of the optimum pitch requires an aggressive design philosophy and therefore a significant amount of design engineering.

As illustrated in Figure 12.23, the *word line pitch* is defined as the width of the word line plus the distance between adjacent word lines; a similar definition is used for the column line pitch. The key significance of the word line pitch is that the layout of the word line decode and buffer circuit must fit in that same pitch (Figure 12.24). A conservative pitch simplifies the design of the word line drivers (and enables a wider ground line to be used, reducing ground shift losses), but at the sacrifice of array density; when multiplied by the number of word lines, this loss in density can be substantial. The column line pitch must also accommodate the column select circuits, as well as the array device width, which is engineered for an access time performance goal. The column line pitch must also

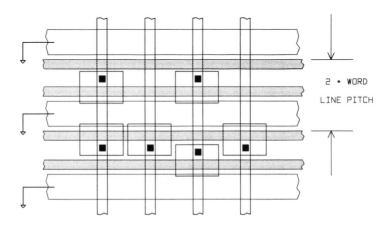

Figure 12.23. Word line pitch for a polysilicon gate ROM.

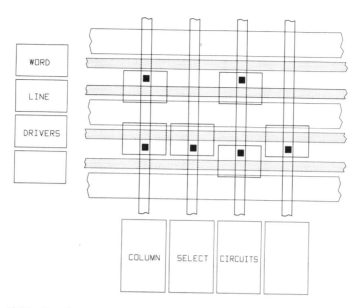

Figure 12.24.   Requirements that the word line driver circuits fit the word line pitch and that the column select circuits must match the column pitch.

permit the placement of the load device (static or clocked precharge) between column lines at the edge of the array.

Once such key parameters are defined for a particular array configuration, the remainder of the background design can be implemented in a relatively straightforward manner; the extreme regularity of the background design does not really necessitate the development of a design system tool to add any but the ROM data personalization shapes.

Such is not the case for a PLA design that has been minimized and folded. The pitches in the two arrays are still key design parameters (and may likewise require resourceful design of bit partitioners or decoders and buffers), but the uniformity of the folded array has been greatly reduced. Therefore, a design system tool is indispensable for the correct construction of the final physical layout. The physical design cells used to generate the final layout include the input bit partitioning circuits, the array load devices, the power distribution, the product term output-to-OR array input connection, and, in particular, the individual array elements; some of these are illustrated in Figure 12.25. The individual physical design cells provide the necessary continuity of interconnections for a single matrix location; the arrays are then constructed for the appropriate number of partitioned inputs, product terms, and outputs by the proper replication of the elemental cells (Figure 12.26). Special physical cell designs may be required to properly cut a folded column or row, as well as around the perimeter of each array where connections to power supply lines, load devices, or decoder or product term outputs must be made. Once the array background has been constructed,

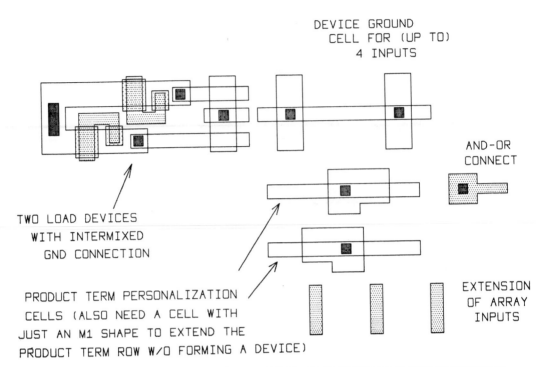

Figure 12.25.   Layouts of some of the elemental cells used in constructing a design system PLA: polysilicon gate segments, array personalization shapes, ground diffusion segments (also stitched in metal), and so on.

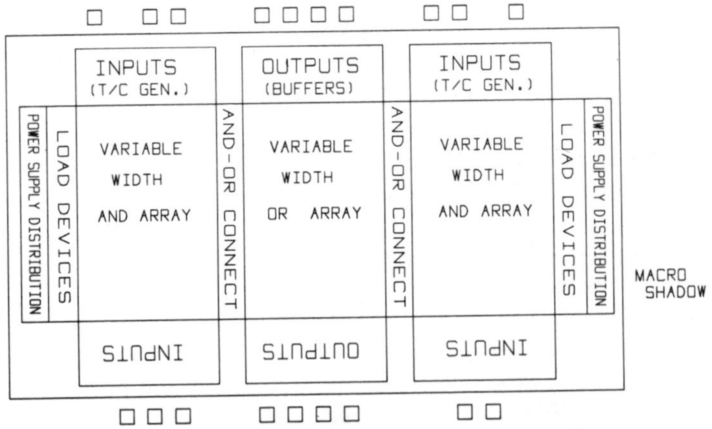

| POWER SUPPLY DISTRIBUTION | LOAD DEVICES | INPUTS (T/C GEN.) VARIABLE WIDTH AND ARRAY | AND-OR CONNECT | OUTPUTS (BUFFERS) VARIABLE WIDTH OR ARRAY | AND-OR CONNECT | INPUTS (T/C GEN.) VARIABLE WIDTH AND ARRAY | LOAD DEVICES | POWER SUPPLY DISTRIBUTION |

INPUTS   OUTPUTS   INPUTS

MACRO SHADOW

Figure 12.26.   Block diagram of a folded design system PLA construction. The intermix of OR array load devices with array outputs and the details of the power supply distribution are not shown.

the personalization of the array involves adding the device personalization physical cell at the calculated coordinate locations of the background for a particular array location.

It should be emphasized that the background design of the predefined ROM library macros and the generated PLA macros must still conform to the overall chip image used by the design system; the connection to the image power supply rails must be aligned correctly, and the placement of the macro input and output pin shapes must lie on the external wiring grid. These restrictions substantially complicate the tool(s) required to be able to generate the PLA layout. As a result, just as the number of ROM configurations offered in the design system library was limited, so too might the flexibility in PLA construction be constrained. Referring again to Figure 12.26, if the number of product term rows is fixed, the height of the resulting physical design for the macro is therefore determined; the appropriate fit on the overall chip image can be assured prior to the actual construction of the array. The number of AND array inputs and their partitioning and OR array outputs could still be permitted to vary from design to design, as long as the required number of product term rows after folding did not exceed the predefined value. The design system library may therefore contain a number of different PLAs, specified in terms of maximum product term rows, which will by construction fit the overall chip image; the width of the macro may vary depending on the final array configuration.

Due to their construction and/or personalization characteristics, PLAs and ROMs are different from other design system macros with regards to physical design by the macro library developer and the logic description provided by the system designer. The physical definition of the background ROM design and the PLA building block cells requires additional information to describe how the personalization cells are to be replicated and mirrored when generating a specific macro design. Recall from the graphics language discussion in Chapter 5 that each cell definition includes a cell header, where the header could conceivably contain a number of different parameters and parameter values; Problems 12.3 and 12.4 ask for the definition of additional parameters to be included in the personalization and background cell header to provide the coordinate information required for macro generation.

In particular, it should be mentioned that although the variable-width PLA should be correct by construction, obviating the need for macro logical-to-physical checking, the need still exists for the design system macro generator tool to produce a macro shadow and macro pin shapes before adding the generated macro to the chip physical design file. This enables the PLA to be included in the global wiring and logical-to-physical checking procedures.

The means of logic entry used to describe the specific ROM or PLA function (and serve as input to the macro generation tool) will likely differ from the means used to describe a combinational network. The array block's logic section in the hierarchical chip model description will likely contain a pointer to a file that contains the array definition, in a convenient, matrixlike format. Problems 12.5

and 12.6 ask for the development of a suitable record format and file description for ROM and PLA personalization files; in addition to the basic array configuration and pin naming information, specific cases to be considered include the address and data representation formats for the ROM and the specification of column and row folding and input bit partitioning to be used in constructing the PLA.

Excellent references regarding the specifics of logic minimization techniques and folding algorithms for PLA design are given at the end of the chapter (references 12.1 through 12.7).

12.4 INTRODUCTION TO RANDOM-ACCESS MEMORIES

Random-access (or read/write) memories allow data to be written and retrieved at comparable rates, usually in the 10- to 500-nanosecond (ns) range, depending on the technology. The semiconductor random-access memory is volatile; if power is removed from the device, the stored information is lost. Semiconductor (and specifically n-channel MOS) random-access memories are commonly divided into two types: static and dynamic. Static memories do not require that the memory chip be accessed with any frequency; dynamic memories, on the other hand, require periodic access to *each* data storage location within a very rigid time interval. This is usually accomplished by a *refresh cycle*, a dummy read and refresh of groups of data storage locations. These refresh cycles are interspersed with the normal data access (read/write) operation cycles of the system; each data storage location must be refreshed within a maximum time interval or else the stored information will potentially be unrecoverable. Nevertheless, dynamic memories are widely used (in considerably larger volumes than static memories) because the cost per bit and power per bit are considerably less than that of comparable static memories.

To a system designer, random-access memories are organized as W words, each of B bits, for a total storage capacity of $W * B$ bits. Storage capacity may also be stated in terms of *bytes*, with $B = 8$ (or 9) bits of a word grouped collectively. Full-chip RAM designs are commonly organized to provide a single data bit for read/write operations (by 1); wider configurations are commonly provided in a design system library for incorporating onto a VLSI chip (by 16, by 32) for use as a *local store* or *register stack*.

The random-access memory provides equal access to each storage location of B bits, independent of the sequence in which the information was stored. A binary-encoded set of input bits provides an *address* that points to a particular group of B bits; the system data will either be read from or written to that address. The access time is independent of the address inputs or their sequence. This is in contrast to a serial access memory, such as a shift register, where the data are available for reading only in the same sequence in which they were stored. Shift-register or recirculating memories will not be discussed in this section.

There are two important specifications for the performance of a random access memory in any technology: *access time* and *cycle time*. Access time equals the time required to read out any word from the memory relative to a valid address strobe or chip select input signal; the cycle time is the reciprocal of the maximum frequency at which words may be accessed for reading or writing without danger of error. For semiconductor RAMs, the cycle time is typically equal to $1 \times$ to $2 \times$ the access time.

Basic Elements

The block diagram of a random-access memory system is shown in Figure 12.27. The basic elements of the memory are the array (where the actual information is stored) and the supporting circuits (which select and either write or sense and amplify the stored signal at a particular array location). It should be evident from the figure that the random-access memory and the read-only memory share many of the same design constraints, although the bit line driver and sense amplifier function is unique to RAMs. In particular, both the RAM and ROM benefit from the use of more than one bit line column in the array of which one is selected (by a column address) to represent the data bit. (In a $\times 1$ RAM configuration, *all* columns refer to the same data bit.) The word line (row) and bit line (column) capacitances have a major effect on the access time performance. Providing an aspect ratio close to one for the array dimension (number of word lines = number of array bit lines) will tend to improve the overall performance considerably. If multiple columns are used for the same data bit, part of the set of address inputs are now used as binary-encoded column select signals, while the remainder of

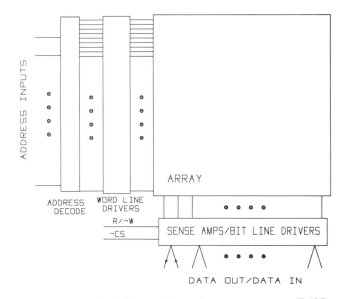

Figure 12.27. Block diagram of a random-access memory (RAM).

the address inputs are decoded to select a particular word line, as for the ROM. As an example, the block diagram of the INTEL 2147 4K $\times$ 1 static RAM is given in Figure 12.28. The memory array has been divided into 64 rows (word lines) $\times$ 64 columns (bit lines). Half of the 12 address inputs are directed to the row decoder, while the remaining half are interpreted by the column decoder. Also note the high-impedance buffer connected to the data output (the column selected by the A_6–A_{11} address inputs). RAMs, whether single-chip designs or embedded within a larger chip, commonly have their data outputs dotted with other output signals. A chip select signal is used to control the dotting capability.

Digression: globally dotted signals on-chip

Globally dotted circuit designs on-chip are really not consistent with a conservative design system approach. Conventionally, all sources to an on-chip bus should be multiplexed (Figure 12.29). This can rapidly become unwieldy as the width of the bus and the number of potential sources onto the bus becomes large. At the expense of additional (potentially manual) checking, various circuit design schemes can be used to synthesize the multiplexing of several sources onto a bus. These techniques differ primarily in how a pull-up device is connected to the bus lines; they essentially all use the same dotted connection circuit design.

    The various dotted sources onto the bus typically use an open drain driver circuit (Figure 12.30), exactly like the open-drain off-chip driver circuit discussed in Section 11.8. The open-drain driver circuit can only *discharge* the bus output; another block is required to either precharge or pull up these lines. Only one pull-up circuit is required per bus line, whereas there may be a potentially large number

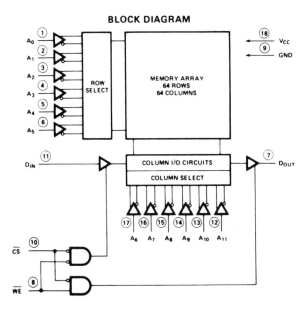

Figure 12.28. Block diagram for the Intel 2147 4K $\times$ 1 static RAM. (From *Memory Components Handbook*. Santa Clara CA: Intel Corp., 1983, p. 3–255. Copyright © 1983 by Intel Corp. Reprinted with permission)

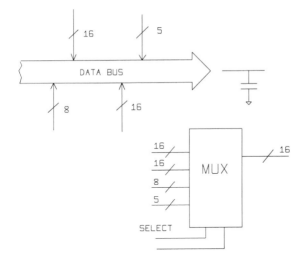

Figure 12.29. Multiplexing of a number of sets of signals connected to a system data bus.

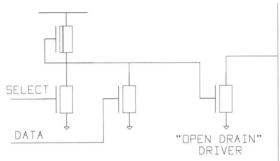

Figure 12.30. Open-drain driver circuit for on-chip dotting to a system bus.

of dotted open-drain driver connections. The select signal is active low; one and only one select line should be active at any time. The bus precharge circuit can be any of the pull-up circuit types illustrated in Figure 12.31; if a precharge pulse is required, it must be generated prior to and *not* overlapping with any select signal.

Two very significant additional constraints on the development of a design system are imposed by the addition of dotted circuits to the design system library (constraints in the sense that additional checking of the *completed* physical design must be performed, which may have to be accomplished manually):

1. It must be verified (after global physical design, when output bus capacitances are explicitly known) that the combination of the pull-up circuit and the precharge pulse duration is indeed sufficient to precharge the output capacitance in the allotted time; likewise, the open-drain driver device must be of sufficient width to discharge the large output capacitance in the appropriate time. This delay information must be calculated and appended to

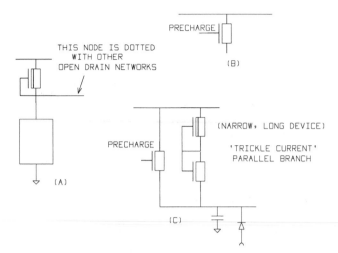

Figure 12.31. Pull-up circuit types for on-chip dotted nets. (a) Static logic circuit with pull-up. (b) Clocked precharge. (c) Clocked precharge with trickle current parallel branch.

the block delay for the simulation model used to represent the open-drain implementation.

2. If the pull-up circuit used with the on-chip open-drain drivers requires that a dc current be sunk by the driver when providing a 0 on the bus, a resistance model of that net should be constructed and the worst-case down-level voltage be evaluated for *all* inputs to which that net is connected. Constructing the resistance model and evaluating the worst-case down-level voltage to all inputs can only be done after the physical design is completed (and requires an accurate model of contact resistance and line resistivity). Figure 12.32 illustrates the model and the necessary checking.

End of digression

Another common approach for the configuration of the memory array is to *split* the array in half around the column decoders and sense amplifiers. This technique is more common among dynamic RAM designs than in static RAM configurations. As will be explained in more detail shortly, dynamic RAM designs depend on the *charge transfer* between the stored charge at a memory location (on the node capacitance of that array location) and the bit line capacitance; the greater the ratio between the storage capacitance and the bit line capacitance, the greater the charge transfer and hence the larger the voltage change on the bit line. A large swing in voltage on the bit line is desirable since that enhances the reliability of the sense amplifier circuitry to which the bit line is connected. (The sense amplifier senses the direction of the voltage swing on the bit line and boosts or amplifies that signal.) As a result, the array configuration is specifically modified to reduce the bit line capacitance as much as possible.

In addition to splitting the memory array to reduce the length of the bit line (by roughly a factor of 2), it is also advantageous to route the bit line on an interconnection layer (diffusion, polysilicon, or metal) with the least capacitance

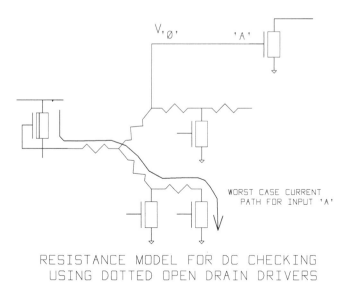

RESISTANCE MODEL FOR DC CHECKING
USING DOTTED OPEN DRAIN DRIVERS

Figure 12.32. DC model for determining the 0 voltage level using an open-drain driver scheme. The global resistance path must be checked to ensure that the 0 voltage level is valid to all gate inputs.

per unit length. Different layout schemes for the dynamic RAM storage cell will be presented later, illustrating some of these design constraints. Depicted in Figure 12.33 is a block diagram of the INTEL 2117, a 16K × 1 bit dynamic memory design. Several features of this design should be highlighted. First, note that the memory array has indeed been split around both sides of the sense amplifier and column decoder function; the 128 × 128 array has been split into two 64 × 128 arrays. One of the seven row address bits is decoded to select the particular half of the array, while the remaining six address bits are input to each of the row decoders. To reduce the package pin count, the seven row addresses and the seven column addresses are multiplexed; thus a 14-bit address can be input to a 16-pin device. The row address strobe ($\overline{\text{RAS}}$) signal is used to latch the row address bits, while the column address strobe ($\overline{\text{CAS}}$) signal latches the subsequent 7-bit column address. The output bit from a read cycle is latched and remains valid at the data output pin as long as the column address strobe ($\overline{\text{CAS}}$) is active.

Refreshing of the storage array locations is performed by sequencing through the 128 row address combinations every 2 msec. The $\overline{\text{CAS}}$ signal may remain active (data present and latched at the output) or inactive (data output in high impedance). The memory array locations are read, sensed, and refreshed a *row* at a time, all 128 array locations in the same row. In particular, note that only 128 sense amplifiers are present, even though there are 256 distinct bit lines (128 for each half of the array). Sense amplifiers are *differential* amplifier circuits, where an input signal to the amplifier is compared to a reference input. Since

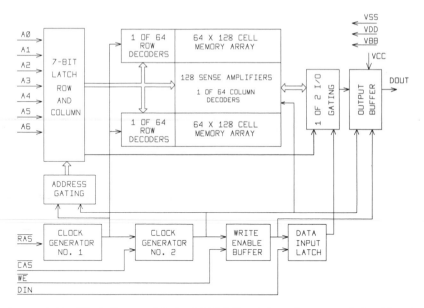

Figure 12.33. Block diagram of the Intel 2117 16K × 1 dynamic memory design. Note the split array around the sense amplifier circuits and the multiplexing of row and columns address inputs on pins *A*0 to *A*6.

only half the array is accessed by the six row address inputs, the other deselected half is used to provide the reference voltage input to the differential amplifier.

To illustrate the definition of the access time and cycle time, the read cycle waveforms and timing specifications for the INTEL 2147 4K × 1 static RAM illustrated earlier are presented in Figure 12.34. Note that the cycle time (minimum) is equal to the access time from address change (maximum). In this particular design, it is possible to continue to change the address inputs while the chip select and read input signals remain active. For completeness, the write cycle timing for the 2147 is sketched in Figure 12.35. The 2147 does not use a common input/output driver for the data input and data output function, but rather uses two separate pins.

The design choice of common I/O data pins versus separate data input and output pins is a very important decision, with extreme significance on the system design, circuit design, and system timing. If the system design implements a bidirectional data bus, the system or chip designer must verify that the appropriate timing between the various data sources is maintained so as to prevent any bus contention problems; that is, data cannot be placed on the bus until the bus has been released by the previous source. Note that in the write cycle example presented for the 2147 (with separate data input and data output pins) the chip select signal going active may *precede* the Read/Write signal going low. A *read* cycle is initiated when chip select ($\overline{\text{CS}}$) goes low and is subsequently interrupted when

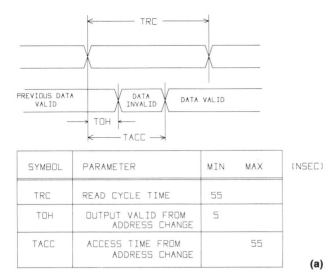

SYMBOL	PARAMETER	MIN	MAX	(NSEC)
TRC	READ CYCLE TIME	55		
TOH	OUTPUT VALID FROM ADDRESS CHANGE	5		
TACC	ACCESS TIME FROM ADDRESS CHANGE		55	

(a)

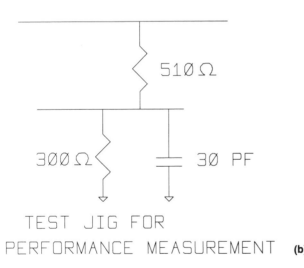

TEST JIG FOR

PERFORMANCE MEASUREMENT   **(b)**

Figure 12.34. (a) Read cycle waveforms and timing specification for the Intel 2147 4K × 1 static RAM. (b) Standardized test circuit used for the performance and voltage level measurements.

the R/$\overline{\text{W}}$ line falls. In the waveform diagram in Figure 12.35, the data-in signal valid condition is shown overlapping the data-out signal transition to high impedance. [The time delay t_{wz} (write cycle initiated to output in high impedance) is specified as 30 ns (maximum) before the chip releases the output bus.] With distinct data in/data out pins, this overlapping condition is certainly feasible. If common I/O pins are used (as with the INTEL 2114 1K × 4 static RAM), the appropriate set of waveform signals for a write cycle is shown in Figure 12.36.

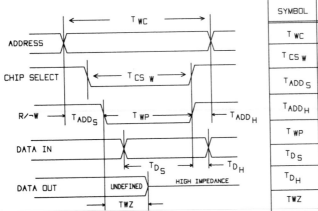

SYMBOL	PARAMETER	(NSEC) MIN	MAX
T WC	WRITE CYCLE TIME	55	
T CS W	CHIP SELECT TO END OF WRITE	45	
T ADD S	ADDRESS SETUP TO WRITE	0	
T ADD H	ADDRESS HOLD TIME	10	
T WP	WRITE PULSE WIDTH	35	
T D S	DATA SETUP TIME	25	
T D H	DATA HOLD TIME	10	
TWZ	WRITE TO OUTPUT IN HIGH IMPEDANCE		30

Figure 12.35. Write cycle timing and waveforms for the Intel 2147. The chip uses separate data in and data out pins, as opposed to a common I/O pin.

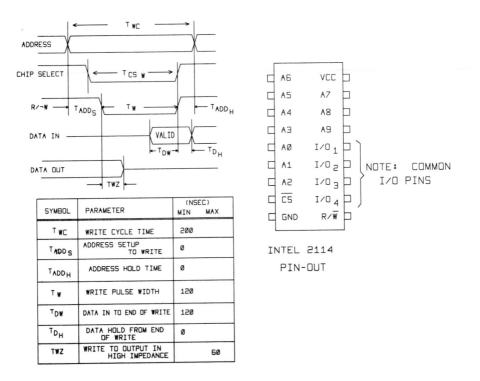

SYMBOL	PARAMETER	(NSEC) MIN	MAX
T WC	WRITE CYCLE TIME	200	
T ADD S	ADDRESS SETUP TO WRITE	0	
T ADD H	ADDRESS HOLD TIME	0	
T W	WRITE PULSE WIDTH	120	
T DW	DATA IN TO END OF WRITE	120	
T D H	DATA HOLD FROM END OF WRITE	0	
TWZ	WRITE TO OUTPUT IN HIGH IMPEDANCE		60

INTEL 2114
PIN-OUT

Figure 12.36. Write cycle timing and waveforms for the Intel 2114 1K × 4 static RAM, using common I/O pins.

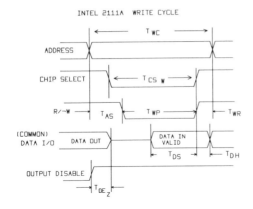

Figure 12.37. Write cycle timing and waveforms for the Intel 2111A 256 × 4 static RAM, which uses an Output Disable signal to provide direct control over the direction of data flow on the common I/O pins.

SYMBOL	PARAMETER	(NSEC) MIN MAX
T_{WC}	WRITE CYCLE TIME	170
T_{CS_W}	CHIP SELECT TO END OF WRITE	150
T_{AS}	ADDRESS SETUP TO WRITE	20
T_{WR}	WRITE RECOVERY TIME (ADDRESS HOLD)	0
T_{WP}	WRITE PULSE WIDTH	150
T_{DS}	DATA SETUP TIME	150
T_{DH}	DATA HOLD TIME	0
T_{OE_Z}	OUTPUT DISABLE SETUP TIME	20

An alternative approach to handling the common I/O data bus contention problem is illustrated by the INTEL 2111A 256 × 4 static RAM, which includes an Output Disable signal to provide more direct control over the direction of data flow on the common I/O pins. The write cycle for the 2111A is illustrated in Figure 12.37; note that control of the common I/O data pins is now the function of the Output Disable signal.

One final design consideration that pertains in particular to the embedded (on-chip) RAM circuit design is the possibility of a *multiport* access, the ability to read from more than one address in the same cycle.

Summary

Alternative RAM design approaches have unique advantages and disadvantages with respect to density, cost, power dissipation, performance, ease of interfacing, complexity of internal circuit design, pin count, and reliability (susceptibility to data misreads). Some of these RAM design options are the following:

1. Static versus dynamic (external refresh) versus quasi-static (internal refresh).
2. Fully decoded address inputs (2^n word lines) versus separate row and column decoders (multiple bit lines per data bit).
3. Latched versus buffered data outputs.
4. Single array versus split subarrays (split around sense amplifiers and column decoders).
5. Differential versus single-ended signal sensing.
6. Common I/O versus separate data input/data output lines.
7. Bidirectional Output Enable data bus control.
8. Single versus multiport addressing.
9. Access time and cycle time performance specifications, specifically:
   a. Read and write cycles
   b. Data/address setup and hold times
   c. Refresh cycles and refresh timing overhead
   d. Access time relative to $\overline{RAS}$, $\overline{CAS}$, and $\overline{CS}$ input signals
   e. Multiplexed address inputs

Many of these different design options are obviously very closely related. The design approach(es) best suited for a particular application should hopefully become evident after the following discussion on static and dynamic memory cell designs and peripheral (support) circuit design.

Before describing these circuits, one final comment should be made about the physical design constraints on the circuit design and layout of an embedded RAM design. Various RAM designs use different interconnect layers for implementing the word lines and bit lines running orthogonally throughout the array; the wiring algorithm for routing global nets must input the macro blockage map in preparation for global wiring. Although the RAM macro's utilization of second-level metal may be sparse, there may be wiring blockage considerations with regard to the additional capacitance of a second-level metal ground plane or capacitive coupling from global switching second metal interconnections; large area blockage maps may require manual macro placement to avoid wireability problems. In addition, the macro may require unique power supply distribution; the array circuit design and read/write operations may be sensitive to transients and/or dc voltage drops on the VDD and ground supply lines (which would result from other circuits connected to the same rails). Separate supply rails, not shared with any other circuits, may be required; separate *chip pads* (VDD and ground for the RAM only) may likewise be used. The embedded RAM should be able to be sufficiently isolated from the remainder of the chip for more efficient testing after fabrication (by enabling the tester to directly access the RAM macro input/output signals). A multitude of testing strategies can be used for the RAM macro. What kind of pattern sensitivity is to be expected in the array physical design? To what extent will the RAM be sensitive to the sequence of 1s and 0s written to and read from a storage location or perhaps to the pattern of 1s and 0s written to and read from *adjacent* storage locations? These considerations and questions are an integral part of the development of a RAM macro for a design system application.

12.5 STATIC RAM DESIGN

A fully static memory cell is a bistable circuit design, capable of being driven into one of two states. After the driving stimulus is removed, the circuit must remain in the applied state, usually held by the feedback from output to input. The two most frequent static cell implementations are the cross-coupled inverter and the Schmitt trigger. (*n*MOS technology circuit examples will be given.)

Cross-coupled Inverter Pair

The *n*-channel MOS enhancement + depletion implementation of the cross-coupled inverter pair is commonly known as the *six-device* cell. Figure 12.38 illustrates this implementation. This cell design has several interesting features:

(1) This design allows for the array configuration illustrated in block diagram form earlier in Figure 12.27, orthogonal bit lines and word lines with a large number of memory locations dotted onto the bit lines and a wide number of locations accessed in parallel by the same word line. For a polysilicon gate FET technology, the two orthogonal interconnection layers used for the word line and bit line are commonly polysilicon and metal, respectively.

(2) The *pass* transistors connect and disconnect the memory location from access (read or write) to the bit lines and allow for cells to be dotted together. When the word line is driven high, the cell is being accessed; one and only one

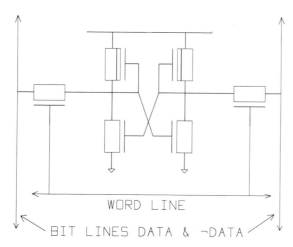

Figure 12.38. Six-device static RAM cell design using *n*-channel enhancement- and depletion-mode devices. The bit lines and word line are illustrated as being orthogonal, as in the block diagram of Figure 12.27.

WORD LINE

BIT LINES DATA & ¬DATA

word line should ever be active at any time (for each pair of bit lines). Since a pass transistor is used, the maximum voltage read from or written to the cell of Figure 12.38 is $\max(VDD, V_{\text{word line}} - V_T)$. If the word line voltage is not bootstrapped above VDD, the ratio of device ratios in the cell must be modified accordingly to provide a good down level with the reduced write input voltage. If a bootstrapped word line is used, then some additional design constraints arise:

(a) The bootstrapped word line voltage must be below the thick (field) oxide threshold voltage so as to avoid turning on any parasitic devices.

(b) To accommodate the bootstrapped driver's precharge portion of the cycle, the memory cycle time will differ from the access time. Consecutive read operations cannot be initiated by simply changing the input address. Specific relative timings between the various control signals ($R/\overline{W}$, addresses, $\overline{CS}$, OE, etc.) will now be required.

(3) The six-device cell illustrated in Figure 12.38 is a relatively high power dissipation implementation; one of the depletion-mode load devices is always dissipating power, since one-half of the cross-coupled circuit is always on. (CMOS static RAM implementations dissipate significantly less internal array power.)

(4) Since both polarities of the stored data value are available, *two* bit lines are routed along the memory cell, one representing the true and the other the complement of the stored valued. When writing into the cell, driving the inputs to the cell in opposite directions from *both* sides will flip the state of the latch faster, reducing the write cycle time. Likewise, when reading from the cell, driving the bit lines in opposite directions from the cell allows a differential sense amplifier to be easily incorporated to boost the transition rates of the two bit lines.

The cross-coupled inverter pair (with dotted word line transistors) also makes a multiport access feature possible. By adding another pair of word line transistors (and separate sets of bit lines and word lines), another I/O port can be provided. A multiport circuit design is shown in Figure 12.39. Some additional system and/or circuit level design constraints need to be imposed regarding any attempt to access the same cell locations through more than one port.

An alternative for the depletion-mode load transistors in the six-device static RAM cell can potentially provide greater density (less cell area) at the expense of additional process complexity. In a conventional polysilicon-gate technology, the polysilicon is heavily doped with impurities to reduce its resistivity and improve its attractiveness as an interconnect material. [The polysilicon is a deposited layer produced from the thermal decomposition of a gas containing silicon (typically silane, SiH_4) and the subsequent deposition and crystallization of silicon onto the wafer surface. As with the single crystal substrate, the resistivity of the material is reduced by the addition of substitutional impurities into the crystalline material.] If a source of impurities is omitted from the polysilicon deposition step, the resulting layer is a relatively *high* resistivity material. If part of the polysilicon layer can be masked off from a subsequent doping step, it is possible for the

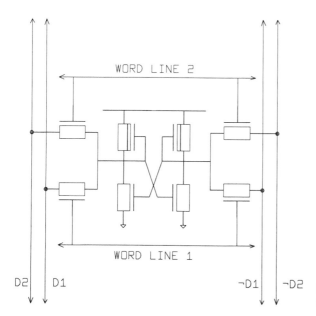

Figure 12.39.   Two-port static RAM cell design.

circuit designer to use the polysilicon layer in two fashions: (1) as a low-resistivity interconnect and gate material (where the impurities were introduced locally), and (2) as a high-resistivity layer that can be patterned to make very high value resistors for load devices and leakage pull-ups (where the impurities were masked off). Whereas the sheet resistivities of polysilicon interconnect layers are typically in the neighborhood of 20 to 30 ohms (Ω) per square, sheet resistivities of 10 MΩ per square are feasible for the high-resistivity layer. A considerably lower power implementation of the cross-coupled inverter pair can be realized by replacing the depletion-mode load devices with polysilicon resistors (Figure 12.40). With a

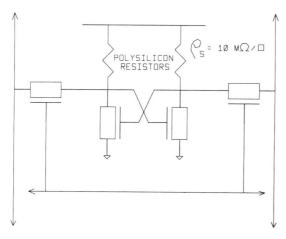

Figure 12.40.   Low-power static RAM cell implementation using high-resistivity polysilicon load resistors.

large load resistance, the gain of the inverters is quite high. An input voltage only millivolts above the device input threshold voltage is required to provide a sufficiently low output voltage to hold off the other inverter (thereby retaining data). This feature permits a very low power battery backup/standby mode of operation where the supply voltage to the cell could be reduced from +5 V to say +2 V in standby mode.

The high sheet resistivity polysilicon resistors provide an additional design constraint in that they can provide only leakage currents (on the order of nano-amperes) and *not* a capacitive charging current. It is therefore necessary to fully precharge the bit lines to a high voltage and then selectively discharge one or the other when the cell is accessed during a read cycle. The cycle time now becomes longer than the access time in order that the bit lines can be precharged between accesses.

Schmitt-trigger Memory Cell

In addition to the bistable cross-coupled inverter pair, another circuit design that can provide two stable states is the Schmitt trigger; the particular implementation of a Schmitt-trigger memory cell to be described is illustrated in Figure 12.41, as presented in reference 12.8.

The two stable states of this memory cell circuit design can be illustrated as follows: note that most proofs of this sort assume the circuit can exist in either one of the two states and indeed that both of those states are stable. Referring to Figure 12.41, first assume that the voltage at node *A* is low. DC current flows in the left half of the circuit only through devices *Q*1, *Q*2, and *Q*4. The device sizes must be designed so that the majority of the supply voltage is dropped across device *Q*1 (providing a low voltage at node *A* as assumed); the device width-to-length ratios given in the figure are from the referenced paper and indicate that,

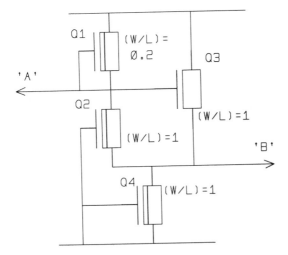

Figure 12.41. Static Schmitt-trigger memory cell (from reference 12.8). Note that *Q*1, *Q*2, and *Q*4 are all depletion-mode devices.

in fact, the impedance of $Q1$ is significantly higher due to its smaller width-to-length ratio. Device $Q1$ limits the current and therefore the power dissipation of the cell in this state.

Second, assume that the voltage at node A is high (equal to the supply voltage). Device $Q3$ is on, and the voltage at node B dropped across device $Q4$ with current flowing through $Q3$ and $Q4$ is designed to be large enough so that the gate-to-source voltage of device $Q2$ (equal to $-V_B$) is more negative than the threshold voltage of device $Q2$; in other words, if V_B is sufficiently large, then device $Q2$ will be off. If device $Q2$ is off, load device $Q1$ will charge node A to the supply voltage as assumed. Current flows in the right side of the circuit only, through $Q3$ and $Q4$. At this qualitative level, the circuit design seems relatively straightforward; however, some very interesting features deserve mention, particularly to illustrate the insightful means that the authors used to describe the circuit's behavior and to demonstrate the importance of thorough simulation.

Consider the two experiments illustrated in Figure 12.42 to determine the circuit's external behavior. These experiments involve application of a dc voltage source to either node A or node B and measuring (from a simulation) the input current drawn from the external source. Two fundamental criteria can be applied to this approach:

1. When the external input current is *zero*, the applied voltage (be it to either node A or B) represents one of the stable voltage states. In other words, when the circuit is in one of its stable states, the dc current flowing in the circuit is at equilibrium; no current is required from the external supply. (Any external current disrupts this equilibrium.) Since this circuit has two stable states, there are three points on the graph of $I_{ext,\,B}$ versus V_B (or $I_{ext,\,A}$ versus V_A) where the source current is equal to zero, the two stable node voltages and the switch point voltage.

2. Since the plot $I_{ext,\,B}$ versus V_B (or $I_{ext,\,A}$ versus V_A) has three zeros (and since the graph is a continuous function), the graph must have local extrema between the zeros. The magnitude of each extremum is an indication of the stability of the particular state; the larger the amount of current required to initiate a transition from a stable state (through the switch point to the other stable state), the *less* likely the probability that a switching transient of short energy or duration will disrupt the stable state.

The static-state plots of $I_{ext,\,B}$ versus V_B (and $I_{ext,\,A}$ versus V_A) for the previously described Schmitt-trigger circuit are sketched in Figure 12.43. The curves indicate the current flowing from the external source if the applied external voltage differs from one of the stable voltage levels. If the external current is positive, this implies that the memory circuit will sink additional current in an attempt to increase the voltage drop across the impedance of the external source and restore the node voltage to a stable value. If the external current is negative, the memory circuit is sourcing current to the external source to reduce the voltage

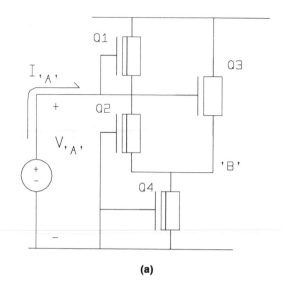

(a)

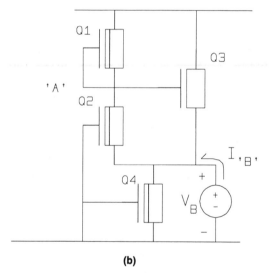

(b)

Figure 12.42. Simulation experiments to measure external $V-I$ characteristics (external perturbation) at the cell's two access nodes. (a) External $V-I$ characteristics at node A. (b) External $V-I$ characteristics at node B.

drop across the source impedance and likewise to return to a stable node voltage value. In other words, with this external current the circuit is tending to force itself back toward one of the stable voltages. The greater the current, the greater the stability of the state (i.e., the faster a displacement from the stable state will return to equilibrium). Past the switch point (the intermediate zero of the I versus V curve), the source current changes sign and the circuit tends to force itself to the opposite state. Comparing the magnitudes of the external currents, it is evident

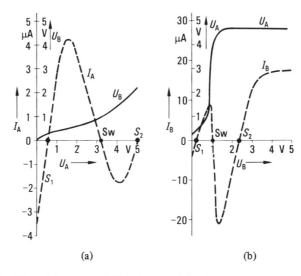

Figure 12.43. Plot of the external *V–I* characteristics from the circuits in Figure 12.42. (From L. Schrader and G. Meusberger, "A New Circuit Configuration for a Static Memory Cell with an Area of 880 μm^2." *IEEE Journal of Solid-State Circuits*, Vol. SC-13, No. 3 (June 1978), p. 346. Copyright © 1978 by IEEE. Reprinted with permission)

that the circuit opposes deviations from its stable voltage states more strongly if node *B* is driven than if node *A* is driven.

Between the two steady-state voltages at nodes *A* and *B* (V_A, V_B = 0.6, 0.2 and V_A, V_B = 5.0, 2.3) is the third zero current point (V_A, V_B = 3.2 V, 0.95 V). This third, or *metastable*, state represents a potential circuit malfunction; the outputs of the cell are not at valid logic levels (see Section 14.7 for a discussion of metastable circuit behavior).

Another means of illustrating circuit behavior, in addition to the static-state diagram, is the *dynamic-state* diagram. If external voltages are applied simultaneously to both nodes *A* and *B* and then disconnected, leaving the two nodes floating, the dynamic-state diagram plots the trajectory of the coordinate pair (V_A, V_B) as these voltages return to their stable states with time. Figure 12.44 depicts the dynamic-state diagram for the Schmitt-trigger circuit; each circle on a trajectory curve represents an equal interval of circuit simulation time. There are several points to note about the dynamic-state diagram (all of which are consistent with the static-state diagram presented earlier):

1. Below the dashed line (called the *separatrix* line), the node voltages will move along a trajectory toward the lower stable state. Likewise, above the separatrix line, the trajectory terminates at the upper stable state.
2. Each trajectory moves quickly toward a single line; these trajectories describe an initial rapid change in V_B with a smaller change in V_A. After reach-

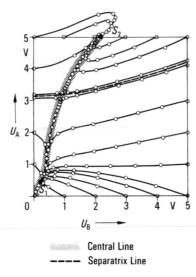

Figure 12.44. Dynamic state diagram for the Schmitt-trigger memory cell. (From L. Schrader and G. Meusberger, "A New Circuit Configuration for a Static Memory Cell with an Area of 880 μm^2." *IEEE Journal of Solid-State Circuits,* Vol. SC-13, No. 3 (June 1978), p. 346. Copyright © 1978 by IEEE. Reprinted with permission)

ing the common central line, the time rate of change of the voltage V_B decreases, limited by the time rate of change of voltage V_A. This is consistent with the static-state diagram shown earlier; the current I_B available to force voltage V_B to its proper stable state is much larger in magnitude than I_A. This smaller current value (and the larger node capacitance at node A) limits the speed with which a stable state is reached.

The Schmitt-trigger circuit differs from a conventional bistable circuit, like the cross-coupled inverter pair circuit described previously, in the following ways:

1. Nodes A and B are *both* high or low simultaneously, rather than at opposite logical values.

2. The dynamic-state diagram for this Schmitt-trigger memory cell differs greatly from that of a symmetric bistable circuit in that the separatrix line is almost horizontal, rather than at a 45° angle, as would be expected for the cross-coupled inverter pair.

3. There are perhaps some unexpected trajectories in the dynamic state diagram, particularly for some of the trajectories above the separatrix line. A few trajectories to the left of the central line overshoot *VDD* at node A, while some to the right of the central line indicate a significant decrease in V_A initially, followed by an increase in V_A as the stable point is approached. Recall that V_B is changing rapidly initially due to the large current I_B. These trajectories indicate that when V_B decreases quickly V_A also drops, and when V_B increases, V_A increases (overshoots); this would tend to indicate that some capacitive coupling between nodes A and B is responsible. Recall that, at the high stable state, device $Q2$ is held off by the voltage at node B. A

valid model for the circuit in this state is shown in Figure 12.45, with the gate-to-source (coupling) capacitance between nodes A and B drawn explicitly. However, when approaching the other (low) stable state, this coupling is *not* observed, since device $Q2$ is on, effectively shunting any capacitive coupling.

Since the two node voltages A and B are simultaneously at the same logic state, a different peripheral memory circuit design is required than that for the cross-coupled inverter pair. The cross-coupled inverter pair can access two bit lines, which are driven in opposite directions by the data stored in the cell during a read cycle. In this Schmitt-trigger design, only one bit line can be used; the option exists to access the bit line from either node A or node B (Figure 12.46).

To write into the memory cell, the bit line is driven to the appropriate value and the word line is raised. To read from the cell, the bit line is precharged to a reference voltage and then left floating (disconnected) prior to raising the word line. The operation of the circuit differs whether the bit line is connected to node A or node B, as described next.

Write Operation: Node A.

When writing a 0 into the memory cell (Figure 12.47), the voltage level of node A falls quickly; device $Q3$ turns off, and the capacitance at node B discharges rapidly through $Q2$ and $Q4$. When writing a 1 into the cell (Figure 12.48), node A initially charges quickly; the device currents through $Q2$, $Q3$, and $Q4$ all increase. As node B charges, the current through device $Q2$ decreases, and eventually $Q2$ turns off. If the word line voltage is not bootstrapped above the supply voltage, the access transistor turns off, leaving only the high-impedance load device $Q1$ to continue to pull node A to the supply. Node B follows this transition as device $Q3$ acts as a source follower for node A.

Write Operation: Node B.

When writing a 0 into the cell using an access transistor connected to node B, the device holding the bit line at 0 must sink the discharge current from node B and node A and the load device current from device

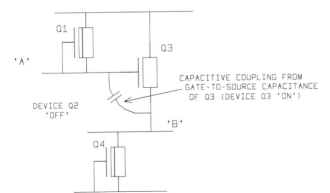

Figure 12.45. Circuit model for the Schmitt-trigger memory cell in its high stable state, used to illustrate the capacitive coupling in the dynamic state diagram of Figure 12.44.

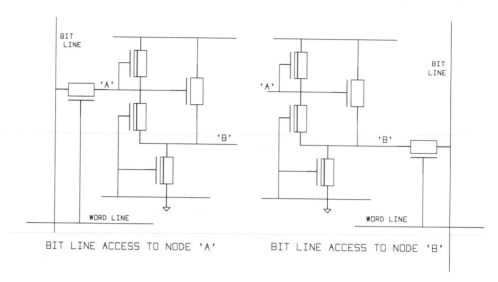

Figure 12.46. Schematic diagrams illustrating the two possible modes of access to the Schmitt-trigger memory cell.

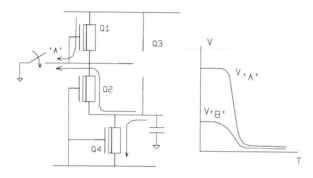

Figure 12.47. Writing a 0 into the Schmitt-trigger memory cell using access at node *A*.

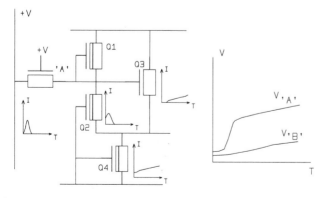

Figure 12.48. Writing a 1 into the Schmitt-trigger memory cell using access at node *A*.

$Q1$ (Figure 12.49). While node B falls rapidly, node A slews more slowly. When writing a 1 to the cell (Figure 12.50), with $+V$ on the bit line, node B charges quickly. Device $Q2$ conducts (in the direction shown in the figure) to charge node A quickly to the magnitude of the threshold voltage of $Q2$ ($V_{GS,Q2} = -V_A$), whereupon device $Q2$ turns off. Now the node capacitance at A is charged through load device $Q1$ only, and A rises at a much reduced rate.

In both cases, writing a 1 into the cell requires a sufficient time for high-impedance load device $Q1$ to charge node A. This affects the memory *cycle* time more than the memory write pulse width, since after the voltage at node A crosses the separatrix voltage, the circuit will continue to force itself to the proper state.

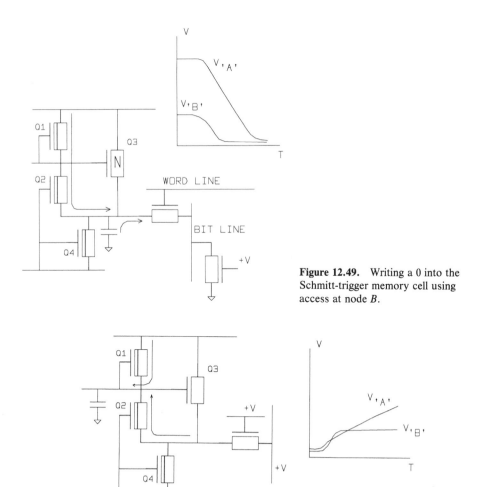

Figure 12.49.   Writing a 0 into the Schmitt-trigger memory cell using access at node B.

Figure 12.50.   Writing a 1 into the Schmitt-trigger memory cell using access at node B.

Read Operation.   When reading from the cell, the current required for the cell circuit to source or sink to the bit line capacitance *must be less* than the extremum values for the two states, as determined from the steady-state diagram. The bit line reference voltage (to which the bit line is precharged) must be chosen to keep these current values within acceptable limits or the operation of reading from the cell may cause the circuit to switch to the opposite state.

Node A.   When reading a value of 1, the circuit of Figure 12.51 applies. Any load current through device $Q1$ to the load capacitance connected to the access transistor will cause V_A to fall; to keep the (V_A, I_A) values within acceptable limits, an approximate bit line capacitance value is calculated and an appropriate *minimum* precharge voltage is determined.

When reading a value of 0, the circuitry of Figure 12.52 pertains. Again, a reference voltage (*maximum*) must be calculated (with the assumed bit line capacitance) to keep the necessary discharge current below unstable limits. This reference voltage must be greater than the previously calculated minimum reference voltage so that a suitable design *window* is present (Figure 12.53).

Node B.   When reading a value of 0, the circuit model of Figure 12.54 should be used for analysis. Again, the discharge current must be below the static diagram limits, which for node B are greater than for node A.

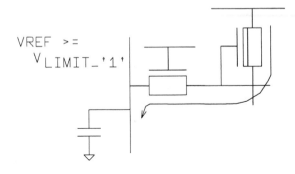

Figure 12.51.   Reading a 1 at node A; V_{ref} is the bit line precharge voltage.

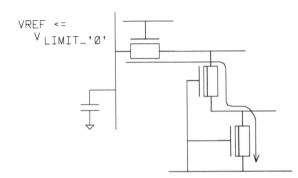

Figure 12.52.   Reading a 0 at node A.

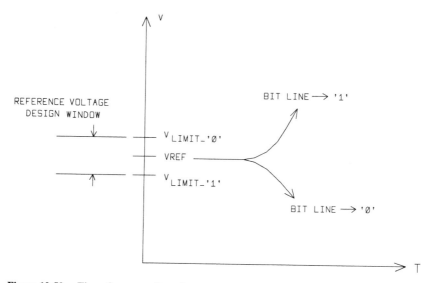

Figure 12.53.   The reference voltage for access at node *A* must be chosen so that it satisfies the 0 and 1 read current restrictions. The cell design must present a suitable design window for the reference voltage.

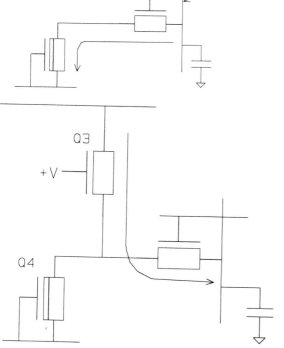

Figure 12.54.   Reading a 0 using access at node *B*.

Figure 12.55.   Reading a 1 using access at node *B*.

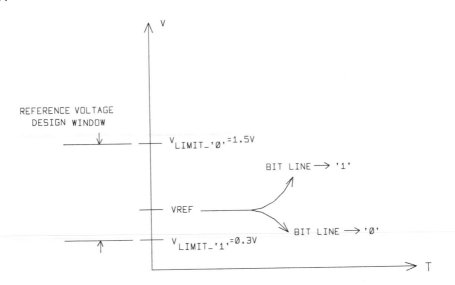

Figure 12.56. Reference voltage design window for accessing the cell at node *B*.

When reading a 1, the enhancement device *Q*3 provides the capacitive charging current (Figure 12.55). The resulting reference voltage window using node *B* as the access node is quite wide (Figure 12.56), making the node *B* access configuration the more attractive of the two options.

Summary

In summary, the design of a static RAM macro requires implementation decisions, detailed circuit design and analysis, and tight (aggressive) physical layout to minimize the cell area and increase the overall macro density. The typical implementation decisions were discussed in Section 12.4. The macro circuit design and analysis involves the following:

1. Choice of cell implementation and type of load device.
2. Developing device design dimensions [constrained by power dissipation, bit line (external) current limits, and cell area].
3. Detailed capacitance modeling.
4. Selecting the mode of the cell access to the bit line(s): access transistors, number of ports, bootstrapped word lines.
5. Design and analysis of precharge and sense amplifier circuitry.

The physical layout phase requires particular resourcefulness to optimally achieve the following:

1. Distribute the orthogonal bit and word lines in a manner so as to minimize the cell area and line capacitance.

2. Utilize the available interconnection levels for word lines, bit lines, and power supply distribution.

3. Exploit the large symmetry of the array by sharing power supply contacts, bit line contacts, and word line contacts between adjacent cells to enhance circuit density.

These design and analysis tasks, and several others, are also part of dynamic RAM array macro development, to be discussed next.

12.6 DYNAMIC RAM DESIGN

By far, the most common static RAM cell design is the cross-coupled inverter six-device cell design described in Section 12.5. This section describes several dynamic RAM cell designs, modifications to the six-device cell design that address trade-offs in density and power/performance. These circuit designs utilize the charge stored on a capacitive node to represent a stored logical value. Each node also has a reverse-biased junction leakage current that modifies the magnitude of the charge stored with time. Before the magnitude of the stored charge has changed sufficiently (so that the stored value can still be accurately and reliably sensed and amplified during a read cycle), it is necessary to replenish the lost charge. Denoted as a *refresh cycle,* the procedure for replenishing the lost charge is to perform an artificial read/sense/amplify/rewrite cycle to *each* array location periodically. Many array locations are commonly refreshed in parallel (simultaneously) to reduce the overall impact on the system memory availability. As a result, the clocking requirements for a dynamic RAM design are considerably more complicated than is typically required for a static RAM design. Part of that additional complexity will be evident throughout this section and likewise in Section 12.8 on circuit designs for functions peripheral to the memory array.

Six-device Clocked-static Cell Design

Although using the same number of devices as the *n*MOS static memory cell and therefore *not* a great savings in cell area, a reduction in power dissipation is possible with the clocked-static six-device cell. The *n*-channel depletion-mode load devices in the static circuit design are replaced by enhancement-mode load devices, whose gate inputs are connected to a read/refresh clock line. These enhancement-mode load devices are usually held off by this clock line and are only occasionally (necessarily) pulsed on. A circuit schematic for the clocked-static, six-device cell is given in Figure 12.57.

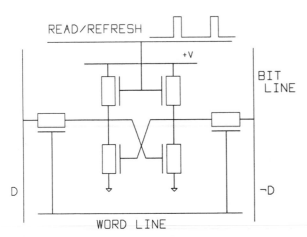

READ/REFRESH

+V

BIT
LINE

D

¬D

WORD LINE

Figure 12.57. Six-device clocked-static memory cell design.

When the load devices are off, the cell dissipates no dc power. As was discussed in Section 12.5, the bit lines must be precharged to the appropriate reference voltage inside the design window such that the access currents do not upset the state of the latch during a read operation. This requires modeling the cell, the transfer devices, and the bit line capacitances accurately; it also involves some design constraints regarding the operating word line and read/refresh line voltages.

The clocked-static memory cell is really a dynamic design; the information is retained dynamically as the charge stored on the capacitance of the off node of the cross-coupled devices (Figure 12.58). It is reasonable to assume that the gate input leakage of a device is negligible; however, since the nodes of the cross-coupled devices are partially diffused areas, the reverse-biased pn^+ junction leakage current eventually results in the loss of the stored information. As a numerical example, assume the following physical and electrical parameters to be representative of the circuit in Figure 12.58:

1. Diffusion node area: 240 μm^2 (20 μm × 12 μm)
2. Junction capacitance: $C_j = C_{area} + C_{perim}$

$$C_{area} = 0.1 \text{ fF}/(V_{reverse} + 0.6 \text{ V})^{1/2} \text{ per } \mu m^2$$

$$C_{perim} = 0.3 \text{ fF}/(V_{reverse} + 0.5 \text{ V})^{1/3} \text{ per } \mu m$$

3. Device dimensions (wafer level): $W = 12 \ \mu m$, $L = 2 \ \mu m$
4. Oxide thickness: $t_{ox} = 300 \text{ Å}$

$$C_{\text{gate input, device on}} = \frac{\epsilon_{ox} * \text{area}}{t_{ox}} = 0.03 \text{ pF}$$

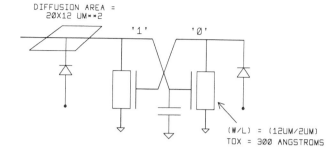

DIFFUSION AREA =
  20X12 UM**2

(W/L) = (12UM/2UM)
TOX = 300 ANGSTROMS

Figure 12.58. Storage node capacitance for the six-device memory cell of Figure 12.57, used to calculate the cell's dynamic behavior.

Assuming the reverse bias voltage across the diffusion node junction is roughly $+5$ V,

$$C_{\text{junction}} = C_{\text{area}} + C_{\text{perim}} = 0.02 \text{ pF}$$

$$C_{\text{total}} = C_{\text{junction}} + C_{\text{gate}} = 0.05 \text{ pF}$$

$$Q_{\text{total}} = 0.05 \text{ pF} * 5 \text{ V} = 0.25 \text{ pC} = 1.5 \times 10^6 \ e^-$$

For the junction leakage current at room temperature, the following coefficients will be assumed:

$$\text{Area:} \quad 0.25 \text{ fA}/\mu\text{m}^2; \quad \text{Perimeter:} \quad 4.0 \text{ fA}/\mu\text{m}$$

$$I_{\text{leak}} = 0.3 \text{ pA}, \qquad C_{\text{total}} = 0.05 \text{ pF}$$

$$\frac{dV}{dt} = \frac{I_{\text{leak}}}{C} = (150 \text{ msec/V})^{-1}$$

Typical dynamic RAM refresh requirements are specified to be in the range of 2 to 4 ms for a cell location. This specification seems extremely conservative when compared against the dV/dt result calculated previously. The significant functional dependence that was not accounted for in the calculation is that the junction leakage current is an exponential function of temperature; at 85°C (a common upper junction temperature limit for commercial hardware), the junction leakage currents *increase* by a factor of 60 over the room temperature values (area and perimeter). The new dV/dt value is now less than $(3 \text{ ms/V})^{-1}$, consistent with the value specified for the required cell refresh time.

While the load devices of the clocked-static six-device cell are on, the cell does dissipate power; however, since the width of the pulse used to refresh the nodes is considerably less than the time between refreshes, duty cycles on the order of 10^{-4} can be used when calculating the total power dissipation.

Four-device Dynamic RAM Cell

If the enhancement-mode load devices are removed from the six-device clocked-static cell, the four-device dynamic RAM cell results (Figure 12.59). As with the previous design, this cell is dynamic, relying on the charge stored on the two cell

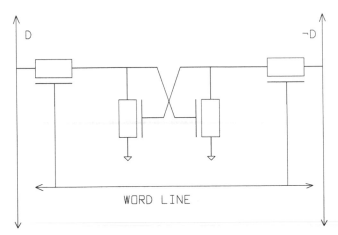

WORD LINE

Figure 12.59. Four-device dynamic RAM cell design.

nodes. Refreshing the data stored in the cell is performed by exercising a *dummy* read cycle; when the word line is raised during a refresh or read cycle, the data on the bit lines is sensed, amplified, and rewritten back into the cell. The amplification and rewrite restores the *full* charge to the node storing a logical 1. The benefits of this design over the six-device cell arise from the elimination of the two load devices, a power supply line, and the read/refresh signal line (representing a reduction of approximately one-third to two-thirds of the cell area), at the expense of the additional refresh cycle generation circuitry.

Three-device Dynamic RAM Cell

Since the means for storing information in the array of a dynamic RAM is primarily charge storage on a capacitive node in the cell, it is *not* necessary to have two devices in the cell in a latch configuration, one storing the true data value and the other its complement. Eliminating one of the cross-coupled devices in the four-device cell results in the three-device cell (Figure 12.60).

The cell is written into by placing the appropriate data value on the data bit line and raising the write select line. The stored data value is retained as charge on capacitance C_{node} after the write select is removed. In reading the cell, the read select line is raised; the read select device and the storage device are effectively in series. The storage device is on or off, depending on the presence or absence of charge at the storage node. The read line will have been precharged; if the data stored (on C_{node}) is a 1, the discharge of the read line results from both devices being on. Note that the complement of the data stored is sensed on the read bit line. Refreshing the cell is performed by reading the data stored in the cell (its complement, actually) and then subsequently placing the appropriate value on the write line and raising the write select line (Figure 12.61).

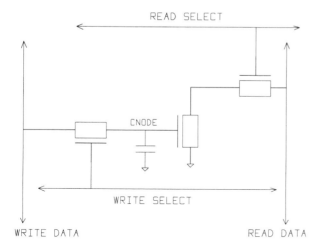

Figure 12.60. Three-device dynamic RAM cell design, removing the cross-coupled device of the previous examples.

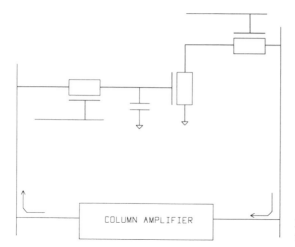

Figure 12.61. Refreshing scheme for the three-device dynamic memory cell.

Several features of the three-device cell should be noted:

1. The area of the cell is reduced considerably since fewer devices are required. In addition, removing the cross-coupled device eliminates one of the major cell design constraints; the cross-coupled device must be sufficiently ratioed (relative to the word line device and the load device) so that when the on device is conducting (read, write, or refresh) a 0 output level is maintained sufficient to hold the other device off. In the three-device cell (and the others to be introduced shortly), the constraints on the cell device design include storage capacitance, read current, amplifier sensitivity, and others, but not

a ratio to another device size. Depending on the targeted performance, these ratioless circuits can often use devices of much smaller dimensions.

2. Unlike the single-device cell design yet to be discussed, reading from the three-device cell does not involve any charge transfer from the storage node to the read bit line.

3. Writing into the three-device cell requires changing only the voltage at a single node; cross-coupled designs require a discharge transient through the on device *and* the (overlapping, but nevertheless still sequential) pull-up transient. Reading from the three-device cell presents a design problem similar to that of any NAND/AND type circuit; if the storage device is off (data stored = 0) and the intermediate node between the two devices in series has been previously discharged, this intermediate node must be charged at the beginning of the read cycle (Figure 12.62). This charging current for the intermediate node should *not* inadvertently be interpreted as a stored cell value of 1.

4. There is a major design consideration relative to the word select voltage used to enable data to be written or refreshed into the cell. In this three-device cell design, rather than using the stored charge at a node to directly represent a data value, the node voltage is used to control the operation of a device. The 1 input voltage to that device is given by $\max[V_{\text{precharge,data line}}, (V_{\text{word select}} - V_T)]$. To maximize the 1 storage node voltage and thereby reduce the on resistance of the storage device, the word select voltage is typically bootstrapped.

Other configurations of the three-device cell that attempt to simplify the cell design at the expense of additional clocking complexity are possible. One possible simplification is to combine the two separate bit lines into a single bit line (Figure 12.63). The read and write cycles can proceed much like before. However, since the data value read from the storage location is actually the complement of the

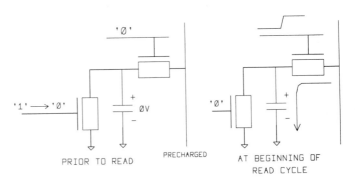

Figure 12.62. Intermediate (NAND) node charging current in the read cycle of the three-device memory cell.

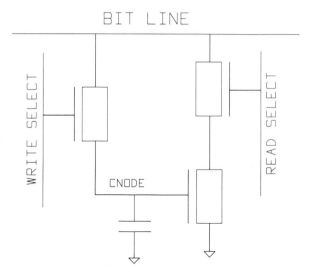

Figure 12.63. Modified three-device memory cell with a common bit line for reading and writing into the cell.

stored data bit, the read/amplify/write refresh cycle must be altered considerably; the bit line must be driven to both data values (and the read select and write select signals must *not* overlap) during a single refresh cycle.

An alternative scheme for reducing the complexity of the three-device cell is to provide both the read select and write select signals on a single line (Figure 12.64). The operation of this configuration is similar to the original three-device cell, except for the following difference: when reading from the storage location, it is mandatory that an immediate write into the cell from the data bit line not occur. Only after the read bit line has been sensed, inverted, and amplified, and the data bit line properly conditioned should a rewrite/refresh into the storage location occur. It would initially appear that using a common select line for both the read select and write select devices imposes an impossible set of constraints to satisfy. However, by providing *two* voltage levels on the select line (a read select voltage and a subsequent write select voltage), proper operation can be

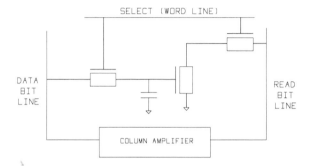

Figure 12.64. Modified three-device memory cell with a common (read and write) word select line.

maintained. To better analyze the circuit operation, consider the following two examples:

1. Storage node = 1; write data bit line precharged to +V (Figure 12.65):
   a. Since $V_{\text{data bit line}} = V_{\text{storage node}}$, no write contention problem is present. To turn the write select device on requires a (word line) select voltage greater than $V_{\text{storage node}} + V_T$.
   b. The read select device will turn on for a select input voltage greater than $V_{T,n}$ (since the intermediate node has been discharged to 0 V).
2. Storage node = 0; write data bit line precharged to +V (Figure 12.66):
   a. If the write select device input is above V_T, the write select device will conduct. However, to effectively change the state of the data bit stored in the cell it is necessary that the enhancement storage device go from a nonconducting to a conducting condition; this requires a write select voltage of at least $2 * V_T$.

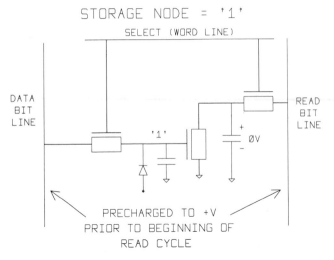

Figure 12.65. Schematic for the analysis of the common select line three-device memory cell when the storage node = 1.

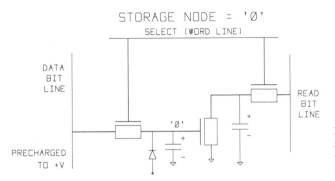

Figure 12.66. Schematic for the analysis of the common select line three-device memory cell when the storage node = 0.

b. If the intermediate node is charged to VDD, the read select device will be off for a read select voltage up to $VDD + V_T$; the read bit line correctly remains at $+V$. Even if the intermediate node has been previously discharged, the read select device current will go to zero (allowing proper sensing on the read bit line) when the intermediate node voltage charges to $V_{\text{read select}} - V_T$.

Points a and b together imply that a read select voltage between V_T and 2 $* V_T$ could be used at the beginning of a read/refresh cycle without a write contention problem (and still provide a proper value on the read bit line). For this particular memory cell to function properly, an intermediate read/refresh select voltage is used at the beginning of a read/refresh cycle; after the write data bit has been appropriately set by the column amplifier, the bootstrapped select voltage is applied to ensure proper full writing into the cell (Figure 12.67).

Single-device Dynamic RAM Cell

Undoubtedly, the most pervasive dynamic RAM cell design among commercially available parts is the single-device cell. It provides a high density, low cost per bit memory at moderate performance levels. The remainder of this section will discuss the single-device cell in detail, illustrating a variety of layout and technology options for its implementation. A good summary of single-device cell design approaches is presented in reference 12.9, from which some of the following layouts and cross sections are used for illustration.

The single-device cell consists of a transistor (an enhancement-mode device) and a capacitor to store either the presence or absence of charge, representing the two logical states. If one node of the storage capacitor is a diffused region (electrically equivalent to one of the nodes of the device), then the storage capacitor is comprised of two components, an oxide capacitance and a junction capacitance, effectively in parallel. This situation is illustrated schematically in Figure 12.68. The oxide capacitance is not a function of voltage, only layout geometry and topography; the junction capacitance is a function of voltage. The oxide capacitance is by design typically an order of magnitude larger than the junction capacitance.

Depending on the process technology options, the storage node can be implemented in a variety of ways: as a diffusion node, as an inversion layer (induced

Figure 12.67.   Word line select waveform for the common select line three-device cell.

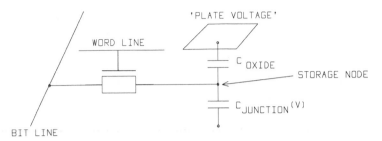

Figure 12.68. One-device cell dynamic RAM.

or implanted), or as a polysilicon plate. These three different approaches are given in cross section in Figure 12.69.[2]

The single-device cell benefits in density from the reduced number of transistors, control lines, *and* contact holes. Its primary disadvantages are related to the means of charge storage and transfer:

1. Unlike the three-transistor cell, which relies on charge storage on a gate capacitance, a separate capacitive node must be included in the design and physical layout.

2. The readout of the data (charge) stored in a cell relies on the charge transfer from the storage node capacitance to the bit line capacitance, rather than controlling the operation of a device, as in the three-device cell. As a result, the readout of the data stored is *destructive*; the amount of charge transferred into or out of the cell disrupts the original quantity of charge stored. In addition, since the cell storage capacitance is typically but a small fraction of the total bit line capacitance, the voltage swing on the bit line when a cell is read is typically small; very sensitive read/refresh amplifiers are required for each bit line. This is illustrated schematically in Figure 12.70. The transfer gate is diagrammed as a switch that is closed when the word line is raised; during a read cycle, there is a transfer of charge between the bit line and the storage node. The direction of charge transfer depends on the difference between the bit line precharge voltage and the cell storage capacitance voltage. The magnitude of the voltage swing is given by the following expression:

$$\Delta v_{\text{bit line}} = \frac{C_{\text{node}}}{C_{\text{node}} + n * C_{\text{bit line}}} * (V_{\text{node}} - V_{\text{precharge}})$$

where $C_{\text{bit line}}$ is the bit line capacitance per row of cells, and n is the number of rows (i.e., the number of dotted cells on the same bit line).

The ratio $C_{\text{node}}/(C_{\text{node}} + n * C_{\text{bit line}})$ is called the *charge transfer ratio* and typically ranges from 5% to 20%. The design and layout of the storage cell are

---

[2] More recent storage node implementations utilize three-dimensional topography, or *trench* capacitor cross sections, in order to try to maximize array density.

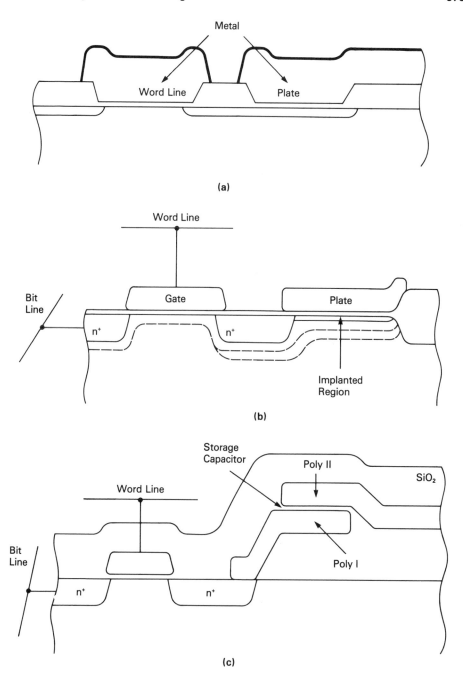

Figure 12.69. Three possible implementations of the storage node plate. (a) Diffusion node (metal gate technology). (b) Implanted or induced inversion layer (polysilicon gate technology). (c) A polysilicon plate (two polysilicon interconnection planes are utilized).

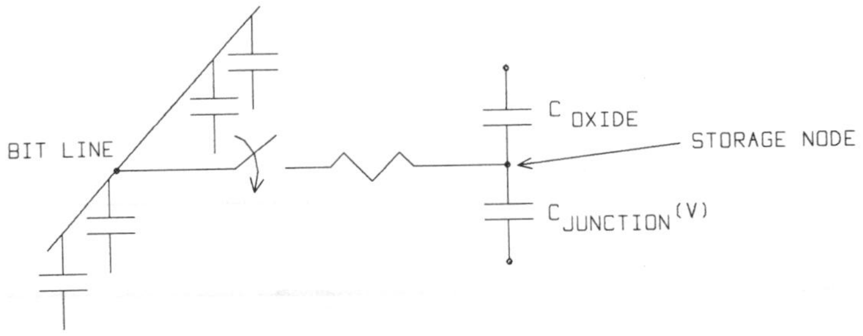

$$\triangle V_{BIT\_LINE} = \left(\frac{C_{NODE}}{C_{NODE} + N*C_{BIT\_LINE}} \right) * (V_{NODE} - V_{BIT\_LINE})_{PRECHARGE}$$

Figure 12.70. Charge transfer between the bit line and the accessed cell; the charge transfer ratio indicates the magnitude of the voltage swing on the bit line after access.

driven to minimize cell area, while at the same time maximizing C_{node} (and minimizing $C_{bit\ line}$). Ideally, the parameter n should be as large as possible (i.e., as large as the sense amplifier sensitivity will allow). Another point to note is the representation of the bit line capacitance in Figure 12.70; in actuality, the bit line is best modeled as a distributed *RC* network. The position of the accessed row in the array will affect the measured transient on the bit line at the sense amplifier input; any clocking requirements internal to the sense amplifier should account for differences between accessing the closest versus the most distant row.

The differences in various single-device cell implementations relate directly to the differences in MOS technologies, particularly in the number and capacitance of different interconnect layers. These different implementations depend to a large degree on the availability in the fabrication process of zero, one, two, or three polysilicon layers; one or two metal layers; enhancement only, enhancement + depletion, or enhancement + depletion + zero (natural) threshold voltage devices; and polysilicon-to-diffusion buried contacts.[3] Increasing the manufacturing complexity, although more costly, can facilitate the design of storage node structures with increased capacitance per unit area; this enables a smaller cell area to be used, which therefore provides increased density, a smaller die size, and increased yield, resulting in an overall lower cost (outweighing the increased process complexity). The single-device cell examples to be presented follow this evolution of fabrication processes toward the ever-present goal of increased array density, while maintaining sufficient charge transfer to ensure reliable sensing of stored values.

[3] A zero or natural threshold voltage device does not receive an enhancement threshold-shifting impurity implant, and thus will have a threshold voltage near zero.

Single-device Cell Types and Layouts

(I) Metal-gate MOS technology

Metal Gate Word-Line Device/Metal Storage Plate/Diffused Bit Line/Diffused Storage Node.   The original single-device cell, as proposed by Dennard in 1968 (reference 12.10), used the given combination of interconnect levels to implement the memory array in a simple four-mask process (diffusion, gate oxide, contact, metal) using planar field oxide to isolate adjacent cells (Figure 12.71). One advantage of this particular combination is its simplicity; no contact holes are required in the layout. The disadvantages with this particular implementation are as follows:

1. The diffused bit line has a large junction capacitance and series resistance.

2. The metal gate process, not being a self-aligned process, typically has a large gate-to-source and gate-to-drain overlap capacitance, increasing the word line-to-cell transient coupling and overall load capacitance on the word line driver.

3. The metal word line and metal plate for the storage capacitor must be routed so as not to intersect; the two lines must pass alongside one another and therefore will consume additional array area.

(II) Polysilicon-gate MOS technology

The vast majority of dynamic RAM designs include polycrystalline silicon as one of the interconnection layers. Including a polycrystalline layer in the processing technology has the distinct advantage that the MOS device in the dynamic RAM cell is self-aligned (i.e., the gate-to-source/drain overlap capacitances are significantly reduced by using polysilicon as the gate material). Figure 12.72 presents a cross section and layout of a single polysilicon layer process with diffused bit line and metal word line (connected by a contact to the polysilicon gate, not shown in the figure). There are several important features to note about this figure:

1. The surface (field oxide) is not planar, but rather is recessed into the original silicon crystal substrate. A single shape on a recessed oxide mask level distinguishes the recessed field oxide regions from the planar (device + storage node + n^+ node) regions. The area of the surface that results in n^+ regions is the area that lies inside the recessed oxide shape but outside any polysilicon shape(s) that intersect(s) the recessed oxide shape. (The polysilicon effectively blocks the implantation of n-type dopant impurities into the crystal; such is the nature of the self-alignment of the source and drain nodes to the gate material.)

2. A dielectric insulator is present between the polysilicon layer and the metal word line that crosses over to electrically isolate the lines. This dielectric

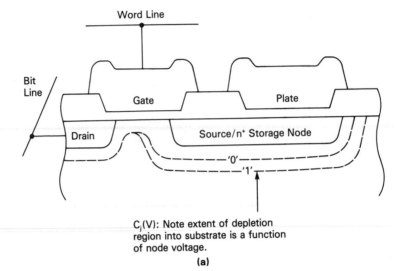

C$_j$(V): Note extent of depletion
region into substrate is a function
of node voltage.

(a)

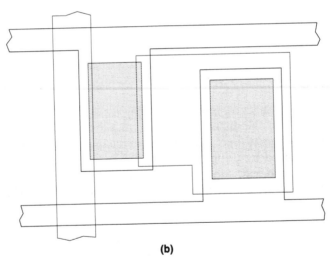

(b)

Figure 12.71.   (a) Cross section and (b) layout of a metal-gate MOS technology single-device cell.

is typically silicon dioxide, usually a combination of thermally grown oxide and deposited silicon dioxide.

3. Without a separate, additional implant step in the fabrication sequence between the recessed oxide and polysilicon masks, the overlap of polysilicon and recessed oxide shapes inhibits the addition of n-type impurities into the substrate; an MOS device channel, rather than an n^+ diffusion node, is formed. For the storage node of the cell in Figure 12.72, the inversion layer

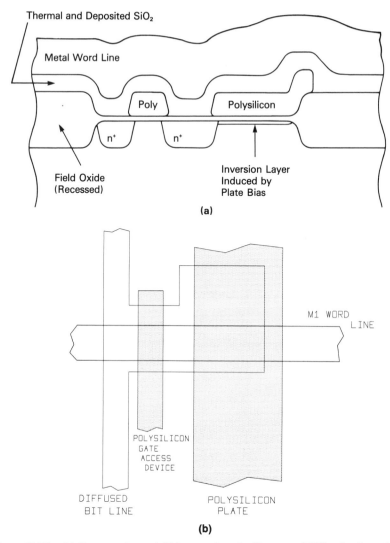

Figure 12.72.   (a) Cross section and (b) layout of a polysilicon gate MOS technology single-device cell (uses polysilicon gate access device and polysilicon upper storage plate).

for the lower plate is *induced* by biasing the polysilicon plate with a large positive voltage, significantly larger than the enhancement device threshold voltage. If a 0 is written into this cell by the presence of a low voltage on the bit line, the storage node inversion layer is likewise taken to a low voltage. The capacitor comprised of the polysilicon plate and the storage node is charged; the potential well of the storage node is filled with electrons (Figure 12.73a). If a 1 is written into the cell, with a high voltage on the bit line, electrons are removed from the induced inversion layer (current flows

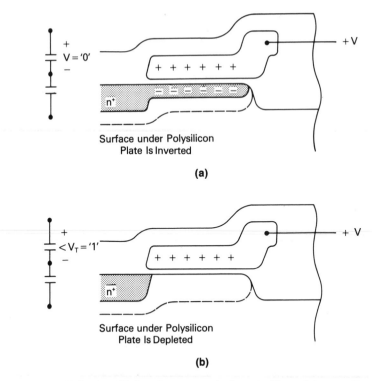

Figure 12.73. Behavior of induced storage node for polysilicon plate capacitor. (a) Data value = 0, device formed by polysilicon upper plate is inverted. (b) Data value = 1, device formed by polysilicon upper plate is depleted.

into the storage node). The voltage across the storage capacitor is now *reduced*; without an additional depletion-mode implant, the surface will be depleted (Figure 12.73b). The potential difference across the storage capacitor nodes is greater when storing a 0 than when storing a 1.

4. The connection between a word line and a polysilicon gate is made using a contact hole etched in the oxide layer over the polysilicon (prior to metal deposition and patterning). The word line contact for this cell design is shown in Figure 12.74, which depicts the layout for two adjacent memory cells. In particular, note the following:

   a. The metal word line that crosses over a cell is *not* the word line for that cell.

   b. A single polysilicon plate is used for many cells.

   c. The layout of each cell is identical; an adjacent cell results from a mirroring operation about both the *x*- and *y*-axes.

   d. The additional area required for the contact is a significant fraction of the total cell area.

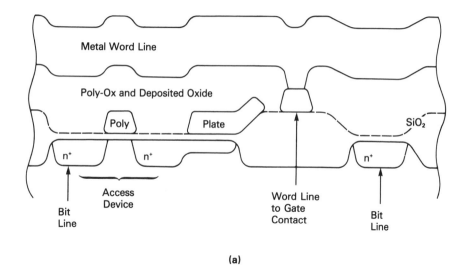

(a)

(b)

Figure 12.74. (a) Cross section and (b) layout of two adjacent memory cells of Figure 12.72, illustrating placement of word line to polysilicon gate contact.

Polysilicon Word Line, Metal Bit Line. Another possible implementation, without any additional masking or interconnect layers, is to use a polysilicon word line and a metal bit line. The metal bit line essentially stitches together the n^+ nodes of the access devices to be dotted. The characteristics of this implementation are as follows:

1. The RC time constant of the bit line has been reduced considerably over the diffused bit line implementation; however, this is at the expense of an increased $R * C$ time constant for the polysilicon word line (due to the relatively large sheet resistivity of the polysilicon layer, typically 20–30 Ω per square). Current processing enhancements under development are attempting to reduce the polysilicon resistivity by forming a metal silicide (low resistivity) material to enhance the word line performance substantially.

2. The surface area of the bit line *diffused* nodes has been reduced considerably, reducing the junction leakage current and bit line precharge voltage droop. However, no change is implemented in the characteristics of the storage node.

3. The pitch between bit lines can be varied considerably, depending on the orientation of the layout. The bit lines must be spaced sufficiently in order that the sense amplifier circuit may be placed between adjacent bit lines (i.e., the sense amplifier layout must fit the bit line pitch). Two layouts for the metal bit line, polysilicon word line cell are presented in Figures 12.75 and 12.76; in particular, note the allowed pitch for the sense amplifier layout. In the preferred layout of Figure 12.76, a contact is shared between adjacent cells in the same column; reducing the number of contact holes leads to a substantial array area savings.

Twin Cells.   Recall from the discussion on the six-device cell static RAM that two bit lines were routed for each column of cells, one carrying the true and the other the complement of the stored data value. The advantage of providing both values was primarily that the sense amplifier design was simplified. (The sense amplifier, being a differential amplifier, requires the generation and application of a reference voltage to the other side of the amplifier if only a single bit line is present.) There are other process and design advantages in having two bit lines available, advantages strong enough to merit the commercial design of twin cells.

Twin cells are exactly as their name implies, two single-device cells per memory array location, storing the true and complemented value of the data bit. Both cells use the same word line, but two separate bit lines are routed in tandem to the edge of the array and the sense amplifier; both bit lines connect to opposite inputs of the same sense amplifier.

Note that the use of two cells side by side and the routing of two bit lines changes some of the constraints on the array configuration; commonly, single-device-cell RAMs use a split array and shared sense amplifiers, with the sense amplifiers between the two array halves. Instead, with the return of two bit lines per cell, the sense amplifiers are returned to the edge of the array. In addition, the problem of fitting the sense amplifier to the array is relaxed by the increase in effective bit line pitch.

There are other advantages in using twin cells, some of which are process related (and which apply to the six-device cell as well):

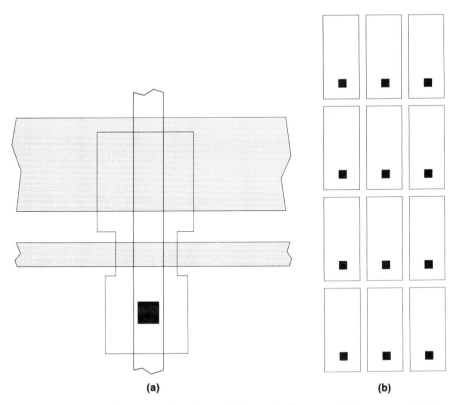

(a)   **(b)**

Figure 12.75.   Possible orientation of metal bit line, polysilicon word line memory cell. (a) Layout. (b) Array configuration. This very narrow cell may present problems in fitting the array sense amplifier circuitry into the tight metal bit line pitch.

1. The close proximity of the two cells implies that electrically the devices and storage node capacitances will track very closely. (Recall that parameter tracking implies that the mismatch tolerance in the value of a particular process parameter increases radially from any point on the chip. To reduce the variation between two devices or circuits, they should be located in close proximity and in identical orientation to one another.) The process parameters (threshold voltage, oxide thickness, impurity distribution, carrier mobility, etc.) are nearly identical for the twin adjacent cells. In the usual single-device cell implementation using a split array, a cell in one-half of the array is compared against a reference (or dummy) cell in the other half during the sensing portion of the read/refresh cycle. The possible variation in the process parameters between two devices or storage nodes in opposite halves of the array can be significant and must be accounted for in such a design.

2. In addition to the device electrical parameters mentioned previously, the masking alignment and process biases for adjacent devices should track very

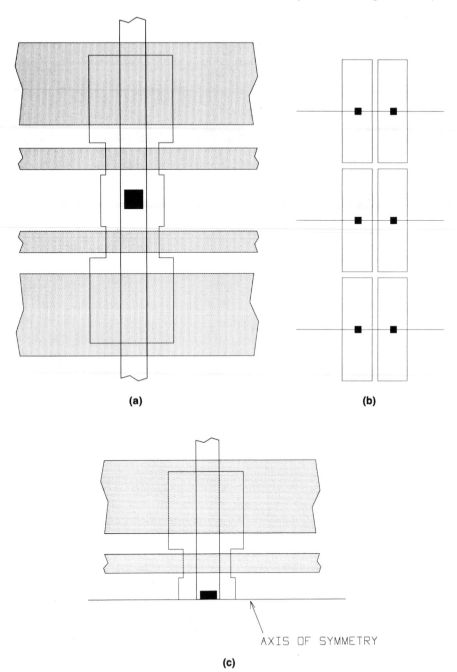

(a)   **(b)**

AXIS OF SYMMETRY

(c)

Figure 12.76.   Preferred layout of metal bit line, polysilicon word line memory cell. This orientation permits the metal bit line contact to be shared between adjacent cells. (a) Layout. (b) Array configuration. (c) A single cell, illustrating the design axis of symmetry.

closely. Specifically, the process bias of the recessed oxide mask dictates the final device width (and the storage node capacitance); the process bias of the polysilicon shapes dictates the final device length (and similarly, the storage node capacitance). The mask level-to-mask level alignment has a *strong* effect on the value of the storage node capacitance (Figure 12.77). It is evident from the figure that twin cells should be oriented in the same direction in order that the physical characteristics be as close as possible.

3. It was mentioned previously that the absolute location of the memory cell and its distance to the sense amplifier can have a significant effect on the overall design; this is particularly true of implementations that use a diffused bit line, which has a significant (distributed) $R * C$ time constant. In the twin-cell design approach, the two cells are an equal distance away from the sense amplifier along the bit line. This reduces the susceptibility to noise on the bit lines since a noise source will typically have an equal effect at the adjacent cell locations. The use of a differential sense amplifier, with

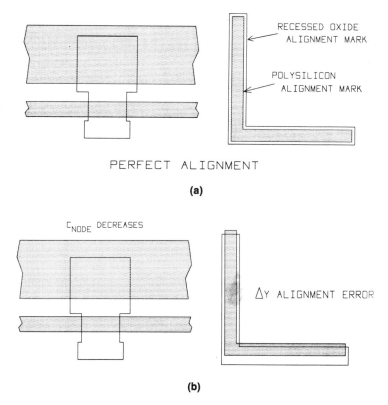

Figure 12.77.   Effects of polysilicon-to-ROX node misalignment on the storage node capacitance. (a) 'Perfect' alignment. (b) Polysilicon-to-diffusion node alignment error. The storage node capacitance of twin cells will track closely.

inputs driven in opposite directions, will provide for good common-mode signal rejection.

The obvious disadvantage of the twin-cell approach is the increase in array area per data bit by essentially a factor of 2. However, this does not imply that the total chip or macro area increases by the same factor; typically, the array area is roughly 50% of the total macro area, the remainder being occupied by sense amplifiers, clock generators, drivers/receivers, and address (row and column) decoders and line drivers. Therefore, the increase in chip or macro area may be more on the order of 25%. In addition, if the complexity of the peripheral circuits can be reduced, additional area may be saved.

Figure 12.78 illustrates the use of the metal bit line, polysilicon word line layout implemented as twin cells. Problem 12.8 requests that the single-cell array technique of *folded* bit lines be researched, a technique that shares many of the advantages of twin cells without the additional array area.

Extra Masking Operations. The attractiveness of MOS technology is due to the few masking steps required; few masks imply reduced processing costs and potentially higher yield (assuming defect density increases with a larger number of masking layers). Additional steps and the increased cost of processing may be economically justifiable if the circuit density and/or performance enhancements merit. Examples follow of additional masking operations that may be included.

Field Implant. Surface leakage currents can be reduced by locally increasing the background impurity concentration under the field oxide (between adjacent cells).

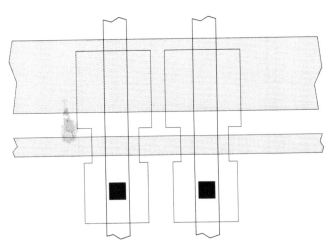

Figure 12.78. Metal bit line, polysilicon word line twin-cell layout.

Additional Polysilicon Layers.   There may be significant density and performance advantages in including an additional polysilicon layer; a second polysilicon layer can be used as a word line or a plate line.

Buried Contacts.   A significant density advantage can be provided by including a contact level between the n-type polysilicon line and the n^+-type source and drain diffusion nodes. Since the polysilicon is reoxidized prior to metallization, the direct polysilicon-to-diffusion contact is usually referred to as a *buried* contact. Buried contacts are very common in the implementation of n-channel depletion-mode logic and peripheral circuits, since a large number of the internal circuit connections can be made without blocking global metal interconnections.

Depletion-Mode Device Implant.   In a polysilicon gate technology, the polysilicon layer is deposited and patterned prior to the incorporation of impurities for the n^+ source and drain regions. The remaining polysilicon acts to block the addition of n-type impurities; as a result, the polysilicon plate does *not* have an n-type diffusion node underneath. As mentioned earlier, an inverted region must be induced by biasing the plate to a large positive voltage. Alternatively, an implant step may be used to locally introduce n-type impurities into the substrate prior to the deposition of polysilicon, to create an inverted layer underneath what will become the polysilicon plate. Figure 12.79 illustrates the layout of a polysilicon plate cell that includes a depletion-mode implant.

Additional single-device cell implementations with extra masking operations

Polysilicon Bit Line, Polysilicon Plate, Metal Word Line (with Polysilicon Gate).   The metal word line, diffused bit line cell (using polysilicon gate MOS devices) can be modified to utilize a polysilicon bit line instead, if buried contacts between the polysilicon bit line and the diffused node of the MOS device are available. If the per unit length capacitance of the minimum-width polysilicon line is less than that of a diffused junction, replacing the diffused bit line with a polysilicon one may reduce the bit line capacitance and improve the charge transfer ratio. A layout of the polysilicon bit line memory cell is shown in Figure 12.80. There are several disadvantages to the polysilicon bit line approach, which have been strong enough to make this approach commercially unattractive. The most significant disadvantages are the following:

1. The process design rule that determines the spacing between the bit line and the access transistor has changed for the worse. The diffused line to polysilicon gate spacing (for the diffused bit line cell) is typically less than the polysilicon to polysilicon line spacing now required for the polysilicon bit line cell (Figure 12.81).

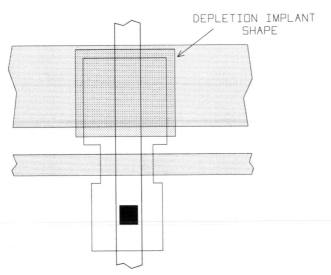

Figure 12.79. Layout of a polysilicon gate technology cell design with an implanted lower plate, introduced during the depletion-mode implant step.

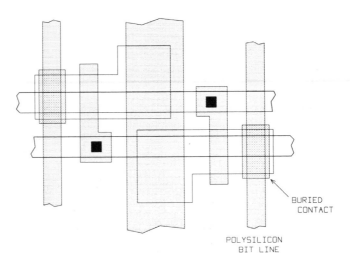

Figure 12.80. Polysilicon bit line, polysilicon gate and metal word line single-device cell layout; the polysilicon bit line uses a buried contact to connect the device node to the polysilicon bit line.

2. The buried contact etch bias must be tightly controlled to avoid encroachment on the device channel of the access transistor (Figure 12.82). In addition, since the presence of a polysilicon line blocks the incorporation of *n*-type impurities underneath, the continuity of the polysilicon-to-access device node contact depends on the outdiffusion of *n*-type impurities in the polysilicon into the substrate to form a continuous *n*-type node (in subse-

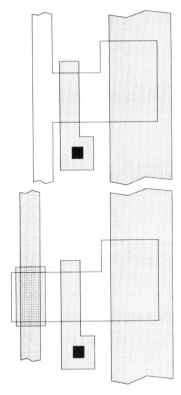

Figure 12.81. Layout comparison between the diffused bit line and polysilicon bit line cell designs; the process design rule for the diffusion to polysilicon space is typically much smaller than the buried contact to polysilicon space.

quent high-temperature processing steps). Due to the characteristics of the materials and the contact formation step, buried contacts tend to be relatively high resistivity contacts and also somewhat nonuniform.

Double polysilicon cells

A significant array density increase can be achieved by including a second polysilicon layer, electrically isolated from the first. A double polysilicon memory cell is illustrated in Figure 12.83. Where the two polysilicon gate layers overlap, a series device channel-to-storage plate connection can be implemented *without* an intervening diffusion node area. A thick dielectric is desirable between the two layers (for reduced capacitance and increased reliability); however, if the dielectric is too thick, the device channels under the two gates will not be continuous. This dielectric layer on top and on the sides of the first polysilicon layer is formed by a thermal oxidation step (which simultaneously provides the gate oxide for the polysilicon-II device). The thickness of the dielectric layer is thus dependent on the parameters of the gate-II oxide and is not independently controlled.

Another characteristic of the double polysilicon cell design in Figure 12.83 is that the length of the access device now depends on an increased number of process parameters; with a single polysilicon technology, the length of a device

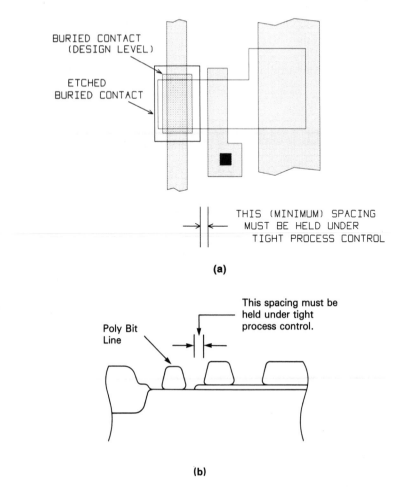

BURIED CONTACT
(DESIGN LEVEL)

ETCHED
BURIED CONTACT

THIS (MINIMUM) SPACING
MUST BE HELD UNDER
TIGHT PROCESS CONTROL

(a)

This spacing must be
held under tight
process control.

Poly Bit
Line

(b)

Figure 12.82. Processing requirement that the etch bias of the buried contact to the polysilicon line be kept under tight process control. The resulting exposed surface from the
etched contact opening must be separated from what will become the channel area for the
access device. (a) Layout, illustrating etch bias on buried contact. (b) Buried contact cross
section, shown after polysilicon deposition and patterning.

depends strictly on the etch bias of the polysilicon line, whereas the length of the
polysilicon-II access transistor is a function of the mask-to-mask alignment of
polysilicon-II to polysilicon-I, the polysilicon-I etch bias, and the polysilicon-II
etch bias.

In addition, the introduction of a second polysilicon layer presents some
potential topography (step coverage) problems. Where the edges of a polysilicon-
I and polysilicon-II shape coincide, a large change in topography results (Figure
12.84; also Figure 12.83b).

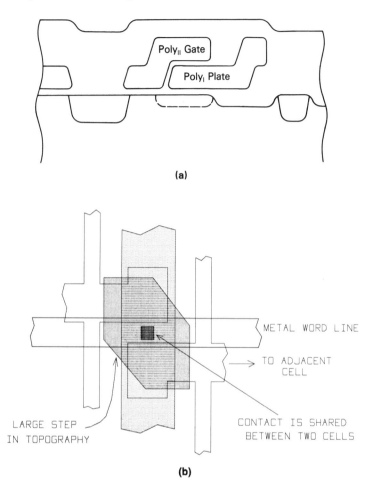

(a)

(b)

Figure 12.83.   Cross section (a) and layout (b) of a double-polysilicon cell design; the second polysilicon level is used as the gate material for the access device, while the first polysilicon layer is the upper capacitor plate. The metal word line to polysilicon II device contact is shared between two cells.

Despite the potential processing problems, it is evident from Figure 12.83 that the cell density can be quite high, particularly if an *interleaved* cell design is used (Figure 12.85). The majority of the area savings comes from the ability to overlap the storage plate (polysilicon-I) with the device gate material (polysilicon-II), eliminating the diffusion node between the device and the storage layer. Additionally, the polysilicon-II shape can be used as the gate input for two adjacent cells (on the same word line) with a single contact, which can be conveniently located over the polysilicon-I storage plate (recall the additional cell area that was required for the metal-to-gate contact in the single polysilicon technology with

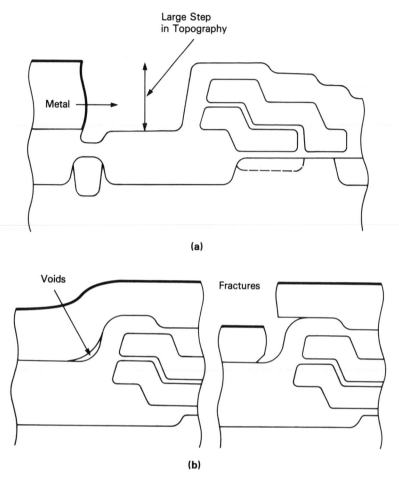

Figure 12.84. Reliability problems (and yield problems) introduced by a large step in wafer topography for metal coverage. (a) Cross section after polysilicon II. (b) Reliability and yield problems with large step coverage.

metal word line and diffused bit line). However, diffused bit line implementations have several disadvantages as compared to metal bit line approaches, which suggests the following alternative cell design.

A metal bit line approach provides perhaps the best combination of cell density, increased charge transfer ratios (reduced bit line capacitance), and simplicity of design. Two possible cell layouts are illustrated in Figure 12.86. (One of the layouts again utilizes an overlap between the polysilicon-II gate and word line and the polysilicon-I plate to eliminate the n^+ node in series with the device and storage capacitor, while the other includes a diffused region between device and storage node.) As before, the contact between metal bit line and the diffused node of the device is shared among two adjacent (mirrored) cells.

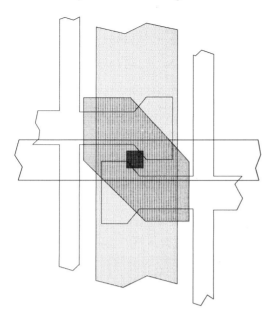

Figure 12.85. Double polysilicon cell with interleaved diffusion storage nodes (similar to Figure 12.83, but with a plate-diffusion area that is interleaved between adjacent cells).

Similarly to polysilicon-I, the second polysilicon layer is a relatively high resistivity material; the distributed $R * C$ time constant of long polysilicon-II word lines may be long, and the design should include the appropriate word line drivers in an attempt to improve the overall performance.

A polysilicon bit line implementation is also feasible, replacing a diffused bit line, if buried contacts are available as a processing option. Note that either polysilicon-I or polysilicon-II may be used as the bit line interconnection layer (Figure 12.87), although there may be an advantage to the choice of implementing the second polysilicon layer as the bit line. If the first polysilicon layer were used for buried contacts, the mask sequence would not facilitate polysilicon-II to polysilicon-I contacts. If the buried contact mask were to be incorporated between polysilicon-I and polysilicon-II deposition and patterning steps, then both polysilicon-II to diffusion and polysilicon-II to polysilicon-I contacts could be realized.

Cell Pitch.   The cell dimensions dictate the bit line and word line pitch; bit line sense amplifiers and word line decoders and drivers must be able to be laid out to fit the pitch dimension. The polysilicon-I word line, metal bit line cell is probably the best example of the constraints the cell design can impose on the peripheral circuits; for this particular example, it is difficult to fit the sense amplifier width to the metal bit line pitch. Several potential options are the following:

1. Put more storage capacitor area between cells, pushing the cells (and the bit lines) farther apart.
2. Stagger the position of the sense amplifiers.

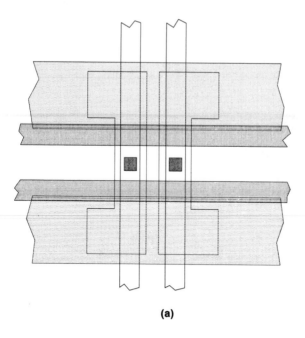

(a)

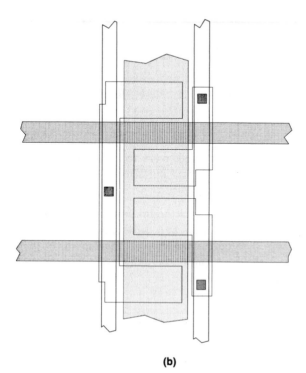

(b)

Figure 12.86. Metal bit line, double polysilicon single-device cell layouts. (a) Polysilicon II to polysilicon I overlap between word line and plate is utilized. (b) Polysilicon I plate is routed orthogonally to the polysilicon II word line.

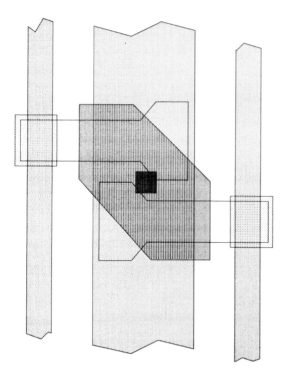

Figure 12.87. Double polysilicon cell with polysilicon bit line. (Polysilicon I is illustrated as the bit line layer; however, polysilicon II could conceivably be used.)

3. Include buried contacts in the sense amplifier layout in order to reduce the need for metal wiring in the peripheral areas.

Cell Orientation and Placement.   The placement and orientation of the memory cell, plus the ability to mirror or interleave cells and share contacts, has a strong effect on the bit line capacitance per cell. In addition, alternating the cell orientation between adjacent bit lines changes the sensitivity to mask-to-mask misalignment; for example, the reference cell on the other side of the sense amplifier should be of the same orientation as the accessed cell in order that mask-to-mask misalignment will track.

Enhanced Processing and Design Features
Doubly Doped or High-*C* Storage Capacitance.   To increase the charge transfer ratio of the cell (without significantly increasing the cell area), it is necessary to increase the storage capacitance. However, the oxide dielectric component of the storage capacitance is essentially fixed by the area and the thickness of the oxide between the storage plate and the storage node. It is possible to increase the second component of the storage node capacitance, the junction capacitance, by locally increasing the substrate background concentration (in the vicinity of the storage node only). This is typically accomplished by the deep implantation of *p*-type impurities under the storage plate area only; the impurities introduced should not affect the threshold voltage of the access transistor. Since the additional

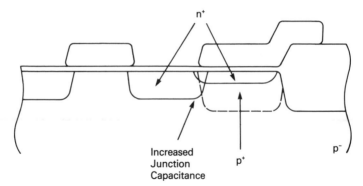

Figure 12.88. Cross section illustrating the high-C storage capacitance cell design (single-polysilicon technology).

p-type impurities raise the threshold voltage of the storage plate gate, either a more positive plate voltage or a shallow *n*-type implant should be included. This high-*C* cell can be used with either single polysilicon or double polysilicon cell designs (Figure 12.88).

12.7 ALPHA PARTICLE INDUCED SOFT ERRORS IN DYNAMIC RAM IMPLEMENTATIONS

When testing a RAM chip or macro, a variety of test approaches can be used to verify the correct operation of each cell in the array; some of the more elaborate test pattern sequences check for a pattern sensitivity. In addition to verifying that into each cell a 0 and 1 can be written and subsequently read, it is also determined if the behavior of the cell is a function of the particular pattern of 1s and 0s stored in adjacent cells. As a result, a variety of test pattern sets are often applied (depending on tester time availability and cost). A list of some of these approaches is given in Figure 12.89 along with a factor indicating the number of patterns required.

A *hard fail* or *hard error* indicates that a particular array location (or a row or column of array locations) repeatedly fails to output the correct data value(s) previously written into the location(s). There are a number of potential sources of hard fails, some of which merit further investigation by the design, process, and test engineering team, as some may be due to causes other than a random physical defect. By and large, the majority of hard fails are random; that is, a random distribution of array (address) locations is hard fails for the failing chip sites across the wafer. Since a relatively large percentage of the area of a dynamic RAM design is gate oxide area, defects in the oxide layer (e.g., pinholes) will contribute substantially to the overall failure rate. Crystalline defects such as *dislocations* and *slip planes* can likewise lead to random hard fails (Figure 12.90).

MEMORY PATTERN TESTING

(N = NUMBER OF BITS)

PATTERN NAME	COMPLEXITY
DIAGONAL	$2 * N$
PARITY	$4 * N$
MARCHING I/O	$8 * N$
WALKING I/O	$2 * N^2$
GALLOPING I/O	$4 * N^2$
WALKING COLUMNS	$2 * N^{3/2}$
MULTIPLE WALKING COLUMNS	$N^{3/2}$
WALKING COMPLEMENTARY COLUMNS	$2 * N^{3/2}$
GALLOPING DIAGONAL	$3 * N^{3/2}$

ADDITIONAL PATTERN TYPES:
CHECKERBOARD, '0' SURROUND, '1' SURROUND

Figure 12.89. Various memory test pattern sequences and their relative complexities. The different test pattern sequences attempt to isolate particular pattern sensitivities in evaluating the RAM. For more information about the specifics of these sequences, consult reference 12.11.

The presence of systematic hard fails, for which an address correlation *can* be found, suggests an alternative failure mode, one that requires a more explicit determination of cause. The first place to look is at the tester itself to verify if any of the following is the culprit:

1. Improper relative timing between applied signals (pulse widths, setup and hold times, etc.)
2. Cabling problem between test card and wafer prober
3. Faulty or intermittent probe pin contact on the chip pads
4. Improper substrate contact
5. Incident light generating excess leakage currents
6. Improper termination on input cables producing ringing at chip inputs (or outputs)

In particular, the mask set used for fabrication should also be inspected for missing shapes and defects.

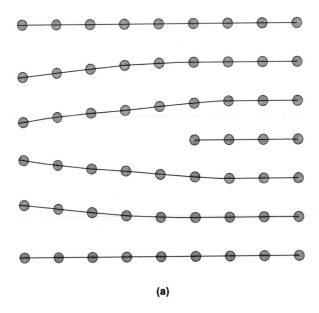

(a)

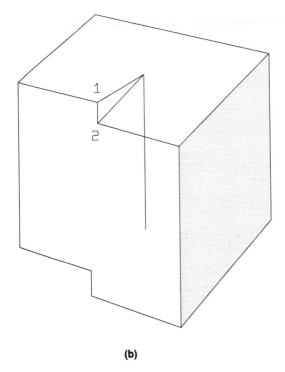

(b)

Figure 12.90. Crystalline defects used to illustrate the source of hard fails in memory array testing. (a) Line dislocation, directed normal to the paper due to the extra half-plane of atoms present. (b) Screw dislocation, about dotted line axis; points 1 and 2 would be coincident if no dislocation were present. The shaded surface area indicates the region of slippage.

A common means to give some test measurement feedback to the design group and to check the adequacy of their design margins is to execute on the tester a "shmoo plot" of the chip functionality versus applied voltage(s). A go/no go measurement of the actual read access time versus the chip specification value is made while the supply and substrate voltages are varied (well beyond their minimum and maximum tolerance values). By knowing over what range of voltage conditions the design succeeds or fails to operate within specification, the design group can rerun their circuit simulation models to verify the degree of model accuracy and perhaps modify the design for a subsequent hardware pass accordingly. An example of a shmoo plot is given in Figure 12.91.

In addition to hard (repeatable) fails measured at wafer level test, *soft fails* or *soft errors* are also a significant problem due to their impact on system reliability. A soft error is a single, nonrecurring read error (on a single bit) of a memory array design; the array location read error is not a permanent error in that no physical process defect is the cause. A subsequent write (then read) cycle into an array location that previously was in error has no greater nor lesser probability of being in error again than any other array location.

The impact of soft errors on system design is tremendous; the system designer must address including parity generation and checking and/or error detecting and correcting codes in the system to maintain a high reliability level at the customer's location (at additional system cost).

In a landmark paper (reference 12.12), a physical failure mode associated with soft errors was identified. (Soft errors can also be caused by supply voltage noise, inadequate design margins, sense amplifier imbalance, etc., but the failure mode to be described is strictly physical in nature and may ultimately curtail the increase in dynamic RAM chip density that has been characteristic of the past years.) As described in detail earlier, dynamic RAMs store a bit of information as the presence or absence of charge (typically electrons, using an *n*-channel device technology) on a node of a storage capacitor. This node can also be pictured

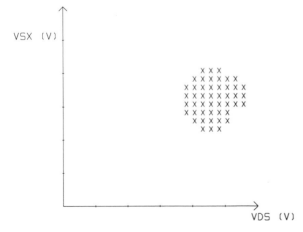

Figure 12.91. Shmoo plot indicating the design margin of the dynamic RAM evaluation over the ranges of applied voltage values. ×s indicate a chip that passes the access time specification for the RAM at the applied voltage conditions.

as a potential well for electrons separated from the *p*-type substrate by a high electric field depletion region. The number of electrons that distinguishes an empty well from a full one (i.e., differentiates between logical 1 and 0) is referred to as the *critical charge*, Q_{crit}, and is usually on the order of 10^6 electrons.

When the temperature is elevated or when light is incident, the generation rate of free hole–electron pairs in the substrate increases significantly. An empty well that fills up with these generated electrons will result in a soft error. These errors can also be caused by incident, collected ionizing radiation. Alpha particles generate free electron–hole pairs as the incident particle transfers energy to the silicon crystal as a result of collisions. Electrons generated in the vicinity of the depletion region between the storage node and the substrate are accelerated by the depletion region electric field and are collected at the storage node. If the fraction collected multiplied by the total number generated exceeds the critical charge, a soft error results.

Alpha particles are He^{2+} nuclei (doubly charged; two protons, two neutrons) emitted from (high atomic number) radioactive elements during decay. Their emitted energy is typically in the range of 8 to 9 MeV. Figure 12.92 illustrates the range of alpha particles in silicon as a function of their incident energy. (The ranges in silicon dioxide and aluminum are comparable in magnitude.) Alpha particles, being charged, interact strongly with the crystalline structure. For example, an incident alpha particle with energy of 4 MeV is traveling at one-twentieth the speed of light and loses energy at the initial rate of 150 keV/μm of travel. The number of free electron–hole pairs generated is determined primarily by the incident energy and the average energy absorbed per pair, which for energetic alpha

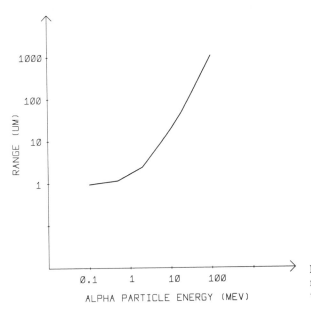

Figure 12.92. Log–log plot of the range of alpha particles in silicon versus alpha particle energy.

particles in silicon has been measured to be 3.6 eV per electron–hole pair. Other pertinent features about alpha-particle radiation are the following:

1. The trajectory of the particle is a straight line.
2. The spread or straggle in distribution of the final stopping point (the range) is small, on the order of 0.02 * range.
3. In the radioactive decay process, alpha particles are emitted at discrete energies. However, alpha particles emitted below the surface of the emitting material lose some of that energy and are effectively incident over a continuum of energies.
4. Alpha-particle emissions are nuclear (probabilistic) events and, as a result, are not a function of environment.
5. The transfer of energy to the silicon crystal (and therefore the generation of electron–hole pairs) is localized to the track of the alpha particle, within 2 to 3 μm.

The number of electron–hole pairs generated and the range of alpha particles have resulted in an increasingly important error mechanism since the critical storage charge and device dimensions have been reduced (in current-generation dynamic memory designs) to comparable values. Other forms of radiation may or may not present a mechanism for soft errors, depending on the energy loss rates in silicon:

Beta: high-speed electron emitted in radioactive decay
Gamma: electromagnetic radiation emitted in radioactive decay (energies comparable to high energy x-rays)

These two types of radiation exhibit low energy loss rates in silicon and are not expected to be a significant source of soft errors.

Cosmic radiation: extraterrestrial radiation consisting of a combination of protons, alpha particles, electrons, and photons that enters the atmosphere and produces secondary radiation of electrons, gamma radiation, and the like

Some evidence exists that at sea level cosmic rays can lead to increased SERs. The vast majority of soft errors can be attributed to alpha particle radiation emitted from trace amounts of radioactive elements in device packaging materials.

Using the energies of alpha particle radiation and the average energy per free electron–hole pair, roughly 2×10^6 electron–hole pairs can be generated per alpha particle. Free electrons and holes generated in the top surface chip passivation layers are not collected; the thickness of these top surface layers is typically on the order of 2 to 4 μm, and therefore only a fraction of the original incident energy is lost. Electron–hole pairs generated in depletion layers separating n^+ diffusion nodes from the substrate are swept by the electric field (electrons into the storage nodes and holes into the substrate). The *collection efficiency* in the depletion region is close to 1 (the ratio of carriers generated to carriers collected). Electrons and holes generated farther down in the bulk silicon diffuse through the substrate. Electrons that reach the edge of the depletion region before recombining are swept into the storage node by the depletion region field. The entire collection process spans a duration of time in the microsecond range. A schematic of the collection process (from reference 12.12) is given in Figure 12.93. (The potential wells are drawn for illustrative purposes only; the actual geometry is an induced inversion and depletion layer.)

The electron collection volume consists of the depletion region plus a minority (electron) carrier diffusion length. The soft error rate is a function of the following:

1. Flux and energy of the incident radiation
2. Collection volume, the collection efficiency, and the recombination rate
3. Q_{crit} of the cell design
4. Cell layout

Experimental Results (from reference 12.12)

Assumptions
1. Errors should be random, single-bit, and nonrecurring.
2. Errors should only occur when an empty storage node collects electrons.
3. The error rate should be proportional to the flux of alpha particles: (alpha particles/area)/time.
4. The error rate is a strong function of the critical charge of the array cell design.

Procedure.   A memory array was written with 1s and 0s in separate halves of the array and then exposed to an intense flux of alpha particles.

Results.   On subsequent memory read cycles, only the half of the array written with 1s (an empty well of electrons) exhibited soft errors. The soft error rate dependence on the alpha particle flux and the critical charge of the dynamic RAM cell design is illustrated in the experimentally measured plot in Figure 12.94. The error rate is linearly dependent on the alpha particle flux over eight decades of alpha particle intensity.

It is also conceivable that other, nonstorage nodes could be subject to soft errors. Bit lines that are precharged and then left floating for tens (or hundreds) of nanoseconds could potentially collect a number of free electrons, disrupting the precharge voltage and potentially leading to a read error. By design, the time

EFFECTS OF AN ALPHA PARTICLE

'0'   '1'

- POTENTIAL WELL FILLED
  WITH ELECTRONS

- P-TYPE SILICON IN
  "INVERSION"

- ~ ONE MILLION ELECTRONS

- POTENTIAL WELL EMPTY

- P-TYPE SILICON IN "DEEP
  DEPLETION"

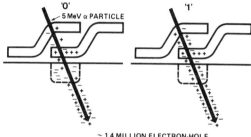

'0'   '1'

5 MeV α PARTICLE

~ 1.4 MILLION ELECTRON-HOLE
PAIRS GENERATED TO A DEPTH
OF ~ 25 μ

- NATURAL ALPHAS UP TO 8 MeV IN ENERGY

- A TYPICAL 5 MeV α: − 25 μ RANGE IN Si
  − 1.4 x 10^6 e-h PAIRS (3.5 eV/e-h PAIR)

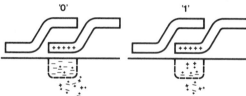

'0'   '1'

- ELECTRON-HOLE PAIRS GENERATED DIFFUSE.

- ELECTRONS REACHING DEPLETION REGION ARE
  SWEPT BY ELECTRIC FIELD INTO WELL. HOLES
  ARE REPELLED.

- "COLLECTION EFFICIENCY" = FRACTION OF
  ELECTRONS COLLECTED.

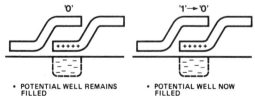

'0'   '1' → '0'

- POTENTIAL WELL REMAINS
  FILLED

- NO APPRECIABLE
  COLLECTION

- POTENTIAL WELL NOW
  FILLED

− IF (COLLECTION EFF.) x (# ELECTRONS GENERATED) > CRITICAL
  CHARGE, A "SOFT ERROR" RESULTS.

− A SINGLE ALPHA CAN CAUSE AN ERROR.

− NO PERMANENT DAMAGE RESULTS.

Figure 12.93. Collection process of free carriers generated by an incident alpha particle. The process illustrates that the cell is sensitive to the alpha particle hit when the storage node is depleted, which is when the cell is storing a 1 for the double polysilicon cell shown. (From T. May and M. Woods, "Alpha-Particle-Induced Soft Errors in Dynamic Memories." *IEEE Transactions on Electron Devices,* Vol. ED-26, No. 1 (January 1979), p. 4. Copyright © 1979 by IEEE. Reprinted with permission)

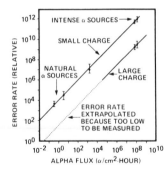

Figure 12.94. Soft error rate (for two test cells with different critical charges) versus alpha particle flux; a linear dependence is indicated over eight decades of alpha particle flux. (From T. May and M. Woods, "Alpha-Particle-Induced Soft Errors in Dynamic Memories." *IEEE Transactions on Electron Devices,* Vol. ED-26, No. 1 (January 1979), p. 4. Copyright © 1979 by IEEE. Reprinted with permission)

duration when the bit line is floating is small, but the collection area is quite large. Depending on the bit line capacitance, the number of collected electrons could put the bit line voltage outside the design margin window. The experimental evidence indicating that the soft errors are specific to an array value of 1, however, would indeed indicate that it is the storage node area that has the greatest sensitivity to alpha particle hits during array operation.

Procedure. The experimental setup to determine that alpha particles (as opposed to other forms of radiation) are indeed the cause of soft errors is depicted in Figure 12.95. Three packaging configurations were used, two with hot lids (trace amounts of radioactive elements), one with a cold lid. One of the hot lids was separated from the chip surface by a piece of 0.005 in. (125 μm) thick masking tape, thick enough to block alpha particles, but not the other forms of radiation.

Results. Only the hot lid *without* the masking tape demonstrated an increased soft error rate.

Note about experimental procedure. Using intense radioactive sources (polonium 210 or americium 241) enhances the soft error rate to accelerate the data-collection process when measuring and comparing the SERs for different package and cell designs, but the emission energies differ from the more common

Figure 12.95. Experimental setup to measure if alpha particles are indeed the source of soft errors; the masking tape separating the hot lid and the chip will block incident alpha particles.

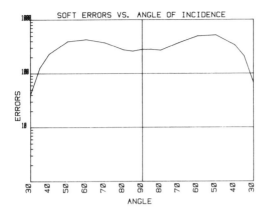

Figure 12.96. Experimentally measured soft error rates versus alpha particle angle of incidence. (From T. May and M. Woods, "Alpha-Particle-Induced Soft Errors in Dynamic Memories." *IEEE Transactions on Electron Devices,* Vol. ED-26, No. 1 (January 1979), p. 4. Copyright © 1979 by IEEE. Reprinted with permission)

radioactive trace elements found in packaging materials, uranium and thorium (which emit alpha particles with slightly larger energies):

$$U^{238} \rightarrow Pb^{206} + 8 \text{ alpha particles}$$

$$Th^{230} \rightarrow Pb^{206} + 6 \text{ alpha particles}$$

with emission energies of 4 to 9 MeV (the chip sees a continuous range of energies). Using an intense flux also increases the probability of a multiple hit at a storage node between refresh cycles. A multiple hit may cause a soft error, whereas a more common single hit may not. After exposure to an intense source for an extended period of time (about 1 hour), a degradation in device characteristics (threshold voltage shifts due to capture of energetic electrons in trap states in the gate oxide) will also increase the measured, experimental failure rate above the naturally occurring value. There is also a dependence on the angle of incidence of the particle; as the angle of incidence varies, so does the depth of penetration and the effective collection area of a storage node.

Figure 12.96 illustrates an experimentally measured soft error rate versus angle of incidence of a collimated alpha-source beam. Of particular importance is the falloff of the soft error rate at acute angles of incidence since some of the most intense alpha particle sources are trace constituents of package sealing materials and thus emit alpha particles at 15° to 30° incident angles.

12.8 PERIPHERAL CIRCUITS FOR DYNAMIC RAM DESIGN

This final section on dynamic RAM design illustrates some of the design alternatives for the circuits peripheral to the memory array, particularly the sense amplifier. The evolution of dynamic RAM design has been fueled by innovation in technology, cell design, and especially the design of the circuits peripheral to the array.

The system designer expects to be able to interface to the RAM with a minimum amount of impact on system timing and with minimal additional hardware required. As a result, memory designs should include the following:

1. Appropriate buffering of address and clock inputs, minimizing the capacitive load presented to the system bus.
2. Internal timing generation of the appropriate sequence of clocking signals.
3. Parallel (or fast serial) byte or nibble mode where a number of stored data values can be accessed in rapid succession (when only a subset of the address inputs are changed).
4. Internally latched address input signals to minimize the overall address valid time interval.

As the amount of stored charge used to distinguish 1 from 0 has been reduced (to accommodate smaller area cells and higher density arrays), the sensitivity required of the sense amplifier has increased. For reliable memory read cycles, the charge transfer to the bit lines must be sufficient to provide a voltage swing on the bit line such that a 1 or 0 can be distinguished at the sense amplifier input. This may require the generation of a reference voltage (swing) at the other of the amplifier's differential inputs.

The most common organization for a dynamic RAM array is to provide a balanced sense amplifier located in the middle of a split array, with memory cells equally divided on either side. The particular implementation of sense amplifier design that is consistent with this organization (and therefore the most common, with several distinguished variations) is the *gated flip-flop*. The advantages of this general class of designs are as follows:

1. Each of the two input nodes of the flip-flop can be connected to half of a bit line, effectively doubling the number of storage array locations per amplifier (consistent with the split memory array configuration).
2. The gated flip-flop can also be viewed as a differential amplifier, with a relatively small offset voltage between 0 and 1. This voltage offset (a strong function of device geometries, threshold voltages, and flip-flop clocking) is further reduced by the close tracking of process and physical parameters between adjacent devices in the amplifier layout.
3. A precharged bit line voltage can easily be provided.

Several commercial implementations of the gated flip-flop sense amplifier circuit design are presented in the following figures. Although all these examples use basically the same regenerative circuit, they differ widely with respect to clocking requirements and complexity, bit line precharge voltage, means of bit line precharge, and reference voltage (swing) generation.

The majority of dynamic RAM designs use a dummy cell (one on both sides of the sense amplifier) to provide a reference voltage swing at the input to the sense amplifier corresponding to the half of the array *not* being addressed during a read access. These dummy cells add two columns to the total array configuration. The design and utilization of these dummy cells differs widely among different sense amplifier designs.

A divided bit line, gated flip-flop sense amplifier is illustrated in Figure 12.97. This sense amplifier implementation operates in the following manner:

When ϕ_P is active: The two bit line halves (A and B) are connected by the cross transistor; the two nodes of the cross-coupled flip-flop reach a common voltage approximately equal to $V_{T,n}$. Simultaneously, a reference voltage is applied directly to the storage capacitance of the two dummy cells, one on both sides of the sense amplifier. During this time, the load clock ϕ_L is inactive (low) and all decoded address and dummy word lines are also low.

Upon initiation of a read access: Initially, clock line ϕ_P is dropped, disconnecting the two bit lines. The address inputs are decoded and a single word line (one out of 2^n for n row address inputs) is raised. The selected cell transfers charge to the bit line corresponding to the stored data value. The dummy word line on the *opposite* side of the flip-flop as the selected cell is raised, whereas the dummy cell on the *same* side of the flip-flop as the selected cell remains isolated; the reference charge is transferred from the dummy cell to the bit line half on the side of the flip-flop opposite to that of the selected cell. The most significant bit of the row address is used to select the appropriate dummy cell. A voltage differential is now present at the two cross-coupled nodes of the flip-flop. The load

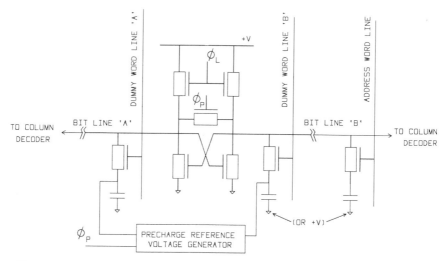

Figure 12.97.   Divided bit line, gated flip-flop sense amplifier; used in Texas Instruments 4K × 1 dynamic RAM (circa 1973).

clock, ϕ_L, is now raised, enabling the two load transistors. Assuming the load devices are balanced, both nodes of the flip-flop rise together until the more positive node begins to turn on the opposite transistor. The regenerative action of the two cross-coupled devices amplifies the oppositely directed signals to full logical 1 and 0 voltage levels, providing the *restoration* of the full charge back into the accessed cell. (Recall that readout from an array location is destructive and that a full restore of the original charge is required.) The selected data bit value passes through its column decoder to the array output.

During a refresh cycle: A refresh cycle is similar to a read cycle in that an entire column of cells is accessed by raising a row address decoded word line, reading from one cell per column and restoring the full amount of charge back into the cell.

During a write cycle: In a write operation, an external data bit value is applied through one of the bidirectional bit line column decoders to the appropriate amplifier; the data and data values drive the amplifier in differential fashion, directing the flip-flop output (and selected storage location) to the proper value. The remaining cells on the raised word line are refreshed. Several features to note about this sense amplifier design are the following:

1. The load clock ϕ_L and the precharge clock ϕ_P are internally generated and must adhere to stringent timing constraints. The precharge clock ϕ_P must be dropped prior to the raising of any decoded address (or dummy) word lines or else some of the subsequent bit line voltage differential will be lost. The load clock ϕ_L must be delayed until all decoded address and dummy word lines have been raised and the appropriate charge transfer has occurred. Nevertheless, the number of clocks and the relatively straightforward timing are attractive features.

2. One major disadvantage of this design is the potentially uncertain precharge voltage of the bit lines. During ϕ_P, the two bit lines are connected and the node voltages reach a common potential approximately equal to $V_{T,n}$, at which point the bit line charging and discharging ceases. Since both cross-coupled enhancement devices are off, the two connected bit line halves are essentially floating. As a result, the common bit line precharge voltage falls due to the junction leakage current of the diffusion nodes on the bit line. The actual starting common-mode bit line voltage when the load devices turn on depends on the amount of leakage current (a strong function of diffusion area and temperature) and the duration of time to the next memory access. The variation in starting voltage varies the amount of time required for the load devices to pull up the two bit line halves to the regenerative region of the flip-flop and, more importantly, increases the sensitivity to any *imbalance* in the pull-up current capabilities of the load devices. Any current imbalance in the two devices could overcome the charge-transfer signal differential given a sufficient common-mode signal transition, leading to a read error.

(3) The design of the dummy cell to facilitate a direct connection to the output of the reference voltage generator requires a significant modification to the design of the array cell. As a result, the calculation of the optimum reference voltage to be applied becomes considerably more involved. Ideally, the resulting voltage of a bit line half when a dummy cell is accessed should be halfway between the voltages provided when a 1 or a 0 is read from an array cell. Referring to the charge transfer equation,

$$\Delta v = \frac{C_{\text{node}}}{C_{\text{node}} + C_{\text{bit line}}} * (V_{\text{node}} - V_{\text{precharge}})$$

where the node subscript refers to either the cell or dummy storage nodes, the determination of the proper $V_{\text{node,dummy}} = V_{\text{reference}}$ is complicated by the fact that the capacitance of the dummy storage node is not equal to the capacitance of the storage node of an array cell, due to the additional area of the dummy storage node contact and the reference generator interconnection capacitance. In addition, the dummy storage node capacitance no longer tracks the cell storage node capacitance very closely and is considerably more sensitive to process variations. Area must also be allocated for the reference voltage generator and the necessary interconnection routing. As a result, designing a reference voltage to provide equal sensitivity in reading either a 1 or a 0 is not straightforward; *very* accurate capacitance and process modeling is required.

An alternative implementation for the memory array sense amplifier is shown in Figure 12.98. The reference voltage generator has been eliminated. The ground connection for the cross-coupled enhancement devices has been replaced by the complement of the load clock. Although the drain nodes of the load devices indicate a connection to *VDD*, connecting the pull-up node to the clock signal ϕ_L is also feasible; indeed, connecting the pull-up node to ϕ_L may be preferable since one less line (*VDD*) needs to be routed to the string of sense amplifiers. The operation of this sense amplifier circuit is as follows (read cycle only; the write cycle is similar to the previous example; refer to the timing diagram in Figure 12.98b):

During ($\phi_P = 1$) * ($\phi_L = 1$).   While the precharge and load clocks are simultaneously active, the circuit schematic of Figure 12.99 applies. The cross transistor is on and *both dummy word lines* are also selected. The cross transistor enables the two bit line halves and the storage node capacitances of the two dummy cells to settle to a voltage equal to the switching point voltage of the symmetric, balanced flip-flop.

Once the precharge voltage has been reached, the flip-flop is shut off by dropping ϕ_P and ϕ_L. The source node of the cross-coupled transistors goes to $\overline{\phi_L}$ = *VDD*. The timing diagram in Figure 12.98b indicates that ϕ_L goes inactive first,

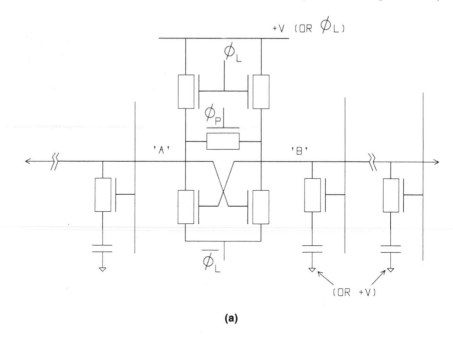

(a)

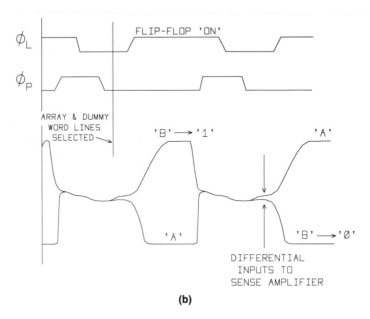

(b)

Figure 12.98.   Implementation of a gated flip-flop sense amplifier, used in Siemens 4K ×
1 dynamic RAM (circa 1972). The dummy cell voltage is developed when ϕ_P is active; it is
equal to the switching voltage of the flip-flop. (a) Circuit schematic. (b) Read cycle timing.

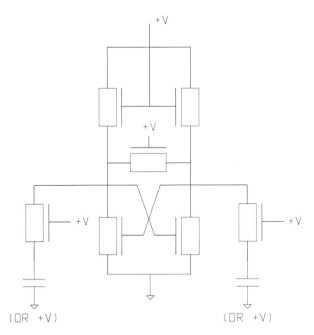

Figure 12.99. Equivalent sense amplifier and dummy cell circuit when clocks ϕ_P and ϕ_I are both active (from Figure 12.98a).

prior to ϕ_P. All the enhancement-mode devices in the sense amplifier are off, and the two nodes are now floating.

The read cycle now proceeds by selecting the decoded address word line and the dummy word line on the opposite side. A voltage differential is developed across the two nodes as before. This voltage differential is amplified by raising ϕ_L, turning the flip-flop on. After restoration of the cell charge, the cycle is terminated by lowering the word lines. This design implementation has an advantage over the example in Figure 12.97 in that a separate reference voltage generator is not required and, consequently, the dummy cell design may be identical to the memory array cell design (and therefore track very closely).

Since the bit line and dummy cells are precharged to the same voltage, it may appear (from a glance at the charge transfer equation) that the dummy cell is unnecessary (i.e., redundant); indeed, the next sense amplifier design to be discussed does not incorporate dummy cells. However, incorporating a dummy cell provides for a feature that (on paper, at least) simplifies some of the design considerations: During a read access, adding the dummy cell capacitance to the bit line capacitance on the other half of the array keeps the total node capacitance (bit line plus cell) on *both* sides of the sense amplifier initially equal; any charge coupled from the transient on the selected word line onto the bit line or into the cell is compensated for by the same coupling on the dummy side.

Connecting the source nodes of the cross-coupled devices to $\overline{\phi_L}$ (rather than ground) increases the sensitivity to imbalance in the load transistors and to the slopes of the ϕ_L and $\overline{\phi_L}$ signals. Other examples to be presented incorporate alternative means for turning on the sense amplifier at a controlled rate to amplify the node voltage differential.

The sense amplifier design in Figure 12.100 illustrates another means of providing the precharge reference voltage to the bit lines and does not use dummy array cells. For this design, the following control and clock signals are used:

$\overline{CE}$: complement of the memory chip enable input signal

ϕ_X: clock signal that enables the decoded address word line to become active (internally generated)

ϕ_S: clock signal (also internally generated) that enables the source discharge transistor (one per sense amplifier) to begin to discharge the flip-flop nodes

ϕ_L: clock signal (internally generated) that enables the load transistors (independently of the source transistor) to pull up the off amplified node to $VDD - V_T$, restoring the charge into the accessed cell

Briefly, the timing sequence for a read access is as follows:

1. When the memory is not selected, the complement of the chip enable signal ($\overline{CE}$) precharges the bit lines to the reference voltage V_{ref}.
2. When the chip is enabled, $\overline{CE}$ goes low, unclamping the bit lines.
3. ϕ_X sequences first, enabling the selected address word line; charge transfer between the cell and the bit line results in a voltage swing at the flip-flop node.
4. ϕ_S is raised next, turning on the source transistor and discharging the source node of the cross-coupled devices toward ground. As this common source node falls, the device whose gate input is at the higher voltage turns on first. The lower of the two flip-flop nodes falls toward ground. while the higher flip-flop node falls only slightly before the device discharging this node turns off.
5. The load clock ϕ_L is now raised; the low node will settle at a voltage determined by the ratio of device (W/L) ratios for the two enhancement devices; the high node will charge relatively slowly, as with most of these sense amplifier designs, to $VDD - V_T$.

This design implementation does not use dummy cells, based on the argument that the balance provided by the dummy cells can be otherwise included (i.e., designed out) by a suitable choice of reference voltage. However, this raises the questions of how such a reference voltage is to be generated and what the sensitivity of the read operation is to the tolerance and noise of this supply.

In this implementation, the clocking signal ϕ_L and its complement $\overline{\phi_L}$ have been replaced by a separate source-clocking transistor (one per sense amplifier) and two separate clocking signals, ϕ_S and ϕ_L. It is feasible to use a single clock as the input to both the source and load devices; with the common-node source-

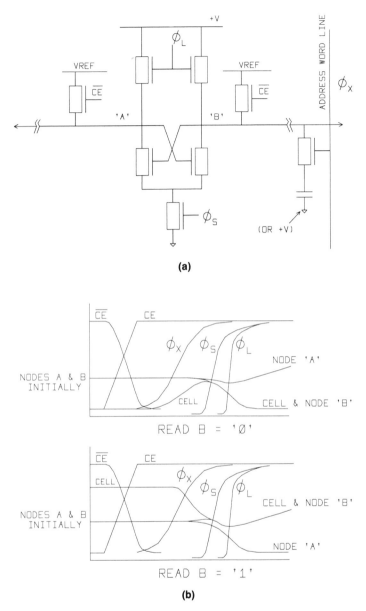

Figure 12.100.   Alternative sense amplifier design that precharges the bit lines using the complement of the Chip Enable signal. (a) Schematic. (b) Read cycle timing waveforms.

clocking transistor connected to ground, this pull-down device will turn on prior to the load devices on the rising edge of the clock signal, as described earlier. However, using separate clock signals (one derived from the other) allows the designer to fine-tune the amplifier operation by independently designing the rise time of the two signals (Figure 12.100b). ϕ_L should have a large (fast) rising slope to enhance the restoration of the high bit line to $VDD - V_T$. The slope of ϕ_S, on the other hand, should be more gradual (without severely degrading the memory access time). The distributed resistance and capacitance of the diffused bit line prevents the sense amplifier from "seeing" all the charge from a cell at the far end of the bit line initially; a slow rise time for ϕ_S minimizes this effect. In addition, the sensitivity of the gated flip-flop as a differential amplifier is increased if the common source voltage is initially high (Figure 12.101):

$$\text{Sensitivity of amplifier:} \quad S = \left. \frac{I_A - I_B}{I_A} \right|_{t=0^+}$$

Using the Sah device model saturated current expression,

$$I = \tfrac{1}{2} * \gamma * \frac{W}{L} * (V_G - V_S - V_T)^2$$

$$S = \left. \frac{\left(\frac{\gamma}{2}\right) * \left(\frac{W}{L}\right)_A (V_B - V_{\text{common}} - V_T)^2 - \left(\frac{\gamma}{2}\right) * \left(\frac{W}{L}\right)_B * (V_A - V_{\text{common}} - V_T)^2}{\left(\frac{\gamma}{2}\right) * \left(\frac{W}{L}\right)_A (V_B - V_{\text{common}} - V_T)^2} \right|_{t=0^+}$$

Assuming all devices track perfectly, the expression for the sensitivity S reduces to approximately

$$S = \left. \frac{2(V_B - V_A)}{V_B - V_{\text{common}} - V_T} \right|_{t=0^+}$$

The higher the value of V_{common} initially, the greater the amplifier sensitivity. One possibility for achieving the proper transition on the common node voltage is to provide a *variable-resistance* path to ground. Initially, a large resistance (a high

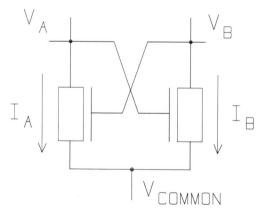

Figure 12.101. Circuit schematic for analyzing the sensitivity of the sense amplifier to the transition at node V_{common}.

V_{common}) is present for maximum sensitivity followed by a reduced resistance for improved performance and proper dc signal levels when clock signal ϕ_L is raised. Figure 12.102 illustrates such a common node design technique used in a Texas Instruments 16K-bit dynamic RAM.

Other alternatives exist for developing the bit line precharge voltage; designs illustrated up to this point have used $V_{T,n}$ (floating), $V_{switch\ point}$, and V_{ref}, and all have suffered from relatively poor performance in charging the 1 bit line to $VDD - V_T$ through the load device clocked by ϕ_L. A performance improvement can be realized by precharging *both* the bit lines completely to VDD and then selectively discharging one or the other. In a diffused bit line implementation, this has the added advantage that the junction bit line capacitance is reduced at the larger initial reverse bias. The problem arises in developing a reference charge on the deselected side of the array; the charge transferred by the dummy storage cell must still result in a voltage differential across the flip-flop nodes for both a 1 or a 0 read from an array cell on the opposite side. One possible solution is to use a *half-area* cell capacitor for the dummy storage node and precharge the dummy storage capacitance to ground. The quantity of charge transferred by the dummy cell $[Q = (C/2) * V]$ will therefore lie between the 1 charge transfer and that for a logical 0. This implementation is illustrated in Figure 12.103 (note the notation for the double polysilicon cell).

Nevertheless, the generation of a half-charge dummy (reference) level from a half-area dummy cell is being squeezed by newer generations of commercial

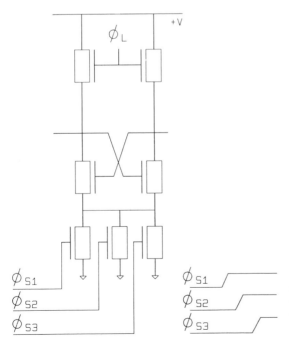

Figure 12.102. Sense amplifier design using staggered sense amplifier clocks to obtain fine control over the transition at the V_{common} node.

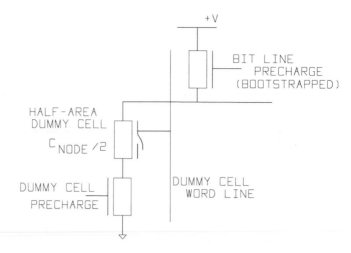

HALF-AREA DUMMY CELL IMPLEMENTATION

Figure 12.103. This implementation is used with a fully precharged bit line array for providing a half-swing on the bit line opposite the accessed cell. The half-area cell is initially precharged to ground. (Note the circuit schematic notation for the double polysilicon cell.)

designs. The design *and control* of accurate half-area cells is increasingly difficult as device or cell design dimensions are reduced. In addition, precharging all the bit lines to *VDD* produces a significant coupling transient current into the substrate, which is potentially problematic for an on-chip substrate voltage generator.

Column Decoding

There are significant benefits in splitting the input address set into separate row and column address inputs: better chip area and aspect ratio, reduced bit line capacitance, refreshing efficiency. The word line decoding mentioned in the previous examples corresponds to decoding the row or *X* address inputs. The column or *Y* decoding is conceptually simple, although potentially difficult to incorporate into the array configuration. One of the most common implementations is illustrated in Figure 12.104, for the split memory array.

The particular column decoder signal that is active is determined by the combinational logic decode of the column address inputs and internal timing signals; either static or dynamic clocked decode logic circuits can be used. Clocked load device circuits may be preferable over passive load options for reduced power dissipation, as all column decoder outputs (except one) will be at a logical 0 after decoding; however, this requires additional clock generation circuitry to provide the correct decoder sequencing. The use of clocked decoder circuitry also eliminates the need for internally latched address inputs; once the clocking sequence has transpired, a change on the address inputs will not alter the decoded output;

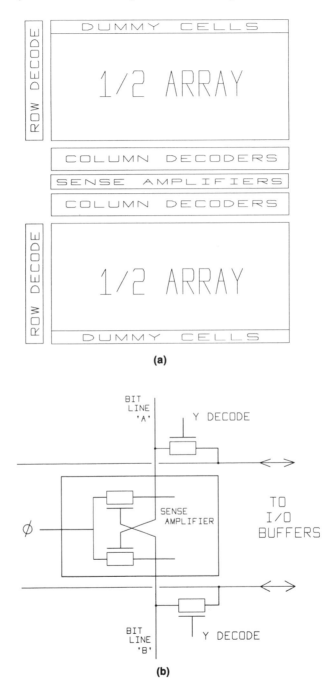

Figure 12.104.   (a) Array configuration illustrating the location of the column decoder circuits; (b) the column decode gating between sense amplifier and data lines.

minimum setup and hold times on the address inputs relative to the address strobe signal must still be maintained.

The lower power dissipation of CMOS technology logic circuits enhances the attractiveness of a static column decode, which simplifies clock generation and enables successive (fast) reads from sense amplifier (row) data values after an initial read access cycle by changing only column addresses.

Row Decoding

Much like column address decoding, row address decoding is commonly implemented with a combinational logic network of address inputs (and the appropriate chip select and/or row address strobe signals), buffered to drive the high-capacitance word lines. Static or clocked logic circuits may be used, the choice again depending on the constraints of power dissipation, performance, latched or non-latched address inputs, internal clock generation, and system address timing.

PROBLEMS

12.1. Research and describe the function of field programmable logic arrays (FPLAs, PALs ™). Specifically, describe the means of array personalization. (PAL is a registered trademark of Monolithic Memories, Inc.)

12.2. Research and describe the function of various customer programmable ROM technology options, specifically, fusible-link PROMs, ultraviolet light erasable PROMs, and electrically alterable PROMs. Contrast the different means of device utilization, data storage, and array organizations. Also, for the reprogrammable ROM options, compare the various manufacturer's specifications on the number of programming cycles and the data retention time.

12.3. The automatic generation of a PLA macro from a suitable logic description requires the assimilation of the various constituent (physical layout) cells in the proper number and orientation to construct the two arrays, connect the array inputs to the signal sources, connect load devices to the array logic signals, add buffers to the OR array outputs, and route power distribution both within the macro and to macro pins. It was discussed in the chapter that to reduce the complexity of the PLA macro generation tool and to ensure that a suitable fit with the overall chip image results, it is reasonable to restrict the variation in PLA macros to a limited number of fixed-height designs (fixing the number of product terms). The width of the macro is nevertheless allowed to vary with the number of macro I/Os. Adopting this limitation permits the background design of the PLA to be better defined and eases the problem of calculating the coordinates of the personalization cells. Using the block diagram of Figure 12.26 as a guide, this problem requests the development of a scheme for the description of the PLA background. Problem 12.5 discusses the techniques to personalize this background.

(a) Complete the block diagram of Figure 12.26, illustrating where the output load devices for the OR array columns should be located, and how the ground and

+ V power distribution should be routed throughout the arrays and to macro pins.

(b) For the fixed-height, variable-width array, what cell designs can be included as part of the background design?

(c) The physical cell description for the PLA background cell must contain the necessary additional information about the design to permit the PLA macro generation tool to calculate the coordinates of the various personalization cell layouts. Define and describe the additional set of keywords and parameters to include in the background cell description to be able to completely calculate the array personalization coordinate information; assume an AND–OR–AND array configuration with folded rows and folded columns, no sequential latch circuits, and no 2-bit input signal partitioning.

12.4. This problem requests many of the same tasks as Problem 12.3, only with regard to fixed array configuration ROM macro generation.

(a) Using Figures 12.16 and 12.19 as a guide, develop a complete block diagram of a ROM macro, including power supply distribution, buffer, and decode circuits. Assume static load devices. The input/output pin shapes must lie on the wiring grid periodicity.

(b) What cells or layouts can be included as part of the background design of the ROM macro?

(c) Define and describe the necessary set of keywords and parameters to include in the background cell definition to be able to calculate the ROM array personalization coordinate information.

12.5. **(a)** Develop a data file format for the general logic specification of the Boolean functions to be implemented in a PLA.

(b) Develop a matrix or array scheme for internally representing a general PLA personalization; provide the appropriate notation to be able to include folded input/output columns and folded product term rows. Assume the AND–OR–AND array configuration, no 2-bit partitioning of the inputs, and no sequential functions in the PLA.

(c) Using one of the algorithms presented in the chapter references (or one of your own design), code a program tool that will input a PLA logical description file, as defined in part (a), and fold the PLA, storing the result in the data structure of part (b).

12.6. Develop a data file format for the specification of the array values to be stored in a ROM macro; compare the format developed with that used by various custom in-house ROM manufacturers.

12.7. **(a)** Compare the array organization, I/O pin utilization, and cycle timing of commercial 64K bit, 256K bit, and 1M bit dynamic RAMs with those presented in this chapter.

(b) What is the specific nature of a read–modify–write timing cycle? From a systems architecture perspective, why are RMW cycles of such extreme importance? (Consider the flexibility required in large data path architectures to be able to perform byte, word, and double-word data manipulations, while maintaining error-correcting code bits in the system memory configuration.)

12.8. Research and describe the implementation known as *folded* bit lines for dynamic RAM chip designs, specifically addressing array configuration and cell area.

12.9. Research and describe the RIPPLEMODE™ operation of current dynamic RAMs, which use CMOS peripheral circuit designs. (RIPPLEMODE is a trademark of Intel Corporation.)

12.10. Research and describe each of the memory-testing techniques listed in Figure 12.89. Include a description of the failure mechanism each test is attempting to isolate. (A good starting point is the IEEE Press book, *Tutorial on LSI Testing*, EHO 122-2, reference 12.11.)

12.11. Develop a PLA matrix description of the following sequential logic network: Design a circuit that receives as input a 1-Hz clock signal and provides hour and minute time-of-day output through four seven-segment common-anode LEDs. The block diagram of the network is shown in Figure 12.105 (roughly in PLA form). The time of day ranges from 1 00 to 12 59; the only logic signal input to the PLA is a reset signal (active high), which is to reset the LED display to 12 00. Assume the PLA is a NOR–NOR configuration with inverting output buffers (to sink the LED current). The latches to be used with the PLA are master-slave data latches; no set/reset function is available.

12.12. Choose several commercial single-device cell dynamic RAM parts that have been well documented in the technical literature; reverse-engineer the sense amplifier and dummy cell circuits used by these designs. (Select older single-level polysilicon, single-level metal designs with coarser lithography to simplify the analysis.) What layout measures were taken to enhance the device tracking between the devices in the sense amplifier? Between the dummy cells and the array storage nodes? Are the dummy cells located adjacent to or the most distant from the sense amplifier differential inputs? Can you determine how the sense amplifier is being clocked?

12.13. Several current RAM architectures include additional or redundant cell locations in the array. These locations are used to replace defective sections of the array, as measured from failing address/bit location maps at manufacturing test. Research

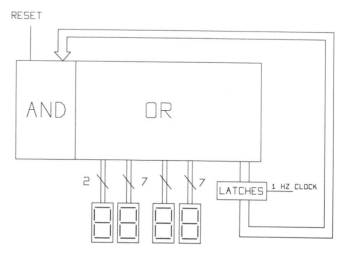

Figure 12.105.   Block diagram of PLA design for Problem 12.11.

the various redundancy techniques; compare and contrast these techniques in several RAM designs, specifically with respect to (1) the number and organization of the additional cells, (2) the means for replacing array locations with the redundant cells, and (3) the additional percentage array and chip area allocated to the redundant cells and the related peripheral circuitry.

12.14. Research the recent techniques being implemented to produce three-dimensional or trench storage node capacitor topographies in an effort to maximize array density. Compare and contrast these designs with regards to cell capacitance, leakage currents, and alpha-particle-induced free electron collection volume.

12.15. Research the current techniques being implemented in memory chip packaging and chip passivation that are attempting to reduce the incident alpha-particle flux to the array storage nodes.

REFERENCES

12.1 Wood, Roy, "A High Density Programmable Logic Array Chip," *IEEE Transactions on Computers,* C-28:9 (September 1979), 602–608.

12.2 Hachtel, G., Newton, A. R., and Sangiovanni-Vincentelli, A., "An Algorithm for Optimal PLA Folding," *IEEE Transactions on Computer-Aided Design of Integrated Circuits and Systems,* CAD-1:2 (April 1982), 63–76.

12.3 Cox, D., and others, "High Density Logic Array," U.S. Patent No. 3,987,287, October 19, 1976.

12.4 Golden, R., and others, "Design Automation and the Programmable Logic Array Macro," *IBM Journal of Research and Development,* 24:1 (January 1980), 23–31.

12.5 Soutschek, E., and others, "PLA Versus Bit Slice: Comparison for a 32-bit ALU," *IEEE Journal of Solid-State Circuits,* SC-17:3 (June 1982), 584–586.

12.6 Cook, P. W., Ho, C., and Schuster, S., "A Study in the Use of PLA-Based Macros," *IEEE Journal of Solid-State Circuits,* SC-14:5 (October 1979), 833–840.

12.7 DeMicheli, G., and Sangiovanni-Vincentelli, A., "Multiple Constrained Folding of Programmable Logic Arrays: Theory and Applications," *IEEE Transactions on Computer-Aided Design of Integrated Circuits and Systems,* CAD-2:3 (July 1983), 151–166.

12.8 Schrader, L., and Meusburger, G., "A New Circuit Configuration for a Static Memory Cell with an Area of 880 μm^2," *IEEE Journal of Solid-State Circuits,* SC-13:3 (June 1978), 345–350.

12.9 Rideout, V. L., "One-Device Cells for Dynamic Random-Access Memories: A Tutorial," *IEEE Transactions on Electron Devices,* ED-26:6 (June 1979), 839–852.

12.10 Dennard, R. H., "Field-effect Transistor Memory," U.S. Patent No. 3,387,286, June 4, 1968.

12.11 *Tutorial on LSI Testing.* New York: IEEE Press, #EH0-122-2, 1978.

12.12 May, T., and Woods, M., "Alpha-Particle-Induced Soft Errors in Dynamic Memories," *IEEE Transactions on Electron Devices,* ED-26:1 (January 1979), 2–9.

Additional References

Arai, E., and Ieda, N., "A 64-kbit Dynamic MOS RAM," *IEEE Journal of Solid-State Circuits*, SC-13:3 (June 1978), 333–338.

Chan, J., and others, "64K Dynamic RAM," *IEEE International Computer Society (Spring) Conference (COMPCON 81S)* (1981), 129–131.

Chatterjee, P., and others, "A Survey of High-Density Dynamic RAM Cell Concepts," *IEEE Transactions on Electron Devices*, ED-26:6 (June 1979), 827–839.

———, ———, "Enhanced-Performance 4K × 1 High-Speed SRAM Using Optically Defined Submicrometer Devices in Selected Circuits," *IEEE Transactions on Electron Devices*, ED-29:4 (April 1982), 700–706.

DeSimone, R., and others, "FET RAM's," *IEEE International Solid-State Circuits Conference* (1979), 154–155, 291.

El-Mansy, Y., and Burghard, R., "Design Parameters of the Hi-C DRAM Cell," *IEEE Journal of Solid-State Circuits*, SC-17:5 (October 1982), 951–956.

Fallin, J., "CHMOS DRAM's in Graphics Applications," *INTEL Solutions* (May/June 1984), 20–27.

Foss, Richard, "The Design of MOS Dynamic RAM's," *IEEE International Solid-State Circuits Conference* (1979), 140–141.

———, and Harland, R., "Peripheral Circuits for One-Transistor Cell MOS RAM's," *IEEE Journal of Solid-State Circuits*, SC-10:5 (October 1975), 255–261.

Hodges, D. A. (ed.), *Semiconductor Memories*. New York: IEEE Press, 1972.

Ieda, N., and others, "Single Transistor MOS RAM Using a Short-Channel MOS Transistor," *IEEE Journal of Solid-State Circuits*, SC-13:2 (April 1978), 218–224.

Kuo, C., and others, "Sense Amplifier Design Is Key to 1-Transistor Cell in 4,096-bit RAM," *Electronics* (September 13, 1973), 116–121.

———, ———, "16-K RAM Built with Proven Process May Offer High Start-up Reliability," *Electronics* (May 13, 1976), 81–86.

Liu, S., and others, "HMOS III Technology," *IEEE Journal of Solid-State Circuits*, SC-17:5 (October 1982), 810–814.

Lynch, W., and Boll, H., "Optimization of the Latching Pulse for Dynamic Flip-Flop Sensors," *IEEE Journal of Solid-State Circuits*, SC-9:2 (April 1974), 49–55.

Madland, P., and others, "CMOS vs. NMOS Comparisons in Dynamic RAM Design," *IEEE International Conference on Computer Design (ICCD 83)* (1983), 379–382.

McPartland, R. J., "Circuit Simulations of Alpha-Particle-Induced Soft Errors in MOS Dynamic RAM's," *IEEE Journal of Solid-State Circuits*, SC-16:1 (February 1981), 31–34.

Ohzone, T., and others, "A 2K × 8-Bit Static MOS RAM with a New Memory Cell Structure," *IEEE Journal of Solid-State Circuits*, SC-15:2 (April 1980), 201–205.

O'Toole, J., and others, "A High Speed 16K × 1 NMOS Static RAM," *IEEE International Computer Society (Spring) Conference (COMPCON 81S)* (1981), 125–128.

Pugh, E., and others, "Solid State Memory Development in IBM," *IBM Journal of Research and Development*, 25:5 (September 1981), 585–602.

Schlageter, J., and others, "A 4K Static 5-V RAM," *IEEE International Solid-State Circuits Conference* (1976), 136–137.

Schroeder, P., and Proebsting, R., "A 16K × 1 Bit Dynamic RAM," *IEEE International Solid-State Circuits Conference* (1977), 12–13.

Smith, F., and others, "A 64 kbit MOS Dynamic RAM with Novel Memory Capacitor," *IEEE Journal of Solid-State Circuits,* SC-15:2 (April 1980), 184–189.

Stein, K., and others, "Storage Array and Sense/Refresh Circuit for Single-Transistor Memory Cells," *IEEE Journal of Solid-State Circuits,* SC-7:5 (October 1972), 336–340.

Tasch, A., and others, "The Hi-C RAM Cell Concept," *IEEE Transactions on Electron Devices,* ED-25:1 (January 1978), 33–42.

Terman, Lewis, "MOSFET Memory Circuits," *Proceedings of the IEEE,* 59:7 (July 1971), 1044–1058.

Varshney, R., and Venkateswaran, K., "Characterization of an MOS Sense Amplifier," *IEEE Journal of Solid-State Circuits,* SC-13:2 (April 1978), 268–270.

Wada, T., and others, "A 64K × 1 Bit Dynamic ED-MOS RAM," *IEEE Journal of Solid-State Circuits,* SC-13:5 (October 1978), 600–606.

White, L., and Chitranhan, R., "Wide-word 64-kbit RAM Expands System Performance," *Electronic Design* (March 18, 1982), 231–238.

Yaney, D., and others, "Alpha-Particle Tracks in Silicon and Their Effect on Dynamic MOS RAM Reliability," *IEEE Transactions on Electron Devices,* ED-26:1 (January 1979), 10–16.

TEST METHODOLOGIES
FOR VLSI

13.1 FAULT MODELING AND DEFINITION

The capacity of VLSI chip technologies to provide high circuit counts and improved performance levels has increased the intensity of the problems associated with test verification. Specifically, these problems include:

1. Modeling the chip design in terms of its test characteristics, highlighting tester clock signals and additional (chip pad) test points.
2. Selecting the criteria for the measurement of test coverage.
3. Generating the set of test pattern stimuli and expected output responses, as well as providing the means for measuring their cumulative effectiveness.
4. Developing the tools to assist with the diagnosis of the cause of a detected failure.
5. Indicating specific performance-related measures to verify at wafer or package test.

Each of these various problems will be discussed briefly in this section; the magnitude of all these tasks increases dramatically as the chip circuit count-to-I/O pad ratio increases.

To reduce the resource required to develop an acceptable level of test verification for each part number, *design for testability* principles and restrictions have been incorporated into design system methodologies. These principles govern many aspects of the design system development and chip design cycles, from restrictions on the structure of logic designs to the suitability of particular circuit types in the design system technology library. In general, design for testability techniques strive to reduce the complexity and intricacy of general sequential networks to more readily manageable combinational logic partitions. Section 13.3 discusses a particular design for testability approach in some detail; this approach, denoted as level-sensitive scan design (LSSD), has a number of associated logic and circuit design restrictions, some of which have been highlighted in previous chapters. The goal of incorporating a DFT approach is to be able to *automatically generate* a near-optimal set of test stimuli, a set that provides a high measure of test coverage and a high confidence level in the functionality and performance of the chips that pass wafer-level testing. An algorithm for generating test patterns to control and observe the internal nets of a combinational network is described in Section 13.2. When used in conjunction with a DFT approach like LSSD, the task of test pattern generation for VLSI chip designs becomes containable.

An alternative approach to test verification adopts the philosophy that unique test pattern stimuli to exercise the chip function are not required. Rather, a subset of the logic simulation input vectors with their associated expected output responses should be used to verify functionality. By this school of thought, no unique test modeling of the design should be necessary. The difficulty in using functional simulation patterns for test stimuli is in calculating a measure of test sufficiency of the pattern set, a measure that correlates closely to the quality level of the parts packaged and shipped. In addition, without a test model for the chip, developing test diagnostics for debug is extremely burdensome. The remainder of this section discusses the considerations associated with generating a test pattern set automatically; comparisons against the use of functional patterns will be included, where appropriate.

Test Modeling

The test pattern count is used by the test engineering group to assess the tester time per chip necessary for verification, and thus the associated testing costs. To minimize these costs, ineffective test patterns (those that exercise logic function previously examined in the test stimuli) should be eliminated. In addition, clock signals used during test should be identified separately; test equipment interfaces commonly permit the definition of periodic signals in a manner so as to efficiently represent the test pattern data file. The chip test model typically contains extensions to the logic simulation model that identify test-specific signal characteristics. For example, as illustrated in Figure 13.1, a data storage register comprised of multiple-port latch pairs (the L1/L2 latch pairs of Section 10.2) may also be connected in shift-register fashion to provide a serial test output of the stored values. The alternating shift clocks, denoted as $A0$ and $B0$ in the figure, are toggled to scan the data out of the shift-register interconnection; the clock $C0$ loads the data in parallel from the driving network. The $A0$ and $B0$ shift clock chip inputs should be uniquely identified to the test pattern generation tool (and subsequently to the

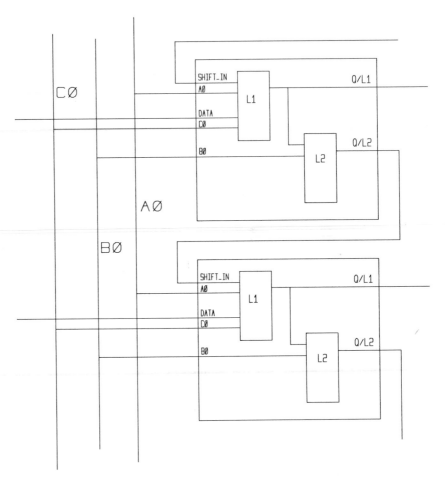

Figure 13.1. Latch pairs connected in shift-register fashion. The nonoverlapping clocks *A*0 and *B*0 implement the shifting of register data through the latch pairs.

tester) in order to efficiently indicate the nature of these repetitive signals during scan. The chip serial Scan In and Scan Out signals should also be identified in the test model to indicate the source of test input stimuli and output responses.

For automatic test pattern generation it is also necessary to substitute a suitable test model for each macro function for which a behavioral simulation model was used. The behavioral was developed for simulation efficiency and to represent and verify the timing constraints of the function; it conveys little or no information about the failure modes internal to the macro to be investigated during chip testing. A test model description from the design system library replaces the behavioral description to form the hierarchical chip test model; note that this step would not be required with the use of logic simulation patterns for chip verification. The test model for the macro should provide what the behavioral cannot, that is, the physical circuit-to-logical fault associations. The nature of faults in the test model of a chip is discussed next.

Criteria for the Measurement of Test Coverage

From a fabrication process engineer's viewpoint, the possible failure modes of a circuit are numerous: an open metal wiring interconnection, shorted metal interconnections, incomplete opening of contact holes, a device source or drain node shorted to the device gate (or chip substrate) due to misaligned or overetched contacts, crystalline or dielectric defects resulting in anomalous device currents, and so on. The development of the design system test methodology must address how to provide a test model for each circuit that associates these physical faults with the resulting incorrect logical behavior. The approach that has become the most widely accepted among the VLSI test engineering community is to assume that physical circuit faults result in stuck-at-0 (s-a-0) or stuck-at-1 (s-a-1) Boolean values at the input or output pins of a block in the test model. For each input and output pin of all blocks in the test model, a s-a-0 and a s-a-1 fault is possible; for *n* pins, 2 * *n* faults is the optimum measure of fault coverage achievable by the test pattern set. The algorithm to be discussed in Section 13.2 begins by assuming that a *single* fault is present on a specific pin in a combinational network; it then works to find a pattern of input signals to the network that will try to assert the correct logic value on that pin and condition the remainder of the network to permit the pin's Boolean value to propagate to an observable output of the network (Figure 13.2a). The propagation of the correct value represents a good machine, while the propagation of the fault is denoted as the bad machine. The collection of network input stimuli and the expected output responses is commonly referred to as a *test pattern* or *test vector*. It should be noted that *input* faults are assigned

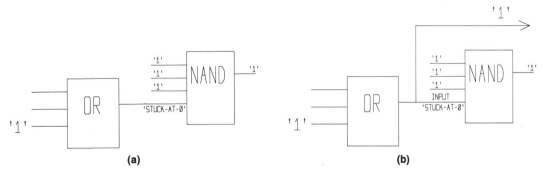

(a)   **(b)**

Figure 13.2.   (a) The signal assertions necessary to provide the opposite value from the assumed stuck-fault value on the pin and condition the remaining network to propagate the actual response to an observable point in the network. (b) Input faults are assigned to block inputs, not to signal nets as a whole; the propagation of a network result to an observable point must be through the affected block (the NAND block in the figure), not on the driving net.

to block *pins* of the test model, rather than to signal nets as a whole; as a result, as illustrated in Figure 13.2b, the propagation of an input pin fault to an observable point cannot use an alternate logic path of the net that contains that pin. Likewise, a stuck-at fault on an output pin is to be regarded as distinct from the possible faults on the input pins in the same net; the propagation of the good or bad value from the output pin may be taken through *any* of the input pins in the net.

A single test pattern will likely cover many possible stuck-at faults of the total fault set; that is, the input test vector to the network usually results in sensitized paths for a number of possible network faults. For the case of an output pin and input pin in the same net discussed just previously, the input pin fault along a sensitized propagation path is covered by the same pattern that detects the output pin fault; indeed, that pattern is suitable for evaluating stuck-at faults on all nodes down the sensitized path that are logically related to the assumed fault value (Figure 13.3). The test pattern generation algorithm to be discussed in Section 13.2 distinguishes the nets that are logically related to the initial assumed fault for a specific test vector, indicating the additional faults that are detected. A recommended strategy is to attempt to find test patterns initially for the longest network paths, starting with faults on the network's primary inputs, to maximize the number of faults detected along the resulting sensitized path early in the test pattern generation procedure. Another algorithm will be described in Section 13.2 that detects the *additional* paths and faults outside the sensitized path that are also covered by the generated test vector.

The following physical faults are commonly *not* incorporated in the stuck-at test model:

1. **Multiple faults:** The condition of more than one stuck-at fault present in the network is not analyzed; the techniques for determining the necessary signals to propagate the initial assumed fault assume that the remainder of the network is error free.

2. **Bridging faults:** A shorted or bridging interconnection between global nets is depicted in Figure 13.4; the resulting correlation of signal values is not a part of the stuck-at fault approach, due in large part to the complexity this

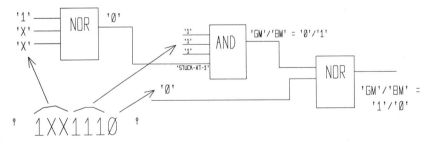

Figure 13.3.   The input pattern 1*xx*1110 is suitable for detecting the faults indicated along the sensitized path; these pins are logically related to the assumed stuck-at fault pin.

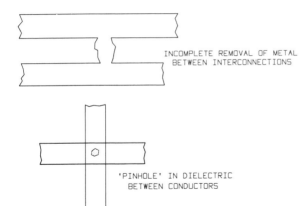

INCOMPLETE REMOVAL OF METAL
BETWEEN INTERCONNECTIONS

'PINHOLE' IN DIELECTRIC
BETWEEN CONDUCTORS

Figure 13.4. Physical faults: bridging between adjacent conductors and shorts between wires on different levels. These physical faults do not directly correspond to input/output pin stuck-at faults.

would introduce into the test model. The test model is an extension to the logic model; no physical design data relative to the location or proximity of signal interconnections are typically provided to the test model or test pattern generation tools. This independence between physical design and test pattern generation permits the two processes to be executed in tandem for the most aggressive design schedules. It also facilitates the identification of testability problems in the logic design relatively early in the design cycle.

3. ac faults: The test modeling of network faults assumes that the bad machine behavior is the result of a static condition that manifests itself at tester data rates; performance-related faults, which cause the chip to malfunction at rated clock frequencies, are therefore not likely to be detected until system-level functional testing. An example of an ac fault is given in Figure 13.5. A shorted interconnection between gate and source nodes of the external depletion-mode load device in a push–pull circuit will increase the circuit's logic transition 0 to 1 delay considerably; the logic output value is nevertheless correct. The circuit would likely pass wafer-level testing, yet may not meet its performance design goals.

One other category of physical faults to consider is the general class of *pattern dependent* failures; that is, the ability to demonstrate that a circuit fault is present depends on the pin's Boolean value just preceding the application of the test pattern of interest. For this class of faults, to distinguish between good machine and bad machine, the previous sequence of patterns must be developed so as to provide a specific starting Boolean value on the net under investigation. An example of this group is the *CMOS open fault* (Figure 13.6). The schematic in the figure depicts an open interconnection between the *p*-channel devices in the AOI circuit. To be able to demonstrate that the connection is indeed present (the good machine behavior) it is necessary that a rising transition on the output be produced by the test pattern, with the charging current flowing through the specific interconnection under analysis. This requirement implies that the starting

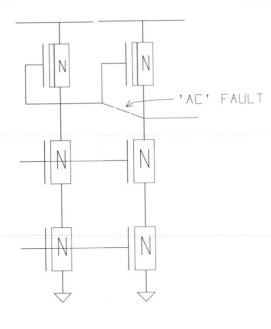

Figure 13.5. An ac or performance fault: if the gate and source nodes of the external depletion-mode device are shorted, the circuit will still operate correctly, but at a considerably reduced performance over the normal push–pull circuit. At tester speeds, this performance-related fault may go undetected.

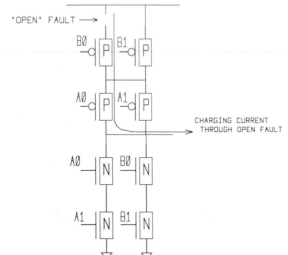

Figure 13.6. CMOS open fault for an AOI circuit. To determine if the connection is indeed present, a rising transition (charging current) on the output through the device connected to input signal $B0$ must be produced by the sequence of test patterns.

value on the circuit output should be a 0 value; the preceding test pattern must discharge the output net in preparation for the rising transition.

The nature of the CMOS open fault is further complicated by the possibilities of dynamic charge transfer between internal circuit nodes and the misinterpretation that might result. Consider again the open fault of the AOI circuit illustrated in Figure 13.6; assume that the internal node capacitance of the stacked p-channel devices has been precharged to the supply voltage during the discharging of the output node by the combination of inputs signals [(A0, A1, B0, B1) = (1, 1, 0, 0)] and that this internal node remains precharged until the application of the pattern (A0, A1, B0, B1) = (0, 1, 0, 1). If the open fault is indeed present, the output node will not rise to *VDD*; however, a charge transfer between the internal node capacitance and the output node capacitive load may result in an output node voltage that could be interpreted as a logical 1 by subsequent circuitry. As a result, the fault may go undetected at test due to the sequence of patterns used.

While on the subject of CMOS faults, another interesting situation is posed by the class of CMOS *shorted device* faults. As illustrated in Figure 13.7, if a CMOS device is indeed shorted, the final test pattern will create a dc current path from supply to ground. In a passive load ratioed circuit design, this would produce the desired bad machine result. However, in this case of ratioless CMOS circuits, it is not clear what output voltage would result. Would an intermediate output voltage ripple through subsequent stages of logic, increasing the dc current substantially? Can a significant dc current be detected as the tester to indicate such a failure? What if the output voltage is such that the presence of the fault goes undetected and the chip passes at the tester? Although the chip functions correctly under tester conditions, will it continue to function correctly over supply voltage and temperature tolerances? The problems of CMOS technology VLSI chip test-

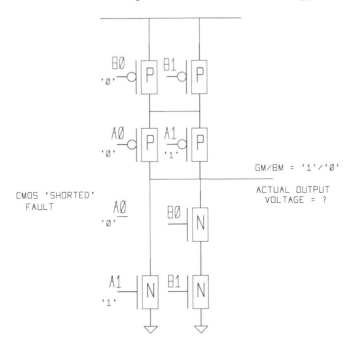

Figure 13.7.   The CMOS shorted fault will result in a dc path from supply to ground with the appropriate test pattern; since CMOS circuits are not ratioed for a dc down level, an intermediate (undefined) output voltage will likely be produced. The dc current that results may potentially be detected at wafer-level test.

ing are severe; input device shorts and device opens translate directly to stuck-at-1 and stuck-at-0 input signal values in NMOS technologies, but not in CMOS. A conventional CMOS stuck-at fault model would assume that the logical input signal to *both* the *n*- and *p*-channel devices is stuck.

It is evident that much work remains in attacking the problems of test pattern generation and analysis for VLSI logic parts, particularly for CMOS technologies. Some of the traditional physical defect-to-test model fault associations of previous technologies may need to be abandoned in favor of other, more efficient approaches. Some of these alternatives are discussed in Section 13.5.

Due to the structure of the logic design itself, the test model may contain *untestable* faults that cannot be suitably controlled and observed by *any* test pattern applied to the combinational logic inputs. This condition arises when parallel (or redundant) logic is present in the network, and the outputs of these redundant blocks reconverge prior to reaching an observable point in the network. These untestable faults should be identified as early as possible in the design cycle in order to assess the possibilities and impacts of changing the logic design, adding more observable test points, and/or accepting the related manufacturing quality level reduction (if a measure of shipped quality level versus stuck-at test fault coverage is provided).

Generation of the Test Pattern Set and Test Diagnostics

The goals of test pattern generation for a combinational network are twofold: (1) to provide an efficient set of patterns that will detect all the assumed failures in the network, and (2) to develop a means of diagnosing the source of a failure when one is detected.

The techniques of finding equivalent faults and additional sensitized paths for a given pattern have already been briefly mentioned; these will result in a reduced test pattern count. In addition, to further reduce the size of the tester data file (TDF), the strategy of combining (or intersecting) existing equivalent patterns may be considered. It is likely that some of the network primary input values in a particular test pattern will remain at an undefined value after test vector generation for a particular fault. In developing sensitized paths for controlling and observing an assumed fault, it may not have been necessary to uniquely specify all network inputs; some may have remained as "don't care" values. Two patterns may be *intersected* in an attempt to further define input values and reduce the overall pattern count. For example, the two input vectors $x01xx0$ and $0x101x$ may be combined into a single pattern: 001010. The main disadvantage with this approach to TDF reduction is the CPU-intensive nature of the search for patterns that may be suitably combined.

The diagnosis of the origin of an observed fault is commonly viewed as a secondary problem to generating the optimal test vector set. A typical approach is to develop a somewhat enhanced (i.e., larger) set of test patterns that will assist in pinpointing the site of the failure. Thus, an efficient pattern set is available for lowest tester-time cost production-level testing, while an enhanced set can be used for debug and failure analysis. The efficient TDF must strive to *cover* the set of circuit faults; the size of this set is reduced by the multiple faults encompassed by individual patterns. The diagnostic set of patterns must strive to provide a *unique combination* of test patterns for each fault. A means of depicting the more stringent requirements of generating test diagnostics is the *fault location table* (Figure 13.8). Each row in the table corresponds to a failure in the network, whereas each column corresponds to a test vector. A mark in a particular array coordinate indicates that the associated test vector produces an observable good or bad machine response for that fault at some output pin of the network. To diagnose a specific single fault, each row must have a unique pattern of marks. By comparing the rows in this table against the measured bad machine patterns at the tester, the fault may potentially be identified. The additional resource invested in generating test diagnostic information can prove to be invaluable in debugging systematic errors measured at wafer-level testing. The conclusions from fault diagnosis are then fed back to the process engineering group to indicate the origin of observed failure(s). A failing net can be traced out into the shapes data present on the masks for the device and/or interconnection lithography levels; a systematic failure is often attributable to a missing mask shape or other mask defect.

Performance Measures and Limitations at Wafer-level Testing

The tester will not typically exercise the parts at their designed performance level; as discussed earlier, the fault modeling used for testability analysis is based on dc test application assumptions. The design system methodology may provide for

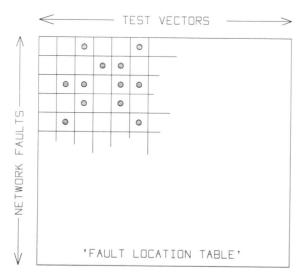

Figure 13.8. Chip diagnostics may be developed from the fault location table if a unique combination of test vectors is provided for each network fault.

the limited measurement of on-chip delay paths for performance screening or sorting.

Besides the test pattern data rates (and the cost of tester time), of additional concern is the environment of the test fixture itself, specifically the following:

1. Range and resolution of applied input signal currents and voltages (i.e., the logic interfaces available)
2. Resolution of measured output voltages and currents
3. Timing resolution and accuracy of applied clock signals
4. Range and resolution of supply voltage(s) and wafer chuck temperature available
5. Cabling noise from the test fixture to the supply voltages and signal measurement points

Some additional parametric measures to be included with the test pattern file were discussed in Section 8.9. These parametric tests also involve requirements for measures of leakage-level currents on input signals and high-impedance output nodes.

To reduce the noise present on applied signals and observed on test measurements, special test fixture designs must be utilized. The probes and cables connecting the chip to the tester act as transmission lines at sufficiently high output signal transition times; as such, special design of impedance-matching networks should be used to prevent spurious reflections from the test measurement jig that might otherwise result in misinterpreted (or potentially destructive[1]) behavior. The inductive $L * (di/dt)$ noise on the tester's supply voltage cabling can also result in substantial shifts in chip *VDD* and *GND* voltage, again possibly leading to failing behavior. The derivative of the supply current with respect to time is at its maximum when a significant number of off-chip driver (OCD) circuits switch simultaneously. For reliable testing, the number of simultaneously switching OCDs for any test pattern should be limited. (This constraint also applies to the distribution of supply voltage and inductive switching noise on the design system chip image itself; refer to the discussion in Section 14.6.) A possible solution to this problem of large transient supply currents at the tester is to stagger the switching of the OCDs for each test pattern with a Test Inhibit signal pad, contacted by the prober during wafer-level test. As illustrated in Figure 13.9, a long *RC* time constant chain may be used to delay the enabling of successive off-chip drivers. The Test Inhibit signal is pulled low prior to the application of the test vector, placing the OCDs in a high-impedance condition. Subsequently, the connection is released and the drivers are enabled as the signal charges to VDD through the distributed *RC* line. After wafer-level test and packaging, the card- and system-level testing may not share the same supply distribution noise prob-

---

[1] See the discussion in Chapter 9 on CMOS latch-up.

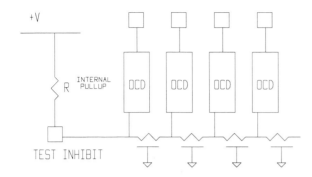

Figure 13.9. A scheme for reducing the number of simultaneously switching off-chip driver circuits at wafer-level testing. The long, resistive interconnection from Test Inhibit will delay the enabling of successive circuits when the Test Inhibit input is no longer being held at 0.

lem; the Test Inhibit pad would therefore not be connected to any package pin, pulled up to its inactive level on-chip.

This section has presented the philosophy that has evolved with respect to testing of VLSI designs, that is, single stuck-at fault modeling. A brief discussion of the design for testability approach was given and will be expanded on in Section 13.3. Fundamentally, this approach separates the testing of sequential and combinational circuitry into partitions and shift-register latch paths for which combinational test pattern generation and analysis techniques suffice. Algorithms to automatically generate test patterns for combinational networks are presented next.

Yet, despite these tools for test pattern generation and analysis, the problems associated with testing VLSI hardware are not abating. The traditional approach is resulting in larger CPU resource requirements to generate the test data file, with long resulting tester times. New technologies and circuit designs have physical faults that do not correlate well to logical stuck-at faults in the circuit's test model. The inability to test at functional speed may also be masking ac faults. Section 13.5 briefly discusses some of the alternative approaches that specifically address these problems.

The best VLSI test solution is the one that results in a high quality level of hardware that successfully exits wafer-level testing, yet that requires the minimum TDF development resource and tester time. Traditionally, the stuck fault coverage was used as a measure of that quality level; the future solution(s) may indeed be some combination of the approaches currently being used and those currently being researched.

13.2 TEST PATTERN GENERATION APPROACHES FOR COMBINATIONAL NETWORKS

Section 13.1 outlined the procedure for automatically generating a test pattern for a specified stuck-at fault in a combinational network, as well as the terminology commonly associated with that process: good machine, bad machine, sensitized path, primary network input (PI), and primary network output (PO). Essentially, the procedure must first determine a suitable path from the fault to a primary output, and then back-trace through the network to develop a corresponding pattern of primary inputs that sensitizes that path to the fault's value.

The iterative (and therefore CPU-intensive) nature of this procedure involves the determination of one among a multitude of possible paths from the pin under analysis to a network PO and the assignment of sensitizing values to internal nets (and PIs) to support the selected path. If an inconsistency in signal values is discovered while in the process of forward propagation or back-tracing, it is necessary to undo the decisions made in determining the chosen path to the point where an alternative assignment of a Boolean value to a net can be made. The test pattern generation procedure then recommences. This process can be modeled by a decision-tree structure, representing the alternative assignments (branches) that may be pursued. Figure 13.10 illustrates this structure, as well as an example of the inconsistencies that may result during assignment of sensitizing values to internal nets. In this example, the forward propagation procedure has encountered an input pin whose previously assigned value conflicts with the current value required to continue the sensitizing path. The assignment $C0 = 1$ in Figure 13.10 was selected when sensitizing the path through block B and has led to a conflict when attempting to sensitize the fault path through block D. To resolve this problem, the decision tree must be retraversed using the alternative assignment $C0 = 0$. This example illustrates that the exclusive-OR (XOR) gate is unique among the set of primitive gate types in that there is no single controlling or sensitizing value to assign to inputs to propagate the fault. It is not surprising then that the early automatic test pattern generation algorithms spent considerable

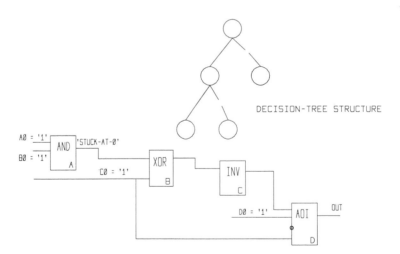

Figure 13.10.   The generation of a test pattern for an assumed fault typically uses a decision-tree structure, where different branches from a root node indicate alternative signal assignments. In the example, the assignment $C0 = 1$ will not lead to a valid test for the assumed fault of the AND gate (A) stuck-at-0.

CPU time in traversing the decision tree and tracing through paths for logic networks that were rich with XOR gates. Parity generator and checker circuits are prime examples of such networks. Refinements to these procedures were conceived to better bound the extent of decision-tree branch traversals before arriving upon a satisfactory test pattern.

The first algorithm to be presented in this section is the D-algorithm for test pattern generation; this process was developed by Roth and others (reference 13.1). An alternative to the D-algorithm will then be discussed, as described by Goel (reference 13.2), who denoted the refined technique as a "Path-Oriented DEcision Making" algorithm (PODEM). Finally, a TEST-DETECT algorithm will be presented that provides an analysis of *all* the faults detected in a combinational network for a given input test vector.

Before discussing the D-algorithm for test pattern generation, the pertinent notation should be introduced. Figure 13.11a depicts a NOR gate and its associated cover (i.e., the minimal truth table that encompasses the logical function of the block). To this Boolean set of prime implicants is added the table of D vectors,

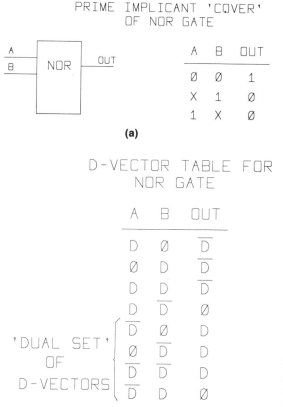

Figure 13.11. (a) A NOR gate and its associated cover of prime implicants. (b) D-vector table for propagating good and bad machine values through the NOR gate.

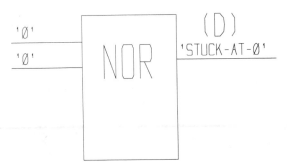

Figure 13.12. A vector is selected
from the prime implicant cover that
provides a good machine output value
opposite from the assumed stuck-at
fault value.

as illustrated in Figure 13.11b; the symbol D may represent either of the two
values 0 or 1, and the value chosen applies to all D symbols in the vector.[2] The
symbol $\overline{D}$ represents the complement of the value assigned to the symbol D. For
completeness, Figure 13.11b also includes the dual set of D vectors for the NOR
gate.

These D values are used to indicate the presence of an assumed fault at a
particular pin, as well as how the good or bad machine value pair would propagate
through the remainder of the network to a primary output. The convention used
is that a D value of 1 (and a $\overline{D}$ value of 0) are assigned so as to represent the good
machine behavior; a measured value of D = 0 or $\overline{D}$ = 1 at the sensitized network
primary output indicates a failing machine. When a pin is to be tested for a stuck-
at-1 (0) fault, it is necessary to try to assert a 0 (1) value on that line. The first
steps in the test pattern generation algorithm are to locate the block driving the
pin with the assumed fault and select an input pattern from the prime implicant
cover of the block type with the good machine output value (Figure 13.12). Once
that pattern has been determined, replace the pin with the assumed fault by the
appropriate D value: $\overline{D}$ for a stuck-at-1 fault (good machine = 0) and D for a
stuck-at-0 fault (good machine = 1). For the case of a stuck-at-0 fault at the output
of the NOR gate in Figure 13.12, the resulting pattern is 00D; for the NOR output
stuck-at-1, one of the two patterns $x1\overline{D}$ or $1x\overline{D}$ may be chosen. For the case of
faults on primary input lines to a network, the initial patterns to propagate are
merely D or $\overline{D}$.

The *intersection* between elements of two D vectors is defined in Figure
13.13. This table of intersections is used to update the current assignment of signal
values in the network when propagating the chain of D (and $\overline{D}$) values to a primary
output. The D vector selected to sensitize a path through a block is *intersected*
with the existing network D vector to describe the new values of network signals
and to ensure that a consistent signal assignment results. For a test pattern to be
valid, at least one of the network POs must have a D (or $\overline{D}$) value, in order to be
able to distinguish the good versus bad network response. The chain(s) of D values

---

[2] Roth's terminology referred to a single vector of Boolean values as a "cube" in the space of
the possible block input/output values.

∧	Ø	1	X	D	D̄
Ø	Ø	NULL	Ø	UNDEFINED	UNDEFINED
1	NULL	1	1	UNDEFINED	UNDEFINED
X	Ø	1	X	D	D̄
D	UNDEFINED	UNDEFINED	D	D	UNDEFINED
D̄	UNDEFINED	UNDEFINED	D̄	UNDEFINED	D̄

Figure 13.13. The intersection of network D-vectors is used to determine the results of signal assignments and implied values in the network; an undefined or null result of the intersection is *not* valid for propagating the D-value toward a network primary output.

in the network are connected to the faulty pin in the sense that (1) the assumed fault has a D (or D̄) value, and (2) for a D (or D̄) value to be present on a block output, at least one of the block inputs must also be D (or D̄).

An example of this intersection process is given in Figure 13.14; note that the network D vector is the concatenation of the 0, 1, x, D, and D̄ values of *each*

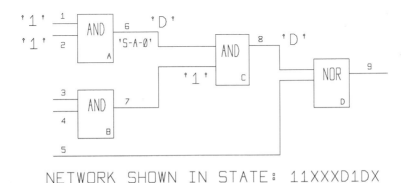

NETWORK SHOWN IN STATE: 11XXXD1DX

Figure 13.14. D-vector intersection procedure for propagating the D-value to a network PO. The network D-vector consists of the values of each signal net concatenated together.

net in the logic circuit. For the block A output stuck-at-0 fault, the following steps to generate a test pattern would be followed:

1. From the cover of the AND gate type of block A, the vector 111 is chosen; the stuck-at-0 output of this block is changed to D, yielding the nine-element vector 11xxxDxxx.

2. The only successor to block A is block C. The D vector for block C selected to propagate the D chain is D1D. Intersecting the two vectors 11xxxDxxx (network) and xxxxxD1Dx (block) yields 11xxxD1Dx (network). During this forward propagation of the D value to a PO, it is necessary to maintain a record of the signal net values that were assigned to 0 or 1, as net 7 in the figure was assigned a value of 1 by the chosen D-vector. These implied signal assignments must be realized when back-tracing to determine the required network PI values. This back-trace involves intersecting the network D vector with prime implicants of the blocks whose output values were assigned. The intersection of the network D vector with block prime implicants is the consistency portion of the test pattern generation algorithm. In this example, the assignment of net 7 to a value of 1 must be pushed to a stack for the subsequent consistency operation.

3. The selected D vector for block D is intersected with the network vector to propagate the D value to a PO: $xxxx0xxD\overline{D} \cap 11xxxD1Dx = 11xx0D1D\overline{D}$. The network primary output has been reached with a value of $\overline{D}$.

4. The consistency operation begins by examining the stack of assigned internal net values; in this case, net $7 = 1$. Examining the prime implicant table for the AND gate type of block B results in the vector $xx11xx1xx$. When intersected with the network D vector, this yields a result of $11110D1D\overline{D}$. Since the intersection did not yield an undefined or null result and since the stack of assigned internal net values is now empty, this final vector represents a test pattern for the original assumed fault.[3]

This example was particularly straightforward in the sense that no alternative paths existed for propagating the fault (each signal had a single fan-out), and no alternative assignments during the consistency back-tracing needed to be considered. In the general case, as discussed next, the test pattern generation algorithm must be able to retrace its decision tree when an inconsistent assignment is uncovered.

The algorithmic description that follows summarizes that of reference 13.1, which provides a detailed description of its implementation. The algorithm entails selecting the initial D vector that represents the assumed fault and attempting to successively intersect the network D vector with block D vectors (propagating

---

[3] Since values have been assigned to all network PIs, this must also conclude the test pattern generation step; either a valid test has been found or an inconsistent situation is uncovered, which would necessitate the selection of an alternate path for propagating the fault.

the fault forward) and with block prime implicants (back-tracing to confirm the consistency of assignments). During the test pattern generation procedure, a list of blocks is maintained; these blocks are on the "frontier" of the connected D chains in the goal of reaching a primary output. To be present on this activity list at any point in the algorithm, a block has one or more Ds (or $\overline{D}$s) on its input(s), an undefined value of x on one or more inputs and (therefore) on its output, and whose possible propagation of the D chain has not already led to an inconsistent result; in short, these are the blocks that are candidates for propagating the D chain, with an input (or inputs) yet to be assigned from the block's D-vector table.

Among the blocks in the activity list, the algorithm must prioritize the sequence in which a block is processed. Assume that the blocks are levelized; that is, each block (and block output) is assigned a level number equal to one plus the highest level number of any block input signal. (Network primary inputs have a level of zero. Since it is assumed a priori that the network is strictly combinational, this levelization is well defined; no feedback paths are present in the network.) One possibility for selecting blocks for assignment from the activity list is to choose the one with the highest level number; in this manner, a rapidly protruding frontier toward a PO would develop. Alternatively, an orderly progression to a PO would result from selecting the block in the activity list with the lowest level number. With either technique, once a network output is reached by the connected D chain, the processing of the activity list ceases; since any network PO is assumed to be an observable point, it is not necessary to try to extend other D chains toward other POs.

It is desirable to determine the ramifications of block signal assignments in the selected D vector as quickly as possible in order to eliminate unnecessary execution time pursuing what could eventually be an inconsistent result. Therefore, once a decision is made in connecting the D chain through a block, the implications of resulting line assignments (from x to 0, 1, D, or $\overline{D}$) are calculated. This means that all other signal assignments that are a *forced* consequence of the totality of current network values are investigated. The calculation of implied network values proceeds immediately after each block vector selection to as large an extent as possible. This is in contrast to the example given earlier in this section, which simply saved the Boolean signal assignment in a stack for later evaluation. When evaluating all the signal assignments directly resulting from the selection of a particular D vector for a block in the activity list, the following changes to other block signal values can be examined:

1. The assignment of a D (or $\overline{D}$) value to the block output extends to other block inputs in the fan-out of the activity list block.
2. The assignment of a value in the D-vector to an activity list block input extends to the driving block output and other block inputs in the net.
3. Additional values on block inputs and outputs may be implied from the signal assignments in item 2.

4. Immediate propagation of the D chain may result from the signals assigned in items 1 and 2 and those calculated in item 3.

The steps of the generalized D-algorithm for each network fault may be outlined as follows:

Step 1.   Begin with a network D vector consisting of undefined values (x's) for all signals. Levelize the network description; ensure that this logic description is strictly combinational and contains no feedback paths. It will be useful to initialize the following data structures (Figure 13.15):

1. A list of lists to store the fan-in and fan-out block numbers for each block in the network.
2. The activity list; blocks with some, but not all D (or $\overline{D}$) input values, and undefined outputs.
3. A list of block numbers whose *outputs* have been defined since the last arbitrary decision was made regarding the selection of a D vector or prime implicant. This list will be examined to determine the possibility of implied block input values. A pointer into this list is used to select the next block to evaluate for back-tracing.
4. A list of block numbers, similar to the one in list 3, whose *inputs* have been defined since the last arbitrary decision; these are candidates for direct calculation of further forwarding signal values.
5. A list of block numbers that have been *added* to the activity list since the last arbitrary decision.
6. A list of block numbers that have been *deleted* from the activity list since the last arbitrary decision.
7. A structure to represent the current decision tree path of alternatives (Figure 13.10).

Lists 5 and 6, along with list 3, will be used to undo the changes in the network status back to the point of the last arbitrary decision in the decision tree if an inconsistent assignment is uncovered. Lists 5 and 6 will restore the activity list to its previous condition, while all the blocks in list 3 would be returned to x output values in the network vector. The information in structure 7 determines the alternative assignment to be selected.

Step 2.   If the assumed fault is on a block output pin, select an element from the prime implicant table (i.e., the good machine) that produces a conflicting value to the assumed fault value. Then, replace that output value with D ($\overline{D}$) if the good machine value is 1 (0). For an assumed fault on a block input pin, set the *driving* block's value to the good machine result, and select a suitable element from the table of D vectors for the *driven* block type to propagate the fault to the

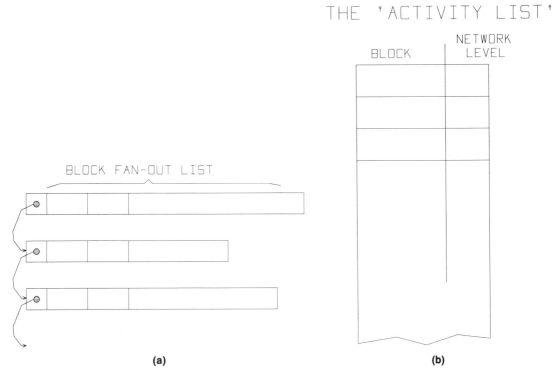

(a)   **(b)**

Figure 13.15.   Data structures used during the test pattern generation D-algorithm. (a) The block fan-out list. (b) The activity list: blocks with D (or $\overline{D}$) inputs and as yet undefined outputs. These are the candidates for propagating the D-chain. (c) The list of blocks whose outputs have changed. These blocks will be examined to see if their currently undefined inputs can be assigned an implied value; the back-tracing implication step is performed frequently in the D-algorithm to detect inconsistent assignments as early as possible. (d) The list of blocks whose inputs have changed. The assignment of a value to a net will add the blocks in the fan-out of that net to this list. These blocks are candidates for further forward calculation and signal assignment. (e) The list of blocks added to the activity list since the last arbitrary decision. (f) The list of blocks deleted from the activity list since the last arbitrary decision *(continues)*.

driven block's output (Figure 13.16). Build a D vector from the resulting signal assignments and intersect the result with the overall network vector. (Since the network vector contains all x's at this point, the assigned signal values become the network values.) Update the block numbers in each of the appropriate lists described in step 1.

The initialization of the network fault condition may already involve an arbitrary decision regarding the prime implicant (output pin fault) or D vector (input pin fault) selected for the block. The decision tree structure should incorporate these potential assignments, if it proves to be necessary to select an alternative.

LIST OF BLOCKS WHOSE   LIST OF BLOCKS WHOSE
OUTPUTS HAVE CHANGED   INPUTS HAVE CHANGED

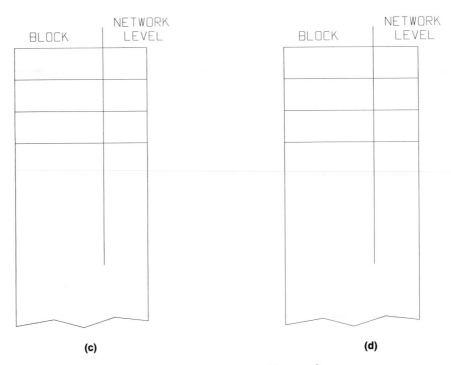

(c)   (d)

Figure 13.15   (*Continued*)

Step 3.   This step begins the fundamental loop within the algorithm. Initiate
the back-tracing of assigned values, referring to the list containing blocks whose
outputs have just changed. Only blocks in this list whose outputs have been assigned Boolean values (0 or 1) need be examined during back-tracing. This list
also includes those assigned to D and $\overline{D}$ to reflect the overall change in network
status. As will be elaborated on shortly, the nature of the algorithm is such that
a D or $\overline{D}$ value on a block output results from a *forcing* input condition; no further
implications of undefined input values are possible. Examine the inputs of each
of these blocks to determine if the currently undefined block input value(s) can
be directly inferred from the new block output value. If this is indeed the case,
then (1) assign these implied values to the signals in the network D vector, (2)
concatenate the driving block numbers of these implied input values to the end
of this same list, and (3) for other blocks in the fan-out of the nets affected, add
their block numbers to the list of blocks whose *input* values have changed (Figure
13.17).

LIST OF BLOCKS ADDED TO
THE ACTIVITY LIST SINCE
THE LAST ARBITRARY DECISION

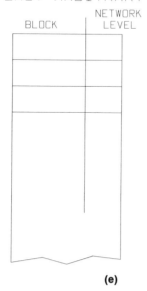

(e)

LIST OF BLOCKS DELETED FROM
THE ACTIVITY LIST SINCE
THE LAST ARBITRARY DECISION

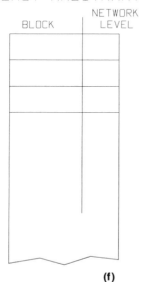

(f)   **Figure 13.15** (*Continued*)

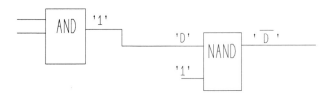

Figure 13.16. If the assumed fault is on a block input pin, the driving block's value must provide the opposite Boolean value from the assumed stuck-at fault value; in addition, an element from the D-vector table of the driven block must be selected so as to propagate the D (or $\overline{D}$) value.

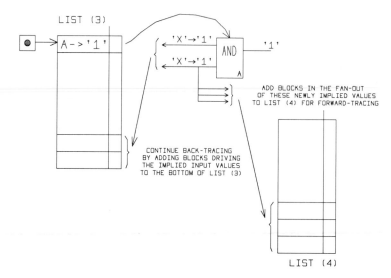

Figure 13.17. When signal values are implied during the back-tracing step of the D-algorithm, the various lists of blocks whose input or output signals have changed need to be updated. The recent assignment of the output of block *A* to 1 has resulted in a *complete*, nonarbitrary assignment of all undefined block inputs. After accepting these values, update the lists of block numbers.

Another possible result of a block's evaluation is that the output value of the block in this list is *inconsistent* with the already defined block inputs; if this is the case, go to step 6.

The last possibility of the block evaluation is that no new input assignment is made (and no inconsistency is uncovered); either the current input values are sufficient to produce the assigned output value or an arbitrary assignment of input values would be required. In this case, the pointer to elements in list (3) is simply incremented.

Continue executing this step until all blocks in this output signal change list have been evaluated. For the sake of execution efficiency, it is only necessary to evaluate in this step the blocks whose level numbers are less than the highest level number previously encountered in propagating the D chain forward. A block in this list that is ahead of the D-chain frontier could not have been put in the list by the process of selecting a block D vector. This block number was therefore added by the step that evaluates the *forward* results of signal assignments to block input pins. This forward calculation process is the next step in the algorithm. If the forward calculation placed this higher-level block number in the output change list, its current inputs already dictate its output value, and no further nonarbitrary block input assignment is possible.

Step 4.   Begin the forward calculation step; refer to the list of block numbers whose inputs have recently changed from x to either 0, 1, D, or $\overline{D}$. For each block in this list, compare the values of the block inputs and output against the prime implicant table and D-vector table of the block type. If the block output has an assigned value and an inconsistency results, go to step 6. If the block output is currently undefined and the input signals are sufficient to determine the output, that output value is set in the network vector and the other blocks in the fan-out of this signal are appended to the end of this same list. The block number that now has a defined output value should be added to the output change list, not for back-tracing but to reflect the cumulative changes in the network condition since the last arbitrary decision.

It is also possible that a *different* input signal to the same block may be uniquely determined (refer to the example in Figure 13.18). When back-tracing to the OR block in the figure as part of step 3, the implications ceased due to the available arbitrary choice; subsequently, the assignment of a block input to 0 in evaluating a forward propagating signal permits the unique assignment of the remaining undefined input. After making this signal assignment, the driving block number should be added to the list of output changes (list 3), and other block fan-in's in the same net should be added to this forward calculation list. If this occurs, the back-tracing block list contains new elements and step 3 may be reexecuted.

The final possibility is that the new block input value results in no change in existing block output (or input) values; this element in the input change list is skipped.

During the execution of this forward calculation step, the activity vector

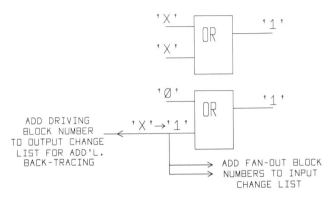

Figure 13.18.   The forward-propagating step may result in the implied assignment of undefined block *inputs*, as indicated for the OR gate. The assignment of the block output to 1 during back-tracing did not result in a signal assignment; the subsequent forward propagating step assigned a 0 to one of the block inputs, dictating the value assignment for the remaining undefined input.

must be updated. If an output value assignment for an activity list block results, that block should be deleted from the activity list (Figure 13.19a). If a D or $\overline{D}$ input change results in no determination of the block output, that block number should be present on the activity list (Figure 13.19b). If a network PO is reached by the D chain during the execution of this forward calculation step, then the current status of the test vector analysis is saved, and the algorithm begins consistency checking in step 7. The current status of the analysis consists of storing

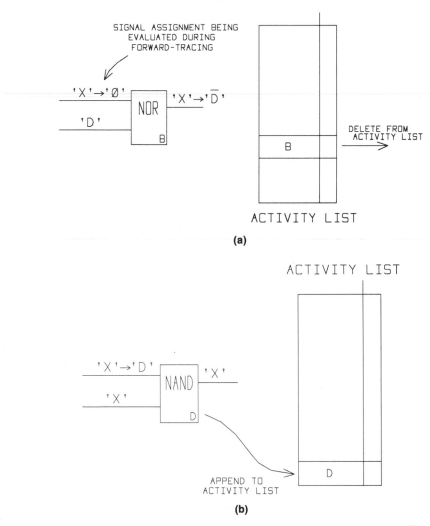

Figure 13.19. (a) When an assignment of a block output (with one or more D or $\overline{D}$ inputs) results from forward-tracing, that block should be deleted from the activity list. (b) If the forward-propagation of a D (or $\overline{D}$) value does not continue past a block due to an undefined signal, that block should be added to the activity list.

the information required to be able to return to the network condition at the point of the last arbitrary decision, in case an inconsistent condition is found. This status includes the cumulative changes to the block activity list plus the newly-defined block output signals (lists 3, 5, and 6), as well as the next alternative of the current decision-tree structure.

The forward calculation and the previous back-tracing steps utilize the lists of blocks whose input and output signals have been assigned or implied from choosing a D vector to propagate the chain through a block. When these lists have been exhausted and a network PO has not been reached, it is necessary to examine the activity list and make a selection on how to further propagate the D chain. If the activity list is empty at this point, the D chain has "died" and no test pattern can result for the initial D vector selected for the assumed fault; in this case, go to step 6. If the activity list is not empty, continue on with step 5.

Step 5.   Select a block from the activity list; in the algorithm developed in reference 13.1, the block with the lowest level number was chosen, effectively trying to extend all branches of the D chain at a constant rate. If the activity list contains more than one block, the current status is saved in order to be able to permit the analysis to return to this point if an inconsistency is subsequently discovered. Although the algorithm may have some means of determining the order of selection (lowest level number, for example) this still reflects an arbitrary decision. The need may subsequently arise to be able to reset the analysis to this point and select a different block from the activity list. Having selected a block from the activity list, it is necessary to assign values to the undefined block inputs and block output in the network D vector; the values to be assigned to undefined block inputs are listed in Figure 13.20 based on the block type.[4] If the selected block is an XOR gate, an arbitrary decision is made initially in the signal assignment to the undefined input of the gate. The network status and nature of this decision are saved in order to be able to return to this block and select the opposite value. The resulting changes in signal values to driving block outputs and driven block inputs are reflected in the appropriate lists; since the back-tracing and forward calculation lists contain new elements for evaluation, return to step 3.

Step 6.   This step is the *backup* process when an inconsistency is found in the test vector being developed. Two situations are possible: no alternative decisions are pending at this point, or one (or more) arbitrary decisions have been made in step 5 and a stack of saved status conditions has been produced.

If no alternative decisions regarding activity list blocks or XOR gate input

---

[4] Using the table of values in Figure 13.20 would preclude assigning a D (or $\overline{D}$) value to a currently undefined input; this option may indeed be useful for extending the D chain through AND, NAND, OR, or NOR gates where logic paths in the network have reconverged from the fault to multiple gate inputs. Roth's selection of a slowly moving D frontier will tend to introduce D values methodically through such a path, precluding the need to investigate assigning a D (or $\overline{D}$) value to an undefined input at the edge of the D frontier.

LOGIC BLOCK TYPE	VALUE TO ASSIGN TO UNDEFINED INPUTS
AND	1
NAND	1
OR	0
NOR	0
XOR	0 / 1

Figure 13.20. After selecting a block from the activity list, undefined block inputs are assigned the values given in the table based on block type. The initial assignment of a Boolean value to an undefined XOR gate input reflects an arbitrary decision, which may need to be reversed if an inconsistency is found.

signal assignments remain, it is necessary to return to steps 1 and 2 to reinitialize the network analysis; when reexecuting step 2 this time, select a different element from the prime implicant table for an output pin stuck-at fault or from the D-vector table for an input pin stuck-at fault. In short, select another top branch of the decision tree structure. If all branches have been exhausted, the fault is indeed untestable. If a different initial choice is possible, reinitialize the block lists and network vector and continue on from that point to step 3.

If a recent arbitrary decision in step 5 means that an alternative is available, it is first necessary to restore the analysis to its most recently saved status. If the previous decision involved propagation through an exclusive-OR block for the first time, now select the other Boolean signal assignment for the undefined XOR block input (Figure 13.21). If the decision to select a particular block from the activity list proved to be problematic, select instead one of the remaining blocks from that list; also, delete the problematic block from the activity list. If the activity list still contains more blocks, save the current status at this point to be able to return to yet another alternative. For this newly selected activity list block, assign the values to the undefined inputs and outputs as given earlier in Figure

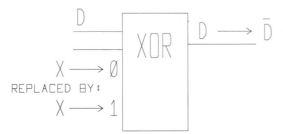

Figure 13.21. If an arbitrary decision has previously been made in propagating the D-value through an XOR gate, select the other Boolean signal assignment for the undefined block input when executing the decision-tree alternative step in the D-algorithm.

13.20; update the necessary lists for the blocks whose signal values have changed and return to step 3.

Step 7.   This step involves the consistency checking of the test vector status once a PO is reached by the D chain; it requires constructing and subsequently evaluating the list of blocks whose outputs have been assigned values without the input values yet assigned so as to *force* this result.[5] Blocks with undefined outputs are not included initially, but may get added to this list by the back-tracing of consistency signal implications. Each block in this dynamic list will eventually be evaluated to determine if an overall consistent assignment of signals can be found; if all the signal implications can be completed without an inconsistency (i.e., when the consistency list becomes empty), a valid test pattern has been obtained. The consistency procedure begins by selecting a block from the newly created list and then searching for a prime implicant for the block that agrees with the current network signal definitions. For this block, a number of prime implicant alternatives may be available to produce the desired output value (consistent with the block's current input definition); selecting from among these alternatives represents an arbitrary decision and requires that the consistency network status be saved. (At least two implicants must initially be available, or else the back-tracing procedure in step 3 would have completed this block.) Once a prime implicant is selected, the block is deleted from the consistency list, and the back and forward signal assignments that result are evaluated immediately, just as with steps 3 and 4. The lists of blocks maintained and the stack of decision tree status results produced are similar in nature to those described in earlier steps. The consistency list is analogous to the activity list used to propagate the D chain; the lists 3, 5, and 6 described in step 1 are likewise maintained for the consistency procedure in order that changes in the network can be undone (back to the previous arbitrary decision), if required. The activity list is not altered.

The consistency procedure continues by back-tracing signal assignments toward PIs, using the list of output signal changes resulting from the prime implicant selected. The back implications of this assignment are pursued as fully as possible, until no more forced implications are evident. During this process, blocks whose outputs are defined but whose input assignments now require an arbitrary decision are *added* to the working consistency list. Blocks whose *inputs* receive signal assignments are listed separately and will be evaluated in the following step.

Step 8.   After step 7 progresses to the point where an arbitrary decision is again required from the consistency list, this step first examines the blocks whose input values have been assigned. This forward examination may result in further input signal implications; refer again to the example in Figure 13.18. A completed

---

[5] Using the table of values in Figure 13.20 to assign to undefined inputs, it will not be the case that a D (or $\overline{D}$) value will appear on a block output without a forward calculation based on the input condition; only blocks whose outputs are Boolean values need be evaluated during the consistency operation.

input signal assignment in this step results in the deletion of this block from the consistency list (the inputs are now defined for the block output) and the addition of the block driving the implied signal to the output value change list for further potential back-tracing.

For the block implicant selected, the back-tracing and forward implication consistency steps 7 and 8 iterate until no further signal assignments result. When this occurs, the next block in the consistency list is extracted, a prime implicant is selected, and the back-tracing of step 7 is restarted. As the prime implicant selection for a new block from the consistency list will be arbitrary, the consistency status of the network is pushed to the stack of alternatives.

When an inconsistency arises, alternative possibilities are examined at the last arbitrary decision; this involves either (1) selecting another prime implicant for the current consistency list block, or (2) if all suitable prime implicants have been examined, backing up the decision tree to the previous consistency block to select a different prime implicant. In either case, the consistency status prior to the arbitrary decision is restored. If the decision tree of alternative prime implicant selections is completely empty, the consistency procedure cannot complete a network signal assignment to force the values that are present in the network D vector. It is therefore necessary to leave the consistency portion of the algorithm and return to the status of the last arbitrary decision made when propagating the D chain, selecting the next alternative (step 6).

This concludes the general description of the D-algorithm for test pattern generation for an assumed stuck-at fault in a combinational network.

It was mentioned previously that for networks with a high percentage of XOR gates, the D-algorithm decision tree may get to be very large, and the execution time may grow considerably. A different approach to test pattern generation for an assumed fault was proposed by Goel (reference 13.2), who replaced the alternating back-tracing and forward-propagating steps of the D-algorithm with a forward-implication process from the combinational network's primary inputs. Basically, this algorithm searches the possible combinations of PI values for a test pattern that will detect the assumed fault at a network PO. The algorithm proceeds by continuing to assign values to primary inputs (all initially at x) until it is discovered that (1) a test pattern has indeed been found, or (2) the current assignment of PIs cannot possibly result in a test for the fault, regardless of the values chosen for the remaining undefined PIs. If the latter condition is determined, the last previous arbitrary PI assignment is reversed, and the resulting network behavior is reevaluated. In the end, either a test will be produced, or the exhaustive examination of all 2^n network input combinations will have been performed, with the fault declared untestable (n being the number of network PIs).[6]

_____

[6] For large n, it would be advisable to abandon the test pattern generation step after a particular number of alternative decisions have been made, rather than using the execution time to fully investigate all 2^n patterns.

This "Path-Oriented DEcision Making" algorithm (PODEM) is flow-charted in Figure 13.22. The steps are as follows:

Step 0.   Initialize all network PIs to x, set the assumed fault node to its D (or $\overline{D}$) value, and initialize the decision tree stack.

Step 1.   Determine if a test for the assumed fault is still a possibility, with the current PI assignment; if not, go to step 5. Developing a test for the assumed fault is still possible, unless either (1) the PI assignment sets the Boolean value of the faulty net to the assumed stuck-at value, or (2) it is determined that there exists *no* signal path from a net with a D (or $\overline{D}$) value to a PO with all x's on that path. The use (and rules for propagation) of D values in this algorithm is the same as used by Roth.

Step 2.   Since a test is still possible, assign a value to an undefined PI; place this assignment on the decision stack, indicating that the opposite Boolean value has not yet been examined.

Step 3.   Propagate throughout the logic the results of the current PI assignment.

Step 4.   If a D (or $\overline{D}$) reaches a PO as a result of this assignment, a test has been found and the algorithm terminates. Otherwise, return to step 1 to determine if a test remains a possibility with this assignment.

Step 5.   It has been determined that the current PI assignment cannot lead to a suitable test. Therefore, an alternative decision regarding a PI value is required. Examine the decision stack; if it is empty, all 2^n possibilities have indeed been examined and the fault is therefore untestable (box 9 in Figure 13.22).

Step 6.   If the stack is not empty, determine if the PI on the top of the stack has been assigned one or both of its possible Boolean values. If only one value has been examined, go to step 8.

Step 7.   Both values for this PI have been evaluated and no test has been found; it is therefore necessary to pop this input off the stack, returning to the previous input assignment. After popping the unsuccessful PI off the stack, reinitialize its value to x. Go to step 5.

Step 8.   Assign the opposite value to the PI at the top of the stack; indicate that both values have now been evaluated. Go to step 3 to determine the implications of this assignment.

To improve algorithm efficiency, PODEM uses several criteria in steps 1 and 2 for selecting among the remaining undefined primary inputs the particular

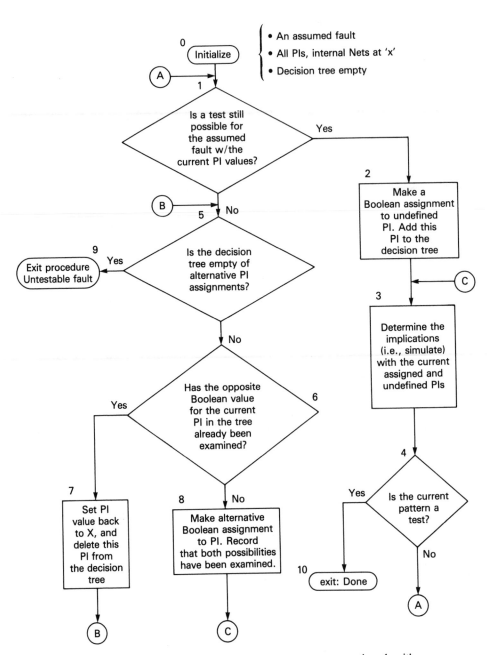

Figure 13.22. Flowchart of the PODEM test pattern generation algorithm.

signal (and Boolean value) that will best meet the goal of generating a test pattern for the assumed fault. Fundamentally, several steps are taken for each new PI assignment to determine an *internal* net objective (and objective value) to help propagate the D chain toward a PO; using this net objective value, a heuristic back-tracing procedure is then executed to reach a network PI with an initial assignment value. (This back-tracing is similar to the D-algorithm consistency check, but is much simplified; consider the example in Figure 13.23. The heuristic nature of the back-tracing process involves the selection of the particular blocks that are intended to form a sensitized path back to a PI; the Boolean assignment objectives for the blocks in this path are based upon the block type. For more details, refer to Goel's paper.)

Fault Simulation

After the test pattern for the assumed fault has been generated, it is necessary to fault-simulate the pattern to determine the set of *all* network faults detected by the pattern. In this manner, the list of untested network faults yet to be assigned for test pattern generation is reduced and the fault diagnostic table is updated for this particular test pattern. Several techniques have been developed for fault-simulating a logic network given the input test pattern; among those that strive for high execution efficiency are the techniques of *parallel fault simulation* and *deductive fault simulation* (Problem 13.1).

The technique presented here is the TEST-DETECT algorithm described by Roth (reference 13.1); the algorithm assumes that all logic signal values in the good machine are well defined (0 or 1) or, equivalently, that all primary inputs in the test pattern are 0 or 1; none remain at an undefined *x* value. (This will commonly *not* be the case; many test patterns, as generated by the methods

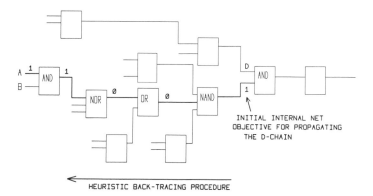

Figure 13.23.   Heuristic procedure used to select the next PI value among the undefined network inputs; an internal net objective is selected to help propagate the D-chain toward a network PO. Then an abbreviated back-tracing procedure is used to determine a suitable PI (and PI value) consistent with that internal net objective.

previously illustrated, will have one or more PI values still at x. The efficiency of individual patterns in detecting multiple network faults is typically increased if PIs at x are indeed assigned a value. Test pattern compaction techniques have been developed to provide this assignment in an effort to reduce the overall size of the test file; two such compaction techniques will be discussed shortly.)

The TEST-DETECT algorithm begins with the set of all good machine logic signal values, as determined from the test pattern of 1s and 0s. Individual stuck-at faults on block pins are then assumed (faults that differ from their good machine values), and the affected block output is subsequently replaced with D (or $\overline{D}$, as appropriate). Using the remaining good machine values to propagate the D chain, the algorithm then determines if a D (or $\overline{D}$) value reaches a network PO. The simulation of the results of this *fault insertion* is greatly simplified over the forward-propagation step of the D-algorithm since *all* remaining block input values are known; the evaluation of the fault insertion can simply proceed level by level through the affected gates toward the network outputs. If no D (or $\overline{D}$) value reaches a network PO for the inserted fault (i.e., the D-chain dies), then the pattern is not a test for that particular fault.

To improve the efficiency of this fault insertion and execution procedure, the algorithm inserts faults for the higher-level nets first (closer to network POs). A list of blocks for which the test pattern detects the opposite stuck-at fault output value is developed during this procedure. When evaluating lower-level nets, it may not be necessary to simulate the fault inserted for that net fully throughout the network; if the propagation of the D-chain of values at any point in the simulation is ever reduced to a single block output, the inserted fault is detected only if this higher-level block output is also detected. Starting with the highest-level blocks and maintaining the detected fault block output list will permit the simulation for subsequent faults to terminate early, if the D-frontier at any point consists of but a single block output.

It is likely that the pattern that results from the test pattern generation algorithm for an assumed fault will leave one or more network primary inputs unassigned; it is beneficial to subsequently attempt to give values to those remaining inputs in order to increase the number of faults detected by the single pattern and reduce the test file size. Two such compaction techniques will be described: static and dynamic compaction (reference 13.3).

The process of static compaction refers to the following steps:

1. Generate independent test patterns for a number of individual stuck-at faults in the combinational network.
2. From that list of patterns, continue to compact two or more patterns together.

Two patterns can be compacted if there are no conflicting 1 and 0 assignments to any primary input; an undefined PI in one pattern combined with an assigned Boolean value in the other requires that the Boolean value be present for the PI in the compacted result.

3. When step 2 has completed its iterative process, fault-simulate the compacted set of tests to determine the network fault coverage (and to generate diagnostic lists for each fault); if additional network faults remain to be analyzed, return to step 1.

The fault simulation is performed after the test compaction so that fewer patterns need to be simulated, reducing the execution time costs. However, as mentioned in Section 13.1, the compaction search process among the existing patterns is itself a time-consuming step. Alternatively, a dynamic compaction technique can be used.

Briefly, the dynamic compaction technique involves expanding on a partial test pattern (one with some undefined primary inputs) until all PIs receive Boolean values. Once an initial pattern for an assumed fault has been generated, a constraint is imposed that the currently assigned PI values are fixed. The first step in the dynamic compaction procedure is to select a *new* stuck-at fault objective for generating a test pattern, using the previously assigned PI values while assigning values to the remaining undefined PIs. (Since only a single network fault is assumed to be present at any time, the D-chain values from the original test pattern generation step are replaced with their calculated Boolean values.) Suitable criteria for selecting this new fault objective include the following:

1. The fault has not yet been tested by previous patterns in the test pattern set.
2. The output of the block associated with the faulty pin remains at x after the logic network has been evaluated with the partial pattern at the primary inputs.

The second of these two criteria is not absolute; for example, there is no guarantee that the new objective fault will indeed be tested by some assignment of the remaining PIs just because its block output is currently undefined. Likewise, it would be possible to select a new stuck-at fault objective for a block pin not at x but with a defined Boolean value (the fault being of the opposite value). This may indeed result in the assignment of primary inputs that detects this (and possibly more) network faults. Rather, criterion 2 is used to select new fault objectives for which the existing primary input values do not constitute a test for this fault (since the pin is at x), new PI values will need to be assigned to detect this fault, and the probability is high that this fault will be testable even with the existing PI values. Once a new fault objective has been selected, use the test pattern generation algorithm to determine if a test can be found, permitting Boolean assignments to only those PIs currently undefined. If the process was successful and undefined PIs remain, select another fault objective and repeat the procedure.

When the selected fault objective cannot be tested with the existing PI values (and there are one or more primary inputs that remain unassigned), the dynamic compaction algorithm presented in reference 13.3 suggests that an *assessment* step be executed. In short, it is suggested that an evaluation be made to determine the benefits and costs of continuing to select new fault objectives to complete the assignment of undefined PIs. If the percentage of primary inputs still at x is small and/or the execution time for completing the test pattern generation step for a new fault objective has grown significantly, it is recommended that the fault objective selection and test pattern generation loop be terminated; the costs of continuing to execute the selection and generation steps will rapidly become prohibitive. Otherwise, select a different fault objective and return to the test pattern generation process.

Once all primary input values are assigned by the dynamic compaction procedure, fault-simulate the test pattern to determine the set of additional faults that has been detected. If the assessment step described previously exits (due to high costs) with one or more primary inputs still undefined, *randomly* assign 1s or 0s to those remaining PIs before fault simulation with the goal of picking up a few additional faults.

Summary

This section has described test pattern generation, test compaction, and fault-simulation techniques used to detect the presence of stuck-at-faults in combinational networks. Section 13.3 will address a design for testability methodology for testing sequential circuits, as well as the application of test patterns to and the observation of responses from combinational networks internal to a VLSI integrated circuit.

Alternative strategies for providing more efficient test pattern sets for VLSI designs are currently being researched; among these techniques are (1) signature analysis, commonly incorporating built-in self-test logic networks, (2) exhaustive testing, (3) testing with patterns developed for logic simulation, and (4) random pattern testing. The first two of these techniques include means of compacting the results of applying a sequence of test patterns to the network into one final vector of output values; the uniqueness of this final output vector should indicate the validity of the network. Section 13.5 briefly discusses these approaches. The last two techniques deserve some special mention. The lowest resource solution to test pattern generation is to use a subset of the logic simulation patterns as the test pattern set. In this manner, the sequence of applied patterns will be more meaningful than an automatically generated set of patterns. However, functional patterns are *not* designed specifically to sensitize paths for propagating fault values to observable nodes; as a result, their fault coverage percentage will commonly be low. Random test pattern generation is a very low resource technique for detecting a substantial percentage of faults; no CPU-intensive test pattern generation algorithm for assumed faults is used. A common test strategy is to *combine* random patterns *with* automatically generated patterns into a single set; the random patterns are produced initially and fault-simulated, whereupon the pattern generation algorithms are exercised against the remaining (more difficult) faults. Last but most certainly not least, the test pattern set (and the network test coverage) can always be enhanced by the addition of manually generated patterns.

13.3 LEVEL SENSITIVE SCAN DESIGN

Section 13.2 described test pattern generation techniques for combinational logic networks. This section will describe a comprehensive design system methodology that has been developed to assist with the application of those techniques to the much larger problem of VLSI testing. The impacts of this methodology on logic design, circuit design, and design system library development are substantial; some of the testability constraints imposed have been discussed in previous chapters. Although several similar methodologies have been developed (see Problem 13.2), this section will concentrate on the level sensitive scan design (LSSD) technique, as first described by Eichelberger and Williams (reference 13.4).

The LSSD methodology encompasses two basic design approaches: (1) all logic networks shall be strictly level sensitive, and (2) all sequential storage elements, with the exception of on-chip memory arrays, should be connected in shift-register (or scan) fashion, in addition to one (or more) data input ports. These two design restrictions will permit combinational logic networks internal to the chip to be controlled and observed to a maximal extent (Figure 13.24). A test pattern is applied to the inputs of the combinational network by a combination of chip input values and/or the values shifted into a shift register string. The output responses of the network are observed at chip outputs and/or through the parallel load plus serial scan of the values in a shift-register string. The shift-register data latches are abbreviated as SRLs. The testing of the shift-register string may be done independently of and prior to the network testing by shifting a sequence of Boolean values through the entire SRL string, from Scan In to Scan Out. The general logic subsystem structure of Figure 13.24 must pervade the entire logic design of the VLSI part in order to most effectively implement this high controllability and observability technique; some measure of design flexibility must be forsaken to stay within the level-sensitive and scan guidelines.

A level-sensitive logic design implies that the steady-state response of a general logic network is a function only of the Boolean values of the network inputs.[7] Specifically, the steady-state network response cannot be dependent on (1) individual gate (or path) delays, (2) signal transition times, or (3) the order in which multiple input changes to a block arrive. A level-sensitive network cannot contain edge-triggered flip-flops, only *clocked, hazard-free* latches. (Several cir-

---

[7] The steady-state response refers to the signal values of all nets after all transitions have ceased (i.e., a sufficient time period after a network input change).

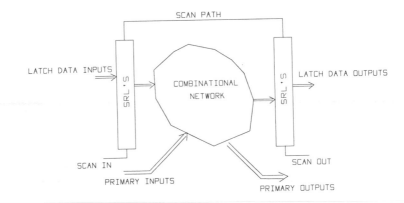

Figure 13.24.   General block diagram of an LSSD test system. Combinational network test pattern inputs are applied via the network primary inputs and via the pattern scanned into the string of shift-register latches. Network responses are observed at the network primary outputs and may be captured in and subsequently scanned out of the SRL string, which has network outputs as latch data inputs.

cuit implementations of these latches were presented in Chapter 9.) No network feedback path providing the ability to store the value of a logic signal is allowed unless the feedback loop is broken by an SRL string with separate nonoverlapping clocks (Figure 13.25). The temporal operation of the logic network is controlled by the frequency and duty cycle of system clock signals to the latches; these signals should be designed to allow the network to reach a steady-state response between clock pulses that update the state of the system (as represented by the values stored in the SRLs). The only delay requirement in a level-sensitive system is that the time between clock pulses be greater than the worst-case delay through the network. In an LSSD chip design, all timing within the chip is effectively controlled by externally generated clock signals.

A block diagram of a shift-register latch is given in Figure 13.26; actually, the SRL consists of *two* latches, denoted as L1 and L2. The L1 latch shown in the figure has two ports; one is the data input port (signals $C0$ and DATA), whereas the other is the input port used to implement the serial shift register function. In its minimum configuration, the L2 latch consists of a single port, whose input comes from the L1 latch output. The clock signal that implements the shift of data from L1 to L2 (denoted as the $B0$ clock in the figure) is distinct from the $A0$ clock that loads the L1 latch from the preceding L2 output in the serial connection. Shifting the latch values through the scan path results from the alternate toggling of *nonoverlapping* $A0$ and $B0$ clock signals connected to L1 and L2 latches, respectively (Figure 13.27). During this time, the system clock $C0$ is held inactive. The overhead associated with incorporating LSSD design for testability tech-

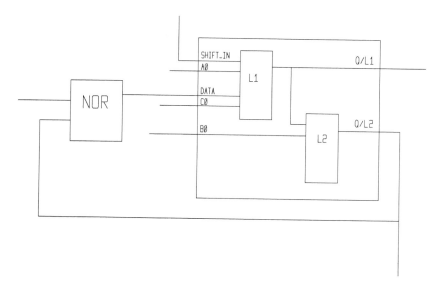

Figure 13.25. In an LSSD structure, no network feedback paths are permitted (other than within the latch designs), unless the feedback loop is broken by the SRL string with non-overlapping clock signals as inputs to the two latches. The design example shown is correct.

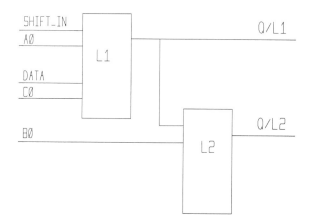

Figure 13.26. Block diagram of a shift-register latch; the SRL consists of two latches, denoted as the L1 and L2, each with a variety of input ports possible. The SRL is an element of a shift-register interconnection when the L2 output is connected to the Scan In input of the succeeding L1 latch in the shift-register string. The A0 and B0 shift clocks to the L1 latches and L2 latches of the entire string should be nonoverlapping for proper shifting operation.

niques on a VLSI chip includes the following:

1. A Scan In and Scan Out pin for each SRL string.
2. A0 clock and B0 clock pins to connect the SRL shift input ports (since the fan-out of the A0 and B0 clocks is commonly quite high, clock powering trees are advisable).
3. The circuit area for the L2 latch (In the shift register of Figure 13.27, the

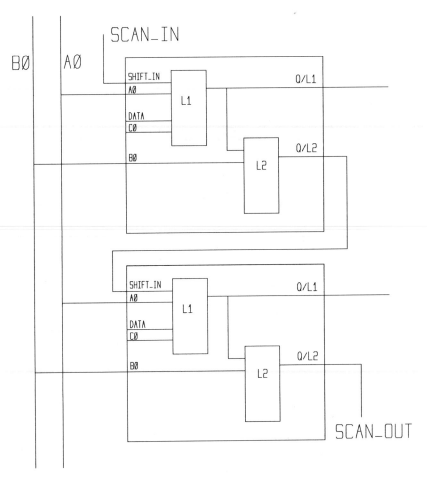

Figure 13.27. Shift-register implementation where the L2 latch output is used only to continue the scan path; the performance requirements of the L2 latch in this implementation are not aggressive, and the additional circuit area occupied by the L2 latch is minimized.

L2 latch output is illustrated as being connected only to the succeeding SRL Scan Input, and not to additional logic gates. As such, the loading on this net is small, and the performance of this signal need only satisfy tester performance requirements; commonly, the L2 latch circuit area can be minimized to reduce the loss in functional circuit density).

An implementation that uses the L2 latch as an independent storage element in the chip design will be illustrated shortly.

In Figure 13.24, it was indicated that the observation of network output values required loading the SRL string through the latch *data ports* and then subsequently toggling the A0 and B0 clocks to shift the captured values out the chip Scan Out pin. Evidently, there must exist design structure rules regarding the system clock inputs to SRLs as well in order to be able to implement this approach at the tester. Examples of these rules follow:

Rule 1.   The output of a latch may feed the data port of another latch, possibly through some combinational network, if and only if the clock that gates this data value into the second latch is *not* a clock of any port of the first latch.

Figure 13.28 can be used to illustrate the reason behind this design rule. The intent of the SRLs is to be able to easily control and observe the behavior of combinational networks to investigate for possible stuck-at faults. If both latches shared the same clock, as depicted in the figure, a new value stored in latch A may desensitize the path in the network used to detect a particular fault and mask the presence of that fault at the input to latch B. Figure 13.29 illustrates why the two latch clocks should be nonoverlapping. If the two clocks overlapped, the value stored in latch B would depend on whether the duration of overlap was greater than the minimum path delay through the network. This delay dependency is not consistent with level-sensitive design principles, which dictate that only *maximum* delay dependencies are permitted.

This rule is difficult to satisfy when strictly L1 latch outputs are being used to represent system data and L2 outputs are used only for scan testing (as in Figure 13.27). Alternatively, the L2 latch outputs may be used to provide system data (in addition to continuing the scan path); Figure 13.30 illustrates such a double-latch design approach. The B0 clock is now synonymous with a system clock input. In this manner, the L1 latch clock design rule is easily satisfied, at the expense of the additional delay through both latches. (Refer also to rule 3 for restrictions on L2 latch output usage.)

Rule 2.   The output of a latch may be used as an input to a block that gates the clock of another latch if and only if the gated clock is *not* an input to the first latch.

This situation is depicted in Figure 13.31; the gating of the clock to latch B requires a 1 output value from latch A. This value would not necessarily be maintained if the gated clock were also an input to a data port of latch A; the resulting clock pulse to latch B would be chopped prematurely.

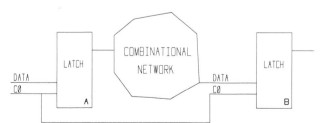

Figure 13.28. Design in violation of LSSD rule 1.

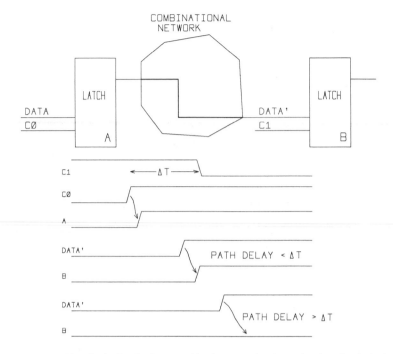

Figure 13.29. The clock signals that provide the test stimulus value into latch A should *not* overlap with the clock signal used to store the network response in latch B, as this introduces a network delay sensitivity. The latch B output value depends on whether the minimum path delay is greater than or less than the overlap $\triangle t$ of the latch clock signals.

Rule 3. System data signals may be from either L1 and/or L2 latch outputs; the L2 output need not strictly be used for test scanning. However, the L1 and L2 latch outputs from the same SRL pair may *not* both be used as inputs to the same combinational network.

The utilization of L1/L2 latch pairs in shift-register fashion permits the loading of an arbitrary pattern of 1s and 0s into either the L1 (or L2) latches *alone*; it is *not* possible to input an arbitrary test pattern into the concatenation of L1/L2 pairs. As Figure 13.32 illustrates, it is not possible to provide the pattern $L1_1 \cdot L2_1 \cdot L1_2 \cdot L2_2 = 1101$ to the combinational network. The only means for an L1 latch to differ from its associated L2 latch is to scan with one more additional $A0$ clock; however, the L1 output is then set equal to the preceding L2 value.

Rule 4. This is the rule that pertains most directly to the parallel loading of network data values into the SRL string for shifting to compare observed versus expected values at the Scan Out pin. Figure 13.33 illustrates a general clock network for an LSSD design. The system clock inputs to all SRLs must be directly

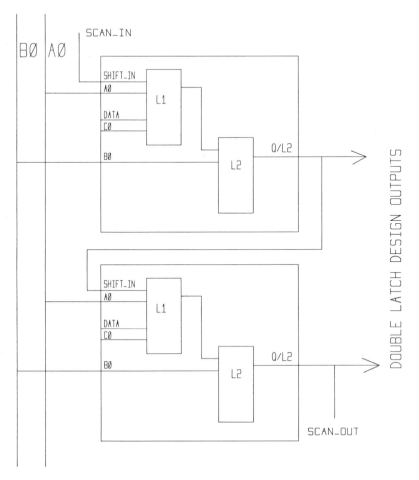

Figure 13.30. Double latch design, where the L2 latch output is used both for continuity of the scan path and for system data storage. The $B0$ clock is now synonymous with a system clock input.

controllable from *chip* inputs in the following manner:

a. An appropriate combination of chip inputs must be provided to enable the scan state:
  1. The $A0$ and $B0$ shift clocks to an SRL may be toggled by toggling the appropriate chip inputs.
  2. All $A0$ and $B0$ clocks to all SRLs may be permitted to be active simultaneously, if all appropriate chip inputs are active.
  3. All system clocks may be held off at SRL inputs during scanning through some combination of input values.

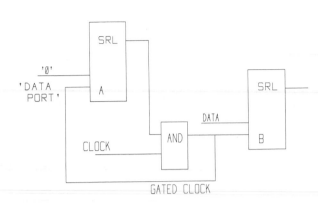

Figure 13.31. Design that violates LSSD design structure rule 2. The output of a latch may be used to produce a gated clock if and only if the gated clock is not an input to the first latch, unlike the situation shown. To provide a clock signal to latch B, a value is scanned into latch A as part of the test pattern to sensitize the gated clock signal; this scanned value may not be the same as the data input value to the port of latch A, which receives this gated clock. As a result, the gated clock pulse to latch B would be chopped prematurely.

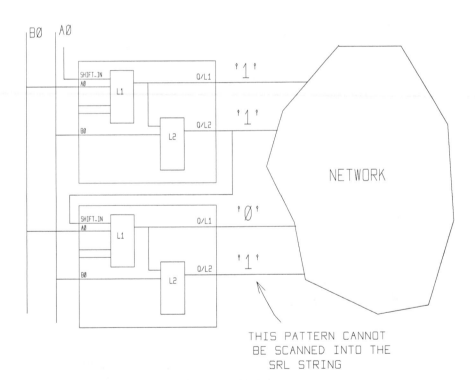

Figure 13.32. Reasons for the imposition of LSSD structure rule 3; L1 and L2 latch outputs may not both be used as inputs to the same combinational network, since a general test pattern vector could not be applied to the network inputs. For example, it is not possible to scan in the vector 1101.

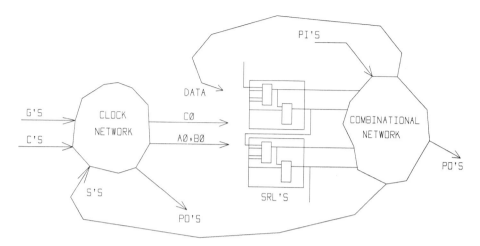

Figure 13.33. General clock network for an LSSD design. The clock system design must incorporate (1) a scan state, (2) clock primary inputs, and (3) a condition of clock gating signals that permits direct toggling of SRL clock input signals from the clock PIs.

b. A set of chip inputs is identified as system clock primary inputs, the Cs in Figure 13.33.

c. These system clock PIs may go through a combinational network; if the clock distribution is more elaborate than a simple fan-out powering tree, the gating path to the SRL clock from a chip input must be able to be sensitized from nonclock chip inputs or preestablished SRL values (the Gs and Ss in Figure 13.33, respectively).

d. Once the gating path is established, the SRL system clock must be able to be toggled by toggling the appropriate chip input, in much the same manner as the $A0$ and $B0$ shift clocks are toggled.

e. Chip clock inputs may *not* be used as SRL data inputs (even if through some combinational network), only as SRL clock inputs or chip outputs.

f. No clock signal may be logically ANDed with the true or complemented value of another clock.

The toggling capability of each SRL clock described in 4a and 4d using only a *single* chip input ensures that each clock is independent of other chip clocks and generally simplifies the generation and application of chip test patterns; at any time, only a single clock signal need be generated to gate data into the latch. Rule 4f is a direct consequence of the desired independence of system clocks. In rule 4a, it was indicated that all $A0$ and $B0$ clocks to all SRLs may be active simultaneously; this condition essentially activates the entire shift-register path from the Scan In pin to the Scan Out pin. A logic value applied to the Scan In pin will flush through the SRL string to the Scan Out pin; a stuck-at fault through the

SRL string may be easily detected. Prior to the application of test patterns to exercise the combinational networks between SRL strings, a flush test of the shift register paths should be performed to verify the functionality of the chip SRLs.

The flush condition may be useful not only for testing purposes, but also during *system* operation. It is common in a VLSI implementation of a synchronous system that an initialization condition be established for the system to perform correctly; this initialization could be viewed, for example, as a reset condition when power to the chip is first applied. During this power-on reset condition, the A0 and B0 clocks could be activated to all SRLs, and the Scan In values for all shift registers established accordingly. Once these clocks return to their inactive values, the initial state of the system, as represented by the collection of all register and latch bit values, is well-defined. To be able to best utilize this feature in a system design environment, it must also be possible to insert logic gates in the shift register path. This design technique is constrained by the following rule.

Rule 5.   Logic gates in the shift-register path (which therefore determine the Scan In value to the succeeding L1 latch) may only be driven by L2 latch outputs and combinational signals directly controlled by chip inputs (Figure 13.34).

In addition, to provide more flexibility in the development of the system's initialization state, the scan path connection may utilize either the positive or negative L2 latch output signal to the succeeding L1 latch; selecting the $-L2$ output will provide a logic inversion in the scan and flush values past that latch.

Design of Clock Distribution to LSSD Shift-register Latches

Clock distribution in LSSD networks is governed by the rules just given; in addition, there will likely be technology rules that must be observed, such as circuit fan-out limitations. (This situation is accentuated by the typically high fan-out of A0 clock and B0 clock signals to all SRLs in a scan path.) The simplest clock distribution is the clock powering tree (Figure 13.35); more complex clock gating networks are also common, but must obey the LSSD rules described earlier. In either case, for system data clocks or shift clocks, the clock must be able to be toggled at chip test by toggling a single chip input.

The nature of a clock gating network is restricted by rule 4f: two clock signals cannot be ANDed together to produce a gated clock as this would require that multiple clocks be active simultaneously. This situation may lead to the performance-sensitive condition illustrated in Figure 13.29. A clock signal may be derived from the AND gating between a clock input and a nonclock input/latch output; a clock signal may also be generated from the OR or two (or more) clock

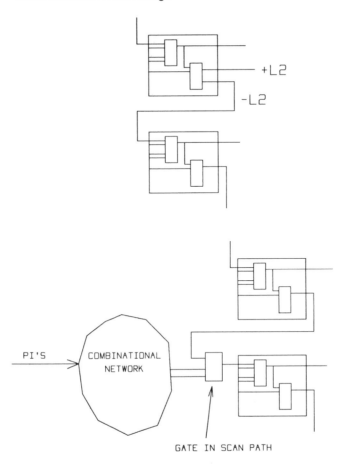

Figure 13.34. LSSD structure rule 5. Logic gates present in the shift-register path may only be driven by L2 latch outputs and combinational signals directly controlled by chip PIs.

inputs, as this would allow latch clock signals to be provided from a single chip input.

Utilization of the L2 Latch

The LSSD design rules regarding the fan-out restrictions of the L2 latch have been given previously; this discussion presents the guidelines for the L2 latch that includes its own *data input port(s)*, in addition to its scan port from the L1 output (Figure 13.36). To distinguish this feature from the conventional SRL latch pair with no independent L2 latch port, this SRL pair is commonly denoted as L1/

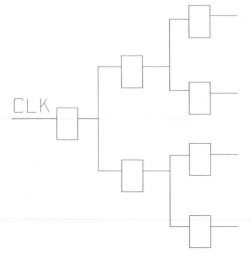

Figure 13.35. Simple clock distribution scheme, the clock powering tree; all gates are buffers.

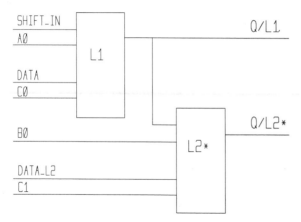

Figure 13.36. The L2 latch may also contain a data input port, other than the shift register input; this latch type is commonly denoted as L2*.

L2* (reference 13.5). The incorporation of a data input port on the L2* latch reduces the circuit density penalty of the previous L1/L2 latch pair, where only the L1 output was used (Figure 13.27), that is, when the L2 latch was included simply to support LSSD testability.

There are no additional LSSD restrictions regarding the system data or clock inputs to the L2* latch; the L1/L2* latch pair may share system data and/or clock signals in common or may be loaded from separate sources with separate clocks. The restriction that prohibits L1/L2 (and now L1/L2*) latch outputs from feeding the same combinational network (rule 3) still applies, so that any general set of pattern values to a network may be scanned into the SRL string.

Incorporating L2* latch designs into a network does present the slight com-

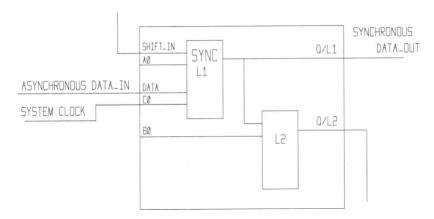

Figure 13.37. The design system library typically contains a special SRL type used for synchronizing an asynchronous input signal to the system clock.

plication that the first shifting clock pulse during serial scan data output will destroy the data values captured in the succeeding latches. For example, after generating signal clocks to load data into the L1 and L2* latches, the first A0 clock pulse will copy the L1 values into the L2*, overwriting those stored test data values. Conversely, if a B0 clock pulse were presented first, the L2* values would write into the following L1 latches. To be able to effectively use the L2* latches for system functionality and LSSD testing, the clock sequencing for test evaluation must be altered to scan the data stored on either the L1 or L2* latches. The algorithms for test pattern generation of the combinational networks between SRL strings remain the same.

Non-LSSD Network-to-LSSD Network Interfacing

A non-LSSD network is one that contains asynchronous sequential logic, edge-triggered storage elements, or, in general, any sequential system that does not adhere to the LSSD rules given previously. For the LSSD design for testability methodology to be successful, there must exist a clean interface between non-LSSD and LSSD networks. The rules that govern that interface follow:

Rule 1.   The asynchronous inputs to an LSSD system are usually fed directly to a synchronizing circuit (Figure 13.37).[8] Transitions of the asynchronous input during the inactive level of the system clock will not be propagated; for the asynchronous signal to be assigned a meaningful value, this signal should not change more than once every system clock cycle.

----

[8] For a more detailed discussion of synchronization, refer to Section 14.7.

Rule 2. The major difficulty encountered in interfacing LSSD and non-LSSD logic relates to the scanning of test pattern data through the shift register latch string to drive the non-LSSD network. Since the non-LSSD network is not level sensitive, the changing latch outputs during scan would likely result in signal transitions on clock inputs that would significantly complicate trying to set the storage elements in the non-LSSD network to a desired state. As a result, the L1/L2 latch outputs cannot be used to drive non-LSSD network inputs; instead, an additional latch is required, a nonshifting storage element, to drive the non-LSSD network (Figures 13.38 and 13.39). To simplify the testing of a non-LSSD network, successive test patterns should present *single input* value changes only.

The nonshifting latch *need not* be present in the signal path for a non-LSSD network driving an LSSD shift-register latch input. The clock signal to the stable, nonshifting latch must also be able to be controlled by chip primary inputs, in the same manner as described previously for L1/L2* clocks. The nonshifting latch clock input may be shared among other nonshifting latches and other L1/L2 latches, provided the design adheres to the clock distribution rules given earlier. In addition, the clock signals that control the nonshifting latches driving the non-LSSD network must be distinct from clock inputs to the SRLs being fed from the network (Figure 13.40). Likewise, for the nonshifting latch to be able to apply a test pattern to the non-LSSD network, the clock input to this latch cannot be used as an input to the non-LSSD network (Figure 13.41).

An important feature of the LSSD design for testability methodology is the straightforward manner in which the design rules can be extended past chip design into higher levels of packaging. Figure 13.42 illustrates the serial scan path connection between chips on a card; the Scan Out of one chip can serve as the Scan In to the succeeding part on the card. In particular, this connection scheme does not require additional pins at the next packaging level (although the card tester

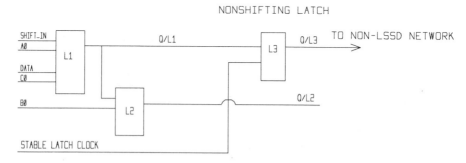

Figure 13.38. For interfacing an LSSD network to a non-LSSD network, a *nonshifting* latch must be used; the latch clock signal for this stable (L3) nonshifting latch must not be used as an input to the non-LSSD network.

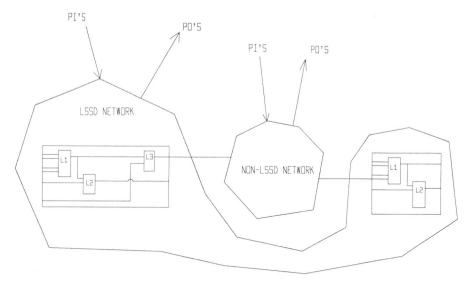

Figure 13.39. Stable L3 latch outputs must be used to feed the non-LSSD network; the outputs of the non-LSSD network may be used directly as SRL latch inputs.

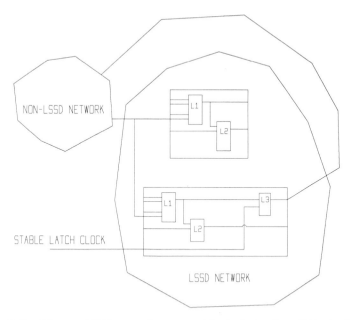

Figure 13.40. The non-LSSD network may provide clock signals to SRLs in the LSSD network; however, the stable latch clock must be distinct from the SRL clock signals provided by the non-LSSD network.

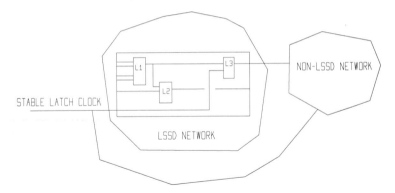

Figure 13.41. Violation of LSSD network-to-non-LSSD network interfacing; the stable latch clock cannot be used as an input to the non-LSSD network.

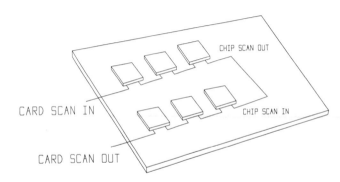

Figure 13.42. The LSSD scan path methodology may be extended to higher levels of system and package design beyond the VLSI chip; as illustrated, a card scan path may be constructed by connecting the Scan Out of individual chips on the card to the Scan In of the successive chip. A minimal overhead in terms of the number of test pins at the card level is maintained.

time is greater). The previously described rules still apply, with references to chip primary inputs changed to card input signals.

Some of the test pattern generation and application techniques that have been developed for logic arrays and memory arrays on VLSI designs will be described in Section 13.4. However, before leaving this discussion, one question remains to be answered: How is the logic design checked for compliance against the aforementioned LSSD testability rules? To implement testability rules checking, a logic tracing procedure, similar to a logic simulation algorithm, can be used (reference 13.6). Briefly, a logic tracing procedure involves initializing all signal values (chip inputs and internal nets) to x; subsequently, one or more inputs are assigned a controlling value, and the propagation of all logic signals affected is simulated. For example, to ensure that all SRL system data ports can be made inactive during scan path shifting, all chip clock inputs are assigned their off values and the design is simulated. Figure 13.43 illustrates that rule violations are indeed detected in this manner if any SRL system clock input is not changed from undefined to inactive. To check rule 4d regarding SRL data port control with a single chip clock input, the level-sensitive design is simulated as before, only with a

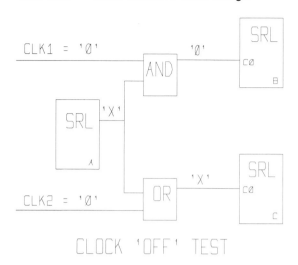

Figure 13.43. The LSSD structure rules are verified by a number of different simulation techniques. To verify that system clocks to internal SRLs are indeed inhibited by an appropriate combination of inactive chip clock PIs, the design is simulated with all internal nets at X and the clock PIs at their off values. In the figure, SRL B is correctly held off, while SRL C is incorrectly designed.

single chip clock input active, one clock input at a time. The simulation of the design under these conditions proceeds by evaluating individual blocks using a controlling logic truth table (Figure 13.44). This controlling truth table differs from a normal simulation table in that the active clock signal is assumed to dominate over *undefined* signal values. This controlling simulation does not seek to determine the other signal values that sensitize the path from chip input to the SRL clock input; the controlling C value simply progresses through the logic gates that are not forced to a Boolean value. During the collective procedure of setting each chip clock input active, all SRL clock inputs in the design must be covered.

The controlling truth table is also used for checking the clock distribution rules pertaining to SRL outputs feeding other SRL clock inputs. After initializing all chip clock inputs to their inactive level and all remaining nets to x, the clock inputs are changed to their active level, as before. However, unlike the previous example, the network simulation *continues after* an SRL clock input is reached; the controlling truth table contains C values for latch outputs when a C value is applied to a latch's clock input. As a result, C values continue to propagate throughout the design. When the propagation of signal changes in the network ceases, the status of all SRL inputs is evaluated. An LSSD clock distribution rule is violated if a C value is present at both clock and data inputs of a latch port or two clock inputs of a latch (Figure 13.45), as this implies both signals are derived from the same original chip clock input. Similar simulation techniques (and logic truth tables) can be used to evaluate the design against the LSSD scan path rules.

The administration of these LSSD design verification procedures requires that the following additional information be provided in the design description to identify the pertinent chip inputs:

1. System clock chip inputs (with their active levels)
2. SRL shift clock chip inputs (with their active levels)

AND	∅	1	C	X
∅	∅	∅	∅	∅
1	∅	1	C	X
C	∅	C	C	C
X	∅	X	C	X

NOR	∅	1	C	X
∅	1	∅	C	X
1	∅	∅	∅	∅
C	C	∅	C	C
X	X	∅	C	X

Figure 13.44. Four-valued logic truth tables using controlling (or clock) values. Note that $\overline{C}$ is not used as a truth table value, only C. Latch outputs are set to C if any of the clock inputs are equal to C.

3. Nonclock chip inputs (and associated Boolean values) that enable the loading of data into the SRL chain to occur

4. Chip Scan In and Scan Out pin (for each shift-register path)

The design compliance against the general LSSD structure rules should be verified as early as possible in the design cycle to avoid major revisions in the design to meet testability and reliability objectives.

A couple of physical design penalties associated with LSSD deserve special mention. The additional loading of the L2 shift input degrades the performance of the L1 latch output; similarly, in a double latch design, the loading of the succeeding L1 latch shift input on the L2 latch output slows the L2 performance. In the first case, the wiring interconnection will be local to the design system library's SRL latch pair book design; in the latter case, however, the wiring interconnection between L2 output and L1 input may be global and may add significantly to the fan-out capacitance on the net. It would be desirable to reduce the testability loading by placing the SRLs in close proximity on the chip image, yet other *system* signal paths may suffer as a result. The question arises, therefore, of what significance to assign the testability interconnections when placing circuits. Alternatively, is it possible to leave the scan path string unspecified in the design until after placement so that an optimum selection of the shortest interconnections between SRLs may be used?

Another physical design methodology option that should be addressed is whether to utilize a single-rail or double-rail scan path, or some combination of both. A single-rail path uses only *one* polarity of the latch output to make the shift register connection, whereas a double-rail path uses both polarities of the latch output (Figure 13.46). If the succeeding latch design includes a data signal inverter circuit, a double-rail interconnection would be advantageous, as it would permit the inverter to be deleted; however, if the double rail path were used for global interconnections, the performance of both latch outputs would degrade

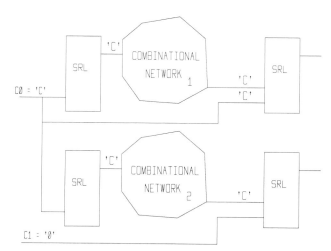

Figure 13.45. A controlling simulation can be used to verify some of the LSSD clock distribution rules; for combinational network 1, the controlling simulation has determined that an SRL data and clock input are both derived from the same clock signal, an LSSD rule violation. The design around combinational network 2 is correct.

(and additional wiring congestion may be introduced). Double-rail paths are therefore most commonly used between L1 and L2 latches in a single SRL pair and between L2 and succeeding L1 latches within a register macro; global nets in the string typically remain a single-rail design.

13.4 ON-CHIP ARRAY TESTING METHODOLOGY

PLA Testing

If a PLA macro on a VLSI chip contains latched outputs for feedback to the PLA inputs, a double-latch LSSD design should be used; the logic functions of the combinational PLA may be tested by using conventional test pattern generation algorithms. This involves developing a detailed equivalent logic PLA test model consisting of the input signal partitioning logic and the NOR circuits of the two arrays.

An interesting technique for general combinational PLA testing is presented in reference 13.7. The unique feature of this method is its universality; that is, the test patterns depend only on the size of the AND–OR arrays and are *independent* of the specific logic functions implemented. Thus, the cost of the test pattern generation for the PLA is reduced considerably.

This universal test method requires augmenting the logic functions of the AND–OR arrays with a set of shift-register latches, an expanded decoder, and a network of XOR gates. In addition, an extra product term row is appended to the AND array while the OR array receives an extra output column (Figure 13.47). The personalization of this extra product term row and output column is selected so that each AND array input column and OR array row contains an *odd* number of connections. The shift register is included to provide the capability of selecting

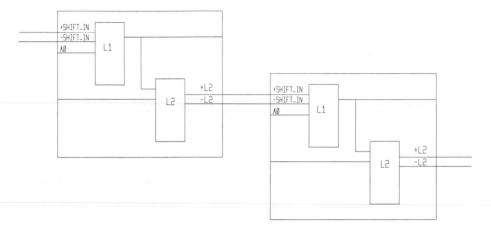

Figure 13.46. A double-rail scan path interconnection is depicted, with both polarities of the L2 latch used to implement the scan path.

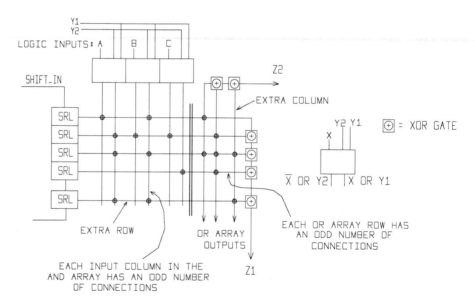

Figure 13.47. Structure of the augmented PLA for enhanced (universal) testability. An extra AND array row and OR array column are included to enable the number of interconnections on each AND array input column and OR array row to be of odd parity. The product term row output is the logical AND of the SRL output value and the AND array minterm. The input decoder includes the additional inputs Y1 and Y2 to be able to force particular values onto the AND array input columns.

a single product term row while forcing all other row outputs to 0 (Figure 13.48). The input decoders use the additional control input signals y_1 and y_2 to be able to select a single AND array input column, rather than complementary values on adjacent columns from the single-bit decode (Figure 13.49). The XOR gates past the product term rows and those past the OR array outputs provide a parity-checker function for a sensitized AND array input column and product term row, respectively. For example, consider the test pattern indicated in Figure 13.50a; an AND array input column is sensitized by the appropriate pattern of shift register outputs and array inputs. By design, the number of product term outputs equal to 0 will be odd, which will dictate the expected output $Z1$ of the XOR chain; a stuck-at-1 fault on the selected decoder output/AND array input is detected. Figure 13.50b illustrates an input pattern that sensitizes a single product term row. The extra OR array column is personalized such that the total number of output columns incorporating this product term is odd; this condition is designed to detect a stuck-at-0 fault on the selected product term output/OR array input, as determined from examining the value of signal $Z2$. The entire PLA test sequence is given in Figure 13.51; for a PLA with n inputs, m product terms, and p outputs, the total number of patterns is $2 * n + 2 * m + 1$ and is independent of the specific logic functions being implemented.

Of particular note is the poor suitability of *random* test pattern generation techniques for PLA stuck-at fault testing; to sensitize a specific product term row, an exact combination of PLA input values is required. For a PLA with a large number of inputs, the likelihood of randomly selecting the input values that correspond to each product term implicant is slight.

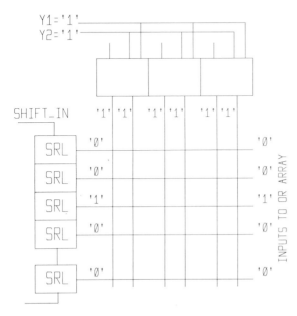

Figure 13.48. The product term row output values are the AND of the SRL outputs and the array minterm; it is therefore possible to force all product term rows except one to 0 with the appropriate condition on the Y1 and Y2 inputs and the proper shift-register vector. This condition will permit checking the parity of interconnections on an OR array input row.

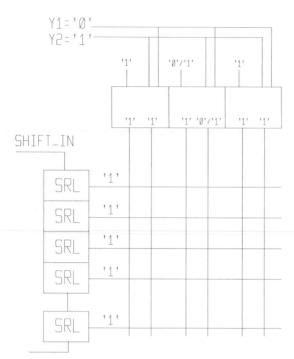

Figure 13.49. The augmented PLA also facilitates the condition where all AND array input columns (inputs *and* their complements) may be forced to 1; likewise, a *single* AND array input column may be selected with the appropriate data inputs.

Memory Array Testing and LSSD Network Interfacing

The ability to incorporate a memory array macro on-chip is a key feature of VLSI technology development; however, the testing of that embedded function (within an LSSD logic environment) presents considerable difficulties. This discussion presents some of the design system rules regarding embedded arrays to facilitate test pattern generation for those functions.

　　A general block diagram for the embedded array is given in Figure 13.52; the array in the figure may be either a RAM or a ROM (this discussion will assume the general case of a RAM macro). There are three unique test generation situations that arise from including the array: (1) the combinational logic whose outputs are used as array inputs will likely require that the good machine/bad machine behavior be propagated through the array to the remaining logic (and subsequently observable outputs), (2) testing the logic fed by the array will require setting the array outputs to specific test pattern values, and (3) all the storage locations within the array itself must be tested for correct behavior. These problems are further complicated by the need to shift in and shift out test pattern results, without disturbing the values previously stored in the array. In Chapter 1, the testing of an embedded array was briefly described; in that discussion, it was recommended that a design for testability rule be established that all embedded array inputs and outputs be directly controllable and observable *at chip I/Os* during the array test

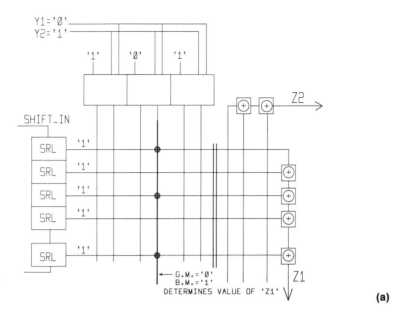

(a)

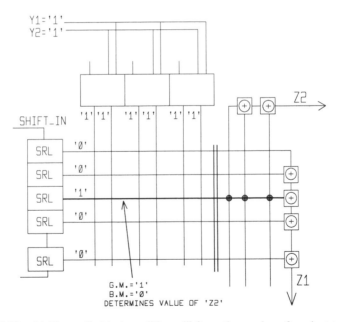

(b)

Figure 13.50. (a) The applied test condition will force the number of product term rows equal to 0 to be odd, which will give the expected value of test signal Z1. If the selected input column is stuck-at-1, the value of the signal Z1 will be opposite that of the good machine value. (b) Selecting a single product term row with the condition Y1 = 1 and Y2 = 1 and a single 1 value in the shift-in string will provide the expected value of the test signal Z2. A stuck-at-0 fault on the selected product term row will result in the opposite (bad machine) value on Z2.

Universal Test Set for Augmented PLA
n Inputs, m Product Terms, p Outputs

#	Data Inputs				Control Inputs		Shift Register Outputs			Parity Outputs	
	1	2	....	n	y_1	y_2	1	2	m	z_1	Z_2
1	x	— x —		x	x	x	0 — 0 — 0			0	0
m	0	— 0 —		0	1	0	1 0 ——— 0 0 1 0 —— 0 ⋮ 0 — 0 1 0 — 0 ⋮			1	1
m	1	— 1 —		1	0	1	1 0 ——— 0 0 1 0 —— 0 ⋮ 0 — 0 1 0 — 0			1	1
2n	0 1 ——— 1 ⋮ 1 — 1 0 1 — 1 ⋮ 1 0 ——— 0 ⋮ 0 — 0 1 0 — 0 ⋮				0 1	1 0	1 ——— 1 1 ——— 1			0: m Odd → 1: m Even	x x
										Total = (2*n) + (2*m) + 1	

Figure 13.51. Universal test set for the augmented PLA.

condition. The block diagram in Figure 13.52 presents the less stringent, more general case of array I/Os also being controlled and observed by the SRL string, in addition to chip I/Os. Testing a large embedded array directly from chip I/Os without requiring repeated scans of SRL values is more efficient by far; however, the remainder of this discussion will assume that the more general structure could suffice.

The following LSSD design rules attempt to reduce the difficulty of test pattern generation for embedded arrays:

Rule 1. Input signals to the array are to be provided from chip inputs and/or SRL outputs, possibly through LSSD combinational logic; the array inputs should not be driven by non-LSSD logic or other array outputs.

Rule 2. Array outputs should be connected to chip outputs or SRL inputs, possibly through an LSSD gating network; as with rule 1, array outputs should not drive non-LSSD networks or other array inputs.

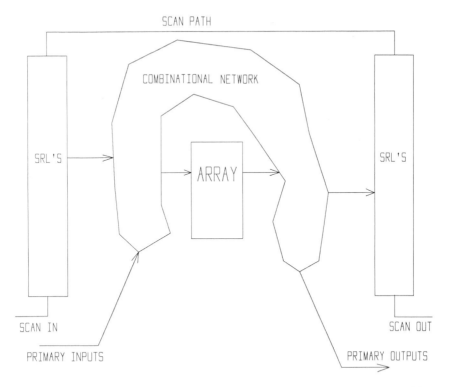

Figure 13.52. General block diagram of the LSSD structure for a chip design containing an embedded storage array.

Rule 3.   To enable the shift-register operation of the SRL string, a condition of the chip input signals must be provided that will prevent writing into the array during this time; this condition is referred to as the *array stable state* (reference 13.8).

To most efficiently apply the necessary pattern of address, data, and control signals to the array, a condition of one-to-one correspondence between the array macro pins and directly controllable and observable nets should be provided; indeed, this design constraint is mandated by further design rules.

Rule 4.   The combinational logic present between the chip inputs, chip outputs, SRLs, and the array must be designed so that there exists some pattern of chip input and SRL output values that establishes a one-to-one correspondence between (1) each array address and data input and an SRL output or chip input, and (2) each array output and an SRL data input or chip output. To be able to apply any pattern of address and data values to the array, the ancillary signals that establish this one-to-one correspondence *cannot* also feed logic that provides the array address and data input values.

Rule 5. The condition of SRL and chip input values specified by the chip designer to satisfy rule 4 for writing into the array is called the *array write state*. In addition, the temporal pattern of *chip input* changes that will actually write data into the array is the *array write sequence*; note that the array write state must enable a *single chip input* to control the array write sequence.

Rule 6. The condition of SRL and chip input values specified by the chip designer to satisfy rule 4 for reading from the array is called the *array read state*; this includes sensitizing the path between the array outputs and SRL latches or chip outputs. The *array read sequence* is the temporal pattern applied to an input during the time when the array read state is active, which completes a read cycle.

As with any LSSD design, the clocking of the SRLs used to provide array inputs and capture array outputs should be insensitive to signal rise and fall times; the same stipulation applies to array clocking signals.

To develop tests for the array itself, the designer should specify the following conditions of chip inputs and SRL values:

1. Array stable state to enable shifting of scan path values without perturbing the array
2. Array read state and array read sequence
3. Array write state and array write sequence
4. One-to-one signal correspondence between array pins and chip/SRL pins during the array read and array write states.

This information, in conjunction with the desired set of array tests, can then be easily translated to the necessary sequences of scan-in, write, read, and scan-out test patterns.

The desired set of array tests should be selected to verify the array behavior, yet not so as to significantly lengthen the tester time; this is particularly important since additional time is required to scan in array stimuli and scan out array results. At a minimum, a test strategy of writing and reading a 1 and a 0 for each storage location could be used. Additional test strategies could be incorporated to investigate specific parameters likely to cause failures in the array. Some of these strategies are as follows:

1. *Address uniqueness:* The address uniqueness test attempts to verify that the cell location retrieved was indeed the cell location addressed; this involves reading each individual cell when all other cells are in the opposite state.

The patterns developed for this test commonly move a single 1 bit value through an array of 0s and then subsequently a single 0 value through a field of 1s.

2. *Address complement:* The address complement test looks for decoder sensitivities by providing an addressing sequence in which all address inputs *except one* are changed each read/write cycle.

3. *Marching data complement:* In this test sequence, a read-write-read cycle is performed at each address, with the value written into the cell being the complement of the data initially read from the cell.

4. *Data sensitivity tests:* A large number of data sensitivity tests have been developed. The application of a particular test to the array should be based on the array architecture and anticipated design weaknesses. These tests commonly attempt to perturb adjacent cells, rows, or columns to the stable cell under evaluation, and then subsequently examine the quiet cell to determine if an upset in data value has occurred.

Fault simulation of the array test patterns will determine the faults covered in the combinational logic network surrounding the array. However, it may be preferable to attempt to verify the logic prior to the lengthy array test sequence; chips that fail the logic test will be abandoned early. Due to the intervening presence of the array, conventional logic test pattern generation algorithms will be difficult to apply. Reference 13.8 suggests that a high percentage of these logic faults may be detected by initializing the values stored in the array and then performing a limited number of *random* write/read cycles. (As mentioned in Section 13.3, random patterns applied initially will typically cover a high percentage of network faults, with a minimum of test pattern generation effort.) Simulating the faults detected by these random array cycles will leave a much reduced set of combinational network faults for which to develop specific test pattern sequences. The nature of these remaining pattern sequences varies depending on whether the assumed fault propagates to an array address input, an array data input, or is located past the array and requires a stimulus pattern from the array outputs. If the array is a ROM, the test pattern generation for the surrounding logic is further complicated by the inability to initialize an array location with a desired value.

The embedded array test methodology described in this section does not easily facilitate the testing of the array's performance, a standard practice with stand-alone VLSI memory chips. The timing specification of the array read sequence and array write sequence may provide some limited detection of defects other than dc stuck-at faults; depending on the controllability and observability afforded by chip inputs and chip outputs, respectively, some access time data may be gathered.

13.5 SELF-TESTING AND SIGNATURE ANALYSIS

The discussion provided in the previous sections of this chapter should not give the impression that VLSI testing is a solved problem, even if design for testability guidelines like LSSD are observed. There is considerable research ongoing to address the problems of tester time, test pattern generation resource, test data volume, and machine speed performance testing, to name but a few. This section briefly discusses one of the most active areas of VLSI test methodology research, the self-testing of the design at machine speeds, utilizing on-chip test stimulus and test response compression circuitry.

A block diagram of a design that includes self-test circuitry is given in Figure 13.53; it is fashioned after the self-test circuitry included in the design of a 32-bit VLSI CPU chip.[9] When in self-test mode, special microcode stored in the on-chip ROM of the microprogrammed CPU is executed. For further testability analysis, the microinstruction register itself is part of a scan path that permits access to the machine instruction (and processor status word).

The figure does not include a function providing the means for the compression and/or checking of the responses of the system under test; this function is commonly provided by a special configuration of storage elements known as a linear feedback shift register or signature analyzer. An example of a 6-bit linear feedback shift register (LFSR) is given in Figure 13.54.

Briefly, signature analysis is a means of providing a vector of binary output values that represents a compression of the data stream input into the linear feedback shift register;[10] the LFSR must begin the self-test operation in an initialized state, while the number of data values clocked into the register is defined as part of the test sequence. At the end of the test sequence, the LFSR output value is compared against the expected result to detect if an error in one or more bits of the input stream has occurred. The likelihood that one or more errors in the input stream results in the same pattern as the expected response (and therefore go undetected) is very low, given a suitably long LFSR and an appropriate interconnection of the XOR gates. The result of checking the signature of the LFSR against the expected response provides a green light/red light (or go/no go) condition on the good machine or bad machine behavior; due to the extensive compression of test data, no diagnostic information is readily available.

In addition to the self-test pattern sequence, the LFSR may also be included on-chip. The verification of the final LFSR value could likewise be implemented on-chip by comparing the resulting vector against an expected signature, which

[9] J. W. Beyers and others, "A 32b VLSI CPU Chip," *IEEE International Solid-State Circuits Conference* (1981), 104–105.

[10] In terms of coding theory, the LFSR performs a polynomial division of the input stream in Boolean algebra with the first input bit as the most significant of the dividend polynomial; the vector of latch values at the end of that sequence represents the remainder, or residue, of the division. For more background, refer to a text on binary codes, such as Wakerly, J. F., *Error-Correcting Codes, Self-Checking Circuits and Applications*. Amsterdam: North-Holland Publishing Co., 1978.

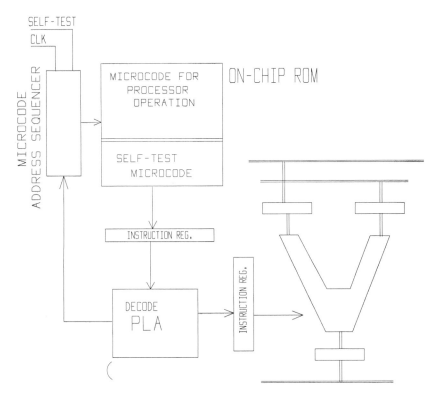

Figure 13.53. CPU chip design containing a microinstruction ROM, which includes microcode to execute in a chip self-test environment.

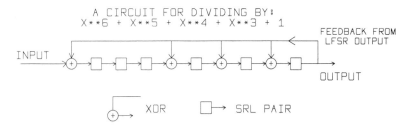

Figure 13.54. Block diagram of a six-bit linear feedback shift register, used to compress self-test data results into a signature. The LFSR performs a polynomial division of the input bit stream in Boolean algebra; the latch values at the end of the input shifting sequence represent the remainder (residue) of the division.

may be stored in an on-chip ROM array location. The multimode LFSR structure to be described simplifies the design of an on-chip test pattern generation and signature analysis function.

The LFSR illustrated in Figure 13.54 uses a relatively inefficient serial data

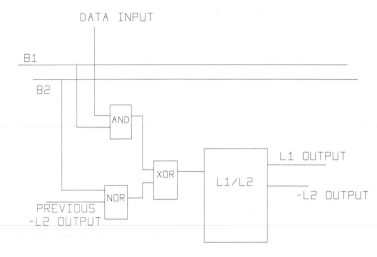

Figure 13.55. Basic element of the multimode shift-register/parallel data input LFSR. Control signals B1 and B2 determine the particular mode of operation of the interconnection of SRLs. This structure is the basic constituent of a BILBO register for pseudorandom test pattern generation and signature analysis. The L1/L2 clock signals are not shown.

input stream. An alternative technique has been developed for utilizing a shift register set of latches in a number of different modes, as data storage latches, as scan path, as pseudorandom test pattern generator, and as a *parallel* data input LFSR (reference 13.9). The basic element of this multimode shift register is given in Figure 13.55; an 8-bit parallel-input LFSR is depicted in Figure 13.56. The values of signals $B1$ and $B2$ control the mode of operation:

— $B1 = 1$, $B2 = 1$:   system data latch
— $B1 = 0$, $B2 = 0$:   scan path connection
— $B1 = 1$, $B2 = 0$:   LFSR operation, with *parallel* system data inputs
— $B1 = 0$, $B2 = 1$:   resets all latches in the register

The implementation of the LFSR in Figure 13.56 differs from that presented earlier in that the feedback taps on the latch outputs are selected according to a prescription for generating pseudorandom output sequences;[11] if the system data inputs are held fixed during LFSR clocking, the LSFR can be used to generate random test patterns to a combinational network. Figure 13.57 illustrates the use of these LFSRs to test a system; when one register is exercised as a random test pattern generator, the other is compressing the network output data values into a signature. At the end of the sequence, the signature vector may be scanned out in serial shift register fashion. Then the roles of the two LFSRs as pattern generator and signature analyzer are reversed.

---

[11] The sequence of register outputs is denoted as pseudorandom as the pattern is *cyclic*, repeating after every $2^n - 1$ cycles, where n is the length of the LFSR.

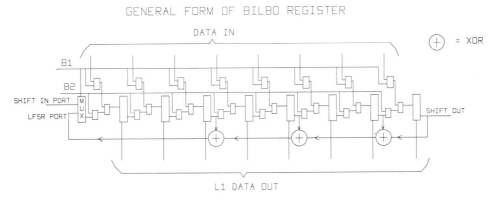

Figure 13.56. Eight-bit parallel-input LFSR utilizing the BILBO element of Figure 13.55. The register data inputs, the shift input and shift output, the mode signals B1 and B2, and the LFSR feedback path are shown; the register clock signals have not been included for clarity.

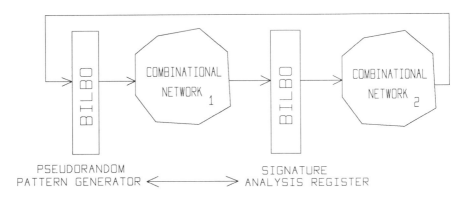

Figure 13.57. General self-test structure utilizing BILBO registers; when one register is used to generate pseudorandom test patterns, the other is operating as a signature analyzer to compress the combinational network responses. At the end of the random pattern testing sequence, the signature is scanned out of the appropriate BILBO register for comparison against the expected response. The feedback from latch outputs is selected in the BILBO LFSR such that a pseudorandom set of test patterns can be generated, if the system data inputs are held fixed.

The overhead for this technique in terms of tester time and test data volume is quite small, due to the on-chip random test pattern generation, LFSR data compression, and the reduced scan requirements. The circuit overhead consists of the additional mode-select and LFSR logic between SRLs; a performance penalty is present in system mode due to the additional logic between the data input and the latch input.

The current research in the area of design for testability and self-test techniques will undoubtedly continue to produce innovative ways of reducing test overhead for evaluating VLSI chip designs. In addition, work is ongoing in applying these concepts to the hierarchical levels of system design. Yet much remains to be investigated in terms of the *quality* of the test pattern set being developed; specifically, how sufficient is the test set that is being applied, and what diagnostic information can be produced to assist with failure analysis? In searching for the answers to these questions, it is important to keep in mind the economics of the situation: the impact on the logic design cycle, the overhead costs of the additional test circuitry, the costs of test pattern development and tester time, and the ultimate costs of servicing and/or replacing a faulty unit in the field. As stated in the beginning of this section, the multifaceted problem of VLSI testing is by no means solved.

PROBLEMS

13.1. Research and discuss the features of the fault simulation techniques known as "parallel fault simulation" and "deductive fault simulation." Compare these alternatives against the TEST-DETECT algorithm presented in this chapter.

13.2. Research and discuss the following design for testability scan techniques: Scan Path, Scan/Set, and Random-access Scan. Compare and contrast these methodologies against LSSD. (A good survey of various design for testability approaches is given in reference 13.10.)

13.3. (a) Construct a flowchart of the D-algorithm from the textual description given in the chapter.

    (b) Develop a test model translation of AOI and OAI functions in terms of basic logic primitives.

    (c) Develop a program to generate test patterns for a combinational network using the D-algorithm. The program should accommodate the following logic gate types: AND, OR, NAND, NOR, XOR, XNOR, AOI, and OAI.

13.4. (a) Develop an algorithm flowchart for an automatic test pattern generation procedure for CMOS technology faults; specifically, this algorithm must precondition the net to the assumed stuck-fault value and then (with a subsequent pattern) attempt to drive the net to the opposite good machine value.

    (b) How do these CMOS technology constraints affect fault-simulation algorithms? Develop the necessary modifications to the TEST-DETECT algorithm to support CMOS technology circuits.

13.5. In a test environment that does not utilize LSSD techniques, the test data file will consist of the test vectors and the gating waveforms to be used for each chip input every tester cycle; in other words, the applied data value at each input is ANDed with one of a number of user-defined timing waveforms to provide the desired temporal sequence of input signal transitions, clocks and otherwise. The chip output values are strobed at a user-defined point in time near the end of each cycle (Figure 13.58).

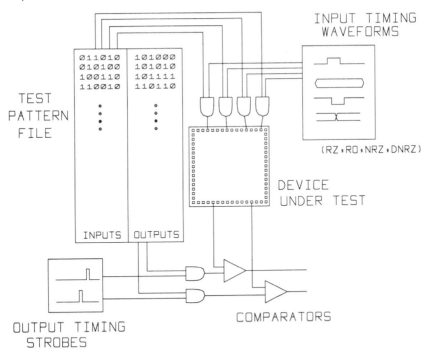

Figure 13.58. Block diagram of general test pattern sequencing and observation of chip response. A number of gating waveform generators are ANDed with test pattern input values and the chip outputs are compared with the expected response at the end of the tester cycle.

(a) Research general chip-testing techniques to determine the following typical parameters:

Number of different input timing waveforms available

Number of patterns that can be efficiently applied from the tester's buffer memory

Number of I/O pins available

Maximum tester cycle frequency

Timing resolution of waveform transitions

Simultaneous switching limits and measures adopted in tester hardware design to reduce switching noise

(b) Indicate the meaning of the testing waveforms denoted as non-return-to-zero (NRZ), delayed non-return-to-zero (DNRZ), return-to-zero (RZ), and return-to-one (RO). Indicate how the test pattern file size can be reduced by using RZ (or RO) timing waveforms for periodic system clock inputs, rather than NRZ transitions.

(c) Describe how the designer would convert the *functional simulation* input stimulus patterns and chip model output responses into a general test pattern file, using the combination of timing waveforms in part (b).

(d) Using the output strobe timing specification within each tester cycle, what degree of performance analysis is afforded? What are the advantages and disadvantages of using more conservative tester cycle times?

REFERENCES

13.1 Roth, J. P., and others, "Programmed Algorithms to Compute Tests to Detect and Distinguish between Failures in Logic Circuits," *IEEE Transactions on Electronic Computers*, EC-16:5 (October 1967), 567–580.

13.2 Goel, Prabhakar, "An Implicit Enumeration Algorithm to Generate Tests for Combinational Logic Circuits," *IEEE Transactions on Computers*, C-30:3 (March 1981), 215–222.

13.3 Goel, P., and Rosales, B., "Test Generation and Dynamic Compaction of Tests," *IEEE International Test Conference* (1979), 189–192.

13.4 Eichelberger, E., and Williams, T., "A Logic Design Structure for LSI Testability," *Journal of Design Automation and Fault Tolerant Computing*, 2:2 (May 1978), 167–178.

13.5 DasGupta, S., and others, "A Variation of LSSD and Its Implications on Design and Test Pattern Generation in VLSI," *IEEE International Test Conference* (1982), 63–66.

13.6 Godoy, H., and others, "Automatic Checking of Logic Design Structures for Compliance with Testability Ground Rules," *14th Design Automation Conference* (June 1977), 469–478.

13.7 Fujiwara, H., and others, "Universal Test Sets for Programmable Logic Arrays," *Digest of the 10th International Symposium on Fault-Tolerant Computing* (1980), 137–142.

13.8 Eichelberger, E., and others, "A Logic Design Structure for Testing Internal Arrays," *3rd USA–Japan Computer Conference* (1978), Session 14-4, 1–7.

13.9 Koenemann, B. and others, "Built-in Logic Block Observation Techniques (BILBO)," 1979 *IEEE Test Conference*, 10/79, p. 37–41.

13.10 Williams, Thomas, and Parker, K., "Design for Testability—A Survey," *Proceedings of the IEEE*, 71:1 (January 1983), 98–112.

Additional References

Bardell, P., and McAnney, W., "Self-testing of Multichip Logic Modules," *Test and Measurement World* (March 1983), 26–29.

Chandramouli, R., "Designing VLSI Chips for Testability," *Electronics Test* (November 1982), 50–60.

Comerford, R., and Lyman, J., "Special Report: Self-Testing," *Electronics* (March 10,

1983): Komonytsky, D., "Synthesis of Techniques Creates Complete System Self-Test," 110–115; Fasang, P., "Microbit Brings Self-testing on Board Complex Microcomputers," 116–119; Kirkland, R., and Flores, V., "Software Checks Testability and Generates Tests of VLSI Design," 120–124.

Eichelberger, E., and Williams, T., "A Logic Design Structure for LSI Testing," *14th Design Automation Conference* (June 1977), 462–468.

Fee, W., "Memory Testing," *IEEE Computer Society Conference (COMPCON)* (1978), 81–88.

Goel, P., and Rosales, B., "PODEM-X: An Automatic Test Generation System for VLSI Logic Structures," *18th Design Automation Conference* (June 1981), 260–268.

IEEE Transactions on Computers, Joint Special Issue on Design for Testability; C-30:11 (November 1981); (also published in *IEEE Transactions on Circuits and Systems*, CAS-28:11 (November 1981) .

McCluskey, E. and Bozorgui-Nesbat, S., "Design for Autonomous Test," *IEEE Transactions on Computers*, C-30:11 (November 1981), 866–875.

Perloff, D., and others, "Microelectronic Test Structures for Characterizing Fine-Line Lithography," *Solid State Technology* (May 1981), 126–129, 140.

Segers, M., "The Impact of Testing of VLSI Design Methods," *IEEE Journal of Solid-State Circuits*, SC-17:3 (June 1982), 481–486.

Wadsack, R., "Fault Coverage in Digital Integrated Circuits," *Bell System Technical Journal*, 57:5 (May–June 1978) 1475–1488.

――― "Fault Modeling and Logic Simulation of CMOS and MOS Integrated Circuits," *Bell System Technical Journal*, 57:5 (May–June 1978) 1449–1474,

Walker, R., and King, D., "Testing Logic Arrays," *Electronics Test* (January 1983), 76–80.

Yau, S., and Tang, Y., "An Efficient Algorithm for Generating Complete Test Sets for Combinational Logic Circuits," *IEEE Transactions on Computers*, C-20:11 (November 1971), 1245–1251.

14

BACK-END DESIGN SYSTEM TOOLS

14.1 INTRODUCTION: GLOBAL AND MACRO LEVEL VERIFICATION

The term *back-end tools* refers to the programs used to verify that the physical chip design satisfies the set of constraints imposed by the design system; these steps include power supply distribution analysis, power dissipation calculations, dc noise margin verification between circuits, block delay calculations, inductive noise modeling, and verification of synchronizing networks. These verification procedures are used to ensure the manufacturability of the design and provide a measure of device failure rates: transient synchronization failures and end-of-life (permanent) failure.

A recurring theme throughout this text has been the distinction made between *global* and *macro*-level design considerations. To best manage the design complexity (and the sheer volume of the design database), the physical layout of circuit functions was partitioned into physical macros, layouts that could be refined, modeled, and verified in a manner so as to efficiently utilize computing resource. Likewise, the description of the logic architecture of a VLSI design was embodied in the definition of logical macros, hierarchically constructed logic networks that were systematically connected to encompass the complete function. At well-defined nodes in this structure, there is a direct logical-to-physical correlation between a macro layout placed on the chip image and its logic and test model representation. The set of design system back-end tools to be provided must support both macro and global design verification tasks; the data developed from the macro analysis (during the design system library development) is stored

694

in *macro files* for later incorporation into global checking tools. For example, the dc power dissipation of a macro's circuits can be calculated and saved (using logic 0 and 1 output duty cycle assumptions). This information can then be extracted during back-end checking to calculate (1) the total chip power dissipation, (2) global power supply distribution voltage shifts, and (3) global dc power supply currents for electromigration analysis.[1] From a systems perspective, data provided by back-end checking tools are used in system and package thermal analysis, noise margin analysis at the card level, and overall system reliability projections.

This chapter discusses some of the common back-end tools used for design verification prior to committing for manufacture; some of these procedures have been introduced in other chapters, while some present new design considerations. In a strict design system environment, where the design is to be *directly released* to a fabrication facility (Section 2.4), the successful completion of these checks is a necessary condition to design acceptance.

14.2 DESIGN RULE CHECKING: GLOBAL AND MACRO

The most computationally intensive task of chip design is the checking of the physical layout against the set of technology rules, as discussed in Section 5.4. Macro-level design rule checking is performed during layout, prior to its incorporation into the macro library. As a consequence of that design effort, macro shadows and pin shapes were provided to assist with global verification, facilitating a vast reduction in the amount of shapes data used as input to global design rule checking.

Global design rule checking, incorporating reduced macro detail and potentially a subset of technology rules, is performed after chip physical design is complete. It is possible that this step may be bypassed if the outputs of automated placement and wiring are not modified by manual editing (i.e., if the design remains correct-by-construction). The manual embedding of incomplete interconnections to finish the design or modifications to macro placements and/or global interconnections to gain performance will require the final physical design to undergo global design rule checking, unless the tool used for editing can unconditionally maintain design correctness.

14.3 LOGICAL-TO-PHYSICAL CHECKING

Logical-to-physical checking involves tracing the myriad of device interconnections in a chip layout to ensure that the physical implementation coincides with a device schematic description of the function. Like design rule checking, it is

---

[1] A modification to this database is required for CMOS logic circuits, where the dc power dissipation is negligible; instead, the chip supply currents and power dissipation are directly related to the clocking frequency.

divided into two levels of detail for increased efficiency when utilizing a chip design system, macro and global.

Macro logical-to-physical (L/P) checking involves comparing the identified nodes and devices in the physical layout with its associated logic and schematic model. In this manner, a high measure of confidence is provided that the function implemented by the layout agrees with the model used for logic block (delay) simulation. The identification of circuit output nodes in the layout is provided by NODE shape statements, which assign a name to a coordinate location and interconnection level in the layout (Section 5.2). This node name must agree with the associated name provided in the schematic model.

Initially, device areas and types are isolated in the layout, using some of the same shape manipulation routines as with design rule checking; subsequently, the gate, source, and drain interconnections for each device are traced out, utilizing a knowledge of the technology's wiring and via lithography levels.[2] The intersection of a NODE identifier with an interconnection shape effectively assigns that node name/number to all shapes that constitute the net, in particular, to the output node of a driving circuit and the device gate inputs of the fan-outs. The path tracing of interconnections within a macro terminates on the macro pins, where *pin shapes* have been included to provide continuity between macro- and global-level design verification; macro-level pin names are also included over the pin shape.

One of the more difficult tasks of macro-level L/P verification is associating the hierarchical block decomposition of the macro's logic simulation model with the single, customized physical layout. The logic simulation model of the macro is composed of a number of branches of a hierarchical structure, each branch with a unique path name; unique signal names are defined for some nets at each level in the structure, replacing the path + pin name string of the signal below. A choice of node names is therefore available to give to the physical layout to identify each circuit output node: the higher-level name used when defining the hierarchy or the fully expanded concatenated path and pin name. The macro's physical layout description consists of a nesting of cell transforms and individual shape statements. To provide the location of a node name in the layout, it is necessary to put all the NODE statements strictly in the top-level macro description and none in the nested cell descriptions (Figure 14.1a); otherwise, if the cell were used in multiple instances, there would be a duplicity of node names. However, considering the large number of signal nets feasible in a macro layout, this approach is somewhat cumbersome and error prone. An alternative would be to provide for *hierarchical* L/P checking, where the node names are included *directly* in the nested cell structure, and the same path identifiers used within the logic structure are also assigned to the physical cell transforms (Figure 14.1b). The

---

[2] Substrate connections provide a unique consideration for logical-to-physical checking; see Problem 14.1.

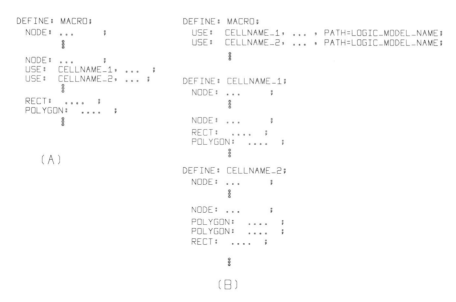

Figure 14.1. Hierarchical L/P checking allows the node name identifiers to be assigned in the nested cell transform (b), as opposed to listing all NODE names in the top-level macro description (a). With hierarchical checking, the path name identifiers in the logic model are also assigned to the physical cell transform; the nesting of the physical design cells must agree exactly with the structure of the logic description model.

NODE statements would no longer be included at the top level of the physical macro description, but rather at the individual circuit level; the logic description path names would be appended to the cell transform statements and concatenated when unnesting the physical design. To successfully implement the node naming convention of hierarchical L/P checking, the nesting of the physical design cells must therefore agree exactly with the logic hierarchy specified for the macro.

A special case would arise for the macro's inputs and outputs, to better facilitate global L/P checking; in this case, *both* the macro pin name and the fully concatenated structural path and pin name would be provided, the NODE statements for the macro pin name in the top-level physical macro description. The presence of two NODE statements in the same physical net should signify a macro pin and provide the necessary continuity between hierarchical macro and global L/P checking steps. The coordinates of the macro NODE statements for its I/O pins should coincide with the desired end point of the automated wiring program line segment, as it is the NODE statements in the macro description that are used for reference by the wiring program.

Problem 14.2 requests that a hierarchical L/P checking methodology be described, including any necessary modifications to the shapes description language of Chapter 5.

Array macros (RAM, ROM, and PLA) are also a special case in that macro-level L/P checking is commonly not executed; no corresponding logic circuit device-level model is commonly available against which to compare these layouts. This is a major factor behind the goal of developing the design system tools to generate these layouts automatically.

The path tracing of signal and power supply interconnections utilizes a list of lithography level associations that provides electrical continuity. The path-tracing step does not typically perform any design rule checking to ensure that the intersections of shapes on these levels satisfies technology rules; after all, that is the function of the macro-level design rule checking procedure. Prior to incorporating the macro layout and its logical model into the technology design system library, *both* L/P and design rule checking procedures must be exited successfully. There is a related function usually associated with the path-tracing step of macro L/P checking, electrical parameter extraction. In identifying the device and interconnection shapes, a wealth of information is readily available for extraction and use within other applications. For example, each device width and length dimension is a key parameter in calculating the total fan-out capacitance on a driving net and the block switching delay due to an input signal transition at the device input. Device dimensions and diffusion series resistance values are likewise an important part of dc checking; interconnection wires between circuits are measured to calculate the appropriate adder to a circuit's capacitive load. The level of detail of the resistive and capacitive networks to be extracted, and the data format in which to store the parameter values, is chosen so as to complement the related design verification applications.

Global L/P checking involves tracing wiring interconnections between physical macro pins (including chip driver and receiver circuits). This step may be skipped if the design was successfully completed by an automated wiring procedure, provided the capacitance parameters of each net are suitably extracted. If the final chip layout required any manual editing of wiring segments, then global L/P checking should be performed. The placement locations of the macros are extracted from the chip design database, and the absolute coordinates of the macro pins calculated from their positions relative to the macro origin. Using these coordinates and the shapes from the global wiring cell of the physical database, the path tracing of global interconnections is executed. As before, the lithography level associations that constitute a continuous interconnection are required. The top levels of the hierarchical logic structure of the chip description (above the physical macro design-associated nodes) are not typically reflected in the global interconnections between layouts on a design system image; as a result, global L/P checking must first unnest the logic design description to the physical macro level, equating higher-level signal names with the full path and macro pin name. Once this unnested description is available, the set of macro pins in each signal net across the logic structure is collected; this list of macro pin names in each net is then used for comparison with the macro pin data extracted from the global physical interconnections.

14.4 DC CHECKING

Chapter 9 discussed the dc circuit design and noise margin considerations for individual logic circuits with passive load devices. To calculate a circuit's worst-case dc logic 0 and 1 output voltage (as both driving and driven circuit) and examine the adequacy of circuit-to-circuit noise margins, the following design system features must be provided:

1. A set of algorithms or equations to calculate these on and off output voltages as a function of device W's and L's, series resistance, shared ground currents, circuit fan-in, and so on.
2. A database where the values of these parameters for each circuit are placed by the parameter extraction procedure.
3. A dc checking tool that incorporates the parametric data, algorithms, and circuit interconnectivity description to provide the noise margin evaluations.

As with the other design system tools discussed in this chapter, the dc verification of physical layouts is most effectively performed in two steps: a detailed analysis at the macro level and a chip-specific verification of global interconnections.

The macro-level dc checking program interrogates the database of circuit parameters for the values to incorporate into the expressions for calculating v_{on} and v_{off} for each circuit. The v_{on} result is then compared to the v_{off} values of all the driven circuit inputs in the macro to calculate the noise margin. As mentioned in Section 9.4, for a circuit with a number of dotted logic functions on the output node, a number of v_{on} calculations are required to determine the worst case output 0 voltage; likewise, the determination of a circuit's v_{off} input voltage requires a consideration of circuit fan-in. When calculating these values for a general usage design system macro, conservative assumptions relative to *global* power supply distribution voltage shifts are made; these assumptions must be subsequently verified for each chip during global dc checking.

One of the additional calculations that can be provided by the macro-level dc checking tool is the logic 0 level dc power dissipation of each circuit; this number can be duty cycled and summed with that of other circuits to produce a macro (and ultimately chip) power dissipation figure. The 0 logic level load device current is used in calculating resistive losses and supply distribution voltage shifts; the dc power dissipation calculation is directly derived from this parameter.

Macro dc checking provides a means to verify that the physical layout has not inadvertently introduced circuit interfacing errors due to resistive voltage losses. Failure to successfully complete macro dc checking will often entail considerable rework to the handcrafted physical layout; it is important that considerations of interconnection sheet resistivity and contact resistance be an integral part of a physical layout from the beginning.

Global dc checking involves two verification steps specific to each chip design: macro pin-to-macro pin noise margin verification and a check to ensure that supply voltage shift assumptions used for internal macro-level checking were indeed sufficient. As mentioned earlier, the voltage shift assumptions used for macro verification are typically conservative; often, the close proximity of circuits within the macro implies that both the power and ground bus structure is largely shared by the two circuits, reducing the voltage difference between them. If none of the internal macro supply distribution is common to the driving and driven circuits, then the supply shift can be calculated from a resistive model of the image power distribution and the macro currents into the branches of that model.

With the adoption of CMOS VLSI technologies, the necessity for macro and global dc verification design system tools is eliminated. However, unlike passive load device technologies, the transient supply current waveforms are now the significant component of the image voltage distribution; this complicates the calculation of chip power dissipation and introduces inductive switching noise concerns (Section 14.6).

14.5 DELAY CALCULATION

A key design system tool for each technology is the *delay calculator*, which utilizes delay equations for different circuit types to determine values for the rising and falling output block delays; the input variables to these equations include the circuit output's capacitive load, the input signal transition time, and the circuit's device dimensions. Block delay variations between circuit input pins can be incorporated into transparent logic *pin* delay adders for more accurate simulation, as discussed in Section 9.5.

The logic simulation model for a network may utilize a mixture of block and behavioral functional descriptions, with an associated variety in the means for which delay values are provided. Block delays are assigned to the individual logic circuits at the lowest level of a macro's hierarchy; the macro's technology library model contains this information. Calculating macro output pin block delay adders requires access to parametric information from the file of global net interconnection capacitances and from other macro files whose pins are in the same net.

The delays calculated and assigned to a chip's logical structure may represent nominal, minimum, and/or maximum values, if corresponding sets of equations are available for the technology. The delay calculation tool should be able to provide the range of delay values appropriate to the simulation application, be it functional verification or timing analysis (Secton 4.8).

An alternative to calculating block level delays for macro models is to utilize behavioral macro simulation routines; the delays through the various circuit paths of the macro are in the behavioral model, while chip-specific capacitive loading and transition time delay information is absorbed into transparent blocks connected to the macro pins. In this case, the delay calculator need only extract the global wiring capacitance data and the parametric information about the macro input/output circuits to calculate the transparent block delay adders; as mentioned in Section 9.5, the necessary input/output pin data can be reduced to a file of tabular delays, based on total output pin loading and input pin transition time. In utilizing behavioral models to provide delays, the efficiencies of the simulator and the delay calculator are enhanced; macro internal circuit transition time delay dependencies are not provided with the behavioral model.

The capacitive load values used for calculating block or macro behavioral pin delays are a combination of interconnection and gate fan-out loading values. The device input capacitance used for totaling the load is commonly the maximum value: $C_{\text{gate}} = (\epsilon_{\text{ox}}/t_{\text{ox}}) * (W/L)$; the variation of this component with input voltage is typically neglected. The interconnection capacitance is based on area and edge-length multipliers for the wiring levels; if minimum-width wires can be assumed, a single per length coefficient may be used instead. It is possible and highly productive to provide *estimated* delay values for global interconnections to use in logic simulations prior to the completion of the chip physical design; that is, the delay calculation program may be run early in the chip design cycle, before beginning chip placement and wiring. The file of global net interconnection capacitance values may use wire lengths that are estimates of the interconnection segments required to complete the chip physical design; these estimates could be based on the number of pins in the net, the total number of circuits and nets in the chip design, the extent to which the net encompasses multiple levels of the logic hierarchy, and so on. A statistical confidence level is commonly given to these estimates to enable the logic designer to evaluate what additional delay margin to provide in critical circuit paths between these estimates and the final chip physical design.

In addition to block and pin delay values, the delay calculator tool may output a file of blocks and pins whose delays are outside designer-specified bounds. This will highlight to the designer specific paths that require immediate simulation analysis. Ideally, the delay calculator would incorporate a power/performance optimization routine; this procedure would examine the list of failing blocks for those corresponding to design system library primitives with multiple performance levels available. The appropriate book power/performance level would replace the default level specified in the chip physical database. It may be the case that several failing blocks in the same network are due to a single sourcing block driving multiple paths; as a result, the power/performance optimization routine should typically examine the blocks in ascending order of levelization in the network.

As discussed in Section 4.8, an additional back-end design system tool, closely tied to the delay calculator and optimizer, is one that performs timing analysis of combinational networks. Logic simulation evaluates paths in a combinational network sensitized by the particular pattern of 1s and 0s provided by the designer's simulation input description. Timing analysis, on the other hand, uses no input stimulus; the tool exhaustively analyzes the delay through all possible paths in a combinational network to verify that designer-specified *network* delay requirements are satisfied, without regard for logic function. In essence, the timing analysis tool takes a combinational network view to delay calculation; it can provide exhaustive analysis of path delays, which may not otherwise have been verified by the designer's simulation cases.

A design system timing analysis tool is applicable to synchronous logic designs, where the arrival of clock signals to storage elements is well defined. The combinational networks between storage elements are levelized, each block and pin having been assigned a rising and falling delay value by the delay calculator. Starting with the blocks at the earliest level and proceeding through the network, a block output signal arrival time is calculated; the maximum output arrival time is determined among the propagation of all input signal arrival times through the block.[3] This analysis results in arrival times at the outputs of the combinational network, which can then be compared against the necessary arrival times to meet setup requirements of the storage elements. Signal arrival times are commonly kept in pairs, corresponding to the rising or falling signal transition; the selection of the block rising or falling delay to calculate the output arrival times is based on the inverting nature of the block function.

The timing analysis tool is not a simulator; except for a block's inverting property, it does not regard its Boolean function. As a result, network outputs denoted as problem paths by the program may be false errors, if the path can never be sensitized or if the path meets special timing criteria.

14.6 NOISE ANALYSIS

The continuing trend in VLSI design applications is to incorporate wider data path architectures on a single chip; these wider data paths result in an increase in the number of simultaneously switching driver circuits. Likewise, the higher bus bandwidth requirements of the VLSI system applications have also led to larger current pulse magnitudes. These current transients generate inductive power supply noise, increased signal transition times, and potentially circuit noise margin failure.

The design system chip image development must incorporate an inductive model analysis of the power supply distribution, specifically that provided to the off-chip driver circuits, due to their large current transients. Two types of noise analysis are required: internal (or self) noise and transmitted noise. The internal noise is that generated in the on-chip power supply distribution and distributed to other on-chip circuits (Figure 14.2); the feedthrough of supply noise directly as logic signal voltages may result in circuit margin failure. The *common-mode* component of the supply noise, due to the shared chip and package supply distribution, is not a contributor to the internal noise failures. Transmitted (off-chip driver) noise refers to the changes in logic signal voltages at receiver inputs of

---

[3] For a fast-path analysis, the minimum output arrival time would be propagated (see Problem 14.3).

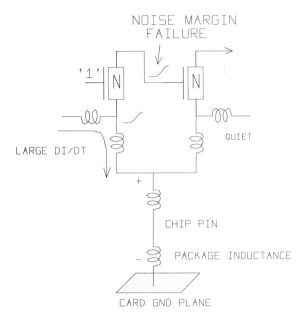

Figure 14.2. The noise generated in the on-chip power supply distribution may be *fed through* as a logic signal voltage, potentially leading to a noise margin failure on a *quiet* line.

other chips originating in the power supply distribution of a quiet driver among a number of simultaneously switching drivers. As depicted in Figure 14.3, the ground noise is transmitted to other receivers on other chips; the external receiver may also have its own self-noise component.

The results of the image power supply distribution analysis should provide guidelines to the system or chip designer regarding the maximum number of simultaneously switching off-chip drivers and any limitations on the relative placement of those circuits. The number of available simultaneously switching drivers

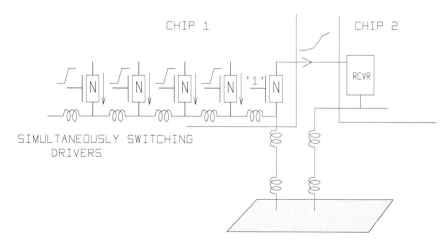

Figure 14.3. Transmitted driver noise between chips.

can be increased by adding more chip and package power pins and reducing the series inductance to a card's power plane. To most effectively utilize these additional supply connections, however, there will likely be placement restrictions imposed on the drivers; the total number should be distributed among each supply and ground pin pair. Manual preplacement of these books at chip I/O sites may be required prior to beginning chip physical design to satisfy inductive noise limitations.

The card-packaging technology also has a dramatic impact on the noise margin analysis of interchip signals. If the driver current pulse has a sufficiently steep transition, the card-level interconnection trace responds as a transmission line; pulse reflections from impedance mismatches and/or discontinuities may result in anomalous driver behavior. A common criterion for selecting to evaluate the interconnection as a transmission line is if the delay along the line (of length l) is greater than one-half the signal transition time. In this manner, the reflected wave is not absorbed in the signal transition, but results in a distinct pulse at an intermediate receiver (Figure 14.4). The example in the figure illustrates the case of a near-end, lightly loaded receiver with a far-end, heavily loaded counterpart; the incident wave passes by the first receiver essentially unmodified by the slight discontinuity, whereas the waveform at the far end is significantly slowed. The reflected wave from the far end may be sufficient in magnitude to cause the near-end receiver voltage to dip below threshold, resulting in circuit failure. Ideally, all receiver waveforms should satisfy an *incident switching* requirement; all waveforms should be such that the receiver will switch values upon the first incident signal, and that no change in value will occur in response to any reflected waves resulting from the incident waveform. The system designer must be cognizant of the transmission line delay (and additional signal delay resulting from increased

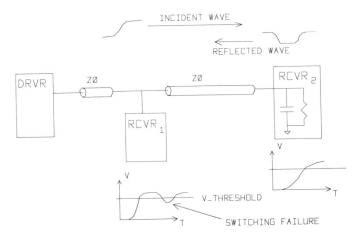

Figure 14.4. If a driver's output transition time is sufficiently steep with respect to the transmission line delay, an intermediate (*near-end*) receiver may be subject to a reflected pulse, which will lead to failure.

transition times); a set of *card-level* delay equations is commonly used. The system designer must also be aware of possible chip-related failures due to reflected waves at the driver output node, which may result in forward-biased junctions, spurious substrate currents, and the potential for CMOS technology latch-up.

14.7 METASTABILITY

Designing for the intercommunication between two computer subsystems that are not based on the same clock commonly requires the implementation of a synchronizing function; that is, the value of an input signal to a synchronous subsystem must be interpreted relative to the clock of that subsystem (*synchronized*). The common solution is to use a flip-flop or latch circuit with the asynchronous signal connected to the data input pin. A synchronizer design is based on the assumption that the output(s) of the flip-flop or latch reach a valid logic voltage level within some maximum *settling* time after the gating clock edge. Since no time relation can be guaranteed between the asynchronous data and system clock signal, the required setup and hold times for the flip-flop or latch will not always be satisfied; the data input signal may be changing in the vicinity of the gating edge of the system clock. As a result, it is possible that the flip-flop or latch circuit (commonly, but erroneously called a *bistable*) may be placed in a third, or *metastable*, state, somewhere between the normal 0 and 1 voltage levels. The data true and complement outputs may hover for an indeterminate time at this intermediate level. It is also possible that both outputs may oscillate *in phase* with each other a number of times between 0 and 1 states before finally settling into an out-of-phase (stable) state. The first condition (metastability) is common of circuits with a small signal delay to rise time ratio (Figure 14.5); the second condition (oscillatory behavior) is common of circuits with a large propagation delay to rise time ratio (Figure 14.6). This discussion pertains strictly to the case of metastability; very little has been published in the technical literature in terms of analysis and/or modeling on the anomalous oscillatory behavior of sequential elements.

Once a storage element is in its metastable state, there is no fixed settling time interval long enough to *guarantee* (with probability *one*) that the outputs will subsequently be at a valid logic state. The metastability problem as defined here is *not* concerned with which particular value the outputs settle out to be, but rather that an intermediate voltage value (one that has not settled) may lead to a subsequent combinational and/or sequential system failure if such values are able to propagate further throughout the system.[4]

The goal of a synchronizing circuit is therefore to reduce the metastability

---

[4] The data input signal is in transition in the neighborhood of the latching clock edge and therefore truly has no defined value; metastability is not related to a correct or incorrect latch value, only intermediate and stable values.

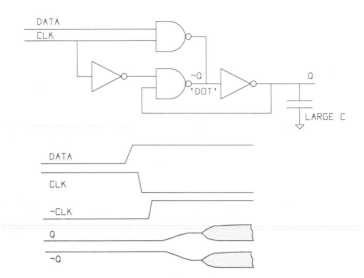

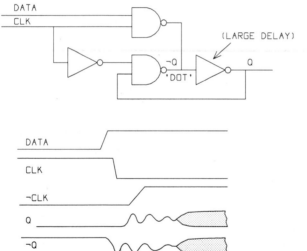

Figure 14.5. Metastability in the synchronizing function (adapted from reference 14.1).

Figure 14.6. Oscillatory behavior in the synchronizing function (adapted from reference 14.1).

failure rate to an acceptable level. This rate signifies the frequency with which sampled asynchronous inputs will result in intermediate latch outputs after a specified settling time. The remainder of this section attempts to briefly summarize some of the analysis, both theoretical and experimental, that has been published to develop models for predicting metastability failure rates (as a function of settling time and the asynchronous data signal frequency), as well as make design and/

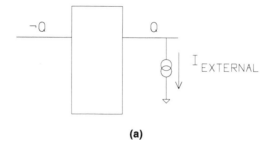

(a)

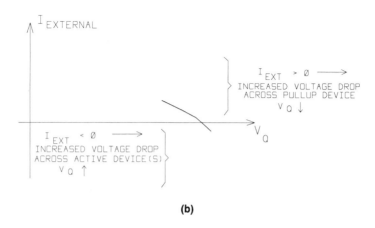

(b)

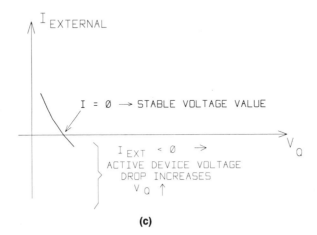

(c)

Figure 14.7.   To indicate the nature of the metastable operating point, an external current requirement is added to a sequential element's output. The perturbations from both 1 and 0 logic states are illustrated.

or characterization recommendations pertinent to a unique design system library synchronizer circuit.

The following heuristic argument is provided to give some measure of proof that all sequential elements must have a metastable state. Consider the black-box circuit design shown in Figure 14.7a with an external (ideal) current source; if the latch state is $(Q, \overline{Q}) = (1, 0)$ an external applied current (either into or out of node Q) will perturb the output voltage either above or below the stable voltage value. This is illustrated in the sketch of I_{ext} versus v_Q in Figure 14.7b. A similar situation applies for perturbations around the state $(Q, \overline{Q}) = (0, 1)$; a positive external current will increase the load voltage drop and reduce v_Q, while a current sinking requirement will increase the active device drop and increase v_Q (Figure 14.7c). Merging the two curves in Figure 14.7b and 14.7c introduces the question of the behavior of this I_{ext} versus v_Q curve in the intermediate region between the two stable logic states; since this curve is continuous, there must be another x-axis intercept between the logical 1 and 0 intercepts. As this point corresponds to a stable ($I_{ext} = 0$) condition, this is the *metastable* operating point. Figure 14.8 illustrates the merged curves with an additional $I_{ext} = 0$ stable point between the logical 1 and 0 intercepts.

To develop a schematic model for the storage element in the metastability region, the design is split into two circuits (Figure 14.9a): a network without a load device/load supply voltage, and the appropriate load connected to the network outputs. From this model, two curves are generated, one representing the v–i characteristics of the network at its output terminals, while the other characterizes the load device at those same terminals (Figure 14.9b). For example,

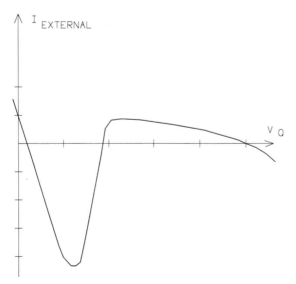

Figure 14.8. External forcing current versus output voltage curve for the circuit configuration of Figure 14.7a; note the third x-intercept ($I_{ext} = 0$) between 1 and 0 states.

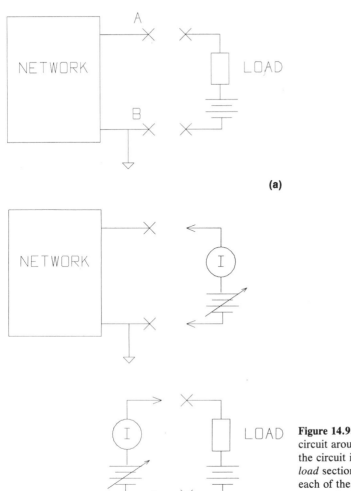

Figure 14.9. To analyze the sequential circuit around the metastability point, the circuit is divided into *network* and *load* sections (a); the characteristics of each of these sections is determined separately (b).

the two curves for an *n*MOS depletion-mode load technology latch circuit are plotted together in Figure 14.10. Two features of Figure 14.10 are of particular note:

1. The load line for the load device intersects the network curve in three points, two stable operating points and the metastable point.
2. The network curve has a region of negative differential resistance of substantial width. (The region of negative small-signal resistance is the region of negative slope on the network curve; that is, applying a *larger* external voltage at the network output terminals *reduces* the external current.)

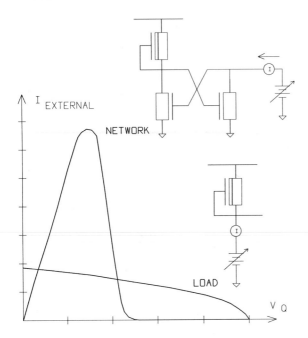

Figure 14.10. Load-line analysis for the modified sequential circuit consisting of network and load.

The process of reaching a stable state from a region surrounding the meta-stable state can be broken down into two parts; the dividing line between these parts depends on the expected magnitude of the circuit noise (reference 14.2). The first part of the output voltage versus time curve consists of the transition within the expected magnitude (or some multiple) of the noise voltage; once the difference between the output voltage and the metastable point significantly exceeds the magnitude of the circuit noise, the time to reach a stable point becomes more deterministic and independent of the effects of noise. The second part is therefore *directed* to one of the stable states. Figure 14.11 illustrates this two-part transition.

Modeling the behavior of the circuit for the initial part of the transition requires developing an equivalent small-signal linear RC model for the network plus load that is valid in the vicinity of the metastable point (Figure 14.12). In parts (b) and (c) of Figure 14.12, the noise current source I_n is assumed to be a *random variable* with zero mean and a normal distribution; the parallel combination of R_{network} and R_{load} has been combined into the equivalent resistance, $-R_{\text{eq}}$. The circuit in Figure 14.12c is a small-signal model valid only around the metastability point. Writing v_{out} in time increments for the network schematic given in Figure 14.12c results in the *approximate solution*:

$$I_n \approx C * \frac{v_{\text{out}_{n+1}} - v_{\text{out}_n}}{\Delta t} + \frac{v_{\text{out}_n}}{-R_{\text{eq}}}$$

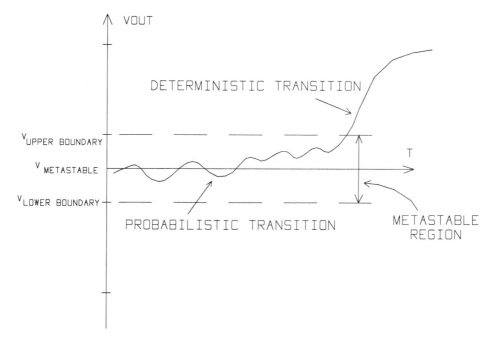

Figure 14.11. *Two-part* transition of the sequential element output voltage in the metastability region.

where $t_{n+1} = t_n + \Delta t$. Rearranging this expression yields

$$v_{\text{out}_{n+1}} \approx \underbrace{v_{\text{out}_n}\left(1 + \frac{\Delta t}{R_{\text{eq}} * C}\right)}_{\substack{\text{no random or} \\ \text{probabilistic nature}}} + \underbrace{\frac{I_n(\Delta t)}{C}}_{\substack{\text{uncertainty due to} \\ \text{the random nature of } I_n}}$$

Considering first just the deterministic term of this initial part of the signal transition:

$$\frac{v_{\text{out}_{n+1}} - v_{\text{out}_n}}{\Delta t} \approx \frac{dv_{\text{out}}}{dt} = \frac{1}{R_{\text{eq}} * C} v_{\text{out}}$$

or

$$v_{\text{out}} \propto e^{t/R_{\text{eq}}C}$$

The deterministic part of the initial transition is characterized by an exponential deviation from the metastable point. The second step is to include (by linear superposition) the effects of the noise term in the solution for the small-signal output voltage. The circuit noise is assumed to be normally distributed with a

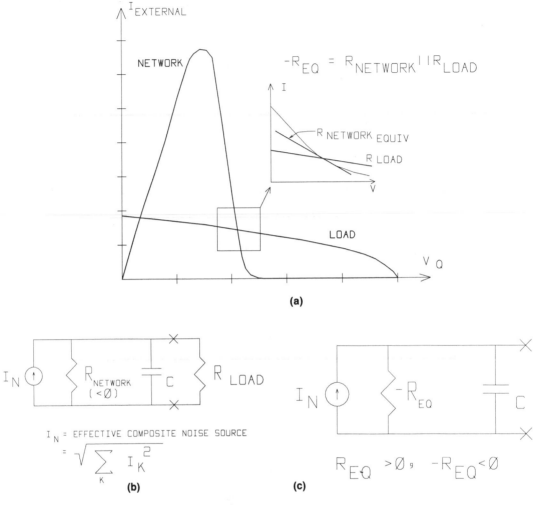

Figure 14.12. (a) Expansion of the load-line analysis for the network and load circuits in the vicinity of the metastability point. (b) Linearized small-signal RC model for the synchronizer circuit in the vicinity of the metastable point. In the expression for the effective composite noise source, I_n, the values i_k represent the noise component of frequency k. (c) Equivalent small-signal circuit.

mean of zero; for small time increments, I_n can be assumed to be a constant over the duration of the time interval. The second term will be referred to as a statistically varying noise voltage: $v_n = I_n * \Delta t / C$. The general analysis for this model proceeds using probability theory for the two, independent summed terms to give a probability density function for v_{out} at each increment of time. This is accomplished by performing a *convolution* of two probability density functions, the output voltage at time t and the noise voltage, to determine the resulting output voltage distribution at each *succeeding* point in time $t + \Delta t$ (Figure 14.13). An example of a repeated set of convolution integrals is shown in Figure 14.13b. The range $v_{\text{lower bound}}$ to $v_{\text{upper bound}}$ represents the interval within which the transition is still subject to the effects of noise; outside this interval, the transition is directed to one or the other stable points.

The probability of exceeding the boundary (terminating the initial portion of the transition) at any time t is found by integrating the area under the probability density function curve outside the boundary at any time t. Once the curve starts to cross the boundary, it is necessary to modify the convolution integral since it has been assumed that the output voltage will not reenter the metastable area; the probability density function used for subsequent calculations should be set to zero outside the boundary. Figure 14.14 is similar to Figure 14.13b except that an initial voltage offset at time $t = 0$ has been assumed. The probability density function $p_{v_{\text{out}}}$ is truncated at the boundary after each time t_i; the remaining area under the curve represents the probability that the output voltage has not escaped the boundary. The key result of this analysis comes from the assumption that the circuit output voltage at time $t = 0$ has an initial uniform probability distribution over the entire boundary range $v_{\text{lower bound}}$ to $v_{\text{upper bound}}$; that is, at time $t = 0$, it is equally likely that the output voltage in metastability can be anywhere in this interval. With this assumption, the probability of escape from the metastable region *up to* time t is given by (reference 14.2)

$$p_e(t) = 1 - e^{-t/R_{\text{eq}} * C}$$

The assumption that the initial output voltage is uniformly distributed over the boundary range implies that the noise effect (with zero average) does not enter the metastability result.

The total settling time is given by the sum of the time in the metastable region plus the deterministic transition time; the deterministic transition time can be determined to a fair accuracy using detailed device-level circuit simulations. The time allotted for an asynchronous input to settle before being sampled, t_{settle}, is a system design criterion and should be maximized to reduce the failure rate to an acceptable level.

The probability that, once in the metastable region, the output voltage has not reached a stable value within t_{settle} is

$$p_{\text{fail}} = 1 - p_{\text{escape}}(t_{\text{settle}} - t_{\text{deterministic}})$$

$$= \exp\left[-\frac{t_{\text{settle}} - t_{\text{deterministic}}}{R_{\text{eq}} * C} \right]$$

This probability should be multiplied by the frequency of synchronization events that will send the synchronizer into the metastable region. This multiplicative factor is much more difficult to attempt to model and calculate analytically; more often it is measured experimentally.

The following description is a composite of some of the analytic models and

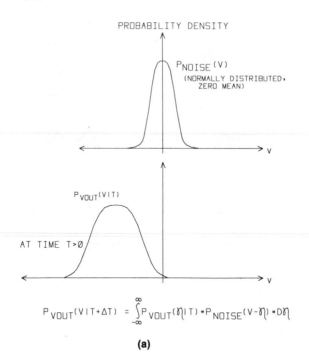

PROBABILITY DENSITY

$P_{NOISE}(V)$

(NORMALLY DISTRIBUTED, ZERO MEAN)

V

$P_{VOUT}(VIT)$

AT TIME T>0

V

$$P_{VOUT}(V|T+\Delta T) = \int_{-\infty}^{\infty} P_{VOUT}(\eta|T) * P_{NOISE}(V-\eta) * D\eta$$

(a)

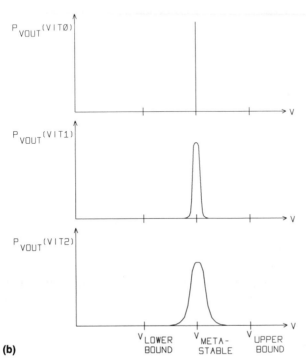

$P_{VOUT}(VIT0)$

V

$P_{VOUT}(VIT1)$

V

$P_{VOUT}(VIT2)$

V

$V_{LOWER \atop BOUND}$ $V_{META- \atop STABLE}$ $V_{UPPER \atop BOUND}$

(b)

Figure 14.13. (a) Convolution of probability distributions for output voltage and noise voltage at time t, which is used to determine the probability distribution of the output voltage at time $t + \Delta t$. (b) Repeated convolutions of the output voltage probability distribution at succeeding time intervals.

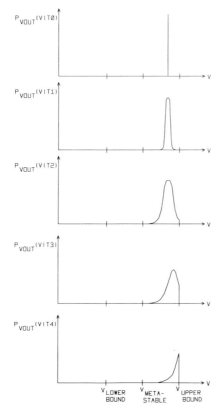

Figure 14.14. Repeated convolutions of the output voltage probability distribution at succeeding time intervals with an *initial voltage offset* (adapted from reference 14.2).

expressions that have appeared in the technical literature for the frequency of synchronization events resulting in metastable behavior:

Notation

f_D: maximum asynchronous data input frequency

f_C: synchronizing clock signal frequency

Δ: time interval in the clock period to which changes in the asynchronous data signal can put the latch into a metastable state (Figure 14.15)

The number of occurrences of a transition of the data input signal within this time window per second is given by

$$n = 2 * \Delta * f_C * f_D = 2 * \Delta * f_D * \frac{1}{T_C}$$

where the factor of 2 is derived from the two transitions on the data signal within the period $1/f_D$. What remains to be defined is a suitable technique to measure the clock pulse metastability window, Δ; it would also be desirable to include in that definition a dependency on the asynchronous input signal rise and fall time.

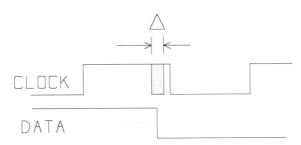

CLOCK

DATA

Figure 14.15. Definition of system clock time window, Δ, in which data transitions will lead to the storage element entering the metastable region. Note that in the figure the input signal edges are assumed to be vertical (no rise or fall time) and that the clock window for metastability need not necessarily include the gating edge of the clock.

It is at this point that most references either (1) treat Δ as a parameter to fit to experimental data, (2) make an assumption about what fraction of the total synchronizer propagation delay should be assigned to Δ (for a particular circuit design), or (3) make a series of circuit simulations to attempt to determine the Δ interval in the clock signal. The first option, to attempt to measure the metastability rates for particular synchronizer designs experimentally, is probably the best. However, gathering accurate data on an *integrated* synchronizer design is not feasible; the measurement procedure itself will very likely perturb the circuit to such a degree so as to invalidate the results. It would seem that the experimental measurement techniques would be limited to discrete designs. To effectively use circuit simulation to measure and/or compare the metastability rates of integrated designs, some standard measurement technique should be used. One such method is to draw a *phase plane* diagram for the circuit design (reference 14.3); an example of such a diagram is given in Figure 14.16. The curves on the diagram are the trajectories of the (v_Q, $v_{\bar{Q}}$) voltage values for different simulation runs with the data input changing at different times in the clock pulse. In this example, the initial state of the latch assumed for all trajectories is (v_Q, $v_{\bar{Q}}$) = (v_1, v_0). (A similar diagram could be generated for the opposite initial state.) Trajectory 1 in the figure corresponds to a data input transition with sufficiently large setup and hold times for the latch. Trajectories 2–5 correspond to the output voltage values for simulation runs with shorter and shorter data setup times. Note that at some point the trajectories do not reach the final value of the data input signal, but rather return to the initial state of the simulation. A 45° line through the origin divides the phase plane into two regions, those trajectories which reach the opposite state from those that return to the original. (The 45° line is a consequence of an assumption of perfect symmetry in two cross-coupled gates; in actuality, this line, called the separatrix line, need not be at 45° to the horizontal nor need it be a straight line.[5]) Note in the phase plane diagram that a couple of the trajectories seem to hover around the separatrix line, moving toward a metastable

---

[5] The phase plane analysis of Figure 14.16 is not to be confused with the *dynamic-state phase diagram* in Section 12.5. In this latter diagram, the assumption that the circuit is in some initial state is removed. In addition, the data and clock signals are absent. Instead, an initial condition at time t = 0 is assumed for the coordinate pair (v_Q, $v_{\bar{Q}}$); at t = 0^+, the circuit is released (with no input stimulus) and the trajectory of the transition to a stable state is plotted. The separatrix line again divides those trajectories that eventually result in opposite stable states.

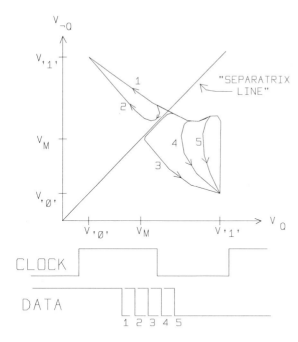

Figure 14.16. Phase plane diagram depicting $(v_Q, v_{\bar{Q}})$ output voltage transitions for differing values of data setup time (adapted from reference 14.3).

operating point designated by the coordinate pair $(v_Q, v_{\bar{Q}}) = (v_M, v_M)$, increasing the time required for the latch to stabilize in one state or another. As mentioned before, this implies that there is a small range of input signal transitions relative to the clock pulse, which will result in an output signal transition time greater than the system settling time. The definition of Δ, as based on the circuit simulations, could be developed in the following manner: Δ is defined by the two least differing data input signals whose simulated trajectories end up in different stable states. The goal of the simulations is therefore to find the narrowest range of data input signals that provide different stable results. To develop as accurate a measure as possible, several precautionary steps should be taken:

1. The circuit simulation step size, absolute error, and relative error values specified should be aggressive and consistent from run to run.

2. The modeling accuracy should reflect the actual physical layout as closely as possible, particularly with regards to junction and wiring capacitances, contact resistances, process level tracking, and the like.

3. The rise time and fall time on the clock and data input signals to the synchronizer should be as close as possible to the expected (chip delay equation) values.

From the circuit simulations that are used to determine the time window Δ, a measure of the worst-case deterministic transition time (the time on the trajectory to move from the separatrix line to a stable point) can also be obtained.

It should be mentioned that the purpose of the transient circuit simulations is to get an estimate for the Δ time window, as well as a measure of the deterministic part of the transition time; any attempt to extract more information from the results is probably unwise. The main functional dependency is present in the negative exponent of the probabilistic factor; it is most important to accurately model the effective small-signal R and C for the network output node around the metastability point.

To summarize the analysis described in this section, the following procedure for developing a metastability failure rate is suggested:

1. Develop an equivalent circuit model for the network + load at the latch output node.

2. Use dc circuit simulations to develop two curves, the (v, i) characteristics of the network at the output node and the load line characteristics; in particular, the two curves should be accurately developed in the vicinity of their third intersection point.

3. From these curves extract the equivalent small-signal resistance of the network and load to develop an equivalent R; from the circuit design layout, determine a value for the nodal capacitance (for voltage-dependent capacitances, evaluate at the metastable node voltage).

4. Using transient circuit simulations, develop a phase plane diagram for the latch at both initial states; determine a suitable time window Δ from the simulations, as well as a measure of the deterministic transition time away from the separatrix line to a stable state.

5. Use the relationship

$$\frac{1}{\text{MTBF}} = \frac{\text{No. of metastability failures}}{\text{year}}$$

$$= 2 * \Delta * f_C * f_D * \exp\left[-\frac{(t_s - t_d)}{R_{\text{eq}} * C}\right] * \left(\frac{\text{POH}}{\text{year}}\right) * (3600)$$

where the term POH/year represents the estimated number of *power-on hours* per year for the system design application.

Recommendations

In addition to the calculation of metastability failure rates, some additional synchronizer design guidelines should be mentioned, some of which are direct consequences of the analysis:

1. The time constant $R_{\text{eq}} * C$ for the output node should be minimized; the circuit design and physical layout should reflect this goal. Specifically, syn-

chronization latch outputs should be buffered to reduce (internal) capacitive loading.

2. The rise and fall times of the clock and data input signals should be kept to a minimum.

3. A common approach is to use a *double-latch* design to effectively multiply two metastable rates together.

4. Perhaps the most intriguing approach to reducing metastability failure rates is based on a comment by Marino (reference 14.3):

It may be possible to intentionally vary the circuit inputs [once in a metastable condition] . . . it may be argued that noise could offset the intentional variations, thus in effect keeping the input constant. But this would not be the case if the intentional variations were made larger in magnitude than the expected noise. . . . it may also be argued that input variations, intentional or otherwise, may simply move the $(v_Q, v_{\bar{Q}})$ state from one invariant to another [along the separatrix line]. But it is not obvious that such behavior could or would occur.

14.8 PHYSICAL DESIGN DATABASE

There are two physical design databases involved with VLSI chip development, the repository of design system library information and the actual shapes and cell transforms that constitute individual chip designs. The design system library physical database consists of the macro and book physical layouts and the file(s) of macro parametric data. Each physical layout macro file is described using the graphics constructs of Chapter 5; in addition to the fabrication description, this file contains the additional information necessary to complete the chip physical design: macro pin locations, macro size, and macro shadows and wiring blockages. The parametric macro files contain the data used for L/P checking, dc noise margin verification, and delay calculation, specifically the following:

1. The macro's hierarchical logic representation for L/P checking

2. The parametric electrical data of each circuit for delay calculation and dc checking

3. The tables of input and output pin delay adders as functions of signal transition time and capacitive loading, respectively

The design system library parametric electrical data can be voluminous, as each macro circuit has a number of uniquely associated resistive, capacitive, and device measures; these data are used strictly for design verification and delay calculation and are not commonly included with the chip database.

The chip design physical database consists of the macro and book layouts and the global wires; in addition, packaging and other related part number information is copied into the chip header.

The complete chip database is the combination of logical, physical, and test descriptions; this collection of data is released to and subsequently managed by manufacturing. Current research efforts are addressing techniques to best coordinate the various facets of the chip design database to ensure that parameters of the physical design are accurately reflected in the logic and test descriptions. These parameters include the following:

1. Signal assignments to chip pads
2. Macro and book placement coordinates
3. Optimized circuit power/performance levels and calculated block delays
4. Translation of logically equivalent pins to physical macro pin names for enhanced wireability

This information is appended to the logic data structure by the physical design tools and becomes part of the chip documentation and diagnostic data. The back-end design system global verification tools (design rule checking, dc checking, and L/P checking) also provide a time-date stamp and completion code to the chip logical description to ensure that the design has successfully passed these audits.

PROBLEMS

14.1. For CMOS technology circuits, logical-to-physical checking also requires verifying that the local well for the fabrication of complementary devices is connected to the correct supply voltage. Briefly describe how the L/P checking methodology can be extended to check CMOS layouts; in particular, describe what additional NODE shape statements are required and how the continuity associations for the well contact are defined.

14.2. Develop a *hierarchical* L/P checking methodology, including necessary modifications to the cell transform statements to be able to build hierarchical node names for verification. Comment on the productivity advantages (or disadvantages) of hierarchical L/P checking, as it pertains to the tasks of composing and checking the layout.

14.3. Describe the uses for a fast-path timing analysis option, where the *minimum* signal arrival times are propagated through the network.

14.4. A system designer has the following synchronizer requirements:

$$t_{\text{settle}} = 25 \text{ nsec}$$

$$f_{\text{clk}} = 14 \text{ MHz}$$

$$f_{\text{data}} = 2.5 \text{ MHz}$$

From an analysis of a synchronizer circuit design and layout, the following parameters were determined:

$$-R_{\text{network}} \parallel R_{\text{load}} = -1.7 \text{ k}\Omega$$

$$C_{\text{node}} = 0.25 \text{ pF}$$

$$\Delta = 100 \text{ psec}$$

$$t_{\text{deterministic}} = 5 \text{ nsec}$$

Calculate the MTBF for this particular application; assume the system is to be operated for 5000 power-on hours per year.

REFERENCES

14.1 Chaney, T., and Molnar, C., "Anomalous Behavior of Synchronizer and Arbiter Circuits," *IEEE Transactions on Computers,* C-22:4 (April 1973), 421–422.

14.2 Couranz, G., and Wann, D., "Theoretical and Experimental Behavior of Synchronizers Operating in the Metastable Region," *IEEE Transactions on Computers,* C-24:6 (June 1975), 604–616.

14.3 Marino, L. R., "General Theory of Metastable Operation," *IEEE Transactions on Computers,* C-30:2 (February 1981), 107–115.

Additional References

Barlow, J., "A New Software Tool for Detecting Problems Caused by Inductively-Generated Switching Noise," *IEEE Test Conference* (1982), 166–169.

Buturla, E., and others, "Finite-Element Analysis of Semiconductor Devices: The FIELDAY Program," *IBM Journal of Research and Development,* 25:4 (July 1981), 218–231.

Catt, I., "Time Loss through Gating of Asynchronous Logic Signal Pulses," *IEEE Transactions on Electronic Computers,* EC-15:1 (February 1966), 108–111.

Chaney, T., "Beware the Synchronizer," *COMPCON '72, IEEE Computer Society Conference* (September 1972), 317–319.

———, "Comments on 'A Note on Synchronizer or Interlock Maloperation'," *IEEE Transactions on Computers,* C-28:10 (October 1979), 802–804.

Elineay, G., and Wiesbeck, W., "A New J-K Flip-Flop for Synchronizers," *IEEE Transactions on Computers,* C-26:12 (December 1977), 1277–1279.

Fleischhammer, W., and Dortok, O., "The Anomalous Behavior of Flip-Flops in Synchronizer Circuits," *IEEE Transactions on Computers,* C-28:3 (March 1979), 273–276.

Hitchcock, Robert, and Goel, N., "Analysis of Cooperative Failures in Digital Computers," *Journal of Digital Systems,* 6:1 (1982), 65–93.

Hu, G., "A Better Understanding of CMOS Latch-up," *IEEE Transactions on Electron Devices,* ED-31:1 (January 1984), 62–67.

Liu, B., and Gallagher, N., "On the 'Metastable Region' of Flip-Flop Circuits," *Proceedings of the IEEE (Letters)*, 65:4 (April 1977), 581–583.

Long, S., "Test Structures for Propagation Delay Measurements on High-Speed Integrated Circuits," *IEEE Transactions on Electron Devices*, ED-31:8 (August 1984), 1072–1076.

Ochoa, A., and others, "Latch-up Control in CMOS Integrated Circuits," *IEEE Transactions on Nuclear Science*, NS-26:6 (December 1979), 5065–5068.

Olson, L., "Application of the Finite Element Method to Determine the Electrical Resistance, Inductance, Capacitance Parameters for the Circuit Package Environment," *IEEE Transactions on Components, Hybrids, and Manufacturing Technology*, CHMT-5:4 (December 1982), 486–492.

Pechoucek, M., "Anomalous Response Times of Input Synchronizers," *IEEE Transactions on Computers*, C-24:6 (June 1975), 604–616.

Ruehli, A., "Inductance Calculations in a Complex Integrated Circuit Environment," *IBM Journal of Research and Development*, 16:5 (September 1972), 470–481.

——, "Survey of Computer-Aided Electrical Analysis of Integrated Circuit Interconnections," *IBM Journal of Research and Development*, 23:6 (November 1979), 626–639.

Russell, P. J., "Physical-to-Logical Checking of FET LSI Chips," *IBM Technical Disclosure Bulletin*, 21:2 (July 1978), 822–824.

Stoll, P. A., "How to Avoid Synchronization Problems," *VLSI Design*, (November/December 1982), 56–59.

Terrill, K., and Hu, C., "Substrate Resistance Calculation for Latchup Modeling," *IEEE Transactions on Electron Devices*, ED-31:9 (September 1984), 1152–1155.

Troutman, R., and Zappe, H., "Layout and Bias Considerations for Preventing Transiently Triggered Latchup in CMOS," *IEEE Transactions on Electron Devices*, ED-31:3 (March 1984), 315–321.

Veendrick, H., "The Behavior of Flip-Flops Used as Synchronizers and Prediction of Their Failure Rate," *IEEE Journal of Solid-State Circuits*, SC-15:2 (April 1980), 169–176.

Wakeman, L., "Silicon-gate C-MOS Chips Gain Immunity to SCR Latchup," *Electronics* (August 11, 1983), 136–140.

Wallmark, J., "Noise Spikes in Digital VLSI Circuits," *IEEE Transactions on Electron Devices*, ED-29:3 (March 1982), 451–458.

Wormald, E., "A Note on Synchronizer or Interlock Maloperation," *IEEE Transactions on Computers (Correspondence)*, C-26:3 (March 1977), 317–318.

——, "Support for T. J. Chaney's Comments on 'A Note on Synchronizer or Interlock Maloperation," *IEEE Transactions on Computers*, C-28:10 (October 1979), 804.

15

INTRODUCTION TO VLSI PROCESS DEVELOPMENT

15.1 CROSS-SECTIONAL DESCRIPTION OF A VLSI PROCESS

This chapter presents a *brief* introduction to integrated-circuit fabrication technology; this section starts with a brief dimensional comparison chart (Figure 15.1), followed by a process description summary (with related device cross sections) of an *n*MOS technology. The technology is that of the chip design system described in reference 3.2.

Process description
1. Initial wafer clean (Figure 15.2): In particular, the following solutions are commonly used:
   (a) H_2SO_4/HNO_3: removes photoresist, other organic materials (oils, greases, waxes, etc.)
   (b) HF: oxide etchant; the top layer of SiO_2 containing any potential contamination is etched away
   (c) NH_4OH/H_2O_2: removes organic contaminants, any remaining fluorine
   (d) HCl/H_2O_2: removes heavy metal contaminants by forming soluble compounds
2. Backside ion implant (Figure 15.3): Providing a region of large crystalline disorder on the backside of the wafer induces the migration of dislocations in the crystal structure to the backside during subsequent high temperature steps. This disorder can be created by implanting the backside with argon, which is not an electrically active dopant.

DIMENSIONAL COMPARISON CHART

METRIC		ENGLISH
	THICKNESS OF MASK	0.090"
	THICKNESS OF WAFER	0.020"
25.4 UM	=	1 MIL = 0.001"
20 UM	WIDTH OF TYPICAL LINE ON 10X RETICLE	
6 UM	THICKNESS OF CHIP SEALING OVERCOAT DIELECTRIC	
2 UM	WIDTH OF (MINIMUM) PHOTORESIST IMAGE LINES AND SPACES	
1.5-2.0 UM	THICKNESS OF M1-M2 INTERLEVEL DIELECTRIC	
1.2-1.5 UM	THICKNESS OF METALLIZATION LAYERS	
1.0 UM = 10^{-6} M = 10000 Å	=	40 U"
5000 Å	(RECESSED) FIELD OXIDE THICKNESS, FINAL	
3500-5000 Å	DEPTH OF SOURCE AND DRAIN JUNCTION NODES, FINAL	
3000-4000 Å	POLYSILICON LINE THICKNESS, FINAL	
750-1000 Å	GATE OXIDE THICKNESS (CIRCA MID-1970'S	
350-600 Å	GATE OXIDE THICKNESS (CIRCA LATE-70'S, EARLY 80'S)	
200-350 Å	GATE OXIDE THICKNESS (MID-1980'S)	

Figure 15.1. Fabrication process technology dimensional comparison chart for a 2-μm technology (circa 1982–1985).

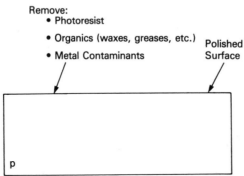

Remove:
- Photoresist
- Organics (waxes, greases, etc.)
- Metal Contaminants

Polished Surface

Figure 15.2. Initial wafer clean.

3. Pad oxide growth, 400 Å (Figure 15.4): The pertinent reactions are:

$$Si + O_2 \rightarrow SiO_2, \qquad Si + 2H_2O \rightarrow SiO_2 + 2H_2$$

The presence of chlorine during the oxidation provides sodium ion passivation and an enhanced growth rate.

4. Pad nitride (Si_3N_4) deposition, 1000 Å (Figure 15.5): Silicon nitride films act as a barrier to the diffusion of oxidizing species to the surface, inhibiting SiO_2 growth. The chemical vapor deposition process for silicon nitride uses

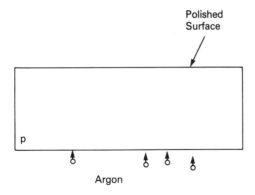

Figure 15.3. Backside ion implant.

Figure 15.4. Pad oxidation.

dichlorosilane (SiH_2Cl_2) and ammonia (NH_3):

$$3SiH_2Cl_2 \text{ (g)} + 4NH_3 \text{ (g)} \rightarrow Si_3N_4 + 6HCl \text{ (g)} + 6H_2$$

5. Anneal

6. Recessed oxide isolation mask definition (Figure 15.6): The photoresist (and Si_3N_4 underneath) will define device area from field isolation oxide regions. The exposed nitride is removed by reactive ion etching from areas not protected by photoresist. The pad oxide remains underneath.

7. Field tailor implant, boron (p-type) (Figure 15.7): This impurity introduction raises the field oxide threshold voltage to reduce leakage between adjacent circuits.

8. Photoresist strip

9. Recessed field oxide growth (Figure 15.8): The silicon nitride acts as a barrier to the diffusion of oxygen to the silicon–silicon dioxide interface. Only in

Figure 15.5. Pad nitride deposition.

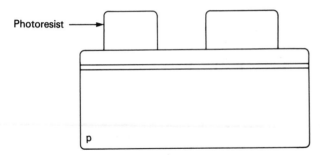

Photoresist

Figure 15.6. Recessed oxide shape patterning into photoresist coat.

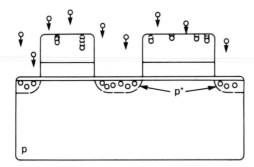

Figure 15.7. Field threshold tailoring implant.

the areas where the nitride is absent does the SiO_2 grow, consuming silicon (locally) from the wafer substrate.

10. (a) Oxynitride etch
    (b) Silicon nitride strip
    (c) Pad oxide etch (Figure 15.9)
11. Gate oxidation, 450 Å: This dielectric layer must be as defect free and contaminant free as possible.
12. Gate oxide anneal
13. Gate 1 enhancement ion implant, boron (p) (Figure 15.10): Boron implanted

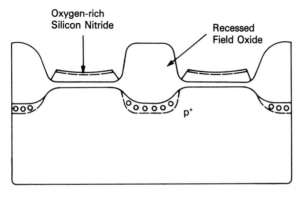

Oxygen-rich Silicon Nitride

Recessed Field Oxide

Figure 15.8. Recessed oxide growth.

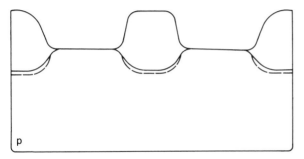

Figure 15.9. Nitride and pad oxide etch removal.

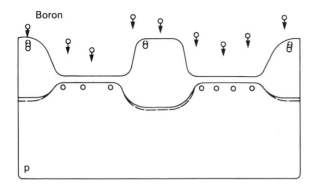

Figure 15.10. Enhancement device threshold ion implant.

under the gate oxide raises the threshold voltage of all devices. This is a blanket implant; the introduction of impurities is not locally masked off.

14. Depletion-mode device mask

15. Depletion device threshold adjust implant, arsenic (n) (Figure 15.11): Depletion-mode device channel areas are locally produced by overcompensation of the enhancement implant with a depletion implant of opposite im-

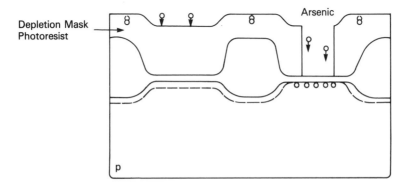

Figure 15.11. Depletion device threshold adjust ion implant.

purity type. The areal density of the *n*-type implant is sufficient to produce a negative threshold voltage.

16. (a) Photoresist strip
    (b) Buried contact mask lithography
    (c) Oxide etch (450 Å) (Figure 15.12)
    (d) Photoresist strip
    It is convenient (but not mandatory) to provide interconnection between polysilicon and diffused areas, utilizing a buried contact. This interconnection saves area and wiring blockages in comparison to a polysilicon-to-first metal-to-diffused area interconnection scheme.

17. Polysilicon deposition (Figure 15.13)

18. Polysilicon doping: Typically, polysilicon is used as the gate material; in previous MOS technologies, metal was used. The polysilicon is doped with impurities to reduce its resistivity (and thereby enhance its attractiveness as an interconnect material). VLSI process research is attempting to further reduce the interconnect resistance by using a metal–silicide compound.

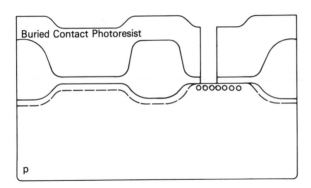

Figure 15.12. Buried contact mask lithography; cross section shown after oxide etch.

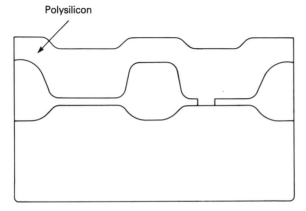

Figure 15.13. Polysilicon deposition.

19. Polysilicon mask:
    (a) Apply, expose, and develop photoresist
    (b) Etch polysilicon
    (c) Oxide etch: removes the exposed thin oxide not under polysilicon (Figure 15.14)
    (d) Strip photoresist
    The sandwich of polysilicon–gate oxide–silicon forms the device channel area. There are two types of devices present, enhancement mode (with a positive threshold voltage) and depletion mode (negative V_T). Natural or zero threshold voltage devices could also be provided by locally masking off *both* enhancement and depletion threshold shifting ion implants.

20. Gate 2 oxidation (450 Å): If a second polysilicon layer were to be included in this technology, it would be deposited and patterned at this point in the fabrication sequence. A single level polysilicon technology will be shown.

21. Gate anneal

22. Source and drain ion implant, arsenic (n) (Figure 15.15): The polysilicon gate defines the edges of the source and drain regions adjacent to the device channel. With self-aligned gates, the gate-to-source node and gate-to-drain node overlap area is greatly reduced (in comparison to metal gate technol-

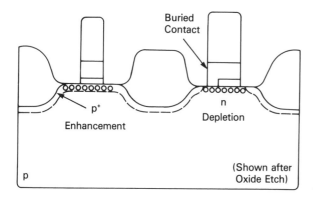

Figure 15.14. Polysilicon patterning.

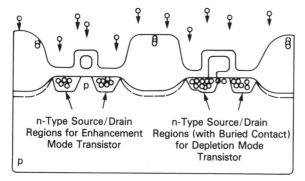

n-Type Source/Drain Regions for Enhancement Mode Transistor

n-Type Source/Drain Regions (with Buried Contact) for Depletion Mode Transistor

Figure 15.15. Device source and drain node ion implant.

ogies), which reduces device input capacitance. Additionally, polysilicon-gate devices eliminate a critical process lithography alignment step present in metal-gate technologies, and facilitate further high-temperature processing steps after gate deposition and patterning.

23. Drive-in oxide, 2700 Å: This high-temperature oxidation step serves two purposes: (1) to drive in the implanted impurities into the source and drain regions, and (2) to grow oxide over the entire device area to seal off the devices (to reduce susceptibility to impurity contamination).

24. Pyrolysis of silicon dioxide, 3000 Å (Figure 15.16): The pertinent reaction is

$$SiH_2Cl_2 \text{ (g)} + 2N_2O \text{ (g)} \rightarrow SiO_2 + 2HCl \text{ (g)} + 2N_2$$

25. Anneal

26. Contact mask (Figure 15.17):
    (a) Apply, expose, develop photoresist
    (b) Etch oxide (note potential undercutting of the photoresist image opening)
    (c) Strip photoresist

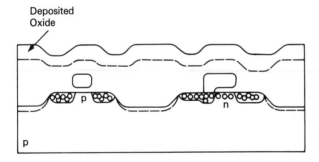

Figure 15.16. Oxide dielectric growth and deposition.

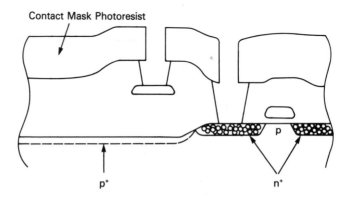

Figure 15.17. Polysilicon and diffusion node contacts (shown after oxide etch).

27. Contact mask, substrate contact:
    (a) Apply, expose, develop photoresist
    (b) Etch oxide

28. Evaporate barrier metal into substrate contact (Figure 15.18)

29. Strip photoresist

30. Anneal

31. First metal mask lithography

32. Metal evaporation, Al/Cu/Si (Figure 15.19)

33. Photoresist strip: The photoresist is dissolved in an agitated bath. The evaporated metal present on this imaged layer is removed from the wafer surface. The photoresist pattern and metal evaporation must produce a clean break between the evaporated metal lines to remain and those to be removed.

34. Anneal

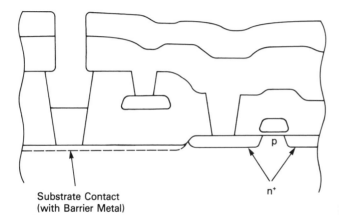

Substrate Contact
(with Barrier Metal)

Figure 15.18.   Substrate contact.

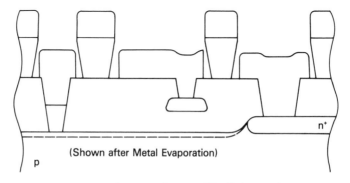

(Shown after Metal Evaporation)

Figure 15.19.   Metal evaporation onto lift-off photoresist pattern.

This concludes the front-end-of-the-line (FEOL) processing. Early characterization data can now be gathered from device kerf test structures. Based on the process monitor results obtained at first metal characterization and test, the run may proceed for back-end-of-line (BEOL) processing or may be halted (and a backup run accelerated). Process monitor wafers are included strictly for individual thin film deposition and implantation steps; it is still necessary to extract product wafers for first metal testing of device structures, specifically investigating the following:

Device characteristics
Open (or nonohmic) contacts
First metal-to-substrate shorts (through diffusion contacts)
Junction leakage currents (to substrate)
Surface leakage currents (between adjacent diffusion nodes)
Sheet resistivities

If any of these measurements provide catastrophic results, the run may be halted and a failure analysis investigation started.

BEOL process description

35. Plasma-enhanced silicon nitride deposition
36. Polyimide coat:
  (a) Preclean
  (b) Polyimide apply
  (c) Cure
  (d) Monochromatic light inspect
37. Via mask: polyimide/nitride composite dielectric via between first and second metal interconnection planes (Figure 15.20)

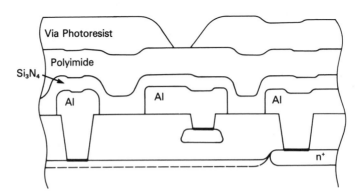

Figure 15.20. M1–M2 via lithography.

38. Reactive ion etch of via (Figure 15.21):

   (a) Polyimide, photoresist: O_2 etch gas chemistry

   (b) Silicon nitride: $CF_4 + O_2$ etch gas chemistry

   The photoresist layer is removed during the polyimide/nitride etch procedure.

39. Second metal mask lithography: Premetal deposition cleaning includes a polyimide ash step followed by aluminum oxide clean steps.

40. Second metal deposition, Al/Cu (Figure 15.22)

41. Second metal photoresist strip

42. Anneal

43. Polyimide coat:

   (a) Preclean, polyimide apply, solvent drive-off, monochromatic light inspect

44. Terminal via mask:

   (a) Apply resist

   (b) Expose

   (c) Develop/etch (Figure 15.23)

   (d) Resist strip

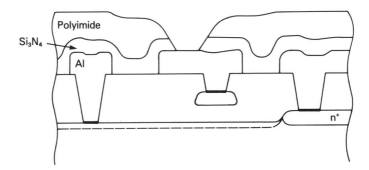

Figure 15.21.   Via etch. The via photoresist profile is transferred into the topography of the dielectric opening.

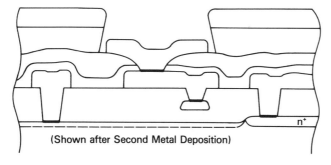

(Shown after Second Metal Deposition)

Figure 15.22.   Second-level metal lithography and evaporation.

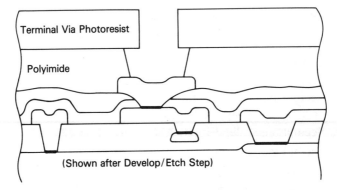

(Shown after Develop/Etch Step)

Figure 15.23. Terminal via mask lithography in preparation for solder ball evaporation.

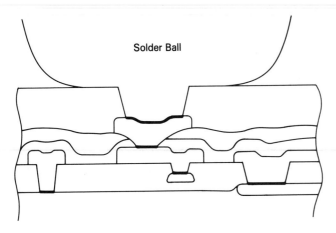

Figure 15.24. Chip to package signal interconnection for this technology is provided through solder balls between second-metal chip pad and package surface wiring trace; the active device surface is mounted face down on the package. (For more details on package attach techniques, refer to Section 15.18.)

45. Polyimide cures

46. Terminal metal (Figure 15.24):
   (a) Evaporate Cr/Cu/Au and Pb/Sn through molybdenum mask
   (b) Reflow cylinders into solder balls

47. Final test → yield

15.2 DESIGN RULES FOR MANUFACTURING

The design rules of a technology represent the commitment from the process engineering group to the physical design group regarding the artwork and wafer level dimensional features achievable in the technology for the manufacturing mask levels. These design rules must be aggressive so as to make the technology competitive in cost and performance and simultaneously conservative so as to guarantee the manufacturability of any part designed to satisfy these rules. It should be noted that the design rules are developed with specific assumptions regarding the following:

Operating voltage

Operating temperature range

Maximum chip size

Current and anticipated processing equipment capabilities

Chip separation and packaging technology

The design rules commonly do not represent the absolute lithography or chemical process limitations of the equipment used in the fabrication facility. In any process, there tends to be a measure of conservatism in the design rules; the main reason for the somewhat relaxed dimensions is the importance of *repeatability*. Once a process reproducibly manufactures parts satisfying the process/performance design window (and truly not until that point) can accurate yield models be developed and representative cost estimates be obtained. Some of the conservatism in design rule development is forsaken, however, in anticipation of enhanced fabrication equipment and techniques. Development of a new process requires time to demonstrate manufacturability; the design rules should be developed so as to encompass the anticipated enhancements during that period.

Designers should use the minimum design rules for widths and spaces of interconnection lines only where the density and performance *leverage* is present. If at all feasible, designers should consider using greater than minimum image and space dimensions in order to make the resulting design as insensitive to process variations as possible. For example, process relief may be achieved if:

1. The spacing between metal lines on the same level is increased.
2. Contact areas between interconnection levels are maximized.
3. All shapes on a contact mask level are of comparable dimension.

(If the etch rate is a function of the contact area, all contacts will have roughly the same etch rate and therefore etch time; conversely, different interconnection overlap design rules may be imposed as a function of contact area.)

Subsets of a process may be available if, for example, a second metal interconnection level or a second polysilicon gate/interconnect level is not required for some application. The design rules should be the same for a potential subset of an overall process.

Linewidth Bias and Measurement

The development of process design rules must account for artwork-to-wafer level etch bias and level-to-level overlay (alignment). In determining the etch bias for

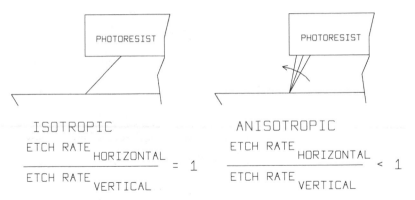

Figure 15.25. Degree of anisotropy of etch process steps, as defined by the ratio of the lateral-to-vertical etch rates.

a particular linewidth, several factors must be considered:[1]

(a) The edge profile of an interconnect line is not vertical. Etching techniques provide some measure of isotropy; the ratio of the lateral etch rate to the vertical etch rate is somewhere between 0 and 1 (Figure 15.25). The definition of the postetch linewidth must therefore include a specification of what elevation in the edge profile is used as the reference.

(b) Some overetch of the material is usually included to ensure the following:

All contact holes are indeed exposing the interconnection below.

All interconnect lines are distinct (with no bridging faults).

No residues or residual areas remain of the material being etched (i.e., no small grains of the material being etched remain on the surface).

Residues are a common problem when etching polysilicon, for example; it is desirable to completely remove the exposed polysilicon layer without overetching to the extent that the etch bias of a polysilicon line increases substantially.

(c) In some cases, the bottom linewidth after patterning is the dimension of interest; in other cases, it is the linewidth at the top. For a polysilicon gate/interconnection layer, the dimension used for capacitance calculations is the bottom linewidth, while contact area design rules must utilize the top.

It is necessary to have an in-line process measurement system to be able to measure photoresist and postetch linewidths (after photoresist removal) to provide

---

[1] Since both additive (lift off) and subtractive process steps may be used for fabricating interconnect lines, a better term would be "process bias," rather than etch bias. In general, the term *bias* represents a difference between the design dimension and a resulting linewidth or contact opening.

immediate feedback on the success of the photoresist expose and develop and etching process steps. Linewidths outside the *process design limits* require the process step to be reworked (or, if rework is not feasible, the process run may have to be scrapped). A block diagram of a typical linewidth measurement system is shown in Figure 15.26. The image of the line to be measured on the mask or wafer is focused in space above the projection options in the plane of the scanning slit. Behind the slit is a photomultiplier light detector. The slit and light detector assembly is driven by a lead screw and stepper motor assembly and is scanned

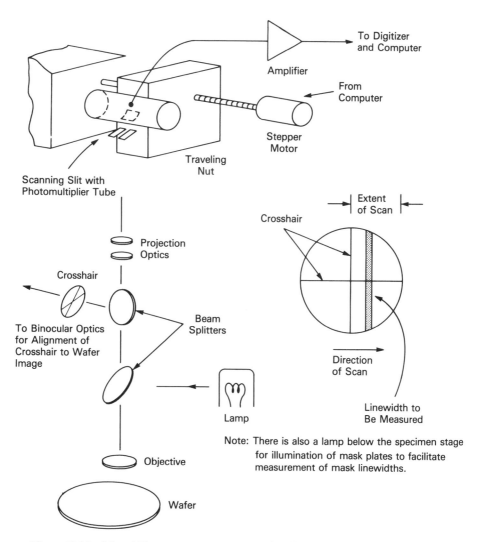

Figure 15.26.   Linewidth measurement system and an illustration of the path across the wafer surface containing the patterned image to be scanned for reflectivity changes.

in many steps across the magnified image of the line. Intensity measurements from the light detector are digitized at each step and stored in the system computer. To develop the design rules and verify that the particular process step remains within the process design limits, the process engineer must evaluate the resulting profile to determine the actual dimension of interest. In particular, the measurement will be sensitive to the amount of defocus of the image being measured; as a result, a threshold can be specified in the reflectance versus position data that defines the linewidth edges. Typical reflectance versus scan plots are illustrated in Figure 15.27.

In addition to an in-line process measurement system, a scanning electron microscope is an invaluable tool in linewidth measurement, particularly during process development (and failure analysis). The accuracy obtained is significantly greater than the optical system just described both in dimensional measurement and edge profile determination.

Overlay

In determining the effect of overlay on the development of design rules, two questions need to be addressed:

1. What are the sources of error in the level-to-level alignment procedure?
2. How do these errors accumulate during the fabrication process?

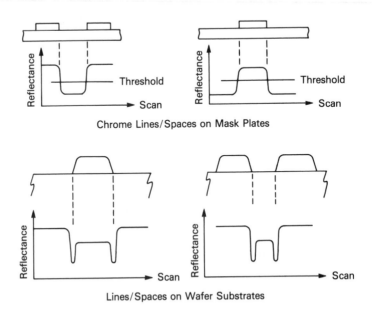

Figure 15.27.   Line-width measurement results and criteria.

The overlay error can be attributed to three sources: mask, operator, and tool. These errors strictly refer to the pattern-imaging procedure and do not include the effects of photoresist sensitivity to exposure, development, and bake conditions, or to the etching process itself:

Mask: magnification
         skew
         chip periodicity
         image placement
Operator: x, y, θ alignment
Tool: temperature
         magnification
         distortion
         mask and wafer warpage (run-out)

The alignment error between a new lithography level and earlier levels accumulates; a quantitative model for these cumulative statistical errors is necessary for the determination of an *alignment strategy* (sequence) for a process. For a technology with n lithography levels, some of the $(n)(n-1)/2$ different level-to-level alignment measures are indeed critical, while the remainder have more relaxed dimensional requirements. Typical assumptions about the alignment process are as follows:

1. The level-to-level error from the sources listed previously is the same for all lithography levels in a given process.

2. The errors measured from a statistically significant sample of level-to-level alignments are *normally distributed*; no favored direction of error is measured. A standard deviation of this distribution can be measured; the lack of a preferred error direction implies that $\sigma = \sigma_x = \sigma_y$. In a random-walk experiment, the most probable distance from the starting point after n steps is given by $R = \sqrt{n} * \sigma$.

Several example of process alignment sequences are shown in Figure 15.28.
    The assumptions that lead to the conclusion that σ is constant are only partially accurate; operator errors will not be constant, wafer warpage will vary during processing [depending on the nature of the film(s) on the wafer surface], and a number of different alignment tools will typically be used during fabrication. The ability to distinguish the features of the alignment marks (i.e., their edge profiles and contrast) will also vary among the different alignment target levels. In general, therefore, a unique σ value may be necessary to characterize a particular level-to-level alignment accurately in order to determine the optimum align-

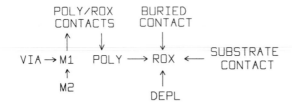

Figure 15.28. Fabrication level alignment sequences.

ment sequence. If this is indeed the case, a σ value for a particular sequence of alignments can be estimated from

$$\sigma_{a-d} = \sqrt{\sigma_{a-b}^2 + \sigma_{b-c}^2 + \sigma_{c-d}^2}$$

$$R_{a-d} = \sqrt{3} * \sigma_{a-d}$$

The goal of alignment strategy development is to analyze potential alignment sequences to determine the optimum. If more than one sequence is satisfactory, additional alignment targets and mask shapes could be provided. This would permit a backup strategy if the preferred target is unusable. A photoresist masking layer for an implant step will leave an alignment target structure intact; other lithography and patterning process steps will commonly add an image on top of the target, thereby prohibiting that target from being used again (if the level is used as a reference for more than one alignment).

It was mentioned previously that process design limits are established for artwork-to-wafer level image bias at each mask level; if the linewidth measurements from a sample of a particular process run are outside the process design limits, some action outside the normal process flow may be required. The same procedure applies for in-line measurements of level-to-level alignment. If an alignment measurement sample indicates that the alignment is sufficiently outside its process design limit, some rework may be required; this decision is made immediately after the photolithographic patterning, *prior* to any etching or deposition step. If the measured alignment between the target level feature and the photoresist pattern of the current level is outside the process design limit, the photoresist may be stripped immediately, recoated, and repatterned, thereby minimizing the amount of rework involved.

It is necessary to be able to accurately measure small displacements between process levels; in addition, the measurement must be made quickly, accurately, and without requiring costly or complex equipment. The overlay measurement is typically made using an optical microscope, viewing two overlapping halves of a *vernier scale*. An example of a vernier scale is given in Figure 15.29a. Each vernier half consists of an array of structures on either side of the reference axis. The difference in periodicity of the male and female arrays is 0.1 μm. The inside edges of the female array are stepped (tapered) in order to ease reading the vernier (Figure 15.29b).

Resolution and Step Size

Two important parameters in the development of the design rules for an integrated circuit process are the resolution and the step size:

Resolution: the minimum design linewidth of a process, that is, the minimum dimension necessary to be resolved photolithographically during fabrication (space or line) on a level.

Step size: the unit of measure used in the definition of design rules for widths, spaces, overlaps, and the like; the minimum length of any edge of

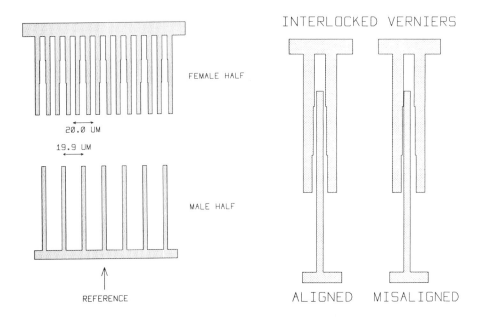

Figure 15.29.   Verniers used to facilitate optical measurement of level-to-level alignment. (a) Complete vernier (male and female) patterns. (b) Vernier detail.

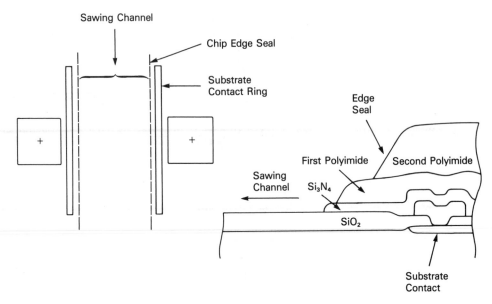

Figure 15.30. The dielectric overcoat must seal the internal chip structures; process design rules are provided to ensure a good seal and a sufficient dicing channel.

any shape permitted; the unit of measure used to distinguish the relative placement of two shapes on different masking levels.

A set of process design rules may be developed using a step size that is comparable in dimension to the resolution; for example, the step size could be equal to one-half of the resolution dimension. Such a design rule set is commonly denoted as lambda based; the intent of this approach is to keep the technology design rules and layout sufficiently general so as to continue to be applicable over an evolution of technologies, as more aggressive resolution dimensions are achievable.[2] By avoiding minute variations in process design rule features, the circuit layout becomes less technology dependent and more likely to remain suitable in future technologies with new step-size dimensional multipliers. The lambda-based design rules also tend to simplify the task of circuit layout, as the variations among individual design rules are reduced. However, the use of a lambda-based set of design rules does not utilize the alignment accuracy of current mask aligners, which are capable of better overlay than resolution by a factor of 4 to 8. Thus, the design rule step size can be tightened considerably with respect to the resolution, permitting enhanced circuit density.

[2] The use of the term lambda originates from the original theoretical work on MOS scaling, where a scaling factor λ was used to predict how a set of process parameters and design features could be made more aggressive.

Chip Pads and Edge Seal Design Rules

In addition to circuit layout design rules, a supplementary set of design rules is necessary to describe the chip pad-to-package interface. The chip pad configuration should be such that the routing of the package wires between a pad and its pin should be straightforward. This will impose constraints on the use of interior pads, multiple rows of perimeter pads, and pad size and spacing. Design rules regarding the size of the pad opening in the dielectric overcoat are also provided, as well as rules defining the extent of the overcoat seal outside the chip (Figure 15.30). Any probeable process monitor structures in the kerf must be sufficiently separated from this chip edge seal. The kerf process evaluation shapes and alignment targets are subsequently merged with the chip design data in preparation for mask manufacture.

Despite the flexibility that may be provided by the chip pad and kerf design rules, a standard set of chip "footprints" is commonly adopted. This permits the development of a standard set of package wiring designs, ultimately reducing the package engineering cost.

15.3 MATERIALS: DIELECTRICS, INTERCONNECTS, SUBSTRATES, PACKAGES

The materials used in integrated-circuit manufacture can be categorized into one of the following groups: dielectrics, interconnects, substrates, and packages. This brief section describes some of the prominent physical characteristics to consider when selecting a material for a particular application and developing the process steps for a technology incorporating that material.

Dielectrics

- Deposition rate and temperature: sensitivity of this rate to tolerances in ambient temperature and pressure, partial pressure of reactants

For thermal oxides, the *growth* rate is the parameter of interest; the sensitivity of this rate to its controlling parameters is likewise of extreme importance. For thermal oxides, the variations in film thickness over regions of different impurity type and concentration can be significant to the overall process development. For deposited films, some measure of planarity may be available; this may be achieved by the use of organic polymers (which are applied in viscous form to the wafer using a spin-coating technique) or by an inorganic film composition with subsequent high-temperature processing (e.g., chemically vapor deposited SiO_2 with a few mole percent phosphorus will *reflow* at elevated temperatures).

- Hydrophilic nature of the film surface

The absorption of water vapor by a dielectric film may result in poor photoresist adhesion in a subsequent photolithography step or ultimately to a metal interconnection corrosion failure.

  - Film etchants (wet): nature of acid solution (or base, in the case of organic polymers) used for etching

The use of liquid etchants for removing a dielectric film must include considerations of buffering and dilution for better etch rate control, extent of attack on other layers, introduction of contaminants, and especially the overall safety in use and disposal.

Interconnects

- Deposition rate and grain size control
- Adhesion to underlying dielectric films
- Temperature limitations in subsequent process steps after deposition of interconnect

Polysilicon (and refractory metal silicides) are used as a gate material and initial interconnection layer due in large part to the availability of subsequent high temperature (oxidation and diffusion) process steps; aluminum interconnects impose strict limitations on the temperatures of succeeding steps.

  - Ductile nature of interconnect (over steps in topography); susceptibility to microcracking

Over steep steps in the surface topography (e.g., into contact holes), the deposited metal may contain microcracks in thinned regions, potentially leading to an electromigration failure. Metal alloys may be selected to reduce this failure mechanism; increased difficulties may arise in the development of a suitable etching procedure, however.

  - Diffusivity of metal into silicon

The junction depth (and ohmic contact alloying step) must be selected so as to avoid the spiking of metal through the device node, resulting in an electrical short to the substrate.

- Growth of metal oxide surface layer

The presence of a metal oxide film after exposure to the atmosphere will require an oxide cleaning step to reduce the likelihood of metal-to-metal contact resistance problems.

- Susceptibility to corrosion and attack by water vapor

The chip-to-package attach metallurgy will potentially be exposed to high temperature and high humidity environments; the corrosion of this metallurgy will lead to early failure.

Substrates

The initial steps in the process description of Section 15.1 provide the wafer substrate preparation; these steps included oxidizing and removing organic residues, as well as introducing backside crystalline damage. In addition to these process steps, substrate characteristics of extreme importance are the following:

- Surface crystalline orientation

The surface crystalline orientation has a major impact on the electrical device characteristics (reflected in the carrier mobility and Q_{ox} terms) and also strongly influences the process development. The introduction of impurities into the crystal by ion implantation relies on the collisions with atoms at lattice sites to slow the impinging ion. If the crystalline orientation is such that the ion is incident on a low density of sites, an anomalously long stopping distance may result; the ion is effectively channeled deeper into the substrate. To reduce this effect, a screen oxide should be provided by the process sequence; this thin layer will tend to provide a more random distribution of incident angles between the path of the ion and the substrate surface. The surface orientation also strongly affects the thermal oxidation growth rate.

- Wafer diameter, wafer thickness, and surface flatness

The mechanical and chemical polishing of the active device surface after wafer sawing will enhance the wafer flatness; nevertheless, the tolerance on this specification will affect the ability to keep a 1:1 mask image in focus when exposing a photoresist-coated wafer. This problem is exacerbated by the use of larger-diameter wafers, which makes a tight flatness specification more difficult to achieve. The defocusing of the image of a large field at the wafer surface is also aggravated by the presence of a dielectric film holding the surface in tension or compression; as a result, the wafer will warp. (This bowing also can make it difficult to pull a vacuum on the wafer backside for holding the wafer onto the chuck of spin-coating and aligning equipment.) The wafer thickness must be increased at larger diameters to reduce this effect.

An additional substrate concern that strongly affects the chip yield is the density of surface crystalline defects; several demarcation techniques are available for accentuating the extent of the defect to obtain an areal count. This defect density is also of importance for epitaxially grown surface layers, as discussed in Section 15.14.

Packages

The two common chip packaging materials are ceramic and plastic; epoxies are used for chip attach to the package cavity prior to wire bonding and for some package lid sealing operations. Briefly, some of the packaging material concerns include the following:

Permeability to water vapor at high temperatures

Thermal conductivity

Outgassing in plastics at elevated temperatures (and the resulting impact on the water vapor permeability)

Thermal expansion coefficient (and mismatch between expansion coefficients of package and chip interconnection)

The materials selected for use in IC fabrication are evolving to meet the demands on process control, dimensional tolerances, and topography coverage. Current VLSI technologies are utilizing recent developments in materials science and chemistry to address these constraints: refractory metal silicides, organic polymer dielectrics, and thin-film metal alloys.

15.4 DIFFUSION

A key design consideration of any fabrication process is the impurity profiles of all junction nodes; the targeted junction depth affects the design rules of device length and node-to-node spacing, while the surface concentration must be sufficient to provide ohmic contacts to metal interconnect. The integral of the profile determines the electrical sheet resistivity; the shape of the profile in the vicinity of the junction depth determines the junction capacitance. The impurities introduced into the substrate will redistribute during the fabrication process by diffusing into regions of lower concentration. The modeling of the process should include calculations of the sequence of impurity redistribution steps to provide an expression for the resulting profile.

The kinetics of impurity diffusion in the crystal are described by Fick's first law:

$$\vec{J} = -D(\vec{\nabla}N)$$

where $\vec{J}$ is the particle current density (flux), D is the diffusion coefficient for the impurity in the substrate crystal, and N is the concentration of impurities. In general, the concentration is a function of all three spatial coordinates and time; $N = N(x, y, z, t)$. The diffusion coefficient is a strong function of temperature and impurity concentration [$D = D(T, N)$]. The diffusion coefficient will increase in regions of high impurity concentration due to straining of the elastic properties of the crystal lattice; this effect is particularly prevalent in compensated impurity regions. The diffusion coefficient of the compensated dopant is enhanced by the higher concentration of the compensating impurity.

Once the impurity enters the crystal, it will become ionized; the mobile carrier contributed by the ion will also diffuse into the substrate (due to its concentration gradient), but with a vastly enhanced diffusion coefficient. A built-in electric field results; this field will be oriented so as to *assist* the diffusion of the ion, according to the relation

$$\vec{J} = D\,\vec{\nabla}N + \mu_{\text{ion}} * \vec{\mathscr{E}} * N$$

The significance of the field-aided diffusion component in this expression is related to the relative magnitudes of N (the additional mobile carrier concentration) and n_i (the intrinsic, thermally generated hole–electron concentration). If $N \ll n_i$, the field-aided term is negligible; if $N \gg n_i$, the field-aided effect should be included. Recall that n_i is an exponentially increasing function of temperature: $n_i \approx 1 \times 10^{19}/\text{cm}^3$ at 1050°C.

The divergence of the impurity current density at a differential volume element is equal to the rate of loss of impurity at the point:

$$\vec{\nabla}\cdot\vec{J} = -\frac{\partial N}{\partial t}$$

$$\frac{\partial J_x}{\partial x} + \frac{\partial J_y}{\partial y} + \frac{\partial J_z}{\partial z} = -\frac{\partial N}{\partial t}$$

or, in one dimension,

$$\frac{\partial J_x}{\partial x} = -\frac{\partial N}{\partial t}$$

Assuming that the diffusion coefficient is independent of concentration (and the field-aided component is small), the general diffusion relationship is

$$\frac{\partial N}{\partial t} = -\vec{\nabla} \cdot \vec{J} = D(T) * \vec{\nabla} \cdot \vec{\nabla} N = D(T) * \nabla^2 N$$

where D is indicated as a function of temperature only.[3]

The diffusion problems to be discussed will assume uniformity over the junction area, reducing the problem to a one-dimensional differential equation describing the junction profile into the substrate. The presence of a field oxide layer around the exposed surface junction node area is assumed to inhibit the introduction of impurities into substrate regions underneath the oxide; the diffusion coefficient of impurities in oxide is typically much less than that in silicon (the process field oxide is sufficiently thick to serve as a masking layer). In general, however, an analysis of the extent of the lateral impurity outdiffusion is of extreme importance; in the presence of surface oxides, the spatial dependence of the diffusion coefficient must be described as a tensor. The boundary conditions used to solve for the impurity profile typically fall into two categories, corresponding to the process steps *predeposition* and *drive-in*.

The predeposition process step is one in which the product lot wafers are loaded vertically into a slotted quartz wafer carrier (called a *boat*) and inserted into an open-end high-temperature furnace tube. Although the wafer can be laid horizontally on the carrier or oriented such that the wafer is parallel to the direction of gas flow, the common orientation is perpendicular to the tube's axis. The dopant may be introduced from a solid, liquid, or gaseous source; examples are phosphine (PH_3, gas), phosphorus oxychloride ($POCl_3$, liquid), boron tribromide (BBr_3, liquid), phosphorus pentoxide (P_2O_5, solid), and boron trioxide (B_2O_3, solid). A gaseous source is introduced at the source end of the tube. A liquid source is held in a temperature-controlled quartz bubbler, over or through which a carrier gas is passed. A solid source powder is heated to vaporization in a section of the diffusion tube distant from the wafers. In the first two cases, the wafers are inserted into an inert ambient prior to impurity flow; a solid dopant source is loaded into the furnace tube prior to (or possibly in conjunction with) the wafers. Gaseous and liquid sources are by far the most common, due in large part to the ease and controllability of their introduction. The dopant compound is transported down the quartz furnace tube by a carrier gas, which is inlet to the source end of the tube. The inert carrier gas is commonly nitrogen and is often

---

[3] The diffusion equation is applicable in several areas of mathematical physics (e.g., the heat conduction in a homogeneous body). Several physical systems are described by differential equations utilizing the Laplacian operator (∇^2); the most familiar of these is the wave equation:

$$\nabla^2 \varphi = \frac{1}{c^2} * \frac{\partial^2 \varphi}{\partial t^2}$$

where c is the velocity of propagation. The diffusion equation differs from the wave equation by the order of the time derivative; the resulting solutions to the respective equations differ drastically.

mixed with oxygen. A quartz diffuser or baffle is commonly placed upstream in the furnace tube to provide some measure of turbulent gas flow down the tube.

The presence of oxygen in the tube enables the impurity to reach the wafer surface as an oxide compound, which has been found to provide the most reproducible introduction of impurities into the wafer; the source molecule is rapidly converted to the stable oxide upon heating in oxygen. Upon absorption of the oxide molecule, the resulting glassy surface layer serves as the impurity source for the silicon area beneath:

$$P_2O_5 + Si \leftrightarrow P + SiO_2, \qquad B_2O_3 + Si \leftrightarrow B + SiO_2$$

The surface concentration of the impurity at the silicon surface will be limited to the solid solubility of the impurity in the silicon substrate at the predeposition temperature; this time-invariant surface concentration is the boundary condition characteristic of the predeposition step:

$$N((x = 0), t) = N_{\text{solid solubility}} * u(t)$$

where $u(t)$ is the ideal time-step function at $t = 0$.

It is assumed that throughout the predeposition process step the partial pressure of the gaseous dopant oxide compound remains sufficiently high so as to ensure the saturation of the silicon surface concentration at the solid solubility limit; this will tend to eliminate variations in the amount of impurity introduced during predeposition due to small variations in the partial pressure. It is likewise assumed that the gas flow in the furnace tube is sufficiently turbulent so that the dopant molecules are transported between the wafer-to-wafer spacing and are absorbed uniformly over the wafer surface. To further reduce variations that may exist between wafers as a function of their front-to-back position in the boat, dummy wafers are typically added to the product lot to absorb any end effects.

While on the subject of gas flow distribution in a furnace tube, it is worthwhile to mention that the impurity introduction process is in actuality a composite of the individual processes of gas phase mass transport, chemical absorption and decomposition, and chemical reaction; this is true for all reactants and products present in the tube. (These same considerations apply to the oxidation and chemical vapor deposition processes, to be discussed shortly.) The attention required to give to the details of gas flow distribution is strongly dependent on the transport rate of gas molecules to the wafer surface as compared to the chemical reaction rate. If the chemical reaction rate is small, it becomes the controlling factor; such a process is denoted as *kinetically limited*. If the diffusivity of the reactants to the wafer surface from the bulk environment is small in comparison to the reaction rate, the process is *transport limited*; this latter condition requires detailed consideration of the gas flow distribution throughout the tube in order to guarantee wafer-to-wafer uniformity. (The same transport-limited situation may arise for the volatility of the products of a reaction and their diffusivity away from the wafer and toward the bulk.) For the diffusion processes discussed here, it is assumed that the gas phase mass transport is *not* a limiting factor and that the glassy surface layer is rich with the dopant atom; the silicon surface concentration remains therefore at the solid solubility limit during the predeposition step.

The solution to the (semi-infinite) one-dimensional diffusion equation[4]

$$\frac{\partial N}{\partial t} = D \frac{\partial^2 N}{\partial x^2}$$

with boundary conditions

$$N(x = 0, t) = N_{\text{solid solubility}} * (u(t)), \qquad N(x = \infty, t) = 0, 0 \le x < \infty$$

and initial condition

$$N(x, t = 0^-)^` = 0$$

is given by the integral expression

$$N = N_{\text{solid solubility}} \left\{ 1 - \frac{2}{\sqrt{\pi}} \int_0^{x/(2\sqrt{Dt})} e^{-\varphi^2} * d\varphi \right\}$$

The expression $(2/\sqrt{\pi}) \int_0^u e^{-v^2} dv$ has a special nomenclature; it is denoted as the error function, and its values are commonly available in tabulated form:

$$\text{erf}(u) = \frac{2}{\sqrt{\pi}} \int_0^u e^{-v^2} dv, \qquad 0 \le \text{erf}(u) < 1 \text{ for } 0 \le u < \infty$$

The term in the preceding impurity profile solution is denoted as the complementary error function:

$$\text{erfc}(u) = \{1 - \text{erf}(u)\} = \left\{ 1 - \frac{2}{\sqrt{\pi}} \int_0^u e^{-v^2} dv \right\}$$

The impurity distribution can therefore be written in shorthand notation as

$$N(x, t) = N_{\text{solid solubility}} * \text{erfc}\left(\frac{x}{2\sqrt{Dt}}\right)$$

The *total* number of impurities introduced into the wafer surface during the predeposition step (per unit area) is the integral of this profile with respect to x:

$$Q_{\text{total}} = \int_0^\infty N(x, t) \, dx = \frac{2}{\sqrt{\pi}} * N_{\text{solid solubility}} * \sqrt{Dt}$$

Assuming that the substrate wafer prior to predeposition contained a uniform background concentration of impurities, N_B, this uniform distribution would be unaffected by a high-temperature step; if the background dopant is of the opposite

---

[4] The assumption that the wafer is semi-infinite extending from the surface is very accurate considering the shallow nature of the predeposition step.

type as the diffusant, the *net* impurity concentration in the material is given by

$$N_{net} = N_{solid\ solubility} * \text{erfc}\ \{x/(2\sqrt{Dt})\} - N_B$$

The metallurgical *pn* junction depth after the predeposition step is the value of x where the net concentration is equal to zero:

$$x_{junction} = 2 * \sqrt{Dt_{predeposition}} * \text{erfc}^{-1} \left(\frac{N_B}{N_{solid\ solubility}} \right)$$

The goal of the predeposition step is to introduce a well-controlled quantity Q_{total} of dopant atoms locally into the exposed junction node areas not masked by a thick layer of thermal oxide. The junction depth, surface concentration, and profile slope can then be modified to their final process target values by one or more subsequent high-temperature drive-in diffusion steps, performed in an ambient such that the quantity of impurities in the wafer is unaltered. This implies that the surface dopant-rich oxide layer is to be etched off after the predeposition, prior to drive-in. Also, as will be discussed shortly, an oxide layer is commonly grown during the drive-in step to seal the exposed wafer surface from loss of impurities.

Predepositions have been replaced in current VLSI processes by ion implantation steps, which can likewise introduce a well-defined quantity of impurities per unit area into the substrate. Thus, an additional high-temperature step is avoided, as are the critical surface cleaning and oxide layer removal steps surrounding the predeposition. Additional flexibilities that are afforded by implanting the impurities (e.g., self-aligning of device nodes to the device channel region) are discussed in Section 15.5.

The drive-in diffusion of impurities in a subsequent high-temperature step redistributes the initial profile, pushing the junction deeper into the substrate. The profile after the predeposition (i.e., the complementary error function profile) would normally be used as the initial condition for the first subsequent drive-in step; however, deriving the resulting solution to the diffusion equation requires a numerical analysis. Rather, an adequate approximation to this initial condition is to assume that a surface impurity concentration per unit area of $Q_{total} * \delta(x)$ is present after predeposition; $\delta(x)$ represents the Dirac delta or unit impulse function and is used to indicate that the concentration is distributed over an infinitesimally small depth into the substrate. The boundary conditions for this case are

$$N(x = \infty, t) = 0 \quad \text{and} \quad \left. \frac{\partial N}{\partial x} \right|_{x=0,\ all\ t>0} = 0$$

This second condition corresponds to the assumption that impurities do not enter nor evaporate from the surface during the drive-in step. With these initial and boundary conditions, the solution to the diffusion equation is given by

$$N_{net} = \frac{Q_{total}}{\sqrt{\pi D_{drive-in} * t_{drive-in}}} * \exp\left(-\frac{x^2}{4D_{drive-in}t_{drive-in}}\right) - N_B$$

The shape of the resulting profile is Gaussian, with the term $2 * \sqrt{Dt}$ as a characteristic length; as the magnitude of this term increases, the extent of the profile into the substrate increases, while the concentration at the $x = 0$ surface decreases.

Subsequent high-temperature process steps (i.e., those where the diffusion coefficient is significant) will continue to redistribute this initial drive-in profile. Assuming the starting profile is Gaussian, as derived previously, the profile after other diffusion steps will remain Gaussian:

$$N_{net}(x, t)$$

$$= \frac{Q_{total}}{\sqrt{\pi(D_1t_1 + D_2t_2 + \cdots + D_nt_n)}} * \exp\left[\frac{-x^2}{4(D_1t_1 + D_2t_2 + \cdots + D_nt_n)}\right]$$

$$- N_B$$

The initial drive-in diffusion after predeposition (or implantation) is commonly performed in an oxidizing ambient; this oxide layer effectively seals the impurities into the substrate and isolates device nodes from sources of contamination. As will be discussed in Section 15.6, the growth of silicon dioxide moves the silicon–silicon dioxide interface into the substrate; the oxide grows from this moving interface as silicon is consumed. This moving interface tends to invalidate the surface boundary assumption used earlier:

$$\left.\frac{\partial N}{\partial x}\right|_{x=0,t} = 0$$

Impurities will be incorporated into the growing oxide and will diffuse away from the silicon–silicon dioxide interface. In equilibrium, the ratio of the interface impurity concentration in silicon to the interface concentration in silicon dioxide is a constant for a particular impurity, denoted as the *segregation coefficient, m*:

$$m = \left.\frac{N_{silicon}}{N_{oxide}}\right|_{Si-SiO_2 \text{ interface}}$$

This concentration ratio is maintained as the interface moves into the substrate during the oxidation. If $m < 1$, the oxide tends to accept the impurity into the layer, effectively leaching the silicon surface concentration; if $m > 1$, the oxide tends to reject the impurity, effectively snow-plowing the impurities at the moving interface, with a resultant increase in surface concentration (Figure 15.31). An

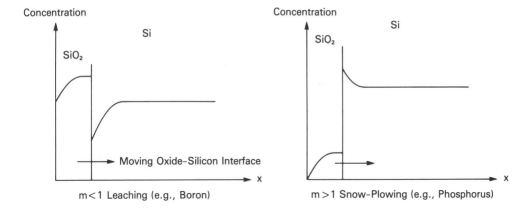

Note: A uniform background impurity concentration is illustrated.

Figure 15.31.   Segregation coefficient at the silicon–silicon dioxide interface, including the characteristics of leaching and snow-plowing.

analysis of the modified diffusion equations describing the silicon dioxide and silicon impurity profiles during oxidation can be found in reference 15.1; in addition to the interface segregation coefficient, the impurity profiles in the oxide and the substrate are functions of the relative rates of oxidation to impurity diffusion and the volumetric ratio of silicon dioxide to silicon (i.e., the ratio of silicon consumed during oxidation to the resulting oxide thickness: $t_{Si,consumed}/t_{SiO_2,grown} = 0.45$). If the oxidation rate is increased (given a fixed diffusion coefficient at the oxidation temperature), the oxide interface impurity concentration will be closer to that of the original impurity profile; as illustrated in Figure 15.32, this will increase the effects of snow-plowing or leaching on the surface impurity concentration.

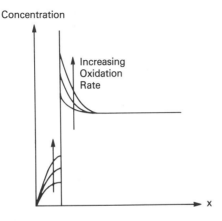

Figure 15.32.   Effects of oxide growth rates on the redistribution of impurities.

15.5 ION IMPLANTATION

Ion implantation is a technique for introducing a precise quantity of impurities into the substrate by accelerating a beam of ionized impurity atoms in vacuum to impinge on the wafer surface. The impurities may be added to the material in selective areas, where a surface masking material is absent. The impurity can be blocked locally from reaching the substrate by a photoresist coat or thermal oxide layer of sufficient thickness. The accelerated ions lose kinetic energy due to atomic collisions and due to coulombic interaction between the ion and the free electron gas in the substrate. Different materials will have varying stopping-power characteristics for these two energy-loss mechanisms.

Ion implantation has effectively replaced predeposition steps for the introduction of impurities into the substrate; in addition to the creation of device junction nodes, this procedure has facilitated the inclusion of fabrication process steps to achieve the following:

1. Increase the field oxide threshhold with a field implant.
2. Adjust the device threshold voltage with a shallow threshold tailoring implant.
3. Provide selective resistivity control for the polysilicon interconnection level.

All this is accomplished without the critical time/temperature requirements of the predeposition and drive-in.

A schematic diagram of an ion implanter is given in Figure 15.33. The ion source must be able to provide a relatively high current level of the desired ion species (in the $10\text{-}\mu A$ to 1-mA range) in order to reduce processing time; the number of different ion species emitted (in addition to the desired ion) and the variation in initial ion energy should both be kept to a minimum. The ions can be produced by any process that imparts sufficient energy to the atom, an energy greater than its first ionization potential energy. The most common means is to provide high-energy free electrons in confinement with a gas of the required atom; collisions between the electron and gas atom can impart sufficient energy to the atom to produce an ion. The *ionization coefficient* describes the number of ionizing collisions produced by an electron passing through a gas (per unit length and per unit pressure); this coefficient is maximized for electron energies in the range of three to five times the ionization energy. The configuration used to provide the necessary ionization is commonly an RF plasma maintained between a hot-filament cathode (source of electrons) and anode. A permanent magnetic field crossed with the anode-to-cathode electric field works to keep the thermionically emitted electrons confined in the plasma (Figure 15.34). The resulting ions that penetrate the screen opposite the cathode are accelerated by the electric field between the screen and extraction electrode. The beam column is evacuated to reduce ion–gas molecule collisions, which could result in unwanted deflection or charge neutralization of the ion.

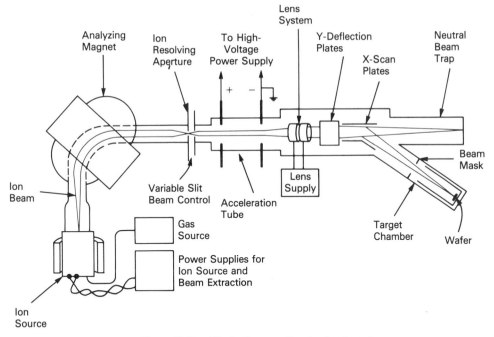

Figure 15.33. Block diagram of an ion implantation system.

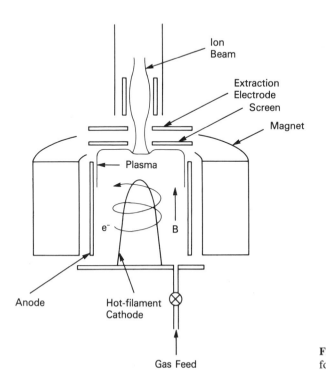

Figure 15.34. RF plasma configuration for the generation of ions.

A particular species of ion is selected by the magnetic mass analyzer (e.g., B^{10+} or B^{11+}). This selection process is based on the circular path of radius R traversed by a charged particle moving in a uniform magnetic field, as given by

$$R = \frac{M * v}{qB} = \frac{1}{B} \sqrt{\frac{2MnV_{\text{extraction}}}{q}}$$

where M is the ion mass, V is the initial accelerating extraction potential, B is the magnetic flux density, and n is the charge state of the ion. A doubly ionized atom can therefore be selected prior to entering the acceleration tube (e.g., P^{2+} accelerated by 200 kV yields an effective implant energy of 400 keV). The ion beam exits the mass analyzer through an aperture, which collects those ions that were deselected (i.e., whose path radius was greater or less than the adjusted target value). A variable slit located past the aperture can be used to adjust the beam intensity over several orders of magnitude.

The final kinetic energy of the ion beam is determined by the potential difference across the acceleration tube, which adds to the initial extraction energy. The beam is then focused and enters a set of xy scanning deflection plates. The beam is scanned in raster fashion to distribute the impinging ions uniformly over the wafer surface.[5] A constant dc voltage is added to the horizontal scan plates to provide a bend in the central axis of the beam; this change in axis is included as a means of separating the ion species from any *neutral* atoms. These neutral atoms may have resulted from a collision with a gas molecule after having exited the bending magnet. Neutral atoms do not undergo this final deflection and are collected outside the wafer scan area. The retrace portion of the beam scan is blocked by an aperture and does not reach the wafer target.

The beam current provides the number of ions per second incident upon the wafer. It is monitored during the implant by measuring the electric current out of the Faraday cage in contact with the wafer backside.[6] The total current received by the wafer is recorded by a charge integrator; when the desired areal dosage of impurities has been accumulated, the beam is diverted away from the wafer to begin the automated unload cycle. The implant dosage is a *surface density* process specification and is calculated from the total integrated beam current incident on the wafer. A major advantage of ion implantation is the accuracy to which this dosage can be controlled over a range of 10^{10} to 10^{17} ions/cm^2.

To enhance the throughput of the implantation of a product wafer lot, the

---

[5] Conversely, several ion implant systems mechanically scan the wafer across a stationary beam.

[6] A *Faraday cage* is a hollow metallic cup that is used to collect a beam of charged particles. Secondary electrons emitted from the wafer as a result of collisions with the ion must not escape in order to maintain an accurate measure of the incoming charged particle current. This will likely require including a biased plate (and/or local magnetic field) to reflect these electrons back to the Faraday cup (Figure 15.35).

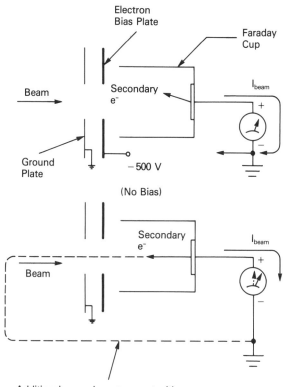

(No Bias)

Additional secondary e⁻ current with no
suppression bias increases current reading.
The wafer will be underdosed.

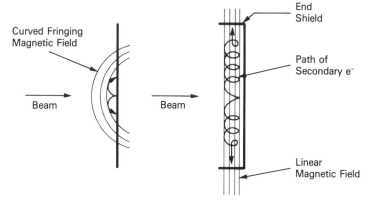

Magnetic Field Suppression of Secondary e⁻

Figure 15.35.   For an accurate measurement of beam current, it is necessary to confine and
retain secondary electrons emitted during implantation.

following equipment options can be contemplated:

    Higher beam currents
    Target fixtures that hold a number of wafers
    Reduced pumpdown time of the target vacuum chamber

Implanters may include a front-end loading station for the automated sequential handling of a product wafer lot; the vacuum pumpdown time for each wafer is managed through a series of interlocks, where the time to evacuate the interlock chambers is pipe-lined with the implant time. In addition to striving for increased throughput, current VLSI process development requires broader control over the location of the impurity profile peak and junction depth; note that the impurity concentration maximum need not be located at the wafer surface, as was the case for the predeposition plus (nonoxidizing) drive-in diffusion steps described in Section 15.4. This broader control over the implanted profile requires higher accelerating voltage equipment; whereas systems of 10 to 200 kV sufficed for LSI technologies, the optimum impurity profiles for the compensated substrate well regions in VLSI CMOS processes utilize accelerating voltages on the order of 0.5 to 2 MV.

    In an amorphous target material, the profile of the implanted impurity is described approximately by a Gaussian function

$$N(d) = N_{\text{peak}} \exp \left[-\frac{(d - R)^2}{2S^2} \right]$$

where d is the distance traveled into the substrate by the incident ion measured in the direction of incidence, R is the mean (or expected value) of the stopping distance, and S is the standard deviation of this profile. Specifically, R is denoted as the *range* of the implant, while S is called the *straggle*.[7] If the implant is performed at a slight (dechanneling) angle with respect to the surface normal, then the projection of the coordinate d onto the surface normal axis is given by $x = d * \cos \theta$. The projected range (R_p) is related to the implant's range by the factor $\cos \theta$. The projected straggle (S_p) is related to the incident and lateral (displacement perpendicular to the incident beam) straggle by the expression

$$S_p^2 = S^2 \cos^2 \theta + \tfrac{1}{2} * S_{\text{lateral}}^2 * \sin^2 \theta$$

The total implant dose is given by the integral of this Gaussian distribution:

$$\frac{\text{Total dose}}{\text{Unit area}} = \int_0^\infty N_{\text{peak}} * \exp \left[-\frac{(x - R_p)^2}{2S_p^2} \right] * dx \cong N_{\text{peak}} * \sqrt{2\pi} * S_p$$

---

[7] Note that the variable d is measured in the direction of incidence; the actual total path length traversed by the ion may be somewhat longer at lower energies due to increased nuclear collisions. A profile expression that incorporates some measure of skew in the distribution can be developed using multiple Gaussians of different standard deviations joined at the peak of the profile.

The range and straggle of the implant are functions of the incident ion kinetic energy, the atomic mass of the ion and target, and the atomic number of the ion and target. For a particular incident kinetic energy, the range of the impurity concentration in silicon decreases with increasing ion mass. A table of range and straggle values versus incident energy in silicon is given in Figure 15.36 for the common implanted dopants; note that these data were generated assuming a de-channeling amorphous silicon layer. The implant profile into a crystalline substrate is subject to channeling when the incident beam is in a crystallographic direction that is relatively open; the magnitude of the stopping distance can be anomalously large. The combination of dechanneled and channeled ions may result in an impurity profile with two peak values. The process sequence should ensure that surface oxide layers are present during implant steps and/or the incident angle of the beam is sufficient (for the particular crystalline surface orientation) to de-channel the ion path. In either case, implant damage extending into the substrate will result from the energy imparted to surface silicon atoms (and potentially oxygen atoms in a surface oxide layer).

The masking of the ion beam can be accomplished by a sufficiently thick surface layer of photoresist, which has been patterned by a previous photolithography step; alternatively, a thick layer of oxide can be patterned prior to implanting. The resist masking layer is advantageous in terms of its simplicity; it is well suited for low-to-medium energy, moderate-dose implant steps. Of concern with photoresist masking for high-energy and high-dosage implants is that the resist temperature may rise to the point where the patterned image is distorted by resist flow. In addition, the organic resist polymer may react so as to cross-link; this reduces the resist solubility and makes the subsequent resist removal after implant more difficult. A high-dosage implant may need to be executed in multiple, smaller-dosage steps.

In addition to channeling, another consequence of implanting into crystalline substrates is that a significant fraction of the impurities will not reside at lattice sites; those that are present substitutionally have displaced the substrate atom, while the remainder will be present interstitially. The interstitial impurities imply

Projected Range and Straggle for Various Ions in Silicon						
Energy (keV)		40	80	120	160	200
Boron	R (μm)	0.14	0.27	0.38	0.47	0.56
	S (μm)	0.04	0.07	0.08	0.09	0.10
Phosphorus	R (μm)	0.049	0.098	0.148	0.200	0.251
	S (μm)	0.016	0.029	0.040	0.051	0.060
Arsenic	R (μm)	0.026	0.047	0.068	0.089	0.110
	S (μm)	0.006	0.010	0.014	0.018	0.022

Figure 15.36.   Ion range and straggle values in Si versus incident energy. [From J. F. Gibbons, *Proc. IEEE*, 56:3 (March 1968), 305.]

that the local electron energy state density will be modified from the band structure described in Chapter 7; rather than adding a mobile carrier to the overall carrier density, a *trap* state may result instead. The extent to which the lattice atoms have been displaced in the implanted region determines the remaining crystalline nature; when the number of displaced atoms is sufficiently large, the material effectively becomes amorphous. The resulting lattice damage may be corrected by annealing the substrate in a subsequent high-temperature processing step; this process will result in an increased electrical activity of the impurity dosage and a reduced electron trap state density, both of which are desirable to achieve the optimum device behavior. This high-temperature annealing step will likewise result in the outdiffusion of the implanted impurity profile, which is undesirable in VLSI technologies, considering the goal of maintaining shallow junctions and a sufficient device punchthrough voltage at shorter channel lengths.

The necessary time and temperature of the annealing step may be determined experimentally by measuring the resulting change in electrical activity of the impurities added to the implanted region. Amorphous implanted layers undergo a recrystallization by a solid-phase epitaxy process, growing from the remaining crystalline interface and incorporating dopant atoms substitutionally. Point defects (i.e., lattice vacancies) migrate at moderately elevated temperatures toward interstitial dopant atoms, allowing them to precipitate to a substitutional lattice site. Extended range defects (i.e., dislocations) require a significantly higher anneal temperature to increase the mobile carrier lifetime and reduce the junction leakage (recombination) current.

Current VLSI process development research is concentrating on various techniques for *rapidly* annealing implanted regions, without introducing substantial modifications to the as-implanted profile. Exposure to high-intensity incoherent radiation (over the near-infrared to near-UV range) and scanned laser annealing are two techniques that have been used to obtain electrical activation without significant impurity redistribution. Another technique is to quickly bring the wafer into close proximity of a high-temperature graphite plate acting as a blackbody radiator.

Current VLSI processes also utilize ion implantation steps for reasons other than the introduction of dopant atoms into the substrate; the damage resulting from a backside argon implant may be used to serve as a gettering location for deep-donor level impurities and lattice defects. A front-side argon implant may be used to locally enhance the etch rate of the damaged surface layer in a subsequent wet chemical or dry plasma etch step, providing for a vertical sidewall topography of the etched opening.

15.6 OXIDATION

The growth of a thermal oxide from the silicon substrate is the fundamental process step of MOS device fabrication; a reproducible, low resulting defect density, contamination-free process technology for gate oxide growth is required. The thermal oxide will serve as the gate dielectric for the device and as the surface-to-interconnection dielectric layer. This section briefly describes a model for the kinetics of the oxidation process and some of the processing alternatives for oxide growth; recall from previous sections that the exposure of the wafer surface to an oxidizing ambient at an elevated temperature will result in the diffusion of existing impurity profiles and an incorporation of impurities into the oxide layer to some degree. The oxidation rate descriptions provided in this section are general in nature and do not include models to reflect dependencies on the specific characteristics of the substrate material, that is, the crystalline orientation and impurity type and concentration.

Oxidation Kinetics[8]

The chemical reactions associated with oxide growth involve either an oxygen (dry) or water vapor (wet) ambient:

$$Si + O_2 \rightarrow SiO_2, \qquad Si + 2H_2O \rightarrow SiO_2 + 2H_2$$

The evolution of hydrogen from the second reaction through the oxide layer out to the ambient will result in some reactivity with oxygen atoms and the formation of hydroxyl (OH) groups; the bridging nature of the oxygen atom is gone, resulting in a more porous film. The dry oxidation process uses high-purity oxygen (or an oxygen–nitrogen mixture); the wet oxidation alternative introduces water vapor into the furnace tube. The water vapor can be provided from the following:

1. A carrier gas (nitrogen or oxygen) forced through a heated water bath (called a *bubbler*) that becomes saturated with water vapor (the temperature of the bath controls the vapor pressure of water).

2. The combustion of a hydrogen and oxygen gas mixture at the furnace tube inlet to form water vapor.

3. A flash system, where water drips onto a heated surface, evaporates, and the resulting water vapor is introduced into the furnace by the flow of a carrier gas.

The ratio of silicon consumed from the substrate to the resulting oxide film thickness is 0.45.

The thermal oxidation of silicon results from the diffusion of the oxidizing species through the existing oxide layer to react at the substrate surface; in a steady-state condition, the following flux densities must be equal:

1. The transport of the reactant from the bulk gaseous ambient to the oxide surface.

---

[8] This section is based on the landmark paper of Deal and Grove, reference 15.2.

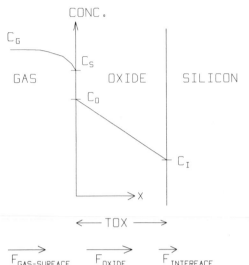

F_{\text{GAS-SURFACE}}$ F_{OXIDE} $F_{\text{INTERFACE}}$

Figure 15.37. Kinetics of the thermal oxidation of Si.

2. The diffusion of the species through the existing oxide layer.
3. The reaction rate at the oxide–silicon interface.

Figure 15.37 illustrates this process, with a concentration of the oxidant described by C_o in the solid at the wafer surface and C_i at the oxide–silicon interface. The three flux densities described are approximated by the following relationships:

$$\text{(a)} \quad F_{\text{gas–surface}} \propto (C_{\text{gas}} - C_S) = h * (C_{\text{gas}} - C_S)$$

where h is the gas-phase mass transfer coefficient, C_S is the oxidant concentration in the gas at the surface, and C_{gas} is the concentration of the oxidant in the ambient, proportional to the partial pressure of the oxidant ($p_G = C_{\text{gas}} * kT$).

$$\text{(b)} \quad F_{\text{oxide}} \propto \frac{dC}{dx} = D_{\text{oxide}} * \frac{C_o - C_i}{t_{\text{ox}}}$$

where D_{oxide} is the diffusion coefficient of the oxidant in silicon dioxide, and t_{ox} is the time-dependent oxide thickness.[9]

$$\text{(c)} \quad F_{\text{interface}} \propto C_i = k_s * C_i,$$

where k_s is the chemical surface reaction rate constant.

---

[9] VLSI process development efforts are continuing to refine the current understanding of the exact nature of the diffusing oxidant. Specifically, electric-field enhanced oxidant diffusion through the existing oxide layer indicates that a charged species (e.g., O^{2-}) may participate. There is also ongoing research investigating the nature of nonreacting O^{2-} moving into vacancy locations in the vicinity of the surface, with a resulting trap state density.

The final relationship required to solve for one of these flux terms relates C_S to C_o and is encompassed by *Henry's law*:

In equilibrium, the concentration of a slightly soluble gas in a solid at the surface is proportional to the partial pressure of that gas at the surface; that is, $C_o \propto p_s = C_S * kT$, or $C_o = H * C_S * kT$, where H is Henry's law constant for the species (and temperature) of the process step.

The solutions for C_o and C_i in terms of the partial pressure of the oxidant in the bulk gas can be found, and therefore the oxidant flux as well:

$$F = \frac{p_G * H * k_s}{1 + (k_s t/D) + (k_s HkT/h)}$$

The flux is the number of oxidant particles per unit area per unit time; the flux of oxidant is equal to the rate at which this reactant molecule is used up to form a new volume of oxide. The oxide thickness growth rate is therefore related to the flux of oxidant reactant by[10]

$$F = (\text{volume density})_{\text{SiO}_2} * \frac{\text{no. of reactant molecules}}{\text{SiO}_2 \text{ molecule}} * \frac{dt_{\text{ox}}}{dt}$$

or

$$\frac{dt_{\text{ox}}}{dt} = \left(\rho_{\text{ox}} * \frac{\text{no. of reactant molecules}}{\text{SiO}_2 \text{ molecule}}\right)^{-1} * \frac{p_G * H * k_s}{1 + (k_s HkT/h) + [k_s t_{\text{ox}}(t)/D]}$$

Separating the variables t_{ox} and t and integrating both sides of the result yields

$$t_{\text{ox}}^2 + A * t_{\text{ox}} = B(t + \tau)$$

where

$$A = 2D\left(\frac{1}{k_s} + \frac{HkT}{h}\right)$$

$$B = 2Dp_G H * \left(\rho_{\text{oxide}} * \frac{\text{no. of reactant molecules}}{\text{SiO}_2 \text{ molecule}}\right)^{-1}$$

$$\tau = \frac{1}{B}(t_{\text{ox}_{\text{init}}}^2 + A * t_{\text{ox}_{\text{init}}})$$

At time $t = 0$, it was assumed that an initial oxide layer of thickness $t_{\text{ox,init}}$ was present; the parameter τ can therefore be regarded as a shift in the oxidation time

---

[10] For a dry O_2 ambient, the transport of one reactant molecule through the oxide per SiO_2 molecule is required; for water vapor, this ratio is 2. The volume density is roughly 2.2×10^{22} SiO_2 molecules/cm^3.

reference due to the initial layer. The oxide thickness as a function of time is given by the solution to the preceding quadratic expression:

$$\frac{t_{\text{ox}}}{A/2} = \sqrt{1 + \frac{t + \tau}{A^2/4B}} - 1$$

For small oxidation times, this expression describes a linear growth rate: $t_{\text{ox}} = (B/A) * (t + \tau)$; for long oxidation times, the thickness is described by a square-root relation: $t_{\text{ox}} = \sqrt{B * t}$. Initially, the reaction is controlled by the oxide–silicon interface rate constant, k_s, and is associated with the bond-breaking energy between silicon atoms in the crystal (and is thus dependent on temperature and crystalline orientation); for extended times, the reaction rate decreases due to the increased time required for the reactant to diffuse to the interface through the growing layer. The gas-phase mass transfer coefficient between the ambient and the surface is typically very large and is therefore not a rate-limiting factor. The nonlimiting value of this coefficient facilitates the simultaneous oxidation of a large number of wafers (oriented perpendicularly to the furnace tube axis) with good wafer-to-wafer thickness uniformity.

Oxidations are often performed in an ambient containing chlorine, from the introduction of HCl. The presence of chlorine during oxidation has been used to reduce the concentration of *mobile, ionic* alkali contaminants (e.g., sodium) in the oxide, detrimental to MOS device characteristics and stability. The presence of chlorine enhances the diffusion coefficient of O_2 and H_2O through the oxide as well as the interface reaction rate constant, k_s; both result in increased oxide growth rates. An alternative procedure to the addition of chlorine to the oxidizing ambient is to utilize a double-walled furnace tube, with HCl flowing in the outer sheath only. In this manner, mobile alkali contaminants from the furnace liner are tied up prior to entering the oxidizing ambient.

A common processing technique for oxidation is to use a dry–wet–dry sequence, utilizing the interface qualities and density of dry oxides and the enhanced growth rates of a wet oxidation.

Oxide Evaluation

The resulting oxide thickness can be predicted from the model described previously, with the values of the rate constants determined by the process conditions; this calculated value can be verified by visual observation. When viewed perpendicularly in white light, the thermal oxide film will produce a characteristic color based on its thickness; the color sequence repeats itself with increasing thickness. The thickness can also be measured from parallel-plate capacitive measurements, the step height from the oxide layer down to a subsequently exposed portion of the wafer surface (determined by the use of a stylus), and by ellipsometry; the advantage of this last technique is the accuracy with which the thickness and dielectric constant can both be determined.

The dielectric strength of the oxide can be determined from the breakdown voltage using parallel-plate capacitance structures; the concentration of mobile ionic oxide contamination is typically found by applying a temperature and voltage stress to the parallel-plate structure and measuring the differences in the resulting C versus V (accumulation–depletion–inversion) curves.

Oxidation Alternatives

The topography step coverage problems of multiple interconnection technologies has been relieved to some extent in current VLSI fabrication processes by replacing planar device fabrication with one incorporating selective recessed field oxides. The growth of the field oxide can be inhibited locally by the presence of a patterned layer in which the diffusion coefficient of O_2 (or H_2O) is small, thereby preventing the reacting species from reaching the silicon surface. Where this inhibiting layer has been removed, the field oxide grows; the fraction of the total film thickness corresponding to the consumption of silicon is the extent to which the field oxide has been recessed from the surface. (The oxidant diffusion at the edges of the inhibiting layer introduces a peculiar variation in surface topography, denoted as the ''bird's beak.'') More recent techniques being evaluated for VLSI processes incorporate the preliminary etching of a nonplanar trench in the field oxide regions of the substrate; the depth of this region of silicon removal is selected so as to result in little or no change in wafer topography after field oxidation.

Other oxidation alternatives address the problems of impurity outdiffusion during a high-temperature oxidation step and the associated loss of circuit density; in particular, executing the oxidaton at elevated pressures (e.g., 25 atmospheres) permits the reduction of furnace temperature to achieve the equivalent oxide layer.

15.7 CHEMICAL VAPOR DEPOSITION

The technique of chemical vapor deposition (CVD) results in the deposition of a stable, solid compound onto the wafer by the thermal decomposition and reaction of gaseous compounds. A variety of CVD reactor designs are available, which provide a number of techniques for gas flow distribution over the wafer surface; in each of these systems, the reactant gas partial pressure, wafer temperature, and chamber wall temperature all strongly affect the resulting film deposition rate. In addition, the deposition may be performed under differing conditions of ambient chamber pressure and with differing means of providing the energy to dissociate the reactant gas molecules.

Chemical vapor deposition can be used to deposit polysilicon, silicon nitride, and silicon dioxide; the inclusion of a gas compound containing a dopant permits the in situ incorporation of that impurity in the deposition of a polysilicon or oxide layer. Chemical vapor deposition of dielectric films over metallization requires reduced temperature processing ($\leqq 450°C$) to prevent cracking of the interconnection at underlying steps in the surface topography.

The deposition rate in a CVD process is dependent on the gas-phase mass transport of the reactants to the wafer surface, the reaction rate at that surface, and the mass transport of the volatile reaction products back to the ambient. Unlike the oxidation and diffusion processes, the CVD process suffers in uniformity from any turbulent mixing of the reactant gases in the tube. At low inlet reactant gas flow rates and atmospheric ambient pressure, the flow patterns are subject to spiraling down the furnace tube axis, produced by the superposition of forced and buoyant gas velocities.[11] As the forced flow rates are increased or the ambient pressure is decreased, the turbulent mixing of the reactant gases is reduced. (These conditions are therefore the most attractive for CVD.) In addition, the deposition uniformity is degraded by the depletion of reactants in the direction from inlet to exhaust.

The model commonly used to develop CVD process characteristics (and CVD reactor design) is based on the high-velocity forced *laminar* flow of a viscous fluid past the deposition surface. A gas flowing in a laminar fashion past a surface forms a boundary layer adjacent to the surface, a layer of reduced gas velocity; the thickness of this boundary layer increases with the distance traveled along the deposition surface. The transport of a reactant from the region of laminar flow to the surface depends on the reactant's diffusivity and the thickness of the boundary layer adjacent to the surface. In addition to the distance along the furnace tube, the boundary layer thickness is a function of the laminar flow region gas velocity. To maintain uniformity within a single product wafer run, it is necessary to minimize the variation in reactant concentrations and boundary layer thickness from front to back of the tube. Atmospheric pressure CVD is performed with the wafer surface oriented parallel to the direction of gas flow; a common atmospheric CVD design utilizes a quartz furnace tube with the wafers situated on a horizontal susceptor. The thickness of the boundary layer may be kept more uniform from front to back if the susceptor is tilted into the laminar stream toward the back of the tube. The wafer-to-wafer deposition rate variation due to the depletion of gas reactants toward the exhaust end of the chamber can be addressed by providing a gradient in furnace temperature from inlet to exhaust or including multiple reactant gas inlets down the length of the tube.

Polysilicon deposition: atmospheric. The deposition of short-range order polycrystalline silicon is commonly achieved by the following reaction:

$$SiH_4(g) + heat \rightarrow Si(s) + 2H_2, \quad T = 600° \text{ to } 700°C, \quad N_2(\text{carrier gas})$$

---

[11] Buoyant forces are present on the gas fluid due to gas density variations associated with the heat-transfer process between furnace and gas ambient.

Silicon dioxide deposition: atmospheric. The deposition of silicon dioxide for dielectric and diffusion masking layers is required where thermal oxidation is not productive or feasible. A typical chemical reaction for the deposited oxide is

$$SiH_4 + 2O_2 \rightarrow SiO_2 + 2H_2O, \quad T = 400° \text{ to } 500°C, \quad N_2 \text{ (carrier gas)}$$

The deposited silicon dioxide layer may contain impurities if the chamber gas composition includes arsene (AsH_3), phosphene (PH_3), or diborane (B_2H_6); in this case, the oxides of the dopants are incorporated into the silicon dioxide film. Impurities are added to the process to serve as gettering sites for subsequent contamination sources, preventing the contaminant from moving through the layer. The concentration of impurity in the oxide layer is typically specified by the mole percentage of the dopant oxide (or by weight percentage of the impurity) in the silicon dioxide film. Phosphorus-doped (and boron + phosphorus-doped) oxides are used in VLSI processes for the ability to reflow the deposited oxide layer at elevated temperatures (e.g., 1100°C). The reflowed layer provides a surface topography with less severe steps over polysilicon interconnection lines. The mole percent of P_2O_5 must be tightly controlled; too low a concentration inhibits reflow, while too high a concentration results in corrosive etching of the subsequent aluminum metallization due to electrolysis between the aluminum and the phosphorus-rich oxide in the presence of atmospheric moisture.

The deposited SiO_2 layer may require a brief high-temperature densification process step to achieve the equivalent physical, electrical, and diffusivity characteristics of thermal oxides.

Silicon nitride deposition: atmospheric. Silicon nitride is used as a dielectric film (of higher dielectric permittivity than SiO_2 and lower diffusivity to contaminants) and as a barrier for the diffusion of oxidant species to locally inhibit (recessed) field oxide growth. Silicon nitride can be deposited using the reaction

$$3SiH_4 + 4NH_3 \rightarrow Si_3N_4 + 12H_2, \quad T = 800° \text{ to } 900°C, \quad N_2 \text{ (carrier gas)}$$

Alternative CVD Techniques

Chemical vapor deposition of films at *low pressure* (e.g., 300 mTorr) can be used to provide layers that are relatively free of contamination, are more uniform stoichiometrically, and provide better coverage over the sidewalls of steps in wafer topography. Low-pressure chemical vapor deposition (LPCVD) utilizes a furnace chamber with a mass flow-controlled gas inlet at one end and a gas exhaust/vacuum pump at the other. LPCVD can be used to deposit the films described earlier, as well as refractory metals (facilitating the use of silicides as gate and low sheet resistivity interconnects). Common LPCVD reactions include the following:

Polysilicon: $SiH_4 \rightarrow Si + 2H_2$,   600° to 700°C

SiO_2: $SiH_2Cl_2 + 2N_2O \rightarrow SiO_2 + 2N_2 + 2HCl$,
       850 to 900°C

$SiH_4 + O_2 \rightarrow SiO_2 + 2H_2$,   400° to 500°C,
       low temperature oxide (LTO)

$\underset{\text{tetraethyloxysilane, TEOS}}{Si(OC_2H_5)_4}$ (l) $\rightarrow SiO_2$ + organic and organosilicon

       products, 650° to 750°C

Si_3N_4: $3SiH_2Cl_2 + 4NH_3 \rightarrow Si_3N_4 + 6HCl + 6H_2$,
       700° to 800°C

Although the deposition rates of LPCVD reactions are lower than that for atmospheric pressure CVD, high effective throughput is maintained by a change in wafer carrier design, permitting a large batch to be processed simultaneously. In an LPCVD process, the wafers are oriented vertically; the high diffusivity (and long mean free path) of the reactant gases at reduced pressure enables a nonlaminar gas flow distribution to be used. The reactants may pass between the wafers and deposit a uniform surface layer provided the wafers are sufficiently spaced. Although the laminar gas flow and boundary layer characteristics of atmospheric CVD are not applicable to LPCVD, considerations of reactant depletion down the tube should be maintained.

In addition to the reflow characteristics of P_2O_5-rich deposited oxide films, another important aspect of CVD oxides for VLSI processes is the *sidewall coverage* of steps in topography. Of particular interest is the extent to which the coverage is conformal to the sidewall of a polysilicon line (Figure 15.38a). The LPCVD reactions of TEOS and dichlorosilane ($SiCl_2H_2$) provide a highly conformal oxide layer; if the deposition is followed by a suitable anisotropic etch process (high vertical-to-lateral etch rates), a sidewall spacer oxide will remain (Figure 15.38b). This sidewall spacer can then serve as an implant mask for the introduction of the source and drain junction node areas of a device; the implant of a *shallow* layer of impurities prior to the LPCVD step (conventionally, using the polysilicon gate as the implant mask) results in a two-step node impurity profile (Figure 15.38c). This structure is denoted as a *lightly doped drain* (or LDD) device and is necessary to maintain a high device punchthrough voltage at short channel lengths. (The drain-to-substrate reverse bias voltage at high VDS results in a channel depletion region that now lies to a larger extent in the lightly doped device node region; the punchthrough voltage, where the two device node-to-substrate depletion regions meet, is therefore increased. Techniques for anisotropic etching are discussed in Section 15.10.)

A major consideration with chemical vapor deposition is the possibility of deposits on the furnace walls flaking off and lodging on the wafer surface. In an

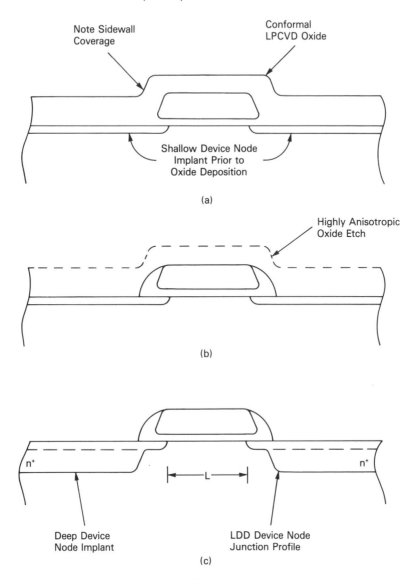

Figure 15.38.   (a) Conformal LPCVD oxide film. (b) Fabrication of gate sidewall spacers. (c) Final LDD device node profile.

atmospheric CVD process, where the wafers are laying horizontally, it is necessary to inhibit the deposition rate on the furnace walls so that flaking does not occur. This reduction can be achieved if the wall temperature is reduced; such a reactor design is denoted as a *cold-wall* system. The furnace walls are forced-air or water-cooled, while the wafer temperature is elevated by the RF inductive heating of a susceptor of suitable composition and coating. (Inductive heating refers to the joule heating resulting from currents induced in a high-resistivity material in the presence of a radio-frequency magnetic field.) For the LPCVD process, the orientation of the wafers normal to the gas flow direction implies that the surfaces will be subject to little resultant contamination from flaking of deposits off furnace walls; as a result, LPCVD depositions are commonly performed in hot-wall systems.

In addition to thermal energy providing the dissociation energy of the reactant compounds, a variety of other energy sources are available to initiate the LPCVD process. One possibility is to utilize a *photo*chemical reaction, with photon energy as the source; the more prevalent option is to perform the reaction in the presence of a plasma. Plasma-enhanced deposition of silicon nitride or silicon dioxide (PECVD) permits the wafer temperature to be reduced considerably (e.g., $T_{deposition} = 300°$ to $400°C$). This facilitates the deposition of dielectrics over metallization. Utilizing an argon plasma, the PECVD reaction for silicon nitride is

$$SiH_4 + xNH_3 \rightarrow SiN_xH_y + H_2$$

The stoichiometry of the deposited film is a strong function of the deposition parameters. Likewise, the PECVD of silicon dioxide using silane and nitrous oxide includes the hydrogen groups SiH, SiOH, and H_2O. The different film compositions result in variations in dielectric strength, hydrogen desorption in subsequent high-temperature steps, and, in particular, residual stress in the deposited film. High tensile stress in a deposited nitride film may result in cracking over steps in topography; as one of the uses of the PECVD film is as an interlevel dielectric between metallization layers, a composite dielectric may be required for enhanced reliability. (Such a composite interlevel dielectric was illustrated in Section 15.1.)

15.8 PHYSICAL DEPOSITION TECHNIQUES

There are three techniques for the deposition of metallization and dielectric films that complement CVD in current VLSI processes. This section briefly describes the uses of evaporation, sputtering, and spin-coating. Excellent reference texts are available to describe the first two of these techniques in more detail.[12]

Evaporation

Aluminum metallization on wafer substrates is commonly deposited by thermal evaporation of a solid source in a high-vacuum chamber with subsequent condensation on the substrate targets. A major consideration in the design of an evaporation system involves that of the wafer holder and the resulting angle of

[12] For example, L. Maissel and R. Glang, *Handbook of Thin Films*, New York: McGraw-Hill Book Co., 1970.

incidence between the evaporant and the wafer substrates. The actual substrate holder configuration employed depends on the metallization coverage characteristics required of the fabrication process; in particular, the two process alternatives to consider are blanket evaporation (with substractive etching) and evaporation onto a patterned photoresist lift-off layer. (The lift-off technique was illustrated in the process description of Section 15.1.) In the case of blanket evaporation, a goal of the evaporation process is to provide good coverage of sidewall steps in the wafer topography to avoid microcracks in a subsequently patterned interconnection wire. To achieve adequate coverage on steep sidewalls, it is necessary to impart incident evaporant atoms over a wide range of angles around the surface normal; blanket evaporation systems place the wafer holder(s) into revolution during evaporation about an axis that is at a considerable angle to the vertical. This planetary orbit system varies the wafer surface angle relative to a point source crucible of evaporant in order to better cover the surface topography. Furthermore, the wafer substrates may be heated during evaporation (to ~300°C) to enhance the surface mobility of the deposited material for less thinning of the film over steps in topography.

For the lift-off process, it is necessary to *avoid* sidewall coverage of the photoresist layer that has been previously coated, patterned, and profiled. Sidewall coverage of the photoresist profile impedes the subsequent dissolution and separation of the photoresist and metal overcoat from the wafer surface. For a lift-off process, normal incidence of the evaporant is required. The substrates are therefore loaded onto a holder of spherical curvature with radius equal to the distance between the holder and the melt. At the base (initial) pressure at which the evaporation is performed (10^{-6} to 10^{-7} torr), the mean free path of the vapor atom is sufficiently large so as to reduce the likelihood of scattering collisions. The lift-off process does *not* permit the wafers to be heated during evaporation as the patterned photoresist image would then reflow and distort; in this case, the evaporated metallization layer is more prone to cracking.

The techniques for vaporizing the source include resistance heating, electron-beam heating, and inductive heating. A refractory metal filament or boat geometry is resistance heated by the passage of dc current through it; the heat transfer to the evaporant source results in its vaporization. The source may be draped over the filament or deposited in the boat carrier. This procedure is inexpensive and requires no radiation sources; its disadvantages are the possible sources of contamination from the heating element, the relatively limited quantity of source material that can be efficiently vaporized, and the inability to deposit a metal alloy (the materials being of different melting temperatures and vapor pressures). This process does broaden the evaporant source area over a point source for better blanket evaporation metallization coverage.

Electron-beam evaporation provides high deposition rates, low contamination, continuous feed of source material, and the possibility of depositing a metal alloy (e.g., Al/Cu). Electrons are thermionically emitted from a filamentary source, shaped into a beam, accelerated by a large potential difference (e.g., 2 A at 5 kV = 10 kW), and focused by a magnetic field onto a crucible containing the evaporant. The bending magnetic field is used to direct the electron beam and screen out massive thermionic impurities emitted by the filament from reaching the crucible. The kinetic energy of the electrons is converted to thermal energy at the melt; the electron beam is scanned continuously across the surface of the melt to maintain a uniform deposition rate. The crucible (or hearth) containing the evaporant is water cooled. The controlled melt volume and source feed provides for the evaporation of a metal alloy, maintaining the stoichiometry of the evaporant during deposition. In the evaporation of an alloy, there will be some dissociation that can be adjusted by varying the composition of the melt.

Electron-beam evaporation results in the emission of a spectrum of radiation in the x-ray wavelengths; this radiation includes both the photon energies characteristic of an electron decay transition from outer to inner shell of the evaporant atom and the continuous Bremsstrahlung spectrum.[13] As an x-ray traverses through the MOS device gate oxide, hole–electron pairs are generated. The relatively large mobility of electrons in the oxide relative to holes produces a net electron transport in the presence of an oxide electric field; the electrons leaving the oxide result in a net positive oxide charge. The effect of this incident radiation is a decrease in $V_{T,n}$ and an increase in $|V_{T,p}|$ ($V_{T,p} < 0$). The MOS device C–V curves are likewise affected. A subsequent metallization anneal process step enhances the ability of the Si to supply electrons to the SiO_2 over the potential barrier at the Si–SiO_2 interface and reduces the magnitude of the device V_T shifts.

Heating of the evaporant can also be produced by the currents induced from a radio-frequency signal applied to a coil surrounding the source crucible (Figure 15.39). Like electron beam evaporation, high deposition rates are available; this technique also has the distinct advantage that no ionizing radiation is produced. As a result, a high-temperature anneal process step is not required. Since the magnetic field energy is coupled directly into the evaporant, it is not necessary for the supporting crucible (typically, boron nitride or ceramic) to be in excess of the vaporization temperatures; this reduces the likelihood of contamination between crucible and source. The extent in the metal to which the energy is coupled is a function of the material skin depth, which is inversely related to the frequency of the RF source. Thus, the effective volume of the evaporant charge that is heated can be varied by varying the frequency; the efficiency of the coupling requires experimentation with the coil diameter, coil-to-crucible spacing, crucible wall thickness, and tuning of the RF supply. The crucible design of Figure 15.39 is thinner at the top in order to thoroughly vaporize any molten metal that has migrated upward before it spills over the brim; the remaining wall thickness is selected to minimize turbulence in the melt, which leads to the undesirable ejection of metal droplets. (The ejection of molten droplets is also a problem with electron

---

[13] Bremsstrahlung radiation is produced by the deceleration of a high-energy electron subject to the electric field of the nucleus of the evaporant atom.

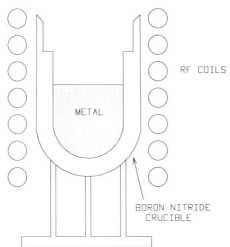

Figure 15.39. Crucible used for RF inductive heating of the evaporant.

beam heating.) As with electron beam heating, dilute metal alloys can be evaporated.

Sputtering

The sputter deposition of a film onto wafer substrates is produced by the impinging of accelerated ions onto a target of the desired film material; a low-energy (e.g., 500 to 1000 eV) ion beam in vacuum or positive ions present in a gas discharge can be used to dislodge the target atom or molecules to deposit onto the substrates. Both metal and dielectric films may be sputtered (although the sputtering of dielectrics utilizing a gas discharge requires a unique means of excitation). The material sputtered from metals is in the form of neutral atoms. Although the physical sputtering mechanisms are extremely complex and not well understood, a model can be developed in which the incident ion effectively imparts sufficient momentum to a surface atom or molecule of the target to result in its ejection from the solid. The momentum of the incident ion is directed inside the target; the elastic constants of the material determine the extent to which the energy imparted to a subsurface target atom upon collision arrives at the surface with momentum reversed in direction. The resulting elastic energy must be greater than or equal to the binding energy of a surface atom for sputtering to occur. Thus, a sputtering threshold of incident ion energies can be defined; Figure 15.40 lists these values for several target metals for a particular experimental evaluation. This threshold energy is dependent on the ratio of the mass of the incident ion to that of the target metal atom, the heat of sublimation of the metal, and the bulk velocity of sound in the metal (which describes the elastic coupling between atoms in the solid).

THRESHOLD ENERGIES FOR SPUTTERING WITH HG
IMPINGING AT NORMAL INCIDENCE ON
DIFFERENT METALS/SEMICONDUCTORS

METAL	SPUTTERING THRESHOLD (EV)
AL	120-140
SI	60-70
CR	60-80
NI	70-90
CU	50-70
MO	80-100
AG	40-50
W	80-100
GE	40-50

Figure 15.40.   Sputtering thresholds for sample target materials. These experimental results support the model that

$$2 * p_{\text{incident ion}} \mid_{\text{threshold}} * \frac{M_{\text{target}}}{M_{\text{target}} + M_{\text{ion}}} = K * \frac{W_H}{v_s}$$

where W_H is the heat of sublimation of the target (kcal/mole), v_s is the bulk sound velocity in the target (cm/sec), and K is a proportionality constant. The term on the left side of the equation describes the momentum of the target atom after an elastic collision with the ion at normal incidence. (For more detail on the model, see G. K. Wehner, *Physical Review*, 93 (1954), 633.)

The sputtering process is characterized by the yield of target atoms released for each incident ion; the deposition rate onto the wafer substrates is proportional to the incident ion density and the sputtering yield. For moderate energies above the threshold, the yield continues to increase with increasing ion energy. Likewise, for a particular accelerating ion potential difference, the deposition rate increases with increasing ionic mass. The sputtering rate also increases with increasing target temperature. The sputtering yield of the target material using ions generated in a gas discharge is reduced due to the increased gas-ejected target atom collisions at the operating gas pressure (e.g., 5 to 40 mTorr) with subsequent redeposition onto the target.

Sputtering is most often performed using the ions produced in a gas discharge present between the target holder and the wafer substrate holder. The characteristics of the discharge (and the production of ions to induce sputtering) are a function of the gas pressure, the discharge path length, the geometry of the holder electrodes, and the applied electrode voltage. The simplest gas discharge configuration for sputtering metals is the diode sputtering system, with plane-parallel target and substrate holder, denoted as the cathode and anode of the diode, respectively (Figure 15.41). A potential difference of sufficient magnitude is initially applied between these plates to cause the gas to break down, resulting in an

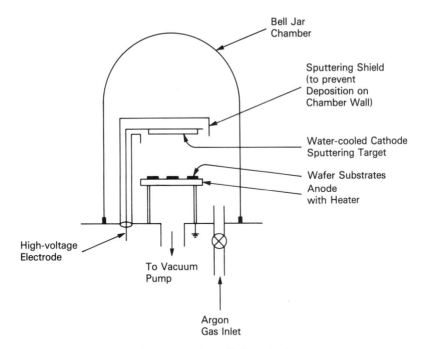

Figure 15.41. A diode sputtering system.

ionization current between anode and cathode. To maintain the gas discharge (at a reduced potential difference), a sufficient number of electrons must be emitted from the target upon impact of an ionized gas atom; these secondary electrons must then produce a suitable concentration of gas ions upon collision to sustain the process. The glow discharge between the plates can be divided into distinct regions (Figure 15.42). This figure denotes the various structures of the gas discharge, and a plot of the potential and light intensity versus position between anode and cathode. The voltage drop across the cathode dark space must provide sufficient electron acceleration to produce positive ions upon collisions with gas atoms. The ions are likewise accelerated by this voltage difference toward the cathode from the edge of the negative glow region. The negative glow region contains a large number of ion–electron pairs and (as evidenced by the plot of the potential) is approximately charge neutral; such a charge neutral region of ionization and excitation is called a plasma.

The v–i characteristic curve of a glow discharge is illustrated in Figure 15.43; to best conceptualize the features of this curve, consider the applied voltage to the diode discharge system to be in series with a ballast resistor. As the supply voltage increases to a critical value, the discharge strikes; the applied voltage necessary to maintain the discharge is less than the striking voltage, as illustrated by the negative differential resistance of the transition region. At relatively low voltages, in the normal glow discharge region, the surface area of the cathode

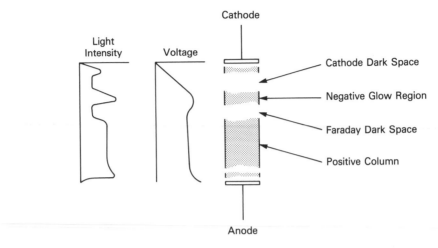

Figure 15.42. Characteristic regions of the glow discharge in a diode sputtering system.

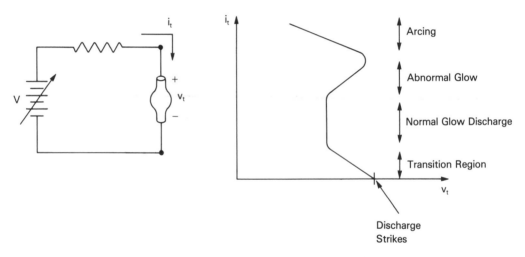

Figure 15.43. $V–I$ characteristic curve of a glow discharge system.

participating in the glow discharge is less than the available target area; a minimum current density is maintained, sufficient to produce enough secondary electrons to sustain the glow discharge. An increase in supply voltage increases the cathode glow area, tending to keep the current density and diode voltage constant. Thus, the total current i_t increases while v_t remains fixed. This current density and voltage difference is maintained with increases in applied power until the abnormal glow region is reached; in this range, the entire cathode surface participates in sputtering and the ion current can be relatively well controlled. Continuing in the abnormal glow region, to achieve the necessary increase in current density from the entire cathode area with increases in applied power, the number of emitted secondary electrons must increase, which requires a greater ion accelerating voltage drop across the cathode dark space. The abnormal glow region is the mode of interest for the deposition of the target material by sputtering, as the ion density and the sputtering yield is large. Further power dissipation increases will raise the target temperature, promoting the thermionic emission of electrons and eventually arcing between electrodes.

The acceleration energy is given up by the secondary electrons in the negative glow region to collisions with gas atoms (both ionization and excitation collisions). Electrons with insufficient remaining ionization or excitation energy are localized at the negative glow–Faraday dark space edge. No excitations or ionizations are occurring as these electrons diffuse toward the anode. In the positive column, a small electric field accelerates the free electrons to the anode; the location of the anode (i.e., the length of the positive column) is not critical to the sputtering process. Indeed, the anode may be sufficiently close to extinguish the positive column and the Faraday dark space without major consequences to the electrical characteristics of the discharge.

An alternative sputtering system for metals utilizes a triode configuration (Figure 15.44). This configuration permits sputtering at lower pressures (e.g., 1 mTorr), which reduces the incorporation of argon in the deposited film and provides less scattering of sputtered atoms (which will therefore impinge on the wafer substrates with higher energy resulting in better adhesion). A hot filament serves as a thermionic source of electrons that are accelerated toward the anode of this configuration. These electrons ionize the gas between the target and substrate; the resulting positive ions are attracted to the negative potential applied to the target. The impedance of the gas discharge, as seen by the target supply, can be independently controlled by varying the thermionic filament current and accelerating anode voltage; high ion currents (and sputtering yields) can be achieved

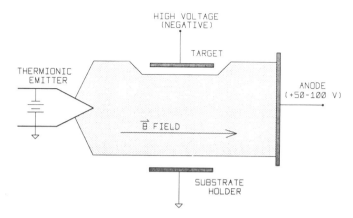

Figure 15.44.   A triode sputtering system.

at lower voltages. For sputtering onto large surfaces, this technique introduces the problem of producing large, uniform thermionic emission and gas ionization.

Sputtering of either metal *or dielectric* films can be produced from radio-frequency excitation of the gas discharge; a RF diode sputtering configuration is illustrated in Figure 15.45. An impedance-matching network is tuned to maximize the power transfer to the diode-sputtering load at the RF supply frequency, essentially matching the 50-Ω output of the generator to the impedance of the gas

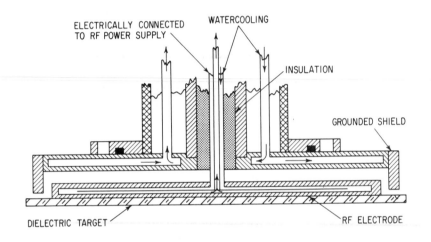

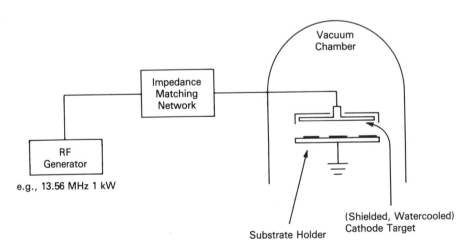

Figure 15.45. RF diode sputtering system. (From F. Maissel, Ed., *Handbook of Thin-Film Technology*. New York: McGraw-Hill, 1979, p. 4–36. Copyright © 1979 by McGraw-Hill Book Company. Reprinted with permission)

discharge (Figure 15.46). In the absence of collisions, a free electron in vacuum subject to the RF field oscillates in simple harmonic motion and absorbs no power from the source. Frequent collisions with gas atoms impart randomness to the electron's path, allowing the electron to absorb electric field energy between collisions. The free electron can build up sufficient energy to make an ionizing collision with a gas atom, liberating additional electrons; the discharge can thus be sustained by the RF source alone, without the need for the negative cathode bias and secondary electrons used in the dc diode system.

For sputtering of the target to begin, the positive ions must impinge on the target with sufficient energy. The positive ions in the discharge are too massive to be effectively accelerated toward the target by the RF field alone; the ions are accelerated to induce target sputtering by the establishment of a dc self-bias at the cathode with respect to ground. The highly mobile free electrons in the gas

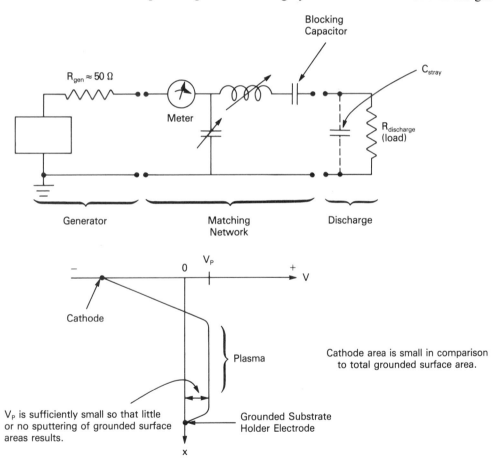

Figure 15.46.   An impedance matching network; voltage distribution in an RF sputtering system.

discharge are readily accelerated by the RF field, as opposed to the more massive positive ions; the external circuit current consists primarily of electron collection current, those electrons that were sufficiently close to the attracting electrode during each half-cycle. The presence of a dc blocking capacitor in the impedance-matching network between RF supply and target electrode implies that the cathode target develops a *negative* pulsating dc potential with respect to the grounded substrate holder anode. The central glow region develops a positive potential with respect to both electrodes. The voltage distribution in the RF glow discharge system is illustrated in Figure 15.46; the equipotential plasma region is positive with respect to both electrodes. The substrate holder and system walls are grounded to provide a single electrode of large area; this will reduce the potential difference between the plasma and the wafer substrate holder, thus minimizing the ion bombardment of this surface. (This assumes that the substrate and deposited material are both conducting; for an insulating substrate, the surface acquires a negative potential just as for the case of an insulating cathode. Grounding the substrate holder plate with insulating substrates produces a locally nonuniform potential distribution and thus nonuniform deposition. In this case, the holder should be allowed to reach a negative floating potential.)

The cathode target self-bias varies with applied RF voltage and gas pressure; the acceleration of positive ions toward the target induces sputtering if the self-bias is of sufficient magnitude. Dielectric materials may also be sputtered, the dielectric target bonded to a conducting back plate; in this case, the dielectric material serves as its own blocking capacitor. The dc voltage on the insulator or capacitor repels free electrons in the glow region, reducing the RF collection current at the cathode to only the most energetic electrons. As no net dc current can flow in the external circuit in steady state through the insulator, the self-bias voltage developed is of a magnitude such that the electron collection and positive ion sputtering currents are equal. The monitoring of the deposition rate using the dc discharge was provided by measuring the cathode current; in an RF system, the control of the deposition rate is commonly provided by monitoring the RF power delivered to the load.

If RF energy is applied to the wafer substrate electrode, the resulting ion bombardment produces *sputter etching*. This technique can be used to remove thin surface oxide layers (e.g., on exposed metal lines at via locations, prior to deposition of the subsequent metal layer). The use of a reactive gas-inert gas mixture (e.g., Ar + O_2) can be used to enhance surface cleaning. A potential problem with sputter etching of substrates is the back-scattering of emitted material; when the substrate surface is a combination of materials (and/or the substrate holder backing plate surface differs from the substrate surface), significant back-scattering will contaminate the surface and locally alter the sputter etching yield. The amount of backscattered material increases with sputtering gas pressure.

As described previously, a floating substrate holder acquires a negative potential, which subjects the substrates to some degree of ion bombardment. A negative supply bias can be applied to the substrate holder to provide some control of the magnitude of substrate resputtering (a dc supply for conducting substrates and deposited film or an RF supply for insulating substrates or films). If the substrate surface is purposefully given a negative potential with respect to the glow discharge (as opposed to reaching a floating potential), the process is denoted as *bias sputtering*. Bias sputtering can produce a high-quality film, relatively free of contaminants. Assuming the surface bond between the deposited material and an incident contaminant is sufficiently weak, the enhanced bias sputtering yield of the contaminant (as opposed to the deposit) keeps the deposited film relatively clean. The small-angle resputtering of the substrate surface material during this process presents possible problems and process advantages. The resputtering of material onto the sidewalls of steps in surface topography complicates the surface cleaning procedure; however, this same bias sputtering characteristic is very attractive in attempting to achieve a measure of surface planarization over topography steps in the deposition of a dielectric film. For example, planar sputtered quartz may be deposited as an interlevel dielectric between first- and second-level metal interconnections using bias sputtering techniques.

A magnetron sputtering system incorporates an applied magnetic field (crossed with the cathode-directed electric field) to confine the free electrons that ionize the gas; the discharge can thus be confined to the face of the target. One geometry commonly used for the magnetic field (and thus the confined gas discharge and target sputtering area) is the planar (or rectangular) magnetron; this configuration is sketched in Figure 15.47. As the deposition area of the target is

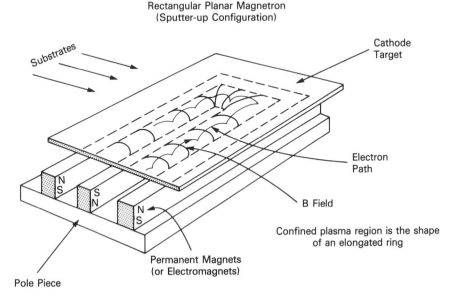

Figure 15.47. Planar magnetron sputtering system.

limited, suitable motion of the substrates past the target is required to achieve film uniformity and step coverage. (For the sputtering of metal films, the initial nucleation of the deposition determines the grain growth characteristics; some variation is introduced due to the different arrival energies and angles of the initial atoms.) Magnetron sputtering can utilize either dc or RF supplies, the latter again required for the deposition of dielectrics. Figure 15.48 illustrates secondary electron capture shields included between target and substrates outside the target magnetron sputtering area. As mentioned earlier, the bombardment of the target with positive ions results in the emission of secondary electrons from the material. The collection of these electrons by the wafer substrates (common in the non-magnetron sputtering configurations) results in wafer heating and possible radiation damage. The secondary electrons emitted in the magnetron-sputtered target area are restrained by the magnetic field (and consequently enhance the ionization in the gas discharge); the water-cooled shields outside the magnetron area reduce the secondary electron substrate bombardment.

Presputtering of targets is commonly performed to clean the target surface prior to film deposition and bring the target temperature up to operating conditions. Surface oxides of metals are therefore removed, and the rise in temperature assists in outgassing the vacuum system. A shutter is placed in front of the substrates. The presputtering operating conditions should be identical to those to be encountered in the subsequent sputtering step.

Sputtering results in a deposited film of the same chemical composition as that of the target material. A multicomponent target may lose the higher sputtering yield constituent first, but a surface layer rich in the other component(s) soon

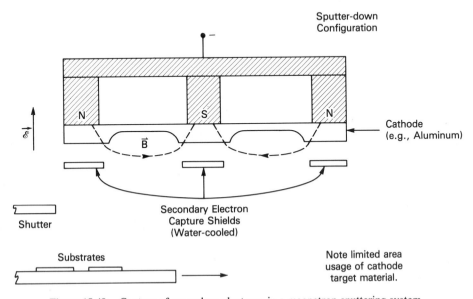

Figure 15.48. Capture of secondary electrons in a magnetron sputtering system.

forms so as to maintain the composition. Alloys can therefore be sputtered, if precautions regarding cathode target temperature and any target surface oxides are observed.

Spin Coating of Organic Polymers

The last physical deposition process to be discussed is the straightforward spin-coating of a viscous polymer resin (polyimide) to be used as an interlevel dielectric. This process uses the conventional dispense/ramp/spin coating equipment used for photoresist application. After coating, the resin is cured and solvents are released; the resulting film thickness is commonly 0.7 to 0.8 of its original value.[14] The polyimide coating provides some degree of planarization over steps in surface topography; multiple coatings (with partial curing between coats) provide enhanced planarization.

15.9 WET ETCHING AND CLEANING SYSTEMS

The patterning of interconnection or dielectric layers using wet chemical etchants has rapidly been replaced in VLSI fabrication processes with dry etch techniques (to be discussed in Section 15.10). The major difficulties with wet etching fabrication steps include the following:

1. The isotropic (i.e., lateral) etch rate under a protective photoresist layer results in severe undercutting of the film, with considerable etch bias between photoresist and final image.

2. The etch rate is typically a strong function of bath temperature, bath volume, and etchant concentration (as reflected by the number of previous wafer lots that have used the same solution).

3. The etch characteristics are also dependent on the degree of agitation of the bath, as etch products (especially hydrogen gas bubbles) must be circulated away from the exposed areas to enable fresh reactants to reach the surface.

4. The etch rate of a film is a function of impurity type and concentration; for example, the etch rate of phosphorus-rich SiO_2 is considerably larger than that of SiO_2 in the same solution.

5. Usually, no in-line etch end-point detection technique is available; most wet etches utilize a predetermined etch rate and estimate of the film thickness to calculate a minimum etch time. (Some degree of overetch is commonly added to accommodate for any variability in film thickness or etch rate across the wafer or from wafer-to-wafer in the lot.)

---

[14] The etching of vias in this dielectric film can be performed either before or after curing, the former using a wet etchant and the latter a dry etch process step. The shrinkage of the film during curing will significantly alter the via if the patterning is performed prior to curing.

6. Any contamination of the etchant bath rapidly spreads to subsequent lots utilizing the same solution. (Special MOS-grade inorganic solutions must be purchased, which specify a low sodium concentration in their assay.)

7. It may be difficult to achieve reproducible wetting of the exposed surface by the etchant; often, a surfactant is added to reduce the surface tension of the solution.

The main advantage of wet etching systems is that the batch immersion process is capable of handling a large number of wafers simultaneously, often utilizing some degree of equipment automation for the immersion and rinse steps. Also, a very high etch selectivity between the exposed film and underlying layers can usually be achieved (i.e., the etch rate of underlying layers is commonly negligible). Therefore, wet etch steps have been retained in VLSI processes for surface cleaning and those steps that require high selectivity (and can tolerate some percentage of undercutting).

A growing concern with the continued utilization of acidic and volatile etchant solutions is one of safety—the safety of the lab technician exposed to etch fumes as well as the safety of chemical etch disposal.

Wet etch solutions are common (indeed, the only practical solution) for periodically cleaning laboratory quartzware. In addition, prior to entering any furnace tube, an aggressive wafer cleaning process (such as listed in Figure 15.49) should be strictly enforced to remove surface contaminants.

Figure 15.50 lists sample solutions used for removing various films. These reactive solutions provide a redox etch process, where the material being etched is converted to a soluble positive ion: $A \rightarrow A^{n+} + ne^-$.[15] Of particular note in

- Immerse in heated H_2O + NH_4OH + H_2O_2 solution, with agitation

(Removes organic contaminants and forms complexes with metal contaminants such as Cu, Ag, Ni)

- Rinse in DI H_2O
- Immerse in heated HCl + H_2O + H_2O_2 solution with agitation

(Removes sodium and heavy metal contaminants and prevents redeposition of those contaminants by the formation of soluble complexes with metal ions)

- Rinse in DI H_2O
- Dry
- Inspect

Assay: 37% HCl, 27% NH_4OH, 30% H_2O_2 (by weight)

Figure 15.49. Cleaning process prior to insertion of product wafers into furnace tube.

[15] Redox is short for oxidation–reduction; in this context, oxidation refers to the loss of electrons from an element in a chemical reaction.

Polysilicon:	HNO_3 + HF
Crystalline Silicon:	65% N_2H_4 (hydrazine) + 35% H_2O (anisotropic, 100× greater etch rate in ⟨100⟩ direction than ⟨111⟩; does not attack SiO_2)
Silicon Polish Etch:	HF + HNO_3 + acetic acid
Silicon Dioxide:	H_2O + HF + NH_4F (buffered HF; a surfactant is often added)
Silicon Nitride:	H_3PO_4 (170°C) (This is a selective etch against an underlying layer of SiO_2; it is also at a very difficult temperature to use with surface photoresist.)
Aluminum:	H_3PO_4 + HNO_3 + acetic acid + H_2O
Aluminum Oxide:	H_2CrO_4 + H_3PO_4 + H_2O (chromic phosphoric)
Photoresist:	90% H_2SO_4 + 10% HNO_3 (90°C)
Assay:	49%–53% HF, 70% HNO_3, 85% H_3PO_4, 37% HCl, 95% H_2SO_4, (glacial) acetic acid (>99.5%)

Figure 15.50. Common etchants.

the figure is the solution for stripping photoresist, after a patterning etch step has been completed; this solution can only be utilized prior to metallization. Also of note is the *anisotropic* etch solution listed for crystalline silicon, a solution whose etch rate is very selective across different crystallographic planes. This technique has been used to fabricate V-grooves in the surface topography for the fabrication of VMOS power devices; patterned SiO_2 serves as the masking layer for this etch.

After etching, a rinse step must be performed. The rinse is executed in a running bath of de-ionized water; the resistivity of the bath overflow is monitored to determine the sufficiency of ionic contaminant removal. A common process step specification is to "rinse to greater than 16 MΩ * cm."

15.10 DRY ETCHING SYSTEMS

The wet etching techniques discussed in the previous section are rapidly being replaced by dry etch processes, utilizing a glow discharge of reactive gases to produce the film removal. Dry etch systems offer the following advantages:

1. The sources of heavy metal and sodium contamination commonly found in solution are eliminated.
2. High wafer-to-wafer reproducibility (with reasonable throughput) is available with small chamber single wafer systems; larger batch systems are also available.
3. The etch process can be readily monitored (through spectroscopy or inter-

ferometry) for a change in etch characteristics, signifying the end point of the etch; with proper design considerations for edge effects, etch uniformity is high and very little overetch time is required.

4. By controlling the flow rate mixing of reactant gases, very good etch rate control is available.

5. Dry etch systems are available that provide good latitude over the anisotropy of the vertical-to-lateral etch rates; very steep edge profiles are achievable for interconnection layers.

6. The use of phosphosilicate glass as a dielectric layer introduces difficulties with wet etch processes; the high etch rate of phosphorus-rich oxides tends to rapidly draw the wet etchant under the photoresist layer by capillary action with subsequent loss of resist adhesion. Dry etch systems typically do not encounter photoresist adhesion problems.

The main disadvantage of a dry etch system is its cost and complexity of operation relative to the simple batch immersion wet etches. Additional considerations for dry etch steps include the etch rates of underlying layers and in particular of the patterned photoresist coating. High selectivity of the exposed film etch rate to that of the photoresist and underlying film is commonly desirable; in cases where the photoresist etch rate is significant, the profiling of the photoresist line width after developing but prior to etching can provide some flexibility in the resulting etch dimension and profile. The removal of photoresist during the dry etch procedure introduces the possibility of polymerization and organic contamination of the system; dry etch photoresist removal is complicated by exposure to elevated wafer temperatures in a previous implant (or other dry etch) step.

Two configurations of dry etch systems are commonly used, with differing etch rate and etch anisotropy characteristics: plasma etching and reactive ion etching (RIE). Ion beam etch techniques are also rapidly evolving, although the relatively poor etch selectivity of this physical process is a disadvantage.

Plasma Etching

Plasma etching is a chemical process utilizing a glow discharge to generate reactant species from a molecular gas constituent of the ambient. This reactant subsequently diffuses to the wafer substrates and etches the exposed surface by formation of products that are volatile. The reaction products are pumped out of the vacuum system. Common inlet gases used for the plasma etching of Si, SiO_2, and Si_3N_4 are CF_4, CHF_3, and C_2F_6. Photoresist is commonly removed utilizing oxygen. A list of the common species present in the plasma is given in Figure 15.51; some of the reactions that produce these species are given in Figure 15.52. Atomic oxygen reacts with the photoresist organic to produce CO, CO_2, and H_2O. Atomic fluorine reacts with Si_3N_4, SiO_2, and Si to form the volatile compound SiF_4. CVD nitride films (typically rich in hydrogen or oxygen: $Si_xH_yN_z$ or $Si_xO_yN_z$) are also removed in a fluorine-containing plasma. $CF_4 + O_2$ gas mixtures are

Input Gases:	CF_4 C_2F_6 H_2 O_2
Neutral Radicals:	CF_3 F SiF_3 COF CF_2 H SiF_2 OH CF O SiF OF C
Stable Products:	C_2F_4 F_2 CO C_2F_2 HF CO_2 SiF_4 H_2O COF_2 CHF_3
Charged Species:	CF_3^+ H_2^+ CF_3^- e^- CF_2^+ H^+ O_2^- CF^+ O_2^+ O^- F^+ O^+

Figure 15.51.   Species present in a plasma etching system. [From M. J. Kushner, *J. Appl. Phys.*, 53:4 (April 1982), 2924]

$$e + CF_4 \rightarrow CF_3 + F + e$$
$$\rightarrow CF_3^+ + F + 2e$$
$$\rightarrow CF_3^- + F$$
$$\rightarrow CF_3 + F^-$$
$$\rightarrow CF_2 + F + F^-$$

$$e + O_2 \rightarrow 2O + e$$
$$\rightarrow O + O^+ + 2e$$
$$\rightarrow O_2^+ + 2e$$
$$\rightarrow O + O^-$$

$$e + SiF_4 \rightarrow SiF_3 + F + e$$

$$e + C_2F_6 \rightarrow 2CF_3 + e$$
$$\rightarrow CF_3^- + CF_3$$
$$\rightarrow CF_3^+ + CF_3 + 2e$$

$$e + CHF_3 \rightarrow CF_3 + H + e$$

$$e + F \rightarrow F^+ + 2e$$
$$\rightarrow F^-$$
$$e + F_2 \rightarrow 2F + e$$
$$\rightarrow F^- + F$$

Figure 15.52.   Electron impact collisions and reactions. [From M. J. Kushner, *J. Appl. Phys.*, 53:4 (April 1982), 2925]

commonly used; the presence of oxygen reduces carbon contamination and increases the concentration of atomic fluorine to enhance the film etch rate.

Two configurations of plasma etch systems are common: planar and barrel. In a planar reactor, two closely spaced parallel electrodes are used (similar to the RF diode sputtering system). The wafers are placed on the grounded electrode, while the RF supply is connected to the top electrode through a suitable impedance-matching network. It should be noted that plasma etching differs considerably from sputtering; the intent of the process is to keep the potential difference between the electrodes and the plasma low (~100 to 200 V) so that little sputtering occurs. This low potential is achieved by using a relatively high gas pressure, typically, 0.2 to 0.5 torr. The design of the electrodes should utilize a material that is not reactive with the plasma species and with a relatively high sputtering threshold. (There is no source of secondary electrons from impinging cathode ions to sustain the discharge, as for dc sputtering; RF energy is used to sustain the plasma and produce the desired chemical etching.) Input gases are fed into the plasma reactor, while etch products are pumped out. A potential advantage of the planar etch system is that the wafer backside is not exposed to the plasma, protecting the back from etching; this is in contrast to the barrel etch system discussed next. A serious disadvantage of the planar configuration is the large area required to achieve high throughput; this introduces a potential for greater nonuniformity in the plasma etch process. Planar plasma etch systems have begun to incorporate unique inlet and exhaust (radial) flow patterns in an attempt to maintain high reproducibility.

A barrel-type plasma reactor utilizes wafers loaded vertically in a holder running down the axis of the vacuum chamber (Figure 15.53). The gas pressure is commonly higher than with a planar system (0.5 to 1.0 torr), with a corresponding reduction in plasma to substrate potential difference (~10 to 20 V). Barrel systems are commonly used for the stripping of the postetch photoresist coat in preparation for further processing. This batch process utilizes the more volume-efficient barrel system rather than a planar reactor; underlying layers are relatively unaffected by the oxygen plasma and the uniformity of the etch rate is not a concern in stripping the resist layer.

To achieve better uniformity in other etch processes, the inlet and exhaust flow patterns should be invariant down the tube, and each product wafer should see a constant environment from front to back; dummy wafers should be included at both ends of the holder. Protecting the wafer backside requires placing each product wafer in contact with a dummy wafer together in a single holder slot.

The system illustrated in Figure 15.53 incorporates an open metal shield or cage onto which the wafers and holder are placed. This cage serves as the floating anode of the RF plasma and confines the discharge to the outside region. The cage is perforated to allow neutral atomic fluorine and oxygen to enter the inner

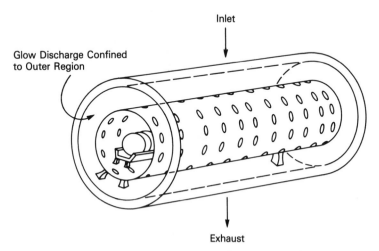

Inlet and exhaust gas flow ports not shown.

Figure 15.53. Barrel configuration of a plasma etch system.

region and react with the wafer surface; ions in the plasma are collected by the RF field shunting cage and do not participate in the etching process.

A consideration in the development of a plasma etch process step is the variability in etch rate with the total active wafer surface area being etched; the greater the loading, the less the etch rate for a given pressure and gas flow parameters, due to the depletion of the reacting species. To minimize this loading variation locally and from run to run, the same number of wafers (each with similar surface coatings) should be used each run. Again, dummy wafers may be added to achieve this consistency. Etch end-point detect techniques can be used to negate some of this concern about the absolute etch rate.

The rise in wafer temperature during plasma etching is a definite concern. The wafer temperature increase is due to the heat generated by the RF power dissipated in the plasma and from the exothermic surface reactions producing the volatile etch products. In addition to adding considerable variability to the etch rates, elevated wafer temperatures result in photoresist reflow and loss of image definition.

Due to the relatively high pressures at which plasma etching is performed, the etch profiles are roughly *isotropic;* plasma etching does not offer a major advantage is this regard. Indeed, the increase in wafer temperature during the etch and the potential change in etch rate at completion (as the loading area is rapidly reduced) tend to introduce significant undercutting of a photoresist image. Preheating the wafers to ~100°C reduces the variability due to temperature changes, while the latter requires careful attention to batch etch uniformity so that overetch time can be minimized once the end point has been detected.

A new application of these plasma systems in VLSI processes is to include procedures for plasma hardening of photoresist after patterning.[16] Brief exposure (at low RF power) to a CF_4 plasma reacts with the positive resist to produce a surface layer with an increased reflow temperature; this layer thus serves as a shell for each shape of the patterned resist film and maintains the as-developed image during exposure to elevated temperatures. Plasma hardening is therefore particularly applicable prior to ion implantation, lift-off metallization, and reactive ion etching process steps, where the energy absorbed by the photoresist-coated wafer leads to increased temperatures.

Reactive Ion Etching

In addition to the high-pressure (barrel or planar) plasma etch systems, the other common technique for the dry etching of films is reactive ion etching (RIE). This process utilizes the configuration of a sputter etching system (as described in

---

[16] For an example of a plasma hardening process description, refer to W. H-L., Ma, "Plasma Resist Image Stabilization Technique," *Technical Digest of the International Electron Devices Meeting* (1980), p. 574.

Section 15.8); the RF energy is applied to the substrate holder (through an impedance-matching network and dc blocking capacitor), while the remaining electrode area is grounded. RIE differs from sputter etching, however, in that the argon gas discharge is replaced with the gas chemistry utilized for *chemical* plasma etching described previously. The RIE process is still primarily the result of a chemical reaction that forms volatile etch products; the effective *physical* sputtering rate of ions in the discharge accelerated toward the substrate is low. Yet, due to the accelerating dc voltage present on the substrate electrode, it is indeed reactive *ions* (as opposed to atomic neutrals) that are the major participating species in the chemical reaction.[17] The accelerating voltage and lower operating pressure imply that the incident reactive ions will be directed normal to the wafer surface; as a result, exposure of photoresist protected areas to the reactant is minimized and highly anisotropic etching of the unprotected areas of the film can be produced. Compared to plasma etching, an additional advantage of RIE is that for a particular set of process parameters the etch rate is relatively insensitive to variations in substrate temperature, film impurity concentration, or crystallinity (for polysilicon and silicon etches); the chemical reaction energy differences associated with these variations are small in comparison to the available incident energy of the reactive ion. RIE also demonstrates less of a loading effect etch rate variation than does plasma etching. The capabilities of RIE to produce patterned films with little undercutting is of key significance in future process development.

The disadvantages of RIE parallel those of the physical sputter etching process, that is, device-related effects of ion bombardment, increase in wafer temperature, and back-scattering of emitted material. In comparison to plasma etching, RIE etch rates are significantly less. Also, the selectivity of SiO_2-to-Si film etch rates is commonly worse for RIE than for plasma etching.

A relatively new dry etching equipment configuration attempts to provide increased control over etch rates and the final edge profile by combining plasma etching and RIE etching features (Figure 15.54). This flexible diode system incorporates the RIE geometry with an additional isolated counterelectrode to which RF energy may be independently applied (either simultaneously or sequentially to the RIE power source).

The discussion of dry etching to this point has concentrated on the gas chemistries used for etching of silicon and silicon compounds in fluorinated discharges and photoresist in oxygen and fluorinated plasmas. The dry etching of metal films (in an RIE or planar plasma configuration) introduces several process complications. The dry etching of interconnection metals (i.e., Al, Al + Si, Al + Cu, and Al + Cu + Si) requires a discharge containing chlorine to produce the volatile products $AlCl_3$ and $SiCl_4$ (at somewhat elevated temperatures). The volatility of AlF_3 is low, precluding the use of a fluorine-containing gas. CCl_4 is

---

[17] The contribution of reactive ions versus neutrals can be further enhanced by a gas chemistry mixture that depletes the concentration of free neutrals.

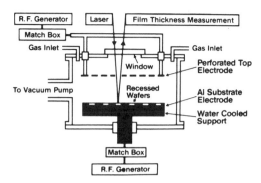

Figure 15.54. Flexible diode RIE/ plasma etching system. (From L. Ephrath, "Reactive Ion Etching for VLSI." *IEEE Transactions on Electron Devices*, Vol. ED-28, No. 11 (November 1981), p. 1316. Copyright © 1981 by IEEE. Reprinted with permission)

a suitable gas for use with aluminum etching, although an unsaturated chloro-carbon polymer can be deposited on the wafers and system walls. The use of BCl_3 avoids this polymerization problem, yet can result in undesirable contamination from boron residues. Adding Cl_2 to the inlet gas may be used to reduce the concentration of unsaturated chlorine compounds.

A surface aluminum oxide film may locally retard the etching of the aluminum layer; an inert gas (e.g., Ar or Xe) can be added to the gas flow to help remove this surface oxide by physical sputtering. The local variation in thickness of this oxide layer and its lower etch rate results in nonuniformities in the etched aluminum film. Oxygen or water vapor in the system reacts with aluminum to inhibit the etch process; the vacuum system must be specifically designed (using low-volume loading stations between runs) to exclude or trap water vapor. In addition, the etch product $AlCl_3$ readily absorbs moisture; on exposure to the atmosphere, the etch chamber increases in water concentration from $AlCl_3$ absorbed on the chamber walls, which leads to etch rate variability in subsequent runs.

The incorporation of a small percentage of copper in the metallization film introduces the problem that no volatile compound is formed by reaction with chlorine; copper or copper chloride residues normally remain after etching, unless a sufficient degree of sputtering of the copper is incorporated into the etch process. A postetch treatment can also potentially be used to remove the residues. (A subsequent water rinse may be sufficient, as copper chloride compounds are soluble.)

The aluminum (or aluminum–copper) metallization lines after the etch are directly susceptible to corrosion by exposure to water in the presence of residual chlorine compounds (thereby producing acidic solutions). These compounds are present on the wafer surface, etched line sidewalls, and in the protective photoresist layer. To reduce this susceptibility to corrosion, the wafers can be subjected to a fluorine-containing plasma immediately after etching before exposure to the atmosphere. The fluorine displaces the residual chlorine and results in compounds that do not readily absorb moisture.

When working with chlorine plasmas, safety precautions must be observed due to the corrosiveness and toxicity of compounds that may be generated in a CCl_4 discharge, such as HCl and $COCl_2$ (phosgene). Extreme care must be taken to purge the chamber of these reaction products to avoid exposure during system unloading and maintenance. The corrosive nature of chlorine gases also strongly affects the design and maintenance of the chamber vacuum system.

The utilization of a lift-off metallization patterning technique avoids some of the aforementioned problems with dry etching of aluminum and aluminum alloys. Lift-off also does not require the patterning of VLSI resolution photoresist lines over a highly reflective (scattering) surface metallization layer. It does have the distinct disadvantage that the metal must be deposited onto cold wafer substrates to prevent photoresist reflow. This implies that the deposit will have a low surface mobility after impinging onto the wafer surface; the resulting metal lines will exhibit poor step coverage, as indicated by the thinning of the film over steep sidewalls. The blanket deposition of a metal film onto heated substrates (with subsequent dry etch patterning) reduces step coverage failure mechanisms (e.g., sidewall voids, microcracks, and fractures). These two metallization process alternatives will continue to compete for acceptance in VLSI technologies; the continued progress in metallization dry etch characteristics and alternatives will help to make the blanket deposition and subtractive removal technique increasingly attractive.

End-point Detection

The determination of the etch time for a dry etch process differs considerably from that of the typical wet etch process; the ratio of the film thickness to the etch rate is not adequate, due to the possible fluctuations in wafer temperature and plasma reactant and product concentrations during the etch. In addition, as the etch selectivity between different film layers is typically less with dry etch processes, controlling the overetch time (added to compensate for batch wafer-to-wafer nonuniformities) is also crucial. Several techniques of etch end-point detection (EPD) have been incorporated into dry etch systems.

One alternative for EPD monitoring is to utilize laser interferometry; a monochromatic beam is directed at the surface of a wafer in the system chamber and the reflected intensity from the etch surface is recorded. The reflected intensity from the etched film surface and from the surfaces of the underlying layers interfere; as the top layer is removed, a series of constructive and destructive interference patterns is produced. The change in the remaining film thickness between signal peaks is given by $t = \lambda/(2 * n)$, where n is the index of refraction of the film. Rather than utilizing this measure, however, a graphic depiction of end point is provided by the change in signal when the etched film–underlying layer interface is exposed. The main disadvantage of this technique is that it is *not* a batch measurement, but only local to the wafer used for monitoring. (It also has the disadvantage that a relatively large exposed area must be provided on a wafer from the product run and aligned under the beam.) To compensate for wafer-to-wafer etch uniformity variation, some overetch time should be included past end point.

Another option for EPD is to monitor the intensity of light radiation emitted from the discharge at a *specific* spectroscopic wavelength characteristic of one of the etch reactants or products. This technique assumes that the intensity of light emitted by excited atoms or molecules of a particular species at a characteristic wavelength is proportional to the concentration of that species in the discharge. This concentration changes when the end point of the reaction is reached; the concentration of a reactant increases, while that of a product decreases. The percentage change in product concentration is usually considerably larger than that of the reactants, and, as such, it is preferable to select an emission wavelength of one of the products. (This assumes that the product concentration is sufficiently high during the etch to be monitorable; this will require some consideration of the total exposed area being etched.) It is also necessary that the characteristic wavelength selected for the gas discharge reactant or product be sufficiently distinct from those of other species present to be resolvable.

Other techniques for EPD include monitoring changes in the impedance of the discharge during the etch and sampling the effluent gas stream from the reactor to the pumping system. In the latter case, a mass spectrometer is used to analyze the stream for products of the etch process.

15.11 CONVENTIONAL PHOTOLITHOGRAPHY

Mask Manufacture

The progress in photolithographic processing technology has been the primary factor enabling the exponential growth in density and performance of modern integrated circuitry. The cornerstone of photolithography is the mask (the series of masks, actually) that accurately defines the patterns that must be transferred into the wafer surface at the appropriate steps in the overall product manufacturing process. The correct pattern registration between each mask plate is one of the most critical problems encountered in mask manufacture. Cleanliness is also vital, both in mask manufacture and production use.

As shown in Figure 15.55, several options for mask manufacture and pattern transfer have been developed. They vary widely in accuracy, ease of manufacture, and cost. The pattern data for mask manufacture is typically provided to the mask house in the form of a magnetic tape or on-line postprocessed data. There must be coordination between the circuit designer and the mask manufacturer to ensure compatibility between design data and mask generation equipment.

The common technique used for mask making in LSI technologies involves the direct photocomposition of a $10\times$ reticle using a system developed strictly for mask manufacture. The pattern generator system consists of a camera assembly, a positioning stage, a magnetic tape reader, and the necessary electronics

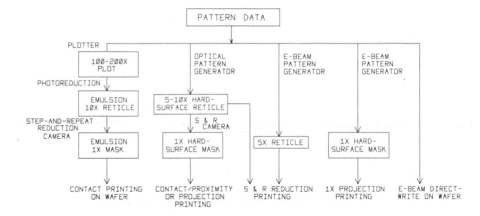

Figure 15.55. Photolithographic techniques.

for overall control. The $10\times$ circuit patterns are produced by assembling rectangular images in building block fashion on a photographic plate; the entire circuit pattern is built up from a series of rectangular exposures. A block diagram of an optical pattern generator was given in Figure 5.55. There are three pattern generator techniques commonly used for the production of reticles, where the pattern data are transferred into a chrome film. (The benefits of a chrome working plate over a photographic emulsion will be discussed shortly.) These approaches are flow-charted in Figure 15.56.

A reticle defect is an undesired interruption or addition to the image pattern that makes the plate unusable. In general, more defects are found on clear field rather than dark field background plates. In addition, in some cases defects can be repaired and a mask salvaged (e.g., pinholes in a dark field background). Some common reticle defects are illustrated in Figure 15.57.

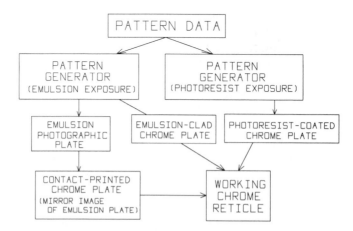

Figure 15.56. Techniques for the manufacture of a working chrome plate.

The use of a pattern generator assumes that the necessary postprocessing of the shapes data has been applied to the internal graphics representation to produce the properly formatted PG input. Each pattern requires five data elements:

1. x coordinate of the center of the rectangle
2. y coordinate of the center of the rectangle
3. Rectangle width
4. Rectangle length
5. Counterclockwise angle of rotation

When evaluating a pattern generator, several machine characteristics should be considered:

1. Exposable area: maximum area that can be exposed on a glass plate, primarily based on the x and y stage travel; does not include glass mounting brackets or the border from the edge of the glass plate to the first exposable pattern
2. Stage speed: note that the maximum stage speed is not a very meaningful specification; the actual mask generation time is more dependent on the stage acceleration and deceleration, the average distance between flashes, and the sorted sequence in which the flash data are presented to the generator
3. Minimum stage increment and addressable resolution in x and y dimensions
4. Positional accuracy and repeatability of the stage movements
5. Orthogonality of x and y stage movements
6. Aperture size range
7. Aperture resolution (smallest aperture increment in size)
8. Aperture speed: the time required to change aperture sizes (varies with the extent of the change)
9. Aperture accuracy and repeatability
10. Rotational speed

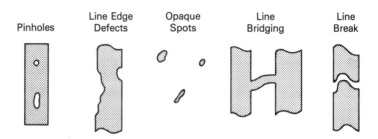

Figure 15.57.   Common reticle defects. Pinholes can potentially be repaired by opaquing.

11. Rotational resolution
12. Rotational accuracy and repeatability
13. Symbol generation: the ability to include mask titling and identification information as keyed in on the PG's terminal
14. English and metric operation
15. Illumination uniformity and exposure control
16. Utilization of emulsion-coated or photoresist-coated plates

After the $10\times$ reticle has been produced, in either emulsion or chrome, a $1\times$ mask may be generated. A $1\times$ mask contains arrays of final size shapes and patterns. To produce the arrays of $1\times$ chip images, a step-and-repeat $10\times$ reduction camera system is used. The step-and-repeat system must satisfy the following requirements:

1. The table used to move the $1\times$ blank during the step-and-repeat operation must be accurate in travel in both x and y directions.
2. Any vertical movement of the table stage plus any curvature in the plate must be less than the depth of focus of the camera system.
3. System vibration isolation is required.
4. The means for changing and aligning $10\times$ reticles in the camera system must ensure that all images produced on the $1\times$ plate (from more than one $10\times$ reticle) are in proper registration with each other. A $1\times$ mask will potentially contain multiple chip designs and test plug sites. When exposing one of multiple $10\times$ plates, the system controls must permit array locations to be skipped. Use of multiple $10\times$ reticles requires alignment of the reticle to the platen of the step-and-repeat camera system, which is perpendicular to the optical axis of the system. Alignment marks on the reticle and on the platen are viewed through an optical microscope; when proper registration is obtained, the reticle can be mechanically clamped and/or held by vacuum to the tube of the reduction optics.

A significant increase in the mask-to-mask registration accuracy for a set of photolithographic plates of a process (as well as the throughput of $1\times$ production) can be realized by the use of multibarrel step-and-repeat camera systems. A number of identical cameras, typically six to ten, are mounted above a single x, y table. While the registration accuracy between masks can be improved due to the simultaneous exposure of many plates, some error can be introduced by any variations in reduction ratios between barrels. In addition, it is likely that not all the plates generated simultaneously will be acceptable due to some of the possible defects illustrated earlier. Defective masks must be remade, potentially with a loss in registration accuracy in comparison to a different set.

Masks containing a pattern in an emulsion (gelatin) film are easily damaged in any contacting operation. To extend the usefulness of an emulsion master ($10 \times$ or $1 \times$), copies are commonly made. The process of copying an emulsion plate involves pressing the master against a blank plate and exposing the blank through the pattern on the master. The submaster now contains the mirrored image of the original emulsion plate. If the original has not been mirrored, another contact printing step from the submaster into a production plate is necessary. The submasters (and production plates, if necessary) may be either emulsion-on-glass or chrome-on-glass types. Prior to exposure, the chrome-on-glass plate is coated with photoresist. After the photoresist is developed, the chrome is etched to produce the final pattern.

The measurement systems for positioning the xy stage of a pattern generator or step-and-repeat camera must be capable of accurately controlling the servomotors that provide the motion. One means for accurate positional measurement is illustrated schematically in Figure 15.58. Two laser interferometers measure x and y motion of the stage. The beam splitter divides the laser light into two beams, one reflecting from a fixed mirror and the other reflecting through a path that varies with the movement of the work stage. By reflecting the two beams back upon themselves and eventually combining on a path to the detector, a series of constructive and destructive interference patterns are produced as the stage trav-

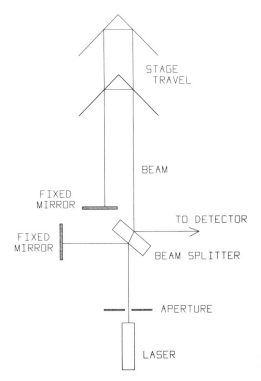

Figure 15.58. Laser interferometer measurement system for stage motion.

els. A digital counter is used to measure the number of intensity pulses. The distance traveled by the stage is given by the relation $d = (n * \lambda)/\gamma$, where d is the distance traveled, n is the number of counts, λ is the laser wavelength, and γ is the path multiplier or sensitivity. For a He–Ne laser of wavelength $\lambda = 6328$ Å and the sensitivity (as illustrated) of 4, the resolution of the interferometer is ~0.15 μm.

It is feasible to use a pattern generator as a step-and-repeat reduction camera as well; this is possible if the construction of the pattern generator permits the variable-sized aperture assembly to be replaced by a reticle holder.

Alternative mask manufacturing techniques that have been developed for VLSI lithography requirements are discussed in Section 15.12.

Mask Materials

The primary materials used in mask manufacture are glass, photoresist, emulsion, and chrome (or chrome oxide). The characteristics of the particular materials used strongly affect the suitability of the resulting mask for manufacture. In particular, the defect densities of the coated films and the defects present in the glass substrates (e.g., voids, inclusions, and surface defects) are all detrimental to mask yield. The need to periodically clean working mask plates also affects the material selection. As always, cost considerations must be included.

One key parameter of the glass mask substrate is its temperature coefficient of expansion. The glass can rise several degrees Celsius above ambient due to absorption of the incident light energy during through-the-mask alignment. The change in lateral dimension is commonly called *run-out* and is illustrated in Figure 15.59 for several glass materials commonly used for mask substrates.

Another key parameter or characteristic of the various glass materials used is the transmission properties of the incident wavelengths used for aligning and exposure. A significant absorption at the wavelength used for exposure necessitates a longer exposure time (to maintain a particular total photoresist exposure energy). Longer exposure times reduce throughput (and increase the sensitivity to vibration with projection aligners). The transmission percentages (for 1.5 and 2.5 mm thickness) of the glass materials illustrated in Figure 15.59 are shown in Figure 15.60. Highlighted on the plot are the wavelengths of mercury vapor arcs, commonly used for ultraviolet photoresist exposure. In particular, note that in an effort to improve resolution by using a shorter exposing wavelength source, major modifications are required to the following:

1. The photoresists used, so that their spectral sensitivity is high at the selected wavelength.
2. The lens system, which must be redesigned for operation at shorter wavelengths.
3. The glass substrate material, to maintain a high transmission percentage.

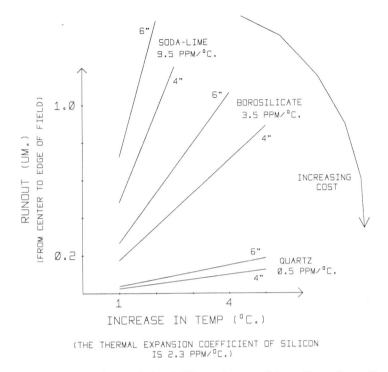

Figure 15.59.   Run-out characteristics of different glass materials used in mask manufacture.

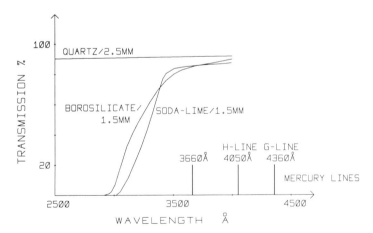

Figure 15.60.   Transmission percentages for 60- and 90-mil thick glass materials for masks; the common UV exposure spectral line wavelengths of mercury vapor arcs are also illustrated.

Yet another key parameter that involves both the choice of glass material and the aligning equipment used is glass flatness and sag. The glass flatness is a function of manufacture and polishing. The gravitational sag of a plate in use is dependent on the thickness of the plate, the plate area, and the fixturing and support provided to the plate during use. The most common and best means of mask support is a vacuum ring that completely surrounds the internal pattern area. This ring can be applied to the top or the patterned side (more commonly, the patterned or working side).

The masking material coated onto the glass substrate must:

Bond strongly to the glass.

Be able to be coated uniformly onto the glass prior to patterning.

Be easily patterned using photoresist-compatible etchants.

Have high absorption coefficients in the ultraviolet wavelengths (and, potentially, some transmissivity in the optical red and green wavelengths).

The thickness of the masking material should be chosen so that any transmitted ultraviolet energy is reduced by destructive interference of multiple reflections through the masking material (Figure 15.61).

Chrome is a commonly used (hard-surface) mask material. Its properties, both good and bad, are as follows:

(+) It is a hard film, with high durability and potentially long mask life.

(+) It is resistant to mask cleaning procedures.

(−) It is opaque at visible wavelengths, which potentially makes alignment of dark-field mask areas more difficult.

(−) It has a high reflectivity.

When transferring a pattern from an emulsion master into a photoresist- and chrome-coated working plate, the reflectivity of the chrome film underneath can

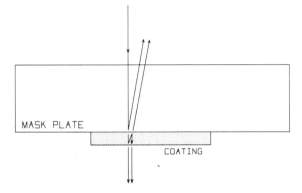

Figure 15.61. Multiple reflections of UV radiation through the opaque mask coating should destructively interfere to minimize transmitted intensity.

significantly alter the resulting line width by exposing adjacent photoresist areas. Antireflective chrome oxide (a chrome film with a thin chrome oxide layer) can reduce the magnitude of this problem.

An emulsion layer (a gelatin layer containing grains of a silver–halogen compound) on a mask plate is inexpensive, relatively thick (and therefore somewhat insensitive to mask surface irregularities but also resolution limited), and relatively soft (and therefore offers only a very limited mask life). Its main advantage is its high sensitivity to longer-wavelength light, which implies that patterning can be done directly without a photoresist coating and ultraviolet exposure. An emulsion coating is thus ideally suited for pattern generator use and the making of masters that are contact printed into chrome working plates.

Mask cleaning is a vital procedure to perform periodically to remove particulate contamination, which adheres to the plate by Coulomb and van der Waals (molecular) forces. Various cleaning approaches are advocated by a variety of manufacturers in a variety of combinations:

> Soft-hair bristle mechanical brushes
>
> Air flow (with bursts)
>
> Solvent cleaners (methanol, acetone)
>
> Water rinsing and high velocity water jets
>
> Resist strippers (for contact masks)

Mask–Wafer Alignment Strategies

The suitability of a particular mask material and the necessary requirements of the mask manufacturing equipment are strongly related to the means of mask–wafer alignment and wafer photoresist exposure. In particular, the following alignment strategies will be briefly described: contact and proximity, projection, and step-and-repeat (with reduction). An alternative strategy for potential use in future technologies is described in Section 15.12.

Specific areas that are used to evaluate the various alignment schemes are as follows:

> Achievable linewidth resolution
>
> Exposing wavelength (as it affects photoresist sensitivity)
>
> Achievable overlay accuracy (plus any compensation for mask-to-wafer runout variations)
>
> Illumination uniformity and efficiency
>
> Exposure times and energy

Some of the less technical specifications, yet significant nevertheless, are those that will ultimately affect the final chip cost:

1. Aligner cost, including maintenance and service, unique environment requirements (power, temperature and humidity control, size, operator interface).

2. Modularity (ability to upgrade to a different exposing wavelength system or a larger wafer size).

3. Aligner yield-detracting characteristics, specifically the amount of wafer and mask handling required.

4. Aligner throughput (wafers/hour), as governed by:
   a. Alignment and exposure time.
   b. Time required for mask change and acclimation.
   c. Machine setup time (the time required and necessary frequency of test for focus, registration, wafer stage movement orthogonality and precision, wafer chuck flatness, illumination intensity).

5. Capability to efficiently handle multiple parts per wafer or split lots (multiple part numbers per lot).

All these specifications and cost factors need to be considered when weighing the alternative alignment options. In addition, the number and volume of part numbers produced by a fabrication facility affect the alignment strategy considerably. For example, in a high-volume, single part number fabrication area (e.g., DRAMs), the dedication of one aligner to each mask level would provide for maximum throughput and eliminate much of the need for mask handling (and potential contamination). Such an approach requires that the aligners for different levels be suitably matched in their characteristics. It is also very common to utilize varied alignment techniques for the different levels of the same process. If the level-to-level overlay requirements for the masking steps vary substantially (a few critical, others more relaxed), different alignment techniques could be used for different levels in an attempt to trade off resolution and overlay specifications versus machine cost. If the fabrication facility falls into the pilot-line category, support for a number of different alignment schemes may be required.[18]

Contact and proximity alignment.   These two alignment strategies are commonly discussed in combination due to their similarities; indeed, the aligners in this category can often be programmed to operate in either contact or proximity

---

[18] A pilot line, as opposed to a fab area, is described by the following characteristics:

1. A large number of part numbers are produced.
2. Low volumes are provided per part number, typically consisting of initial prototype design hardware and low volume special application-specific master slice designs.
3. A number of processes are commonly supported.
4. Advanced process development, experimentation, and implementation are provided.
5. Most importantly, for designers who require early hardware evaluation, *fast* turn-around time is offered.

mode. Briefly, the method of contact alignment involves the positioning of a photoresist-coated wafer relative to a fixed mask while in the off-contact mode. The alignment of the mask to the wafer is accomplished by viewing the pattern in the wafer surface through the mask using an optical microscope. When proper alignment is achieved, the wafer and mask are placed in contact and the shutter opened, allowing ultraviolet radiation to expose the areas of photoresist under the transparent areas of the mask. Contact masking techniques result in progressive deterioration of the mask (particularly an emulsion-coated mask) through physical damage and particulate contamination. As a result, it is necessary to make several copies of a mask and institute thorough, periodic mask inspection and cleaning procedures.

Proximity printing involves exposing the photoresist-coated wafer with the mask and wafer separated by a distance just sufficient to avoid physical contact. The exposing optical system for proximity printing must be modified from that for contact printing to produce an acceptable photoresist image when the mask and wafer are not in contact, primarily due to *diffraction* effects. The exposing intensity distribution spreads in extent (under opaque mask areas) as the separation distance increases. A mask commonly consists of a periodic pattern of transparent and opaque lines. In the case of a multislit (or *grating*) pattern, the intensity distribution is of the form shown in Figure 15.62 (for five slits). For a particular line, the pertinent contrast in exposure is given by the intensity modulation function:

$$\text{Modulation} = \frac{I_{\max} - I_{\min}}{I_{\max} + I_{\min}}$$

where a value of 1 is equivalent to perfect contrast and 0 provides no contrast. For a particular (linewidth plus space) pitch and exposing wavelength, as the

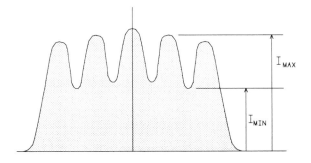

Figure 15.62.   Intensity distribution versus position for five openings with a proximity alignment scheme.

mask-to-wafer separation increases, the intensity modulation (contrast) decreases. Similarly, photoresist has a contrast specification defined as (minimum exposure to maintain solubility)/(maximum exposure to maintain nonsolubility), as for a positive resist. This contrast ratio is a strong function of the photoresist film thickness. To reduce these diffraction effects, shorter wavelengths are required; in addition, low reflectivity masks can reduce the amount of light backscattered into unexposed areas. The gap between the mask and wafer should be as small as possible, given the tolerances on mask and wafer flatness and the precision of the mechanical stage.

The major advantages of contact and proximity printing are high throughput and low cost (e.g., $160K for a proximity aligner and $50K to $100K for a contact aligner). The high defect density tends to limit their usefulness and applicability at VLSI densities and dimensions.

Projection printing (1×). A 1:1 projection printing aligner images the pattern in a 1× mask onto the photoresist-coated wafer through an optical system with the mask and wafer distant from each other. Projection printing offers two main advantages over the contact and proximity systems: longer mask life and less wafer contamination, permitting increased yields. However, the machine cost is considerably greater ($200K to $500K).

The most prevalent optical system for 1:1 projection aligners is a reflecting system; as compared to a refractive optical system, a reflective system results in reduced light scattering and introduces no source of wavelength-dependent aberrations. As a result, no light filtering at the lamp source is required and the entire ultraviolet spectrum of the mercury lamp can be used for exposure. This system is the heart of Perkin–Elmer's "Micralign"™ series of projection aligners.

Fundamentally, the optical system consists of two concentric spherical reflecting surfaces (Figure 15.63). Light from a point on the mask is reflected from the primary mirror to a secondary mirror and back to the opposite side of the primary. The final reflection in the optical path images the desired point onto the wafer surface. Only a small zone (or slit) of the mask surface is illuminated in order to reduce the area on the mirror surfaces (and in the overall system), which must be free of abberation. There are two items of particular note in Figure 15.63:

1. There is no lefthand-to-righthand image inversion; the wafer image exactly resembles the mask image. As a result, working plates generated by contact printing would not be used for projection printing.

2. Coverage of the entire mask and wafer surface is achieved by scanning the mask and wafer in synchronism across the illumination slit. Using the previously illustrated optical system, this requires that the mask and wafer be moved in *opposite* directions, a mechanically awkward and undesirable situation.

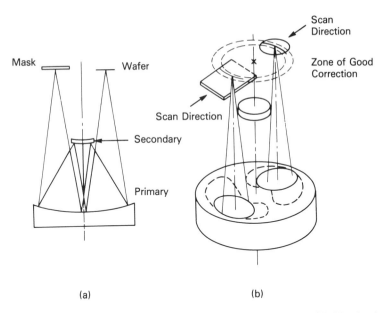

Figure 15.63. (a) A 1:1 projection alignment reflective system. (b) Slit illumination and exposure. (Adapted from D. Markle, "A New Projection Printer." *Solid State Technology*, June 1974, p. 50. Reprinted with permission from *Solid State Technology*, published by Technical Publishing, a Company of Dun & Bradstreet)

These two obstacles are eliminated by the use of a folded mirror projection carriage (to which the mask and wafer are held by vacuum after alignment). Figure 15.64 illustrates this folded reflective system. Exposure of the wafer surface is accomplished by scanning the carriage across the illuminated slit. The carriage is pivoted on a flexure blade bearing operating well below its elastic limit. The primary concern for the operation of the 1:1 projection aligner is that the contour of the slit be precisely focused onto the mask surface.

The light source for the Micralign projection aligner consists of a 1000-W mercury vapor lamp in a small capillary tube (2-mm inside diameter) with two tungsten electrodes approximately 1 inch apart. Air is drawn through the lamp housing by a fan to extract heat.

The resolution of the exposing system can be traded off against the speed of exposure by varying the width of the slit image at the mask surface. The usable width typically ranges from 1.0 mm (best resolution) to 4.0 mm (for coarse geometries). Changes in output intensity over the life of the lamp are compensated for by maintaining a scan speed proportional to lamp intensity.

For alignment, the operator views the surface of the wafer illuminated by the image of the mask pattern. A split-field binocular microscope is used for viewing the wafer surface; the two halves of the split field are normally positioned

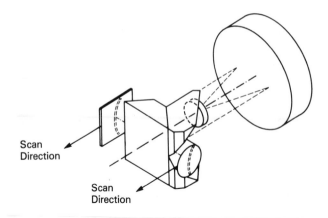

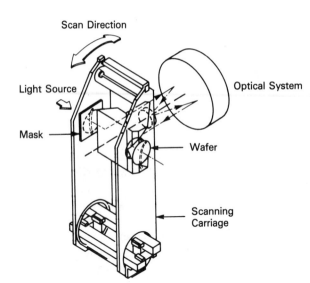

Figure 15.64. Folded reflective system. (Adapted from D. Markle, "A New Projection Printer." *Solid State Technology*, June 1974, p. 50. Reprinted with permission from *Solid State Technology*, published by Technical Publishing, a Company of Dun & Bradstreet)

near the top and bottom of the wafer, but can be moved along the slit. Alignment at any point across the wafer surface is possible by scanning the carriage normal to the slit. There is also a wide field viewing system used for initial (coarse) alignment.

The magnification of the reflective system at any point should conceptually be exactly equal to 1. Some distortion is introduced if the pivoting axis of the scan carriage is not parallel to the line normal to the mask and wafer. This source of error can be reduced by adjusting the tilt in two directions of the folded mirror assembly. For the Micralign system, the depth of focus (i.e., the total out-of-flatness tolerance for both mask and wafer) is ± 5.5 μm for resolving 2-μm lines. As a result, the mask flatness is a very significant fabrication parameter (and potentially a very costly one). In addition, as with all $1 \times$ alignment alternatives, time-varying mask temperature and process-induced wafer distortion can introduce run-out errors. Some of the focus and run-out-related alignment error can be eliminated by using direct-step-on-wafer (DSW) alignment machines, to be discussed next.

Direct-step-on-wafer projection and reduction aligners. The DSW technique involves projecting the mask pattern of a $10 \times$ master (or $5 \times$ master) through a reduction optics lens system onto the wafer surface and stepping and repeating the reduced $1 \times$ image in an array pattern over the wafer surface. As with a step-and-repeat camera used for mask manufacture, the DSW aligner combines a photolithographic optical system with the appropriate stage motions, controls, and software. In particular, the technique of exposing one chip site at a time allows for the software and controls to make real-time corrections for focus, run-out (interpolated), and, in particular, site-by-site alignment. Several technological developments were necessary before DSW could become a production tool. Specifically, these developments include a laser-interferometer xy stage positioning system, an autofocus technique, and a higher-intensity light source.

A representative industrial DSW tool is the GCA Mann 4800™ aligner. The rest of this discussion describes some of the features of this machine. The improved light source was necessary to keep the wafer throughput at an acceptable level (since the total wafer expose time is a large multiple of the individual site exposure time). The light source for the Mann 4800 consists of a mercury lamp source surrounded by four sets of collection optical systems. The collected light energy from different quadrants is transmitted into a four-branched fiber-optic bundle that terminates just prior to the optics in front of the mask (object) plane (Figure 15.65). Typical exposure times for 1-μm lines in 1-μm-thick AZ1370 positive photoresist are reduced to 0.4 sec. Stage motion must also speed up to maintain high throughput. A 9-mm translation occurs in 0.45 sec; the shutter is not opened until the stage has settled at the desired $(x\ y)$ location to within ± 0.1 μm. Exposing one chip site at a time allows each individual exposure to be focused automatically. A photoelectric sensor monitors the change in height of the wafer at the center of the exposure and shifts the optical column accordingly.

The wafer alignment approach is perhaps the most unusual feature of the Mann 4800 aligner. An off-axis alignment technique is used. The wafer is viewed directly through an alignment microscope, *not* through the mask and lens system. The mask is initially aligned to the axis of exposure through an optical microscope to alignment marks on the platen of the optical column, as with the step-and-repeat camera described earlier in the section on mask making. The wafer alignment microscope contains two viewing objectives that are spaced by a considerable distance. A high-resolution video camera and remote monitor are used for remote control of the alignment procedure and to isolate the operator from the

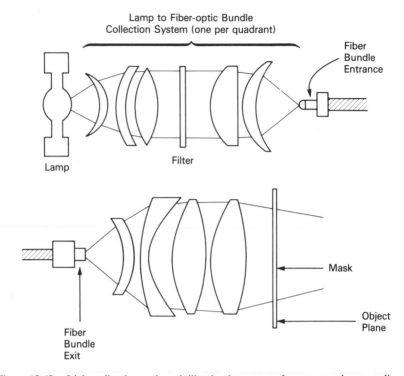

Figure 15.65.   Light collection and mask illumination system for a step-and-repeat aligner.

aligner's environmental chamber. The wafer is aligned to alignment marks in the off-axis microscope itself, *not* to any alignment marks present on the mask. One of the alignment objectives is used to allow the operator to correctly position the (x, y) location of the wafer relative to its alignment marks; the other is used to control the rotational alignment of the stage (Figure 15.66a). The electronic controls and system software must monitor the magnitude of the stage movement from the initial reference position until alignment is complete. That wafer displacement from its original position will be maintained throughout, as the wafer is moved under the optical axis in preparation for step-and-repeat exposure. The periodicity of the exposures and the pattern of the exposures across the wafer surface, including any sites to be skipped, are also software controlled.

Referring to Figure 15.66a, it should be noted that when the x, y, and θ alignment are completed, it is possible that the wafer alignment mark used for rotational alignment may not coincide exactly with the (x, y) lines in the off-axis optical microscope alignment cross. Because of wafer run-out (due to thermal processing between the level of the wafer alignment target and the current level to be exposed), some displacement of the θ alignment mark may be observed (Figure 15.66b). The operator may be able to compensate for wafer runout; after completing (x, y, θ) alignment, a special alignment mode is entered. The alignment

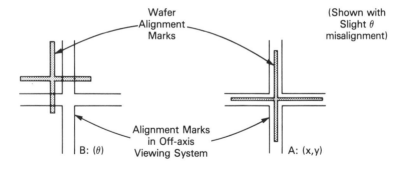

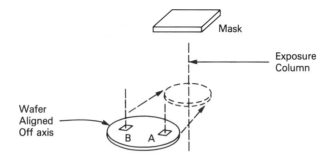

Wafer
Aligned
Off axis

Alignment Procedure:
- Align wafer for (x,y) positioning through objective A
- Align wafer for θ rotation through objective B
- Move wafer under optical column
- Expose

(a)

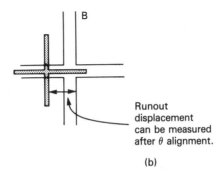

(b)

Figure 15.66.   Off-axis alignment procedure: (a) Off-axis optical alignment viewing system. (b) Indication of run-out displacement as viewed in the alignment mark.

displacement is saved for subsequent step-and-repeat exposure and the additional lateral displacement required to bring only the rotational alignment mark into (x, y) alignment is recorded. Dividing this runout displacement by the distance between objectives, a per-site measure can be determined to add to the chip stepping periodicity (essentially linearly interpolating between distant chip sites).

Using an off-axis alignment technique, the alignment target area on the wafer is viewed through a separate optical system and *not* through the mask; the mask itself may be either transparent or opaque in the alignment target area. All alignments are essentially bright field alignments, even with a dark field mask. The mask may be designed so that the wafer alignment targets remain protected by photoresist after the exposure and development process steps and therefore are not modified by a subsequent etching step. After photoresist removal, the alignment target may potentially be used again.

Two major limitations to the system previously described should be highlighted. Pertaining to the use of the off-axis alignment approach, it is critical to be able to measure with extreme accuracy the distance between the alignment and exposing optical axes; it is this software-controlled distance that the stage travels to initiate exposure after alignment (adjusting for the specific wafer alignment modifications). It is necessary to periodically measure for any *drift* in the baseline displacement and make appropriate corrections to the axis-to-axis measure.

The other *major* disadvantage of the direct step-on-wafer technique is the catastrophic effect of mask defects and/or contamination. On a $1\times$ mask, a dust particle (or photoresist residue or other mask defect) will ruin but a single chip site, whereas on a step-and-repeat system, *every* chip site will contain an image of the defect. A mask defect that gets printed into every chip site of each wafer in the lot could destroy any possibility of yield if it goes undetected. As a result, several process procedures must be adopted:

Extensive, periodic mask-cleaning procedures
Minimal mask handling and changing
Mask pellicles

A *pellicle* is a frame plus thin, protective film (nonlight absorbing) that is attached to the patterned surface of the mask plate to effectively seal the mask surface from contamination. The pellicle puts dust distant enough from the object plane mask surface to sufficiently defocus the image of the particle at the wafer.

Optical inspection of photoresist image of mask on dummy wafer

After each mask change, a dummy photoresist-coated wafer is exposed, developed, and *exhaustively* inspected for any photoresist image defects. If any are found (related to mask contamination) that in the judgment of the technician or process engineer would be critically detrimental to the chip yield, the photoresist step is halted temporarily to clean the mask and rerun a dummy for inspection. This is a time-consuming and judgmental task and can be detrimental to throughput and turnaround time. Three aspects of the step-and-repeat exposure tend to reduce the iterative mask clean and dummy inspect process:

1. Defects are commonly present primarily around the mask perimeter (chip kerf). As a result, if a defect is detected in the kerf area, processing typically proceeds and a note is made that any kerf structures affected should not be tested.
2. The contrast between transparent and opaque areas of the mask is great, whereas the contrast between a dust particle and transparent areas is typically less; using the high exposure intensities commonly found in DSW systems, a dust particle present on a transparent area of the mask may not end up being imaged in the photoresist.
3. A 10:1 reduction of the mask pattern will also reduce the image size and therefore reduce the likelihood that the image of the contaminant will even be resolved in the photoresist.

Photoresist

Two general classifications of photoresist are used in IC manufacture: negative and positive. Negative photoresist (a mixture of resin and photosensitizer) cross-links upon absorption of ultraviolet radiation; the resulting high-molecular-weight polymer chains are insoluble in the negative photoresist developer solvent solution. Transparent areas of the mask correspond to areas on the wafer where the protective photoresist coat of negative photoresist remains. Positive photoresist is developed away in areas where it receives ultraviolet exposure; transparent areas of the mask correspond to areas on the wafer where the protective photoresist coat is removed. Positive resists consist of a mixture of resin, solvent, and develop-inhibiting photoactive compound. When exposed, this photoactive compound chemically decomposes into a carboxylic acid, which is readily soluble in a base solution (e.g., KOH). Negative resists photopolymerize upon exposure, while positive resists are photosolubilized when exposed. In both cases, the polymer resin is the component that provides the etch resistance upon exposure to acidic wet etching solutions.

In VLSI fabrication processes, positive photoresist is utilized almost exclusively. Negative photoresists suffer from the following disadvantages, which make them undesirable at finer lithography dimensions:

1. The necessary cross-linking of long polymer chains to inhibit dissolution by the developer limits the fundamental resolution of negative resists.
2. The negative photoresist developer is largely composed of a solvent to remove the low-molecular-weight, noncross-linked polymer resin; this solvent is absorbed to some degree by the exposed negative resist, which swells the

remaining resist film. (A postdevelop bake is used to assist in driving off this solvent.) The positive resist developer is a solution, which does not result in image swelling.

3. In step-and-repeat aligners, a high-intensity, short-duration exposure is necessary to maintain high effective throughput. The *reciprocity* of a photosensitive medium is the extent to which the intensity * time product necessary for exposure (for a given thickness) remains constant over a range of intensities. In the range of short-duration exposure times, positive resists are indeed reciprocal, while negative resists are not. Thus, the determination of the proper exposure time (using a light integrator, for example) is more difficult for negative resists.

4. Negative resists are also more sensitive to low levels of exposure energy than positive resists; in other words, negative resists have less contrast. The reflectivity and scattering of the incident light energy from the wafer film below (e.g., an aluminum metallization layer) tends to cross-link (and therefore maintain) resist in areas outside the original image.

The general procedure for a photolithography process step is as follows:

(1) Apply resist adhesion promoter to wafers: For most surface films that are to be coated with resist (e.g., SiO_2, Si_3N_4, polysilicon) an adhesion promoter is initially applied. A solution is applied to the wafer surface by a spin-coating technique, using a high-rpm, high-acceleration-rate wafer chuck; the wafer is held by vacuum. (Vapor coating can also be employed.) The most common adhesion promoter is hexamethyldisilazane (HMDS, an organosilane). Prior to application of the adhesion promoter, a surface cleaning step is commonly included; in particular, a bake at 170° to 180°C is often included to drive off absorbed moisture from a hydrophilic surface layer.

(2) Application of resist: Again using a spin-coat wafer chuck system, a precise amount of resist is dispensed onto the wafer surface, allowed to dwell momentarily, and then spun at high rpm to throw excess resist off the wafer into a collecting bowl-shaped shroud. The chuck spin speed, spin time, and acceleration rate are selected to produce a particular resist film thickness and uniformity. The resist is filtered during application to remove particulate contamination.

(3) Prebake: The resist coat is exposed to elevated temperature (~85° to 95°C, 30 min convection bake oven) to drive off solvents; a high residual solvent concentration would otherwise change the photosensitivity and develop rate of the resist. In addition to placing a cassette containing the wafer lot into a forced air flow oven, a number of additional heating sources can be used, often incorporated directly into the spin-coating equipment (e.g., conveyors with infrared heaters, vacuum hot plates, and microwave energy sources). The vacuum hot plate has the distinct advantage that the heat flow is radiant outward from the resist–wafer surface interface; this enhances the drive-off of solvents and reduces the residual concentration.

(4) Resist exposure: After alignment (using a nonactinic wavelength for illumination), the photoresist coat is exposed to ultraviolet through the transparent and opaque mask areas. The selection of the proper exposure time and focus is typically made from inspection of the resulting patterned image; on a step-and-repeat reduction aligner, a focus-and-exposure matrix is commonly produced initially on a nonproduct wafer and the optimum developed image selected after inspection to determine the exposure characteristics. (The wafer used for focus-and-exposure experimentation should be included with the product lot processing in order that it receive thin-film depositions of the same quality and thickness as the product wafer lot.) Often, a measure of a reference photoresist linewidth in the mask kerf is made to assist in selecting the optimum conditions.

Of particular concern in establishing the correct exposure parameters is the reflectivity of underlying layers on the wafer surface; the scattering of incident light off sidewalls over topography steps can distort the resulting developed image. In addition, standing waves can be produced in resist sidewalls due to constructive and destructive interference of incident plus reflected radiation in the resist film. The periodic variation in exposure leads to variations in the rate of attack by the resist developer and a resulting wavelike resist edge profile. These effects can be reduced by reducing the reflectivity of the underlying layer (e.g., a metal oxide film on top of a metallization layer).

(5)   Resist develop: The exposed resist film is then subjected to a developer solution to produce the latent resist pattern. The dissolution rate ratio between exposed and unexposed areas is commonly large (positive photoresist), but the develop time and temperature must nevertheless be controlled quite accurately. The quenching of the developer action by a subsequent rinse follows immediately. (The dilution of the developer 1:1 with water is typical for positive resists to slow the develop rate and gain process latitude.)

Several techniques can be used for developing the resist, the simplest being batch immersion in a developer bath with agitation. The most common technique utilizes the single wafer vacuum chuck and spin mechanism characteristics of resist coaters; as the wafer is spun, a developer spray is directed back and forth from wafer center to circumference. In this manner, fresh developer is continually being introduced to the wafer surface. The rinse follows immediately after the develop time expires while the wafer continues to spin.

A brief bake after exposure but prior to development may be added to improve resist adhesion during developing and further reduce the rate of developer attack, increasing process latitude in developer time.

(6) Resist postbake: The final resist treatment step (prior to etching of the exposed film) is a postbake at elevated temperatures (e.g., 30-min bake at 120°C in an oven). This step increases the etch resistance of the coating, enhances the resist adhesion to the underlying layer, and drives off remaining solvents.

Careful consideration must be given to the temperatures and times of postbaking so that the plastic flow of the resist film does not severely alter the development image; in some cases, a slight reflow of the resist may be used for sidewall profile control in the resulting etched film, if the resist and film are removed at comparable etch rates.

In critical dimension (dry etching) process steps, the postbake may be deleted altogether or may be followed by a brief O_2 plasma etchback procedure.

This sequence of steps describes the conventional photoresist processing technique. Alternative steps are taken to produce resist sidewall profiles with a slope suitable for metallization lift-off procedures.

Photoresist Removal

The complete removal of the photoresist coat after etching is a key process step to eliminate contamination in subsequent processing. Resist removal is complicated by exposure to a high-temperature ion implantation or plasma etch procedure.

In addition to O_2 plasma stripping mentioned earlier, a number of organic and inorganic solutions are commonly used. The strongly oxidizing nature of sulfuric acid and nitric acid mixtures at elevated temperatures makes this solution the preferred choice for difficult resist removal steps, prior to metallization. Organic strippers are used when aluminum is present. Unlike the acids, which oxidize the resist carbon polymer, these solvents break the carbon-chain bonds of the polymer.

15.12 ADVANCED PHOTOLITHOGRAPHY

In the ongoing effort to provide reduced linewidth resolution and increased process control, alternatives to deep ultraviolet lithography are currently being pursued. The relative cost, throughput, availability, and reliability of these alternatives are major considerations in the decision to evolve to these lithographic techniques. Currently, both direct-write electron beam exposure and x-ray illumination exposure are vying for the emphasis and investment required to move from research to production facilities. The fundamentals of these two techniques will be briefly discussed in this section.

Electron Beam Exposure

A focused electron beam (e-beam) may be scanned across a resist-coated wafer surface in vacuum using a deflection and blanking system; this direct-write technique produces the desired resist exposure pattern by computer control of the beam intensity. The beam is blanked from areas that are not to be exposed, while the exposure of shapes in close proximity to one another requires some degree of modulation of the beam intensity. This is due to the proximity effect of direct-write e-beam exposure; incident electrons (of 20 to 25 keV for a 1-μm resist film) undergo scattering due to collisions with resist and wafer substrate atoms. Electrons that are backscattered from the substrate may be absorbed in adjacent resist areas, leading to undesired resist image widening and overexposure of images in close proximity. It is therefore necessary to modify the calculation of the average electron dose to be received by each exposed image based on the nature of adjacent shapes.

The beam is typically a focused spot (of Gaussian intensity around the center); the diameter of the beam is commonly on the order of one-fourth to one-fifth the minimum required linewidth to provide suitable definition of the edges and corners of exposed shapes. The beam is displaced by one-half the diameter on subsequent passes of an area to provide for more uniform illumination. To increase the throughput of exposure, considerable effort is ongoing in developing variable-shaped beam systems, capable of exposing larger (rectangular) areas much like an optical pattern generator system (reference 15.3).

Two systems are used for scanning the beam across the wafer surface: raster scan and vector scan. The area of scan is curtailed considerably to avoid defocusing of the beam. After the area is completed, an xy wafer stage table is used to bring the next area into the active region.

The use of electron beams for imaging has required modifications to resist formulations to increase their sensitivity to electron absorption (specified as microcoulombs/cm^2, at a particular thickness and incident electron energy). Negative e-beam resists cross-link when exposed (as with optical resists). Positive e-beam resists undergo bond breaking upon electron absorption. For both resist types, the resulting low-molecular-weight polymer chains are dissolved by the developer. A current technique used to achieve fine lithographic resolution using e-beam exposure incorporates multiple resist coatings or a resist plus etch-masking material composite layer. In the case of multiple resist layers, both layers can be simultaneously exposed to the mask pattern. The resist materials can be selected such that they utilize disjoint solvents (i.e., the developer for one of the resist layers does not attack the other). In this manner, the top layer may be developed, without attacking the bottom layer. The developer is then changed and the bottom layer removed, utilizing the pattern in the top layer as a mask. The surface planarity of the thick bottom layer improves the uniformity of coating and exposure of the thin top layer. Rather than using two dissimilar resists with unique developers, it is also possible to use two layers of a common resist material in a single develop step; a substantial difference in the develop rates of the layers can be produced from a variety of pretreatments. If the top layer develops much more slowly than the bottom layer, the resist removal will proceed slowly until the interface is reached; subsequently, the bottom layer will be quickly dissolved, without excessive etch bias in the top layer pattern. The resulting resist profile for such a procedure is often of a pedestal nature, where the bottom resist layer has been undercut; such resist profiles are required for lift-off metallization.

It is also possible that the two layers may be exposed sequentially; the top layer would be e-beam exposed and developed as before. The bottom layer may then be exposed to deep ultraviolet radiation to which it is sensitive and the top layer is opaque. This second exposure may be a flood exposure and thus not a serious detractor to photolithography throughput.

An alternative technique for multilayered lithography patterning uses three layers: a thick layer of organic polymer on the bottom, a thin photoresist layer on top, and an intermediate layer that serves as the masking layer for the patterning of the polymer. The resist is exposed and developed as before and the image pattern transferred to the intermediate layer in a dry etching step. For example, the intermediate layer may be a plasma-enhanced chemically vapor deposited SiO_2 layer, which would then be etched after resist development using a CF_4 plasma. Having transferred the image to this SiO_2 layer, an O_2 reactive ion etch step is used to define the pattern in the thick polymer coat; this anisotropic etch also removes the top-layer photoresist but not the underlying oxide. A thin oxide layer is sufficient protection for the RIE of the polymer. The anisotropic nature of the RIE etch can result in a high aspect ratio resist pattern (height to width), necessary for finer dimension lithography. Of particular concern in the development of these techniques utilizing thin resist plus intermediate masking layers is the susceptibility to pinhole defects and the resulting fabrication yield impacts.

The utilization of e-beam lithography (or a hybrid e-beam plus optical lithography approach) requires careful attention to the alignment technique. The chip kerf area contains an isolated alignment mark used for site-by-site alignment. The alignment mark is initially overscanned by the electron beam; the coordinates of the alignment mark are determined from a measure of the backscattered incident electrons or secondary electrons emitted from the scanned surface. The collected signal during overscan is a function of the e-beam energy, alignment mark geometry, photoresist thickness and topography, and the angle of collection spanned by the detector. The selection of the alignment mark target material also affects the collected signal due to the differences in electron reflectivity.

The throughput limitations combined with the cost of direct-write e-beam systems have impeded their acceptance in production environments. However, electron beam exposure is rapidly gaining acceptance in mask-making approaches to VLSI technologies. Wafer topographies require thick resist coatings (e.g., 1.0 to 2.0 μm) for adequate topography step coverage; the flat mask surface and relatively mild chrome etch process permit thinner resist coatings (and thus lower-energy electrons) to be used for exposure. As a result, proximity effects are reduced. In addition, adequate edge delineation is not as difficult to achieve with a $5\times$ or $10\times$ projection reticle as is required with a $1\times$ wafer exposure.

X-Ray Lithography

The major problems posed by x-ray exposure systems for enhanced lithography over ultraviolet or deep ultraviolet exposure techniques are fabrication of high-quality masks, development of sensitive resists (able to withstand RIE process steps), and the design and construction of uniform, stable x-ray sources (over a large wafer field). X-ray wavelengths commonly range from 8.3 Å (aluminum line) to 4.4 Å (palladium line) for metal targets excited by an incident high energy electron beam; an alternative x-ray source is the radiation from the radial acceleration of electrons moving in a synchrotron storage ring. In the former case, the diameter of the incident electron beam on the metal target effectively creates a limited area radiation source; a ray diagram from this source through the mask to the wafer indicates that there is a lateral displacement of images (which increases proportionately with distance off axis) and some degree of defocusing around the edges of mask shapes. To simplify the task of compensating for these effects, the mask and wafer are placed in very close proximity (e.g., 40-μm mask–wafer separation for a 3-mm-diameter x-ray spot and 50-cm x-ray source–wafer separation over a 3-in field). At these dimensions, exposed shapes are 80 ppm greater than the mask image; their lateral displacement increases with off-axis distance. Special wafer chuck and mask holder design considerations are used to make this mask–wafer separation as uniform as possible. The synchrotron radiation source can provide a high-intensity collimated beam over a range of wavelengths (10 to 50 Å). The synchrotron radiation is emitted in the plane of orbit of the accelerated electrons; the intensity falls off very sharply with the displacement from this plane. The geometrical shadowing distortion present with the metal target source is thus eliminated. The option exists to provide scanning of the mask plus wafer assembly or to scan the beam (utilizing a wobbling mirror). Current research efforts are investigating the potential for x-ray focusing systems that may lead to step-and-repeat reduction printing; complementary efforts are pursuing the development of lower cost storage ring x-ray sources.

The x-ray source using a metal target is a high-vacuum region, consisting of an electron gun, electrostatic focusing shields, and the water-cooled target. The x-rays produced from the incident beam leave the chamber and enter the aligner column through a beryllium, low atomic number, low x-ray absorptance window. The x-ray resist is exposed in much the same manner as an electron-beam resist layer; absorbed x-rays result in the release of a photoelectron from the resist material. The cross-linking (negative resist) or polymer-scission (positive resist) of the layer is the same result as achieved by e-beam exposure. The absorption efficiency of e-beam resist films is commonly quite low (a few percent for conventional resist thicknesses), leading to long exposure times. The resist sensitivity can be increased by the incorporation of elements whose absorption coefficients are high at the specific x-ray wavelength. Chlorinated resists are used with Pd (4.4 Å) sources. The x-ray wavelength is selected so as to minimize the sources of imaging error; the blurring of the mask image by resist photoelectrons is reduced at shorter wavelengths, while diffraction of mask features increases with shorter wavelengths.

The fabrication of an x-ray lithography mask presents some very difficult challenges:

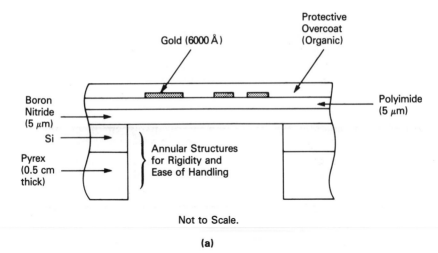

(a)

I. After e-beam resist exposure/develop, the exposed top surface Ta is removed in a CF_4 plasma.

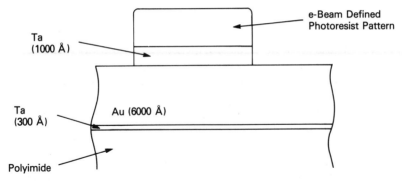

II. The top surface Ta protects the Au layer from removal during RF sputter etching.

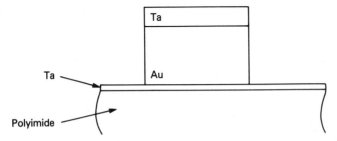

III. The bottom Ta layer is patterned in the alignment mark areas (not shown); these window areas are etched to enhance optical transparency.

Figure 15.67.   Mask manufacture for an x-ray exposure system.

1. The mask substrate must be rigid enough to support the patterned opaque layer, yet thin enough to provide minimum attenuation of the x-ray radiation to be transmitted.

2. The opaque layer must have a very high x-ray absorption coefficient; a patterned gold film of approximately 0.7-μm thickness is commonly used.

3. In addition to dimensional stability, the defect density of the thin support and absorbing films must be small.

An example of a composite x-ray mask is given in Figure 15.67a; the supporting mask substrate is a layer of CVD boron nitride plus polyimide. These films are coated onto a Si wafer; the exposed backside of the wafer is etched away to the boron nitride layer. A set of sputtered thin-film metals is then deposited. The pattern in the gold masking layer is produced by e-beam resist exposure and develop, followed by RIE of the exposed tantalum (Ta) layer and sputter etching of the exposed gold (Au). The patterning is completed by selective etching of windows in the bottom Ta layer, in order to enhance the optical transparency of the mask in alignment mark areas (Figure 15.67b).

The selection of the mask substrate materials and thicknesses also involves a consideration of the alignment strategy, which is performed using conventional optical techniques (and thus requires a transparent layer). After alignment, the mask plus wafer stage is transported under the column and the shutter opened for the required time of x-ray exposure. Since the mask and wafer are separated, a unique objective lens design is required to allow high resolution viewing of both mask and wafer. Also, the temperature rise in the absorbing Au layer during exposure to the high x-ray energy density requires exceptional adhesion strength and uniform tensile stress in the substrate support layers.

The potential for x-ray lithography also involves a consideration of the pos-

	OPTICAL STEP & REPEAT	FULL WAFER X-RAY	DIRECT-WRITE E-BEAM
RESOLUTION (R) (UM)	1.0	0.25	0.1
THROUGHPUT (N) (WAFERS/HR.)	40	75	10
SYSTEM COST (C) (K$)	700	225	3000
CLEAN ROOM AREA (A) (SQ. FT.)	50	12	65
FIGURE OF MERIT: $\dfrac{N * 10^3}{(C + 10*A)*R^2}$	33	3480	274

Figure 15.68. Comparison of different photolithography alternatives (circa 1981). [From A. Zacharias, *University/Government/Industry Microelectronics Symposium* (1981), II-15.]

sible process modifications and reliability impacts introduced with regard to the effects on device behavior. Additional annealing may be required to eliminate drift in device threshold due to capture of energetic electrons in the gate oxide.

A figure of merit comparison between different photolithography alternatives (circa 1981) is given in Figure 15.68.

15.13 ANNEALING

VLSI processes are motivated to reduce the number of high-temperature steps, to minimize the outdiffusion of impurity junction nodes. In previous sections, the need for high-temperature treatments was mentioned to repair ion implantation crystalline damage, to densify deposited films, and to reflow impurity-rich oxide layers. The potential for rapid thermal annealing was indicated in order to provide the requisite energy of formation without a large time–diffusion coefficient product.

An additional requirement for an annealing process step is to form a resistive ohmic contact between metal and device and substrate nodes. The ohmic contact is a result of the incorporation of a high-impurity concentration at the semiconductor side of the junction and the formation of a metal–silicon alloy at the interface. The anneal furnace temperature must be well below the temperature at which a molten solution would be formed under equilibrium conditions; the eutectic temperature for the binary silicon–aluminum system is 577°C. The solubility of aluminum in silicon is high; it is therefore critical to control the depth of alloying to avoid exceeding the desired contact region. Aluminum is also the preferred choice for metalization due to its ability to penetrate the surface oxide layer that inevitably forms in the contact windows.

Contact annealing steps are typically performed at 400° to 450°C for 30 min with appropriate preheat and cool-down cycles. During this step, the furnace carrier gas typically consists of a hydrogen–nitrogen mixture (known as *forming gas*). The hydrogen is introduced to stabilize device characteristics; it has been experimentally determined that the exposure of the substrates to hydrogen is effective in reducing the density of unoccupied electron energy states (traps) at the $Si–SiO_2$ interface through chemical reaction (reference 15.4). Postoxidation annealing steps, in addition to postmetallization annealing, are often included with the goal of reducing the interface trap density.

15.14 PROCESS ALTERNATIVES

The chemical vapor deposition of polycrystalline silicon films was discussed in Section 15.7; this section will briefly introduce the VLSI processing alternatives that utilize an *epitaxial* silicon layer grown onto the starting crystalline substrate material. Bipolar device IC technologies have long been dependent on the growth of an epitaxial layer *during* device fabrication; the increasing attractiveness of CMOS technology for VLSI applications has led to the specification that an epitaxial layer be present on the substrates prior to beginning fabrication.

The impurity concentration in the epi layer is relatively independent of the background concentration of the wafer substrate. The epi layer can therefore be a high-resistivity layer with a low-resistivity bulk region below. The high-resistivity epi is attractive for reducing junction capacitance and threshold voltage body effect coefficients, not to mention the requirements of impurity compensation to form the CMOS technology well regions. The low-resistivity substrate is selected to reduce the localized substrate resistance and substantially reduce the current gain of one of the parasitic bipolar transistor types. The likelihood of CMOS transient latch-up is thus reduced.

During the well drive-in diffusion step, the impurities in the low-resistivity substrate outdiffuse into the epitaxial layer. This *autodoping* reduces the effective epi layer thickness; the initial deposited thickness should be selected appropriately.

The deposition of an epitaxial layer can be performed in a cold-wall system with the wafers lying horizontally on an inductively heated susceptor, in a similar manner to atmospheric CVD. Unlike the chemical vapor deposition of polysilicon, however, it is necessary that the silicon atoms deposited on the wafer surface be sufficiently mobile so as to allow for incorporation into a crystalline lattice site. Thus, the wafer temperature must be higher and the deposition rate slower; too large a deposition rate results in a polycrystalline layer.

Epitaxial silicon can be grown using any of the following reactions, although each has its own distinct advantages and disadvantages (reference 15.5):

$$SiCl_4 + 2H_2 \rightleftharpoons Si \text{ (s)} + 2HCl \text{ (g)}, \quad 1150°\text{--}1250°C, \quad 0.4\text{--}1.5 \ \mu m/min$$

$$SiHCl_3 + H_2 \rightleftharpoons Si \text{ (s)} + 3HCl \text{ (g)}, \quad 1100°\text{--}1200°C, \quad 0.4\text{--}2.0 \ \mu m/min$$

$$SiH_2Cl_2 \rightleftharpoons Si \text{ (s)} + 2HCl \text{ (g)}, \quad 1050°\text{--}1150°C, \quad 0.4\text{--}3.0 \ \mu m/min$$

$$SiH_4 \rightarrow Si \text{ (s)} + 2H_2 \text{ (g)}, \quad 950°\text{--}1050°C, \quad 0.2\text{--}0.3 \ \mu m/min$$

The doping of the deposited epitaxial layer is accomplished by the addition of phosphine (PH_3), diborane (B_2H_6), or arsine (AsH_3) to the gas flow; the incorporation of dopant is roughly proportional to the partial pressure of the dopant gas constituent.

The presence of HCl as a reaction product (in all but the silane reaction) serves to provide some measure of surface etching during deposition; the reversibility of the reactions can be controlled by varying the excess hydrogen carrier gas. Some surface etching is desirable during the deposition to reduce the defect density of spikes in the resulting film. *Spikes* are large precipitates of deposited material due to contaminants or foreign silicon material falling onto the wafer surface. Crystalline vacancies are also present in the deposited film, which can lead to locally enhanced diffusion coefficients for subsequent impurity introduction steps (also known as diffusion *pipes*).

Unlike bipolar technology fabrication, the utilization of epitaxial layer material for CMOS processing does not involve mask alignment to any previous lithography in the starting material. Thus, the potential wash-out or shift of an original pattern is not a concern. In addition, since the impurity type is the same for the epi and substrate materials, the redistribution of impurities by autodoping is again not as critical to control as with bipolar device geometries. The added material cost for specifying an initial epi layer is a consideration; in particular, the epi layer defect densities may have a considerable impact on device yields, as described previously.

A potential future option for device fabrication is the use of a crystalline layer on an *insulating* substrate. This choice can effectively result in the elimination of the area component of device node junction capacitance. Alternatives for a silicon-on-insulator (SOI) technology include the following:

1. The deposition of a silicon epitaxial layer onto a crystalline material whose lattice parameters are sufficiently similar to that of the desired surface orientation [e.g., (100) silicon on ($1\bar{1}02$) sapphire (Al_2O_3)]. Potential problems with this technology option are aluminum autodoping, thermal expansion coefficient mismatch, and film defect density (a strong function of the distance from the silicon–insulator interface).
2. The deposition of polycrystalline silicon onto a noncrystalline (amorphous) substrate, with subsequent recrystallization of the deposited film.

The energy for recrystallization can potentially be provided by a laser or a strip heater, where a molten zone is slowly scanned across the wafer surface; upon cooling, the film recrystallizes. The noncrystalline substrate may be quartz or potentially a silicon wafer onto which an insulating oxide or nitride film has been deposited. For the case of the wafer with insulating film, a window in the dielectric would be opened prior to polysilicon deposition; the recrystallization scanning would then start at this point, allowing the underlying substrate crystalline orientation to serve as a seed for the crystalline layer to be formed. SOI technologies are not actively being used in production of VLSI hardware, as the yield impacts of their defect densities (and other cost-related factors) are presently prohibitive.

15.15 TEST STRUCTURES FOR PROCESS AND YIELD EVALUATION

Throughout the fabrication of a single product wafer lot, it is necessary to obtain process characteristics of individual, critical steps (e.g., gate oxide thickness and dielectric strength, ion implant dosage, etch biases of linewidth and contact images). The importance of these individual measurements is magnified not only by the potential yield impacts on the run itself but also by the identification of immediate equipment and process modifications required to enhance the yields of subsequent runs. Since subsequent runs will likely reach critical process steps before a product run is completed for testing, it is necessary to provide very rapid turnaround on the individual step characterization data.

The useful process characterization data to be gathered from a run generally fall into two categories: those that require separate wafer monitors and those that can be measured in the chip kerf and test site areas on the product wafers by probing after metallization.

Process monitor wafer blanks are added to the product wafer lot at specific points in the process and then subsequently withdrawn, after having received the desired sequence of steps. The number of monitor wafers present at any one time should be kept to a minimum, as equipment and wafer carrier considerations will provide a bound on the total number of wafers in the run; the monitors are thus displacing product wafers. However, a single monitor for a particular step may not suffice. For example, if it is important to measure the variability in deposited or grown film thickness in a batch process over the product lot, several monitors may be required; in addition, the *positioning* of these monitors within the furnace tube or vacuum chamber is significant to gather these data. Three gate oxide monitor wafers might be added, located at the front, middle, and back of the furnace tube.

After the monitor has received the desired subset of processing steps, it is extracted from the lot and a new one potentially takes its place. The monitor may require additional unique processing to facilitate probing of the wafer, commonly metallization, and patterning. Film thickness/dielectric strength/defect density monitors of thermally grown oxide films commonly use a metal dot pattern on the oxide surface; capacitance versus voltage (C–V) measurements of these films are made to evaluate the dielectric. The data gathered from C–V measurements typically include oxide leakage currents, mobile contamination densities (under temperature and voltage stress), and a pulsed C–V relaxation time measurement to investigate generation and recombination rates for a surface region in deep depletion. The dots are of various sizes in order to gather sufficient statistics to calculate an areal defect density. Another technique for evaluation of the permittivity and index of refraction of dielectric layers is ellipsometry. The sheet resistance of ion implanted layers can be measured by four-point probe techniques (see Problem 15.1). Junction node depths can be measured by angle-lap and stain techniques.

In addition to the use of distinct monitor wafers, a large amount of electrical characterization and defect density data can be gathered from the postmetallization probing of test structures present in the kerf area or in separate test sites merged with the step-and-repeat pattern of chip sites. These data can include the following:

Electrical: Device characteristics: process width and length biases, threshold
voltage, carrier mobility, subthreshold behavior
Contact resistance (usually using a long, series chain of contacts)
Sheet resistance (using a van der Pauw cross structure)
Memory cell and array circuit behavior
Thick field oxide threshold voltage (polysilicon and aluminum)

Defect: Line continuity over severe topography
Parallel-plate capacitor structures

The layouts that provide the structures for these measurements *need not* satisfy
the design rules of the technnology used for chip design; indeed, the process
engineer gains a considerable amount of insight into process latitudes if many of
the test sites include *more* aggressive device, contact, and linewidth dimensions
than would be present on the design.

It is not necessary (nor particularly desirable) to extract many wafers from
the lot for postmetal probing; the goal is merely to gather sufficient data from the
run to confirm that there is no gross failure present in device fabrication and to
be able to later ascertain the cause of perturbations from the anticipated yield.
The collection of a statistically significant sample of data on a few wafers requires
the use of automated probing equipment with related instrumentation programming. (The duration for which the run is held prior to beginning back-end-of-line
processing should be kept to a minimum.) The characterization structures should
therefore be designed to utilize a standard probe pattern, if at all possible.

One set of process data that does not lend itself well to automation, but
which is nevertheless critical to record, is the in-line optical measurement of
linewidths and alignment tolerances. After developing an exposed resist pattern,
a measure of the photoresist linewidth (and/or space) of a representative target
should be made; an immediate decision to strip and rework the photoresist layer
may be required if the linewidth measure is sufficiently off-spec to require action.
The target linewidth dimension in the patterned film after etch and resist removal
should likewise be recorded so that the etch bias of the process step can easily
be calculated. As VLSI lithography dimensions continue to be reduced, this etch
bias measure becomes especially important in determining the adequacy of dry
etch parameters. Product lot wafers should be used to obtain these measures, so
that the underlying film thicknesses and reflectivity (which help to produce the
final resist image) are representative. In addition to linewidth measurement targets, a number of alignment verniers should be included in the kerf area to enable
an optical measurement of level-to-level alignment. The lithography levels required for the verniers are based on the alignment strategy for the particular
technology. Alignment tolerance measurements should be gathered at multiple
sites across the surface of the product wafers selected to determine the extent of
effects such as process-induced wafer distortions.

There will likely be additional defect structures that are related to the back-end-of-line metallization levels, which are to be measured prior to sending the product lot for chip testing.

Prior to the application of test patterns to the chip design, a set of parametric tests is applied at appropriate chip I/O pads; failure to pass this group of tests usually indicates that the particular site need not be further evaluated. These tests commonly include the following:

Forward-bias junction characteristics at all signal pins

Substrate leakage current

Input pin gate leakage

Output high-impedance leakage current

Total chip supply current

15.16 IMPACTS OF TOPOGRAPHY ON VLSI PROCESSING AND DESIGN RULES

Improved linewidth and etch bias control is only one aspect of VLSI fabrication technology development; another major facet to implementing a new technology in production is the requirement that each level of chip interconnect be able to traverse the surface topography of the dielectric and interconnect layers below. The process characteristics and layout design rules must both be developed with consideration for the topography to be covered and the changes in topography produced. Some of the process and design parameters affected are as follows:

Process: Interconnect line thickness and sidewall angle
Dielectric thickness and contact or via sidewall angle
Design: Minimum extent of contact and via overlap by interconnections
Terraced via design rules for use with multiple dielectrics
Terracing design rules for coincident edges of underlying multilevel interconnections
Stacked via layout rules (if such a via is indeed permitted)

The design rules referred to previously are best explained by illustration. Figure 15.69 provides a layout and resulting cross section illustrating the requirement for interconnect extensions past a via shape; the steep via sidewall terminating on the first metal line edge introduces a difficult topography to cover with a second metal line. This can lead to microcracking or a void area where contamination can reside, potentially leading to accelerated corrosion. Some processing technologies provide a two-layer dielectric film between interconnection planes; if the resulting step height of a single via sidewall through both dielectrics is too great for reliable coverage, a set of via terracing design rules is required

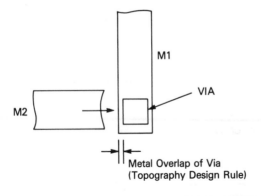

Metal Overlap of Via
(Topography Design Rule)

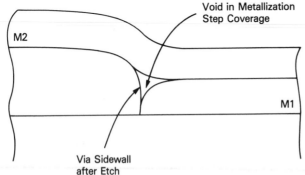

Figure 15.69. Surface topography after via etch, with small M1–VIA design overlap.

(Figure 15.70a). The via-to-via spacing may suffer as a result, although some relief may be gained by merging the outer via shapes (Figure 15.70b); for reliability and yield considerations, the nonvia area with a single dielectric layer between interconnection planes should be minimized. There may also be terracing design rules related to the coincidence of edges of multiple underlying interconnections, which would otherwise produce unacceptable steps in surface topography. Figure 15.71 depicts that coincident metal and polysilicon lines may introduce step coverage difficulties for a second metal line attempting to reach an adjacent via; a design rule may be required to ensure a sufficient terrace in the topography to facilitate step coverage. Similarly, the fabrication technology may not be able to accommodate the coincidence or stacking of a diffusion or polysilicon contact plus first metal via (Figure 15.72). If indeed stacked vias are feasible, this structure will likely have its own unique set of design rules. In all these cases, the process engineer must work closely with the design engineer to determine the proper trade-offs between available process latitudes and required wiring density.

VLSI process engineering has continued to work toward providing less severe topographies for each interconnection layer. The initial processing features that were adopted to help achieve this goal included recessed field oxides and viscous, spin-coated dielectrics. More recent advances include bias-sputtered planar oxides, reflowable silicate glasses, and photoresist profiling prior to dry

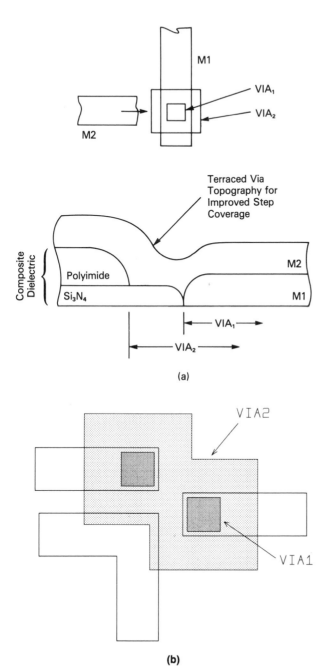

(a)

(b)

Figure 15.70.   (a) Terracing of via topography through multiple dielectric layers. (b) Unioning of outer via shapes to facilitate closer via-to-via spacing.

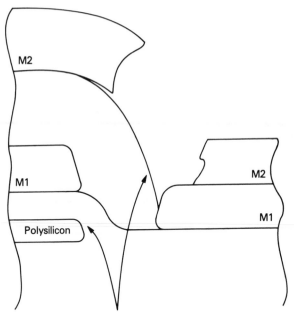

Step Height of Via Sidewall is
aggravated by coincident edges of
adjacent metal and polysilicon lines.

Figure 15.71. Cross section illustrating
step coverage problems arising from
the coincident edges of multiple
underlying interconnection lines.

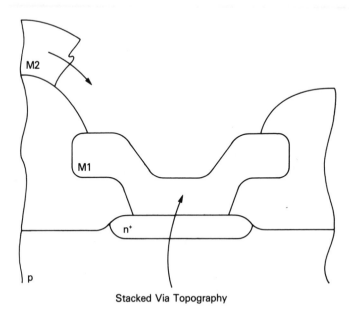

Stacked Via Topography

Figure 15.72. Cross section illustrating the surface topography of a stacked contact plus
first metal via.

etch transfer of the patterned image. Advances currently being pursued include fully recessed (trench) field oxides and studded vias. The former involves etching a trench in the substrate in the field areas prior to LPCVD of oxide, while the latter involves the addition of a metal stud to fill a via hole *prior* to the deposition of the next metallization layer. Although the simplest approach to less severe topographies is just to reduce the interconnect line and dielectric film thickness, process and design engineers must be cognizant of the resulting negative impacts on circuit performance (interconnection resistance and dielectric capacitance), dielectric pinhole density, and interconnect current density and electromigration failure.

15.17 CONTAMINATION AND RELIABILITY CONTROL

Section 15.16 described some of the process restrictions imposed to avoid the contamination and reliability problems related to step coverage failure. This section will briefly discuss some of the additional process and design measures adopted to address contamination and reliability concerns; these measures will be placed under the headings of gettering and corrosion protection.

Gettering

Gettering refers to the technique of trapping fast-diffusing (heavy metal) impurities from reaching device nodes during processing. The presence of these impurities (substitutionally or interstitially) in the substrate provides an enhanced recombination probability for free carriers (due to the introduction of electron energy states between the silicon conductance and valence band edges). The presence of these impurities in junction depletion regions therefore increases reverse-biased leakage currents, a critical parameter for dynamic memory nodes.

Gettering of metal impurities resident in the substrate is accomplished by the introduction of backside wafer damage as an early process step, commonly provided by the implant of an inert gas such as argon. The resulting high concentrations of dislocations and polycrystalline grain boundaries after subsequent high temperature steps serve as gettering sites for mobile metal impurities.

VLSI processes commonly incorporate a phosphorus-rich deposited oxide layer prior to metallization. Phosphorus serves as a getter or sink of mobile heavy metal impurities; metals such as copper or gold form a strong ionic bond with the donor impurity phosphorus as a result of electron transfer.

Corrosion Protection

The permeability of water vapor through the integrated circuit package and upper dielectric layers leads to aluminum corrosion failure. The electrochemical reaction leading to corrosive etching of the aluminum is a result of exposure to ions in the presence of water vapor; these ions may be carried from the package, present on the chip as a residual from incomplete cleaning, or present as part of the fabrication process itself. (A high phosphorus concentration silicate glass can result in aluminum corrosion.) The predominant mechanism is exposure of the aluminum to residual or transported chlorine. (Recall the discussion on the dry etching of aluminum and the precautions required to avoid exposure of the system to the atmosphere in the presence of residual chlorine.) From an IC design and fabrication standpoint, the appropriate measures to follow are to select a low-permeability overcoat chip dielectric and to ensure a suitable edge seal is present around the chip perimeter. Considerable evaluation is nevertheless required of the packaging material.

The top layer dielectric is selected to provide some degree of scratch protection, as well as the aforementioned low permeability to water vapor (at accelerated temperature and humidity). The chip perimeter design includes layout rules pertaining to the minimum extent of the dielectric overcoat between design shapes and the kerf sawing channel.

15.18 PACKAGING ALTERNATIVES

This section only briefly describes some of the packaging options available to the VLSI chip designer; it does not describe the reliability, cost, nor potential for automated chip and card assembly that a more thorough discussion would include. The demand for increased chip I/O has continued to drive packaging engineers toward developing higher pin-count packages; the simultaneous switching requirements of wider bus architectures have likewise made package lead inductance a primary concern. As discussed later, these trends show little signs of abating.

Chip designs are commonly categorized as being either *circuit limited* or *pin limited*. The former term refers to the case where the number of logic circuits present on the chip is sufficiently large such that the die size required readily accommodates the number of pads for chip I/O signals in the design. A pin-limited design implies that the number of chip I/O dictates the particular die size (with pads at minimum spacing); the designer is inclined to find additional function to incorporate on-chip (without requiring additional pads) in order to best utilize the available internal circuit area.

Chip Terminal Design

There are currently two prevalent alternatives for the interconnection of chip pads to the corresponding package pins: wire bonding and raised pad bonding. These techniques differ considerably in package and interconnection material options, reliability, ease of automation, and, ultimately, cost.

Wire-bonded die are mounted face up and utilize gold (or possibly aluminum) wires between chip pad and associated interconnection point on the package lead frame. The attachment of the wires, typically 0.001 to 0.004 in. in diameter, is realized by either thermocompression or ultrasonic bonding techniques.

Thermocompression bonding refers to the method of applying heat and pressure simultaneously to a wire positioned on a terminal pad. The temperature used is well below the melting temperature of the metals (e.g., 300° to 400°C for gold wires on aluminum pads), yet plastic flow and molecular bonding nevertheless result. This temperature can be reached by heating the bonding head, the chip substrate, or both. The bonding head pressure precludes the use of a polymer interlevel dielectric between second metal pad and underlying films; a sputtered oxide dielectric (or large area first metal via) is required. To minimize pad surface irregularities for wire bonding, no device structures (or other topography) should reside under the pad.

Thermocompression bonding has several variants based on the configuration of the bonding head (i.e., ball, wedge, and stitch bonding; Figure 15.73). Ball and

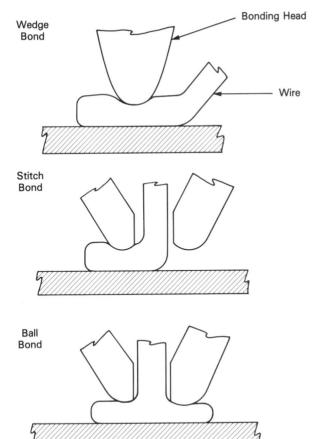

Figure 15.73. Alternative bonding head configurations for thermocompression bonding.

stitch bonding utilize wire that is fed through a capillary section of the bonding head. The name ball bonding refers to the geometry found at the end of the wire prior to contact; the subsequent compression onto the chip pad results in a nail-head cross section. The bond head is then moved to the package pin, where a stitch bond is made. The wire is then cut using a localized source of heat (e.g., a hydrogen flame), and the surface tension of the melt forms the ball for the next chip pad bond. As wedge and stitch bonds reduce the cross-sectional area of the wire in the area of contact, they are usually of weaker separation strength than ball bonds.

Gold is the only metal useful for ball bonding, as a ball-type cross section cannot be formed in this manner with aluminum; either gold or aluminum can be used for wedged or fully stitched bonds. Gold wire bonding to aluminum pads can result in the formation of intermetallic compounds upon extended exposure to elevated temperatures (150° to 200°C). These compounds are mechanically brittle, have poor conductivity, and will accelerate the rate of contact failure; their purplish-black color has led to the use of the term "purple plague." In the case of exposure to elevated temperatures, aluminum wires should be used. To avoid oxidation of the aluminum chip pads during thermocompression bonding, this process is executed in a nitrogen ambient.

Ultrasonic bonding is equivalent to the fusion (or seizure) of nonlubricated surfaces due to friction. The bonding tip vibrates mechanically in a direction parallel to the chip surface; the energy source to the tip comes from the deformation of a magnetostrictive transducer vibrating at 60 kHz. The energy dissipated in the shearing friction produces a localized heating to form the bond; the overall process can be carried out at room temperature. Ultrasonic bonding is attractive for aluminum-to-aluminum interconnections, as the aluminum oxide surface layer on the pad is readily broken down.

The IC die may be attached to the package carrier by one of a variety of techniques, depending on whether a ceramic or plastic package is used and whether the chip substrate voltage is to be applied via a backside connection. For ceramic substrates, the chip is epoxied to the package base with a thermally conductive resin; the die is commonly located in a cavity to minimize the length of the subsequent bonding wires. A metal pad is present on the base to reduce the thermal resistance, effectively increasing the radiant area. A lead frame is attached to the base, which has been patterned in such a manner so as to best route bonding post positions for chip pads to package pins. A ceramic cover is commonly attached by the heating and solidification of a layer of glass (a glass frit) between lid and base.

Plastic packages are attractive from cost considerations, yet provide significantly less resistance to moisture penetration at elevated temperatures than do ceramic packages; the chlorine present in polyvinyl chloride compounds internally increases metal corrosion. The lowest-cost plastic packaging is available with an injection-molded process, where the chip, bonding wires, and lead frame are completely encapsulated by an epoxy resin introduced into a mold at low pressure; economies of scale result from the use of a mold assembly containing multiple lead frames. Figure 15.74 illustrates a cutaway view of a plastic-encapsulated dual-in-line package (DIP). The lead frame consists of a flag, a tie bar, the package leads, and a supporting structure; after encapsulation, the supports are severed and the pins bent accordingly.

New packaging options are rapidly becoming available for VLSI parts, to take advantage of the printed circuit board density achievable if pin-through-plated via hole connections are not required. These surface-mount packages include leaded and leadless chip carriers. Pin-grid array packages are also used with wire-bonded interconnections, utilizing an array of package pins (with 100-mil spacing) outside the chip cavity that are connected to a pattern of wire traces on a package substrate. Despite packaging advances in pin count and printed circuit board mounting technology, wire-bonded package techniques have the distinct disadvantage that the bonding process is sequential in nature. The labor costs associated with manual bonding have made domestic assembly prohibitive; automated wire bonding equipment is helping to increase manufacturing throughput.

Rather than using discrete wire bond interconnections, an alternative technique is to utilize a direct connection between raised chip pads and a mating pattern of package interconnections; all pad bonds could then be formed simul-

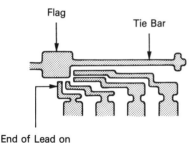

Flag

Tie Bar

End of Lead on
Frame for Bonding

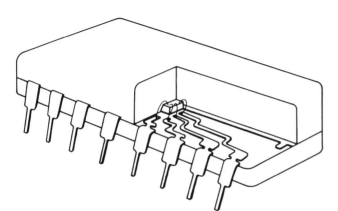

Figure 15.74. Lead frame for chip attach and cutaway view of a plastic encapsulated dual-in-line (DIP) integrated-circuit package.

taneously. In one possible implementation, the interconnections between chip pads and package traces could be provided using intermediate wire leads attached to a supporting polyimide film substrate; a 35-mm tape reel can be used for automation efficiencies. The film area where the die is to be attached has been removed, and the ends of the interconnection leads on the tape are free-standing. Die bonding can be accomplished either by thermocompression or the soldering (formation of a eutectic) between the plated leads and the raised chip pads. The bonding operation requires positioning the sawed wafer die site under the bonding head, transporting and aligning the next 35-mm frame of interconnections on the tape reel, and bonding. The attached die can then be transported with the film to test and package attach stations. Having attached the inner end of the leads, it is now necessary to separate the chip and interconnection leads from the tape carrier, bend the leads appropriately, position the die onto the package substrate, bond all the outer ends of the leads to the mating traces on the substrate, and attach the die to the substrate by elevated temperature curing of an epoxy (or formation of a eutectic). (Rather than using a package base, the die and leads can also be attached to a frame of the type illustrated in Figure 15.74.) This tape-automated bonding (TAB) procedure for chip packaging provides the potential for high throughput; in addition, the accessibility to chip connections after inner lead bonding facilitates immediate testing of the correctness of the bonding process.

The flip-chip package utilizes raised chip pads, as did the TAB packaging technique; however, with this approach, no interconnection leads from chip to package are added during die attach. Rather, the interconnections between chip pads and the array of package pins are previously patterned directly onto a ceramic substrate; the die is oriented face down and all raised chip pads are soldered to the traces simultaneously (Figure 15.75). The raised pad in the figure consists of a ball of Pb/Sn solder, with additional underlying metal films included to provide adhesion to the aluminum terminal pad. Since a soldering technique is used, it is necessary to restrict the flow of the molten solder down the interconnection

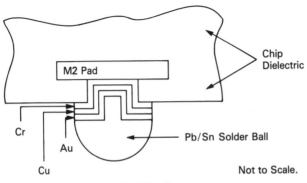

Not to Scale.

Figure 15.75. Cross section of a solder ball used for flip-chip packaging.

trace to avoid the collapse of the chip onto the substrate. A controlled collapse is provided by the addition of a dam near the end of the interconnection or by localizing the interconnection area that has been previously tinned (Figure 15.76). This packaging technology has several unique characteristics:

1. Pb/Sn solder can tolerate considerable strain during thermal cycling, introduced by the mismatch in thermal expansion coefficients between silicon and the ceramic. The amount of strain is a function of the difference in thermal expansion coefficients of the die and ceramic package ($\Delta\alpha$), the change in operating temperature (ΔT), and a geometrical factor (D_N/H):

$$\epsilon = (\Delta\alpha)(\Delta T)\frac{D_N}{H}$$

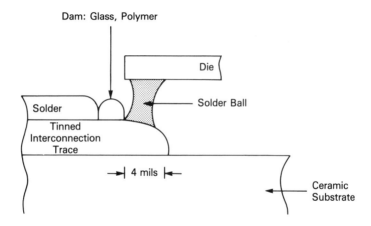

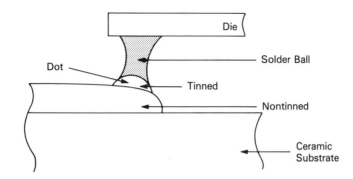

Figure 15.76. Techniques for controlling the collapse of the flip-chip solder ball upon reflow and bonding.

In this expression, D_N is the distance from a particular pad to the chip's neutral point and H is the height of the connection. The limits of the solder pad's compliance dictate the maximum distance to any pad from the chip's center.

2. The surface tension characteristics of the solder pad provide several distinct advantages:

   a. After evaporation of the solder onto each pad, a reflow heating operation is performed; the surface tension draws the solder into a sphere or ball. The reflow operation can be performed repeatedly, each time with the same result. This is particularly important in healing the damage caused by the application of probes during chip testing. (Probing on second metal pads without the solder ball terminals produces irreversible gouging and must be limited to a maximum of two or three passes).

   b. The solder pads must support the weight of the die during the chip-joining process and prevent the collapse of the die onto the ceramic surface; since the solder does not "wet" the dielectric around the periphery of each pad, relatively large solder balls can be used.

   c. Due to surface tension, the chip-joining process is to some degree self-aligning; after coarse chip-to-package trace alignment, the chip will float into the correct registration during reflow.

3. Device-related topography under the second metal pad presents no problem to the chip-joining process.

4. Soldered joints are mechanically stronger than wire-bonded pads.

5. The die may be removed and replaced by reheating the chip or package to the solder melt temperature.

6. Multiple chips on a single ceramic substrate may be joined simultaneously.

7. Signal and power pads need not be confined to the perimeter of the die, as with TAB and wire bonded packages; solder pads may be located throughout the internal chip area if associated routes for the interconnection traces on the ceramic package can be found. For VLSI hardware, minimizing the series resistance and inductance between supply voltage pin and the chip internal distribution is crucial; this packaging technique can accommodate multiple internal chip power pads with wide trace area to package pins.

The primary disadvantages of the flip-chip bonding technique are the inability to utilize low-cost plastic packages to contain the surface interconnection traces, the unique chip-to-package alignment equipment required, and the distance-to-neutral point limitation mentioned earlier. The use of internal pads in the chip footprint also presents unique probe card design requirements.

Figure 15.77 illustrates the final sealed package, including a top seal that covers the die and back seal that prevents contamination from entering between the ceramic package substrate and the crimped metal cap.

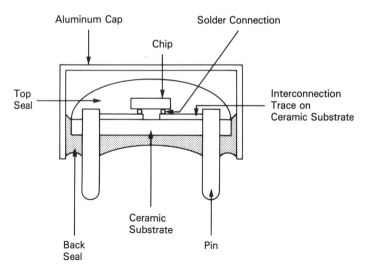

Figure 15.77.   Cross section of the final flip-chip package, illustrating the sealing coats used to provide contamination barriers.

Multichip Packaging

The development of packages that incorporate chip-to-chip interconnections (potentially on multiple wiring planes of a composite substrate) has had a significant impact on the performance achieved in main-frame computer systems. The design objectives of a ceramic or ceramic plus polyimide multichip package are to:

Increase performance and/or reduce power consumption (in off-chip drivers) by reducing the interconnection length between chips.

Accommodate an increase in power density.

Maintain sufficient logic noise margin voltages through lower resistance and inductance supply and ground distribution.

The physical design of the package can be assisted by the same computer-aided design tools as are applied to chip design. This includes the description of pin locations, surface interconnection pattern, VDD and ground distribution, and layer-to-layer via patterns (for a multilayer interconnection package). A design system tool is required that provides a transmission line pulse analysis based on the electrical characteristics of the materials and driver or receiver circuits using the connection lengths determined by the wiring layout. This analysis determines the delay time associated with a chip output transition as well as the magnitude of reflected transitions from impedance mismatches. For optimum performance, the initial pulse arriving at the receiver should be of sufficient magnitude to provide and maintain a logic switching voltage to the receiver.

The design objectives of a multilayer interconnection package are to maximize density and minimize cost by keeping wire-to-wire spacing as small as practical. The constraints on a multilayer design include minimizing cross-talk between switching and quiet signal nets and providing predictable capacitance and inductance characteristics, allowing for accurate delay calculation.

Die Separation Techniques

The physical separation of individual die on a wafer can utilize a variety of techniques, each with distinct advantages and disadvantages. Briefly, these alternatives are diamond scribing, diamond sawing, and laser-heated separation. All three techniques require some consideration of the exposed surface material(s) in the kerf areas after fabrication to ensure a reliable separation; the frame area around the chip on each masking level must be of appropriate polarity and design. Process design rules dictate the width of the kerf and the minimum distance to internal chip circuit structures.

Diamond scribing.   Scribing refers to the mechanical dragging of a diamond stylus through the kerf area between chip sites (often referred to as the *streets*), relying on the tendency of the silicon crystal to cleave along preferred planes. The shape and pressure of the scribe head determine the depth of the groove, which extends only slightly into the surface. The wafer chuck steps by the chip periodicity between the scribe lines and is rotated 90° to provide the orthogonal grid. Note that the exposed crystalline surface is required after fabrication is completed; this prohibits incorporating test devices in the kerf. Diamond scribing is therefore best suited for 1:1 projection lithography, where distinct wafer test sites are provided on the mask. It is also limited to thin wafer material, where the limited depth of the cut can produce an adequate break without fragmenting or cracking. Nevertheless, diamond scribing is the least expensive of the separation techniques, with equipment costs on the order of $10,000. It is a relatively clean process and does not produce the loss of kerf material characteristic of sawing. As a result, the die may be spaced closer together, producing more sites per wafer.

Laser scribing.   Laser scribing involves the focusing of a high-energy beam to vaporize the silicon in the kerf. The depth of the U-shaped groove can be a few mils, which relaxes the dependence on providing stress along a specific crystalline orientation for breaking; as a result, nonrectangular cuts are feasible. Also, the larger wafer thickness required for larger wafer diameters presents less of a problem. The focused laser can utilize a very narrow kerf, much less than required for a saw cut. The silicon material ejected from the kerf results in surface particles that must be subsequently cleaned off. Laser scribing remains an expensive process, with scribers costing on the order of $100,000.

Diamond sawing. Sawing is the most prevalent means of die separation, with a machine expense on the order of $30,000. Rather than a scribe-and-break means of separation, wafers are typically mounted on an adhesive backing and sawed completely through. Unlike the scribing technique, the kerf area may contain device and parametric structures. Sawing is therefore conducive to a step-and-repeat exposure technique. During sawing, a high-pressure water jet is directed at the saw cut to eject particulates. Critical parameters of the sawing process include blade thickness, diamond grit size, and smoothness of blade pressure; concerns with sawing include blade breakage, cracking and chipping underneath the die during sawing, and the undesirability of sawing through polymer dielectric films.

15.19 YIELD MODELING AND RELIABILITY TESTING

Yield modeling is a crucial aspect to providing cost estimates for manufacture; a manufacturing facility must be able to accurately estimate the yield for a new design when requested by a system developer to quote the per piece cost. Yield estimate models can be extremely complex. They typically include empirical defect density data combined with factors pertaining to the following:

Design rule lithography dimensions

Chip circuit area (especially gate oxide area)

Number (and defect sensitivities) of mask levels

Number and combination of different circuit types (since the defect sensitivities of combinational logic circuits, sequential circuits, and dynamic circuits vary considerably)

The defect density information that is gathered ranges from surface crystalline defects present in the starting material to film deposition and growth defects (e.g., pinholes, hillocks) to the average number and size of photoresist coating imperfections (e.g., striations, globules). These statistics applied against the chip and process lithography parameters commonly assume that each individual class of defect is randomly distributed and that a limiting yield percentage can be assigned to individual process steps. (The tendency of defects to cluster together, rather than to be of random distribution, is often not incorporated in yield modeling statistics.) The product of these independent yield factors (each very close to unity) reflects the ultimate manufacturing yield.

Mask-related defects are usually not a part of process yield modeling; it is assumed that a mask has been inspected sufficiently to ensure that the mask itself is free of defects and that the mask pattern indeed accurately represents the design data.

In addition to yield models based on process defect densities and chip defect sensitivities, there is yield loss due to the failure of some wafers in the initial lot to reach completion. There will undoubtedly be wafers (or sections of wafers) lost to breakage and equipment malfunction; there may be entire lots that are scrapped due to gross processing errors. The attention given to both equipment monitoring and technician training will reap benefits in both percentage of wafers reaching completion and the ultimate process yield.

Yield improvement on a high-volume, long-active-life (custom) VLSI part is an ongoing effort. The data collected from kerf testing are correlated to the resulting product yield. Ultimately, a process window is developed; that is, the yield can be specified as a function of the set of parameter values extracted from kerf testing at first metal. This information is used to maintain the process in the center of this multidimensional region in the space of all process variables. This process window and the related parameter distributions are communicated to the chip design group to facilitate design modifications and/or enhance the yield of future chip designs.

Although the VLSI system designer is keenly aware of chip and package cost estimates from manufacturing, an assessment of the part's reliability is also crucial. The failure of a part in a system design results in additional costs associated with service and inventory. Each estimated chip mean time to failure (MTTF) is a crucial parameter in calculating an overall system reliability figure-of-merit. A representative value is often determined from an experimental plot of the failing percentage over time from an initial sample. Rather than using the mean of this distribution, a frequently used parameter is the time required for the cumulative total to reach 50%, denoted as t_{50}. Effectively, this gives the median time to fail.

Two means are used to characterize the reliability of a chip population (Figure 15.78). The *failure density* function, $f(t)$, illustrates the number of failures observed at time t divided by the size of the original population. The function $z(t)$ is the *failure rate* and depicts the number of failures at time t divided by the population remaining at that time. [Both $f(t)$ and $z(t)$ are truly defined over a time interval in the neighborhood of t; a continuous curve is illustrated for both functions, in the limit that the width of the interval is very small.] The overall speed at which failures are occurring is given by $f(t)$; $z(t)$ gives the instantaneous rate of failure. The cumulative plot described to measure t_{50} is the integral of $f(t)$.

The distinctive shape of the $z(t)$ curve makes it the more useful measure of a part's reliability. This bathtub curve can be divided into three regions, as shown in the figure. A high infant failure rate is initially observed, followed by a constant random failure rate, and ultimately a wear-out region. The infant fails are commonly due to a manufacturing process defect that was not evident at wafer- or package-level test. Screening these early fails from distribution will increase t_{50} substantially; a number of different techniques can be used in an attempt to cause

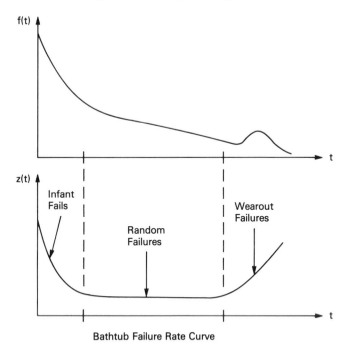

Bathtub Failure Rate Curve

Figure 15.78.   Failure density and failure rate functions, used to characterize a chip population with respect to overall reliability.

infant fails to be revealed. These techniques attempt to accentuate the mechanisms that characterize infant fails (without excessively aging the remaining hardware). The simplest technique would be to subject the chip population to operation for t_1 hours to determine which have failed (burn-in). Alternative methods try to accelerate this process by subjecting the parts to a voltage and/or temperature stress, outside the normal operating environment. For example, gate oxide dielectric weak spots and/or mobile ionic contamination can typically be isolated by applying an overvoltage to the supply pins with the chip at elevated temperatures. (Some subset of the chip test patterns could be applied to exercise the logic, if feasible.) Subsequent retesting after stress may detect an input pin failure due to dielectric breakdown or sufficient parameter drift to introduce a circuit failure.

Having eliminated the infant fails, it is necessary to develop t_{50} models for the remaining population; however, the time required to gather sufficient experimental data from a sample population would be prohibitive. Accelerated stress testing is again required. For example, Figure 15.79 lists experimental results for the measurement of t_{50} from a high-temperature, high-humidity stress test intended to examine the aluminum metallization corrosion sensitivity of a plastic encapsulated integrated circuit; the acceleration factors achieved by performing the

EXPERIMENTAL RESULTS OF ACCELERATED
CORROSION TESTING

TEST CONDITIONS	T (HRS.)	ACCELERATION
85°C./81% R.H.	3300	1
115°C./81% R.H.	446	7.4
130°C./81% R.H.	85.0	38.8
150°C./81% R.H.	12.4	266

Figure 15.79. Experimental acceleration factors achievable by stress testing. [From S. K. Malik, University/Government/Industry Microelectronics Symposium (1981), VIII-51.]

evelution at higher temperatures are also depicted. These experimental data and many other failure rate mechanisms fit an Arrhenius rate model for temperature dependence:

$$t_{50} = A * e^{-(E_A/kT)}$$

where E_A is denoted as an activation energy. By evaluating individual failure mechanisms under accelerated stress, extrapolation of measured t_{50} values down to the expected range of operating conditions provides parameters for a reliability model. The system designer must then input to this model the associated environmental values anticipated for the part:

Number of power-on hours

Number of on–off cycles (thermal cycling)

Industrial versus military versus office usage (as this varies the expected range of ambient temperature and humidity, exposure to strong electromagnetic fields or particle radiation, and the vibrations and mechanical shock to which the part would be subjected)

The system designer must address the trade-offs between reduced device operating temperatures (and therefore enhanced reliability) and the costs associated with improved means of withdrawing the heat generated by the hardware.

PROBLEMS

15.1. Research and describe the four-point probe measurement technique for the determination of the sheet resistivity of surface impurity layers.

15.2. Verify that the complementary error function expression for the predeposition impurity profile satisfies the differential diffusion equation and initial and boundary conditions. Verify that the Gaussian expression(s) for the drive-in impurity profile satisfies the diffusion equation and initial profile and boundary conditions.

15.3. Sketch how the impurity profiles in Figure 15.31 illustrating leaching and snowplowing would be altered if the diffusion coefficient of the impurity in SiO_2 were large.

15.4. Research and describe the techniques in use to provide photoresist sidewalls with a slope suitable for lift-off metallization processing.

15.5. Develop a mask alignment strategy for the process outlined in Section 15.1. Which of the level-to-level alignments were assumed to be *critical*?

15.6. Show how a conformal LPCVD oxide layer followed by an anisotropic etch will result in a sidewall spacer. How does the width of the spacer vary with the angle of the sidewall and the deposited oxide thickness?

15.7. Determine the appropriate names and characteristics of the glow discharge regions not identified in the diode sputtering configuration of Figure 15.42.

15.8. What are the factors that inhibit etching of aluminum in a barrel plasma reactor?

15.9. Research and describe the processing alternatives currently under development for utilizing a metal-silicide gate and interconnection material.

15.10. Research and describe the MOS device fabrication techniques used to provide device channel regions located on the sidewalls of V-grooves etched into the substrate.

15.11. Research and describe the equipment configurations and unique processing steps associated with plasma-enhanced chemical vapor deposition of dielectric films. What design measures are taken to maintain high wafer throughput?

15.12. Research and describe the materials and techniques used for adding pellicles to mask plates.

15.13. Describe in detail the specific nature of the pulsed C–V characterization technique mentioned in Section 15.15.

15.14. Research and describe dielectric film characterization using ellipsometry.

15.15. What are the potential VLSI process enhancements available by implanting oxygen in silicon? What difficulties are presented in terms of dosage, energy, and impact on device characteristics?

15.16. What is the electron flood technique utilized with ion implants? What failure mechanism is being addressed by neutralizing the beam in this manner?

15.17. The discussion of impurity predeposition indicated that a solid source powder is heated to vaporization and transported down the diffusion tube to the product wafers; alternatively, wafers containing a dopant compound can be interspersed with product wafers in the boat to provide the source of impurities. Research and describe this alternative for impurity introduction from a solid source, specifically addressing predeposition procedure, dopant compounds used as solid wafers, and the advantages and disadvantages of this solid source option (e.g., uniformity of deposition, unique handling requirements, cost).

REFERENCES

15.1 Huang, J., and Welliver, L., "On the Redistribution of Boron in the Diffused Layer during Thermal Oxidation," *Journal of the Electrochemical Society*, 117:12 (December 1970), 1577–1580.

15.2 Deal, B. and Grove, A., "General Relationship for the Thermal Oxidation of Silicon," *Journal of Applied Physics*, 36:12 (December 1965), 3770–3778.

15.3 Stickel, W., "Direct Write E-beam Systems in IBM's Integrated Circuit Production," *Proceedings of the University/Government/Industry Microelectronics Symposium* (1981) II-33 to II-44.

15.4 Deal, B., and others, "Characteristics of Fast Surface States Associated with SiO$_2$-Si and Si$_3$N$_4$-SiO$_2$-Si Structures," *Journal of the Electrochemical Society*, 116:7 (July 1969), 997–1005.

15.5 Hammond, M., "Silicon Epitaxy," *Solid State Technology* (November 1978), 68–75.

Additional References

Process development

AMI Staff, "MOS Processes," *IEEE Transactions on Consumer Electronics*, CE-24:2 (May 1978), 155–167.

Fair, Richard, and others, "Modeling Physical Limitations on Junction Scaling for CMOS," *IEEE Transactions on Electron Devices*, ED-31:9 (September 1984), 1180–1185.

Khan, M., and Godejahn, G., "A Self-aligned Contact MOS Process for Fabricating VLSI Circuits," *Journal of the Electrochemical Society*, 128:6 (June 1981), 1333–1335.

Mikkelson, J., and others, "An NMOS VLSI Process for Fabrication of a 32 bit CPU Chip," *IEEE International Solid-State Circuits Conference* (1981), 106.

Shibata, T., and others, "An Optimally Designed Process for Submicrometer MOSFET's," *IEEE Transactions on Electron Devices*, ED-29:4 (April 1982), 531–535.

Su, S., "Low-Temperature Silicon Processing Techniques for VLSIC Fabrication", *Solid State Technology* (March 1981), 72–82.

Yu, K., and others, "HMOS-CMOS—A Low-Power High-Performance Technology," *IEEE Journal of Solid-State Circuits*, SC-16:5 (October 1981), 454–459.

Lithography

Allan, R., "Semiconductors: Toeing the (Microfine) Line," *IEEE Spectrum* (December 1977), 34–40.

Broers, A., "Resolution, Overlay, and Field Size for Lithography Systems," *IEEE Transactions on Electron Devices*, ED-28:11 (November 1981), 1268–1278.

Deckert, C., and Ross, D., "Microlithography—Key to Solid-State Device Fabrication," *Journal of the Electrochemical Society*, 127:3 (March 1980), 45C-55C.

Dill, F., "Optical Lithography," *IEEE Transactions on Electron Devices*, ED-22:7 (July 1975), 440–444.

Lyman, J., "Lithography Chases the Incredible Shrinking Line," *Electronics* (April 12, 1979), 105–116.

Watts, R., and Bruning, J., "A Review of Fine-Line Lithographic Techniques: Present and Future," *Solid State Technology* (May 1981), 99–105.

Mask manufacture

Cast, R., "Computer Controlled Artwork Generation at 10 X," *Solid State Technology* (February 1971), 29–32, 60.

Goeders, T., "Reticles for Wafer Imaging Systems," *Solid State Technology* (May 1980), 91–96.

Henriksen, G., "Reticles by Automatic Pattern Generation," *Society of Photo-optical Instrumentation Engineering, Vol. 100, Semiconductor Microlithography II* (1977), 86–95.

O'Malley, A., "An Overview of Photomasking Technology in the Past Decade," *Solid State Technology* (June 1971), 57–62.

———, "Technological Implications in the Photomasking Process," *Solid State Technology* (June 1975), 40–45.

Richardson, F., and Resor, G., "Computer Aided Design for Photomask Production," *Solid State Technology* (June 1970), 60–64.

Shoho, R., "Fabrication of Microelectronics Reticles," *Solid State Technology* (February 1979), 75–79.

Tong, J., "Mask Manufacture for Integrated Circuits," *Solid State Technology* (July 1968), 19–26.

Alignment systems

Bossung, J., "Projection Printing Characterization," *Society of Photo-optical Instrumentation Engineering, Vol. 100, Semiconductor Microlithography II* (1977), 80–84.

———, and Muraski, E., "Advances in Projection Microlithography," *Solid State Technology* (August 1979) 109–112.

Heim, R., "Practical Aspects of Contact/Proximity, Photomask/Wafer Exposure," *Society of Photo-optical Instrumentation Engineering, Vol. 100, Semiconductor Microlithography II* (1977), 104–114.

Markle, D., "A New Projection Printer," *Solid State Technology* (June 1974), 50–53.

Nakase, M., and Shinozaki, T., "Resolution and Overlay Precision of a 10 to 1 Step-and-Repeat Projection Printer for VLSI Circuit Fabrication," *IEEE Transactions on Electron Devices*, ED-28:11 (November 1981), 1416–1421.

Resor, G., and Tobey, A., "The Role of Direct Step-on-the-Wafer in Microlithography Strategy for the 80's," *Solid State Technology* (August 1979), 101–108.

Roussel, J., "Step-and-Repeat Wafer Imaging," *Solid State Technology* (May 1978), 67–71.

Schneider, W., "Testing the Mann Type 4800DSW™ Wafer Stepper™," *Society of Photo-optical Instrumentation Engineering, Vol. 174, Developments in Semiconductor Microlithography IV* (1979), 6–14.

Stover, H., "Stepping into the 80's with Die-by-Die Alignment," *Solid State Technology* (May 1981), 112–120.

Tobey, A., "Wafer Stepper Steps up Yield and Resolution in IC Lithography," *Electronics* (August 16, 1979), 109–112.

Wittekoek, S., "Step-and-Repeat Wafer Imaging," *Solid State Technology* (June 1980), 80–84.

Photoresist patterning

Allen, R., and others, "Deep U.V. Hardening of Positive Photoresist Patterns," *Journal of the Electrochemical Society*, 129:6 (June 1982), 1379–1381.

Deckert, C., and Peters, D., "Processing Latitude in Photoresist Patterning," *Solid State Technology* (January 1980), 76–80.

Fujimori, S., "Computer Simulation of Exposure and Development of a Positive Photoresist," *Journal of Applied Physics*, 50:2 (February 1979), 615–623.

Hatzakis, M., "Single-Step Optical Lift-Off Process," *IBM Journal of Research and Development*, 24:4 (July 1980), 452–460.

Hiraoka, H., and Pacansky, J., "High Temperature Flow Resistance of Micron Sized Images in AZ Resists," *Journal of the Electrochemical Society*, 128:12 (December 1981), 2645–2647.

Kaplan, L., and Bergin, B., "Residues from Wet Processing of Positive Resists," *Journal of the Electrochemical Society*, 127:2 (February 1980), 386–395.

Peters, D., and Deckert, C., "Removal of Photoresist Film Residues from Wafer Surfaces," *Journal of the Electrochemical Society*, 126:5 (May 1979), 883–885.

Widmann, D., and Binder, H., "Linewidth Variations in Photoresist Patterns on Profiled Surfaces," *IEEE Transactions on Electron Devices*, ED-22:7 (July 1975), 467–477.

Oxidation

Antoniadis, D., and others, "Impurity Redistribution in SiO_2-Si during Oxidation: A Numerical Solution including Interfacial Fluxes," *Journal of the Electrochemical Society*, 126:11 (November 1979), 1939–1948.

Chaudari, P., and others, "Stability of MOSFET Devices with Phosphorus-Doped Oxide as Gate Dielectric," *Journal of the Electrochemical Society*, 124:12 (December 1977), 1897–1900.

Chiu, K., and others, "A Bird's Beak Free Local Oxidation Technology Feasible for VLSI Circuits Fabrication," *IEEE Journal of Solid-State Circuits*, SC-17:2 (April 1982), 166–170.

Deal, Bruce, "Standardized Terminology for Oxide Charges Associated with Thermally Oxidized Silicon," *Journal of the Electrochemical Society*, 127:4 (April 1980), 979–981.

————, "The Current Understanding of Charges in the Thermally Oxidized Silicon Structure," *Journal of the Electrochemical Society*, 121:6 (June 1974), 198C–205C.

DiMaria, D., "The Properties of Electron and Hole Traps in Thermal Silicon Dioxide Layers Grown on Silicon," *IBM Research Report*, RC 7104, April 14, 1978.

Fair, R., "Oxidation, Impurity Diffusion, and Defect Growth in Silicon—An Overview," *Journal of the Electrochemical Society*, 128:6 (June 1981), 1360–1368.

Goodwin, C., and Brossman, J., "MOS Gate Oxide Defects Related to Treatment of Silicon Nitride Coated Wafers Prior to Local Oxidation," *Journal of the Electrochemical Society*, 129:5 (May 1982), 1066–1070.

Ho, C., Plummer, J., and Meindl, J., "Thermal Oxidation of Heavily Phosphorus-Doped Silicon," *Journal of the Electrochemical Society*, 125:4 (April 1978), 665–671.

Hu, S., "New Oxide Growth Law and the Thermal Oxidation of Silicon," *Applied Physics Letters*, 42:10, (May 15, 1983), 872–874.

Ishikawa, Y., and others, "The Enhanced Diffusion of Arsenic and Phosphorus in Silicon by Thermal Oxidation," *Journal of the Electrochemical Society*, 129:3 (March 1982), 644–648.

Lee, H., Dutton, R., and Antoniadis, D., "On Redistribution of Boron during Thermal Oxidation of Silicon," *Journal of the Electrochemical Society*, 126:11, (November 1979), 2001–2007.

Marcus, R., and Sheng, T., "The Oxidation of Shaped Silicon Surfaces," *Journal of the Electrochemical Society*, 129:6 (June 1982), 1278–1282.

————, and others, "Polysilicon/SiO_2 Interface Microtexture and Dielectric Breakdown," *Journal of the Electrochemical Society*, 129:6 (June 1982), 1282–1289.

Monkowski, J., "Role of Chlorine in Silicon Oxidation," *Solid State Technology*, Part I (July 1979), 58–61; Part II (August 1979), 113–119.

Raider, S., and Berman, A., "On the Nature of Fixed Oxide Charge," *Journal of the Electrochemical Society*, 125:4 (April 1978), 629–632.

Revitz, M., and others, "Effect of High-temperature, Postoxidation Annealing on the Electrical Properties of the Si-SiO_2 Interface," *Journal of Vacuum Science Technology*, 16:2 (March/April 1979), 345–347.

Sunami, H., "Thermal Oxidation of Phosphorus-Doped Polycrystalline Silicon in Wet Oxygen," *Journal of the Electrochemical Society*, 125:5 (June 1978), 892–897.

Taft, E., "Index of Refraction of Steam Grown Oxides on Silicon," *Journal of the Electrochemical Society*, 127:4 (April 1980), 993–994.

Wu, C., and others, "Redistribution of Ion-Implanted Impurities in Silicon During Diffusion in Oxidizing Ambients," *IEEE Transactions on Electron Devices*, ED-22:9 (September 1976), 1095–1097.

Plasma processing

Adams, A., "Plasma Planarization," *Solid State Technology* (April 1981), 178–181.

Bell, A., "Abstract: Fundamentals of Plasma Chemistry," *Journal of Vacuum Science Technology*, 16:2 (March/April 1979), 418–419.

Libby, W., "Plasma Chemistry," *Journal of Vacuum Science Technology*, 16:2 (March/April 1979), 414–417.

Tolliver, D., "Plasma Processing in Microelectronics—Past, Present, and Future," *Solid State Technology* (November 1980), 99–105.

Plasma etching

Bergeron, S., and Duncan, B., "Controlled Anisotropic Etching of Polysilicon," *Solid State Technology* (August 1982), 98–103.

Bower, D., "Planar Plasma Etching of Polysilicon Using CCl_4 and NF_3," *Journal of the Electrochemical Society*, 129:4 (April 1982), 795–799.

Busta, H., and others, "Plasma Etch Monitoring with Laser Interferometry," *Solid State Technology* (February 1979), 61–64.

Coburn, J., and Kay, E., "Abstract: Some Chemical Aspects of the Fluorocarbon Plasma Etching of Silicon and Its Compounds," *Journal of Vacuum Science Technology*, 16:2 (March/April 1979), 407.

——, and Winters, H., "Plasma Etching—A Discussion of Mechanisms," *Journal of Vacuum Science Technology*, 16:2 (March/April 1979), 408–409.

Donnelly, V., and Flamm, D., "Anisotropic Etching in Chlorine-Containing Plasmas," *Solid State Technology* (April 1981), 161–166.

Eisele, K., "SF$_6$, a Preferable Etchant for Plasma Etching Silicon," *Journal of the Electrochemical Society*, 128:1 (January 1981), 123–126.

Enomote, T., "Loading Effect and Temperature Dependence of Etch Rate of Silicon Materials in CF$_4$ Plasma," *Solid State Technology* (April 1980), 117–121.

Ephrath, L., "Dry Etching for VLSI—A Review," *Journal of the Electrochemical Society*, 129:3 (March 1982), 62C–66C.

——, "Etching Needs for VLSI," *Solid State Technology* (July 1982), 87–92.

——, "Reactive Ion Etching for VLSI," *IEEE Transactions on Electron Devices*, ED-28:11 (November 1981), 1315–1319.

——, and DiMaria, D., "Review of RIE Induced Radiation Damage in Silicon Dioxide," *Solid State Technology* (April 1981), 182–188.

Hayes, J., and Pandjumsoporn, T., "Planar Plasma Etching of Polycrystalline Silicon," *Solid State Technology* (November 1980), 71–78.

Hirobe, K., and others, "Some Problems in Plasma Etching of Al and Al-Si Alloy Films," *Journal of the Electrochemical Society*, 128:12 (December 1981), 2686–2688.

Hutt, M., and Class, W., "Optimization and Specification of Dry Etching Processes," *Solid State Technology* (March 1980), 92–97.

Kawata, H., and others, "The Relation between Etch Rate and Optical Emission Intensity in Plasma Etching," *Journal of the Electrochemical Society*, 129:6 (June 1982), 1325–1329.

Kushner, M., "A Kinetic Study of the Plasma-etching Process. I. A Model for the Etching of Si and SiO$_2$ in C$_n$F$_m$/O$_2$ Plasmas; II. Probe Measurements of Electron Properties in an RF Plasma-etching Reactor," *Journal of Applied Physics*, 53:4 (April 1982), 2923–2946.

Mauer, J., and Logan, J., "Reactant Supply in Reactive Ion Etching," *Journal of Vacuum Science Technology*, 16:2 (March/April 1979), 410–413.

Mogab, C., "The Loading Effect in Plasma Etching," *Journal of the Electrochemical Society*, 124:8 (August 1977), 1262–1268.

——, and Harshbarger, W., "Abstract: Plasma-assisted Etching for Pattern Transfer," *Journal of Vacuum Science Technology*, 16:2 (March/April 1979), 408–409.

Reynolds, J., and others, "Simulation of Dry Etched Line Edge Profiles," *Journal of Vacuum Science Technology*, 16:6 (November/December 1979), 1772–1775.

Schwartz, G., and Schaible, P., "Reactive Ion Etching of Silicon," *Journal of Vacuum Science Technology*, 16:2 (March/April 1979), 404–406.

Stach, J., and Woytek, A., "Dry Etching Systems Expand IC Device Processing Arsenal," *Industrial Research and Development*, (July 1982), 107–110.

Tuck, J., "Plasma Etching," *Circuit Manufacturing* (July 1982), 69–77.

Viswanathan, N., "Simulation of Plasma-etched Lithographic Structures," *Journal of Vacuum Science Technology*, 16:2 (March/April 1979), 388–390.

Wang, D., and Maydan, D., "Reactive-ion Etching Eases Restrictions on Materials and Feature Sizes," *Electronics* (November 3, 1983), 157–161.

Chemical vapor deposition

Baudrant, A., and Sacilotti, M., "The LPCVD Polysilicon Phosphorus Doped in Situ as an Industrial Process," *Journal of the Electrochemical Society*, 129:5 (May 1982), 1109–1115.

Brown, W., and Kamins, T., "An Analysis of LPCVD System Parameters for Polysilicon, Silicon Nitride, and Silicon Dioxide Deposition," *Solid State Technology* (July 1979), 51–57.

Dong, D., and others, "Preparation and Some Properties of Chemically Vapor-Deposited Si-Rich SiO_2 and Si_3N_4 Films," *Journal of the Electrochemical Society*, 125:3 (May 1978), 819–823.

Hammond, M., "Introduction to Chemical Vapor Deposition," *Solid State Technology* (December 1979), 61–65.

Kamins, T., "Oxidation of Phosphorus-Doped Low Pressure and Atmospheric Pressure CVD Polycrystalline-Silicon Films," *Journal of the Electrochemical Society*, 126:5 (May 1979), 838–844.

————, "Resistivity of LPCVD Polycrystalline-Silicon Films," *Journal of the Electrochemical Society*, 126:5 (May 1979), 833–837.

Mattson, B., "CVD Films for Interlayer Dielectrics," *Solid State Technology* (January 1980), 60–63.

Rosler, R., "Low Pressure CVD Production Processes for Poly, Nitride, and Oxide," *Solid State Technology* (April 1977), 63–70.

Sinha, A., "Plasma Deposited Silicon Nitride Films," *Solid State Technology* (April 1980), 133–136.

van de Ven, E., "Plasma Deposition of Silicon Dioxide and Silicon Nitride Films," *Solid State Technology* (April 1981), 167–171.

Ion implantation

Ahmed, H., and Charpentier, A., "Selective Ion Implantation to Reduce Power Consumption in MOS Integrated Circuits," *IEEE Transactions on Electron Devices*, ED-25:5 (May 1978), 547–548.

Gibbons, James, "Ion Implantation in Semiconductors," *Proceedings of the IEEE*: "Part I—Range Distribution Theory and Experiments," 56:3 (March 1968), 295–319; "Part II—Damage Production and Annealing," 60:9 (September 1972), 1062–1096.

Lee, D., and Mayer, J., "Ion-Implanted Semiconductor Devices," *Proceedings of the IEEE*, 62:9 (September 1974), 1241–1255.

Macdougall, J., "Ion Implantation Equipment and Processes for Semiconductor Device Manufacture," *Solid State Technology* (October 1971), 46–50.

Prussin, S., "Ion Implantation Gettering: A Fundamental Approach," *Solid State Technology* (July 1981), 52–54.

Reddi, V., and Yu, A., "Ion Implantation for Silicon Device Fabrication," *Solid State Technology* (October 1972), 35–41.

Interconnections and dielectrics

Blech, I., "Electromigration in Thin Aluminum Films on Titanium Nitride," *Journal of Applied Physics*, 47:4 (April 1976), 1203–1208.

Brown, G., "Reliability Implications of Polyimide Multilevel Insulators," *IEEE International Reliability Physics Symposium (RELPHYS 81)* (1981), 282–286.

Cohen, S., and others, "Al-0.9% Si/Si Ohmic Contacts to Shallow Junctions," *Journal of the Electrochemical Society*, 129:6 (June 1982), 1335–1338.

Mohammadi, F., "Silicides for Interconnection Technology," *Solid State Technology* (January 1981), 65–72, 92.

Mukai, K., and others, "Planar Multilevel Interconnection Technology Employing a Polyimide," *IEEE Journal of Solid-State Circuits*, SC-13:4 (August 1978), 462–467.

Sato, K., and others, "A Novel Planar Multilevel Interconnection Technology Utilizing Polyimide," *IEEE Transactions on Parts, Hybrids, and Packaging*, PHP-9:3 (September 1973), 176–180.

Schreiber, H., and Grabe, R., "Electromigration in Sputtered Aluminum Films," *IEEE Transactions on Electron Devices*, ED-28:3 (March 1981), 351–353.

Crystalline growth

Domey, Kenneth, "Computer Controlled Growth of Single-Crystal Ingots," *Solid State Technology* (October 1971), 41–45, 58.

Heil, R., "Advances in Zone Refining," *Solid State Technology*, (January 1968), 21–28.

Law, J., "Silicon Wafer Technology—State of the Art," *Solid State Technology* (January 1971), 25–29.

Manufacturing process control

Mills, A., "Reverse Osmosis for Purification of Water," *Solid State Technology*: Part 1 (July 1970), 41–45, Part 2 (August 1970), 71–75.

Morrison, P., "Selection of Clean Environments for Product Manufacture," *Western Electric Engineer*, 8:4 (October 1964), 14–18.

Pink, W., "Clean Environments for Photo-Resist Manufacturing Operations," *Western Electric Engineer*, 8:4 (October 1964), 19–23.

Rott, C., "Clean Rooms," *Solid State Technology*; (July 1970), 35–40.

Savola, W., and Wallace, J., "Deionized Water for Integrated Circuit Fabrication," *Solid State Technology* (November 1973), 47–49.

Schuegraf, K., "Advances in Wafer Process Control," *Solid State Technology* (February 1980), 87–94.

Taubenest, R., and Ubersax, H., "Ultrapure Water in Semiconductor Manufacturing," *Solid State Technology* (June 1980), 74–79.

Vacuum system considerations for fabrication

Duval, P., "Vacuum Problems in Today's Integrated Circuit Manufacturing Systems: Part I," *Solid State Technology* (August 1982), 110–116.

Lam, D., and Koch, G., "Vacuum System Considerations for Plasma Etching Equipment," *Solid State Technology* (September 1980), 99–101.

Peterson, J., and Steinherz, H., "Vacuum Pump Technology: A Short Course on Theory and Operations—Part I," *Solid State Technology* (December 1981), 82–86.

Chip–package interconnection

Budnick, A., "Manufacturing the Plastic Dual in-Line Integrated Circuit," *Solid State Technology* (August 1968), 37–42.

Miller, L. F., "Controlled Collapse Reflow Chip Joining," *IBM Journal of Research and Development* (May 1969), 239–250.

Characterization and fabrication measurements

Coates, V., "Computerized Optical Systems for Linewidth and Film Thickness Measurements on Microelectronic Circuits," *Society of Photo-optical Instrumentation Engineering, Vol. 174, Developments in Semiconductor Microlithography IV* (1979), 184–192.

Gegenwarth, R., and Laming, F., "Effect of Plastic Deformation of Silicon Wafers on Overlay," *Society of Photo-optical Instrumentation Engineering, Vol. 100, Semiconductor Microlithography II* (1977), 66–73.

Hornstra, J., and van der Pauw, L., "Measurement of the Resistivity Constants of Anisotropic Conductors by Means of Plane-Parallel Discs of Arbitrary Shape," *Journal of Electronic Control*, 7:2 (August, 1959), 169–171.

Irene, E., and Dong, D., "Ellipsometry Measurements of Polycrystalline Silicon Films," *Journal of the Electrochemical Society*, 129:6 (June 1982), 1247–1353.

MacIver, B., and Puzio, L., "A Method for Observing Micrometer-Size Defects in Resists," *Journal of the Electrochemical Society*, 129:10 (October 1982), 2384–2385.

McMillan, L., "MOS C-V Techniques for IC Process Control," *Solid State Technology* (September 1972), 47–52.

Pliskin, W., and Conrad, E., "Nondestructive Determination of Thickness and Refractive Index of Transparent Films," *IBM Journal of Research and Development* (January 1964), 43–51.

Test and Measurement World Staff, "Optical Microscopes—Inspection and Measurement for the Electronics Industry," *Test and Measurement World* (March 1983), 31–42.

van der Pauw, L., "A Method of Measuring Specific Resistivity and Hall Effect of Discs of Arbitrary Shape," *Philips Research Reports*, 13:1 (February 1958), 1–9.

———, "Determination of Resistivity Tensor and Hall Tensor of Anisotropic Conductors," *Philips Research Reports*, 16:2 (April 1961), 187–195.

Wiley, John, and Miller, G., "Series Resistance Effects in Semiconductor CV Profiling," *IEEE Transactions on Electron Devices*, ED-22:5 (May 1975), 265–272.

Zaininger, K., and Heiman, F., "The C-V Technique as an Analytical Tool," *Solid State Technology*: Part I (May 1970), 49–56; Part II (June 1970), 46–55.

General texts on integrated circuit fabrication

Burger, R. M., and Donovan, R. P., *Fundamentals of Silicon Integrated Device Technology*. Englewood Cliffs, N.J.: Prentice-Hall, Inc., 1967.

Elliott, D., *Integrated Circuit Fabrication Technology*. New York: McGraw-Hill Book Co., 1982.

Gise, P., and Blanchard, R., Fairchild Corp., *Semiconductor and Integrated Circuit Fabrication Techniques*. Reston, VA.: Reston Publishing Co., 1979.

Runyan, W. R., *Silicon Semiconductor Technology*. New York: McGraw-Hill Book Co., 1965.

Wolf, H. F., *Silicon Semiconductor Data*. Elmsford, N.Y.: Pergamon Press, 1969.

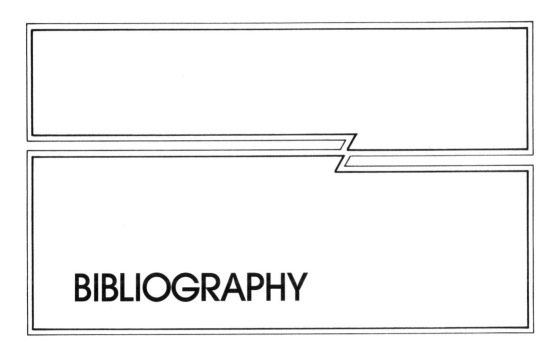

BIBLIOGRAPHY

GENERAL TEXTS ON INTEGRATED-CIRCUIT DESIGN AND TECHNOLOGY

AMI Engineering Staff, *MOS Integrated Circuits*. New York: Van Nostrand Reinhold, 1972.

Camenzind, H., *Electronic Integrated Systems Design*. New York: Van Nostrand Reinhold, 1972.

Carr, W., and Mize, J., *MOS/LSI Design and Application*. New York: McGraw-Hill Book Co., 1972.

Cobbold, R., *Theory and Applications of Field-Effect Transistors*. New York: Wiley-Interscience, 1970.

Glaser, A., and Subak-Sharpe, G., *Integrated Circuit Engineering*. Reading, Mass.: Addison-Wesley Publishing Co., 1977.

Gray, P., and Meyer, R., *Analysis and Design of Analog Integrated Circuits*. New York: John Wiley & Sons, Inc., 1977.

Grebene, A., *Analog Integrated Circuit Design*. New York: Van Nostrand Reinhold, 1972.

Mavor, J., Nervyn, J., and Denyer, P., *Introduction to MOS LSI*. Reading, Mass.: Addison-Wesley Publishing Co., 1983.

Mead, C., and Conway, L., *Introduction to VLSI Systems*. Reading, Mass.: Addison-Wesley Publishing Co., 1980.

Muller, R., and Kamins, T., *Device Electronics for Integrated Circuits*. New York: John Wiley & Sons, Inc., 1977.

Muroga, S., *VLSI System Design*. New York: Wiley-Interscience, 1982.

853

Rice, Rex (ed.), *Tutorial: VLSI—The Coming Revolution in Applications and Design*. New York: IEEE Computer Society Press, 1980.

Richman, P., *MOS Field-Effect Transistors and Integrated Circuits*. New York: Wiley-Interscience, 1973.

SPECIAL ISSUES OF TECHNICAL JOURNALS ON INTEGRATED CIRCUIT DESIGN AND TECHNOLOGY

IEEE Transactions on Electron Devices

Special Issue on Pattern Generation and Microlithography, ED-22:7 (July 1975).

Special Issue: Historical Notes on Important Tubes and Semiconductor Devices, ED-23:7 (July 1976).

Special Issue on Device Reliability, ED-26:1 (January 1979).

Joint Special Issue on Very Large Scale Integration, ED-26:4 (April 1979); also *IEEE Journal of Solid-State Circuits*, SC-14:2 (April 1979).

Special Issue on Semiconductor Memory, ED-26:6 (June 1979).

Joint Special Issue on Very Large Scale Integration, ED-27:8 (August 1980): also *IEEE Journal of Solid-State Circuits*, SC-15:4 (August 1980).

Special Issue on Hot Electron Effects in Short-Channel Devices, ED-28:8 (August 1981).

Special Issue on High-Resolution Fabrication of Electron Devices, ED-28:11 (November 1981).

Joint Special Issue on Very Large Scale Integration, ED-29:4 (April 1982); also *IEEE Journal of Solid-State Circuits*, SC-17:2 (April 1982).

Joint Special Issue on Numerical Simulation of VLSI Devices, ED-30:9 (September 1983).

Joint Special Issue on Devices and Technologies for Custom Integrated Circuits, ED-31:2 (February 1984); also *IEEE Journal of Solid-State Circuits*, SC-19:1 (February 1984).

Proceedings of the IEEE

Special Issue on Micron and Submicron Circuit Engineering, 71:5 (May 1983).

Special Issue on Computer-aided Design, 69:10 (October 1981).

Special Issue on VLSI Circuit Design, 71:1 (January 1983).

Journal of Solid-State Circuits

Special Issue on Semiconductor Memory and Logic, SC-11:5 (October 1976).

Special Issue on Digital Circuits, SC-15:5 (October 1980).

Special Issue on Digital Circuits, SC-16:5 (October 1981).

IBM Journal of Research and Development

Semiconductory Memory Technology, 24:3 (May 1980).
VLSI Circuit Design, 25:2 and 3 (May 1981).
25th Anniversary Issue, 25:5 (September 1981).
Semiconductor Manufacturing Technology, 26:5 (September 1982).

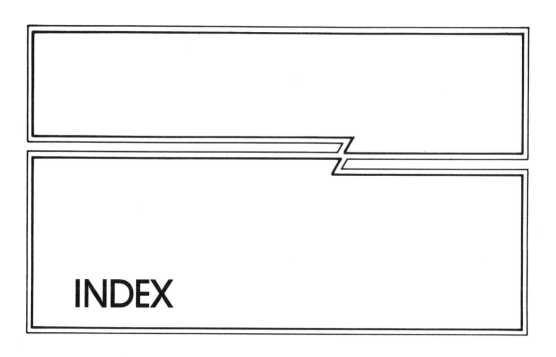

INDEX